Adenosine

Susan Masino • Detlev Boison
Editors

Adenosine

A Key Link between Metabolism and Brain Activity

Springer

Editors
Susan Masino
Trinity College
Hartford, CT, USA

Detlev Boison
Legacy Research Institute
Portland, OR, USA

ISBN 978-1-4614-3902-8 ISBN 978-1-4614-3903-5 (eBook)
DOI 10.1007/978-1-4614-3903-5
Springer New York Heidelberg Dordrecht London

Library of Congress Control Number: 2012943605

Printed on acid-free paper

Springer is part of Springer Science+Business Media (www.springer.com)

Preface

For many years the "retaliatory metabolite" and neuromodulator adenosine has been recognized as an endogenous anticonvulsant and neuroprotective molecule. As the core molecule of ATP and of nucleic acids, adenosine forms a unique link among cell energy, gene regulation, and neuronal excitability. Adenosine has long been a highly coveted therapeutic target, and its actions at the A_1 receptor subtype hold well-established and profound therapeutic potential for conditions such as stroke, brain injury, pain, and epilepsy, among others. To date, receptor-based strategies to augment the therapeutic influence of adenosine have been unable to harness its clinical potential, primarily due to side effects at identical peripheral adenosine receptors.

Emerging evidence ignites new hope for adenosine-based therapies in the central nervous system. Parallel and converging lines of research include studies on the ongoing regulation of adenosine by nonpathological, physiologically relevant stimuli, by metabolism and astrocytes and also by exciting new cell-based approaches. All of these modes of regulation have detailed associated mechanisms established by basic research and, accordingly, proven or predicted clinical implications.

In this book we assemble diverse recent research on regulating the influence of adenosine in the central nervous system. We outline clinical implications and highlight emerging adenosine-based therapies. For example, the relationship between adenosine and the highly altered metabolic state during hibernation has important predictions for neuroprotection. As part of a major paradigm shift, we now understand that the regulation of extracellular adenosine by astrocytes is critical for normal sleep and could be pivotal for the prevention and treatment of epilepsy. In addition, there is an established functional relationship where adenosine and dopamine receptors form receptor heteromers and extend the clinical reach of the regulation of adenosine directly into the realm of dopaminergic disorders and neuropsychiatry. Finally, a ketogenic (high-fat, low-carbohydrate) diet is a metabolic therapy used to treat pediatric and medically refractory epilepsy for nearly 100 years. Although its efficacy is well established, the key neural mechanisms underlying its success remained unclear, hindering development of analogous pharmacological solutions. Converging evidence points to a critical link between ketogenic metabolism and increased adenosine.

Given its ubiquitous presence in the extracellular fluid, and the central role of adenosine as both a metabolic sensor and an effector, it is not surprising that adenosine function and dysfunction in the central nervous system appears to be a common link between diverse neurological and neuropsychiatric conditions. The broad therapeutic implications of understanding the regulation of adenosine extend to acute and chronic neurological disorders as diverse as brain injury, epilepsy, pain, neurodegenerative disorders, and dopamine-related disorders. Here we combine cellular and clinical evidence in an attempt to strengthen the conceptual underpinnings of this important research area and foster a broad and interdisciplinary perspective regarding the therapeutic potential and mechanisms associated with adenosine in the central nervous system.

We would like to express our sincere thanks to all the authors for their excellent contributions and commitment for making this book possible. We thank them for their patience and adherence to editorial timelines. We thank the many experts who contributed to the peer-review process and the success of this publishing endeavor. We are indebted to Sue Crawford who provided invaluable assistance. Finally we wish to thank Simina Calin and Ann Avouris at Springer for their continued support and guidance.

Hartford, CT, USA Susan Masino
Portland, OR, USA Detlev Boison

Contents

Contributors

Richard Allen Department of Neurology, Johns Hopkins School of Medicine, Baltimore, MD, USA

Rogelio O. Arellano Instituto de Neurobiología, Universidad Nacional Autónoma de México, Ciudad Nezahualcóyotl, Mexico

Vânia Batalha Neurosciences Unit, Instituto de Medicina Molecular, Lisbon, Portugal

David Blum Inserm U837, Lille Cedex, France

UDSL, Jean-ierre Aubert Research Centre, Institut de Médecine Prédictive et de Recherche Thérapeutique, Université Lille-Nord de France, Lille, France

CHRU-Lille, Lille, France

Detlev Boison Legacy Research Institute, Portland, OR, USA

Steven A. Brown Institute of Pharmacology and Toxicology, University of Zurich, Zurich, Switzerland

Luc Buée Inserm U837, Lille Cedex, France

UDSL, Jean-ierre Aubert Research Centre, Institut de Médecine Prédictive et de Recherche Thérapeutique, Université Lille-Nord de France, Lille, France

CHRU-Lille, Lille, France

Sylvie Burnouf Inserm U837, Lille Cedex, France

Jean-Pierre Aubert Reserach Centre, Institut de Médecine Prédictive et de Recherche Thérapeutique, Université Lille-Nord de France, Lille, France

CHRU-Lille, Lille, France

Kelly A. Cavanaugh Department of Psychology and the Brain Research Institute, University of California, Los Angeles, CA, USA

Xuesong Chen Department of Pharmacology, Physiology and Therapeutics, School of Medicine and Health Sciences, University on North Dakota, Grand Forks, ND, USA

Merce Correa Division of Behavioral Neuroscience, Department of Psychology, University of Connecticut, Storrs, CT, USA

Department of Psicobiology, Universitat Jaume I, Spain

Francesca Corti Department of Preclinical and Clinical Pharmacology, University of Florence, Florence, Italy

Sofia Cristóvão-Ferreira Faculty of Medicine and Unit of Neurosciences, Institute of Pharmacology and Neurosciences, Institute of Molecular Medicine, University of Lisbon, Lisbon, Portugal

Balázs Csóka Department of Surgery, New Jersey Medical School, University of Medicine and Dentistry of New Jersey, Newark, NJ, USA

Arpád Dobolyi Semmelweis University and the Hungarian Academy of Sciences, Budapest, Hungary

Alexander S. Doney Department of Medicine and Therapeutics, Ninewells Hospital and Medical School, University of Dundee, Dundee, Scotland

Kelly L. Drew Alaska Basic Neuroscience Program, Department of Chemistry and Biochemistry Institute of Arctic Biology, University of Alaska, Fairbanks, AK, USA

Chris G. Dulla Neuroscience Department, Tufts University Medical School, Boston, MA, USA

Christopher J. Earley Department of Neurology, Johns Hopkins School of Medicine, Baltimore, MD, USA

Inge-Marie Eigsti Psychology Department, University of Connecticut, Farmington, CT, USA

Sergi Ferré Intramural Research Program, Department of Health and Human Services, National Institute on Drug Abuse, National Institutes of Health, Baltimore, MD, USA

Lucia Frau Department of Biomedical Sciences - Toxicology, University of Cagliari, Cagliari, Italy

Bertil B. Fredholm Department of Physiology and Pharmacology, Karolinska Institutet, Stockholm, Sweden

Bruno G. Frenguelli School of Life Sciences, University of Warwick, Coventry, UK

Jonathan D. Geiger Department of Pharmacology, Physiology and Therapeutics, School of Medicine and Health Sciences, University on North Dakota, Grand Forks, ND, USA

María Victoria Sánchez-Gómez CIBERNED and Departamento de Neurociencias, Universidad del País Vasco (UPV/EHU), Leioa, Spain

Estibaliz González-Fernández CIBERNED and Departamento de Neurociencias, Universidad del País Vasco (UPV/EHU), Leioa, Spain

Xavier Guitart Intramural Research Program, Department of Health and Human Services, National Institute on Drug Abuse, National Institutes of Health,, Baltimore, MD, USA

Seema Gulyani Department of Neurology, Johns Hopkins School of Medicine, Baltimore, MD, USA

Radhey S. Gupta Department of Biochemistry and Biomedical Sciences, McMaster University, Hamilton, ON, Canada

Malika Hamdane Inserm U837, Lille Cedex, France

UDSL, Jean-ierre Aubert Research Centre, Institut de Médecine Prédictive et de Recherche Thérapeutique, Université Lille-Nord de France, Lille, France

CHRU-Lille, Lille, France

György Haskó Department of Surgery, New Jersey Medical School, University of Medicine and Dentistry of New Jersey, Newark, NJ, USA

Department of Medical Chemistry, Medical and Health Science Center, University of Debrecen, Debrecen, Hungary

Philip G. Haydon Department of Neuroscience, Tufts University School of Medicine, Boston, MA, USA

Dustin J. Hines Department of Neuroscience, Tufts University School of Medicine, Boston, MA, USA

Liang Hui Department of Pharmacology, Physiology and Therapeutics, School of Medicine and Health Sciences, University on North Dakota, Grand Forks, ND, USA

Edwin K. Jackson Department of Pharmacology and Chemical Biology, University of Pittsburgh, Pittsburgh, PA, USA

Tulasi R. Jinka Alaska Basic Neuroscience Program, Department of Chemistry and Biochemistry, Institute of Arctic Biology, University of Alaska, Fairbanks, AK, USA

David L. Kaplan Department for Biomedical Engineering, Tufts University, Medford, MA, USA

Masahito Kawamura Jr. Department of Pharmacology, Jikei University School of Medicine, Tokyo, Japan

Patrick M. Kochanek Department of Critical Care Medicine, School of Medicine, University of Pittsburgh, Pittsburgh, PA, USA

Balázs Koscsó Department of Surgery, New Jersey Medical School, University of Medicine and Dentistry of New Jersey, Newark, NJ, USA

Zsolt Kovacs Department of Zoology, University of West Hungary, Szombathely, Hungary

Pierre Krystkowiak Department of Neurology, Amiens University Hospital, Amiens, France

CNRS, Amiens, France

Cyril Laurent Inserm U837, Lille Cedex, France

UDSL, Jean-ierre Aubert Research Centre, Institut de Médecine Prédictive et de Recherche Thérapeutique, Université Lille-Nord de France, Lille, France

Antoine Leboucher Inserm U837, Lille Cedex, France

UDSL, Jean-ierre Aubert Research Centre, Institut de Médecine Prédictive et de Recherche Thérapeutique, Université Lille-Nord de France, Lille, France

Catherine Ledent IRIBHM, Université Libre de Bruxelles, Brussels, Belgium

Luísa V. Lopes Neurosciences Unit, Instituto de Medicina Molecular, Lisbon, Portugal

Laura Lopez-Cruz Department of Psicobiology, Universitat Jaume I, Castelló, Spain

Alberto Martire Section of Central Nervous System Pharmacology, Department of Therapeutic Research and Medicine Evaluation, Istituto Superiore di Sanità, Viale Regina Elena, Rome, Rome, Italy

Susan A. Masino Trinity College, Hartford, CT, USA

Carlos Matute CIBERNED and Departamento de Neurociencias, Universidad del País Vasco (UPV/EHU), Leioa, Spain

Alessia Melani Department of Preclinical and Clinical Pharmacology, University of Florence, Florence, Italy

Thomas R. Minor Department of Psychology and the Brain Research Institute, University of California, Los Angeles, CA, USA

Micaela Morelli Department of Biomedical Sciences - Toxicology, University of Cagliari, Cagliari, Italy

CNR, Institute of Neuroscience, Cagliari, Italy

Centre of Excellence for Neurobiology of Dependence, University of Cagliari, Cagliari, Italy

Christine Muheim Institute of Pharmacology and Toxicology, University of Zurich, Zurich, Switzerland

Kwaku D. Nantwi Department of Anatomy and Cell Biology, Wayne State University, School of Medicine, Detroit, MI, USA

Stephanie zur Nedden School of Life Sciences, University of Warwick, Coventry, UK

Molecular and Cellular Biophysics, Institute of Science and Technology Austria, Klosterneuburg, Austria

Eric J. Nunes Division of Behavioral Neuroscience, Department of Psychology, University of Connecticut, Storrs, CT, USA

Marco Orrú Intramural Research Program, Department of Health and Human Services, National Institute on Drug Abuse, National Institutes of Health, Baltimore, MD, USA

Marta Pardo Department of Psicobiology, Universitat Jaume I, Castelló, Spain

Jaoek Park Department of Biochemistry, McMaster University, Hamilton, QC, Canada

Florence Pasquier CHRU-Lille, Lille, France

Department Neurology, CHRU-Lille, Lille, France

Felicita Pedata Department of Preclinical and Clinical Pharmacology, University of Florence, Florence, Italy

Annalisa Pinna Department of Biomedical Sciences - Toxicology, University of Cagliari, Cagliari, Italy

CNR, Institute of Neuroscience, Cagliari, Italy

Traci N. Plumb Department of Psychology and the Brain Research Institute, University of California, Los Angeles, CA, USA

Patrizia Popoli Section of Central Nervous System Pharmacology, Department of Therapeutic Research and Medicine Evaluation, Istituto Superiore di Sanità, Rome, Italy

Tarja Porkka-Heiskanen Institute of Biomedicine, Biomedicum Helsinki, University of Helsinki, Helsinki, Finland

Eleanor M. Pritchard Department for Biomedical Engineering, Tufts University, Medford, MA, USA

Anna Maria Pugliese Department of Preclinical and Clinical Pharmacology, University of Florence, Florence, Italy

César Quiroz Intramural Research Program, Department of Health and Human Services, National Institute on Drug Abuse, National Institutes of Health, Baltimore, MD, USA

Patrick A. Randall Division of Behavioral Neuroscience, Department of Psychology, University of Connecticut, Storrs, CT, USA

Jong M. Rho Departments of Pediatrics and Clinical Neurosciences, Alberta Children's Hospital, University of Calgary, Calgary, AB, Canada

Joaquim A. Ribeiro Faculty of Medicine and Unit of Neurosciences, Institute of Pharmacology and Neurosciences, Institute of Molecular Medicine, University of Lisbon, Lisbon, Portugal

David N. Ruskin Neuroscience Program and Department of Psychology, Trinity College, Hartford, CT, USA

Bernard Sablonnière Inserm U837, Lille Cedex, France

Jean-Pierre Aubert Reserach Centre, Institut de Médecine Prédictive et de Recherche Thérapeutique, Université Lille-Nord de France, Lille, France

CHRU-Lille, Lille, France

John D. Salamone Division of Behavioral Neuroscience, Department of Psychology, University of Connecticut, Storrs, CT, USA

Ursula Sandau Legacy Research Institute, Portland, OR, USA

Jana Sawynok Department of Pharmacology, Dalhousie University, Halifax, Nova Scotia, Canada

Ana M. Sebastião Faculty of Medicine and Unit of Neurosciences, Institute of Pharmacology and Neurosciences, Institute of Molecular Medicine, University of Lisbon, Lisbon, Portugal

Nicola Simola Department of Biomedical Sciences - Toxicology, University of Cagliari, Cagliari, Italy

Philipp Singer Legacy Research Institute, Portland, OR, USA

Sarah R. Sterlace Department of Psychology and the Brain Research Institute, University of California, Los Angeles, CA, USA

Julia Svedova Neuroscience Program/Psychology Department, Trinity College, Hartford, CT, USA

Jonathan D. Verrier Department of Pharmacology and Chemical Biology, University of Pittsburgh, Pittsburgh, PA, USA

Amy K. Wagner Department of Physical Medicine & Rehabilitation, University of Pittsburgh, Pittsburgh, PA, USA

Benjamin K. Yee Legacy Research Institute, Portland, OR, USA

Beth Zupec-Kania Ketogenic Seminars, The Charlie Foundation, Santa Monica, CA, USA

Part I
Introduction

Chapter 1
Adenosine and Metabolism—A Brief Historical Note

Bertil B. Fredholm

Abstract Research on adenosine has a long history, and since the beginning more than 80 years ago there has been a link to energy metabolism. This is due to the fact that levels of adenosine increase under conditions of relative energy deficiency and that adenosine can have actions that limit this: increase respiration, increase blood flow, and reduce cellular work. There are also effects on intermediary metabolism and on overall energy homeostasis. This minireview provides a brief and personal historical background to these developments.

1.1 The Earliest History

The first report on adenosine actions by Drury and Szent-Györgyi (Drury and Szent-Gyorgyi 1929) mainly considered the profound cardiac actions. A second paper (Bennet and Drury 1931) reported many more actions including a marked lowering of body temperature in guinea pigs. They also discussed the possibility that different forms of tissue trauma could release adenosine and/or AMP that could cause intra-arterial (i.a.) vasodilatation. Nevertheless, the focus was on cardiac actions (Drury 1932) and remained so for some time. This focus was also due to interest in the therapeutic uses of a preparation, "Lacarnol," containing i.a. adenylic acid developed by the German firm I.G. Farben. Rothmann reported on the effects of Lacarnol in humans (Rothmann 1930). One case in his paper was a patient with mitral valve disease in heart failure, having cyanosis, tachypnea, and atrial fibrillation at a ventricular rate of 150/min. The injection of 5 ml of a 1 % solution of Lacarnol immediately caused transient ventricular asystole due to complete heart block, but after

B.B. Fredholm (✉)
Department of Physiology and Pharmacology,
Karolinska Institutet, Stockholm S-171 77, Sweden
e-mail: bertil.fredholm@ki.se

S. Masino and D. Boison (eds.), *Adenosine: A Key Link between Metabolism and Brain Activity*, DOI 10.1007/978-1-4614-3903-5_1,

that the patient recovered and his cardiac condition improved. Later studies, however, failed to confirm this positive outcome and the use of the preparation was gradually stopped.

1.2 Adenosine and Shock

Partly because of the huge number of young men who died in shock due to trauma and surgery during the Second World War there was a resurrection of interest in the use of adenine derivatives in the clinic. Much work was done by Green and Stoner (Stoner and Green 1945a, b). They found protective effects of administration of such derivatives, despite the fact that blood pressure was not increased. This was actually a part of the paradigm shift where the attempts to maintain blood pressure levels with vasoconstrictors was shifted to the use of i.a. vasodilators to improve peripheral tissue perfusion, but this would actually take many years more to come about completely. It was also shown that there was a relationship between the severity of the trauma and the magnitude of the loss of tissue adenine nucleotides in trauma and the release of breakdown products (Kalckar and Lowry 1947; Kovach et al. 1952; Ninomiya et al. 1961a, b). The reports on the protective effect of adenine nucleotides continued to appear and one dominant concept was that the administration of high energy phosphate compounds was able to replace the diminishing tissue stores by some kind of energy transfer or even by the direct uptake of the ATP into the energy depleted cell (Sharma and Eiseman 1966). However, by the mid-60s other results had become available and in one study it was concluded that "The protective effect of adenosine triphosphate in the prevention of irreversibility may be attributed to vasodilation, energy transfer, anticoagulation, or a combination of these mechanisms (Massion et al. 1965)." I will not further comment on the reports that adenine nucleotides altered "coagulation," but it is now clear that the mechanism was that ATP was rapidly broken down to adenosine that acted on adenosine receptors on platelets to reduce their activation and hence their contribution to intravascular coagulation. By contrast, the third alternative, vasodilatation, needs to be considered more.

1.3 Adenosine and Vasodilatation

It took a long time until the suggestion in the paper by Bennet and Drury was seriously followed up. Two independent papers (Berne 1963; Gerlach et al. 1963), demonstrated that relative energy lack by hypoxia caused the breakdown of myocardial adenine nucleotides and that the breakdown products were able to cause coronary vasodilatation. This was the beginning of a very attractive hypothesis that *adenosine is a (or the) mediator of hypoxic vasodilatation*. The hypothesis is intellectually satisfactory in that *adenosine could, by increasing oxygen supply to an energy depleted tissue, limit its own formation*—a classical tenet of homeostasis and negative feedback. These studies were preceded by several investigations demonstrating that adenine compounds cause vasodilatation (Duff et al. 1954; Frohlich 1963; Lindner and Rigler 1931). There was also precedence for the interest in hypoxia and adenine compounds in coronary circulation (Gercken 1961; Piorkowski 1950) and even in the result that actions of adenine nucleotides could be mimicked by adenosine (Rowe et al. 1962). Nevertheless, it was only with the studies of Berne and Gerlach that the concepts were generally accepted and serious research started—not least in their own laboratories. The work depended on careful development of methods to adequately determine the levels of adenine nucleotides and adenosine in body fluids and tissues, and to establish correlations (Katori and Berne 1966; Rubio and Berne 1969; Rubio et al. 1969), see also (Schutz et al. 1981). There were also attempts to broaden the concept to other tissues including brain (Rubio et al. 1975). The studies on adenosine effects in the heart were also very important in demonstrating that methylxanthines such as theophylline and the theophylline derivative aminophylline acted as adenosine antagonists (Curnish et al. 1972; Mustafa et al. 1975a). They also established the use of adenosine uptake inhibitors such as dipyridamole as tools in adenosine research (Mustafa et al. 1975b). What is often referred to as the Berne hypothesis has been extensively studied since the proposal in 1963 (Feigl 2004). It seems fair to say that in its most extended version—namely that adenosine is a major regulator of normal physiological regulation of coronary flow—the hypothesis has not held up well. It does contribute to basal coronary tone, but plays little role in the increase during exercise (Duncker and Merkus 2007). By contrast, adenosine appears to be a very important factor in pathophysiological conditions including ischemia (Feigl 2004). It is also a very important factor in preconditioning (Eckle et al. 2007; Lankford et al. 2006).

The kidney provides an interesting twist of the concept that adenosine acts to reduce its own rate of formation via effects on the circulation. It was early recognized that here adenosine could lead to reductions in blood flow (Harvey 1963; Scott and Falconer 1965) and that adenine compounds could alter renal autoregulation (Harvey 1964) even though systemic administration usually caused renal vasodilatation(Houck et al. 1947). The vasoconstrictor effect also involved enhancement of the constriction caused by sympathetic nerve stimulation (Hedqvist and Fredholm 1976), even though adenosine at the same time caused a reduction of the release of the transmitter adenosine. Important studies demonstrated that methylxanthines abolished the renal vascular effects of adenosine in the kidney (Osswald 1975) and the tubuloglomerular feedback (Schnermann et al. 1977), and presynaptic control of noradrenaline release (Hedqvist et al. 1978) thereby implicating endogenous adenosine as the underlying mechanism. Energy deficiency was shown to reduce adenine nucleotide levels and increase adenosine and inosine levels (Fredholm and Hedqvist 1978; Osswald et al. 1977). Now, how does this vasoconstrictor effect of adenosine tally with a feed-back restoration of energy deficiency? The transport work in the kidney, especially the reuptake of salts, is directly related to the amount of plasma ultrafiltered in the glomeruli and hence to renal cortical blood flow. A way to limit the transport work, and hence ATP degradation leading to adenosine formation, is therefore to reduce glomerular blood flow (Vallon et al. 2006).

The careful work on adenine nucleotide and adenosine levels in the heart and vessels led to the realization that the interconversion of these compounds is exceedingly rapid. The kidney was a major organ to study adenosine metabolism and adenine nucleotide breakdown given the profound importance of ATPases in this tissue (Zydowo 1959), and this naturally leads us over to the next important chapter.

1.4 Adenosine Formation and Degradation

The history of adenosine starts with Justus Liebig and his interest in a meat extract that could be a nutritional supplement. It obviously contained a myriad of substances, and composition varied with exact preparation. From it he isolated a barium precipitate of inosinic acid (what we now call IMP). A long and arduous process of structural identification took place and it was only in the laboratory of Emil Fischer that the structures of the purine bases were finally conclusively demonstrated and proven with synthesis. The next important step was the identification of adenylic acid from fresh muscle by Gustav Georg Embden (Embden and Zimmermann 1927). Although he did recognize the extremely rapid degradation of higher phosphate compounds by his extract he missed the discovery of ATP. He examined the deamination of AMP to IMP by a deaminase with formation of ammonia and assumed that this played an important part in muscle contraction.

Further progress was greatly helped by the improved analytical methods for phosphate containing compounds developed by Fiske and Subba-Row. They used them to discover creatine phosphate, which they perceived as an energy source in muscle contraction. In 1928 Karl Lohmann discovered large amounts of free pyrophosphate in freshly prepared muscle, and in the following year Fiske and Subba-Row discovered an adenylic compound that contained the pyrophosphate, and also that in really fresh muscle the amounts of adenylic acid were very low. They failed to make the connections to physiology and to name the compound. Lohmann could confirm and extend the findings, particularly with regard to the physiology in studies together with Meyerhof. Finally in 1935 Lohmann showed the structure of ATP and its chemical or enzymatic degradation to AMP.

Despite these initial findings it took a very long time until it was appreciated how very fast nucleotides and nucleosides were metabolized. A particular source of contention was the rapidity of conversion of ATP to adenosine, which meant that some effects of the adenine nucleotides could be ascribed to adenosine. Unfortunately for a long time strength of personal conviction was allowed to dominate over solid data and an extremely unfortunate rift between research focusing on adenine nucleotides and research focusing on adenosine developed. Only recently with a new generation of scientists has the interdependence of the two areas been very profitably pursued.

The methods to synthesize nucleosides and nucleotides took off particularly as a result of the efforts of Sir Alexander Todd, who received the 1957 Nobel Prize in Chemistry, partly for his synthesis of ATP. The synthesis of these compounds was very important as it freed scientists from the use of often imperfectly purified compounds from natural sources. This did, however, continue to be a real problem and we may just remind ourselves of the remarkable contribution of a minor contamination by GTP in Martin Rodbells ATP preparations used for adenyl cyclase experiments, and which was to be the starting point of his discovery of what is now called G-proteins. During the war there was a great effort to develop new chemical entities to combat infections. One avenue that was pursued then and later was nucleic acid derivatives. Such derivatives were also developed for cancer chemotherapy. Some of the most successful developers of this strategy were Gertrude Elion and George Hitchings who received the 1988 Nobel Prize in Physiology or Medicine (shared with Sir James Black).

1.4.1 Elimination of Adenosine

The breakdown of adenosine was followed by the appearance of ribose. The "enzyme" responsible was called a nucleosidase. Now we know that there is no enzyme doing this and hence, what was studied was a mixture of enzyme activities. The first true adenosine degrading enzyme discovered was *adenosine deaminase* by Schmidt in 1928 (Schmidt 1928). He had earlier discovered AMP deaminase, and could show that there were two separate enzyme activities. A few years later the enzyme splitting inosine into hypoxanthine and ribose was discovered by Klein (Klein 1935), and the "nucleosidase" activity was apparently accounted for. However, the enzyme, now known as purine nucleoside phophohydrolase, actually generates ribose-1-phosphate from inosine and phosphate and in order to generate ribose further, chemical or enzymatic hydrolysis is required. There was much interest in adenosine deaminase when it was realized that a deficiency in this enzyme was associated with severe immune deficiency as well as other disorders (Tritsch et al. 1985). As noted elsewhere in this article the use of the enzyme itself to remove adenosine, or of inhibitors thereof to raise the levels, were increasingly used to assess the biological role of the nucleoside. However, it was becoming clear that adenosine deaminase was not the major route of adenosine removal, because the enzyme had a rather high Km (about 40–60 μM) for adenosine and actually appeared predominantly concerned with removing 2-deoxy-adenosine (Fredholm and Lerner 1982; Smith and Henderson 1982).

Hypoxanthine is further broken down by *xanthine oxidoreductase*. This enzyme converts purines to what is the end point in man, namely, uric acid. The enzyme normally acts as a NAD-dependent *dehydrogenase*, but can be converted to an *oxidase* in various ways, often associated with tissue damage or oxidative stress. For a long time there was interest in the possible pathophysiological role of the enzyme, but this is less obvious today.

Another pathway that plays some, but often minor, role in regulating adenosine levels is *S-adenosyl homocysteine hydrolase* (see Fig 1.1). Accumulation of adenosine will block the hydrolysis of S-adenosyl homocysteine (SAH), because the enzyme can drive the reaction in either direction (Hershfield and Krodich 1978; Kredich and Martin 1977). By blocking SAH removal the transmethylation reactions from SAM will also be blocked and this can lead to toxicity. Interestingly increasing the concentration of L-homocysteine will also drive the reaction towards SAH. This will then lead to a lowering of intracellular adenosine (Schrader et al.

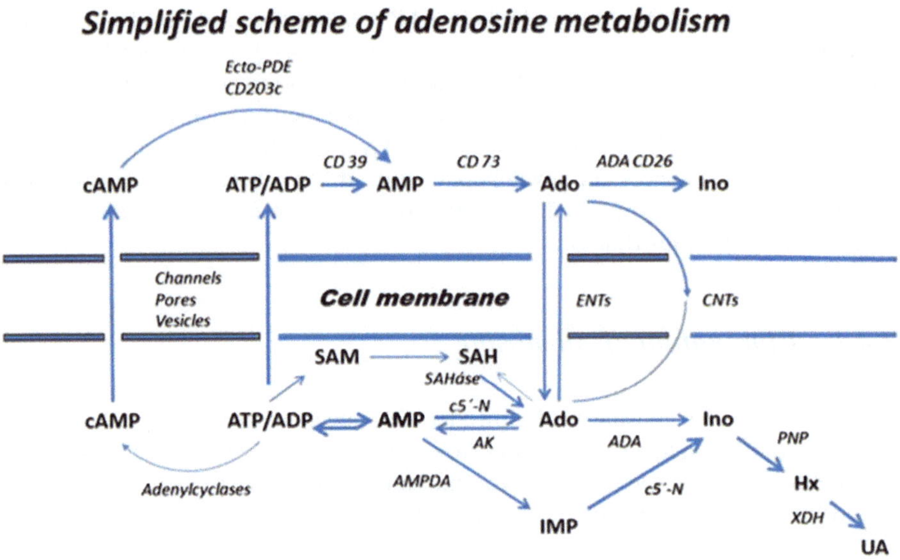

Fig. 1.1 Important pathways of ATP degradation in man. *AC* adenylate cyclase, *ADA* adenosine deaminase, *AK* adenosine kinase, *AMPD* AMP deaminase, *eN* ectophosphatase(s), *CD39* ectonucleotide phosphohydrolases, CD203c, ecto-cyclic nucleotide phosphohydrolase, *c5′-N and CD73* cytosolic and ecto-5′-nucleotidases, respectively, *ENT* equilibrative nucleoside transporter(s), *MTs* S-adenosylmethionine-dependent methyl transferases, *NPPCs* nucleotide-permeable pores and channels, *PNP* purine nucleoside phosphohydrolase, *SAH* S-adenosyl-homocysteine hydrolase, *SAM* S-adenosylmethionine, *XDH* xanthine dehydrogenase. Redrawn and slightly modified from R.A. Olsson (unpublished)

1981). This method of "trapping" intracellular adenosine in the form of SAH could also be used to show that much of the adenosine generated by hypoxia is probably intracellular both in heart (Schrader et al. 1981) and brain (Lloyd et al. 1993). Although the SAH pathway may contribute significantly to the basal adenosine production in some tissues and cells it did not appear to be critically involved in adenosine increases in conditions of stress or in removal. Another important point is that a part of total intracellular adenosine is bound to SAHhydrolase, which means that measurement of total tissue adenosine may not give an adequate estimate of the actual free intra- or extracellular adenosine level. The SAHase pathway was used to estimate the free intracellular adenosine concentration (Deussen et al. 1988a, b). Interestingly that agrees well with the concentration of extracellular adenosine estimated by microdialysis (Ballarin et al. 1991).

When neither ADA nor SAH hydrolase contributed much to removal during physiological conditions, it instead became clear that the physiologically more important mechanism of adenosine removal was *uptake into cells* (Arch and Newsholme 1978; Wohlhueter and Plagemann 1980). The transport was shown to require some form of transporter (Cass and Paterson 1975; Oliver and Paterson 1971). The driving force for the transport was, in most studied cells, the intracellular phosphorylation of adenosine by *adenosine kinase (AK)* that is able to convert adenosine to AMP using ATP as a source of phosphate (Caputto 1951). This enzyme has a very high affinity for adenosine (often estimated to be in the high nanomolar range). AK is critically important in maintaining the physiological levels of adenosine low, and also in maintaining depots of adenine nucleotides (Boison 2006). The critical role of adenosine kinase in maintaining intracellular concentration of adenosine (and thereby the extracellular level) was dramatically shown by genetic and pathophysiological manipulation of adenosine kinase amounts and the resulting effects on neuronal firing (Li et al. 2007; Masino et al. 2011).The transport machinery itself was symmetrical, implying an equilibrative transporter (Wohlhueter and Plagemann 1982). The transport could be blocked by some structural analogues (Cass and Paterson 1975) as well as other drugs (Paterson et al. 1981) including dipyridamole and dilazep that had uses as platelet inhibitors and/or vasodilators. Now we know that purine transport is mediated by several equilibrative as well as concentrating (Na-dependent) transporters with an uneven tissue distribution (Parkinson et al. 2011).

1.4.2 Formation of Adenosine

Although, as noted above, degradation of SAH to yield ADO + homocysteine could yield adenosine the major pathway was hydrolysis of AMP. The work from Berne's and Gerlach's groups noted above confirmed the earlier result that AMP could be broken down either by deamination to IMP, a pathway favored in skeletal muscle, or by dephosphorylation to adenosine, which predominates in heart (Conway and Cooke 1939; Gerlach and Deuticke 1966; Imai et al. 1964). The enzyme activity responsible for adenosine formation from AMP was 5′-nucleotidase, known for a long time (Gulland and Jackson 1938). The distribution of the enzyme activity did not agree very well with areas of known adenosine formation (Ferrans et al. 1969; Gordon et al. 1967; Gordon and Torack 1967; McManus et al. 1952; Pearse and Reis 1952; Reis 1950). In particular, the bulk of the enzyme activity detected appeared to be an ectoenzyme, whereas the enzyme needed to explain many of the findings had to be intracellular. We now know that the ecto-5′-nucleotidase—also known as CD 73—is extremely important in a variety of cells, tissues, and conditions, because it generates adenosine from extracellular nucleotides (see Fig 1.1), but these events and findings are too recent to be covered in this chapter. It was realized that the hydrolysis of extracellular adenine nucleotides to adenosine close to the cell membrane allowed adenosine to be rapidly taken up into cells before it was metabolized by other extracellular enzymes, and this was even described as "vectorial transport" following actions of ecto-5′-N (Frick and Lowenstein 1978), a concept that now has low credibility. In 1967 a cytosolic form of the enzyme was discovered (Itoh et al. 1967), and such forms were soon observed in many tissues. Through work from several laboratories, in particular that of Newby, the critical role of this enzyme in cardiac adenosine formation during energy deficiency was established (Newby et al. 1985). Later it was realized that there was more than one form of c-5′-N (Burgess et al. 1985), with some forms preferentially acting on pyrimidines, and one of the two purine selecting inosine and guanosine, whereas the others preferred AMP or deoxyAMP.

As mentioned above scientists interested in adenosine for a long time disregarded evidence that adenosine might be formed from extracellular adenine nucleotides, or at least that this was a physiologically very meaningful process. It was always recognized that even minor trauma could damage the integrity of cell membranes and lead to the release of ATP, and that this given the high intracellular concentrations (3–5 mM), would significantly increase adenosine levels, because of the rapid breakdown of nucleotides via ectoenzymes. It was also recognized that this was a major source of error in many experiments in vivo and in vitro. Another well known factor of concern was that platelets readily release ADP, which is converted to adenosine. This fact plagued the possibility of adequately measuring blood or plasma adenosine levels for a long time and I personally believe that a perfectly adequate way around this problem still does not exist.

Attempts were made to examine the role of intracellular vs. extracellular adenosine formation by means of blockers of ect-5′-nucleotidase. However, the compounds used (α,β-methylene-ADP and GMP) were neither very potent nor specific. Therefore, there was no really convincing evidence until genetic methods became more readily available. The knockouts of CD39 and CD73 have yielded very interesting and important data (Atkinson et al. 2006; Beldi et al. 2008; Hart et al. 2008; Resta et al. 1998; Yegutkin 2008). The other important aspect is that there is now hard evidence for a number of different physiological mechanisms by which adenine nucleotides can be released. To mention a few examples: It can be released by exocytosis, both when there is complete fusion between vesicle and cell membrane, and when it is transient, so-called "kiss-and-run" release (MacDonald et al. 2006); it can be released through connexin hemichannels (Eltzschig et al. 2006), and it is released from cell membranes subjected to stretch (Okada et al. 2006).

1.5 Adenosine as a Retaliatory Metabolite

As already alluded to above there is compelling evidence from work on the heart that with increasing work load intracellular adenosine formation is related to indices such as oxygen consumption. There is also good evidence that increased energy consumption (or decreased energy supply) in brain tissue causes adenosine levels to rise. Using thin slices of brain tissue McIlwain and coworkers showed that electrical field stimulation led to a marked increase in cAMP levels (Kakiuchi et al. 1969), and that this was largely due to release of adenosine which in turn stimulated cAMP formation (Pull and McIlwain 1972, 1973). Somewhat similar results were obtained in synaptosomal preparations (Kuroda and McIlwain 1974). Then it was shown that release of neurotransmitters could be reduced by adenosine (Fredholm 1974; Fredholm and Dunwiddie 1988; Fredholm and Hedqvist 1980; Hedqvist and Fredholm 1976; Ribeiro and Dominguez 1978). Thus, adenosine also in the central nervous system has the ability to limit its own formation.

The studies on adenosine in ischemia, in cardiovascular control and in the nervous system thus argued that adenosine could be released in conditions of metabolic emergency and that the released adenosine, in different ways, tended to limit the emergency and its own formation. This general concept was very clearly formulated in an influential review by Newby where he coined the term that adenosine was a "retaliatory metabolite" (Newby et al. 1985).

1.5.1 Adenosine and Fat and Carbohydrate Metabolism

It was shown already by Vincent Dole that adenosine had insulin like action in that it decreased fatty acid mobilization from fat (Dole 1962). When Martin Rodbell had described his method to isolate fat cells in 1964 (Rodbell 1964), a number of scientists became interested. Since the fat cells were perhaps the first nonblood cells that could be reproducibly purified and cultured in acute primary culture it was interesting to use for studies on signal transduction (Butcher et al. 1965). It also became a favorite for scientists interested in insulin action and diabetes. At this time the recent discovery of cyclic AMP as a mediator of hormone action was in focus. A tool in the study of its role was theophylline, which was considered to be a phosphodiesterase inhibitor—pure and simple. Theophylline was a good stimulator of lipolysis and this was taken as another proof of the critical role of cyclic AMP (Hynie et al. 1966). After a few years it was shown by several researchers that adenosine (and adenosine analogues) could potently inhibit lipolysis in rat fat cells (Fain 1973a, b; Fain et al. 1972; Vapaatalo et al. 1972) as well as in vivo in dogs (Fredholm 1974). An important additional result from the latter study was that adenosine could also block the release of noradrenaline triggered by sympathetic nerve stimulation. Despite the fact that the adenosine antagonistic effect of methylxanthines had been realized in heart and brain it took several years before it was realized that the effect of methylxanthines might be related to adenosine antagonism. An important result was the demonstration that removal of adenosine by the enzyme adenosine deaminase could stimulate lipolysis and cAMP accumulation just like methylxanthines (Fain and Wieser 1975). It was also shown that adenosine was released from adipose tissue by lipolytic stimuli (Fredholm 1976).

The fat cell was also critically important in the experiments that led to the definition of two receptor subtypes denoted R_i and R_a (Londos et al. 1980). A subdivision that was later proven to be identical with that defined by work on cultured brain cells (van Calker et al. 1979) and leading to the definition of two receptors denoted A_1 and A_2. Indeed, the fat cell is very richly endowed with A_1 receptors and adipose tissue responds to already very low levels of adenosine. Furthermore, the degree of lowering of cAMP formation in fat cells to reduce lipolysis is very small (Fain 1973a, b; Fain et al. 1972; Hjemdahl and Fredholm 1976; Vapaatalo et al. 1972, 1975b). Therefore, there is a huge "receptor reserve" in adipose tissue which means that weak partial agonists can reduce at cell lipolysis without inducing many other potentially A_1 dependent effects (Dhalla et al. 2003). A nonredundant role of adenosine as a tonic lipolysis-reducing agent in fat that acts together with insulin has been confirmed in recent studies using A_1 receptor knockout animals (Johansson et al. 2007, 2008).

Less is known about a role for adenosine in regulation of carbohydrate metabolism. The interest in the topic of adenosine receptors as potential regulators of glucose metabolism has been boosted by studies on caffeine (Estler 1976; Estler et al. 1978) and in particular by the findings of several large epidemiological studies that caffeine reduces the risk for Type II diabetes (Lee et al. 2005). The effect is dose-dependent, and risk reduction is marked at higher doses (risk ratio between 0.6 and 0.7 with more than 4 cups/day). Glucose homeostasis is partly linked to lipid metabolism. There is evidence that activation of A_1 receptors could significantly enhance insulin sensitivity. Indeed, as mentioned above this is certainly true in adipose tissue. However, this may not be physiologically very important as the glucose uptake in adipose tissue (Johansson et al. 2008) and skeletal muscle (Johansson et al. 2007) is unaltered in A_1–/– mice vs. control animals, but A_{2B} receptors may be more important. Reducing ATP conversion to adenosine results in reduced hepatic insulin sensitivity (Enjyoji et al. 2008), but it is not known to what extent adenosine and its receptors are responsible.

It has, however, been known that adenosine can influence insulin secretion (Chapal et al. 1981), but for a long time the interest was focused on adenine nucleotides and cAMP metabolism. Only comparatively late were adenosine analogues tested and a role for inhibitory adenosine receptors established (Hillaire-Buys et al. 1987). We found that plasma insulin levels were significantly increased in A_1–/– mice compared to their wild-type controls after a glucose challenge (Johansson et al. 2007), but there was no difference in basal insulin levels. Using the perfused pancreas a glucose infusion (raising levels from 4 to 16 mM) caused the expected rapid increase in insulin release, followed by a rapid return to a low steady level of insulin secretion in pancreata from wild-type mice (Johansson et al. 2007). A_1 receptors regulate pulsatility of insulin release (Salehi et al. 2009), but it is not known if this is important in humans.

1.6 Adenosine and Central Temperature and Metabolic Control

Already Drury and Szent-Györgyi noted marked central effects of peripheral adenosine administration. The site of action was however unclear until Feldberg and Sherwood (Feldberg and Sherwood 1954) injected adenosine icv and noted a marked hypnogenic effect. Later this was confirmed and it was shown that adenosine analogues were powerful somnogens probably via actions on receptors (Radulovacki et al. 1984). There is a close relation between sleep and energy metabolism (Horne 2009; Saper 2006). One of the reasons why we sleep is probably a conservation of energy. Furthermore these processes show similarities in how they are regulated. As an example hypocretins or orexins (Hcrt/Orx), hypothalamic neuropeptides that are synthesized by hypothalamic neurons, are involved in the regulation of feeding, thermoregulation, as well as the sleep–wakefulness cycle (Nunez et al. 2009). Numerous other factors also regulate these processes in an apparently concerted manner, and one of these is adenosine. It has been shown that adenosine increases during sleep deprivation and that induces sleep by blocking a number of wakefulness-inducing neuronal pathways (Porkka-Heiskanen et al. 1997).

A very important factor in regulating energy consumption and in the sustained need for energy is the maintenance of body temperature (Kayser and Malan 1963). Heat is generated as a by-product of ATP production and ATP utilization. When temperature falls thermogenic processes are turned on: increased muscle activity that may turn into shivering and heat production in brown adipose tissue (as well as in specialized white adipocytes) that depends on uncoupling of oxidative phosphorylation. The process is regulated by heat and cold sensitive neurons both peripherally and centrally. By contrast when the supply of energy is very low (or when the prospects for future energy supply are bleak) animals may go into hibernation involving deep sleep and reduced body temperature. This may develop further into torpor. Torpor involves a remarkable drop in whole body metabolism down to 1–2 % of basal values in awake small animals and shows some deep similarities to sleep (Heller and Ruby 2004). The body temperature may also fall in response to hypoxia. Given that a reduction in body temperature increases the area surviving an infarct this could be an important endogenous mechanism to increase survival after a cerebral insult (Drew et al. 2007). Adenosine is a possible mediator of this response.

It has long been known that systemic and intracerebroventricular adenosine and derivatives produces a marked fall in body temperature and that methylxanthines can increase it (Thithapandha et al. 1972; Vapaatalo et al. 1975a). Pharmacological studies have mainly implicated A1 receptors, and this has been supported by the finding that the temperature drop can be substantially reduced in mice lacking A_1 receptors (Yang et al. 2007). Thus, the biological roles of adenosine in regulation of metabolism may be very wide, and it seems likely that a careful examination of adenosine physiology from a phylogenetic perspective could prove rewarding, but this clearly is not history but future, and this may provide a suitable point at which to end this brief overview.

Acknowledgments In writing this I have repeatedly consulted an unpublished review by R.A. Olsson, *A History of Adenosine Research*. I am most grateful to him for letting me read it. I am also grateful to numerous coworkers.

References

Arch JR, Newsholme EA (1978) Activities and some properties of 5′-nucleotidase, adenosine kinase and adenosine deaminase in tissues from vertebrates and invertebrates in relation to the control of the concentration and the physiological role of adenosine. Biochem J 174:965–977

Atkinson B, Dwyer K, Enjyoji K, Robson SC (2006) Ecto-nucleotidases of the CD39/NTPDase family modulate platelet activation and thrombus formation: potential as therapeutic targets. Blood Cells Mol Dis 36:217–222

Ballarin M, Fredholm BB, Ambrosio S, Mahy N (1991) Extracellular levels of adenosine and its metabolites in the striatum of awake rats: inhibition of uptake and metabolism. Acta Physiol Scand 142:97–103
Beldi G, Enjyoji K, Wu Y, Miller L, Banz Y, Sun X, Robson SC (2008) The role of purinergic signaling in the liver and in transplantation: effects of extracellular nucleotides on hepatic graft vascular injury, rejection and metabolism. Front Biosci 13:2588–2603
Bennet DW, Drury AN (1931) Further observations relating to the physiological activity of adenine compounds. J Physiol 72:288–320
Berne RM (1963) Cardiac nucleotides in hypoxia: possible role in regulation of coronary blood flow. Am J Physiol 204:317–322
Boison D (2006) Adenosine kinase, epilepsy and stroke: mechanisms and therapies. Trends Pharmacol Sci 27:652–658
Burgess FW, el Kouni MH, Parks RE Jr (1985) Nucleotidase activities of human peripheral lymphocytes. Biochem Pharmacol 34:3061–3070
Butcher RW, Ho RJ, Meng HC, Sutherland EW (1965) Adenosine 3′,5′-monophosphate in biological materials. II. The measurement of adenosine 3′,5′-monophosphate in tissues and the role of the cyclic nucleotide in the lipolytic response of fat to epinephrine. J Biol Chem 240:4515–4523
Caputto R (1951) The enzymatic synthesis of adenylic acid; adenosinekinase. J Biol Chem 189:801–814
Cass CE, Paterson AR (1975) Inhibition by nitrobenzylthioinosine of uptake of adenosine, 2′-deoxyadenosine and 9-beta-D-arabinofuranosyladenine by human and mouse erythrocytes. Biochem Pharmacol 24:1989–1993
Chapal J, Loubatieres-Mariani MM, Roye M (1981) Effect of adenosine and phosphated derivatives on insulin release from the newborn dog pancreas. J Physiol (Paris) 77:873–875
Conway EJ, Cooke R (1939) The deaminases of adenosine and adenylic acid in blood and tissues. Biochem J 33:479–492
Curnish RR, Berne RM, Rubio R (1972) Effect of aminophylline on myocardial reactive hyperemia. Proc Soc Exp Biol Med 141:593–598
Deussen A, Borst M, Kroll K, Schrader J (1988a) Formation of S-adenosylhomocysteine in the heart. II: a sensitive index for regional myocardial underperfusion. Circ Res 63:250–261
Deussen A, Borst M, Schrader J (1988b) Formation of S-adenosylhomocysteine in the heart. I: an index of free intracellular adenosine. Circ Res 63:240–249
Dhalla AK, Shryock JC, Shreeniwas R, Belardinelli L (2003) Pharmacology and therapeutic applications of A1 adenosine receptor ligands. Curr Top Med Chem 3:369–385
Dole VP (1962) Insulin-like actions of ribonucleic acid, adenylic acid, and adenosine. J Biol Chem 237:2758–2762
Drew KL, Buck CL, Barnes BM, Christian SL, Rasley BT, Harris MB (2007) Central nervous system regulation of mammalian hibernation: implications for metabolic suppression and ischemia tolerance. J Neurochem 102:1713–1726
Drury AN (1932) Nucleic acid derivatives and the heart beat. J Physiol 74:147–155
Drury AN, Szent-Gyorgyi A (1929) The physiological activity of adenine compounds with especial reference to their action upon the mammalian heart. J Physiol 68:213–237
Duff F, Patterson GC, Shepherd JT (1954) A quantitative study of the response to adenosine triphosphate of the blood vessels of the human hand and forearm. J Physiol 125:581–589
Duncker DJ, Merkus D (2007) Exercise hyperaemia in the heart: the search for the dilator mechanism. J Physiol 583:847–854
Eckle T, Krahn T, Grenz A, Kohler D, Mittelbronn M, Ledent C, Jacobson MA, Osswald H, Thompson LF, Unertl K et al (2007) Cardioprotection by ecto-5′-nucleotidase (CD73) and A2B adenosine receptors. Circulation 115:1581–1590
Eltzschig HK, Eckle T, Mager A, Kuper N, Karcher C, Weissmuller T, Boengler K, Schulz R, Robson SC, Colgan SP (2006) ATP release from activated neutrophils occurs via connexin 43 and modulates adenosine-dependent endothelial cell function. Circ Res 99:1100–1108
Embden G, Zimmermann M (1927) Z Physiol Chem 167:137

Enjyoji K, Kotani K, Thukral C, Blumel B, Sun X, Wu Y, Imai M, Friedman D, Csizmadia E, Bleibel W et al (2008) Deletion of cd39/entpd1 results in hepatic insulin resistance. Diabetes 57:2311–2320

Estler CJ (1976) Stoffwechsel Einzelner Organe. In: Eichler O (ed) Kaffee und Coffein. Springer-Verlag, Berlin, pp 183–213

Estler CJ, Ammon HP, Herzog C (1978) Swimming capacity of mice after prolonged treatment with psychostimulants. I. Effects of caffeine on swimming performance and cold stress. Psychopharmacology (Berl) 58:161–166

Fain JN (1973a) Biochemical aspects of drug and hormone action on adipose tissue. Pharmacol Rev 25:67–118

Fain JN (1973b) Inhibition of adenosine cyclic 3′, 5′-monophosphate accumulation in fat cells by adenosine, N6-(phenylisopropyl) adenosine, and related compounds. Mol Pharmacol 9: 595–604

Fain JN, Pointer RH, Ward WF (1972) Effects of adenosine nucleosides on adenylate cyclase, phosphodiesterase, cyclic adenosine monophosphate accumulation, and lipolysis in fat cells. J Biol Chem 247:6866–6872

Fain JN, Wieser PB (1975) Effects of adenosine deaminase on cyclic adenosine monophosphate accumulation, lipolysis, and glucose metabolism of fat cells. J Biol Chem 250:1027–1034

Feigl EO (2004) Berne's adenosine hypothesis of coronary blood flow control. Am J Physiol Heart Circ Physiol 287:H1891–H1894

Feldberg LA, Sherwood PD (1954) Injection of drugs into the lateral ventricle of the cat. J Physiol Lond 123:148–167

Ferrans VJ, Hibbs RG, Buja LM (1969) Nucleoside phosphatase activity in atrial and ventricular myocardium of the rat: a light and electron microscopic study. Am J Anat 125:47–85

Fredholm BB (1974) Vascular and metabolic effects of theophylline, dibuturyl cyclic AMP and dibuturyl cyclic GMP in canine subcutaneous adipose tissue in situ. Acta Physiol Scand 90: 226–236

Fredholm BB (1976) Release of adenosine-like material from isolated perfused dog adipose tissue following sympathetic nerve stimulation and its inhibition by adrenergic alpha-receptor blockade. Acta Physiol Scand 96:122–130

Fredholm BB, Dunwiddie TV (1988) How does adenosine inhibit transmitter release? Trends Pharmacol Sci 9:130–134

Fredholm BB, Hedqvist P (1978) Release of 3H-purines from [3H]-adenine labelled rabbit kidney following sympathetic nerve stimulation, and its inhibition by alpha-adrenoceptor blockage. Br J Pharmacol 64:239–245

Fredholm BB, Hedqvist P (1980) Modulation of neurotransmission by purine nucleotides and nucleosides. Biochem Pharmacol 29:1635–1643

Fredholm BB, Lerner U (1982) Metabolism of adenosine and 2′-deoxy-adenosine by fetal mouse calvaria in culture. Med Biol 60:267–271

Frick GP, Lowenstein JM (1978) Vectorial production of adenosine by 5′-nucleotidase in the perfused rat heart. J Biol Chem 253:1240–1244

Frohlich ED (1963) Local effect of adenosine mono-, di-, and triphosphate vessel resistance. Am J Physiol 204:28–30

Gercken G (1961) Glucose, lactic acid, ATP and ADP concentrations in the arterial and venous coronary blood of dogs in oxygen deficiency. Pflugers Arch Gesamte Physiol Menschen Tiere 272:323–335

Gerlach E, Deuticke B (1966) Comparative studies on the formation of adenosine in the myocardium of different animal species in oxygen deficiency. Klin Wochenschr 44:1307–1310

Gerlach E, Deuticke B, Dreisbach RH, Rosarius CW (1963) On the behavior of nucleotides and their dephosphorylation degradation products in the kidney in ischemia and short-term postischemic re-establishment of blood circulation. Pflugers Arch Gesamte Physiol Menschen Tiere 278:296–315

Gordon GB, Price HM, Blumberg JM (1967) Electron microscopic localization of phosphatase activities within striated muscle fibers. Lab Invest 16:422–435

Gordon JS, Torack RM (1967) Inhibition of cerebral adenosinetriphosphatase activity by various aldehyde fixatives. J Neurochem 14:1155–1160
Gulland JM, Jackson EM (1938) 5-Nucleotidase. Biochem J 32:597–601
Hart ML, Henn M, Kohler D, Kloor D, Mittelbronn M, Gorzolla IC, Stahl GL, Eltzschig HK (2008) Role of extracellular nucleotide phosphohydrolysis in intestinal ischemia-reperfusion injury. FASEB J 22:2784–2797
Harvey RB (1963) Characteristics of blood flow in branches of the renal artery. Am J Physiol 205:977–981
Harvey RB (1964) Effects of adenosinetriphosphate on autoregulation of renal blood flow and glomerular filtration rate. Circ Res 15(Suppl):178–182
Hedqvist P, Fredholm BB (1976) Effects of adenosine on adrenergic neurotransmission; prejunctional inhibition and postjunctional enhancement. Naunyn Schmiedebergs Arch Pharmacol 293:217–223
Hedqvist P, Fredholm BB, Olundh S (1978) Antagonistic effects of theophylline and adenosine on adrenergic neuroeffector transmission in the rabbit kidney. Circ Res 43:592–598
Heller HC, Ruby NF (2004) Sleep and circadian rhythms in mammalian torpor. Annu Rev Physiol 66:275–289
Hershfield MS, Krodich NM (1978) S-adenosylhomocysteine hydrolase is an adenosine-binding protein: a target for adenosine toxicity. Science 202:757–760
Hillaire-Buys D, Bertrand G, Gross R, Loubatieres-Mariani MM (1987) Evidence for an inhibitory A1 subtype adenosine receptor on pancreatic insulin-secreting cells. Eur J Pharmacol 136:109–112
Hjemdahl P, Fredholm BB (1976) Cyclic AMP-dependent and independent inhibition of lipolysis by adenosine and decreased pH. Acta Physiol Scand 96:170–179
Horne J (2009) REM sleep, energy balance and 'optimal foraging'. Neurosci Biobehav Rev 33: 466–474
Houck CR, Bing RJ et al (1947) Renal hyperemia in the dog after intravenous infusion of adenosine, adenylic acid, or adenosinetriphosphate. Fed Proc 6:132
Hynie S, Krishna G, Brodie BB (1966) Theophylline as a tool in studies of the role of cyclic adenosine 3',5'-monophosphate in hormone-induced lipolysis. J Pharmacol Exp Ther 153:90–96
Imai S, Riley AL, Berne RM (1964) Effect of Ischemia on Adenine Nucleotides in Cardiac and Skeletal Muscle. Circ Res 15:443–450
Itoh R, Mitsui A, Tsushima K (1967) 5′-nucleotidase of chicken liver. Biochim Biophys Acta 146: 151–159
Johansson SM, Lindgren E, Yang JN, Herling AW, Fredholm BB (2008) Adenosine A1 receptors regulate lipolysis and lipogenesis in mouse adipose tissue-interactions with insulin. Eur J Pharmacol 597:92–101
Johansson SM, Salehi A, Sandstrom ME, Westerblad H, Lundquist I, Carlsson PO, Fredholm BB, Katz A (2007) A1 receptor deficiency causes increased insulin and glucagon secretion in mice. Biochem Pharmacol 74:1628–1635
Kakiuchi S, Rall TW, McIlwain H (1969) The effect of electrical stimulation upon the accumulation of adenosine 3′,5′-phosphate in isolated cerebral tissue. J Neurochem 16:485–491
Kalckar HM, Lowry OH (1947) The relationship between traumatic shock and the release of adenylic acid compounds. Am J Physiol 149:240–245
Katori M, Berne RM (1966) Release of adenosine from anoxic hearts. Relationship to coronary flow. Circ Res 19:420–425
Kayser C, Malan A (1963) Central nervous system and hibernation. Experientia 19:441–451
Klein W (1935) Experimentelle Studien über den Nucleinstoffwechsel. XXXVII. Über Nucleosidase. Hoppe-Seyler's Z Physiol Chem 231:125–148
Kovach AG, Bagdy D, Balazs R, Antoni F, Gergely J, Menyhart J, Iranyi M, Kovach E (1952) Traumatic shock and adenosine triphosphate. Acta Physiol Hung 3:330–344
Kredich NM, Martin DV Jr (1977) Role of S-adenosylhomocysteine in adenosinemediated toxicity in cultured mouse T lymphoma cells. Cell 12:931–938
Kuroda Y, McIlwain H (1974) Uptake and relase of (14 C)adenine derivatives at beds of mammalian cortical synaptosomes in a superfusion system. J Neurochem 22:691–699

Lankford AR, Yang JN, Rose'Meyer R, French BA, Matherne GP, Fredholm BB, Yang Z (2006) Effect of modulating cardiac A1 adenosine receptor expression on protection with ischemic preconditioning. Am J Physiol Heart Circ Physiol 290:H1469–H1473

Lee S, Hudson R, Kilpatrick K, Graham TE, Ross R (2005) Caffeine ingestion is associated with reductions in glucose uptake independent of obesity and type 2 diabetes before and after exercise training. Diabetes Care 28:566–572

Li T, Lan JQ, Fredholm BB, Simon RP, Boison D (2007) Adenosine dysfunction in astrogliosis: cause for seizure generation? Neuron Glia Biol 3:353–366

Lindner F, Rigler R (1931) Über die Beeinflussung der Weite der Herzkranzgefässe durch Produkte des Zellkernstoffwechsels. Pflugers Arch 226:697–708

Lloyd HG, Lindstrom K, Fredholm BB (1993) Intracellular formation and release of adenosine from rat hippocampal slices evoked by electrical stimulation or energy depletion. Neurochem Int 23:173–185

Londos C, Cooper DM, Wolff J (1980) Subclasses of external adenosine receptors. Proc Natl Acad Sci U S A 77:2551–2554

MacDonald PE, Braun M, Galvanovskis J, Rorsman P (2006) Release of small transmitters through kiss-and-run fusion pores in rat pancreatic beta cells. Cell Metab 4:283–290

Masino SA, Li T, Theofilas P, Sandau US, Ruskin DN, Fredholm BB, Geiger JD, Aronica E, Boison D (2011) A ketogenic diet suppresses seizures in mice through adenosine A1 receptors. J Clin Invest 121:2679–2683

Massion WH, Talaat SM, Schilling JA (1965) Protective effects of adenosine triphosphate administration in irreversible hemorrhagic shock. Anesth Analg 44:104–109

McManus JF, Lupton CH, Harden G (1952) Histochemical studies of 5 nucleotidase. I. Method and specificity. Lab Invest 1:480–487

Mustafa SJ, Berne RM, Rubio R (1975a) Adenosine metabolism in cultured chick-embryo heart cells. Am J Physiol 228:1474–1478

Mustafa SJ, Rubio R, Berne RM (1975b) Uptake of adenosine by dispersed chich embryonic cardiac cells. Am J Physiol 228:62–67

Newby AC, Worku Y, Holmquist CA (1985) Adenosine formation. Evidence for a direct biochemical link with energy metabolism. Adv Myocardiol 6:273–284

Ninomiya H, Buxton RW, Michaelis M (1961a) A note on adenosine and adenylate deaminases in brain and quadriceps muscle of the rat in Noble-Collip drum shock. Bull Sch Med Univ Md 46: 61–64

Ninomiya H, Buxton RW, Michaelis M (1961b) Observations on rat brain dehydrogenases in tumbling shock. Enzymologia 23:78–84

Nunez A, Rodrigo-Angulo ML, Andres ID, Garzon M (2009) Hypocretin/Orexin neuropeptides: participation in the control of sleep-wakefulness cycle and energy homeostasis. Curr Neuropharmacol 7:50–59

Okada SF, Nicholas RA, Kreda SM, Lazarowski ER, Boucher RC (2006) Physiological regulation of ATP release at the apical surface of human airway epithelia. J Biol Chem 281:22992–23002

Oliver JM, Paterson AR (1971) Nucleoside transport. I. A mediated process in human erythrocytes. Can J Biochem 49:262–270

Osswald H (1975) Renal effects of adenosine and their inhibition by theophylline in dogs. Naunyn Schmiedebergs Arch Pharmacol 288:79–86

Osswald H, Schmitz HJ, Kemper R (1977) Tissue content of adenosine, inosine and hypoxanthine in the rat kidney after ischemia and postischemic recirculation. Pflugers Arch 371:45–49

Parkinson FE, Damaraju VL, Graham K, Yao SY, Baldwin SA, Cass CE, Young JD (2011) Molecular biology of nucleoside transporters and their distributions and functions in the brain. Curr Top Med Chem 11:948–972

Paterson AR, Kolassa N, Cass CE (1981) Transport of nucleoside drugs in animal cells. Pharmacol Ther 12:515–536

Pearse AG, Reis JL (1952) The histochemical demonstration of a specific phosphatase (5-nucleotidase). Biochem J 50:534–536

Piorkowski G (1950) Adenylphosphoric acid in the treatment of disorders of the coronary circulation. Ther Ggw 89:314–315

Porkka-Heiskanen T, Strecker RE, Thakkar M, Bjørkum AA, Greene RW, McCarley RW (1997) Adenosine: A mediator of the sleep-inducing effects of prolonged wakefulness. Science 276: 1265–1268
Pull I, McIlwain H (1972) Output of (14 C)adenine derivatives on electrical excitation of tissues from the brain: calcium ion-sensitivity and an accompanying re-uptake process. Biochem J 127:91P
Pull I, McIlwain H (1973) Output of (14 C)adenine nucleotides and their derivatives from cerebral tissues. Tetrodotoxin-resistant and calcium ion-requiring components. Biochem J 136:893–901
Radulovacki M, Virus RM, Djuricic-Nedelson M, Green RD (1984) Adenosine analogs and sleep in rats. J Pharmacol Exp Ther 228:268–274
Reis JL (1950) Studies on 5-nucleotidase and its distribution in human tissues. Biochem J 2:21–22
Resta R, Yamashita Y, Thompson LF (1998) Ecto-enzyme and signaling functions of lymphocyte CD73. Immunol Rev 161:95–109
Ribeiro JA, Dominguez ML (1978) Mechanisms of depression of neuromuscular transmission by ATP and adenosine. J Physiol (Paris) 74:491–496
Rodbell M (1964) Metabolism of isolated fat cells. I. Effects of hormones on glucose metabolism and lipolysis. J Biol Chem 239:375–380
Rothmann H (1930) Der Einflub der Adenosinphosphosaure auf die Herztatigkeit. (The effect of adenosine phosphoric acid on cardiac activity.). Naunyn Schmiedeberg's Arch E xp Path Pharmakol 155:129–138
Rowe GG, Afonso S, Gurtner HP, Chelius CJ, Lowe WC, Castillo CA, Crumpton CW (1962) The systemic and coronary hemodynamic effects of adenosine triphosphate and adenosine. Am Heart J 64:228–234
Rubio R, Berne RM (1969) Release of adenosine by the normal myocardium in dogs and its relationship to the regulation of coronary resistance. Circ Res 25:407–415
Rubio R, Berne RM, Bockman EL, Curnish RR (1975) Relationship between adenosine concentration and oxygen supply in rat brain. Am J Physiol 228:1896–1902
Rubio R, Berne RM, Katori M (1969) Release of adenosine in reactive hyperemia of the dog heart. Am J Physiol 216:56–62
Salehi A, Parandeh F, Fredholm BB, Grapengiesser E, Hellman B (2009) Absence of adenosine A1 receptors unmasks pulses of insulin release and prolongs those of glucagon and somatostatin. Life Sci 85:470–476
Saper CB (2006) Staying awake for dinner: hypothalamic integration of sleep, feeding, and circadian rhythms. Prog Brain Res 153:243–252
Schmidt G (1928) Über fermentative desaminierung im Muskel. (On enzymatic deamination in muscle.). Hoppe-Seyler's Z Physiol Chem 179:243–282
Schnermann J, Osswald H, Hermle M (1977) Inhibitory effect of methylxanthines on feedback control of glomerular filtration rate in the rat kidney. Pflugers Arch 369:39–48
Schrader J, Schutz W, Bardenheuer H (1981) Role of S-adenosylhomocysteine hydrolase in adenosine metabolism in mammalian heart. Biochem J 196:65–70
Schutz W, Schrader J, Gerlach E (1981) Different sites of adenosine formation in the heart. Am J Physiol 240:H963–H970
Scott NS, Falconer IR (1965) Enzymic assay of adenosine 3',5'-monophosphate. Anal Biochem 13:71–73
Sharma GP, Eiseman B (1966) Protective effect of ATP in experimental hemorrhagic shock. Surgery 59:66–75
Smith CM, Henderson JF (1982) Deoxyadenosine triphosphate accumulation in erythrocytes of deoxycoformycin-treated mice. Biochem Pharmacol 31:1545–1551
Stoner HB, Green HN (1945a) The action of magnesium and calcium on the enzymic breakdown of certain adenine compounds. Biochem J 39:474–478
Stoner HB, Green HN (1945b) Experimental limb ischaemia in man with especial reference to the role of adenosine triphosphate. Clin Sci 5:159–175
Thithapandha A, Maling HM, Gillette JR (1972) Effects of caffeine and theophylline on activity of rats in relation to brain xanthine concentrations. Proc Soc Exp Biol Med 139:582–586

Tritsch GL, Paolini NS, Bielat K (1985) Adenosine deaminase activity associated with phagocytic vacuoles. Cytochemical demonstration by electron microscopy. Histochemistry 82:281–285

Vallon V, Muhlbauer B, Osswald H (2006) Adenosine and kidney function. Physiol Rev 86: 901–940

van Calker D, Muller M, Hamprecht B (1979) Adenosine regulates via two different types of receptors, the accumulation of cyclic AMP in cultured brain cells. J Neurochem 33:999–1005

Vapaatalo H, Bieck P, Westermann E (1972) Actions of various cyclic nucleotides, nucleosides and purine bases on the synthesis of corticosterone in vitro. Naunyn Schmiedebergs Arch Pharmacol 275:435–443

Vapaatalo H, Neuvonen PJ, Bieck P, Westermann E (1975a) Actions of various methylxanthines and papaverine on the synthesis of corticosterone in vitro. Arzneimittelforschung 25:1265–1269

Vapaatalo H, Onken D, Neuvonen PJ, Westermann E (1975b) Stereospecificity in some central and circulatory effects of phenylisopropyl-adenosine (PIA). Arzneimittelforschung 25:407–410

Wohlhueter RM, Plagemann PG (1982) On the functional symmetry of nucleoside transport in mammalian cells. Biochim Biophys Acta 689:249–260

Wohlhueter RM, Plagemann PGW (1980) The role of transport and phosphorylation in the uptake of nutrients in cells in culture. Int Rev Cytol 64:171–240

Yang SC, Chiu TH, Yang HW, Min MY (2007) Presynaptic adenosine A1 receptors modulate excitatory synaptic transmission in the posterior piriform cortex in rats. Brain Res 1156: 67–79

Yegutkin GG (2008) Nucleotide- and nucleoside-converting ectoenzymes: Important modulators of purinergic signalling cascade. Biochim Biophys Acta 1783:673–694

Zydowo M (1959) Deaminases of adenylic acid and adenosine in rat kidney. Nature 184(Suppl 21):1641–1642

Part II
Molecular Biology, Pharmacology, and Cell Biology of Adenosine

Chapter 2
Adenosine Metabolism, Adenosine Kinase, and Evolution

Jaoek Park and Radhey S. Gupta

Abstract The enzyme adenosine kinase (AdK, ADK, or AK) plays an important role in regulating the intracellular as well as extracellular concentrations of adenosine and hence its diverse physiological actions. In view of the enormous pharmacological potential of adenosine, there has been much interest in studying adenosine kinase over the past few decades. This chapter summarizes the wealth of information that has accumulated concerning its structure and function. The aspects that are reviewed include the enzymological aspects of ADK including its reaction mechanism and ionic requirement; insights provided by the crystal structure of the enzyme; a brief overview of work on identification and development of ADK inhibitors; novel aspects of the ADK gene structure; tissue distribution and subcellular localization of the two ADK isoforms; novel information provided by mammalian cells harboring mutations of ADK; and lastly the evolutionary relationship of ADK to other related proteins. Despite enormous progress several important gaps exist in our knowledge regarding ADK, particularly concerning the cellular functions of the two isoforms and how their relative amounts in different tissues are regulated, that need to be understood in order to fully realize the therapeutic potential of increased local concentration of adenosine by modulation of this key enzyme.

Keywords Adenosine kinase • Long and short isoforms • Protein structure • AK gene structure • Chinese hamster cell mutants • Differential localization of AdK isoforms • Pentavalent ion dependency

---

J. Park
Department of Biochemistry, McGill University, Montreal, QC, Canada

R.S. Gupta (✉)
Department of Biochemistry and Biomedical Sciences,
McMaster University, Hamilton, ON, Canada
e-mail: gupta@mcmaster.ca

S. Masino and D. Boison (eds.), *Adenosine: A Key Link between Metabolism and Brain Activity*, DOI 10.1007/978-1-4614-3903-5_2,

2.1 Adenosine Kinase: Function and Role in Adenosine Metabolism

2.1.1 Adenosine

The purine nucleoside adenosine with neuromodulating properties produces a broad range of physiological responses in mammalian tissues via interaction with adenosine receptors. Much interest in adenosine metabolism has arisen from the discovery that adenosine receptor-mediated signals are involved in the protection of cells and tissues undergoing ischemic stress. Under normal oxygenated conditions, the rates of ATP catabolism and anabolism are equal, and this maintains stable cytosolic concentrations of both ATP and ADP (5–10 mM and 40–60 μM in the heart, respectively) (Gard et al. 1985; Decking et al. 1997). This high ratio of [ATP] and [ADP] is accompanied by a low level of AMP, whose increase in concentration indicates a threat to the ATP supply. Under conditions of tissue stress or trauma, such as ischemia or hypoxia, the oxygen supply is inadequate for oxidative phosphorylation of ADP to regenerate ATP, and the [ATP]–[ADP] ratio is compromised (Headrick and Willis 1990). The increasing concentration of ADP leads to adenylate kinase activity (i.e., 2 ADP↔ATP+AMP) in the direction of AMP formation to maintain a stable [ADP]–[AMP] ratio (Ballard 1970). There is a subsequent net increase in the concentration of AMP (Headrick and Willis 1990), which in turn is hydrolyzed to adenosine (Decking et al. 1997; Gustafson and Kroll 1998; Bak and Ingwall 1998). Due to the increased cytosolic concentration, adenosine is released into the extracellular space, where it can bind to adenosine receptors (see Fig. 2.1).

Adenosine receptor-mediated responses (receptor subtypes: A_1, A_{2A}, A_{2B}, and A_3) can have profound protective effects throughout the entire human body, including the nervous, cardiovascular, gastrointestinal, renal, muscular, and immune systems (Mubagwa and Flameng 2001; Stone et al. 2009; Wojcik et al. 2010; Zylka 2011). For this reason, adenosine has been variously called as "signal of life," "retaliatory metabolite," and "body's natural defense"(Engler 1991; Mullane and Bullough 1995; Cohen and Downey 2008; Boison 2008a). Naturally, there has been an enormous research interest in studying the therapeutic potential of adenosine. The list of prospective areas for its medical application is expanding rapidly; because these topics are covered in other chapters, only a brief mention of them is made here. Adenosine has been shown to provide protection against ischemia–reperfusion injuries in the heart (Mubagwa and Flameng 2001), brain and spinal cord (Phillis and Goshgarian 2001; Williams-Karnesky and Stenzel-Poore 2009), and kidney (Bauerle et al. 2011). It also exhibits potent pain-relieving effects in diverse preclinical models of chronic pain (Zylka 2011). Adenosine is a powerful anticonvulsant against epilepsy (Boison and Stewart 2009), and it participates in the pathology of several neuronal and neurodegenerative disorders such as schizophrenia, Alzheimer's disease, Huntington's disease, amyotrophic lateral sclerosis, and multiple sclerosis (Boison 2008a; Stone et al. 2009; Amadio et al. 2011; Boison et al. 2011). Recent work by

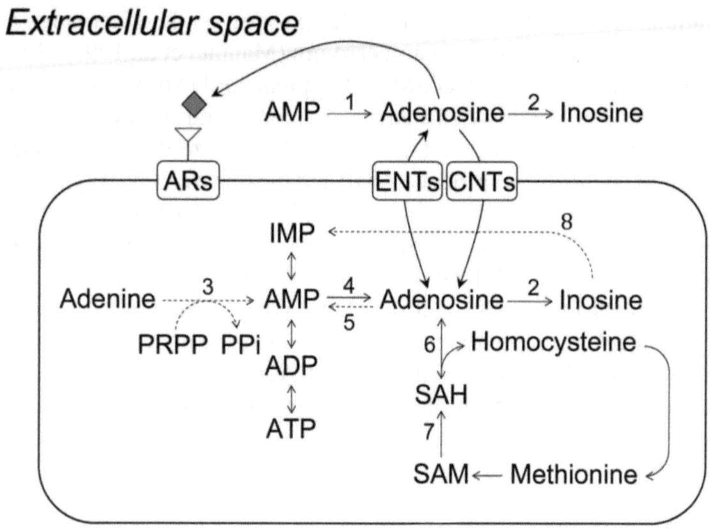

Fig. 2.1 Adenosine metabolism, transport, and receptor binding: a general overview. Enzymes: 1, ecto-5′-nucleotidase (5′-NT); 2, adenosine deaminase (ADA); 3, adenine phosphoribosyltransferase (APRTase); 4, 5′-nucleotidase (cN-I); 5, adenosine kinase (ADK); 6, *S*-adenosyl-L-homocysteine hydrolase; 7, various methyl-transferases and 8, inosine-guanosine kinase. *AR* adenosine receptor, *ENT* equilibrative nucleoside transporter, *CNT* concentrative nucleoside transporter, *PRPP* 5-phosphoribose-1-pyrophosphate, *PPi* inorganic pyrophosphate, *SAH* *S*-adenosyl-L-homocysteine, *SAM* *S*-adenosyl-L-methionine. The salvage pathway reactions are shown as *dotted arrows*. Modified from Park and Gupta 2008

Boison's group establishes key roles for adenosine and ADK in the development of epilepsy in a mouse model (Boison 2008b, 2010). Their work has led to "Adenosine Kinase hypothesis of Epileptogenesis," according to which the increased production of ADK by astrocytes in astrogliosis, leads to a lowering of adenosine concentration in critical brain regions, thereby resulting in the loss of inhibition of neuronal excitation leading to epileptic seizure (Boison 2008b, 2010). Because adenosine augmentation by inhibition of ADK, or by other means such as implantation of encapsulated cells engineered to release adenosine, provided protection from epileptic seizure, it establishes the usefulness of these modalities for development of novel therapies for treatment of epilepsy (Boison and Stewart 2009; Boison 2009a, b, 2011; Theofilas et al. 2011).

2.1.1.1 Adenosine Metabolizing Enzymes

As adenosine initiates signaling cascades that can produce such intense and broad downstream effects, its cellular concentration needs to be tightly regulated. The cytosolic concentration of adenosine under normoxic conditions is estimated to range between 0.01 and 0.1 μM (Kroll et al. 1992; Deussen et al. 1999). Interstitial concentrations of adenosine during hypoxic episodes, on the other hand, have been found as high as 30 μM within the myocardium (Manthei et al. 1998), 30 μM in the brain (Latini et al. 1999), and 100 μM in the spinal cord (McAdoo et al. 2000). The enzymes responsible for lowering adenosine concentration are adenosine kinase (ADK) and adenosine deaminase (ADA), which catalyze the phosphorylation and deamination of adenosine to produce AMP and inosine, respectively (Fig. 2.1). The K_m of ADA for adenosine is in the range of 25–150 μM (depending upon the isoform) (Ford et al. 2000; Singh and Sharma 2000), whereas that of ADK is approximately 1 μM (De Jong 1977; Drabikowska et al. 1985). Because the K_m of ADK for adenosine is much lower than that of ADA, ADK is thought to be the principal enzyme responsible for regulating the level of adenosine under physiological conditions. In proof of this concept, inhibition of ADK, but not of ADA, resulted in increased adenosine concentration and its subsequent release into the interstitial space in rat brain slices (Pak et al. 1994).

2.1.1.2 Adenosine Producing Enzymes

While ADK and ADA are the enzymes responsible for the removal of cellular adenosine, there are three enzymes whose immediate actions result in the formation of adenosine. The most significant source of adenosine during enhanced oxygen demand or metabolic load, quantitatively, is the AMP-selective cytosolic 5′-nucleotidase (cN-I, Fig. 2.1), which hydrolyzes AMP into adenosine and inorganic phosphate. Some studies have suggested that this enzyme is responsible for more than 70 % of adenosine production in mammalian heart cells (Darvish et al. 1996; Garvey and Prus 1999). Another enzyme responsible for the production of adenosine is ecto-5′-nucleotidase (5′-NT, Fig. 2.1), a glycoprotein found on the plasma membrane of eukaryotic cells (Misumi et al. 1990). Although this enzyme carries out the same reaction as cN-I, its sequence bears no similarity to that of the latter enzyme, representing a case of convergent evolution (Sala-Newby et al. 1999). Interestingly, 5′-NT is inhibited by both ADP and ATP (Sullivan and Alpers 1971; Naito and Lowenstein 1985), whereas cN-I is allosterically activated by ADP but not by ATP (Yamazaki et al. 1991; Darvish and Metting 1993). The end result is probably that the relative contribution of each enzyme to the total adenosine pool varies as a function of its tissue-specific expression, as well as the cellular energy status. In addition to these two enzymes, adenosine is also produced in heart by hydrolysis of *S*-adenosylhomocysteine (SAH) via SAH-hydrolase (Fig. 2.1). Although this later mechanism is responsible for the production of majority of basal adenosine in heart under normoxia, its role in other tissues such as brain seems to be quite limited (Pascual et al. 2005; Dulla et al. 2005; Frenguelli et al. 2007). The reaction catalyzed by SAH-hydrolase is reversible, and the direction of its activity is dependent on the local concentrations of adenosine and homocysteine (Loncar et al. 1997). The enzyme, however, generally favors the hydrolysis of SAH, as both adenosine and homocysteine are rapidly metabolized under normal conditions (Deussen et al. 1988).

2.1.1.3 Adenosine Transport

In addition to its role as a signaling molecule that communicates intracellular metabolic events to receptors on the cell surface, adenosine has another important cellular function: a building block for ATP, and ultimately nucleic acids, via the salvage pathway. Adenosine recycling is common and important, as de novo nucleotide synthesis is energetically expensive and highly tissue-specific. Both adenosine signaling and adenosine salvage require efficient transport of the molecule across the plasma membrane (Fig. 2.1). Since adenosine is hydrophilic, this is mediated by specialized membrane-associated proteins. Two gene families have been identified to code for such proteins in mammals: solute carrier families 28 and 29 (SLC28 and SLC29) (Molina-Arcas et al. 2009). SLC28 encodes the concentrative nucleoside transporters, CNT1, CNT2, and CNT3, which carry out unidirectional (inwardly directed), sodium-coupled, energy-dependent active transport (Gray et al. 2004). SLC29 encodes the equilibrative nucleoside transporters, ENT1, ENT2, ENT3, and ENT4, which mediate bidirectional, sodium-independent facilitated diffusion (Baldwin et al. 2004). ENTs are considered as ubiquitous transporters, expressed in a wide range of tissues including the vascular endothelium, skeletal muscles, heart, brain, liver, kidney, and placenta (Griffiths et al. 1997; Pennycooke et al. 2001; Baldwin et al. 2005; Barnes et al. 2006). CNTs, on the other hand, have been found in specialized epithelial cells in the small intestine, kidney, and liver (Felipe et al. 1998; Valdes et al. 2000; Pennycooke et al. 2001), as well as in immune cells (Pennycooke et al. 2001; Soler et al. 2001; Minuesa et al. 2008) and the brain (Anderson et al. 1996; Ritzel et al. 2001; Guillen-Gomez et al. 2004).

Equilibrative Nucleoside Transporters

While all of the four ENTs can mediate the influx and efflux of adenosine, their capacity to transport adenosine and other nucleosides varies. Human ENT1 and ENT2, for example, display a broad selectivity of substrates, transporting most purines and pyrimidines (Crawford et al. 1998; Yao et al. 2002). Both transporters show the highest affinity for adenosine out of all the nucleoside substrates, with the apparent K_m values of 40 and 100 μM, respectively (Molina-Arcas et al. 2009). ENT2, but not ENT1, is able to transport various nucleobases as well (Osses et al. 1996). ENT3, on the other hand, can also transport a wide range of purine and pyrimidine nucleosides but shows relatively low affinity for adenosine ($K_m \cong 2$ mM for the human protein) (Baldwin et al. 2005). This protein contains two endosomal/lysosomal targeting motifs in its N-terminus (Hyde et al. 2001) and shows the optimum pH value of 5.5 (Baldwin et al. 2005), which reflects its location in acidic, intracellular compartments (Molina-Arcas et al. 2009). ENT4, in contrast to the other ENTs, transports adenosine (Baldwin et al. 2004) and serotonin (Engel et al. 2004) but no other nucleosides. Its adenosine transport activity is also optimal at acidic pH (apparent K_m values of 0.78 and 0.13 mM for the human and mouse forms, respectively, at pH 5.5) but absent at pH 7.5 (Barnes et al. 2006). ENT4's abundance in the heart, particularly in the plasma membranes of ventricular myocytes and vascular endothelial cells, suggests its pH optimum is related to the acidotic conditions associated with ischemia (Barnes et al. 2006).

Concentrative Nucleotide Transporters

In comparison to ENTs, CNTs show higher affinity as well as selectivity for their nucleoside substrates (Gray et al. 2004). CNT1 is a pyrimidine-preferring transporter and cannot transport adenosine, which nevertheless binds to the translocation site of the protein with high affinity ($K_d = 14$ μM for the human protein) (Larrayoz et al. 2004). CNT2 is a purine-preferring transporter, although it can also transport uridine (Gray et al. 2004). CNT3, on the other hand, shows broader substrate specificity and translocates both purine and pyrimidine nucleosides (Ritzel et al. 2001). The apparent K_m values of human CNT2 and CNT3 for adenosine have shown to be 8 and 15 μM, respectively (Molina-Arcas et al. 2009). In addition to their substrate specificity and tissue distribution, these transporters also differ in their cotransport requirement. CNT1 and CNT2, for example, transport one sodium ion per nucleoside (Ritzel et al. 1998; Smith et al. 2004), while CNT3 transport two sodium ions (Ritzel et al. 2001). Also, CNT3 shows pH-dependent nucleoside transport, as it can cotransport H^+ with nucleosides as well (in a 1:1 stoichiometry) (Smith et al. 2005). CNT1 and CNT2, on the contrary, are strictly Na^+-dependent. As mammalian tissues express several nucleoside transporters in a single cell type, often combining both the concentrative and equilibrative proteins, the regulation of extracellular adenosine is a complex process. Termination of adenosine signaling is a result of concerted work by these nucleoside transporters as well as the adenosine-metabolizing enzymes, ADK and ADA.

2.1.2 Adenosine Kinase

As one of the most ubiquitous and abundant nucleoside kinases known (Krenitsky et al. 1974; Snyder and Lukey 1982), ADK (EC 2.7.1.20) is the first enzyme in the metabolic pathway of adenosine salvage. ADK functions as an important regulator of extracellular adenosine, as clearly demonstrated in animal models of epilepsy and ischemia (Pignataro et al. 2007, 2008; Li et al. 2008). ADK in addition carries out an essential role in facilitating cellular methylation reactions, and its deficiency in transgenic $ADK^{-/-}$ mice leads to development of acute neonatal hepatic steatosis, which causes early postnatal mortality (Boison et al. 2002). ADK is also responsible for the phosphorylation of other nucleosides and their analogs into the corresponding monophosphates, which include antimetabolites that have profound biological effects (Gupta 1989). After its first isolation from yeast and mammalian tissues 60 years ago (CAPUTTO 1951), an extensive amount of work has been carried out on many different aspects of this important enzyme.

2.1.2.1 Enzymological Aspects of Adenosine Kinase

One of the unique characteristics of ADK reaction (i.e., Adenosine + ATP→ AMP + ADP) is that the phosphoryl accepting substrate adenosine shares the same structural motif with the phosphoryl donor ATP. As a result, both adenosine and its phosphorylated product AMP bind to the ATP site (Pelicano et al. 1997), resulting in enzyme inhibition. As this complicates the inhibition pattern of the enzyme, it has been difficult to interpret the kinetic data and come to a consensus in regards to ADK's reaction mechanism. Chang et al. (1983) suggested a two-site ping-pong mechanism, while several others proposed ordered Bi Bi mechanisms (Henderson et al. 1972; Palella et al. 1980; Mimouni et al. 1994a). The order of substrate binding and product release suggested in each of these studies also differed from one another. The information obtained from the crystal structures of human (Mathews et al. 1998) and *Toxoplasma gondii* (Schumacher et al. 2000) ADK, on the other hand, allowed clear understanding of the reaction mechanism: an ordered Bi Bi mechanism, in which adenosine is the first substrate to bind, and AMP is the last product to dissociate.

Another interesting aspect of ADK, which also complicates its enzymological characterization, is its adenosine-AMP exchange activity. In characterizing ADK from rat liver (Mimouni et al. 1994a, b) and Chinese hamster cells (Hao and Gupta 1996; Gupta 1996), it was discovered that the enzyme could transfer the phosphate from AMP to radioactively labeled adenosine in the absence of ATP. This at first suggested a classical ping-pong mechanism, but formation of a phosphoryl-enzyme intermediate could not be demonstrated. The exchange reaction showed complete dependence on the presence of ADP and Mg^{2+} (Mimouni et al. 1994a; Gupta 1996), and thus could be explained with the backward reaction of ADK. In the exchange reaction, the enzyme first carries out the backward reaction, in which AMP and ADP are converted to adenosine and ATP. Accumulation of ATP then allows the forward reaction, in which both the labeled and nonlabeled adenosine is phosphorylated without discrimination. Overall, therefore, ADP functions as a carrier compound that shuttles a phosphate group from one adenosine to another. This finding also corroborates a Bi Bi reaction mechanism for ADK.

Magnesium Requirement

With respect to the phosphate-donating substrate, ATP and GTP are the preferred nucleotides for ADK, although other nucleotides can also participate in the catalytic reaction with lower efficiency (Miller et al. 1979, 1982; Chang et al. 1983; Rotllan and Miras Portugal 1985). However, the true phosphate-donating substrate of ADK, like most kinases, is the complex of a nucleotide and a divalent metal ion, typically magnesium. In a reaction mixture, ATP exists in several states of ionization depending on the pH of the solution. Mg^{2+} binds to ATP^{4-}, the fully dissociated species of ATP, and forms $MgATP^{2-}$, which then binds to ADK. In support of this concept, absence of magnesium in the reaction resulted in lack of enzyme activity in human ADK, while in the presence of magnesium, optimal activity was observed at pH values where ATP and the divalent cation exist primarily in the chelated form (Palella et al. 1980). The magnesium ion is thought to neutralize the negative charges on the phosphate groups of the nucleotide for its binding to the enzyme (Mildvan 1987).

The effect of free magnesium on ADK should also be noted. Excess magnesium in a reaction results in the saturation of available ATP^{4-}, and thus, presence of free Mg^{2+}. Interestingly, ADK activity increases as the level of free Mg^{2+} in the reaction increases up to an optimal concentration, after which point additional free Mg^{2+} leads to inhibition of the enzyme (Palella et al. 1980; Rotllan and Miras Portugal 1985; Maj et al. 2002a). A similar free magnesium requirement for optimal catalysis has also been observed in other ATP-utilizing enzymes, including 6-phosphofructo-2-kinase (Parducci et al. 2006), phosphoribosylpyrophosphate synthetase (Arnvig et al. 1990), pyruvate kinase (Baek and Nowak 1982), and phosphoenolpyruvate kinase (Lee et al. 1981). The free, catalytic magnesium ion is thought to bind at the active site and induce the transition state of the reaction by increasing the electrophilicity of the γ-phosphorous atom of the nucleotide via its interaction with the oxygen atoms (Parducci et al. 2006). In addition, this ion may induce tighter binding and/or proper orientation of the substrate's functional groups (Rivas-Pardo et al. 2011).

Substrate Specificity

As for the phosphate-accepting, nucleoside substrate, ADK shows remarkably broad specificity. Alteration of the ribosyl moiety of adenosine is well tolerated, although this decreases substrate efficiency by two- to fivefold (Bennett and Hill 1975). One important structural requirement is that the nucleoside must have a 2′-hydroxyl group, and this hydroxyl group should be in the *trans* conformation. If these conditions are met, the 3′-hydroxyl and 4′-hydroxyl-methyl groups may be in either *trans* or *cis* conformation. Substitution or addition of functional groups in the purine base moiety can also be well accepted. In fact, compounds such as 8-azaadenosine, show even higher phosphorylation efficiency than the natural substrate adenosine (Bennett and Hill 1975). The nucleosides and their analogs that are efficiently phosphorylated by ADK generally differ from adenosine at the 2-, 6-, 7-, 8-, and 9-subsitituent positions (Lindberg et al. 1967; Bennett and Hill 1975; Miller et al. 1979).

One of the direct consequences of ADK's broad selectivity for the nucleoside substrate is that the enzyme can activate a large number of antimetabolites. Adenosine analogs, such as fludarabine, are easily transported into the cell and after phosphorylation produce cytotoxic effects by interfering with DNA synthesis (Parker et al. 2004). A promising new area of application for these antimetabolites, in addition to the current anticancer and antiviral applications, is antiparasitic chemotherapy (Datta et al. 2008). Nucleoside kinases are important in parasitic protozoa, such as *Leishmania*, *Trypanosoma*, and *Toxoplasma*, due to their absolute reliance on the salvage of preformed purines from their hosts. ADK in these organisms shows profoundly different substrate specificity compared to the human enzyme (Luscher et al. 2007; Al Safarjalani et al. 2008; Cassera et al. 2011) and thus may allow selective activation of adenosine antimetabolites in the invading parasite but not in the host cells.

Regulation by Inorganic Phosphate

In addition to divalent metal ions, the activity of ADK is also affected by inorganic phosphate, whose precise mechanism of action is not fully understood. This property of ADK, which is referred to as phosphate dependency, was first demonstrated with the Chinese hamster enzyme (Hao and Gupta 1996). In this study, the catalytic activity of ADK was shown to be almost completely dependent on the presence of inorganic phosphate. The addition of inorganic phosphate served to increase the maximum velocity of ADK, as well as to decrease the K_m for its substrate, adenosine. Interestingly, other phosphate analogs, such as arsenate and vanadate, were able to substitute for inorganic phosphate as activators of ADK (Hao and Gupta 1996). Kinetic studies have shown that these ions also function to increase the V_{max} of ADK and to decrease the K_m for adenosine. On the contrary, arsenate, vanadate, and phosphate showed little effect on the K_m for ATP. ADK isolated from different sources, such as human, cow, spinach, yeast, and a protist, *Leishmania donovani*, have all shown phosphate dependency, indicating that this is a conserved property of the enzyme (Maj et al. 2000, 2002; Park et al. 2006).

In addition to ADK, there are many other enzymes which have shown phosphate dependency, including ribokinase (EC 2.7.1.15) (Maj and Gupta 2001), 6-phosphofructo-1-kinase (EC 2.7.1.11) (Hofer et al. 1982), 6-phosphofructo-2-kinase (EC 2.7.1.105) (Laloux et al. 1985), and phosphoribosylpyrophosphate synthetase (EC 2.7.6.1) (Switzer 1969). Like ADK, these enzymes catalyze the transfer of phosphate from ATP to a sugar-derived substrate. Importantly, these enzymes are all involved in the production of essential, high-energy metabolites. It is not surprising that the activities of these enzymes are regulated by inorganic phosphate, whose concentration can function as an index of the cellular energy-status. The concentration of inorganic phosphate measured in guinea pig hearts, for example, was 1.7, 7.1 and 13.9 mM under normal, mildly hypoxic, and severely hypoxic conditions, respectively (Gorman et al. 1997). Inorganic phosphate results in half maximal velocity of ADK at about 2 mM, and maximal velocity at about 10 mM concentration in vitro (Park et al. 2004, 2006). The physiological level of inorganic phosphate, therefore, allows a full range of control for ADK.

The activation of ADK by phosphate has led to an examination of many other phosphorylated compounds and their analogs (Park et al. 2004, 2006). This study has identified a number of additional compounds that can activate mammalian ADK. Kinetic experiments showed that these compounds act like inorganic phosphate, resulting in increased V_{max} of the enzyme and decreased K_m for adenosine. Most of the activators identified are common and important high-energy cellular metabolites, such as acetyl phosphate, carbamoyl phosphate, and phosphoenolpyruvate, and may have a physiologically relevant role as regulators of the phosphate-dependent enzymes. Interestingly, a number of compounds that showed structural homology to the activators instead inhibited ADK in the same study (Park et al. 2004, 2006). The fact that these activators and inhibitors show structural homology indicates that there is a structural basis for phosphate dependency. This thought is further supported by the observation that the inhibitor compounds inhibit ADK in a competitive manner with respect to the activating inorganic phosphate. The hypothesis is that both the activators and inhibitors, sharing similar structure motifs, can bind to the phosphate regulatory site on ADK. Although bound to the regulatory site, however, the inhibitor compounds cannot promote enzyme catalysis due to the differences in their electrochemical properties (Park and Gupta 2008). The inhibitors thus elicit enzyme inhibition by sequestering the regulatory (or activation) site, which phosphate or other activator compounds need to occupy for full enzyme activity.

In this light, it is very interesting that crystal structures of ribokinase, an enzyme evolutionarily related to ADK with a remarkable structural homology, were determined with a bound phosphate in the active site (Sigrell et al. 1998; 1999). This phosphate makes close contacts with the residues Asn187 and Glu190, forming direct and indirect hydrogen interactions. The asparagine and glutamic acid residues at these positions are highly conserved in the ribokinase family, to which both ADK and ribokinase belong, and identify a new sequence motif, NXXE (Maj et al. 2002). Site-directed mutations of these residues (Glu242 and N239) have been introduced in Chinese hamster ADK (Maj et al. 2002). Some of the mutants were completely inactivated as a result, but the ones with remaining activity (E242D and N239Q mutants) showed a significantly altered phosphate requirement, demonstrating the importance of the NXXE motif in the phosphate activation of ADK. These mutants also showed decreased affinity for free Mg^{2+}, indicating that the phosphate regulation of ADK activity is closely related to the catalytic magnesium ion. Mutation of the corresponding Glu190 residue in *E. coli* 6-phosphofructo-2-kinase, another member of the ribokinase family that show phosphate-dependency, also resulted in a protein with an altered Mg^{2+} requirement (Parducci et al. 2006). These results suggest the NXXE motif as the site of phosphate-binding, which in turn is required for proper binding of the catalytic Mg^{2+} at the active site.

Crystallographic Structure

Further insights into ADK's mechanism of action have been provided with the enzyme's crystal structures. The human enzyme was the first to be crystallized and characterized (Mathews et al. 1998). ADK exists as a monomer (Fig. 2.2a shows the overall structure of human ADK), which comprises two distinct domains, a $\alpha\beta\alpha$ three layer sandwich domain and a smaller lid-like domain. The larger domain is formed by a central nine-stranded β-sheet which is flanked by ten α-helices and provides most of the specific binding interactions for the two substrates, adenosine and ATP. The smaller domain contains a five-stranded mixed β-sheet flanked by two α-helices on the side exposed to the solvent and acts as a lid over the active site. Four peptide segments connect these domains, and the substrate adenosine binds in the cleft formed between the two domains. The nucleotide binding site, on the other hand, is located in a groove in the $\alpha\beta\alpha$ domain and is not entirely covered by the smaller lid domain. Important residues involved in the binding of adenosine include Leu16, Leu40, Leu134, Ala136, Leu138, Phe170, and Phe201, which form the hydrophobic pocket for the substrate and provide stacking interaction with its adenine base. The residues

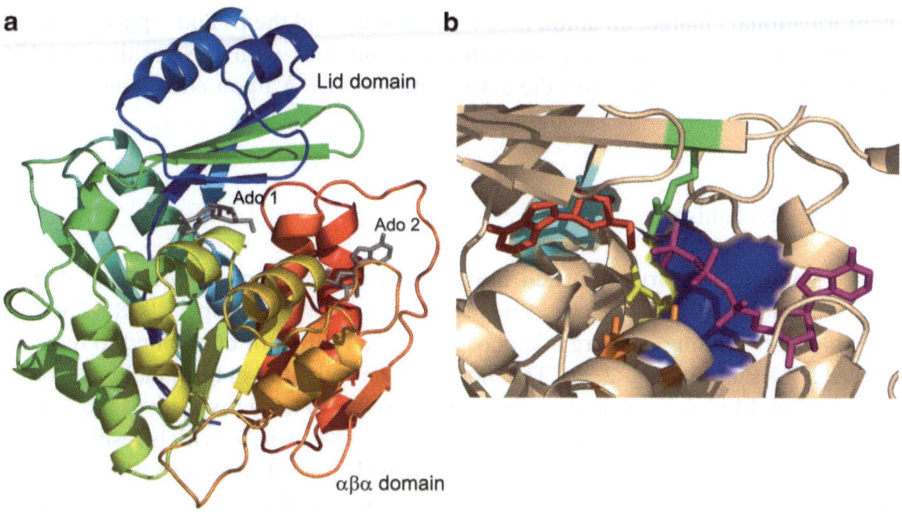

Fig. 2.2 Crystal structures of adenosine kinase. (**a**) A cartoon representation of human adenosine kinase (PDB code: 1BX4) with two bound adenosine molecules. The enzyme is shown in a *rainbow color* scheme (*blue* at the N-terminus and *red* at the C-terminus), whereas the adenosine molecules (Ado 1 and Ado 2) are depicted as *sticks in gray*. The second adenosine ligand (Ado 2) is bound at the ATP-binding site. (**b**) A close-up view of *Toxoplasma gondii* adenosine kinase (PDB code: 1LII) active site. Bound adenosine and AMP-PCP are shown in *red* and *magenta*, respectively. A number of other important structural features are also shown in this representation: i) the surface of the anion hole (*blue*); ii) the surface of the GG switch (*cyan*); Asn223 and Glu226 of the NXXE motif (*orange*); the catalytic base Asp318 (*yellow*); and Arg136 of the lid domain which shows a close contact with the bound nucleotide (*green*)

involved in ATP-binding, on the other hand, include Asn196, Asn223, Glu226, Thr265, Gly267, Val284, and Ile292, which provide fewer hydrophobic interactions (hence higher K_m for the nucleotide) compared to those in the adenosine site.

Additional information on ADK structure and function became available when the crystal structures of the enzyme from *T. gondii*, a parasitic protozoan, were determined (Schumacher et al. 2000). In this study, the structures of both apo enzyme and the substrate-bound forms were solved, and their comparison revealed drastic conformational differences between them. The most striking difference is the 30° hinge bending caused by the binding of adenosine, which brings the large and small domains together. This conformational change can be conveniently described as an opening and closing of the lid domain, if the αβα domain is taken as a point of reference. In this case, the apo structure takes the "open" conformation, exposing the adenosine-binding site to the surrounding solvent environment. The substrate-bound structure, on the contrary, takes the "closed" conformation, in which the bound adenosine is almost completely buried under the lid domain. This conformational change is likely brought about by residues Gly68 and Gly69, called GG switch (Fig. 2.2b), which undergo large torsional changes as they interact with adenosine. Subsequent ATP (AMP-PCP in the crystal structure) binding to the adenosine–enzyme complex induces a less dramatic but nonetheless important conformational change: an anion hole is created around the β- and γ-phosphates of the ATP (Fig. 2.2b). These phosphates also interact with the residue Arg136 (Fig. 2.2b), which moves into the active site by 13.7 Å upon adenosine-induced lid closing.

Catalytic Mechanism

Based on both the structural and biochemical observations, a detailed catalytic mechanism of ADK has been recently proposed (Park and Gupta 2008). The first step involves binding of inorganic phosphate or an activator compound to the conserved NXXE motif (Fig. 2.3a). The binding of an activator, in turn, facilitates binding of a free Mg^{2+} as well as the substrate adenosine to the active site of the "open" enzyme. The binding of adenosine turns on the GG switch, which results in the hinge-bending and the subsequent burial of the substrate. The closing of the lid then increases the enzyme affinity for MgATP, as seen with the translocation of Arg136 (Arg132 in human ADK) (Fig. 2.3b). MgATP binding induces formation of the anion hole, which stabilizes the pentacovalent transition state typical of an inline S_N2 displacement reaction. The catalytic magnesium ion enhances the electrophilicity of the γ-phosphorous in ATP by withdrawing charge via its interaction with the nonbridging oxygen atoms. The bound inorganic phosphate, on the other hand, may have a similar interaction with the β-phosphate of ATP. Due to the increased electronegativity of the γ- and β-phosphorous atoms, the electrons on the bridging oxygen are drawn away, making the covalent bonds weak and elongated. In the mean time, the catalytic base Asp318 (Asp300 in human ADK) deprotonates the 5′-hydroxyl end of adenosine, which subsequently attacks the positive center of the γ-phosphate. In the final step (Fig. 2.3c), the γ-phosphate is transferred to adenosine, and the products are released in the order of ADP and AMP as the adenosine-binding site is more deeply buried.

2.1.2.2 Inhibitors of Adenosine Kinase as Therapeutic Agents

The recent interest in ADK function and structure is largely due to its potential as a pharmacological target. As mentioned previously, adenosine shows a broad range of tissue-protective effects as a neuromodulator. In tissues under adverse conditions, the intracellular production of adenosine is dramatically increased (Williams and Jarvis 2000). Increased cytosolic concentration results in the net release of adenosine into the extracellular space via bidirectional ENTs (Griffith and Jarvis 1996). By activating various adenosine receptors on the cell surface, adenosine acts to limit tissue damaging and restore normal function (Ralevic and Burnstock 1998). Utilization of exogenous adenosine for pharmacological purposes, however, is difficult to achieve due to its short half-life in the interstitial fluid, which is on the order of seconds (Moser et al. 1989). Systematic introduction of adenosine receptor agonists have also proven limited, due to the detrimental side effects caused by the

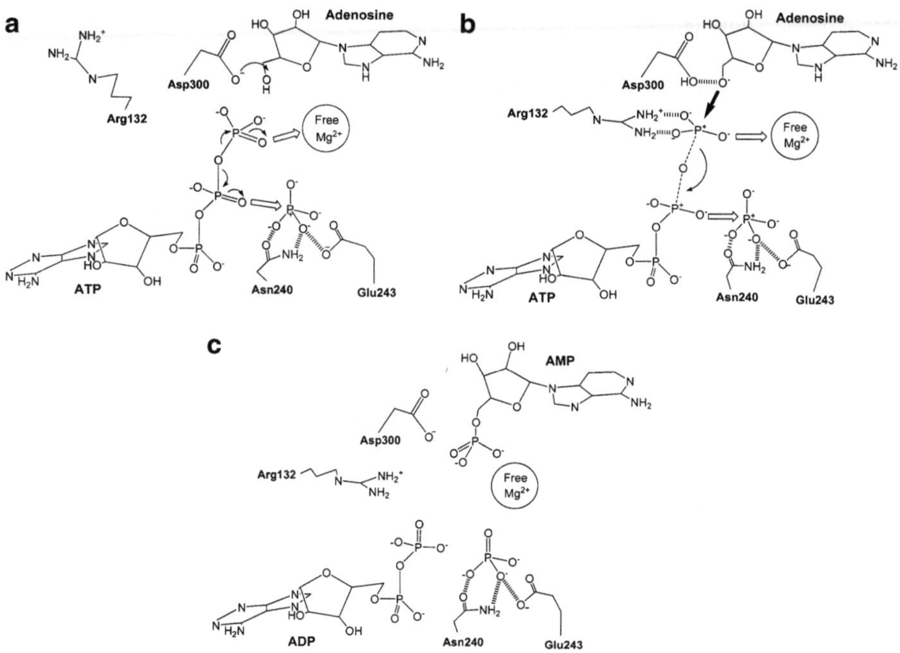

Fig. 2.3 Catalytic mechanism of adenosine kinase. Binding of inorganic phosphate facilitates binding of free Mg^{2+} as well as the substrate adenosine to the active site. This induces closing of the lid domain, which in turn increases the enzyme affinity for the second substrate ATP. The Mg^{2+} ion complexed with ATP is not shown in the scheme. *Hatched lines* indicate hydrogen-bonds. (**a**) Once ATP is bound, the phosphate molecule enhances the electrophilicity of the β-phosphorous atom by withdrawing charge via its interaction with the nonbridging oxygen of the β-phosphate. The catalytic magnesium ion, on the other hand, may have a similar interaction with the γ-phosphate of ATP. Due to the increased electronegativity of the β- and γ-phosphorous atoms, the electrons on the bridging oxygen are drawn toward the phosphorous atoms, making the bonds between the β- and γ-phosphate weak and elongated. In the mean time, the catalytic residue Asp300 deprotonates the 5′-hydroxyl of adenosine. (**b**) The positive center of the γ-phosphate attracts the negative charge developed on the 5′-hydroxyl of adenosine after proton abstraction by the catalytic base Asp300. Arg132 helps stabilize the negative charges on the nonbridging oxygen atoms of the γ-phosphate. (**c**) The γ-phosphate is transferred to adenosine. Arg132 flips back, and the products ADP and AMP are released

activation of adenosine receptors in nontarget tissues (Williams 1996). Inhibition of ADK, which can help maintain the elevated level of adenosine in the extracellular compartment (Pak et al. 1994), has been proposed as a promising alternative to exploit the adenosine receptor-mediated tissue protection. Since the release of endogenous adenosine is highly localized to the tissues under stress and trauma, the effects of ADK inhibition may only be pronounced at these specific sites (Kowaluk and Jarvis 2000). In view of this potential as a pharmacological agent, possibly without the adverse side effects associated with direct-acting adenosine receptor agonists, there has been a great interest and effort to identify and optimize ADK inhibitors.

Nucleoside Analogs

One of the most extensively characterized groups of ADK inhibitors consists of structural analogs of adenosine, such as 5′-amino,5′-deoxyadenosine (5′NH_2dAdo), 5-iodotubercidin (5IT) and 5′-deoxy,5-iodotubercidin (5′d-5IT). With IC_{50} values on the order of several nanomoles, these compounds are highly potent in inhibiting ADK in vitro (Kowaluk and Jarvis 2000). Although these compounds have shown positive effects in several animal models of pain, seizure, and ischemia (Miller et al. 1996; Jiang et al. 1997; Kowaluk et al. 1999; Wiesner et al. 1999), their pharmacological utility is largely compromised due to short plasma half-lives, poor bioavailability, lack of target selectivity, and the potential to form cytotoxic metabolites (Cottam et al. 1993; Wiesner et al. 1999). The therapeutic limitations are due to the fact that these adenosine analogs can also function as the substrates of ADK and/or agonists of the adenosine receptors. Additionally, these compounds can also potentially inhibit the transport or movement of adenosine into or to the outside of the cells.

In an effort to optimize the aforementioned ADK inhibitors for better in vivo properties, a large series of 5′NH_2dAdo and 5IT derivatives were synthesized and tested (Ugarkar et al. 2000; Bookser et al. 2005). Rounds of structure–activity relationship studies identified a number of diaryltubercidin analogs as potent inhibitors of ADK. Compounds such as GP790 (Ugarkar et al. 2003) and GP3966 (Boyer et al. 2005), for example, showed inhibition of human recombinant ADK at low nanomolar concentrations. When tested for anticonvulsant, anti-inflammatory, and analgesic properties in animal models, these compounds showed high efficacy with relatively mild adverse effects, which included decreased locomotor activity and muscular flaccidity. The traditional ADK inhibitors and adenosine receptor agonists, on the contrary, generally exert profound hemodynamic effects at equieffective doses and have much lower LD_{50} values (Malhotra and Gupta 1997; Wiesner et al. 1999). However, the clinical trial for GP3966 was stopped due to a preliminary report of CNS hemorrhage in rats and dogs (McGaraughty and Jarvis 2006). GP3269, a similar ADK inhibitor, on the other hand, has been studied in man (Phase I), but the results have not been disclosed (Kowaluk and Jarvis 2000). In spite of the promising preclinical data, the efficacy and safety of this class of compounds in clinical settings have yet to be verified.

Nonnucleoside Inhibitors

In addition to optimizing nucleoside analog ADK inhibitors, there has been a great effort in identifying nonnucleoside ADK inhibitors. The rationale for such an endeavor is that with compounds that are effectively different from adenosine, side effects due to the systemic activation of adenosine receptors would be minimal. Some of the most promising nonnucleoside ADK inhibitors belong to two different classes, the pyridopyrimidine and the alkynylpyrimidine derivatives. These two groups of compounds were initially derived from a single high-throughput screening lead (Jarvis et al. 2000), and have been subjected to rigorous rounds of structure–activity relationship studies (Lee et al. 2001; Matulenko et al. 2007). Compound ABT-702, an example of pyridopyrimidine with the IC_{50} value of 1.7 nM (Jarvis et al. 2000), is as comparably efficient as morphine for pain (Jarvis et al. 2000; Kowaluk et al. 2000) and has shown to attenuate cardiac hypertrophy (Fassett et al. 2011) in animal models. An alkynylpyrimidine compound, 5-(4-(dimethylamino)phenyl)-6-(6-morpholin-4-ylpyridin-3-ylethynyl)pyrimidin-4-ylamine, which shows a similar level of efficacy as ABT-702, has been crystallized with human ADK (Muchmore et al. 2006). Some of these pyridopyrimidine and alkynylpyrimidine derivatives, in animal studies, showed little effects on motor coordination, exploratory locomotor activity, mean arterial pressure, and heart rate, at effective doses for pain relief, likely due to lack of nucleoside-like structural features (Kowaluk and Jarvis 2000; Zheng et al. 2003).

Additionally, a group of pyrrolobenzoxa(thia)zepinones has recently been identified as a new class of nonnucleoside ADK inhibitors (Butini et al. 2011). These compounds are novel in that they function allosterically (i.e., binding at a site away from the active site), in contrast to all of the aforementioned compounds which are competitive (with respect to adenosine) inhibitors. Therefore, they are not likely to act as adenosine receptor agonists. These compounds also showed no affinity for the enzyme ADA and a low cytotoxic potential on cultured murine cells. If proven safe, ADK inhibitors may represent a new group of multi-purpose drugs, with analgesic, anti-inflammatory, and antiepileptogenic properties, as well as cardioprotective effects.

We have also carried out limited work on screening of chemical libraries to identify novel nonnucleoside inhibitors of ADK. These studies have identified a number of compounds not belonging to the pyridopyrimidine or alkynylpyrimidine groups of compounds that inhibited ADK in submicromolar range. All four of these compounds (viz. 2-tert-Butyl-4*H*-benzo[1,2,4]thiadiazine-3-thione; *N*-(5,6-Diphenyl-furo[2,3-d]pyrimidin-4-yl)-propionamide; 3-[5,6-Bis-(4-methoxy-phenyl)-furo[2,3-d]pyrimidin-4-ylamino]-propan-1-ol; and 2-[2-(3,4-Dihydroxy-phenyl)-5-phenyl-1*H*-imidazol-4-yl]-fluoren-9-one) inhibited human ADK both in vitro and in cultured cells, without exhibiting any cellular toxicity in the effective dose range (Park et al. 2007).

2.1.2.3 Isoforms of Adenosine Kinase and Subcellular Localization

There are two isoforms of ADK in various mammalian organisms including humans (Juranka and Chan 1985; Spychala et al. 1996; Singh et al. 1996; McNally et al. 1997). The recombinant proteins of both forms are enzymatically functional and show no differences in their kinetic properties (Sakowicz et al. 2001; Sahin et al. 2004). The two isoforms are identical except at the N-terminus, where the long ADK isoform (ADK-L) contains extra 20–21 amino acids (MAAAEEEPKPKKLK VEAPQAL in human ADK) that replace the first four amino acids (MTSV in human ADK) in the short isoform (ADK-S). Most programs for prediction of cellular localization (e.g., PSORT, BaCelLo) revealed no specific localization of these isoforms, and thus they were both assumed to be present in the cytoplasm (Nakai and Horton 1999; Sakowicz et al. 2001). Recently, it has been shown that ADK-L is localized within the nucleus, whereas ADK-S is found in the cytoplasm (Cui et al. 2009). The extra 20–21 amino acids present at the N-terminus of ADK-L were capable of directing other proteins to the nucleus (Cui et al. 2009), thus qualifying as a new, nonclassical nuclear localization signal.

Tissue Distribution

The two ADK isoforms have shown to express differentially in mammalian tissues (Sakowicz et al. 2001; Sahin et al. 2004; Cui et al. 2011). In the most recent tissue-distribution study (Cui et al. 2011), both the long and short isoforms were shown to express in comparable amounts in rat liver, kidney, pancreas, and lung. The heart, spleen, skeletal muscle, and thymus, in contrast, showed expression of mainly the long isoform. Of the different tissues examined, the brain was the only tissue in which the short isoform was predominantly expressed. The overall level of ADK expression, on the other hand, has shown to be the highest in the liver and decrease in the kidney, lung, heart, brain, and skeletal muscle, in the order of appearance, in both rats and mice (Sakowicz et al. 2001; Sahin et al. 2004; Cui et al. 2011).

Involvement in Methylation Processes

The differential expression of the ADK isoforms and the observed differences in their subcellular localization suggest that they are involved in different physiological processes or functions. Interestingly, the enzyme ADK, in addition to its well-studied role in the purine salvage pathway, also plays an essential role in facilitating various methyltransferase reactions (Boison et al. 2002). All transmethylation reactions catalyzed by *S*-adenosylmethionine (SAM)-dependent methyltransferases result in the production of *S*-adenosylhomocysteine (SAH) (Fig. 2.1). SAH, which is also a powerful inhibitor (product inhibition) of the SAM-dependent methyltransferases, needs to be removed efficiently to ensure the continuation of transmethylation reactions. The enzyme responsible for the removal of SAH is SAH-hydrolase, which converts SAH to adenosine and homocysteine (Fig. 2.1). As SAH-hydrolase action is reversible and adenosine concentration-dependent, rapid removal of the inhibitory end product adenosine by ADK is essential for the SAM-dependent methylation reactions (Kredich and Martin 1977; Boison et al. 2002). The nuclear localization of ADK-L, therefore, suggests that this isoform is involved in DNA methylation and needs to be in close proximity with other methylation machineries (Cui et al. 2009). Other key enzymes involved in DNA methylation, such as DNA methyltransferase and SAH-hydrolase, are also localized in nucleus (Jue et al. 1995; Kloor et al. 2007). In contrast to ADK-L, the cytoplasmically localized ADK-S is likely involved mainly in purine salvage and keeping the adenine nucleotides pool in balance (Cui et al. 2009).

DNA methylation is one of the main epigenetic regulation mechanisms that operate throughout the development of organisms. In animals, it is most active at the embryonic and postnatal stage, during which the promoters of specific genes are turned on/off at strict time points (Reik 2007). In the ADK knockout mouse, the fine tuned epigenetic regulatory process is likely disturbed due to the failure in adenosine removal, and this is one of the possible reasons for its neonatally lethal phenotype (Boison et al. 2002). The importance of the biochemical link between ADK and methylation has been well established in plants as well (Moffatt and Weretilnyk 2001). ADK knockdown (7–65 % of the WT ADK activity) in *Arabidopsis thaliana*, for example, produced dramatic and detrimental developmental abnormalities (Moffatt et al. 2002; Pereira et al. 2006). The essential role of ADK in the methylation reactions also provides a plausible explanation as to why, in contrast to other enzymes in the purine salvage pathway, such as ADA, purine nucleoside phosphorylase, and hypoxanthine-guanine phosphoribosyltransferase, whose deficiency leads to a variety of genetic disorders (Boss and Seegmiller 1982), no mutations or diseases linked to the ADK locus have been identified in humans.

2.1.2.4 Mutants Affected in Adenosine Kinase

Unlike in whole organisms, mutants affected in ADK can be readily obtained in cultured cells by selecting in the presence of toxic concentrations adenosine analogs (Gupta 1989). ADK mutants survive in such conditions, as they are unable to convert the analogs into the toxic, phosphorylated forms. Based upon their chemical structures, these adenosine analogs can be grouped into two broad classes. Compounds such as tubercidin and toyocamycin, like adenosine, contain a carbon–nitrogen bond in the ribosidic linkage and are referred to as N-nucleosides. Other analogs such as formycin A and formycin B, in which a carbon–carbon bond links the ribose sugar to the purine base, are referred to as C-nucleoside (Daves and Cheng 1976).

Mutants of mammalian cells have been selected from several cell types (Gupta and Siminovitch 1978; Gupta and Singh 1983; Mehta and Gupta 1983, 1985; Juranka and Chan 1985; Gupta and Mehta 1986; Cui et al. 2011). Most of these mutants contain no detectable ADK activity in the cell extract and exhibit high levels of cross-resistance to various N- and C-nucleosides. The resistance phenotype of these mutants (Class A) behaves recessively in the cell hybrids formed with sensitive cells, as the WT enzyme from the sensitive cells can phosphorylate and activate adenosine analogs. Another group of mutants (Class B) exhibit cross-resistance to various C-nucleosides while remaining sensitive to N-nucleosides. What is intriguing about the Class B mutants is that, although these cells are sensitive to N-nucleosides, no ADK activity (i.e., phosphorylation of adenosine, which is also an N-nucleoside) is detected in their cell extracts. Further, their drug-resistant phenotype behaves dominantly in the cell hybrids formed with either WT cells or Class A mutant cells, as both types of the hybrids are resistant to C-nucleosides but sensitive to N-nucleosides. Other mutants affected in ADK (Class C) show increased resistance to certain adenosine analogs, of which they exhibit reduced phosphorylation in the cell extracts.

Molecular Characterization of Mutants

Molecular characterization of various ADK mutants has been recently carried out. Many of the Class A mutants (e.g., Tub^R-1, Tub^R-3, Tub^R-5, Toy^R-3, and Toy^R-4, derived from Chinese hamster ovary (CHO) cells) have shown to contain large deletions in the ADK gene, which result in inactive, truncated proteins or no expression of the enzyme at all (Singh et al. 2001; Singh and Gupta 2004; Cui et al. 2011). Some Class A mutants, however, show normal expression of intact, but catalytically inactive ADK, which probably is affected by site-specific mutations. $DrToy^R$-18 (also derived from a CHO cell line), for example, contains a single base substitution (G to A) that changes Glu242 in the conserved NXXE motif to Lys (Cui et al. 2011). As discussed earlier, the NXXE motif is involved in the binding of the activating phosphate and magnesium ions, and the complete lack of activity in the mutant enzyme provides direct evidence for the importance of this motif.

Of the Class B mutants, the molecular alteration in Fom^R-4 (a CHO cell line derived mutant, selected against formycin A) has been characterized (Cui et al. 2011). This mutant contains in ADK a mutation of Ser191 to Phe, which is located near the adenosine-binding pocket. The C-nucleosides, to which the Fom^R-4 mutant exhibits selective resistance, exist predominantly in Syn conformation, while adenosine and other N-nucleosides do so in anticonformation (Daves and Cheng 1976; Gupta 1989). Hence, it is likely that this molecular alteration selectively prevents the binding of C-nucleosides to the mutant enzyme (Cui et al. 2011). This account, however, cannot explain the two puzzling, aforementioned aspects of the mutant: the lack of ADK activity in the cell extract and the dominance of its resistance phenotype in cell hybrids. Although the underlying biochemical and molecular mechanisms for these phenomena are not clear, it is possible that the mutant enzyme interacts and/or forms a complex with a downstream enzyme such as adenylate kinase (Cui et al. 2011). Adenylate kinase, which carries out the reaction $AMP+ATP \leftrightarrow 2\ ADP$, and nucleoside diphosphate kinase, which catalyzes the subsequent phosphorylation of ADP to ATP, are important for the toxicity of adenosine analogs, as it is the di- and triphosphate forms of these compounds that get into and interfere with different metabolic pathways (Gupta 1989). In this context, it should be noted that the S191F mutation is present on the surface of the protein and is replacing a hydrophilic residue with a hydrophobic one. The surface hydrophobic patch created here may be responsible for the proposed protein–protein interaction (Cui et al. 2011).

Mutants VF18 and VF19 (derived from Chinese hamster V79 cells), which also exhibit preferential resistance to C-nucleosides, have been characterized as well (Cui et al. 2011). The molecular alterations observed in these mutants are L188F in VF18 ADK and L188F as well as F221L in VF19 ADK. As these residues are near the adenosine-binding site, it is likely that they also selectively prevent the binding of C-nucleosides to the mutant enzymes. The mutants VF18 and VF19 are also of much interest, because in contrast to all other ADK mutants, they are the only mutants exhibiting increased ADK activity relative to the parental V79 cells in cell extracts. Further, these mutants show higher expression of ADK-S in comparison to ADK-L, whereas the parental V79 cells show higher expression of ADK-L. These observations indicate that VF18 and VF19, in addition to the molecular changes identified in the coding region, also contain additional genetic changes affecting the expression of the two isoforms. To fully understand the functional significance of different changes in the ADK mutants, further characterization of these mutants must be carried out at genetic, molecular, and biochemical levels.

2.1.2.5 Adenosine Kinase Gene Structure

The gene structure of ADK has been at least partially characterized in several organisms (Singh et al. 2001; Singh and Gupta 2004; Cui et al. 2011). In various mammalian species (viz. human, mouse, and Chinese hamster), the gene for ADK-L is linked in a head-to-head manner with the gene for the μ3A subunit of the adaptor protein complex-3 (AP-3), and both these genes are transcribed from a single bidirectional promoter (Singh and Gupta 2004). The AP-3 μ3A adaptor protein is involved in vesicle-mediated protein sorting, and an alteration in this protein has been observed in Hermansky–Pudlak syndrome, a genetic disorder characterized by defective lysosome-related organelles (Dell'Angelica et al. 2000a, b). The possible significance of the close linkage of these two genes and their sharing a common promoter is presently not clear. The promoter region as well as the first exon for ADK-S, on the other hand, lies within the first intron of the ADK-L gene (Cui et al. 2011). The ADK-L and ADK-S isoforms are identical except for the amino acids encoded by their first exons and are thought to arise by differential splicing of their unique first exons and the other common exons. The fact that each of these first exons has their own promoter strongly suggests that the expression of the two ADK isoforms is independently regulated at the transcriptional level. It is of much interest to further characterize the promoter regions as well as various transcriptional regulatory elements of the two isoforms to understand the specific factors or triggers that are responsible for differential expression of these two isoforms in different tissues under various conditions.

Another interesting aspect of ADK genes in various organisms is their extremely large size (Singh et al. 2001; Singh and Gupta 2004). The human ADK is around ~550 kb long, and the genes from rat, mouse, and Chinese hamster are also very large. Both human and Chinese hamster ADK genes have identical structures, consisting of 11 exons, whose lengths range from 36 to 173 bps and total to 1.1 kb. The lengths of the intervening introns in the human genome vary from 4.2 to 128.6 kb, averaging ~50 kb. The ratio of the noncoding to coding sequence for human ADK (i.e., >550) is the highest known for any gene in mammalian or other organisms. Recent analyses (unpublished results) of various animal genomes show that the large size of the ADK gene is a unique characteristic of the amniotes (i.e., tetrapod vertebrates that have a terrestrially adapted egg, including various mammals, birds, and reptiles). On the contrary, ADK genes in other eukaryotic organisms are small (Singh et al. 2001; Singh and Gupta 2004). ADK genes in amphibians and fish, for example, are ~20–25 kb. *Drosophila melanogaster* and *Caenorhabditis elegans*

ADK genes are only 1.5 and 1.3 kb long, containing two and four small introns, respectively. *A. thaliana*, on the other hand, has 10 small introns in its ADK gene, whose total length is 2.4 kb.

The large size of the ADK gene in humans and other amniotes raises many interesting questions about its biological significance. Based on its size, it is estimated that the transcription of the human ADK gene should take about 4 h (Tennyson et al. 1995). It is possible, therefore, that this gene serves as some kind of developmental timer or checkpoint to ensure the completion of some critical steps during development (e.g., transcription of certain genes that may be methylated subsequently). Further, although the human genome (NCBI) database currently indicates that no other genes are present within the 546 kb region covered by the ADK gene, a recent analysis (unpublished) has identified within its introns several large ORFs that exhibit high degrees of sequence similarity to known proteins. The cellular functions of these predicted proteins and their potential involvement in the regulation of ADK remain to be determined.

2.1.2.6 Adenosine Kinase and the Ribokinase Family Proteins

Sequence comparisons with other enzymes have shown that ADK belongs to the ribokinase family of carbohydrate kinases, which phosphorylates the hydroxymethyl groups of a variety of sugar moieties (Wu et al. 1991). The ribokinase family forms one of the three nonhomologous branches of sugar kinases, along with the hexokinase family and the galactokinase family (Bork et al. 1993). At the time of its identification, the ribokinase family consisted of only a small number of evolutionarily related enzymes. Besides ribokinase from *E. coli* and yeast, other members of this group included fructokinase, 1-phosphofructokinase, 6-phosphofructo-2-kinase (PfkB or PFK2), tagatose-6-phosphate kinase, and inosine kinase (INGK) (Bork et al. 1993). As more proteins were sequenced and characterized, a number of other enzymes were soon added to the ribokinase family, including ADK (Spychala et al. 1996; Singh et al. 1996) and 2-dehydro-3-deoxygluconokinase (Hugouvieux-Cotte-Pattat et al. 1994). It should be noted that nucleoside kinases, such as ADK and INGK, are also sugar kinases in that they phosphorylate the sugar moiety of the nucleosides. The proteins of the ribokinase family typically range from 280 to 430 amino acid residues in the total length (Park and Gupta 2008).

The members of the ribokinase family are now identified by signature sequences in the two highly conserved motifs (Fig. 2.4a) (Park and Gupta 2008). The first motif is found in a glycine-rich area located in the N-terminal region of these enzymes. Crystallographic data for ADK from various sources as well as other family members (Mathews et al. 1998; Sigrell et al. 1998; Schumacher et al. 2000; Arnfors et al. 2006; Cabrera et al. 2010), indicate that this motif includes two structurally significant consecutive glycine residues, the GG switch. As described earlier, this GG dipeptide is a part of the hinge between the αβα domain and the lid domain, which undergoes torsional changes upon the sugar substrate binding. The second motif is found in the C-terminal region. Structural evidence suggests that these residues

a

Enzymes		N-terminal sequence		C-terminal sequence
RK (ECOLI)	**41**	GGkGaNQAvaAgR-----SganiafiactG	**249**	DTiAAGDtfnGALI
RK (BACSU)	**38**	GGkGaNQAvaAaR-----LgaqvfmvgkvG	**236**	DTtGAGDtfnAAFA
RK (YEAST)	**37**	GGkGlNQAaaIgKlknpsSRysvrmignvG	**277**	DTtGAGDtflGGLV
RK (HUMAN)	**52**	GGkGaNQCvqSaR-----LgamtsmvckvG	**263**	DTtGAGDsfvGALA
AK (MYCTU)	**47**	GGvGgNMAfaIgV-----LggevalvgaaG	**251**	DPtGVGDafrAGFL
AK (YEAST)	**63**	GGaGqNTArgSaYvl---GagqvvyfgsvG	**287**	DTnGAGDafaGGFM
AK (MOUSE)	**79**	GGsGqNSMkvAqwliqe-PhkaatffgciG	**310**	DTnGAGDafvGGFL
AK (HUMAN)	**80**	GGsGqNSIkvAqwmiqq-PhkaatffgciG	**311**	DTnGAGDafvGGFL
K1PF (BACSU)	**36**	GGkGiNVSrlLkR-----HhvaskalgfvG	**243**	NSvGAGDsvvAGFL
K1PF (ECOLI)	**39**	AGkGiNVAkvLkD-----LgidvtvggflG	**249**	STvGAGDsmvGGLI
K1PF (MYCPN)	**34**	GGkGiNMAivMaL-----FgikptvltflG	**241**	STtGAGDtllGVFL
KHK (MOUSE)	**40**	GGnGsNSCtvLsL-----LgarcafmgslA	**252**	DTlGAGDtfnASVI
KHK (HUMAN)	**40**	GGnGsNSCtvLsL-----LgapcafmgsmA	**252**	DTlGAGDtfnASVI
KDGK (ECOLI)	**27**	GGdGlNTSvyIaRqvdp-AaltvhyvtalG	**258**	DTtAAGDsfsAGYL
LACC (STRPN)	**36**	GGkGlNVTrvLsE-----FgdsvlatglvG	**247**	NPvGSGDstvAGIS
FRLD (ECOLI)	**24**	GGnGvNVAvyCtR-----YgiqpgcitwvG	**218**	DTmGAGDsfiAGFL

b

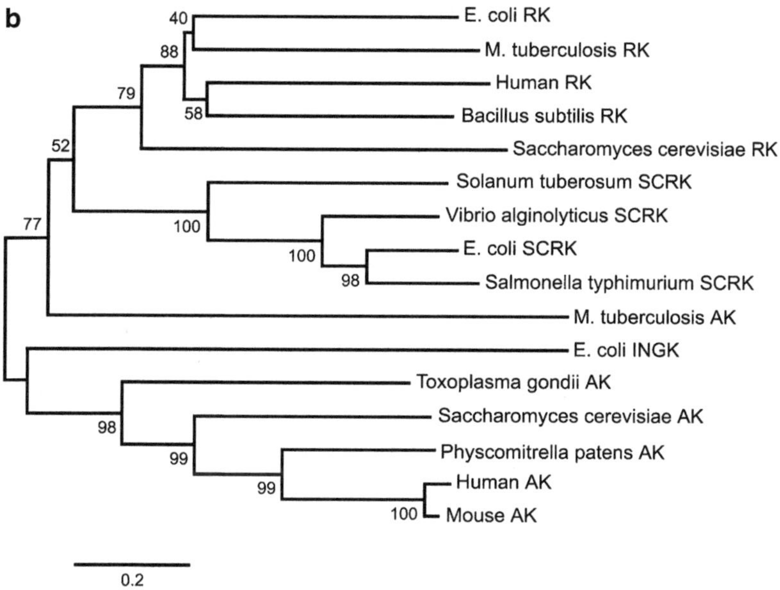

Fig. 2.4 The sequence motifs characteristic of the ribokinase family of proteins. (**a**) Conserved signature sequences of the ribokinase family enzymes. Conserved amino acid residues are shown capitalized in the alignment. Abbreviations for the enzymes: *RK* ribokinase, *ADK* adenosine kinase, *K1PF* 1-phosphofructokinase, *KHK* ketohexokinase, *KDGK* 2-dehydro-3-deoxygluconokinase, *LACC* tagatose-6-phosphate kinase, *FRLD* fructoselysine kinase. Abbreviations for the species: ECOLI *Escherichia coli*, BACSU *Bacillus subtilis*, MYCTU *Mycobacterium tuberculosis*, MYCPN *Mycoplasma pneumoniae*, STRPN *Streptococcus pneumoniae*. The amino acid numbering of mouse and human ADK is for the long isoform. (**b**) A consensus neighbor-joining tree for the ribokinase family of proteins. The numbers on the nodes are bootstrap scores. The abbreviations for the enzymes are: *RK* ribokinase, *SCRK* fructokinase, *ADK* adenosine kinase, *INGK* inosine-guanosine kinase

are involved in ATP binding and the formation of the anion hole. This sequence motif also contains an absolutely conserved aspartic acid residue which serves as the catalytic base (Asp317 in human ADK-L). Although the overall sequence identity between the family members is less than 30 %, the crystal structures of the ribokinase family members are extremely similar, retaining the overall structure composed of the large αβα sandwich domain and the small lid domain (Park and Gupta 2008). Enzymes in this group differ only to a small extent structurally, in the sugar substrate binding moiety and minor peripheral structures. Advances in sequencing technology have led to a rapid increase in the number of enzymes that belong to the ribokinase family. New members include fructoselysine kinase, the bifunctional enzyme D-hepta-D-heptose-7-phosphate kinase/D-hepta-D-heptose-1-phosphate adenosyltransferase, and a number of uncharacterized sugar kinases mostly of bacterial sources (Park and Gupta 2008).

Interestingly, the ability to utilize adenosine as a substrate appears to have arisen at least twice in the ribokinase family. ADK activity until recently was only found in the eukaryotic organisms. However, a recently identified enzyme from *Mycobacterium tuberculosis* has been denoted as the first, and thus far the only, bacterial ADK (Long et al. 2003). Sequence alignment identified some regions in this *M. tuberculosis* enzyme that are conserved in ADK from other sources. Analysis of the aligned sequences suggests that the conserved residues are involved in the binding of the substrates adenosine and ATP (Long et al. 2003). The *M. tuberculosis* enzyme, however, shows greater overall sequence homology with ribokinase and fructokinase (35 %) than it does with other ADKs (less than 24 %). Other ADKs, in contrast, show more than 50 % sequence homology with one another. In a phylogenetic tree of the ribokinase family, *M. tuberculosis* ADK branches with ribokinases and fructokinases, while ADKs from other sources form a separate branch (Fig. 2.4b) (Park et al. 2009). The substrate specificity of this *M. tuberculosis* enzyme is also very different from that of other eukaryotic ADKs (Park et al. 2009). For example, while human ADK favors adenosine approximately 900-fold over ribose as a substrate, *M. tuberculosis* ADK does so only about ninefold. Further, the *M. tuberculosis* enzyme can also phosphorylate fructose as effectively as it can phosphorylate ribose. Based on the results of the sequence alignment and biochemical studies, as well as the fact that ADK is not found in other bacteria, the ADK activity exhibited by the *M. tuberculosis* enzyme appears to have resulted from convergent evolution.

2.1.3 Concluding Remarks

The enzyme adenosine kinase plays a central role in the metabolism of adenosine and in determining its intra- as well as extracellular levels. In the past few decades, although much has been learnt about the structure of ADK, its catalytic mechanism and some aspects of its genomic organization, our understanding of the cellular functions of the two ADK isoforms and how their relative levels in various tissues that exhibit large differences are regulated is entirely lacking. Due to the proven neuroprotective and cardioprotective effects of increased concentration of adenosine, inhibition of ADK is now widely recognized as an important strategy to benefit from this body's endogenous defense mechanism (Wiesner et al. 1999; Kowaluk and Jarvis 2000; Boison 2009a, 2011; Theofilas et al. 2011). However, due to the important role that ADK and adenosine play in different tissues and regulatory pathways, the compounds that are general inhibitors of ADK have proven of limited therapeutic value due to their undesired toxic effects. To develop more specific inhibitors, whose effects are limited to a desired function(s) of the ADK at a specific location (viz. tissue), it is important to understand the cellular functions of the two isoforms and different protein factors that are involved in their tissue-specific expression. The knowledge gained by these studies should prove valuable in designing specific means (inhibitors) to modulate the activity of this enzyme in a highly specific manner to realize the therapeutic potential of increased local concentration of adenosine for treatment of neuronal (viz. epilepsy, schizophrenia) as well as other diseases (Boison 2011; Boison et al. 2011). Another very interesting aspect of ADK that needs to be investigated relates to the unusually large size of its gene in the amniotes and the possible functions of several predicted proteins that are present within its large introns. These observations suggest that ADK might be playing an important role during development that is presently not understood. Lastly, despite the important roles of ADK in a number of different pathways, no genetic condition has been attributed to mutations of ADK thus far. The emerging role of ADK in the development of a number of neurological disorders (Boison 2010; Shen et al. 2010; Boison et al. 2011) suggests that it would be of interest to examine them for possible alteration in the ADK gene.

Acknowledgments This work was supported by a research grant from the Heart and Stroke Foundation of Canada.

References

Al Safarjalani ON, Rais RH, Kim YA, Chu CK, Naguib FN, El Kouni MH (2008) 7-Deaza-6-benzylthioinosine analogues as subversive substrate of Toxoplasma gondii adenosine kinase: activities and selective toxicities. Biochem Pharmacol 76:958–966

Amadio S, Apolloni S, D'Ambrosi N, Volonte C (2011) Purinergic signalling at the plasma membrane: a multipurpose and multidirectional mode to deal with amyotrophic lateral sclerosis and multiple sclerosis. J Neurochem 116:796–805

Anderson CM, Xiong W, Young JD, Cass CE, Parkinson FE (1996) Demonstration of the existence of mRNAs encoding N1/cif and N2/cit sodium/nucleoside cotransporters in rat brain. Brain Res Mol Brain Res 42:358–361

Arnfors L, Hansen T, Schonheit P, Ladenstein R, Meining W (2006) Structure of Methanocaldococcus jannaschii nucleoside kinase: an archaeal member of the ribokinase family. Acta Crystallogr D Biol Crystallogr 62:1085–1097

Arnvig K, Hove-Jensen B, Switzer RL (1990) Purification and properties of phosphoribosyl-diphosphate synthetase from Bacillus subtilis. Eur J Biochem 192:195–200

Baek YH, Nowak T (1982) Kinetic evidence for a dual cation role for muscle pyruvate kinase. Arch Biochem Biophys 217:491–497

Bak MI, Ingwall JS (1998) Regulation of cardiac AMP-specific 5′-nucleotidase during ischemia mediates ATP resynthesis on reflow. Am J Physiol 274:C992–C1001

Baldwin SA, Beal PR, Yao SY, King AE, Cass CE, Young JD (2004) The equilibrative nucleoside transporter family, SLC29. Pflugers Arch 447:735–743

Baldwin SA, Yao SY, Hyde RJ, Ng AM, Foppolo S, Barnes K, Ritzel MW, Cass CE, Young JD (2005) Functional characterization of novel human and mouse equilibrative nucleoside transporters (hENT3 and mENT3) located in intracellular membranes. J Biol Chem 280:15880–15887

Ballard FJ (1970) Adenine nucleotides and the adenylate kinase equilibrium in livers of foetal and newborn rats. Biochem J 117:231–235

Barnes K, Dobrzynski H, Foppolo S, Beal PR, Ismat F, Scullion ER, Sun L, Tellez J, Ritzel MW, Claycomb WC, Cass CE, Young JD, Billeter-Clark R, Boyett MR, Baldwin SA (2006) Distribution and functional characterization of equilibrative nucleoside transporter-4, a novel cardiac adenosine transporter activated at acidic pH. Circ Res 99:510–519

Bauerle JD, Grenz A, Kim JH, Lee HT, Eltzschig HK (2011) Adenosine generation and signaling during acute kidney injury. J Am Soc Nephrol 22:14–20

Bennett LL Jr, Hill DL (1975) Structural requirements for activity of nucleosides as substrates for adenosine kinase: orientation of substituents on the pentofuranosyl ring. Mol Pharmacol 11: 803–808

Boison D (2008a) Adenosine as a neuromodulator in neurological diseases. Curr Opin Pharmacol 8:2–7

Boison D (2008b) The adenosine kinase hypothesis of epileptogenesis. Prog Neurobiol 84:249–262

Boison D (2009a) Adenosine augmentation therapies (AATs) for epilepsy: prospect of cell and gene therapies. Epilepsy Res 85:131–141

Boison D (2009b) Engineered adenosine-releasing cells for epilepsy therapy: human mesenchymal stem cells and human embryonic stem cells. Neurotherapeutics 6:278–283

Boison D (2010) Adenosine dysfunction and adenosine kinase in epileptogenesis. Open Neurosci J 4:93–101

Boison D (2011) Modulators of nucleoside metabolism in the therapy of brain diseases. Curr Top Med Chem 11:1068–1086

Boison D, Stewart KA (2009) Therapeutic epilepsy research: from pharmacological rationale to focal adenosine augmentation. Biochem Pharmacol 78:1428–1437

Boison D, Scheurer L, Zumsteg V, Rulicke T, Litynski P, Fowler B, Brandner S, Mohler H (2002) Neonatal hepatic steatosis by disruption of the adenosine kinase gene. Proc Natl Acad Sci U S A 99:6985–6990

Boison D, Singer P, Shen HY, Feldon J, Yee BK (2011) Adenosine hypothesis of schizophrenia—opportunities for pharmacotherapy. Neuropharmacology 62(3):1527–1543

Bookser BC, Ugarkar BG, Matelich MC, Lemus RH, Allan M, Tsuchiya M, Nakane M, Nagahisa A, Wiesner JB, Erion MD (2005) Adenosine kinase inhibitors. 6. Synthesis, water solubility, and antinociceptive activity of 5-phenyl-7-(5-deoxy-beta-D-ribofuranosyl)pyrrolo[2,3-d] pyrimidines substituted at C4 with glycinamides and related compounds. J Med Chem 48: 7808–7820

Bork P, Sander C, Valencia A (1993) Convergent evolution of similar enzymatic function on different protein folds: the hexokinase, ribokinase, and galactokinase families of sugar kinases. Protein Sci 2:31–40

Boss GR, Seegmiller JE (1982) Genetic defects in human purine and pyrimidine metabolism. Annu Rev Genet 16:297–328

Boyer SH, Ugarkar BG, Solbach J, Kopcho J, Matelich MC, Ollis K, Gomez-Galeno JE, Mendonca R, Tsuchiya M, Nagahisa A, Nakane M, Wiesner JB, Erion MD (2005) Adenosine kinase inhibitors. 5. Synthesis, enzyme inhibition, and analgesic activity of diaryl-erythro-furanosyltubercidin analogues. J Med Chem 48:6430–6441

Butini S, Gemma S, Brindisi M, Borrelli G, Lossani A, Ponte AM, Torti A, Maga G, Marinelli L, La PV, Fiorini I, Lamponi S, Campiani G, Zisterer DM, Nathwani SM, Sartini S, La MC, Da SF, Novellino E, Focher F (2011) Non-nucleoside inhibitors of human adenosine kinase: synthesis, molecular modeling, and biological studies. J Med Chem 54:1401–1420

Cabrera R, Babul J, Guixe V (2010) Ribokinase family evolution and the role of conserved residues at the active site of the PfkB subfamily representative, Pfk-2 from Escherichia coli. Arch Biochem Biophys 502:23–30

Caputto R (1951) The enzymatic synthesis of adenylic acid; adenosinekinase. J Biol Chem 189:801–814

Cassera MB, Ho MC, Merino EF, Burgos ES, Rinaldo-Matthis A, Almo SC, Schramm VL (2011) A high-affinity adenosine kinase from Anopheles gambiae. Biochemistry 50:1885–1893

Chang CH, Cha S, Brockman RW, Bennett LL Jr (1983) Kinetic studies of adenosine kinase from L1210 cells: a model enzyme with a two-site ping-pong mechanism. Biochemistry 22:600–611

Cohen MV, Downey JM (2008) Adenosine: trigger and mediator of cardioprotection. Basic Res Cardiol 103:203–215

Cottam HB, Wasson DB, Shih HC, Raychaudhuri A, Di Pasquale G, Carson DA (1993) New adenosine kinase inhibitors with oral antiinflammatory activity: synthesis and biological evaluation. J Med Chem 36:3424–3430

Crawford CR, Patel DH, Naeve C, Belt JA (1998) Cloning of the human equilibrative, nitrobenzylmercaptopurine riboside (NBMPR)-insensitive nucleoside transporter ei by functional expression in a transport-deficient cell line. J Biol Chem 273:5288–5293

Cui XA, Singh B, Park J, Gupta RS (2009) Subcellular localization of adenosine kinase in mammalian cells: the long isoform of AdK is localized in the nucleus. Biochem Biophys Res Commun 388:46–50

Cui XA, Agarwal T, Singh B, Gupta RS (2011) Molecular characterization of Chinese hamster cells mutants affected in adenosine kinase and showing novel genetic and biochemical characteristics. BMC Biochem 12:22

Darvish A, Metting PJ (1993) Purification and regulation of an AMP-specific cytosolic 5′-nucleotidase from dog heart. Am J Physiol 264:H1528–H1534

Darvish A, Pomerantz RW, Zografides PG, Metting PJ (1996) Contribution of cytosolic and membrane-bound 5′-nucleotidases to cardiac adenosine production. Am J Physiol 271:H2162–H2167

Datta AK, Datta R, Sen B (2008) Antiparasitic chemotherapy: tinkering with the purine salvage pathway. Adv Exp Med Biol 625:116–132

Daves GD Jr, Cheng CC (1976) The chemistry and biochemistry of C-nucleosides. Prog Med Chem 13:303–349

De Jong JW (1977) Partial purification and properties of rat-heart adenosine kinase. Arch Int Physiol Biochim 85:557–569

Decking UK, Schlieper G, Kroll K, Schrader J (1997) Hypoxia-induced inhibition of adenosine kinase potentiates cardiac adenosine release. Circ Res 81:154–164

Dell'Angelica EC, Aguilar RC, Wolins N, Hazelwood S, Gahl WA, Bonifacino JS (2000a) Molecular characterization of the protein encoded by the Hermansky-Pudlak syndrome type 1 gene. J Biol Chem 275:1300–1306

Dell'Angelica EC, Mullins C, Caplan S, Bonifacino JS (2000b) Lysosome-related organelles. FASEB J 14:1265–1278

Deussen A, Borst M, Schrader J (1988) Formation of S-adenosylhomocysteine in the heart. I: An index of free intracellular adenosine. Circ Res 63:240–249

Deussen A, Stappert M, Schafer S, Kelm M (1999) Quantification of extracellular and intracellular adenosine production: understanding the transmembranous concentration gradient. Circulation 99:2041–2047

Drabikowska AK, Halec L, Shugar D (1985) Purification and properties of adenosine kinase from rat liver: separation from deoxyadenosine kinase activity. Z Naturforsch C 40:34–41

Dulla CG, Dobelis P, Pearson T, Frenguelli BG, Staley KJ, Masino SA (2005) Adenosine and ATP link PCO2 to cortical excitability via pH. Neuron 48:1011–1023

Engel K, Zhou M, Wang J (2004) Identification and characterization of a novel monoamine transporter in the human brain. J Biol Chem 279:50042–50049

Engler RL (1991) Adenosine. The signal of life? Circulation 84:951–954

Fassett JT, Hu X, Xu X, Lu Z, Zhang P, Chen Y, Bache RJ (2011) Adenosine kinase regulation of cardiomyocyte hypertrophy. Am J Physiol Heart Circ Physiol 300:H1722–H1732

Felipe A, Valdes R, Santo B, Lloberas J, Casado J, Pastor-Anglada M (1998) Na^{+}-dependent nucleoside transport in liver: two different isoforms from the same gene family are expressed in liver cells. Biochem J 330(Pt 2):997–1001

Ford H Jr, Dai F, Mu L, Siddiqui MA, Nicklaus MC, Anderson L, Marquez VE, Barchi JJ Jr (2000) Adenosine deaminase prefers a distinct sugar ring conformation for binding and catalysis: kinetic and structural studies. Biochemistry 39:2581–2592

Frenguelli BG, Wigmore G, Llaudet E, Dale N (2007) Temporal and mechanistic dissociation of ATP and adenosine release during ischaemia in the mammalian hippocampus. J Neurochem 101:1400–1413

Gard JK, Kichura GM, Ackerman JJ, Eisenberg JD, Billadello JJ, Sobel BE, Gross RW (1985) Quantitative 31P nuclear magnetic resonance analysis of metabolite concentrations in Langendorff-perfused rabbit hearts. Biophys J 48:803–813

Garvey EP, Prus KL (1999) A specific inhibitor of heart cytosolic 5′-nucleotidase I attenuates hydrolysis of adenosine 5′-monophosphate in primary rat myocytes. Arch Biochem Biophys 364:235–240

Gorman MW, He MX, Hall CS, Sparks HV (1997) Inorganic phosphate as regulator of adenosine formation in isolated guinea pig hearts. Am J Physiol 272:H913–H920

Gray JH, Owen RP, Giacomini KM (2004) The concentrative nucleoside transporter family, SLC28. Pflugers Arch 447:728–734

Griffith DA, Jarvis SM (1996) Nucleoside and nucleobase transport systems of mammalian cells. Biochim Biophys Acta 1286:153–181

Griffiths M, Yao SY, Abidi F, Phillips SE, Cass CE, Young JD, Baldwin SA (1997) Molecular cloning and characterization of a nitrobenzylthioinosine-insensitive (ei) equilibrative nucleoside transporter from human placenta. Biochem J 328(Pt 3):739–743

Guillen-Gomez E, Calbet M, Casado J, de LL, Soriano E, Pastor-Anglada M, Burgaya F (2004) Distribution of CNT2 and ENT1 transcripts in rat brain: selective decrease of CNT2 mRNA in the cerebral cortex of sleep-deprived rats. J Neurochem 90:883–893

Gupta RS (1989) Purine nucleoside analogs. In: Gupta RS (ed) Drug resistance in mammalian cells. CRC, Florida, pp 89–110

Gupta RS (1996) Adenosine-AMP exchange activity is an integral part of the mammalian adenosine kinase. Biochem Mol Biol Int 39:493–502

Gupta RS, Mehta KD (1986) Genetic and biochemical characteristics of three different types of mutants of mammalian cells affected in adenosine kinase. Adv Exp Med Biol 195(Pt B): 595–603

Gupta RS, Siminovitch L (1978) Genetic and biochemical studies with the adenosine analogs toyocamycin and tubercidin: mutation at the adenosine kinase locus in Chinese hamster cells. Somatic Cell Genet 4:715–735

Gupta RS, Singh B (1983) Quantitative mutagenesis at the adenosine kinase locus in Chinese hamster ovary cells. Development and characteristics of the selection system. Mutat Res 113:441–454

Gustafson LA, Kroll K (1998) Downregulation of 5′-nucleotidase in rabbit heart during coronary underperfusion. Am J Physiol 274:H529–H538

Hao W, Gupta RS (1996) Pentavalent ions dependency of mammalian adenosine kinase. Biochem Mol Biol Int 38:889–899

Headrick JP, Willis RJ (1990) Adenosine formation and energy metabolism: a 31P-NMR study in isolated rat heart. Am J Physiol 258:H617–H624

Henderson JF, Mikoshiba A, Chu SY, Caldwell IC (1972) Kinetic studies of adenosine kinase from Ehrlich ascites tumor cells. J Biol Chem 247:1972–1975

Hofer HW, Allen BL, Kaeini MR, Pette D, Harris BG (1982) Phosphofructokinase from Ascaris suum. Regulatory kinetic studies and activity near physiological conditions. J Biol Chem 257:3801–3806

Hugouvieux-Cotte-Pattat N, Nasser W, Robert-Baudouy J (1994) Molecular characterization of the Erwinia chrysanthemi kdgK gene involved in pectin degradation. J Bacteriol 176: 2386–2392

Hyde RJ, Cass CE, Young JD, Baldwin SA (2001) The ENT family of eukaryote nucleoside and nucleobase transporters: recent advances in the investigation of structure/function relationships and the identification of novel isoforms. Mol Membr Biol 18:53–63

Jarvis MF, Yu H, Kohlhaas K, Alexander K, Lee CH, Jiang M, Bhagwat SS, Williams M, Kowaluk EA (2000) ABT-702 (4-amino-5-(3-bromophenyl)-7-(6-morpholinopyridin-3-yl)pyrido[2, 3-d] pyrimidine), a novel orally effective adenosine kinase inhibitor with analgesic and anti-inflammatory properties: I. In vitro characterization and acute antinociceptive effects in the mouse. J Pharmacol Exp Ther 295:1156–1164

Jiang N, Kowaluk EA, Lee CH, Mazdiyasni H, Chopp M (1997) Adenosine kinase inhibition protects brain against transient focal ischemia in rats. Eur J Pharmacol 320:131–137

Jue K, Bestor TH, Trasler JM (1995) Regulated synthesis and localization of DNA methyltransferase during spermatogenesis. Biol Reprod 53:561–569

Juranka P, Chan VL (1985) Analysis of adenosine kinase mutants of baby hamster kidney cells using affinity-purified antibody. J Biol Chem 260:7738–7743

Kloor D, Hermes M, Fink K, Schmid H, Klingel K, Mack A, Grenz A, Osswald H (2007) Expression and localization of S-adenosylhomocysteine-hydrolase in the rat kidney following carbon monoxide induced hypoxia. Cell Physiol Biochem 19:57–66

Kowaluk EA, Jarvis MF (2000) Therapeutic potential of adenosine kinase inhibitors. Expert Opin Investig Drugs 9:551–564

Kowaluk EA, Kohlhaas KL, Bannon A, Gunther K, Lynch JJ III, Jarvis MF (1999) Characterization of the effects of adenosine kinase inhibitors on acute thermal nociception in mice. Pharmacol Biochem Behav 63:83–91

Kowaluk EA, Mikusa J, Wismer CT, Zhu CZ, Schweitzer E, Lynch JJ, Lee CH, Jiang M, Bhagwat SS, Gomtsyan A, McKie J, Cox BF, Polakowski J, Reinhart G, Williams M, Jarvis MF (2000) ABT-702 (4-amino-5-(3-bromophenyl)-7-(6-morpholino-pyridin- 3-yl)pyrido[2,3-d]pyrimidine), a novel orally effective adenosine kinase inhibitor with analgesic and anti-inflammatory properties. II. In vivo characterization in the rat. J Pharmacol Exp Ther 295:1165–1174

Kredich NM, Martin DV Jr (1977) Role of S-adenosylhomocysteine in adenosinemediated toxicity in cultured mouse T lymphoma cells. Cell 12:931–938

Krenitsky TA, Miller RL, Fyfe JA (1974) Levels of nucleoside and nucleotide kinases in rhesus monkey tissues. Biochem Pharmacol 23:70–72

Kroll K, Deussen A, Sweet IR (1992) Comprehensive model of transport and metabolism of adenosine and S-adenosylhomocysteine in the guinea pig heart. Circ Res 71:590–604

Laloux M, Van Schaftingen E, Francois J, Hers HG (1985) Phosphate dependency of phosphofructokinase 2. Eur J Biochem 148:155–159

Larrayoz IM, Casado FJ, Pastor-Anglada M, Lostao MP (2004) Electrophysiological characterization of the human Na(+)/nucleoside cotransporter 1 (hCNT1) and role of adenosine on hCNT1 function. J Biol Chem 279:8999–9007

Latini S, Bordoni F, Pedata F, Corradetti R (1999) Extracellular adenosine concentrations during in vitro ischaemia in rat hippocampal slices. Br J Pharmacol 127:729–739

Lee MH, Hebda CA, Nowak T (1981) The role of cations in avian liver phosphoenolpyruvate carboxykinase catalysis. Activation and regulation. J Biol Chem 256:12793–12801

Lee CH, Jiang M, Cowart M, Gfesser G, Perner R, Kim KH, Gu YG, Williams M, Jarvis MF, Kowaluk EA, Stewart AO, Bhagwat SS (2001) Discovery of 4-amino-5-(3-bromophenyl)-7-(6-morpholino-pyridin-3-yl)pyrido[2,3-d]pyrimi dine, an orally active, non-nucleoside adenosine kinase inhibitor. J Med Chem 44:2133–2138

Li T, Ren G, Lusardi T, Wilz A, Lan JQ, Iwasato T, Itohara S, Simon RP, Boison D (2008) Adenosine kinase is a target for the prediction and prevention of epileptogenesis in mice. J Clin Invest 118:571–582

Lindberg B, Klenow H, Hansen K (1967) Some properties of partially purified mammalian adenosine kinase. J Biol Chem 242:350–356

Loncar R, Flesche CW, Deussen A (1997) Determinants of the S-adenosylhomocysteine (SAH) technique for the local assessment of cardiac free cytosolic adenosine. J Mol Cell Cardiol 29: 1289–1305

Long MC, Escuyer V, Parker WB (2003) Identification and characterization of a unique adenosine kinase from Mycobacterium tuberculosis. J Bacteriol 185:6548–6555

Luscher A, Onal P, Schweingruber AM, Maser P (2007) Adenosine kinase of Trypanosoma brucei and its role in susceptibility to adenosine antimetabolites. Antimicrob Agents Chemother 51: 3895–3901

Maj MC, Gupta RS (2001) The effect of inorganic phosphate on the activity of bacterial ribokinase. J Protein Chem 20:139–144

Maj M, Singh B, Gupta RS (2000) The influence of inorganic phosphate on the activity of adenosine kinase. Biochim Biophys Acta 1476:33–42

Maj MC, Singh B, Gupta RS (2002) Pentavalent ions dependency is a conserved property of adenosine kinase from diverse sources: identification of a novel motif implicated in phosphate and magnesium ion binding and substrate inhibition. Biochemistry 41:4059–4069

Malhotra J, Gupta YK (1997) Effect of adenosine receptor modulation on pentylenetetrazole-induced seizures in rats. Br J Pharmacol 120:282–288

Manthei SA, Reiling CM, Van Wylen DG (1998) Dual cardiac microdialysis to assess drug-induced changes in interstitial purine metabolites: adenosine deaminase inhibition versus adenosine kinase inhibition. Cardiovasc Res 37:171–178

Mathews II, Erion MD, Ealick SE (1998) Structure of human adenosine kinase at 1.5 A resolution. Biochemistry 37:15607–15620

Matulenko MA, Paight ES, Frey RR, Gomtsyan A, DiDomenico S Jr, Jiang M, Lee CH, Stewart AO, Yu H, Kohlhaas KL, Alexander KM, McGaraughty S, Mikusa J, Marsh KC, Muchmore SW, Jakob CL, Kowaluk EA, Jarvis MF, Bhagwat SS (2007) 4-amino-5-aryl-6-arylethynylpyrimidines: structure-activity relationships of non-nucleoside adenosine kinase inhibitors. Bioorg Med Chem 15:1586–1605

McAdoo DJ, Robak G, Xu GY, Hughes MG (2000) Adenosine release upon spinal cord injury. Brain Res 854:152–157

McGaraughty S, Jarvis MF (2006) Purinergic control of neuropathic pain. Drug Dev Res 95: 376–388

McNally T, Helfrich RJ, Cowart M, Dorwin SA, Meuth JL, Idler KB, Klute KA, Simmer RL, Kowaluk EA, Halbert DN (1997) Cloning and expression of the adenosine kinase gene from rat and human tissues. Biochem Biophys Res Commun 231:645–650

Mehta KD, Gupta RS (1983) Formycin B-resistant mutants of Chinese hamster ovary cells: novel genetic and biochemical phenotype affecting adenosine kinase. Mol Cell Biol 3:1468–1477

Mehta KD, Gupta RS (1985) Chinese hamster ovary cell mutants specifically affected in the phosphorylation of C-purine nucleosides. Can J Biochem Cell Biol 63:1044–1048

Mildvan AS (1987) Role of magnesium and other divalent cations in ATP-utilizing enzymes. Magnesium 6:28–33

Miller RL, Adamczyk DL, Miller WH, Koszalka GW, Rideout JL, Beacham LM III, Chao EY, Haggerty JJ, Krenitsky TA, Elion GB (1979) Adenosine kinase from rabbit liver. II. Substrate and inhibitor specificity. J Biol Chem 254:2346–2352

Miller RL, Adamczyk DL, Rideout JL, Krenitsky TA (1982) Purification, characterization, substrate and inhibitor specificity of adenosine kinase from several Eimeria species. Mol Biochem Parasitol 6:209–223

Miller LP, Jelovich LA, Yao L, DaRe J, Ugarkar B, Foster AC (1996) Pre- and peristroke treatment with the adenosine kinase inhibitor, 5′-deoxyiodotubercidin, significantly reduces infarct volume after temporary occlusion of the middle cerebral artery in rats. Neurosci Lett 220:73–76

Mimouni M, Bontemps F, Van den BG (1994a) Kinetic studies of rat liver adenosine kinase. Explanation of exchange reaction between adenosine and AMP. J Biol Chem 269:17820–17825

Mimouni M, Bontemps F, Van den BG (1994b) Production of adenosine and nucleoside analogues by an exchange reaction catalyzed by adenosine kinase. Adv Exp Med Biol 370:613–616

Minuesa G, Purcet S, Erkizia I, Molina-Arcas M, Bofill M, Izquierdo-Useros N, Casado FJ, Clotet B, Pastor-Anglada M, Martinez-Picado J (2008) Expression and functionality of anti-human immunodeficiency virus and anticancer drug uptake transporters in immune cells. J Pharmacol Exp Ther 324:558–567

Misumi Y, Ogata S, Ohkubo K, Hirose S, Ikehara Y (1990) Primary structure of human placental 5′-nucleotidase and identification of the glycolipid anchor in the mature form. Eur J Biochem 191:563–569
Moffatt BA, Weretilnyk EA (2001) Sustaining S-adenosyl-L-methionine-dependent methyltransferase activity in plant cells. Physiol Plant 113:435–442
Moffatt BA, Stevens YY, Allen MS, Snider JD, Pereira LA, Todorova MI, Summers PS, Weretilnyk EA, Martin-McCaffrey L, Wagner C (2002) Adenosine kinase deficiency is associated with developmental abnormalities and reduced transmethylation. Plant Physiol 128:812–821
Molina-Arcas M, Casado FJ, Pastor-Anglada M (2009) Nucleoside transporter proteins. Curr Vasc Pharmacol 7:426–434
Moser GH, Schrader J, Deussen A (1989) Turnover of adenosine in plasma of human and dog blood. Am J Physiol 256:C799–C806
Mubagwa K, Flameng W (2001) Adenosine, adenosine receptors and myocardial protection: an updated overview. Cardiovasc Res 52:25–39
Muchmore SW, Smith RA, Stewart AO, Cowart MD, Gomtsyan A, Matulenko MA, Yu H, Severin JM, Bhagwat SS, Lee CH, Kowaluk EA, Jarvis MF, Jakob CL (2006) Crystal structures of human adenosine kinase inhibitor complexes reveal two distinct binding modes. J Med Chem 49:6726–6731
Mullane K, Bullough D (1995) Harnessing an endogenous cardioprotective mechanism: cellular sources and sites of action of adenosine. J Mol Cell Cardiol 27:1041–1054
Naito Y, Lowenstein JM (1985) 5′-Nucleotidase from rat heart membranes. Inhibition by adenine nucleotides and related compounds. Biochem J 226:645–651
Nakai K, Horton P (1999) PSORT: a program for detecting sorting signals in proteins and predicting their subcellular localization. Trends Biochem Sci 24:34–36
Osses N, Pearson JD, Yudilevich DL, Jarvis SM (1996) Hypoxanthine enters human vascular endothelial cells (ECV 304) via the nitrobenzylthioinosine-insensitive equilibrative nucleoside transporter. Biochem J 317(Pt 3):843–848
Pak MA, Haas HL, Decking UK, Schrader J (1994) Inhibition of adenosine kinase increases endogenous adenosine and depresses neuronal activity in hippocampal slices. Neuropharmacology 33:1049–1053
Palella TD, Andres CM, Fox IH (1980) Human placental adenosine kinase. Kinetic mechanism and inhibition. J Biol Chem 255:5264–5269
Parducci RE, Cabrera R, Baez M, Guixe V (2006) Evidence for a catalytic Mg^{2+} ion and effect of phosphate on the activity of Escherichia coli phosphofructokinase-2: regulatory properties of a ribokinase family member. Biochemistry 45:9291–9299
Park J, Gupta RS (2008) Adenosine kinase and ribokinase—the RK family of proteins. Cell Mol Life Sci 65:2875–2896
Park J, Singh B, Maj MC, Gupta RS (2004) Phosphorylated derivatives that activate or inhibit mammalian adenosine kinase provide insights into the role of pentavalent ions in AK catalysis. Protein J 23:167–177
Park J, Singh B, Gupta RS (2006) Inhibition of adenosine kinase by phosphonate and bisphosphonate derivatives. Mol Cell Biochem 283:11–21
Park J, Vaidyanathan G, Singh B, Gupta RS (2007) Identification and biochemical studies on novel non-nucleoside inhibitors of the enzyme adenosine kinase. Protein J 26:203–212
Park J, Singh B, Gupta RS (2009) Mycobacterial adenosine kinase is not a typical adenosine kinase. FEBS Lett 583:2231–2236
Parker WB, Secrist JA III, Waud WR (2004) Purine nucleoside antimetabolites in development for the treatment of cancer. Curr Opin Investig Drugs 5:592–596
Pascual O, Casper KB, Kubera C, Zhang J, Revilla-Sanchez R, Sul JY, Takano H, Moss SJ, McCarthy K, Haydon PG (2005) Astrocytic purinergic signaling coordinates synaptic networks. Science 310:113–116
Pelicano H, Maury G, Elalaoui A, Shafiee M, Imbach JL, Goody RS, Divita G (1997) Study of the substrate-binding properties of bovine liver adenosine kinase and inhibition by fluorescent nucleoside analogues. Eur J Biochem 248:930–937

Pennycooke M, Chaudary N, Shuralyova I, Zhang Y, Coe IR (2001) Differential expression of human nucleoside transporters in normal and tumor tissue. Biochem Biophys Res Commun 280:951–959

Pereira LA, Schoor S, Goubet F, Dupree P, Moffatt BA (2006) Deficiency of adenosine kinase activity affects the degree of pectin methyl-esterification in cell walls of Arabidopsis thaliana. Planta 224:1401–1414

Phillis JW, Goshgarian HG (2001) Adenosine and neurotrauma: therapeutic perspectives. Neurol Res 23:183–189

Pignataro G, Simon RP, Boison D (2007) Transgenic overexpression of adenosine kinase aggravates cell death in ischemia. J Cereb Blood Flow Metab 27:1–5

Pignataro G, Maysami S, Studer FE, Wilz A, Simon RP, Boison D (2008) Downregulation of hippocampal adenosine kinase after focal ischemia as potential endogenous neuroprotective mechanism. J Cereb Blood Flow Metab 28:17–23

Ralevic V, Burnstock G (1998) Receptors for purines and pyrimidines. Pharmacol Rev 50:413–492

Reik W (2007) Stability and flexibility of epigenetic gene regulation in mammalian development. Nature 447:425–432

Ritzel MW, Yao SY, Ng AM, Mackey JR, Cass CE, Young JD (1998) Molecular cloning, functional expression and chromosomal localization of a cDNA encoding a human Na^+/nucleoside cotransporter (hCNT2) selective for purine nucleosides and uridine. Mol Membr Biol 15: 203–211

Ritzel MW, Ng AM, Yao SY, Graham K, Loewen SK, Smith KM, Ritzel RG, Mowles DA, Carpenter P, Chen XZ, Karpinski E, Hyde RJ, Baldwin SA, Cass CE, Young JD (2001) Molecular identification and characterization of novel human and mouse concentrative Na^+-nucleoside cotransporter proteins (hCNT3 and mCNT3) broadly selective for purine and pyrimidine nucleosides (system cib). J Biol Chem 276:2914–2927

Rivas-Pardo JA, Caniuguir A, Wilson CA, Babul J, Guixe V (2011) Divalent metal cation requirements of phosphofructokinase-2 from E. coli. Evidence for a high affinity binding site for Mn^{2+}. Arch Biochem Biophys 505:60–66

Rotllan P, Miras Portugal MT (1985) Adenosine kinase from bovine adrenal medulla. Eur J Biochem 151:365–371

Sahin B, Kansy JW, Nairn AC, Spychala J, Ealick SE, Fienberg AA, Greene RW, Bibb JA (2004) Molecular characterization of recombinant mouse adenosine kinase and evaluation as a target for protein phosphorylation. Eur J Biochem 271:3547–3555

Sakowicz M, Grden M, Pawelczyk T (2001) Expression level of adenosine kinase in rat tissues. Lack of phosphate effect on the enzyme activity. Acta Biochim Pol 48:745–754

Sala-Newby GB, Skladanowski AC, Newby AC (1999) The mechanism of adenosine formation in cells. Cloning of cytosolic 5′-nucleotidase-I. J Biol Chem 274:17789–17793

Schumacher MA, Scott DM, Mathews II, Ealick SE, Roos DS, Ullman B, Brennan RG (2000) Crystal structures of Toxoplasma gondii adenosine kinase reveal a novel catalytic mechanism and prodrug binding. J Mol Biol 296:549–567

Shen HY, Li T, Boison D (2010) A novel mouse model for sudden unexpected death in epilepsy (SUDEP): role of impaired adenosine clearance. Epilepsia 51:465–468

Sigrell JA, Cameron AD, Jones TA, Mowbray SL (1998) Structure of Escherichia coli ribokinase in complex with ribose and dinucleotide determined to 1.8 A resolution: insights into a new family of kinase structures. Structure 6:183–193

Sigrell JA, Cameron AD, Mowbray SL (1999) Induced fit on sugar binding activates ribokinase. J Mol Biol 290:1009–1018

Singh B, Gupta RS (2004) Genomic organization and linkage via a bidirectional promoter of the AP-3 (adaptor protein-3) mu3A and AK (adenosine kinase) genes: deletion mutants of AK in Chinese hamster cells extend into the AP-3 mu3A gene. Biochem J 378:519–528

Singh LS, Sharma R (2000) Purification and characterization of intestinal adenosine deaminase from mice. Mol Cell Biochem 204:127–134

Singh B, Hao W, Wu Z, Eigl B, Gupta RS (1996) Cloning and characterization of cDNA for adenosine kinase from mammalian (Chinese hamster, mouse, human and rat) species. High frequency mutants of Chinese hamster ovary cells involve structural alterations in the gene. Eur J Biochem 241:564–571

Singh B, Lin A, Wu ZC, Gupta RS (2001) Gene structure for adenosine kinase in Chinese hamster and human: high-frequency mutants of CHO cells involve deletions of several introns and exons. DNA Cell Biol 20:53–65

Smith KM, Ng AM, Yao SY, Labedz KA, Knaus EE, Wiebe LI, Cass CE, Baldwin SA, Chen XZ, Karpinski E, Young JD (2004) Electrophysiological characterization of a recombinant human Na^+-coupled nucleoside transporter (hCNT1) produced in Xenopus oocytes. J Physiol 558: 807–823

Smith KM, Slugoski MD, Loewen SK, Ng AM, Yao SY, Chen XZ, Karpinski E, Cass CE, Baldwin SA, Young JD (2005) The broadly selective human Na^+/nucleoside cotransporter (hCNT3) exhibits novel cation-coupled nucleoside transport characteristics. J Biol Chem 280:25436–25449

Snyder FF, Lukey T (1982) Kinetic considerations for the regulation of adenosine and deoxyadenosine metabolism in mouse and human tissues based on a thymocyte model. Biochim Biophys Acta 696:299–307

Soler C, Garcia-Manteiga J, Valdes R, Xaus J, Comalada M, Casado FJ, Pastor-Anglada M, Celada A, Felipe A (2001) Macrophages require different nucleoside transport systems for proliferation and activation. FASEB J 15:1979–1988

Spychala J, Datta NS, Takabayashi K, Datta M, Fox IH, Gribbin T, Mitchell BS (1996) Cloning of human adenosine kinase cDNA: sequence similarity to microbial ribokinases and fructokinases. Proc Natl Acad Sci U S A 93:1232–1237

Stone TW, Ceruti S, Abbracchio MP (2009) Adenosine receptors and neurological disease: neuroprotection and neurodegeneration. Handb Exp Pharmacol 193:535–587

Sullivan JM, Alpers JB (1971) In vitro regulation of rat heart 5′-nucleotidase by adenine nucleotides and magnesium. J Biol Chem 246:3057–3063

Switzer RL (1969) Regulation and mechanism of phosphoribosylpyrophosphate synthetase. I. Purification and properties of the enzyme from Salmonella typhimurium. J Biol Chem 244: 2854–2863

Tennyson CN, Klamut HJ, Worton RG (1995) The human dystrophin gene requires 16 hours to be transcribed and is cotranscriptionally spliced. Nat Genet 9:184–190

Theofilas P, Brar S, Stewart KA, Shen HY, Sandau US, Poulsen D, Boison D (2011) Adenosine kinase as a target for therapeutic antisense strategies in epilepsy. Epilepsia 52:589–601

Ugarkar BG, DaRe JM, Kopcho JJ, Browne CE III, Schanzer JM, Wiesner JB, Erion MD (2000) Adenosine kinase inhibitors. 1. Synthesis, enzyme inhibition, and antiseizure activity of 5-iodotubercidin analogues. J Med Chem 43:2883–2893

Ugarkar BG, Castellino AJ, DaRe JS, Ramirez-Weinhouse M, Kopcho JJ, Rosengren S, Erion MD (2003) Adenosine kinase inhibitors. 3. Synthesis, SAR, and antiinflammatory activity of a series of l-lyxofuranosyl nucleosides. J Med Chem 46:4750–4760

Valdes R, Ortega MA, Casado FJ, Felipe A, Gil A, Sanchez-Pozo A, Pastor-Anglada M (2000) Nutritional regulation of nucleoside transporter expression in rat small intestine. Gastroenterology 119:1623–1630

Wiesner JB, Ugarkar BG, Castellino AJ, Barankiewicz J, Dumas DP, Gruber HE, Foster AC, Erion MD (1999) Adenosine kinase inhibitors as a novel approach to anticonvulsant therapy. J Pharmacol Exp Ther 289:1669–1677

Williams M (1996) Challenges in developing P2 purinoceptor-based therapeutics. Ciba Found Symp 198:309–321

Williams M, Jarvis MF (2000) Purinergic and pyrimidinergic receptors as potential drug targets. Biochem Pharmacol 59:1173–1185

Williams-Karnesky RL, Stenzel-Poore MP (2009) Adenosine and stroke: maximizing the therapeutic potential of adenosine as a prophylactic and acute neuroprotectant. Curr Neuropharmacol 7:217–227

Wojcik M, Zieleniak A, Wozniak LA (2010) New insight into A1 adenosine receptors in diabetes treatment. Curr Pharm Des 16:4237–4242

Wu LF, Reizer A, Reizer J, Cai B, Tomich JM, Saier MH Jr (1991) Nucleotide sequence of the Rhodobacter capsulatus fruK gene, which encodes fructose-1-phosphate kinase: evidence for a kinase superfamily including both phosphofructokinases of Escherichia coli. J Bacteriol 173: 3117–3127

Yamazaki Y, Truong VL, Lowenstein JM (1991) 5′-Nucleotidase I from rabbit heart. Biochemistry 30:1503–1509

Yao SY, Ng AM, Vickers MF, Sundaram M, Cass CE, Baldwin SA, Young JD (2002) Functional and molecular characterization of nucleobase transport by recombinant human and rat equilibrative nucleoside transporters 1 and 2. Chimeric constructs reveal a role for the ENT2 helix 5-6 region in nucleobase translocation. J Biol Chem 277:24938–24948

Zheng GZ, Mao Y, Lee CH, Pratt JK, Koenig JR, Perner RJ, Cowart MD, Gfesser GA, McGaraughty S, Chu KL, Zhu C, Yu H, Kohlhaas K, Alexander KM, Wismer CT, Mikusa J, Jarvis MF, Kowaluk EA, Stewart AO (2003) Adenosine kinase inhibitors: polar 7-substitutent of pyridopyrimidine derivatives improving their locomotor selectivity. Bioorg Med Chem Lett 13:3041–3044

Zylka MJ (2011) Pain-relieving prospects for adenosine receptors and ectonucleotidases. Trends Mol Med 17:188–196

Chapter 3
Adenosine and Energy Metabolism—Relationship to Brain Bioenergetics

Xuesong Chen, Liang Hui, and Jonathan D. Geiger

Abstract The brain is one of the most metabolically active organs in the body, but because it has a limited capacity to store bioenergetic molecules it requires a continuous supply of oxygen and energy. In order to meet the high metabolic needs of the brain, the blood–brain barrier selectively expresses specific transport systems including glucose transporters and monocarboxylic acid transporters that transport lactate and ketone bodies. As a universal energy source, adenosine triphosphate (ATP) drives biological reactions essential for brain functions, and loss of such cellular energy results in profound abnormalities in brain function. The high-energy phosphate bonds of ATP are rather labile and thus energy inherent in ATP is readily released when ATP is hydrolyzed sequentially to ADP, AMP, and finally adenosine. Although each of these molecules serves different functions and can activate different signaling pathways, our focus here is on adenosine and brain energy metabolism. Under basal conditions, brain levels of adenosine are nearly 10,000-fold lower than ATP. Therefore, unnoticeable and possibly physiologically irrelevant decreases in ATP levels can result in dramatic and physiologically relevant rises in adenosine levels. As brain energy levels drop, adenosine levels rise to adjust brain energy supply and to retaliate against an external stimulus that would otherwise cause excessive ATP breakdown. These actions of adenosine are mediated by adenosine receptors located on target cells including neurons, glial cells, and brain endothelial cells. A critical issue in studying brain bioenergetics is the precise and accurate measurement of levels of ATP and its metabolites including adenosine. Because these molecules can be degraded rapidly, it is challenging to make such measurements. Essential components in the correct assessment of brain energetics should include justifying carefully the methodology used and putting the data in the context of what is already known about brain energy metabolism.

X. Chen • L. Hui • J.D. Geiger (✉)
Department of Pharmacology, Physiology and Therapeutics,
School of Medicine and Health Sciences, University of North Dakota,
501 N. Columbia Road, Grand Forks, ND 58202, USA
e-mail: jonathan.geiger@med.und.edu

S. Masino and D. Boison (eds.), *Adenosine: A Key Link between Metabolism and Brain Activity*, DOI 10.1007/978-1-4614-3903-5_3,

Keywords Adenosine • Adenylate energy charge • Blood–brain barrier • Glucose • Glycogen • Ketone bodies • High-energy focused microwave irradiation

3.1 Overview of Sources and Requirements for Brain Bioenergetics

The brain is a complicated and active organ that requires an almost constant supply of energy. While the brain comprises only about 2 % of the body weight, it uses approximately 20 % of total body energy consumption; and that is during a resting state (Rolfe and Brown 1997)! Roughly one-half of the brain's energetic requirements are used to maintain ionic gradients across neuronal plasma membranes and the vast majority of this is used by Na^+/K^+-ATPase, which pumps K^+ into and Na^+ out of neuronal cells: a process with keen requirements for adenosine triphosphate (ATP), a universal and ubiquitous source of cellular energy (Attwell and Laughlin 2001). The brain contains the largest density of Na^+/K^+-ATPase molecules in the body (~11 pmol/g tissue); by comparison the concentration of this enzyme is about 20 times lower in heart, another highly active organ (Clausen 1998). Other than Na^+/K^+-ATPase, about one-half of the brain's energy consumption is used for transport of neurotransmitters and other essential molecules across cell membranes, for transport of substances within cells, and for synthesis of the building blocks of cells including proteins, lipids, and carbohydrates (Attwell and Laughlin 2001; McKenna et al. 2006).

Although metabolically active, the brain has a very limited capacity to store bioenergetic molecules. These energetic stores include ATP, glycogen, ketone bodies, and in some cases lipid-based molecules. Essential for physiological regulation of brain function is the continuous supply of oxygen and energy-yielding nutrients and of course all of this originates from blood. Thus, it is easy to understand why even a very brief interruption (seconds) in blood flow and/or absence of oxygen results in dramatic losses of cellular energy, loss of consciousness, and profound abnormalities in brain function up to and including increased susceptibility to insult-induced neuronal cell death.

Further complicating access of energy substrates from blood into brain is the presence of the blood–brain barrier (BBB), a physical and metabolic barrier that permits selective passage of cells and substances into brain (Abbott et al. 2010). While water and oxygen can readily diffuse across the BBB, other larger particles cannot cross the BBB freely. The restrictive nature of the BBB is due primarily to tight junctions formed between adjacent endothelial cells, a lack of pinocytosis by endothelial cells, tightly coupled end-feet of astrocytes, and less so by the presence of an underlying continuous basement membrane (Abbott et al. 2010). Functionally, these restrictions provide a structural and functional barrier capable of preventing blood-borne toxins and cells from entering brain. However, although the BBB works to protect the brain microenvironment it does limit the entry of energy-yielding nutrients and in doing so the BBB increases the risk of the brain experiencing insult-induced bioenergetic crises. In order to meet the high metabolic needs of the brain, the BBB selectively expresses specific transport systems (Spector 2009): the highly glycosylated 55 kDa form of the glucose transporter (GLUT1) that transports blood glucose into brain parenchyma (Ito et al. 2011; Vannucci et al. 1997) and the monocarboxylic acid transporter (MCT1) that transports monocarboxylic acids including lactate and the ketone bodies acetoacetate and β-hydroxybutyrate (Ito et al. 2011; Vannucci and Simpson 2003).

During early stages of development the brain is more flexible in terms of using various substrates for energy. However, adult brain relies primarily on blood glucose. Indeed, it has been estimated that brain uses about 25 % of total body glucose and 20 % of respired oxygen (Magistretti et al. 1995). Once blood glucose is transported into brain parenchyma via the insulin-insensitive GLUT1 transporters, glucose is taken up by astrocytes via the 45 kDa form of GLUT1, by neurons via the high-capacity GLUT3 transporter, and by microglia via GLUT5. Most of the glucose that enters mitochondria is metabolized to pyruvate and then converted to acetyl-CoA before entering the Kreb's tricarboxylic acid cycle to produce ATP; only about 13 % of glycolytic pyruvate is converted to lactate. In addition to energy production, glucose is also used for amino acid and neurotransmitter biosynthesis, glycogen synthesis, lipid biosynthesis via the pentose phosphate shunt, maintenance of reduced glutathione, and incorporation in glycoproteins and glycolipids (McKenna et al. 2006).

In developing brain, during periods of prolonged fasting when blood glucose levels are low, or during the ingestion of high-fat diets that are extremely restricted in terms of available carbohydrates (the so-called ketogenic diet that mimics fasting/starvation)—ketone bodies provide a key alternative fuel for the brain capable of providing 60–70 % of the energy need of the brain (Owen et al. 1967). Consisting of acetoacetate, β-hydroxybutyrate, and acetone, ketone bodies are formed in hepatic mitochondria as by-products of fatty acid metabolism. Acetoacetate and β-hydroxybutyrate are transported to the brain through MCT1 expressed in the BBB. Once inside brain, these ketone bodies are taken up by neurons via the high-affinity MCT2 transporters, and by glia cells via MCT1 and MCT4 transporters (Bergersen et al. 2002; Rafiki et al. 2003; Vannucci and Simpson 2003). In addition to ketone bodies coming from systemic circulation, astrocytes have been shown to produce ketone bodies from fatty acids and leucine (Auestad et al. 1991; Bixel and Hamprecht 1995; Guzman and Blazquez 2001; Guzman and Blazquez 2004; Lopes-Cardozo et al. 1986). Ketone body production by astrocytes could contribute significantly to brain bioenergetics because astrocytes can outnumber neurons 9:1 and can occupy at least 50 % of the cerebral volume (White and Venkatesh 2011).

In normal adults, blood levels of ketone bodies are very low and cerebral utilization of ketones is negligible. However, when glucose stores are exhausted and the rate of gluconeogenesis is insufficient to meet the needs for glucose by brain, blood levels of ketone bodies rise to low mM concentrations as a result of the rapid fat catabolism (Levitsky et al. 1977; Owen et al. 1967) and the uptake and utilization of ketone bodies increase dramatically (Hasselbalch et al. 1994; Leino et al. 2001). Through the actions of D-β-hydroxybutyrate dehydrogenase, acetoacetate-succinyl-CoA transferase and acetoacetyl-CoA-thiolase ketone bodies in brain can be converted into acetyl-CoA, which then can be used to produce ATP, lipids, and/or proteins (Patel et al. 1975). In contrast to glucose, ketone bodies can bypass glycolysis and enter directly into mitochondria where they are oxidized to acetyl-CoA. Thus, ketone bodies are more efficient in producing ATP than is glucose and can stimulate mitochondrial biogenesis via the upregulation of genes encoding energy metabolism and mitochondrial enzymes (Kashiwaya et al. 1994; Sato et al. 1995; Veech et al. 2001). In addition, metabolism of ketone bodies produces less reactive oxygen species than does glucose (Mobbs et al. 2007). Thus, the utilization of ketone bodies as an alternative energy fuel could exert beneficial effects to brain (Cunnane et al. 2011; LaManna et al. 2009).

In addition to ketone bodies, lactate can be also used for energy in brain (Itoh et al. 2003; Smith et al. 2003). Like ketone bodies, lactate from systemic circulation is transported to brain at the BBB via MCT1. Once inside brain, lactate is uptaken via MCTs and is then oxidized to produce ATP. In addition to peripherally derived lactate, astrocytes have been implicated as lactate sources inside the brain where they are thought to participate in supplying lactate to neurons especially during periods of enhanced synaptic activity (Chih and Roberts 2003; Kasischke et al. 2004).

Glycogen, primarily stored in astrocytes, is the largest energy reserve in brain. When brain glucose levels are very low, glycogenolysis in astrocytes can provide neurons the necessary glucosyl units in the form of lactate to maintain ATP synthesis (Choi and Gruetter 2003). However, glycogen levels in brain are very low (3 mmol/kg), about 100-fold lower than in liver, and this amount of glycogen can only sustain brain energy for a very short period of time (Brown 2004; Cruz and Dienel 2002; Nelson et al. 1968). As such, a continuous supply of glucose and oxygen is necessary to sustain neuronal activity.

For decades, it has been suggested that brain does not use fatty acids as a source of fuel, even though about 60 % of solid matter in the brain is fat. This view is supported by observations that fatty acids that are attached to lipoproteins cannot cross the BBB and brain has no triacylglycerol stores. This point of view might be true for neurons, because enzyme activity for fatty acid oxidization is very low (Yang et al. 1987). However, recent studies on brain energy metabolism have revealed a more intricate picture of fatty acid metabolism in brain and have highlighted the important role of astrocytes. Indeed, certain fatty acids (e.g., arachidonic acid and docosahexaenoic acid) can be rapidly taken up by brain and oxidized for energy production by astrocytes that express enzymes for fatty acid oxidization. In support of this, octanoate, a medium chain fatty acid, was shown to contribute significantly to total brain oxidative energy production (Ebert et al. 2003).

3.2 Adenosine and ATP Metabolism

ATP, consisting of three phosphate groups bound to adenosine, is a universal energy source that drives a huge number of biological reactions essential for life and a variety of cellular functions. ATP is integrally involved in neuronal and glial metabolism—notably the maintenance of transmembrane ion gradients, cAMP formation, kinase-mediated phosphorylation, vesicular storage of neurotransmitters, and glutamate metabolism. The high-energy phosphate bonds of ATP are rather labile and thus energy inherent in ATP is readily released when ATP is hydrolyzed sequentially to ADP, AMP, and finally adenosine. Although each of these molecules serves different functions and can activate different signaling pathways, they do influence directly and indirectly cell activity and brain energy homeostasis. Our focus here is on adenosine and brain energy metabolism.

Under basal conditions, brain levels of adenosine (~100 nM) are nearly 10,000-fold lower than ATP (Delaney and Geiger 1996; Pazzagli et al. 1993; Van Wylen et al. 1986). Therefore, unnoticeable decreases in ATP levels can result in dramatic rises in adenosine levels. In general, adenosine acts to inhibit neuronal activity and in so doing allows target cells like neurons and astrocytes to adjust their energy demands to match better energy supply—this is the basic concept behind adenosine being referred to as a "retaliatory metabolite," for example, retaliating against external stimuli and insults capable of causing excessive ATP breakdown (Meghji and Newby 1990; Newby et al. 1990).

In brain, de novo synthesis of adenine nucleotides is minimal and for adenosine is absent; purine salvage is absolutely key to the maintenance of cellular levels of these purines. For adenosine, its levels are regulated enzymatically by two enzymes that catalyze its synthesis, 5′-nucleotidases and *S*-adenosylhomocysteine hydrolase, and four enzymes that catalyze its removal, *S*-adenosylhomocysteine hydrolase, 5′-nucleotidases, adenosine deaminase, and adenosine kinase. In addition to these enzymes, a whole family of equilibrative and concentrative (sodium-dependent) nucleoside transporters are involved in the regulation of intra- and extracellular levels of adenosine (King et al. 2006; Latini and Pedata 2001; Peng et al. 2005). Equilibrative nucleoside transporters are bidirectional and they facilitate not only the movement of adenosine into cells when its extracellular concentration exceeds its intracellular concentration, but also the efflux of adenosine from cells when its intracellular concentration is increased. Under conditions when Na^+ electrochemical gradients are reversed, concentrative nucleoside transporters facilitate adenosine release from cells. It is important to note that extracellular and intracellular levels of adenosine are usually maintained in the same range because most cells possess highly efficient equilibrative nucleoside transporters (Cass et al. 1999).

In addition to neurons releasing ATP as a neurotransmitter, astrocytes too have been identified as a major source of adenosine formation (Pascual et al. 2005). Adenosine generated from astrocytes originates from vesicular release of ATP (Pascual et al. 2005), direct release of ATP through hemichannels (Iglesias et al. 2009; Kang et al. 2008), and adenosine released via nucleoside transporters (King et al. 2006; Latini and Pedata 2001; Peng et al. 2005) and chloride channels (Darby et al. 2003). Astrocytes may also play an important role in enzyme-linked regulation of adenosine levels because adenosine kinase is mainly expressed in astrocytes and is known to control salvage of adenosine to 5′-AMP (Boison et al. 2010; Studer et al. 2006). The intracellular astrocyte-specific enzyme adenosine kinase plays an important role in re-uptaking extracellular (Boison 2008; Boison et al. 2010), and the removal of adenosine is predominantly regulated by adenosine kinase via conversion into 5′-AMP (Boison 2006; Pak et al. 1994).

As such, astrocytes are uniquely situated to have a major influence on adenosine levels as they are related to brain bioenergetics. Astrocytes contact synapses and coordinate synaptic networks, and astrocytes release ATP and its subsequent degradation into adenosine has a major regulatory function in setting a global adenosine-mediated inhibitory tone within a neuronal network. On the other hand, end feet of astrocytes are an integral part of the BBB, brain endothelial cells express both adenosine A_1 and A_{2A} receptors, and activation of these receptors modulates the permeability of the BBB (Carman et al. 2011). Thus adenosine could have a major role in maintaining BBB integrity, regulating blood flow, and modulating the entry of energy-yielding nutrients into the brain (O'Regan 2005).

Adenosine signaling first starts with it binding to specific cell surface G-protein-coupled receptors (Benarroch 2008; Boison et al. 2010; Fredholm et al. 2005). There are four subtypes of adenosine receptors: A_1, A_{2A}, A_{2B}, and A_3 receptors. The A_1 and A_3 receptors couple to the G_i family of G proteins and inhibit cAMP formation; A_{2A} receptors couple to members of the G_s family, whereas A_{2B} receptors couple to many G proteins including G_s, G_q, and G_{12}. Adenosine is approximately equipotent on A_1 and A_{2A} receptors, with A_1 receptors having the highest affinity, whereas A_3 and A_{2B} receptors require much higher agonist concentrations. Although all four types of adenosine receptors are expressed in brain, the levels of A_1 and A_{2A} receptors are much higher than those of A_{2B} and A_3 receptors. Moreover, manipulation of A_{2B} and A_3 receptors has relatively modest impact on brain function. Thus, the impact of adenosine on brain function mostly depends mainly on the actions of A_1 and A_{2A} receptors.

The actions of adenosine depend on the cellular and subcellular location of adenosine receptors. In neurons, adenosine A_1 receptors are located presynaptically, postsynaptically, and nonsynaptically. Activation of presynaptic A_1 receptors inhibits excitatory synaptic transmission by inhibiting N-type and P/Q-type calcium channels (Ambrosio et al. 1997; Gundlfinger et al. 2007). Postsynaptic A_1 receptors can influence the responsiveness to excitatory stimuli by inhibiting NMDA receptors. Nonsynaptic A_1 receptors can stimulate potassium channels leading to neuronal hyperpolarization. Although A_{2A} receptors are mostly expressed at neuronal synapses, the activation of A_{2A} receptors appears to have limited impact on synaptic transmission but plays a crucial role in controlling synaptic plasticity (Canas et al. 2009; Gomes et al. 2011; Mills et al. 2011). All four types of adenosine receptors are expressed on brain glial cells, and adenosine can modulate glia function in many different ways including energy metabolism, reactive gliosis, glutamate uptake, and immune responses (Boison et al. 2010; Dare et al. 2007; Gomes et al. 2011). A_{2A} receptors are highly expressed in endothelial cells of brain capillaries, where they play an important role in controlling brain microvascular function (Mills et al. 2011; O'Regan 2005; Pelligrino et al. 2010).

3.3 Adenosine and Brain Adenylate Energy Charge

There are many different ways to assess brain energy status including adenylate energy charge (EC). Adenylate energy charge is a measure of the phosphorylating power of the adenylate pool, equal to one-half the average number of phosphate anhydride bonds per adenosine moiety present in the pool. It is defined in terms of the concentrations of AMP, ADP, and ATP in the pool and expressed by the equation $EC = ([ATP] + \frac{1}{2}[ADP])/([ATP] + [ADP] + [AMP])$. The maximum (mostly theoretical) EC value is 1.0, but practically speaking under physiological conditions measured brain adenylate energy charges range to 0.95, meaning that about 95 % of the adenylates are in the form of ATP (Atkinson and Walton 1967). As cellular energy levels drop, ATP levels decrease, and ADP and AMP levels increase; the result is a decrease in EC.

Maintaining brain EC and ATP levels is critical for proper neuronal function, and compromised brain energy levels have detrimental effects on neuronal function, as occur in a variety of acute and chronic neurological and neuropsychiatric disorders such as brain ischemia, sleep deprivation, epilepsy, Huntington's disease, Parkinson's disease, Alzheimer's disease, depression, bipolar disorder, and schizophrenia (Gomes et al. 2011). Because basal brain adenosine levels are nearly 10,000-fold lower than ATP levels, even negligible drops in ATP levels have the potential to produce dramatic rises in adenosine levels. As such, elevated levels of adenosine are observed in all aforementioned neurological and psychological disorders. Although such a rise in adenosine levels could affect the functions of neuron, glia, and endothelial cells in many different ways, here we focus on effects of adenosine on brain bioenergetics. We found that adenosine levels increase as brain energy levels decrease following a variety of excitatory stimuli (Shepel et al. 2005). However, this relationship was not stoichiometric, but rather occurred in three distinct phases, each defined by a linear relationship with a unique slope.

In phase 1, when there is no significant external excitatory stimulation, brain adenosine levels increase only moderately while brain adenylate energy levels decrease only modestly within a physiological range. For example, moderate increases in neuronal activity can increase the release of glutamate that then can trigger the release of ATP from astrocytes and the rapid formation of adenosine. Such moderate increases in adenosine will preferentially activate adenosine A_1 receptors, inhibit voltage-gated calcium channels, inhibit excitatory synaptic transmission, inhibit NMDA receptors, and stimulate potassium channels. As such, neuronal activity will be dampened and ATP levels will be conserved.

In phase 2, when there is significant external excitatory stimulation, brain adenosine levels increase rapidly as brain adenylate energy charges drop below the threshold of the physiological range. For instance, when brain energy levels and ATP levels are compromised, as might occur during brain ischemia, sleep deprivation, and seizures, a cascade of harmful excitotoxic events may be initiated. During excitotoxicity, released glutamate at synapses perpetuates neuronal depolarization through activation of AMPA and kainate receptors, alters intracellular calcium signaling through activation of NMDA receptors, and limits its removal by reversing glutamate uptake. Under such conditions, not only ATP expenditure is increased, but the production of ATP can be also impaired as a result of calcium overload in mitochondria. As such, adenosine levels rise more rapidly and broadly, not just at synaptic terminals but also in glial cells and at the BBB. Unlike in phase 1 where adenosine A_1 receptors at synapses are preferentially activated, in phase 2 as occurs under pathological conditions adenosine A_{2A} receptors are upregulated and may mediate the actions of adenosine. At the BBB, activation of A_{2A} and possibly A_{2B} receptors induces vasodilation, which increases cerebral blood flow and brings more oxygen and glucose to the activated brain regions. Activation of adenosine A_{2A} receptors can also increase GLUT1 expression at the brain capillary endothelial cell and thus promote glucose transport into brain as a fuel (Takagi et al. 1998). In addition to its effects on glutamate clearance, controlling neurotrophic factors, and neuro-inflammatory effects, short-term activation of adenosine A_{2A} induces glycogenolysis in astroctyes (Sorg and Magistretti 1991) and long-term activation of adenosine A_{2B} receptors promotes glycogen synthesis in astrocytes (Allaman et al. 2003), thus affecting brain energy status. Activation of neuronal adenosine A_{2A} receptors promotes glycolysis and mitochondria metabolism (Hammer et al. 2001), thus increasing ATP production. In contrast to adenosine A_1 receptors, activation of neuronal A_{2A} receptors increases glutamate release (Marchi et al. 2002) and activates NMDA receptors (Rebola et al. 2008), as occurs when neurons are highly activated. As a consequence, neuronal ATP expenditure might increase as a result of activation of neuronal adenosine A_{2A} receptors. Such an activation of neuronal adenosine A_{2A} receptors could trigger plastic changes in excitatory synapses, defining the threshold for induction of plastic changes in excitatory synapses (Cunha et al. 2008).

In phase 3, when brain is exposed to drastic insults such as global ischemia, brain adenosine levels increase dramatically as brain adenylate energy charge drops below 0.5, a critical level between life and death (Atkinson and Walton 1967). Indeed, drastic insults such as global ischemia and ischemia associated with decapitation reduce brain adenylate energy charge levels to the 0.2–0.3 range (Delaney and Geiger 1996; Onodera et al. 1986; Verhaegen et al. 1995). Under such conditions, widespread activation of NMDA receptors could facilitate a cascade of excitotoxic events including intracellular calcium dysregulation, free radical production, and inhibition of mitochondrial function, which inhibits energy production, resulting in rapid adenosine formation. Such drastic rises in levels of adenosine may not overcome the loss of brain energy levels, but rather activate low-affinity adenosine A_3 receptors and mobilize tissue repair mechanisms including apoptosis and astrocyte proliferation (Abbracchio et al. 1997; von Lubitz et al. 1999).

In short, whenever brain energy levels drop adenosine levels rise to adjust brain energy supply and to retaliate against an external stimulus that would cause excessive ATP breakdown through activation of different adenosine receptors located on target cells including neurons, glial cells, and brain endothelial cells.

3.4 Difficulty in Accurately and Precisely Measuring Levels of ATP and Adenosine

Precisely and accurately measuring levels of ATP and its metabolites including adenosine in discrete brain regions is a critical issue in studying brain bioenergetics. The main challenge in making such measurements is that even brief periods of hypoxia/ischemia cause rapid degradation of ATP and precipitous increases in levels of adenosine. Therefore, in order to measure ATP precisely and accurately, brain samples must be obtained either from well-ventilated anesthetized or conscious animals, or from postmortem brains where the enzymes responsible for production and metabolism of adenosine have been instantaneously and completely inactivated.

Studies dating back to the 1950s demonstrated the rapid loss of ATP following decapitation. Kratzing and Narayanaswami (Kratzing and Narayanaswami 1953) obtained small pieces of the cortex of young guinea pigs by either killing the animals and dissecting out tissue as quickly as possible or by rapidly freezing the brain with liquid air. Values from the rapidly frozen brains were around 3.0 mmol/kg whereas those from the brain slices were 0.73 mmol/kg. In order to minimize such a loss in ATP various freezing methods were employed and it was found that most of the decrease in ATP levels occurred in the first minute following the ultimate hypoxic/ischemic insult: decapitation. In comparing ATP values from mouse brains rapidly dissected (20 s) and then frozen in liquid nitrogen versus brains obtained from mouse heads that were dropped immediately into liquid nitrogen the levels of ATP were 0.38 mmol/kg for the dissected brains and 1.4 mmol/kg for the rapidly frozen brains (Mandel and Harth 1961). When the heads of decapitated mice were dropped into liquid Freon 12 at intervals ranging from immediately to 10 min the highest values of ATP obtained were 2.5 mmol/kg and were one-half that value with an interval delay of about 6 s and reached a low rather constant level with an interval delay of about 1 min (Lowry et al. 1964). In attempting to obviate these rather dramatic postmortem decreases in ATP, a method to freeze the brain in situ was devised. Using this method, liquid nitrogen was poured directly onto the skulls of anesthetized rats that had received implanted thermocouples so that the freeze front could be timed and the cessation of blood flow could be measured (Ponten et al. 1973). At a depth of only 3 mm, blood flow continued for 5 to 10 s at which time the tissue had reached 20 °C and it took another 5–6 s for the tissue to freeze. Thus, one would infer that decapitating animals directly into liquid nitrogen would involve 10–15 s of ischemia even at the most superficial layers of the brain. Another way of inactivating enzymes responsible for adenine nucleotide breakdown and adenosine production is the "freeze-blowing" technique (Veech et al. 1973), in which high-pressure air is blown into one side of the cranial cavity via a sharp hollow steel probe and the supratentorial brain expelled via another tube on the contralateral side onto a liquid N_2-cooled aluminum plate, thus immediately halting metabolic reactions at defined time points on the order of seconds after an intervention. The above-mentioned freezing techniques, while to varying degrees are successful at providing accurate measures of purine levels, do not allow for the precise and accurate measurement for ATP and endogenous adenosine levels in discrete brain regions.

An innovative way to measure brain energetics precisely and accurately is instantaneously inactivating the catabolic enzymes with high-energy focused microwave irradiation (Delaney and Geiger 1996). Using this technique, brain tissue temperature is brought up to an enzyme-inactivating temperature of about 85 °C within a few seconds depending on the size of the animal and the power setting of the head-focused high-energy microwave. They found that a power level of 10 kW heated rat brain tissue to this temperature in 1 s, 6 kW in 2 s, and 3.5 kW in 3 s. The highest irradiation power level used (10 kW for 1.25 s) maintained the brain adenylate energy charge closest to the theoretical maximal value of 1.0. The levels of ATP increased and levels of adenosine decreased as irradiation power increased. In addition, they sacrificed rats and extracted tissues using techniques such as decapitation, immersion of the decapitated head in liquid nitrogen, and in situ freezing of the brain tissue with liquid nitrogen. The decapitation into liquid nitrogen yielded values around 0.3 mmol/kg; the decapitation followed by dissection and then freezing yielded values around 0.06 mmol/kg; and the brains that were exposed to 10 kW microwave radiation yielded values around 3.5 mmol/kg. Because the data reported by Delaney and Geiger were reported in mg/protein, the data showing here were translated, assuming that the protein content of the tissue was 12 % (Sagar et al. 1987).

Even such an extreme method as high-energy focused microwave irradiation creates nonuniform heating throughout the brain with the basal structures of brain heating up before the dorsal structures. As a result, purinergic enzymes would not be denatured as quickly in areas of brain that do not reach 85 °C and therefore ATP levels would be lower and adenosine levels would be higher. Another drawback of the high-energy focused microwave irradiation method is its inability to differentiate intracellular and extracellular levels of ATP and its metabolites, because brain regions are homogenized and analyzed as a whole.

Alternatively, levels of ATP and its metabolites can be measured in vivo in well-ventilated anesthetized or conscious animals. The first of such measurements are the brain microdialysis and cortical cup methods, which allow for direct measurement of ATP and adenosine from brain interstitial fluid (Lutz and Kabler 1997; Melani et al. 2005; Pazzagli et al. 1993; Van Wylen et al. 1986). Such methods can also inform as to the temporal sequence of release events and whether ATP is released before adenosine or vice versa. The disadvantages of such techniques are the tissue trauma caused by the implantation of the dialysis cannula and poor temporal resolution in detecting ATP and adenosine levels. The development of adenosine and ATP biosensors allows for direct and real-time measurements under physiological condition, which provide a higher temporal and spatial resolution in detection of adenosine and ATP levels (Dale et al. 2000; Frenguelli et al. 2003; Frenguelli et al. 2007; Llaudet et al. 2003, 2005). However, purine levels measured by those biosensors are not true concentrations of adenosine and ATP, but rather relative values.

In addition to direct measurement of ATP and its metabolites to assess brain energy status, it is also critical to measure accurately and precisely the brain energy reserve, glycogen. Glycogen can be visually identified in brain sections using a cytochemical staining method that relies on the interaction of glycogen with osmium tetroxide (Cataldo and Broadwell 1986; Koizumi 1974; Wender et al. 2000) or the use of periodic acid–Schiff's base (Kong et al. 2002). The absolute glycogen content can be measured using quantitative methods, which have traditionally been carried out using a biochemical assay on postmortem tissue. However, glycogen degrades extremely rapidly, and such degradation must be halted in order to measure glycogen content accurately and precisely. Similar freezing methods as those used to inactivate enzymes responsible for adenine nucleotide breakdown and adenosine production have been used for quantitative glycogen measurement. Such methods include immersion of whole animals or decapitated heads in liquid nitrogen (Passonneau and Lauderdale 1974) or Freon-12 (King et al. 1967), freezing brains in situ using liquid nitrogen poured directly on the skull (Cruz and Dienel 2002), and rapidly removing brain tissue at 0 °C followed by immediate freezing (Wender et al. 2000). The high-energy focused microwave irradiation, which denatures enzymes responsible for glycogen metabolism preserving glycogen content at the time of irradiation, has also been used (Sagar et al. 1987). Another critical step in accurate measurement of glycogen is to minimize animal handling prior to assay (Cruz and Dienel 2002). The most accurate method for measuring glycogen is debranching glycogen into glucose followed by standard biochemical assay of glucose (Passonneau and Lauderdale 1974), with glycogen being expressed as glucosyl units, as the debranching enzymes yield one glucose molecule per glucosyl unit of the glycogen molecule. The manner in which glycogen content is expressed can be confusing, since glycogen is often expressed as glucosyl units either relative to a normalized protein concentration (pmol glycogen/μg protein) or relative to the wet or dry weight of the tissue (μmol glycogen/g wet tissue). Assuming that wet brain tissue contains about 12 % protein content (Sagar et al. 1987), and that dry brain tissue contains about half the protein of wet brain tissue, a simple calculation could allow interconversion.

Alternatively, glycogen content can be estimated in vivo using a noninvasive nuclear magnetic resonance (NMR) method, which detects the ^{13}C glucose incorporation into the brain glycogen. In such a method, ^{13}C-labeled glucose is injected systemically and the labeled glucose is slowly incorporated into the brain glycogen pool producing a detectable glycogen signal that can be continuously monitored (Choi et al. 1999). The advantage of this noninvasive technology is that it could be used in humans to record glycogen measurements under both pathological and physiological conditions. However, the observed increase in ^{13}C signal over time might not necessarily be reflecting changes in glycogen concentration alone, but also an increased incorporation of the ^{13}C label. A new approach was recently developed that allows the measurement of in vivo brain glycogen concentration more accurately (Morgenthaler et al. 2008). In this method, the influence of glycogen turnover, a potential confounding factor in the ^{13}C NMR, was minimized by prelabeling the animals and achieving a near steady-state isotopic enrichment using long-term [1-^{13}C]-glucose administration.

In summary, it is not an easy task to measure brain energy status accurately and precisely. Essential components in assessing correctly brain energetics should include justifying carefully the methodology used and putting the data in the context of what is already known about brain energy metabolism.

Acknowledgments The work currently conducted in our laboratories is supported by 2P20RR0017699 from the NCRR, a component of the NIH, and by R01NS069597.

References

Abbott NJ, Patabendige AA, Dolman DE, Yusof SR, Begley DJ (2010) Structure and function of the blood-brain barrier. Neurobiol Dis 37:13–25

Abbracchio MP, Ceruti S, Brambilla R, Franceschi C, Malorni W, Jacobson KA, von Lubitz DK, Cattabeni F (1997) Modulation of apoptosis by adenosine in the central nervous system: a possible role for the A_3 receptor. Pathophysiological significance and therapeutic implications for neurodegenerative disorders. Ann N Y Acad Sci 825:11–22

Allaman I, Lengacher S, Magistretti PJ, Pellerin L (2003) A_{2B} receptor activation promotes glycogen synthesis in astrocytes through modulation of gene expression. Am J Physiol Cell Physiol 284:C696–C704

Ambrosio AF, Malva JO, Carvalho AP, Carvalho CM (1997) Inhibition of N-, P/Q- and other types of Ca^{2+} channels in rat hippocampal nerve terminals by the adenosine A1 receptor. Eur J Pharmacol 340:301–310

Atkinson DE, Walton GM (1967) Adenosine triphosphate conservation in metabolic regulation. Rat liver citrate cleavage enzyme. J Biol Chem 242:3239–3241

Attwell D, Laughlin SB (2001) An energy budget for signaling in the grey matter of the brain. J Cereb Blood Flow Metab 21:1133–1145

Auestad N, Korsak RA, Morrow JW, Edmond J (1991) Fatty acid oxidation and ketogenesis by astrocytes in primary culture. J Neurochem 56:1376–1386

Benarroch EE (2008) Adenosine and its receptors: multiple modulatory functions and potential therapeutic targets for neurologic disease. Neurology 70:231–236

Bergersen L, Rafiki A, Ottersen OP (2002) Immunogold cytochemistry identifies specialized membrane domains for monocarboxylate transport in the central nervous system. Neurochem Res 27:89–96

Bixel MG, Hamprecht B (1995) Generation of ketone bodies from leucine by cultured astroglial cells. J Neurochem 65:2450–2461

Boison D (2006) Adenosine kinase, epilepsy and stroke: mechanisms and therapies. Trends Pharmacol Sci 27:652–658

Boison D (2008) The adenosine kinase hypothesis of epileptogenesis. Prog Neurobiol 84:249–262

Boison D, Chen JF, Fredholm BB (2010) Adenosine signaling and function in glial cells. Cell Death Differ 17:1071–1082

Brown AM (2004) Brain glycogen re-awakened. J Neurochem 89:537–552

Canas PM, Porciuncula LO, Cunha GM, Silva CG, Machado NJ, Oliveira JM, Oliveira CR, Cunha RA (2009) Adenosine A_{2A} receptor blockade prevents synaptotoxicity and memory dysfunction caused by beta-amyloid peptides via p38 mitogen-activated protein kinase pathway. J Neurosci 29:14741–14751

Carman AJ, Mills JH, Krenz A, Kim DG, Bynoe MS (2011) Adenosine receptor signaling modulates permeability of the blood-brain barrier. J Neurosci 31:13272–13280

Cass CE, Young JD, Baldwin SA, Cabrita MA, Graham KA, Griffiths M, Jennings LL, Mackey JR, Ng AM, Ritzel MW et al (1999) Nucleoside transporters of mammalian cells. Pharm Biotechnol 12:313–352

Cataldo AM, Broadwell RD (1986) Cytochemical identification of cerebral glycogen and glucose-6-phosphatase activity under normal and experimental conditions. II. Choroid plexus and ependymal epithelia, endothelia and pericytes. J Neurocytol 15:511–524

Chih CP, Roberts EL Jr (2003) Energy substrates for neurons during neural activity: a critical review of the astrocyte-neuron lactate shuttle hypothesis. J Cereb Blood Flow Metab 23: 1263–1281

Choi IY, Gruetter R (2003) In vivo ^{13}C NMR assessment of brain glycogen concentration and turnover in the awake rat. Neurochem Int 43:317–322

Choi IY, Tkac I, Ugurbil K, Gruetter R (1999) Noninvasive measurements of [1-^{13}C]glycogen concentrations and metabolism in rat brain in vivo. J Neurochem 73:1300–1308

Clausen T (1998) Clinical and therapeutic significance of the Na^+, K^+ pump*. Clin Sci (Lond) 95: 3–17

Cruz NF, Dienel GA (2002) High glycogen levels in brains of rats with minimal environmental stimuli: implications for metabolic contributions of working astrocytes. J Cereb Blood Flow Metab 22:1476–1489

Cunha RA, Ferre S, Vaugeois JM, Chen JF (2008) Potential therapeutic interest of adenosine A_{2A} receptors in psychiatric disorders. Curr Pharm Des 14:1512–1524

Cunnane S, Nugent S, Roy M, Courchesne-Loyer A, Croteau E, Tremblay S, Castellano A, Pifferi F, Bocti C, Paquet N et al (2011) Brain fuel metabolism, aging, and Alzheimer's disease. Nutrition 27:3–20

Dale N, Pearson T, Frenguelli BG (2000) Direct measurement of adenosine release during hypoxia in the CA1 region of the rat hippocampal slice. J Physiol 526(Pt 1):143–155

Darby M, Kuzmiski JB, Panenka W, Feighan D, MacVicar BA (2003) ATP released from astrocytes during swelling activates chloride channels. J Neurophysiol 89:1870–1877

Dare E, Schulte G, Karovic O, Hammarberg C, Fredholm BB (2007) Modulation of glial cell functions by adenosine receptors. Physiol Behav 92:15–20

Delaney SM, Geiger JD (1996) Brain regional levels of adenosine and adenosine nucleotides in rats killed by high-energy focused microwave irradiation. J Neurosci Methods 64:151–156

Ebert D, Haller RG, Walton ME (2003) Energy contribution of octanoate to intact rat brain metabolism measured by ^{13}C nuclear magnetic resonance spectroscopy. J Neurosci 23:5928–5935

Fredholm BB, Chen JF, Cunha RA, Svenningsson P, Vaugeois JM (2005) Adenosine and brain function. Int Rev Neurobiol 63:191–270

Frenguelli BG, Llaudet E, Dale N (2003) High-resolution real-time recording with microelectrode biosensors reveals novel aspects of adenosine release during hypoxia in rat hippocampal slices. J Neurochem 86:1506–1515

Frenguelli BG, Wigmore G, Llaudet E, Dale N (2007) Temporal and mechanistic dissociation of ATP and adenosine release during ischaemia in the mammalian hippocampus. J Neurochem 101:1400–1413

Gomes CV, Kaster MP, Tome AR, Agostinho PM, Cunha RA (2011) Adenosine receptors and brain diseases: neuroprotection and neurodegeneration. Biochim Biophys Acta 1808:1380–1399

Gundlfinger A, Bischofberger J, Johenning FW, Torvinen M, Schmitz D, Breustedt J (2007) Adenosine modulates transmission at the hippocampal mossy fibre synapse via direct inhibition of presynaptic calcium channels. J Physiol 582:263–277

Guzman M, Blazquez C (2001) Is there an astrocyte-neuron ketone body shuttle? Trends Endocrinol Metab 12:169–173

Guzman M, Blazquez C (2004) Ketone body synthesis in the brain: possible neuroprotective effects. Prostaglandins Leukot Essent Fatty Acids 70:287–292

Hammer J, Qu H, Haberg A, Sonnewald U (2001) In vivo effects of adenosine A_2 receptor agonist and antagonist on neuronal and astrocytic intermediary metabolism studied with ex vivo ^{13}C MR spectroscopy. J Neurochem 79:885–892

Hasselbalch SG, Knudsen GM, Jakobsen J, Hageman LP, Holm S, Paulson OB (1994) Brain metabolism during short-term starvation in humans. J Cereb Blood Flow Metab 14:125–131

Iglesias R, Dahl G, Qiu F, Spray DC, Scemes E (2009) Pannexin 1: the molecular substrate of astrocyte "hemichannels". J Neurosci 29:7092–7097

Ito K, Uchida Y, Ohtsuki S, Aizawa S, Kawakami H, Katsukura Y, Kamiie J, Terasaki T (2011) Quantitative membrane protein expression at the blood-brain barrier of adult and younger cynomolgus monkeys. J Pharm Sci 11:3939–3950

Itoh Y, Esaki T, Shimoji K, Cook M, Law MJ, Kaufman E, Sokoloff L (2003) Dichloroacetate effects on glucose and lactate oxidation by neurons and astroglia in vitro and on glucose utilization by brain in vivo. Proc Natl Acad Sci U S A 100:4879–4884

Kang J, Kang N, Lovatt D, Torres A, Zhao Z, Lin J, Nedergaard M (2008) Connexin 43 hemichannels are permeable to ATP. J Neurosci 28:4702–4711

Kashiwaya Y, Sato K, Tsuchiya N, Thomas S, Fell DA, Veech RL, Passonneau JV (1994) Control of glucose utilization in working perfused rat heart. J Biol Chem 269:25502–25514

Kasischke KA, Vishwasrao HD, Fisher PJ, Zipfel WR, Webb WW (2004) Neural activity triggers neuronal oxidative metabolism followed by astrocytic glycolysis. Science 305:99–103

King AE, Ackley MA, Cass CE, Young JD, Baldwin SA (2006) Nucleoside transporters: from scavengers to novel therapeutic targets. Trends Pharmacol Sci 27:416–425

King LJ, Lowry OH, Passonneau JV, Venson V (1967) Effects of convulsants on energy reserves in the cerebral cortex. J Neurochem 14:599–611

Koizumi J (1974) Glycogen in the central nervous system. Prog Histochem Cytochem 6:1–37

Kong J, Shepel PN, Holden CP, Mackiewicz M, Pack AI, Geiger JD (2002) Brain glycogen decreases with increased periods of wakefulness: implications for homeostatic drive to sleep. J Neurosci 22:5581–5587

Kratzing CC, Narayanaswami A (1953) The enzymic determination of energy-rich phosphates in brain. Biochem J 54:317–323

LaManna JC, Salem N, Puchowicz M, Erokwu B, Koppaka S, Flask C, Lee Z (2009) Ketones suppress brain glucose consumption. Adv Exp Med Biol 645:301–306

Latini S, Pedata F (2001) Adenosine in the central nervous system: release mechanisms and extracellular concentrations. J Neurochem 79:463–484

Leino RL, Gerhart DZ, Duelli R, Enerson BE, Drewes LR (2001) Diet-induced ketosis increases monocarboxylate transporter (MCT1) levels in rat brain. Neurochem Int 38:519–527

Levitsky LL, Fisher DE, Paton JB, Delannoy CW (1977) Fasting plasma levels of glucose, acetoacetate, D-beta-hydroxybutyrate, glycerol, and lactate in the baboon infant: correlation with cerebral uptake of substrates and oxygen. Pediatr Res 11:298–302

Llaudet E, Botting NP, Crayston JA, Dale N (2003) A three-enzyme microelectrode sensor for detecting purine release from central nervous system. Biosens Bioelectron 18:43–52

Llaudet E, Hatz S, Droniou M, Dale N (2005) Microelectrode biosensor for real-time measurement of ATP in biological tissue. Anal Chem 77:3267–3273

Lopes-Cardozo M, Larsson OM, Schousboe A (1986) Acetoacetate and glucose as lipid precursors and energy substrates in primary cultures of astrocytes and neurons from mouse cerebral cortex. J Neurochem 46:773–778

Lowry OH, Passonneau JV, Hasselberger FX, Schulz DW (1964) Effect of ischemia on known substrates and cofactors of the glycolytic pathway in brain. J Biol Chem 239:18–30

Lutz PL, Kabler S (1997) Release of adenosine and ATP in the brain of the freshwater turtle (Trachemys scripta) during long-term anoxia. Brain Res 769:281–286

Magistretti PJ, Pellerin L, Martin J-L (1995) Brain energy metabolism: an integrated cellular perspective. In: Bloom FE, Kupfer DJ (eds) Psychopharmacology: the fourth generation of progress. Raven Press, New York, pp 657–670

Mandel P, Harth S (1961) Free nucleotides of the brain in various mammals. J Neurochem 8: 116–125

Marchi M, Raiteri L, Risso F, Vallarino A, Bonfanti A, Monopoli A, Ongini E, Raiteri M (2002) Effects of adenosine A_1 and A_{2A} receptor activation on the evoked release of glutamate from rat cerebrocortical synaptosomes. Br J Pharmacol 136:434–440

McKenna MC, Rolf Gruetter R, Sonnewald U, Waagepetersen HS, Schousboe A (2006) Energy metabolism of the brain. In: Siegel G, Albers RW, Brady S, Price DL (eds) Basic neurochemistry. London, Elsevier, pp 531–557

Meghji P, Newby AC (1990) Sites of adenosine formation, action and inactivation in the brain. Neurochem Int 16:227–232

Melani A, Turchi D, Vannucchi MG, Cipriani S, Gianfriddo M, Pedata F (2005) ATP extracellular concentrations are increased in the rat striatum during in vivo ischemia. Neurochem Int 47:442–448

Mills JH, Alabanza L, Weksler BB, Couraud PO, Romero IA, Bynoe MS (2011) Human brain endothelial cells are responsive to adenosine receptor activation. Purinergic Signal 7:265–273

Mobbs CV, Mastaitis JW, Zhang M, Isoda F, Cheng H, Yen K (2007) Secrets of the lac operon. Glucose hysteresis as a mechanism in dietary restriction, aging and disease. Interdiscip Top Gerontol 35:39–68
Morgenthaler FD, van Heeswijk RB, Xin L, Laus S, Frenkel H, Lei H, Gruetter R (2008) Non-invasive quantification of brain glycogen absolute concentration. J Neurochem 107:1414–1423
Nelson SR, Schulz DW, Passonneau JV, Lowry OH (1968) Control of glycogen levels in brain. J Neurochem 15:1271–1279
Newby AC, Worku Y, Meghji P, Nakazawa M, Skladanowski AC (1990) Adenosine: a retaliatory metabolite or not? Physiology 5:67–70
O'Regan M (2005) Adenosine and the regulation of cerebral blood flow. Neurol Res 27:175–181
Onodera H, Iijima K, Kogure K (1986) Mononucleotide metabolism in the rat brain after transient ischemia. J Neurochem 46:1704–1710
Owen OE, Morgan AP, Kemp HG, Sullivan JM, Herrera MG, Cahill GF Jr (1967) Brain metabolism during fasting. J Clin Invest 46:1589–1595
Pak MA, Haas HL, Decking UK, Schrader J (1994) Inhibition of adenosine kinase increases endogenous adenosine and depresses neuronal activity in hippocampal slices. Neuropharmacology 33:1049–1053
Pascual O, Casper KB, Kubera C, Zhang J, Revilla-Sanchez R, Sul JY, Takano H, Moss SJ, McCarthy K, Haydon PG (2005) Astrocytic purinergic signaling coordinates synaptic networks. Science 310:113–116
Passonneau JV, Lauderdale VR (1974) A comparison of three methods of glycogen measurement in tissues. Anal Biochem 60:405–412
Patel MS, Johnson CA, Rajan R, Owen OE (1975) The metabolism of ketone bodies in developing human brain: development of ketone-body-utilizing enzymes and ketone bodies as precursors for lipid synthesis. J Neurochem 25:905–908
Pazzagli M, Pedata F, Pepeu G (1993) Effect of K^+ depolarization, tetrodotoxin, and NMDA receptor inhibition on extracellular adenosine levels in rat striatum. Eur J Pharmacol 234:61–65
Pelligrino DA, Xu HL, Vetri F (2010) Caffeine and the control of cerebral hemodynamics. J Alzheimers Dis 20(Suppl 1):S51–S62
Peng L, Huang R, Yu AC, Fung KY, Rathbone MP, Hertz L (2005) Nucleoside transporter expression and function in cultured mouse astrocytes. Glia 52:25–35
Ponten U, Ratcheson RA, Salford LG, Siesjo BK (1973) Optimal freezing conditions for cerebral metabolites in rats. J Neurochem 21:1127–1138
Rafiki A, Boulland JL, Halestrap AP, Ottersen OP, Bergersen L (2003) Highly differential expression of the monocarboxylate transporters MCT2 and MCT4 in the developing rat brain. Neuroscience 122:677–688
Rebola N, Lujan R, Cunha RA, Mulle C (2008) Adenosine A_{2A} receptors are essential for long-term potentiation of NMDA-EPSCs at hippocampal mossy fiber synapses. Neuron 57:121–134
Rolfe DF, Brown GC (1997) Cellular energy utilization and molecular origin of standard metabolic rate in mammals. Physiol Rev 77:731–758
Sagar SM, Sharp FR, Swanson RA (1987) The regional distribution of glycogen in rat brain fixed by microwave irradiation. Brain Res 417:172–174
Sato K, Kashiwaya Y, Keon CA, Tsuchiya N, King MT, Radda GK, Chance B, Clarke K, Veech RL (1995) Insulin, ketone bodies, and mitochondrial energy transduction. FASEB J 9:651–658
Shepel PN, Ramonet D, Stevens P, Geiger JD (2005) Purine level regulation during energy depletion associated with graded excitatory stimulation in brain. Neurol Res 27:139–148
Smith D, Pernet A, Hallett WA, Bingham E, Marsden PK, Amiel SA (2003) Lactate: a preferred fuel for human brain metabolism in vivo. J Cereb Blood Flow Metab 23:658–664
Sorg O, Magistretti PJ (1991) Characterization of the glycogenolysis elicited by vasoactive intestinal peptide, noradrenaline and adenosine in primary cultures of mouse cerebral cortical astrocytes. Brain Res 563:227–233
Spector R (2009) Nutrient transport systems in brain: 40 years of progress. J Neurochem 111:315–320

Studer FE, Fedele DE, Marowsky A, Schwerdel C, Wernli K, Vogt K, Fritschy JM, Boison D (2006) Shift of adenosine kinase expression from neurons to astrocytes during postnatal development suggests dual functionality of the enzyme. Neuroscience 142:125–137

Takagi H, King GL, Aiello LP (1998) Hypoxia upregulates glucose transport activity through an adenosine-mediated increase of GLUT1 expression in retinal capillary endothelial cells. Diabetes 47:1480–1488

Van Wylen DG, Park TS, Rubio R, Berne RM (1986) Increases in cerebral interstitial fluid adenosine concentration during hypoxia, local potassium infusion, and ischemia. J Cereb Blood Flow Metab 6:522–528

Vannucci SJ, Maher F, Simpson IA (1997) Glucose transporter proteins in brain: delivery of glucose to neurons and glia. Glia 21:2–21

Vannucci SJ, Simpson IA (2003) Developmental switch in brain nutrient transporter expression in the rat. Am J Physiol Endocrinol Metab 285:E1127–E1134

Veech RL, Chance B, Kashiwaya Y, Lardy HA, Cahill GF Jr (2001) Ketone bodies, potential therapeutic uses. IUBMB Life 51:241–247

Veech RL, Harris RL, Veloso D, Veech EH (1973) Freeze-blowing: a new technique for the study of brain in vivo. J Neurochem 20:183–188

Verhaegen M, Iaizzo PA, Todd MM (1995) A comparison of the effects of hypothermia, pentobarbital, and isoflurane on cerebral energy stores at the time of ischemic depolarization. Anesthesiology 82:1209–1215

von Lubitz DK, Ye W, McClellan J, Lin RC (1999) Stimulation of adenosine A_3 receptors in cerebral ischemia. Neuronal death, recovery, or both? Ann N Y Acad Sci 890:93–106

Wender R, Brown AM, Fern R, Swanson RA, Farrell K, Ransom BR (2000) Astrocytic glycogen influences axon function and survival during glucose deprivation in central white matter. J Neurosci 20:6804–6810

White H, Venkatesh B (2011) Clinical review: Ketones and brain injury. Crit Care 15:219

Yang SY, He XY, Schulz H (1987) Fatty acid oxidation in rat brain is limited by the low activity of 3-ketoacyl-coenzyme A thiolase. J Biol Chem 262:13027–13032

Chapter 4
Adenosine and Autocrine Metabolic Regulation of Neuronal Activity

Masahito Kawamura Jr. and David N. Ruskin

Abstract The cause-and-effect relationships between altered metabolism and neurological conditions are not fully understood. However, some metabolic conditions including ischemia/hypoxia, fasting and ketogenic diet therapy are known to modulate processes in the nervous system such as neuroprotection and seizures, indicating that altered metabolism can regulate neuronal excitability. Adenosine is one of the agents linking metabolism to neuronal activity. In this chapter, we discuss purinergic signaling via ATP release and subsequent activation of adenosine receptors, revealing a CNS pathway for metabolic autocrine regulation. ATP is the final product of brain energy metabolism and it can be an indicator of brain metabolic changes. ATP can be released to the extracellular space through sites including gap junction hemichannels, vesicles, and chloride channels. After subsequent dephosphorylation of ATP by ectonucleotidases, the resulting metabolite adenosine can activate adenosine receptors to limit neuronal excitability. All the proteins for purinergic signaling (ATP releasing sites, ectonucleotidases, and adenosine receptors) may be expressed in the same cell, allowing for autocrine regulation. We give one example: our finding of adenosinergic autoregulation through a cascade involving pannexin-1 channels, adenosine A_1 receptors, and ATP-sensitive potassium channels in hippocampal CA3 pyramidal neurons. These neurons can react to changes in ATP and glucose levels with altered electrical activity through this purinergic signaling pathway. This pathway may underlie effects seen in ischemic conditions or during ketogenic diet treatment. Adenosinergic autocrine regulation might have a significant role in neuroprotection and seizure regulation in the CNS.

M. Kawamura Jr. (✉)
Department of Pharmacology, Jikei University School of Medicine,
3-25-8 Nishi-shimbashi, Minato-ku, Tokyo 105-8461, Japan
e-mail: m.kawamura@jikei.ac.jp

D.N. Ruskin
Neuroscience Program and Department of Psychology, Trinity College,
300 Summit St, Hartford, CT 06106, USA
e-mail: david.ruskin@trincoll.edu

S. Masino and D. Boison (eds.), *Adenosine: A Key Link between Metabolism and Brain Activity*, DOI 10.1007/978-1-4614-3903-5_4,

Keywords ATP • Adenosine • ATP release • Adenosine A_1 receptors • Autocrine regulation • Pannexin-1 channels • ATP-sensitive potassium channels • Hippocampal CA3 pyramidal neurons

4.1 Purinergic Signaling Links Metabolism to Neuronal Activity

4.1.1 Altered Metabolism with Ischemia/Hypoxia, Fasting, or Dietary Therapy Modulates Neuronal Activity

Neuronal ATP is generated by oxidative phosphorylation in mitochondria. In the brain, mitochondria use pyruvate from cytoplasmic glycolysis of glucose for producing NADH and $FADH_2$ by enzymes of the tricarboxylic acid cycle. NADH and $FADH_2$ transfer energy from the tricarboxylic acid cycle to the electron transport chain and then the electron transport chain forms ATP from ADP and phosphate with ATP synthase (Hertz et al. 2007). ATP production is dependent on mitochondrial function and oxygen supply and thus is sensitive to altered metabolism. Altered metabolism is known to be one of the hallmarks of various neurological conditions including neurodegenerative disorders (Sas et al. 2007) and psychiatric disorders (Rezin et al. 2009). Although it has been reported that some patients with neurological disorders have mitochondrial dysfunction (Rezin et al. 2009), the cause-and-effect relationship between energy metabolism and these neurological disorders is not understood. Ischemia/hypoxia, fasting, and dietary therapy are some of the exceptions in which we know the cause and the effect. Ischemic/hypoxic stress is well determined to cause neuroprotection by reducing neuronal excitability (Ballanyi 2004; Dunwiddie and Masino 2001; Johansson et al. 2001; Miller and Hsu 1992; Rudolphi et al. 1992; Zhang and Krnjević 1993), although ischemia/hypoxia of sufficient magnitude can certainly overcome this neuroprotective effect to produce damage. Fasting has been reported to suppress seizure activity (Greene et al. 2003) and to be neuroprotective (Davis et al. 2008). These phenomena indicate that some alterations in metabolism can limit neuronal activity. The most convincing evidence might be anticonvulsant dietary therapy, such as the ketogenic diet or simple caloric restriction. The ketogenic diet (low carbohydrate, high fat) mimics the metabolic condition of fasting (Wilder 1921) and has been used successfully to treat pediatric (Freeman et al. 2007) and medically refractory epilepsy (Hallböök et al. 2007). In rodents, ketogenic diet (Bough et al. 2006; Hartman et al. 2008; Rho et al. 1999; Uhlemann and Neims 1972) or caloric restriction (Greene et al. 2001) also reduces seizure activity. These reports strongly argue that manipulations of brain metabolism caused by dietary therapies can limit neuronal activity for an anticonvulsant effect; some pathway must connect metabolism to regulation of neuronal excitability. The mitochondrial metabolites including purines, NADH, and $FADH_2$ and mitochondrial calcium signaling are reported to be metabolic links for regulating neuronal activity (Kann and Kovacs 2007). Adenosine is thought to be one of the important links between the metabolic modulation of neurons and the effects of ischemia/hypoxia, fasting and anticonvulsant diets.

4.1.2 Adenosine Links Metabolism to Neuronal Activity

Adenosine is the core molecule of ATP and a well-known neuromodulator with receptor-mediated neuroprotective and anticonvulsant effects (Dunwiddie and Masino 2001; Fredholm 1997; Stone 2002; Wardas 2002). These receptors are characterized into four subtypes: A_1, A_{2A}, A_{2B}, and A_3 (Fredholm et al. 2000; Ralevic and Burnstock 1998). Adenosine A_1 receptors are widely distributed throughout the whole brain (Goodman and Snyder 1982), and are functionally expressed in various neuronal cell types (Dunwiddie and Masino 2001; Fredholm 1995) with coupling by G-proteins to modulate activity through calcium or potassium channels (Schulte and Fredholm 2003). Adenosine A_{2A} receptors are located mainly in the basal ganglia and olfactory tubercle (Rosin et al. 2003) and are known to form receptor heteromers with dopamine D_2 receptors (Ferré et al. 2007). Adenosine A_{2B} and A_3 receptors are less expressed in the CNS, but are functionally expressed in astrocytes (Chen et al. 2001; Kawamura and Kawamura 2011; Peakman and Hill 1994).

Through altered metabolism, adenosine can link ischemia/hypoxia, fasting and the effects of the ketogenic diet to neuronal activity for neuroprotective and/or anticonvulsant outcomes:

1. Activation of adenosine A_1 receptors reduces neuronal activity by decreasing excitatory synaptic transmission (Dunwiddie and Hoffer 1980; Masino et al. 2002; Masino et al. 2001; Thompson et al. 1992) and/or causing hyperpolarization (Pan et al. 1995; Thompson et al. 1992), thereby augmenting neuroprotective (Fredholm 1997; Sebastião et al. 2001; Stone 2002; Wardas 2002) and anticonvulsant effects (Anschel et al. 2004; Dunwiddie 1999).
2. Ischemic or hypoxic conditions increase extracellular adenosine concentrations in the CNS (Fowler 1993; Frenguelli et al. 2003; Frenguelli et al. 2007; Latini and Pedata 2001; Parkinson et al. 2005), and the anticonvulsant effect of fasting or the ketogenic diet is thought to be caused in part by adenosine (Masino and Geiger 2008; Masino et al. 2009; Masino et al. 2011).
3. One source of extracellular adenosine is the breakdown of extracellular ATP (Pascual et al. 2005). There are various mechanisms for ATP efflux to the extracellular space (see Sect. 2), and ATP release sites have been reported to colocalize with ectonucleotidases (which dephosphorylate ATP into adenosine (Joseph et al. 2003)): dephosphorylation of ATP to adenosine occurs rapidly ($t_{1/2}$ ~200 ms) in the extracellular space (Dunwiddie et al. 1997).
4. There is also evidence that, during hypoxia or ischemia, adenosine can be directly released into the extracellular space (Frenguelli et al. 2007; Lloyd et al. 1993). This release might be largely from astrocytes (Martín et al. 2007).

5. Most importantly, ATP is a final product of energy metabolism. The most crucial role of ATP is not as a source of extracellular adenosine, but as a source for intracellular energy. ATP concentration is easily changed by modulating mitochondrial activity via altered metabolism. For instance, the ketogenic diet (DeVivo et al. 1978; Nakazawa et al. 1983; Nylen et al. 2009) and potentially Huntington's disease (Oláh et al. 2008) are known to increase brain ATP levels.

Given the above, it is plausible that changes in cellular ATP will be reflected in changes in extracellular adenosine, which can then regulate neuronal activity via its receptors. This adenosine receptor-activated purinergic cascade links metabolism to neuronal activity and might play a significant anticonvulsant or neuroprotective role in neurological conditions or dietary therapy.

4.2 Regulation of Extracellular Adenosine Levels in the CNS

4.2.1 ATP Release as a Source of Extracellular Adenosine

Classically, ATP is thought to be released with global and synchronized cell death (Inoue 1998). Local thermal brain injury has been reported to cause ATP release from astrocytes in the cortex (Davalos et al. 2005). These types of ATP release, however, involve transient efflux and an irreversible stimulus. A more physiological source of extracellular adenosine is extracellular ATP efflux through ATP releasing sites. In the CNS, ATP releasing sites are well established in astrocytes (Coco et al. 2003). Table 4.1 shows the cell types in the CNS reported to release ATP into the extracellular space. ATP releasing sites are of three different types: connexin/pannexin-based channels, vesicular release, and chloride channels. It has been reported that astrocytes release ATP from all these types. Hippocampal astrocytes express one of the subtypes of gap junction proteins (connexin-43), and hemichannels made of this connexin are permeable to ATP (Kang et al. 2008). With low concentrations of extracellular divalent cations, connexin-43 hemichannels release ATP to the extracellular space from astrocytes (Stout et al. 2002). Pannexins are connexin-related proteins that form channels but not gap junctions (MacVicar and Thompson 2009); they also release ATP with ischemic stress of cortical astrocytes (Iwabuchi and Kawahara 2011). This ATP release is regulated with negative-feedback mechanisms via $P2X_7$ receptors activated by releasing ATP through the pannexin-1 channels (Iwabuchi and Kawahara 2011). The role of ischemia-induced ATP release through pannexin channels is still controversial. Oxygen-glucose deprivation-induced ATP release in the hippocampal slices is not inhibited by one of the broadly selective gap junction blockers (Frenguelli et al. 2007). It has been reported that ischemic stress causes opening of pannexin-1 channels with anoxic depolarization (Thompson et al. 2006), but another group has described that ischemia-induced anoxic depolarization is insensitive to gap junction blockade (Madry et al. 2010). Astrocytic exocytosis of

Table 4.1 ATP releasing sites and mechanisms in various brain cells

ATP releasing sites		Cell types	References
Connexin-43 hemichannels		Hippocampal astrocytes and C6 glioma cells	Kang et al. (2008)
		Spinal astrocytes	Garré et al. (2010)
		Astrocytes (mixed glial cultures) and C6 glioma cells	Stout et al. (2002)
Connexin-36 hemichannels		Cortical neurons	Schock et al. (2008)
Pannexin-1 channels		Cortical astrocytes	Iwabuchi and Kawahara (2011)
		Spinal astrocytes	Garré et al. (2010)
		Hippocampal neurons	Kawamura et al. (2010)
Exocytosis (vesicular release)		Hippocampal astrocytes	Coco et al. (2003); Pascual et al. (2005)
		Spinal astrocytes	Garré et al. (2010)
Chloride channels	Maxi anion channels	Cortical astrocytes	Liu et al. (2008a); Liu et al. (2008b)
	CFTR	Spinal microglia	Liu et al. (2006)
	Anion channels (VRAC?)	Cortical astrocytes	Anderson et al. (2004); Darby et al. (2003)
	VAAC	Dorsal root ganglion axons	Fields and Ni (2010)

CFTR cystic fibrosis transmembrane conductance regulator, *VRAC* volume-regulated anion channels, *VAAC* volume-activated anion channels

ATP is now thought to maintain extracellular adenosine concentrations in hippocampus (Pascual et al. 2005). Interestingly, spinal cord astrocytes release ATP from vesicles, pannexin-1 channels, and connexin-43 hemichannels with different time courses and different mechanisms involving ATP-induced ATP release via activation of $P2X_7$ receptors, suggesting that each releasing site plays a different role (Garré et al. 2010). ATP-induced ATP release also occurs through anion channels opened by P2 receptors in cortical astrocytes (Anderson et al. 2004). Together with $P2X_7$ receptor negative or positive feedback regulation in astrocytes, one of the roles of P2 receptors might be the regulation of ATP release. Cortical astrocytes also release ATP from anion channels with hypoosmotic stress (Darby et al. 2003; Liu et al. 2008b) and ischemia (Liu et al. 2008a). Another glial cell type, microglia, releases ATP via one of the chloride channels, cystic fibrosis transmembrane conductance regulator (CFTR; Liu et al. 2006). Hippocampal astrocytes can also release adenosine directly during hypoxia (Martín et al. 2007).

ATP is also effluxed from neurons. Neurons in several brain regions release ATP from synaptic vesicles (Pankratov et al. 2006). Cortical neurons release ATP through hemichannels formed of connexin-36 during hyperexcitability in high KCl concentrations (Schock et al. 2008). Hippocampal neurons release ATP from pannexin-1 channels in hypoglycemic conditions ((Kawamura et al. 2010); also see Sect. 3). Dorsal root ganglion axons release ATP through volume-activated anion channels (VAAC) with action potentials (Fields and Ni 2010).

4.2.2 Other Sources of Extracellular Adenosine

It is known that the adenosine transporter regulates extracellular adenosine levels (Dunwiddie and Diao 2000). Adenosine transporters work to salvage extracellular adenosine into cells for use in ATP synthesis. They are functionally characterized into two categories: equilibrative nucleoside transporter and concentrative nucleoside transporter (Latini and Pedata 2001). Equilibrative nucleoside transporters move purine and pyrimidine nucleosides across cell membranes dependent on their concentration gradient, and have been cloned from rat and human into four subtypes: ENT1-4 (Baldwin et al. 2004). Concentrative nucleoside transporters carry nucleosides driven by the transmembrane sodium gradient and have been cloned into three subtypes: CNT1-3 (Gray et al. 2004). These nucleoside transporters are bidirectional for moving the purines to both inside and outside the cell (Latini and Pedata 2001).

Inhibitors of equilibrative nucleoside transporters usually increase extracellular adenosine concentrations with blocking adenosine uptake (Ballarin et al. 1991; Fredholm et al. 1994) and facilitate activation of adenosine receptors (Dunwiddie and Diao 2000). However, it is also reported that these inhibitors attenuate direct adenosine release with electrical field stimulation or high potassium stimulation from the equilibrative nucleoside transporter (Jonzon and Fredholm 1985; Sweeney 1996). Ischemia has also been reported to cause direct adenosine release. Oxygen-glucose deprivation stress increases extracellular adenosine from astrocytes in the hippocampus (Martín et al. 2007). This ischemia-induced adenosine release has an earlier onset than ATP release (Frenguelli et al. 2007) and is insensitive to inhibitors of ectonucleotidase or blockers of ATP releasing sites, indicating that it is direct adenosine release not secondary to ATP release (Frenguelli et al. 2007; Lloyd et al. 1993). However, the mechanisms of ischemia-induced direct adenosine release are not fully determined yet.

Another source of extracellular adenosine is breakdown from cAMP. Extracellular cAMP is known to activate adenosine receptors after metabolism to adenosine (Brundege et al. 1997), and it has been reported that activation of adrenergic receptors induces cAMP release (Rosenberg et al. 1994) from an anion efflux pump (Henderson and Strauss 1991).

4.2.3 ATP Breakdown into Adenosine is the Main Pathway for Metabolic Autocrine Regulation

One of the major pathways for purinergic signaling linking metabolism and neuronal activity is ATP breakdown into adenosine and buildup of extracellular adenosine. Just like ATP releasing sites, adenosine receptors are found in both glial and neuronal cells in many brain areas. Adenosine A_1 receptors especially are strongly expressed in the cortex, hippocampus, and spinal cord (Dixon et al. 1996; Fredholm 1997; Mahan et al. 1991) in both astrocytes (Biber et al. 1997) and neurons (Dunwiddie and Masino 2001). Ectonucleotidase activity is also high in these regions (Goodman and Snyder 1982) and is colocalized in the same cells as ATP release sites (Joseph et al. 2003). In hippocampal synaptosomes, ATP-derived extracellular adenosine inhibits transmitter release (Cunha et al. 1994). This evidence indicates that all the players for a metabolic purinergic signaling cascade (ATP releasing site, ectonucleotidase, adenosine receptors) may be expressed in individual CNS cells and might integrate autocrine/paracrine regulation.

4.3 Autocrine Metabolic Regulation in Hippocampal CA3 Pyramidal Neurons

4.3.1 Autocrine Regulation in CA3 Pyramidal Neurons via Adenosine A_1 Receptors: An In Vitro Model of the Ketogenic Diet

We recently described autocrine metabolic regulation in CA3 pyramidal neurons using whole-cell patch-clamp methods in hippocampal slices (Kawamura et al. 2010). Extracellular glucose and intracellular ATP are both main factors regulating CNS metabolism. We measured the effects of reducing extracellular glucose concentration from 11 to 3 mM and of different concentrations of ATP in the intracellular solution (between 0.5 and 5 mM). When intracellular ATP concentrations were low (0.5 mM), reduced extracellular glucose caused transient inward currents in CA3 pyramidal neurons. With high or sufficient intracellular ATP concentrations and reduced glucose, outward current (or hyperpolarization in current-clamp mode) was produced. The outward current was dose-dependent on intracellular ATP concentration between 0.5 and 2 mM, suggesting that they were caused by autocrine modulation of the recorded neuron. Interestingly, 5 mM intracellular ATP concentration solution reduced the amplitude of this current. The reversal potential of the reduced glucose-induced outward current was similar to the reversal equilibrium potential of potassium, and blockers of ATP-sensitive potassium channel inhibited the outward current, suggesting that this current was caused by opening of ATP-sensitive potassium channels. Furthermore, involvement of ATP-sensitive potassium channels explains the amplitude reduction of outward current with 5 mM ATP, as these channels are blocked by high concentrations of intracellular ATP. Reduced glucose-induced outward current was abolished by an adenosine A_1 receptor antagonist and was not present in adenosine A_1 receptor knockout mice. Adenosine A_1 receptors couple to ATP-sensitive potassium channels (Andoh et al. 2006) through reduction of high-ATP blockade (Andoh et al. 2006; Hu et al. 1999) and our data strongly argue that the outward currents are caused by activation of adenosine A_1 receptors. Gap-junction blockers and a peptide blocker specific for pannexin-1 channels inhibited the outward currents, suggesting that the source of extracellular adenosine was ATP release from pannexin-1 channels. Together, these results suggest

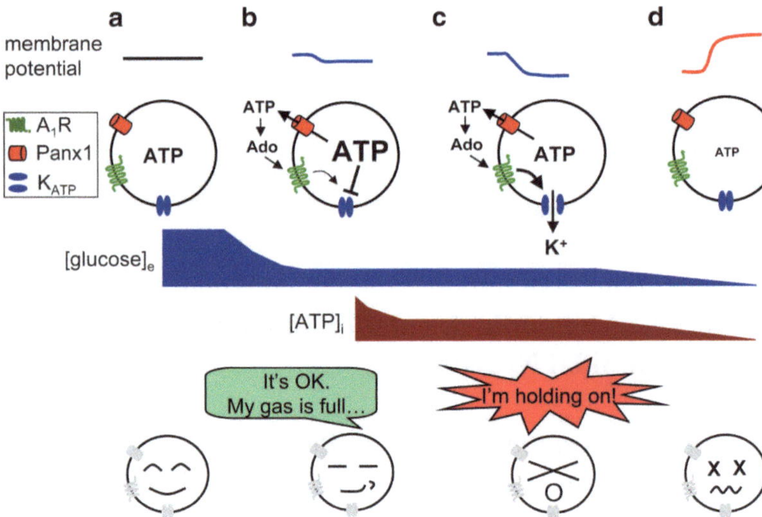

Fig. 4.1 An interpretation of the role of metabolic autocrine regulation via purines in hippocampal CA3 pyramidal neurons. At *top*, schematic traces showing the changing membrane potential with four different metabolic conditions (**a–d**), and, immediately below, illustrations of the corresponding state of the purinergic metabolic autoregulatory pathway. Below the illustrations, schematic traces of extracellular glucose concentrations ($[glucose]_e$) and intracellular ATP concentrations ($[ATP]_i$) and their changes from **a** to **d**. *At bottom*, corresponding visual representations of neuronal health using the Happy-to-Unhappy Cellular Health Scale (HUCHS; MK, Jr). (**a**) The baseline neuronal state. (**b**) With ketogenic diet feeding, extracellular glucose is low and intracellular ATP is high. Released ATP activates an autocrine inhibition through KATP channels, which, however, is largely blocked by high intracellular ATP. (**c**) As intracellular ATP runs down, the autocrine inhibition is unmasked, producing neuroprotective and antiseizure effects. (**d**) With sufficient lost of intracellular ATP, the autocrine adenosinergic mechanism fails and, lacking energy sources, the neuron enters a pathological depolarization. A_1R adenosine A_1 receptors, *Panx1* pannexin-1 channels, K_{ATP} ATP-sensitive potassium channels

that in conditions of high or sufficient intracellular ATP concentration and reduced extracellular glucose, CA3 pyramidal neurons hyperpolarize via direct ATP release through pannexin-1 channels with the subsequent activation of adenosine A_1 receptors and opening of ATP-sensitive potassium channels (Fig. 4.1c).

As mentioned in Sect. 1, dietary therapy is one of the specific models illustrating that changing the brain's metabolism will affect neuronal activity, and the ketogenic diet is a well-known metabolic therapy that reduces epileptic seizures. Our experimental condition of reducing extracellular glucose and increasing intracellular ATP concentration mimicked the ketogenic diet in vitro. Ketogenic diets increase brain concentrations of ATP in rodents (DeVivo et al. 1978; Nakazawa et al. 1983; Nylen et al. 2009), and in humans, Pan et al. (Pan et al. 1999) showed ketogenic diet increased the phosphocreatine to ATP ratio. These reports suggest that increased ketone body levels might accelerate intracellular ATP production. Blood glucose levels drop with the ketogenic diet (Bough et al. 2006; Huttenlocher 1976) and mild reduction of blood glucose has been reported to control seizures in both animal models (Greene et al. 2001; Mantis et al. 2004) and clinical trials (Muzykewicz et al. 2009). During ketogenic diet feeding, CA3 pyramidal neurons may experience increased intracellular ATP concentrations and reduced extracellular glucose concentrations, and these metabolic changes should cause hyperpolarization through adenosinergic autocrine regulation. Hyperpolarizing CA3 pyramidal neuron should have a strong anticonvulsant effect, and this metabolic autoregulation through purines might be an important mechanism underlying the success of ketogenic diets. Recently, the therapeutic implications of ketogenic diet have extended to other neurological disorders that involve ATP dysregulation. It is known that traumatic brain injury (Signoretti et al. 2001) and psychiatric disorders (Rezin et al. 2009) cause mitochondrial dysfunction and reduce levels of ATP in the brain. The ketogenic diet has been reported to improve traumatic brain injury ((Deng-Bryant et al. 2011) but only in young rats) or psychiatric disorders such as mood disorder (El-Mallakh and Paskitti 2001) and schizophrenia (Kraft and Westman 2009). The regulation of adenosine may be one likely focus for such a ketogenic diet efficacy (Masino et al. 2009).

4.3.2 An Interpretation of the Role of Adenosinergic Autocrine Modulation in CA3 Pyramidal Neurons

Adenosinergic autocrine regulation of hippocampal pyramidal neurons may be the mechanism underlying a ketogenic diet's effectiveness; however, the condition produced by a ketogenic diet (increased ATP concentration and mild hypoglycemia) is a particular and artificial situation. What is the physiological or pathophysiological role of autocrine metabolic modulation in CA3 pyramidal neurons? Figure 4.1 shows our interpretation of the pathophysiological role for this regulatory cascade in four different metabolic conditions. (1) With normal glucose concentration (euglycemia), no specific changes are caused in CA3 pyramidal neurons (Fig. 4.1a). (2) In response to hypoglycemia, pannexin-1 channels release ATP. When intracellular ATP is abundant (e.g. over 5 mM), this level of ATP inhibits the opening of ATP-sensitive potassium channels, and does not change or only slightly changes the membrane potential (Fig. 4.1b): if the reduction of glucose is transient, these neurons do not need to reduce activity because damage is not likely and priority is given to normal CNS activity and function. (3) When hypoglycemia continues and the intracellular ATP levels are likely to be decreased (by 1–2 mM) from depletion of the glucose supply and/or anoxia (the combination of which is ischemia), the neuron hyperpolarizes with adenosine A_1 receptor activation of ATP-sensitive potassium channels to dramatically reduce activity for self-protection (Fig. 4.1c). In this stage the neuron remains in a state of minimal energy expenditure until the metabolic abnormality is resolved. (4) If the hypoglycemia and/or anoxia persists for an extended time leading to a lack of intracellular ATP for release, the neuron can no longer use this mechanism to resist the metabolic challenge, and switches into neuronal necrosis (Kaminogo et al. 1998) with anoxic depolarization (Fig. 4.1d) via activation of glutamate receptors (Rossi et al. 2000) and/or opening of pannexin channels (Thompson et al. 2006).

Glucose and ATP are major players in metabolism. Through this adenosinergic autocrine pathway, hippocampal CA3 pyramidal neurons can react to changes in these molecules by modulating their activity using pannexin-1 channels, adenosine A_1 receptors, and ATP-sensitive potassium channels. A similar phenomenon has been reported in dorsal vagal neurons (Kulik et al. 2000) and hippocampus (Hansen et al. 1982; Zhang and Krnjević 1993). The depletion of extracellular glucose alone caused hyperpolarization through changing potassium conductance in dorsal vagal neurons; however, the full ischemic condition causes membrane depolarization (Kulik et al. 2000). In the hippocampus, brief anoxia hyperpolarizes CA1 or CA3 pyramidal neurons, while prolonged anoxia eventually depolarizes them (Zhang and Krnjević 1993). These reports show that neurons have protective mechanisms for dealing with altered metabolism (anoxic or ischemic conditions) and our finding suggests that one of the neuroprotective mechanisms against ischemia is generated by autoregulation using purinergic signaling.

4.4 Conclusions

Here we note that altered brain metabolism links to neuronal activity via adenosinergic autocrine regulation with dephosphorylation of ATP into adenosine. In the CNS, most gray matter structures have various cell types including neurons, astrocytes and microglia. Unlike in peripheral tissues, these cell types have different roles in brain functions even if they are neighbors. Each cell in the CNS, thereby, needs to sense metabolic changes and regulate its own activity individually. Autocrine regulation is a self-modulating system of releasing neurotransmitters/neuromodulators and sensing them with receptors expressed by the same cell to control its own activities. ATP is a final product of energy metabolism and it might be a signal for altered metabolism. ATP can efflux to the extracellular space via various release sites and, after breakdown into adenosine, activates adenosine receptors. The actors in this signaling pathway are expressed in individual CNS cells and link altered metabolism to neuronal activity. Delineation of this autoregulatory cascade has shed new light on the yin-and-yang relationship of ATP and adenosine, and described a powerful mechanism by which neurons can protect themselves against an adverse metabolic environment. Conversely, dietary treatments effective against epilepsy or against brain damage due to ischemic or physical insults (e.g., fasting, caloric restriction, ketogenic diet) might engage this very mechanism to produce their beneficial outcomes.

Acknowledgments We acknowledge the support of Grant-in Aid for Scientific Research (KAKENHI) for Young Scientists (B) (No. 23790303 to M.K.Jr.), Takeda Science Foundation (to M.K.Jr.) and of the National Institute of Neurological Disease and Stroke (No. NS-065446 to D.N.R.).

References

Anderson CM, Bergher JP, Swanson RA (2004) ATP-induced ATP release from astrocytes. J Neurochem 88:246–256

Andoh T, Ishiwa D, Kamiya Y, Echigo N, Goto T, Yamada Y (2006) A_1 adenosine receptor-mediated modulation of neuronal ATP-sensitive K channels in rat substantia nigra. Brain Res 1124:55–61

Anschel DJ, Ortega EL, Kraus AC, Fisher RS (2004) Focally injected adenosine prevents seizures in the rat. Exp Neurol 190:544–547

Baldwin SA, Beal PR, Yao SY, King AE, Cass CE, Young JD (2004) The equilibrative nucleoside transporter family, SLC29. Pflugers Arch 447:735–743

Ballanyi K (2004) Protective role of neuronal K_{ATP} channels in brain hypoxia. J Exp Biol 207: 3201–3212

Ballarin M, Fredholm BB, Ambrosio S, Mahy N (1991) Extracellular levels of adenosine and its metabolites in the striatum of awake rats: inhibition of uptake and metabolism. Acta Physiol Scand 142:97–103

Biber K, Klotz KN, Berger M, Gebicke-Härter PJ, van Calker D (1997) Adenosine A_1 receptor-mediated activation of phospholipase C in cultured astrocytes depends on the level of receptor expression. J Neurosci 17:4956–4964

Bough KJ, Wetherington J, Hassel B, Pare JF, Gawryluk JW, Greene JG, Shaw R, Smith Y, Geiger JD, Dingledine RJ (2006) Mitochondrial biogenesis in the anticonvulsant mechanism of the ketogenic diet. Ann Neurol 60:223–235

Brundege JM, Diao L, Proctor WR, Dunwiddie TV (1997) The role of cyclic AMP as a precursor of extracellular adenosine in the rat hippocampus. Neuropharmacology 36:1201–1210

Chen Y, Rathbone MP, Hertz L (2001) Guanosine-induced increase in free cytosolic calcium concentration in mouse astrocytes in primary cultures: does it act on an A3 adenosine receptor? J Neurosci Res 65:184–189

Coco S, Calegari F, Pravettoni E, Pozzi D, Taverna E, Rosa P, Matteoli M, Verderio C (2003) Storage and release of ATP from astrocytes in culture. J Biol Chem 278:1354–1362

Cunha RA, Ribeiro JA, Sebastião AM (1994) Purinergic modulation of the evoked release of [^{3}H] acetylcholine from the hippocampus and cerebral cortex of the rat: role of the ectonucleotidases. Eur J Neurosci 6:33–42

Darby M, Kuzmiski JB, Panenka W, Feighan D, MacVicar BA (2003) ATP released from astrocytes during swelling activates chloride channels. J Neurophysiol 89:1870–1877

Davalos D, Grutzendler J, Yang G, Kim JV, Zuo Y, Jung S, Littman DR, Dustin ML, Gan WB (2005) ATP mediates rapid microglial response to local brain injury in vivo. Nat Neurosci 8: 752–758

Davis LM, Pauly JR, Readnower RD, Rho JM, Sullivan PG (2008) Fasting is neuroprotective following traumatic brain injury. J Neurosci Res 86:1812–1822

Deng-Bryant Y, Prins ML, Hovda DA, Harris NG (2011) Ketogenic diet prevents alterations in brain metabolism in young but not adult rats after traumatic brain injury. J Neurotrauma 28: 1813–1825

DeVivo DC, Leckie MP, Ferrendelli JS, McDougal DB Jr (1978) Chronic ketosis and cerebral metabolism. Ann Neurol 3:331–337

Dixon AK, Gubitz AK, Sirinathsinghji DJ, Richardson PJ, Freeman TC (1996) Tissue distribution of adenosine receptor mRNAs in the rat. Br J Pharmacol 118:1461–1468

Dunwiddie TV (1999) Adenosine and suppression of seizures. Adv Neurol 79:1001–1010

Dunwiddie TV, Diao L (2000) Regulation of extracellular adenosine in rat hippocampal slices is temperature dependent: role of adenosine transporters. Neuroscience 95:81–88

Dunwiddie TV, Diao L, Proctor WR (1997) Adenine nucleotides undergo rapid, quantitative conversion to adenosine in the extracellular space in rat hippocampus. J Neurosci 17:7673–7682

Dunwiddie TV, Hoffer BJ (1980) Adenine nucleotides and synaptic transmission in the in vitro rat hippocampus. Br J Pharmacol 69:59–68

Dunwiddie TV, Masino SA (2001) The role and regulation of adenosine in the central nervous system. Annu Rev Neurosci 24:31–55

El-Mallakh RS, Paskitti ME (2001) The ketogenic diet may have mood-stabilizing properties. Med Hypotheses 57:724–726

Ferré S, Ciruela F, Woods AS, Lluis C, Franco R (2007) Functional relevance of neurotransmitter receptor heteromers in the central nervous system. Trends Neurosci 30:440–446

Fields RD, Ni Y (2010) Nonsynaptic communication through ATP release from volume-activated anion channels in axons. Sci Signal 3:73

Fowler JC (1993) Purine release and inhibition of synaptic transmission during hypoxia and hypoglycemia in rat hippocampal slices. Neurosci Lett 157:83–86

Fredholm BB (1995) Purinoceptors in the nervous system. Pharmacol Toxicol 76:228–239

Fredholm BB (1997) Adenosine and neuroprotection. Int Rev Neurobiol 40:259–280

Fredholm BB, Arslan G, Halldner L, Kull B, Schulte G, Wasserman W (2000) Structure and function of adenosine receptors and their genes. Naunyn Schmiedebergs Arch Pharmacol 362:364–374

Fredholm BB, Lindström K, Wallman-Johansson A (1994) Propentofylline and other adenosine transport inhibitors increase the efflux of adenosine following electrical or metabolic stimulation of rat hippocampal slices. J Neurochem 62:563–573

Freeman JM, Kossoff EH, Hartman AL (2007) The ketogenic diet: one decade later. Pediatrics 119:535–543

Frenguelli BG, Llaudet E, Dale N (2003) High-resolution real-time recording with microelectrode biosensors reveals novel aspects of adenosine release during hypoxia in rat hippocampal slices. J Neurochem 86:1506–1515

Frenguelli BG, Wigmore G, Llaudet E, Dale N (2007) Temporal and mechanistic dissociation of ATP and adenosine release during ischaemia in the mammalian hippocampus. J Neurochem 101:1400–1413

Garré JM, Retamal MA, Cassina P, Barbeito L, Bukauskas FF, Sáez JC, Bennett MV, Abudara V (2010) FGF-1 induces ATP release from spinal astrocytes in culture and opens pannexin and connexin hemichannels. Proc Natl Acad Sci U S A 107:22659–22664

Goodman RR, Snyder SH (1982) Autoradiographic localization of adenosine receptors in rat brain using [^{3}H] cyclohexyladenosine. J Neurosci 2:1230–1241

Gray JH, Owen RP, Giacomini KM (2004) The concentrative nucleoside transporter family, SLC28. Pflugers Arch 447:728–734

Greene AE, Todorova MT, McGowan R, Seyfried TN (2001) Caloric restriction inhibits seizure susceptibility in epileptic EL mice by reducing blood glucose. Epilepsia 42:1371–1378

Greene AE, Todorova MT, Seyfried TN (2003) Perspectives on the metabolic management of epilepsy through dietary reduction of glucose and elevation of ketone bodies. J Neurochem 86: 529–537

Hallböök T, Köhler S, Rosén I, Lundgren J (2007) Effects of ketogenic diet on epileptiform activity in children with therapy resistant epilepsy. Epilepsy Res 77:134–140

Hansen AJ, Hounsgaard J, Jahnsen H (1982) Anoxia increases potassium conductance in hippocampal nerve cells. Acta Physiol Scand 115:301–310

Hartman AL, Lyle M, Rogawski MA, Gasior M (2008) Efficacy of the ketogenic diet in the 6-Hz seizure test. Epilepsia 49:334–339

Henderson GB, Strauss BP (1991) Evidence for cAMP and cholate extrusion in C6 rat glioma cells by a common anion efflux pump. J Biol Chem 266:1641–1645

Hertz L, Peng L, Dienel GA (2007) Energy metabolism in astrocytes: high rate of oxidative metabolism and spatiotemporal dependence on glycolysis/glycogenolysis. J Cereb Blood Flow Metab 27:219–249

Hu K, Li GR, Nattel S (1999) Adenosine-induced activation of ATP-sensitive K^+ channels in excised membrane patches is mediated by PKC. Am J Physiol 276:H488–495

Huttenlocher PR (1976) Ketonemia and seizures: metabolic and anticonvulsant effects of two ketogenic diets in childhood epilepsy. Pediatr Res 10:536–540

Inoue K (1998) ATP receptors for the protection of hippocampal functions. Jpn J Pharmacol 78: 405–410

Iwabuchi S, Kawahara K (2011) Functional significance of the negative-feedback regulation of ATP release via pannexin-1 hemichannels under ischemic stress in astrocytes. Neurochem Int 58:376–384

Johansson B, Halldner L, Dunwiddie TV, Masino SA, Poelchen W, Gimenez-Llort L, Escorihuela RM, Fernandez-Teruel A, Wiesenfeld-Hallin Z, Xu XJ et al (2001) Hyperalgesia, anxiety, and decreased hypoxic neuroprotection in mice lacking the adenosine A1 receptor. Proc Natl Acad Sci U S A 98:9407–9412

Jonzon B, Fredholm BB (1985) Release of purines, noradrenaline, and GABA from rat hippocampal slices by field stimulation. J Neurochem 44:217–224

Joseph SM, Buchakjian MR, Dubyak GR (2003) Colocalization of ATP release sites and ecto-ATPase activity at the extracellular surface of human astrocytes. J Biol Chem 278: 23331–23342

Kaminogo M, Suyama K, Ichikura A, Onizuka M, Shibata S (1998) Anoxic depolarization determines ischemic brain injury. Neurol Res 20:343–348

Kang J, Kang N, Lovatt D, Torres A, Zhao Z, Lin J, Nedergaard M (2008) Connexin 43 hemichannels are permeable to ATP. J Neurosci 28:4702–4711

Kann O, Kovács R (2007) Mitochondria and neuronal activity. Am J Physiol Cell Physiol 292: C641–C657

Kawamura M Jr, Kawamura M (2011) Long-term facilitation of spontaneous calcium oscillations in astrocytes with endogenous adenosine in hippocampal slice cultures. Cell Calcium 49:249–258

Kawamura M Jr, Ruskin DN, Masino SA (2010) Metabolic autocrine regulation of neurons involves cooperation among pannexin hemichannels, adenosine receptors, and K_{ATP} channels. J Neurosci 30:3886–3895

Kraft BD, Westman EC (2009) Schizophrenia, gluten, and low-carbohydrate, ketogenic diets: a case report and review of the literature. Nutr Metab (Lond) 6:10

Kulik A, Trapp S, Ballanyi K (2000) Ischemia but not anoxia evokes vesicular and Ca^{2+}-independent glutamate release in the dorsal vagal complex in vitro. J Neurophysiol 83:2905–2915

Latini S, Pedata F (2001) Adenosine in the central nervous system: release mechanisms and extracellular concentrations. J Neurochem 79:463–484

Liu GJ, Kalous A, Werry EL, Bennett MR (2006) Purine release from spinal cord microglia after elevation of calcium by glutamate. Mol Pharmacol 70:851–859

Liu HT, Sabirov RZ, Okada Y (2008a) Oxygen-glucose deprivation induces ATP release via maxi-anion channels in astrocytes. Purinergic Signal 4:147–154

Liu HT, Toychiev AH, Takahashi N, Sabirov RZ, Okada Y (2008b) Maxi-anion channel as a candidate pathway for osmosensitive ATP release from mouse astrocytes in primary culture. Cell Res 18:558–565

Lloyd HG, Lindström K, Fredholm BB (1993) Intracellular formation and release of adenosine from rat hippocampal slices evoked by electrical stimulation or energy depletion. Neurochem Int 23:173–185

MacVicar BA, Thompson RJ (2009) Non-junction functions of pannexin-1 channels. Trends Neurosci 33:93–102

Madry C, Haglerod C, Attwell D (2010) The role of pannexin hemichannels in the anoxic depolarization of hippocampal pyramidal cells. Brain 133:3755–3763

Mahan LC, McVittie LD, Smyk-Randall EM, Nakata H, Monsma FJ Jr, Gerfen CR, Sibley DR (1991) Cloning and expression of an A_1 adenosine receptor from rat brain. Mol Pharmacol 40:1–7

Mantis JG, Centeno NA, Todorova MT, McGowan R, Seyfried TN (2004) Management of multifactorial idiopathic epilepsy in EL mice with caloric restriction and the ketogenic diet: role of glucose and ketone bodies. Nutr Metab (Lond) 1:11

Martín ED, Fernández M, Perea G, Pascual O, Haydon PG, Araque A, Ceña V (2007) Adenosine released by astrocytes contributes to hypoxia-induced modulation of synaptic transmission. Glia 55:36–45

Masino SA, Diao L, Illes P, Zahniser NR, Larson GA, Johansson B, Fredholm BB, Dunwiddie TV (2002) Modulation of hippocampal glutamatergic transmission by ATP is dependent on adenosine A_1 receptors. J Pharmacol Exp Ther 303:356–363

Masino SA, Geiger JD (2008) Are purines mediators of the anticonvulsant/neuroprotective effects of ketogenic diets? Trends Neurosci 31:273–278

Masino SA, Kawamura M, Wasser CA, Pomeroy LT, Ruskin DN (2009) Adenosine, ketogenic diet and epilepsy: the emerging therapeutic relationship between metabolism and brain activity. Curr Neuropharmacol 7:257–268

Masino SA, Latini S, Bordini F, Pedata F, Dunwiddie TV (2001) Changes in hippocampal adenosine efflux, ATP levels, and synaptic transmission induced by increased temperature. Synapse 41: 58–64

Masino SA, Li T, Theofilas P, Sandau US, Ruskin DN, Fredholm BB, Geiger JD, Aronica E, Boison D (2011) A ketogenic diet suppresses seizures in mice through adenosine A_1 receptors. J Clin Invest 121:2679–2683

Miller LP, Hsu C (1992) Therapeutic potential for adenosine receptor activation in ischemic brain injury. J Neurotrauma 9(Suppl 2):S563–577

Muzykewicz DA, Lyczkowski DA, Memon N, Conant KD, Pfeifer HH, Thiele EA (2009) Efficacy, safety, and tolerability of the low glycemic index treatment in pediatric epilepsy. Epilepsia 50:1118–1126

Nakazawa M, Kodama S, Matsuo T (1983) Effects of ketogenic diet on electroconvulsive threshold and brain contents of adenosine nucleotides. Brain Dev 5:375–380

Nylen K, Velazquez JL, Sayed V, Gibson KM, Burnham WM, Snead OC 3rd (2009) The effects of a ketogenic diet on ATP concentrations and the number of hippocampal mitochondria in *Aldh5a1-/-* mice. Biochim Biophys Acta 1790:208–212

Oláh J, Klivényi P, Gardián G, Vécsei L, Orosz F, Kovacs GG, Westerhoff HV, Ovádi J (2008) Increased glucose metabolism and ATP level in brain tissue of Huntington's disease transgenic mice. FEBS J 275:4740–4755

Pan JW, Bebin EM, Chu WJ, Hetherington HP (1999) Ketosis and epilepsy: ^{31}P spectroscopic imaging at 4.1T. Epilepsia 40:703–707

Pan WJ, Osmanović SS, Shefner SA (1995) Characterization of the adenosine A_1 receptor-activated potassium current in rat locus ceruleus neurons. J Pharmacol Exp Ther 273:537–544

Pankratov Y, Lalo U, Verkhratsky A, North RA (2006) Vesicular release of ATP at central synapses. Pflugers Arch Eur J Physiol 452:589–597

Parkinson FE, Xiong W, Zamzow CR (2005) Astrocytes and neurons: different roles in regulating adenosine levels. Neurol Res 27:153–160

Pascual O, Casper KB, Kubera C, Zhang J, Revilla-Sanchez R, Sul JY, Takano H, Moss SJ, McCarthy K, Haydon PG (2005) Astrocytic purinergic signaling coordinates synaptic networks. Science 310:113–116

Peakman MC, Hill SJ (1994) Adenosine A_{2B}-receptor-mediated cyclic AMP accumulation in primary rat astrocytes. Br J Pharmacol 111:191–198

Ralevic V, Burnstock G (1998) Receptors for purines and pyrimidines. Pharmacol Rev 50: 413–492

Rezin GT, Amboni G, Zugno AI, Quevedo J, Streck EL (2009) Mitochondrial dysfunction and psychiatric disorders. Neurochem Res 34:1021–1029

Rho JM, Kim DW, Robbins CA, Anderson GD, Schwartzkroin PA (1999) Age-dependent differences in flurothyl seizure sensitivity in mice treated with a ketogenic diet. Epilepsy Res 37: 233–240

Rosenberg PA, Knowles R, Knowles KP, Li Y (1994) Beta-adrenergic receptor-mediated regulation of extracellular adenosine in cerebral cortex in culture. J Neurosci 14:2953–2965

Rosin DL, Hettinger BD, Lee A, Linden J (2003) Anatomy of adenosine A_{2A} receptors in brain: morphological substrates for integration of striatal function. Neurology 61:S12–18

Rossi DJ, Oshima T, Attwell D (2000) Glutamate release in severe brain ischaemia is mainly by reversed uptake. Nature 403:316–321

Rudolphi KA, Schubert P, Parkinson FE, Fredholm BB (1992) Adenosine and brain ischemia. Cerebrovasc Brain Metab Rev 4:346–369

Sas K, Robotka H, Toldi J, Vécsei L (2007) Mitochondria, metabolic disturbances, oxidative stress and the kynurenine system, with focus on neurodegenerative disorders. J Neurol Sci 257:221–239

Schock SC, Leblanc D, Hakim AM, Thompson CS (2008) ATP release by way of connexin 36 hemichannels mediates ischemic tolerance *in vitro*. Biochem Biophys Res Commun 368: 138–144

Schulte G, Fredholm BB (2003) Signalling from adenosine receptors to mitogen-activated protein kinases. Cell Signal 15:813–827

Sebastião AM, de Mendonca A, Moreira T, Ribeiro JA (2001) Activation of synaptic NMDA receptors by action potential-dependent release of transmitter during hypoxia impairs recovery of synaptic transmission on reoxygenation. J Neurosci 21:8564–8571

Signoretti S, Marmarou A, Tavazzi B, Lazzarino G, Beaumont A, Vagnozzi R (2001) N-Acetylaspartate reduction as a measure of injury severity and mitochondrial dysfunction following diffuse traumatic brain injury. J Neurotrauma 18:977–991

Stone TW (2002) Purines and neuroprotection. Adv Exp Med Biol 513:249–280

Stout CE, Costantin JL, Naus CC, Charles AC (2002) Intercellular calcium signaling in astrocytes via ATP release through connexin hemichannels. J Biol Chem 277:10482–10488

Sweeney MI (1996) Adenosine release and uptake in cerebellar granule neurons both occur via an equilibrative nucleoside carrier that is modulated by G proteins. J Neurochem 67:81–88

Thompson RJ, Zhou N, MacVicar BA (2006) Ischemia opens neuronal gap junction hemichannels. Science 312:924–927

Thompson SM, Haas HL, Gähwiler BH (1992) Comparison of the actions of adenosine at pre- and postsynaptic receptors in the rat hippocampus *in vitro*. J Physiol 451:347–363

Uhlemann ER, Neims AH (1972) Anticonvulsant properties of the ketogenic diet in mice. J Pharmacol Exp Ther 180:231–238

Wardas J (2002) Neuroprotective role of adenosine in the CNS. Pol J Pharmacol 54:313–326

Wilder RM (1921) The effects of ketonemia on the course of epilepsy. Mayo Clin Bull 2: 307–308

Zhang L, Krnjević K (1993) Whole-cell recording of anoxic effects on hippocampal neurons in slices. J Neurophysiol 69:118–127

Chapter 5
Physiologic and Metabolic Regulation of Adenosine: Mechanisms

Chris G. Dulla and Susan A. Masino

Abstract Adenosine is the purine nucleoside core of adenosine triphosphate (ATP), a molecule that allows biological energy storage, and changes in intracellular adenosine levels are thought to reflect the energy state of the cell: during net ATP consumption, adenosine accumulates. Via adenosine A_1 receptors, increases in extracellular adenosine can decrease neuronal excitability rapidly. Accordingly, the interplay between energy use and neuronal activity provides a feedback system by which increased energy utilization can decrease energy demand. This is, however, a simplistic view of the regulation of adenosine in the nervous system. Advances in adenosine measurement and genetic tools to dissect the sources of adenosine have lent insight into a dynamic physiological regulation of adenosine and a bevy of cellular changes which alter adenosine tone. Here we examine recent work probing specific physiologic and metabolic changes which influence adenosine (and thus synaptic transmission), focusing primarily on the hippocampus. We summarize the current understanding of how tonic extracellular adenosine levels are set and highlight unanswered questions. Regarding dynamic physiological regulation, we outline how stimuli such as pH, neuronal activity, altered metabolism, and hypoxia can modulate adenosine levels. Using specific examples, we describe how changes in adenosine may act to normalize or restore physiologic and metabolic balance.

C.G. Dulla (✉)
Neuroscience Department, Tufts University Medical School, Boston, MA, USA
e-mail: chris.dulla@tufts.edu

S.A. Masino
Psychology Department and Neuroscience Program,
Trinity College, Hartford, CT, USA

S. Masino and D. Boison (eds.), *Adenosine: A Key Link between Metabolism and Brain Activity*, DOI 10.1007/978-1-4614-3903-5_5,

Keywords Epilepsy • Ketogenic diet • Hypoxia • Adenosine receptors • ATP • Astrocytes

5.1 Introduction

At its most fundamental level, adenosine is the purine nucleoside core of adenosine triphosphate (ATP), the energy currency of the cell. ATP is an energy storage molecule critical to cellular survival and it has numerous effects via ATP receptors which fall outside the scope of this review. Adenosine itself is much more than the core of ATP; among other biochemical roles and signaling roles, it is a ubiquitous neuromodulator in the brain, providing tonic inhibition of excitatory neurotransmission in most brain regions (Masino and Dunwiddie 2001).

In this chapter we will focus on adenosine acting at the widely distributed A_1 receptor subtype (A_1R), which mediates the bulk of adenosine's inhibitory effects. A_1Rs activate a G-protein coupled signaling cascade that leads to the activation of postsynaptic potassium channels and the inhibition of presynaptic calcium channels (Dunwiddie and Fredholm 1985). Enhanced potassium channel activation decreases intrinsic neuronal excitability; reduced calcium channel function decreases the probability of neurotransmitter release. Via these two ion channels, A_1Rs have the potential to completely attenuate evoked and spontaneous neurotransmission in the hippocampus (Winn et al. 1979, 1980; Fowler 1993; Dale et al. 2000; Masino and Dunwiddie 1999). The effects of A_1R activation are neuroprotective (Johansson et al. 2001) and anticonvulsant (Dunwiddie 1999) even in pharmacoresistant models of epilepsy (Gouder et al. 2003). Because of the powerful inhibitory effects of A_1Rs during physiological and pathological conditions, understanding the regulation of extracellular adenosine levels is critically important.

Although there are many mechanisms involved in adenosine homeostasis, the amount of adenosine present in the extracellular space of the brain can fluctuate (Latini and Pedata 2001). For instance, during wakefulness adenosine levels build up in the extracellular space and then decrease with the onset of sleep (Porkka-Heiskanen et al. 1997). Hypoxia (oxygen deprivation) (Winn et al. 1980; Dale et al. 2000), ischemia (oxygen and glucose deprivation) (Winn et al. 1979), hypoglycemia (glucose deprivation) (Fowler 1993), temperature changes (Masino and Dunwiddie 1999), and seizures (Dunwiddie 1999) all cause increased levels of extracellular adenosine. *N*-methyl-D-aspartate (NMDA) receptor activation also has been shown to release adenosine (Manzoni et al. 1994). Evidence suggests that exercise can increase brain adenosine (Dworak et al. 2007), and recent studies suggest that metabolic changes associated with a ketogenic diet result in A_1R-dependent inhibition of neurotransmission (Masino et al. 2011b). Here we will present the current understanding of the regulation of extracellular adenosine, with an emphasis on the hippocampus, and examine recent studies that aim to understand how changes in pH, neuronal activity, metabolism, and hypoxia alter extracellular adenosine tone.

5.2 Tonic Extracellular Adenosine Comes from the Breakdown of ATP Released by Astrocytes

Understanding the control of extracellular adenosine levels has been a long-standing, enigmatic problem. Adenosine sits at the intersection of multiple metabolic and cell signaling cascades; thus parsing apart the molecular players controlling extracellular adenosine levels has been challenging. For many years it has been known that adenosine can enter and exit the cell via the nucleoside transporter, which equilibrates the intracellular and extracellular concentrations of adenosine by allowing adenosine to flow down its concentration gradient (Thorn and Jarvis 1996). In the hippocampus, the normal concentration of free adenosine ranges from low nanomolar (Brundege and Dunwiddie 1997) to hundreds of nanomolar (Dunwiddie and Diao 1994), whereas ATP levels inside cells are much higher (at least 3 mM). Because of this, it is thermodynamically favorable for adenosine to enter neurons via the nucleoside transporter (Thorn and Jarvis 1996). When the nucleoside transporter is blocked pharmacologically, adenosine accumulates in the extracellular space, consistent with adenosine normally moving into cells via equilibrative transport.

Adenosine is the building block of many important cellular molecules, including free adenosine, bound to one (AMP), two (ADP), or three (ATP) high-energy phosphates, and cyclic AMP (cAMP). Each form has a unique role in different cellular processes (Cunha 2001). Adenosine kinase is the main intracellular enzyme that phosphorylates adenosine into adenosine mono-phosphate (AMP). This astrocyte-based intracellular enzyme is a key regulator of extracellular adenosine: once phosphorylated to AMP, adenosine can no longer exit the cell via equilibrative transport. In this sense adenosine kinase acts to “lock” adenosine into the cell and helps clamp a low level of free intracellular adenosine. AMP is further phosphorylated into adenosine diphosphate (ADP) by adenylate kinase. ADP can then be phosphorylated into ATP by creatine kinase (Lloyd and Fredholm 1995). More often, however, ADP enters the mitochondrial inner matrix via adenine nucleotide translocase to undergo its final phosphorylation into ATP via ATP synthase, a molecule that harnesses a pH gradient created by the electron transport chain (Nicholls and Budd 2000) and can alter the production of ATP from ADP (Whittingham et al. 1989).

Because adenosine normally moves into the cell via the nucleoside transporter, and because adenosine is not known to be packaged into vesicles for release, for many years there was no clear mechanism for adenosine to exit the cell other than efflux through the transporter (which would require high intracellular adenosine, at least locally). An important step forward in determining the extracellular source of adenosine came when Tom Dunwiddie’s laboratory discovered that extracellular application of adenine nucleotides including ATP resulted in activation of A_1Rs. When A_1R-dependent potassium currents were measured in hippocampal CA1 pyramidal cells, the kinetics of activation showed that adenosine itself evoked the most rapid activation of an A_1R-dependent potassium current; direct application of ATP evoked the same current with a temporal delay. These studies indicated that extracellular ATP is converted to adenosine within hundreds of milliseconds and

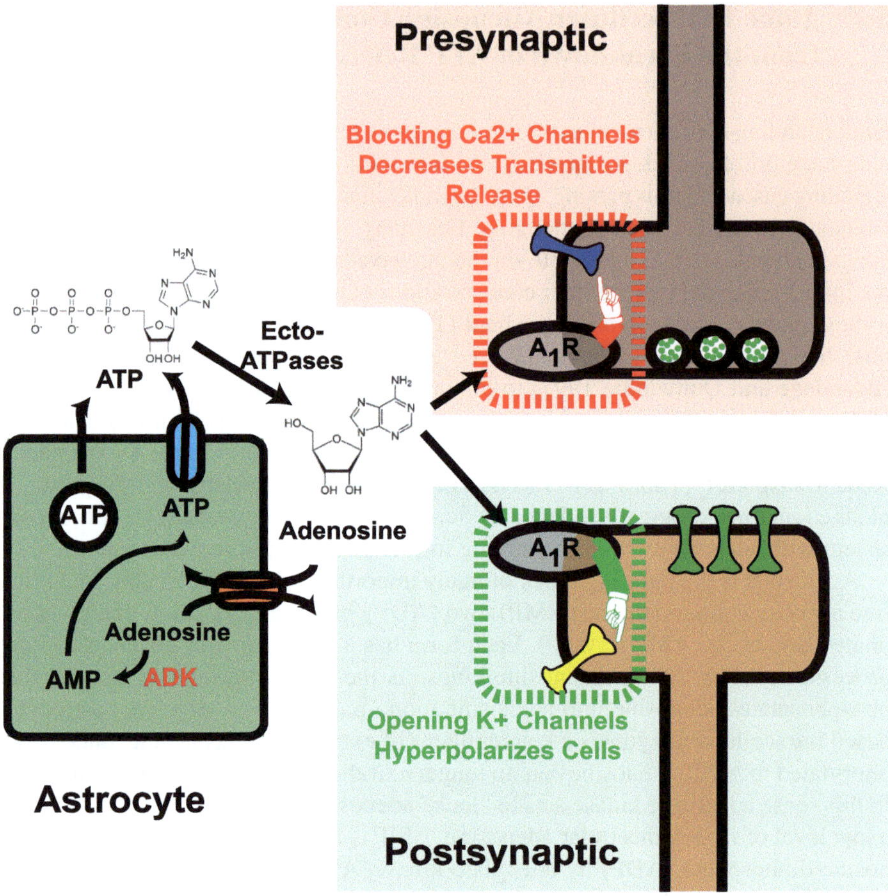

Fig. 5.1 Major synaptic mechanisms of adenosine regulation and consequences of adenosine A1 receptor (A_1R) activation. Adenosine in the synaptic space under normal conditions appears to be derived perhaps exclusively from dephosphorylation of vesicular ATP released from astrocytes (*upper left*). However, ATP can also enter the extracellular fluid via channels (similarly dephosphorylated to adenosine by ecto-ATPases/ectonucleotidases) and adenosine can enter directly via equilibrative transporters. For simplicity, these mechanisms are illustrated on the astrocyte at *left*, but they are also present in neurons. Once in the synaptic space, adenosine binds to pre- and postsynaptic A_1Rs, which close calcium channels and open potassium channels, respectively. Adenosine is removed from the synaptic space via transporters and rephosphorylated by astrocyte-based adenosine-kinase (ADK)

activates A_1Rs (Dunwiddie et al. 1997). cAMP was also shown to activate A_1Rs in the hippocampus but only following prolonged exposure; brief application of cAMP did not lead to A_1R activation (Brundege et al. 1997). These studies combined suggested that the release of ATP and its conversion to adenosine in the extracellular space was a likely mechanism by which extracellular adenosine levels could be maintained dynamically (Fig. 5.1).

The extracellular breakdown of ATP into adenosine is mediated by ectonucleotidases, a family of enzymes which includes ecto-ATPase, ecto-apyrases, and ecto-5′ nucleotidase (Dunwiddie et al. 1997). Ecto-ATPases are expressed by neurons and glia and can be anchored to the extracellular membrane or can be free in the extracellular space. Interestingly, extracellular ATPases require either calcium or magnesium ions to function. Ecto-apyrases cleave both ATP and ADP to AMP and are found extracellularly in both neuron and astrocyte membranes and may be enriched in hippocampal postsynaptic densities (Wang and Guidotti 1998). Ecto-5′ nucleotidase has both soluble and membrane-bound forms, exists intracellularly and extracellularly, and is expressed by both neurons and glia. ATP can reach the extracellular space by being co-released with neurotransmitters (glutamate, noradrenaline, and acetylcholine) (Cunha et al. 1996; Juranyi et al. 1997; Sperlagh et al. 1998; Brager and Thompson 2003), released via channels such as connexin hemichannels (Cotrina et al. 1998) or pannexin channels (Sosinsky et al. 2011; Kawamura et al. 2010; Thompson et al. 2008) or released from astrocytes via soluble NSF attachment protein receptor (SNARE)-dependent vesicular release (Pascual et al. 2005). Regarding the latter, genetic manipulation of the astrocytic vesicular release machinery has shown that in the hippocampus, basal extracellular adenosine arises from astrocyte-released ATP.

In 2005, Pascual et al. showed that CA1 fEPSPs recorded in transgenic animals that express a dominant-negative version of the SNARE (dn-SNARE) protein do not respond to antagonists of A_1Rs, while control litter-mates showed approximately a 60 % increase in evoked hippocampal fEPSP amplitude. This showed that a genetic manipulation that blocked astrocytic vesicular release abolished tonic activation of A_1Rs. This group, led by Phil Haydon, went on to show that in normal animals, astrocytes release ATP which is converted to adenosine (Pascual et al. 2005), in accord with results reported by Dunwiddie and colleagues almost a decade earlier (Dunwiddie et al. 1997). In short, this landmark study showed that dn-SNARE animals lacked the vesicular release of ATP from astrocytes and thus had no tonic activation of A_1Rs, and presumably no tonic extracellular adenosine. Studies of dn-SNARE animals show disruption in sleep (Fellin et al. 2009; Halassa et al. 2009) and nociception (Foley et al. 2011), suggesting potential roles for astrocyte-derived adenosine in multiple behavioral pathways. These experiments added a critical piece of the puzzle regarding adenosine's regulation in the synapse, and thus its regulation of synaptic transmission. Therefore, based on these studies, we now know that astrocytes tonically release ATP via SNARE-dependent vesicular release (Fig. 5.1). This ATP is converted rapidly to adenosine by ectonucleotidase. This astrocyte-derived pool of adenosine provides tonic activation of A_1Rs and resting adenosine-mediated inhibition.

The pathways that remove extracellular adenosine are fairly well understood. Adenosine is removed from the extracellular space via equilibrative transport, and astrocyte-based adenosine kinase plays a major role in determining intracellular (and thus extracellular) adenosine. As previously mentioned, adenosine kinase phosphorylates adenosine to AMP intracellularly, thus creating a driving force for adenosine influx. It is important to note that adenosine kinase is almost exclusively expressed in astrocytes (Studer et al. 2006) and is critical to maintaining extracellular adenosine levels (Boison 2006). The extracellular breakdown of adenosine does not significantly contribute to adenosine levels as inhibitors of adenosine deaminase do not influence extracellular adenosine levels (Zhu and Krnjevic 1994; Pak et al. 1994). Together these studies highlight the importance of astrocytes as both the "source" and "sink" responsible for setting resting extracellular adenosine levels in the hippocampus.

5.3 Neuronal Activity Increases Extracellular Adenosine Levels

Although adenosine levels rise following electrical activity in the hippocampus, the exact mechanisms by which adenosine increases occur are still not understood completely. The development of high-speed enzymatic adenosine biosensors by Dale and Frenguelli at the University of Warwick has been a major advance in facilitating our ability to quantify and understand activity-dependent changes in extracellular adenosine levels (Dale et al. 2000; Frenguelli et al. 2003). The same group has recently published a review of activity-dependent release of adenosine which provides a historical perspective and comprehensive reevaluation of activity-dependent purine release (Wall and Dale 2008). Recently other promising methods for real-time detection of adenosine have been introduced, including fast-scan voltammetry using carbon fiber microelectrodes (Huffman and Venton 2009; Cechova and Venton 2008). Here we examine how extracellular adenosine levels rise in response to neuronal activity and evaluate potential mechanisms which may mediate these changes.

As with tonic extracellular adenosine levels, activity-dependent increases in adenosine likely come from multiple adenosine stores. Studies show that electrical stimulation of neural tissue evokes increases in extracellular purine levels (reviewed in Dale and Frenguelli 2009; Pankratov et al. 2006), but the specific composition of released purines and their mechanism of release is not well understood. One of the first convincing studies showing activity-induced elevation of extracellular adenosine was performed by Mitchell et al., who demonstrated that tetanic stimulation of the Schaffer collateral pathway resulted in A_1R-dependent depression of fEPSPs evoked in a separate, nontetanized pathway (Mitchell et al. 1993). This effect was potentiated by blockade of the nucleoside transporter and by blockade of adenosine catabolism. The mechanism of release, however, was not determined. Frank seizure activity is known to increase extracellular adenosine levels (Etherington and Frenguelli 2004; Schrader et al. 1980) which shortens seizure duration dramatically. The specific mechanism by which seizure activity increases extracellular adenosine levels, however, again is unclear.

A candidate mechanism for activity-dependent adenosine release was identified by Manzoni et al. in the early 1990s, who showed that NMDA receptor activation led to A_1R-dependent suppression of fEPSPs in the hippocampus (Manzoni et al. 1994). Further studies showed that NMDA receptor activation decreased mEPSC frequency in an A_1R-dependent way (Brambilla et al. 2005). Manzoni et al. suggested that perhaps enkephalinergic interneurons were the source of NMDA-dependent adenosine release (Manzoni et al. 1994) but this suggestion has not been confirmed nor did it provide any specific molecular insight into how NMDA receptors on enkephalinergic neurons might lead to adenosine release.

A possible source of increased extracellular adenine nucleotides following neuronal activity is ATP co-released in neurotransmitter-filled vesicles. In the hippocampus, ATP may be co-released with glutamate (Mori et al. 2001). At other central synapses ATP can be co-released with noradrenaline (hypothalamus) (Sperlagh et al. 1998), and acetylcholine (cortex) (Richardson and Brown 1987), GABA (hypothalamus) (Jo and Role 2002), and dopamine (limbic system) (Krugel et al. 2003). Based on studies of basal extracellular adenosine, one might assume that the activity-dependent release of ATP leads solely to its catabolism to adenosine and subsequent activation of A_1Rs. This model, however, is too simple. First, the catabolism of ATP to adenosine is dependent on extracellular ectonucleotidases and occurs with a time delay based on their enzymatic properties. In parallel, a diverse family of membrane-bound P2 (ATP) receptors exists, and the immediate and complete conversion of ATP to adenosine would obviate active signaling at these receptors—an outcome which is perhaps unrealistic. Therefore, to understand the contribution of co-released ATP to extracellular adenosine levels one must consider the location of ATP release relative to ectonucleotidases and A_1Rs and ATP receptors. It remains to be seen if ATP is packaged in all glutamate-containing vesicles in the hippocampus (Larsson et al. 2011), and how the spatial distribution of ATP release (Pankratov et al. 2006) aligns with the relative localization of catabolic enzymes (Belcher et al. 2006) and A_1R and P2 receptors (Rebola et al. 2003; Tonazzini et al. 2007). However, evidence suggests that there is a spatial "channeling" from the ectoenzymes to A_1Rs (Cunha et al. 1998).

A study looking at the balance between ATP and adenosine and the frequency of neuronal activity underscores the complexity of these systems and our in complete understanding. In the hippocampus, high (100 Hz) but not low (5 Hz) frequency electrical stimulation of the Schaffer collateral pathway leads to ATP release and its subsequent catabolism to adenosine (Cunha et al. 1996). In contrast, with low frequency stimulation, extracellular adenosine levels do rise, but in a nucleoside transporter-dependent manner. While extracellular ATP breakdown unquestionably contributes to extracellular adenosine levels under certain circumstances, this study also shows that neuronal activity can itself lead to efflux of adenosine via equilibrative transport.

A recent study supplementing hippocampal slices with D-ribose and adenine demonstrated that this manipulation more rapidly restored and then elevated ATP levels in acute slices, which undergo a major loss of energy charge after the slicing procedure. While no changes were observed in baseline recording conditions, there was a significant activity-dependent increase in the influence of adenosine (observed during induction of long-term potentiation; (zur Nedden et al. 2011)). Another report focused on conditions of sufficient or high intracellular ATP in CA3 pyramidal neurons in acute hippocampal slices. During these ATP conditions, a moderate decrease in glucose mobilized adenosine-dependent inhibition (Kawamura et al. 2010).

Hypoglycemia—typically tested using a glucose-free medium—has consistently been shown to increase extracellular adenosine (Fowler 1993; Masino and Dunwiddie 2001).

Changes in intracellular adenosine levels dictate the directionality of flow of adenosine through the equilibrative transporter. Intracellular adenosine levels appear to be controlled mainly by adenosine kinase and/or the rapid consumption of ATP. Indeed, blockade of adenosine kinase results in a large increase in extracellular adenosine levels as measured by enzymatic adenosine biosensors (Etherington et al. 2009). Seizure activity, which puts a large metabolic demand on neurons, increases extracellular adenosine levels and depresses synaptic transmission (Etherington et al. 2009). Massive decreases in hippocampal ATP levels follow kainate-induced seizures (Sugiura et al. 2011) and blockade of nucleoside transporters partially blocked seizure-induced increases in adenosine (Etherington et al. 2009) consistent with this model. Furthermore, seizure activity can rapidly, but transiently, decrease adenosine kinase levels, which would lead to increased extracellular adenosine levels (Gouder et al. 2004). However, another form of activity-dependent increase in extracellular adenosine must also be considered. Some evidence indicates that adenosine is directly released from neurons following activity which does not depend on extracellular ATP catabolism or efflux via nucleoside transporters (Brager and Thompson 2003). This suggests that an unidentified mechanism may exist that allows for the direct release of adenosine, perhaps via vesicular release.

To postulate a preliminary model linking intra- and extracellular adenosine levels, one would predict that limiting adenosine kinase function and ensuring high intracellular ATP levels would be critical to both maintaining extracellular adenosine levels and the ability to have activity-induced adenosine release. Preliminary evidence suggests that such a mechanism may be mobilized by the metabolic switch from predominantly glucose-based to ketone-based metabolism. This metabolic switch is precipitated by maintaining a high-fat, low-carbohydrate (ketogenic) diet (Masino et al. 2011a; Masino and Geiger 2008), an established treatment for epilepsy (Lee and Kossoff 2011). However, the relationship between purinergic metabolism and neuronal excitability is complex. Sufficient adenosine kinase activity is likely crucial to providing metabolic precursors for ATP production. Overabundant adenosine kinase activity, however, will lower extracellular adenosine levels and drive neurons into a hyperexcitable state (Boison 2010). We hope that future studies combining genetic manipulation of potential adenosine release pathways with high-speed adenosine biosensors will shed new light on how metabolism and/or neuronal activity are linked to extracellular adenosine levels.

5.4 pH, Neuronal Excitability, and Extracellular Adenosine

pH has long been linked to neuronal excitability. Overall, pH is necessarily tightly regulated in vivo, and global deviations from physiological pH can have catastrophic clinical consequences. Nevertheless, neuronal activity has clearly been shown to cause measurable intracellular acidosis and extracellular alkalosis (reviewed in (Chesler 2003)), and changes in pH are certainly exaggerated in distinct microdomains of the neuron (Pantazis et al. 2006). The detailed mechanisms by which activity alters pH are diverse, the subject of intense study, and fall outside of the scope of this review. Here we will highlight work that contributes to our understanding of the physiologic and metabolic relationship among pH, neuronal excitability, and extracellular adenosine.

Overall, alkalosis can increase neuronal excitability and induce seizures (Aram and Lodge 1987; Helmy et al. 2011), and acidosis can decrease neuronal excitability (Balestrino and Somjen 1988) and cause stupor and coma (Guisado and Arieff 1975). Seizure-like events in vitro appear to begin when pH levels rise above a certain threshold and stop when pH drops (Xiong et al. 2000). These studies suggest that excitability may be regulated directly by pH-dependent changes in mechanisms such as ion channels, enzymes kinetics, or metabolic balances. Manipulating pH has multiple effects on a wide variety of cellular processes; by no means do we suggest that all changes in excitability are mediated by adenosine. For example, acidosis decreases NMDA receptor activation (Traynelis and Cull-Candy 1990); ASIC channels are activated by acidosis and regulate excitability (Ziemann et al. 2008). A multitude of other cellular changes are linked to changes in pH.

In vivo, respiration and blood flow are the most global regulators of pH. pH buffering is controlled largely by a bicarbonate buffering system strongly linking CO_2 levels and pH. pH is maintained locally by a family of carbonic anhydrases which catalyze the hydration of CO_2 (Supuran and Scozzafava 2007; Supuran 2007; Shah et al. 2005). A long-used neurological tool has been to have patients suffering from absence epilepsy to hyperventilate (thereby decreasing blood CO_2 levels and increasing pH) to induce an absence seizure for EEG diagnosis. How this change in pH/CO_2 leads to altered excitability has long been enigmatic. In keeping with these historical observations, recent findings show that inhalation of an increased level of CO_2 (5 %) can act as a potent anticonvulsant in multiple species, including humans (Tolner et al. 2011). As discussed in detail below, our work and others' have suggested that pH-dependent changes in extracellular adenosine may be involved.

After noting that many stimuli associated with adenosine release also alter pH, we postulated that changes in pH may be a proximal mechanism in regulating adenosine, and may thus explain pH-induced changes in excitability. For example, as mentioned above, NMDA receptor activation was one of the first specific examples of a stimulus that increases extracellular adenosine levels (Manzoni et al. 1994). NMDA receptor activation also causes intracellular acidification via a calcium-dependent process (Irwin et al. 1994); increased neuronal activity in general causes intracellular acidification and increases extracellular adenosine levels (Chesler 2003; Dale and Frenguelli 2009; Pankratov et al. 2006). Hypoxia is known to increase extracellular adenosine levels (Dale et al. 2000) and also cause intracellular acidification (Diarra et al. 1999; Sheldon and Church 2002). Finally, in hippocampal slices, increased temperature increases extracellular adenosine (Masino and Dunwiddie 1999) and decreases intracellular pH (Swain 1983). Taken together, this body of evidence suggested that pH and adenosine levels may be linked.

We first tested the relationship between intracellular pH and adenosine by decreasing intracellular pH using a weak organic acid (Dulla et al. 2005). This decrease in intracellular pH did not increase extracellular adenosine (measured using enzymatic adenosine biosensors) nor did it alter hippocampal fEPSPs consistently. However, when we altered both intracellular and extracellular pH, by either increasing CO_2 (hypercapnia/acidification) or decreasing CO_2 (hypocapnia/alkalinization) we did find significant concomitant changes in extracellular adenosine and neuronal excitability (Dulla et al. 2005). Specifically, when CO_2 levels were increased (and thus intracellular pH decreased), extracellular adenosine increased and fEPSPs decreased in an A_1R- and ATP receptor-dependent manner. Furthermore, CA3 network bursting in hippocampal slices (an in vitro model of epilepsy) was attenuated by hypercapnia in an A_1R-dependent manner. This relationship among CO_2, adenosine and excitability was bidirectional: when CO_2 levels were decreased, adenosine levels dropped and fEPSPs increased in an A_1R- and ATP receptor-dependent manner. In CA3, hypocapnia increased burst frequency, and this increase could be blocked by combined blockade of A_1Rs and ATP receptors. These findings were reminiscent of clinical observations that CO_2 can modulate neuronal excitability; here the modulation was determined to be via adenosine and ATP receptors.

Based on previous work by Tom Dunwiddie and others, we investigated the role of extracellular ATP breakdown by ectonucleotidases in mediating these pH-dependent changes in extracellular adenosine levels. Not surprisingly, we found that the blocking ectonucleotidases led to a decrease in extracellular adenosine levels similar to that caused by hypocapnia. Importantly, pretreating slices with ARL-67156, an ectonucleotidase antagonist, completely attenuated hypocapnia-induced decreases in extracellular adenosine. These experiments suggest that decreased CO_2 increases extracellular pH, which inhibits ectonucleotidases and slows or blocks the conversion from ATP to adenosine. These results suggest a plausible mechanism by which changes in CO_2 mediates alterations in neuronal excitability, and may explain why hyperventilating can trigger seizures (Dulla et al. 2005).

In the above series of experiments we did not, however, pinpoint the mechanism by which decreased pH leads to increased extracellular adenosine levels. Thus, there are many potential mechanisms which remain unexplored and deserve examination in future experiments. ATP production occurs mainly via mitochondrial oxidative phosphorylation (Nicholls and Budd 2000). This process is driven by an electrochemical and proton gradient across the mitochondria membrane. Therefore, increased intracellular proton concentration may decrease the driving force for ATP production and lead to accumulation of intracellular adenosine (Whittingham et al. 1989). As mentioned previously, increased intracellular adenosine can be translated into extracellular adenosine via the nucleoside transporter. Although this is a plausible mechanism, evidence indicates that the electrochemical gradient, not the pH gradient, across the mitochondrial membrane is essential for rapid ATP production (Nicholls and Budd 2000). Therefore, alterations in intracellular pH may not translate into slowed ATP production.

Another plausible but unconfirmed possibility is the action of pH on adenosine kinase function. As previously mentioned, NMDA receptor activation causes intracellular acidification and increases extracellular adenosine. In 2003 a group headed by Paul Rosenberg showed that NMDA receptor activation inhibited adenosine kinase function as well as decreased intracellular ATP levels (Lu et al. 2003). While they did not investigate the role of pH in their studies, it is a potential mechanism by which NMDA receptor activation could be linked to changes in adenosine level and it suggests that adenosine kinase is a molecule which is modulated by pH, as other studies have shown (Yamada et al. 1980). In this study, however, inhibition of adenosine kinase did not preclude NMDA receptor-dependent adenosine release, again indicating the complexity of adenosine regulatory pathways.

A potential but thus for unstudied mechanism by which adenosine levels could be linked to changes in pH is via proton-dependent modulation of the nucleoside transporter. If decreased pH inhibited equilibrative transport, adenosine could accumulate extracellularly. While no specific evidence exists in the brain to demonstrate this potential mechanism, it is known that binding of antagonists to the nucleoside transporter can be pH-dependent, and a novel cardiac nucleoside transporter, CNT4, is activated at low pH (Barnes et al. 2006). Blockade of equilibrative transport itself by protons is an untested but intriguing possibility. We hope that future experiments will be able to dissect the link between pH and extracellular adenosine in order to define specific molecules which alter adenosine and might be targeted for therapeutic uses.

5.5 pH_i, pH_e, and Neuronal Activity: What's the Right Mix to Increase Extracellular Adenosine?

As mentioned above, changing intracellular pH alone using weak organic acids did not alter either extracellular adenosine levels or hippocampal fEPSPs consistently. However, we did find that intracellular acidification may be sufficient during specific physiologic or metabolic conditions, such as increased neuronal activity. To determine this, we increased excitability in evoked fEPSPs in hippocampal slices (by blocking $GABA_A$ receptors with picrotoxin), and then applied a weak organic acid to cause intracellular acidification. Under these conditions intracellular acidification caused a significant increase in extracellular adenosine levels and caused A_1R-dependent inhibition of fEPSPs. Consistent with this interpretation, adenosine release and fEPSP inhibition caused by intracellular acidification during $GABA_A$ receptor blockade were both attenuated by increasing intracellular bicarbonate buffering capacity (Dulla et al. 2009).

We further addressed the relationship between intracellular pH, excitability, and adenosine during a more physiological change in excitability—persistent bursting in CA3 induced via tetanic stimulation in a modified extracellular buffer. The frequency of bursting is dependent on synaptic strength and neuronal excitability, and this approach using synaptic plasticity promotes increased excitability without blocking $GABA_A$ receptors. Intracellular acidification via application of weak organic acid increased extracellular adenosine levels and completely attenuated spontaneous bursting; blocking A_1Rs prevented effects of intracellular acidification on bursting (Dulla et al. 2009).

Together these experiments indicated that changes in intracellular pH were specifically mediating changes in adenosine and excitability. At the same time, these studies raise additional questions. Why does intracellular acidification alone cause adenosine release during increased excitability but not under control conditions? Are we establishing a global change in excitability in vitro that may happen locally in vivo to regulate adenosine? Certainly, there are a number of possibilities which could act either locally or globally. First, when activity increases, intracellular pH drops. Perhaps the combination of activity-induced acidification with weak organic acid-induced acidification causes a summated pH change which crosses a pH threshold sufficient to induce adenosine release. Furthermore, because increased activity causes intracellular acidification, perhaps endogenous intracellular buffering capacity is compromised. As additional possibilities, adenosine kinase is known to be pH-dependent (Yamada et al. 1980); perhaps combined activity-dependent and weak organic acid induced acidification drive enzyme activity to an extremely low level.

An equally plausible possibility is that a combination of activity-induced calcium signaling and intracellular acidification is required to increase extracellular adenosine levels. It is known that blockade of $GABA_A$ receptors increases intracellular calcium levels during stimulus-evoked excitation (Vogt et al. 2011). Perhaps induction of a calcium-dependent process sensitizes metabolic pathways to release adenosine following intracellular acidification. As in picrotoxin experiments, spontaneous bursting of CA3 increases intracellular calcium levels (Takahashi et al. 2010) and likely induces activity-dependent intracellular acidification.

Alternatively, is there sufficient cross talk between calcium and proton signaling to establish a pH-dependent form of adenosine release that is not present under low calcium conditions? In skeletal muscle it is known that changes in pH can affect calcium binding and unbinding kinetics of processes that control muscle relaxation (Metzger and Moss 1990). In vertebrate photoreceptors it is well known that protons modulate calcium release from internal stores (Krizaj et al. 2011). A number of studies suggest direct links between proton and calcium signals, but intracellular acidification does not appear to potentiate activity-induced intracellular calcium transients in hippocampal neurons (Kelly and Church 2006). This suggests that protons do not directly potentiate currents in neurons, but poses a complex and difficult-to-test possibility: that protons and calcium ions interact directly in shaping enzymatic properties which modulate adenosine metabolism.

5.6 Can Altered Metabolism Increase Adenosine?

As we try to understand the regulation and effects of adenosine in vivo, a body of work has found that global physiological changes can modulate adenosine levels (Fig. 5.2). These conditions include pathological conditions such as brain injury, seizures, stroke, and hypoxia (discussed in more detail below), as well as normal

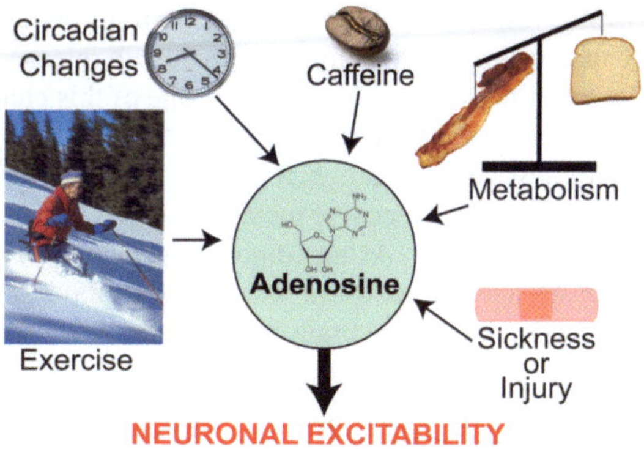

Fig. 5.2 In vivo changes that have been shown to influence adenosine. Diverse global changes in physiology, which occur on a regular or even a daily basis, have been shown to influence adenosine or its actions at A_1Rs. Tom Dunwiddie, PhD (1951–2001), shown here telemark skiing, published many seminal papers on adenosine

circadian changes (Dunwiddie and Masino 2001). Here we focus on physiological conditions that increase adenosine or adenosine-mediated inhibition primarily because of altered metabolism; therefore, the metabolic change translates to altered neuronal activity. For example, there is emerging evidence that conditions such as intense exercise (Dworak et al. 2007), and more recently, a ketogenic diet (Masino et al. 2011b), may alter adenosine levels or signaling. As a more global concept, these studies highlight a dynamic metabolic interplay between ATP and adenosine which affects synaptic transmission. This relationship is in addition to the clear signaling contributions of the rapid conversion of extracellular ATP to adenosine and the net dephosphorylation of ATP and adenosine efflux found during a pathological condition such as hypoxia or ischemia.

In general the relationship between metabolism and neuronal activity is generating a lot of interest. It is recognized increasingly that mitochondrial and metabolic dysfunction accompanies neuronal dysfunction (Knott et al. 2008). Nevertheless, in most cases, the cause-and-effect relationship between neuronal and metabolic dysfunction has not been established. As one exception, the ketogenic diet is a metabolic therapy used to treat epilepsy—a condition with concurrent neuronal and metabolic dysfunction (Pan et al. 2008). In the case of the ketogenic diet the cause-and-effect of the treatment is known: the diet's high fat, low carbohydrate formula is effective at reducing seizures in the majority of patients; some patients become seizure-free (Lee and Kossoff 2011).

With respect to metabolic regulation of adenosine, a ketogenic diet has been shown to increase mitochondrial gene expression (Bough et al. 2006) and increase ATP (Nakazawa et al. 1983), and recent work suggests that A_1R activation is a key component of the ketogenic diet's anticonvulsant effects (Masino et al. 2011b). Interestingly, initial evidence suggests that a ketogenic diet may also reduce adenosine kinase (Masino et al. 2011b). These data showing decreased adenosine kinase, coupled with previous evidence showing that a ketogenic diet increases ATP

(Nakazawa et al. 1983), fulfills conditions for maintaining extracellular adenosine levels alongside a capacity for activity-induced adenosine release. A full discussion of the ketogenic diet and adenosine is outside the scope of this chapter, but a dietary therapy or analogous metabolic strategies, such as those that enhance ATP levels, may yield insights into new cellular targets for the regulation of adenosine.

Exercise as a way to increase adenosine metabolically is an intriguing possibility which deserves further exploration. As noted above, previous work has shown that intense exercise increases adenosine. More recent studies have shown that exercise alters adenine nucleotide hydrolysis in brain (Siqueira et al. 2010) and heart (Roque et al. 2011). Importantly, recent studies demonstrate that exercise not only increases mitochondria in muscle, but also precipitates mitochondrial biogenesis in brain (Steiner et al. 2011; Little et al. 2011). In parallel, for years evidence has been growing steadily that exercise is protective, preventative or restorative in many neurological disorders. These recent brain-specific findings may help explain some of the broad-based beneficial effects of exercise and, like the ketogenic diet, provide a therapeutic strategy for linking metabolism to neurological function.

If brain mitochondrial biogenesis can increase adenosine (which remains to be demonstrated directly) then this is likely a very relevant mechanism for the effects of the ketogenic diet as well as other burgeoning metabolic therapies which, like adenosine, could offer antiseizure and neuroprotective effects. The potential for disease-modifying metabolic therapies is enormous. Increased mitochondrial biogenesis and increased adenosine may be some of the key effects of ketogenic diet therapy, which is now being studied for a host of acute and chronic neurological conditions from brain injury to Alzheimer's disease (Freeman and Kossoff 2010). Metabolic strategies to increase bioenergetics and increase adenosine have the potential to prevent disease and restore homeostasis (Boison et al. 2011).

5.7 Does Adenosine Release During Pathological Insult Give Insight into Release Mechanisms?

Throughout this chapter we have highlighted tools and concepts that have helped to parse apart the molecular pathways that control extracellular adenosine concentration. While enhanced adenosine detection techniques and genetic approaches to modulate specific adenosine-metabolizing enzymes have been extremely useful, more traditionally a great deal of our insight into the regulation of adenosine levels comes from pathological states which produce significant increases in extracellular adenosine, such as ischemia and hypoxia. Here we focus briefly on studies that utilize hypoxia to identify sources of adenosine and discuss potential molecular candidates which these studies suggest might be important for controlling adenosine levels.

Overall, studies of hypoxia have been incredibly useful is understanding the modulation of extracellular adenosine, and during an insult like hypoxia, adenosine levels can increase over 100-fold—more than enough to abolish synaptic transmission. Adenosine levels rise progressively during brief episodes of hypoxia (Dale et al. 2000; Dale and Frenguelli 2009; Pearson et al. 2001), suggesting that adenosine is accumulating in the extracellular space due to (1) either a new steady-state production of adenosine without the equivalent presence of an "adenosine sink" (2) or the loss of an "adenosine sink" with the normal production of adenosine. Although extracellular adenosine levels surge following hypoxic insults (Pearson et al. 2006), the mechanism of this surge is unknown; a remaining paradox is why glutamatergic transmission can return while extracellular adenosine levels remain high.

A recent study using hypoxia as an adenosine-releasing stimulus highlighted the importance of nucleoside transporter expression in controlling adenosine levels. Overexpression of the human nucleoside transporter, ENT1, reduced resting extracellular adenosine levels and, unexpectedly, decreased the adenosine-mediated inhibition caused by hypoxia (Zhang et al. 2011). Based on these studies, the relative expression, and presumably activity, of nucleoside transporters can directly modulate extracellular adenosine levels.

The measurable release of extracellular adenosine becomes progressively less with repeated hypoxic episodes (Pearson et al. 2001), and the progressive reduction in released adenosine does not appear to result from increased reuptake of extracellular adenosine. This study suggests that there is a depletable pool of adenosine, and therefore, repeated hypoxic episodes evoke less adenosine-mediated inhibition of neurotransmission. Are there compartmentalized subsets of ATP (or adenosine) that have different access to the extracellular space? Are there mechanisms whereby cells can limit the amount of adenosine they release into the extracellular space to ensure sufficient precursors for ATP production? How does this relate to preconditioning (an adenosine—based phenomenon whereby an initial modest insult offers neuroprotection against a subsequent insult) (Pugliese et al. 2003; Heurteaux et al. 1995)? These are unanswered but tantalizing questions.

5.8 Conclusion

Understanding the regulation of adenosine is complex. Multiple molecular control sites control extracellular adenosine, and multiple physiologic and metabolic stimuli influence adenosine—which itself has broad and interacting effects. Indeed, stimuli that classically evoke adenosine release (hypoxia and ischemia) consist of multiple changes that alter an enormous number of biological pathways. It may be naïve to think that single targeted perturbations will disrupt adenosine metabolism in a way sufficient enough to increase extracellular adenosine concentrations, as adenosine is designed to provide a dynamic feedback between metabolism and neuronal activity. Conceptualizing adenosine more broadly as a homeostatic bioenergetic network regulator (Boison et al. 2011) may be more relevant as we try to understand its concurrent regulation by physiology and metabolism.

With these caveats in mind, however, we are optimistic about the growing number of genetic tools to modulate biochemical pathways, an ever-evolving ability to quantify extracellular adenosine levels with high temporal resolution, and new insights into bioenergetic regulation of adenosine. Perhaps new insights will also more clearly delineate the contributions of neurons versus glia during different conditions. Together these approaches will make it possible to identify key aspects of the regulation of adenosine, and, more importantly, take advantage of its therapeutic potential.

Acknowledgments We thank the National Institutes of Health, National Science Foundation, Tufts University School of Medicine and Trinity College for their support.

References

Aram JA, Lodge D (1987) Epileptiform activity induced by alkalosis in rat neocortical slices: block by antagonists of N-methyl-D-aspartate. Neurosci Lett 83:345–350

Balestrino M, Somjen GG (1988) Concentration of carbon dioxide, interstitial pH and synaptic transmission in hippocampal formation of the rat. J Physiol 396:247–266

Barnes K, Dobrzynski H, Foppolo S, Beal PR, Ismat F, Scullion ER, Sun L, Tellez J, Ritzel MW, Claycomb WC, Cass CE, Young JD, Billeter-Clark R, Boyett MR, Baldwin SA (2006) Distribution and functional characterization of equilibrative nucleoside transporter-4, a novel cardiac adenosine transporter activated at acidic pH. Circ Res 99:510–519

Belcher SM, Zsarnovszky A, Crawford PA, Hemani H, Spurling L, Kirley TL (2006) Immunolocalization of ecto-nucleoside triphosphate diphosphohydrolase 3 in rat brain: implications for modulation of multiple homeostatic systems including feeding and sleep-wake behaviors. Neuroscience 137:1331–1346

Boison D (2006) Adenosine kinase, epilepsy and stroke: mechanisms and therapies. Trends Pharmacol Sci 27:652–658

Boison D (2010) Adenosine dysfunction and adenosine kinase in epileptogenesis. Open Neurosci J 4:93–101

Boison D, Masino SA, Geiger JD (2011) Homeostatic bioenergetic network regulation—a novel concept to avoid pharmacoresistance in epilepsy. Expert Opin Drug Discov 6:713–724

Bough KJ, Wetherington J, Hassel B, Pare JF, Gawryluk JW, Greene JG, Shaw R, Smith Y, Geiger JD, Dingledine RJ (2006) Mitochondrial biogenesis in the anticonvulsant mechanism of the ketogenic diet. Ann Neurol 60:223–235

Brager DH, Thompson SM (2003) Activity-dependent release of adenosine contributes to short-term depression at CA3-CA1 synapses in rat hippocampus. J Neurophysiol 89:22–26

Brambilla D, Chapman D, Greene R (2005) Adenosine mediation of presynaptic feedback inhibition of glutamate release. Neuron 46:275–283

Brundege J, Dunwiddie TV (1997) Role of adenosine as a modulator of synaptic activity in the central nervous system. Adv Pharmacol 39:353–391

Brundege JM, Diao L, Proctor WR, Dunwiddie TV (1997) The role of cyclic AMP as a precursor of extracellular adenosine in the rat hippocampus. Neuropharmacology 36:1201–1210

Cechova S, Venton BJ (2008) Transient adenosine efflux in the rat caudate-putamen. J Neurochem 105:1253–1263

Chesler M (2003) Regulation and modulation of pH in the brain. Physiol Rev 83:1183–1221

Cotrina ML, Lin JH, Alves-Rodrigues A, Liu S, Li J, Azmi-Ghadimi H, Kang J, Naus CC, Nedergaard M (1998) Connexins regulate calcium signaling by controlling ATP release. Proc Natl Acad Sci U S A 95:15735–15740

Cunha RA (2001) Adenosine as a neuromodulator and as a homeostatic regulator in the nervous system: different roles, different sources and different receptors. Neurochem Int 38:107–125

Cunha RA, Vizi ES, Ribeiro JA, Sebastiao AM (1996) Preferential release of ATP and its extracellular catabolism as a source of adenosine upon high- but not low-frequency stimulation of rat hippocampal slices. J Neurochem 67:2180–2187

Cunha RA, Sebastiao AM, Ribeiro JA (1998) Inhibition by ATP of hippocampal synaptic transmission requires localized extracellular catabolism by ecto-nucleotidases into adenosine and channeling to adenosine A1 receptors. J Neurosci 18:1987–1995

Dale N, Frenguelli BG (2009) Release of adenosine and ATP during ischemia and epilepsy. Curr Neuropharmacol 7:160–179

Dale N, Pearson T, Frenguelli BG (2000) Direct measurement of adenosine release during hypoxia in the CA1 region of the rat hippocampal slice. J Physiol 526(Pt 1):143–155

Diarra A, Sheldon C, Brett CL, Baimbridge KG, Church J (1999) Anoxia-evoked intracellular pH and Ca^{2+} concentration changes in cultured postnatal rat hippocampal neurons. Neuroscience 93:1003–1016

Dulla CG, Dobelis P, Pearson T, Frenguelli BG, Staley KJ, Masino SA (2005) Adenosine and ATP link PCO2 to cortical excitability via pH. Neuron 48:1011–1023

Dulla CG, Frenguelli BG, Staley KJ, Masino SA (2009) Intracellular acidification causes adenosine release during states of hyperexcitability in the hippocampus. J Neurophysiol 102:1984–1993

Dunwiddie TV (1999) Adenosine and suppression of seizures. Adv Neurol 79:1001–1010

Dunwiddie TV, Diao LH (1994) Extracellular adenosine concentrations in hippocampal brain slices and the tonic inhibitory modulation of evoked excitatory responses. J Pharmacol Exp Ther 268:537–545

Dunwiddie TV, Fredholm BB (1985) Adenosine modulation of synaptic responses in rat hippocampus: possible role of inhibition or activation of adenylate cyclase. Adv Cyclic Nucleotide Protein Phosphorylation Res 19:259–272

Dunwiddie TV, Masino S (2001) The role and regulation of adenosine in the central nervous system. Annu Rev Neurosci 24:31–55

Dunwiddie TV, Diao L, Proctor WR (1997) Adenine nucleotides undergo rapid, quantitative conversion to adenosine in the extracellular space in rat hippocampus. J Neurosci 17:7673–7682

Dworak M, Diel P, Voss S, Hollmann W, Struder HK (2007) Intense exercise increases adenosine concentrations in rat brain: implications for a homeostatic sleep drive. Neuroscience 150: 789–795

Etherington LA, Frenguelli BG (2004) Endogenous adenosine modulates epileptiform activity in rat hippocampus in a receptor subtype-dependent manner. Eur J Neurosci 19:2539–2550

Etherington LA, Patterson GE, Meechan L, Boison D, Irving AJ, Dale N, Frenguelli BG (2009) Astrocytic adenosine kinase regulates basal synaptic adenosine levels and seizure activity but not activity-dependent adenosine release in the hippocampus. Neuropharmacology 56:429–437

Fellin T, Halassa MM, Terunuma M, Succol F, Takano H, Frank M, Moss SJ, Haydon PG (2009) Endogenous nonneuronal modulators of synaptic transmission control cortical slow oscillations in vivo. Proc Natl Acad Sci U S A 106:15037–15042

Foley JC, McIver SR, Haydon PG (2011) Gliotransmission modulates baseline mechanical nociception. Mol Pain 7:93

Fowler JC (1993) Purine release and inhibition of synaptic transmission during hypoxia and hypoglycemia in rat hippocampal slices. Neurosci Lett 157:83–86

Freeman JM, Kossoff EH (2010) Ketosis and the ketogenic diet, 2010: advances in treating epilepsy and other disorders. Adv Pediatr 57:315–329

Frenguelli BG, Llaudet E, Dale N (2003) High-resolution real-time recording with microelectrode biosensors reveals novel aspects of adenosine release during hypoxia in rat hippocampal slices. J Neurochem 86:1506–1515

Gouder N, Fritschy JM, Boison D (2003) Seizure suppression by adenosine A1 receptor activation in a mouse model of pharmacoresistant epilepsy. Epilepsia 44(7):877–885

Gouder N, Scheurer L, Fritschy JM, Boison D (2004) Overexpression of adenosine kinase in epileptic hippocampus contributes to epileptogenesis. J Neurosci 24:692–701

Guisado R, Arieff AI (1975) Neurologic manifestations of diabetic comas: correlation with biochemical alterations in the brain. Metabolism 24:665–679

Halassa MM, Florian C, Fellin T, Munoz JR, Lee SY, Abel T, Haydon PG, Frank MG (2009) Astrocytic modulation of sleep homeostasis and cognitive consequences of sleep loss. Neuron 61:213–219

Helmy MM, Tolner EA, Vanhatalo S, Voipio J, Kaila K (2011) Brain alkalosis causes birth asphyxia seizures, suggesting therapeutic strategy. Ann Neurol 69:493–500

Heurteaux C, Lauritzen I, Widmann C, Lazdunski M (1995) Essential role of adenosine, adenosine A1 receptors, and ATP-sensitive K^+ channels in cerebral ischemic preconditioning. Proc Natl Acad Sci U S A 92:4666–4670

Huffman ML, Venton BJ (2009) Carbon-fiber microelectrodes for in vivo applications. Analyst 134:18–24

Irwin RP, Lin SZ, Long RT, Paul SM (1994) N-methyl-D-aspartate induces a rapid, reversible, and calcium-dependent intracellular acidosis in cultured fetal rat hippocampal neurons. J Neurosci 14:1352–1357

Jo YH, Role LW (2002) Coordinate release of ATP and GABA at in vitro synapses of lateral hypothalamic neurons. J Neurosci 22:4794–4804

Johansson B, Halldner L, Dunwiddie TV, Masino SA, Poelchen W, Gimenez-Llort L, Escorihuela RM, Fernandez-Teruel A, Wiesenfeld-Hallin Z, Xu XJ, Hardemark A, Betsholtz C, Herlenius E, Fredholm BB (2001) Hyperalgesia, anxiety, and decreased hypoxic neuroprotection in mice lacking the adenosine A1 receptor. Proc Natl Acad Sci U S A 98:9407–9412

Juranyi Z, Orso E, Janossy A, Szalay KS, Sperlagh B, Windisch K, Vinson GP, Vizi ES (1997) ATP and [3H]noradrenaline release and the presence of ecto-Ca(2+)-ATPases in the capsule-glomerulosa fraction of the rat adrenal gland. J Endocrinol 153:105–114

Kawamura M Jr, Ruskin DN, Masino SA (2010) Metabolic autocrine regulation of neurons involves cooperation among pannexin hemichannels, adenosine receptors, and KATP channels. J Neurosci 30:3886–3895

Kelly T, Church J (2006) Relationships between calcium and pH in the regulation of the slow afterhyperpolarization in cultured rat hippocampal neurons. J Neurophysiol 96:2342–2353

Knott AB, Perkins G, Schwarzenbacher R, Bossy-Wetzel E (2008) Mitochondrial fragmentation in neurodegeneration. Nat Rev Neurosci 9:505–518

Krizaj D, Mercer AJ, Thoreson WB, Barabas P (2011) Intracellular pH modulates inner segment calcium homeostasis in vertebrate photoreceptors. Am J Physiol Cell Physiol 300:C187–C197

Krugel U, Schraft T, Kittner H, Kiess W, Illes P (2003) Basal and feeding-evoked dopamine release in the rat nucleus accumbens is depressed by leptin. Eur J Pharmacol 482:185–187

Larsson M, Sawada K, Morland C, Hiasa M, Ormel L, Moriyama Y, Gundersen V (2011) Functional and anatomical identification of a vesicular transporter mediating neuronal ATP release. Cereb Cortex 22(5):1203–14

Latini S, Pedata F (2001) Adenosine in the central nervous system: release mechanisms and extracellular concentrations. J Neurochem 79:463–484

Lee PR, Kossoff EH (2011) Dietary treatments for epilepsy: management guidelines for the general practitioner. Epilepsy Behav 21:115–121

Little JP, Safdar A, Benton CR, Wright DC (2011) Skeletal muscle and beyond: the role of exercise as a mediator of systemic mitochondrial biogenesis. Appl Physiol Nutr Metab 36:598–607

Lloyd HG, Fredholm BB (1995) Involvement of adenosine deaminase and adenosine kinase in regulating extracellular adenosine concentration in rat hippocampal slices. Neurochem Int 26:387–395

Lu Y, Chung HJ, Li Y, Rosenberg PA (2003) NMDA receptor-mediated extracellular adenosine accumulation in rat forebrain neurons in culture is associated with inhibition of adenosine kinase. Eur J Neurosci 17:1213–1222

Manzoni OJ, Manabe T, Nicoll RA (1994) Release of adenosine by activation of NMDA receptors in the hippocampus. Science 265:2098–2101

Masino SA, Dunwiddie TV (1999) Temperature-dependent modulation of excitatory transmission in hippocampal slices is mediated by extracellular adenosine. J Neurosci 19:1932–1939

Masino SA, Dunwiddie TV (2001) Role of purines and pyrimidines in the central nervous system. In: Abbracchio MP, Williams M (eds) Purinergic and pyrimidinergic signalling I. Springer, Berlin, pp 251–288

Masino SA, Geiger JD (2008) Are purines mediators of the anticonvulsant/neuroprotective effects of ketogenic diets? Trends Neurosci 31:273–278

Masino SA, Kawamura M Jr, Ruskin DN, Geiger JD, Boison D (2011a) Purines and neuronal excitability: links to the ketogenic diet. Epilepsy Res Epub Aug 29

Masino SA, Li T, Theofilas P, Sandau US, Ruskin DN, Fredholm BB, Geiger JD, Aronica E, Boison D (2011b) A ketogenic diet suppresses seizures in mice through adenosine A receptors. J Clin Invest 121:2679–2683

Metzger JM, Moss RL (1990) pH modulation of the kinetics of a Ca2(+)-sensitive cross-bridge state transition in mammalian single skeletal muscle fibres. J Physiol 428:751–764

Mitchell JB, Lupica CR, Dunwiddie TV (1993) Activity-dependent release of endogenous adenosine modulates synaptic responses in the rat hippocampus. J Neurosci 13:3439–3447

Mori M, Heuss C, Gahwiler BH, Gerber U (2001) Fast synaptic transmission mediated by P2X receptors in CA3 pyramidal cells of rat hippocampal slice cultures. J Physiol 535:115–123

Nakazawa M, Kodama S, Matsuo T (1983) Effects of ketogenic diet on electroconvulsive threshold and brain contents of adenosine nucleotides. Brain Dev 5:375–380

Nicholls DG, Budd SL (2000) Mitochondria and neuronal survival. Physiol Rev 80:315–360

Pak MA, Haas HL, Decking UK, Schrader J (1994) Inhibition of adenosine kinase increases endogenous adenosine and depresses neuronal activity in hippocampal slices. Neuropharmacology 33:1049–1053

Pan JW, Williamson A, Cavus I, Hetherington HP, Zaveri H, Petroff OA, Spencer DD (2008) Neurometabolism in human epilepsy. Epilepsia 49(Suppl 3):31–41

Pankratov Y, Lalo U, Verkhratsky A, North RA (2006) Vesicular release of ATP at central synapses. Pflugers Arch 452:589–597

Pantazis A, Keegan P, Postma M, Schwiening CJ (2006) The effect of neuronal morphology and membrane-permeant weak acid and base on the dissipation of depolarization-induced pH gradients in snail neurons. Pflugers Arch 452:175–187

Pascual O, Casper KB, Kubera C, Zhang J, Revilla-Sanchez R, Sul JY, Takano H, Moss SJ, McCarthy K, Haydon PG (2005) Astrocytic purinergic signaling coordinates synaptic networks. Science 310:113–116

Pearson T, Nuritova F, Caldwell D, Dale N, Frenguelli BG (2001) A depletable pool of adenosine in area CA1 of the rat hippocampus. J Neurosci 21:2298–2307

Pearson T, Damian K, Lynas RE, Frenguelli BG (2006) Sustained elevation of extracellular adenosine and activation of A1 receptors underlie the post-ischaemic inhibition of neuronal function in rat hippocampus in vitro. J Neurochem 97:1357–1368

Porkka-Heiskanen T, Strecker RE, Thakkar M, Bjorkum AA, Greene RW, McCarley RW (1997) Adenosine: a mediator of the sleep-inducing effects of prolonged wakefulness. Science 276: 1265–1268

Pugliese AM, Latini S, Corradetti R, Pedata F (2003) Brief, repeated, oxygen-glucose deprivation episodes protect neurotransmission from a longer ischemic episode in the in vitro hippocampus: role of adenosine receptors. Br J Pharmacol 140:305–314

Rebola N, Pinheiro PC, Oliveira CR, Malva JO, Cunha RA (2003) Subcellular localization of adenosine A(1) receptors in nerve terminals and synapses of the rat hippocampus. Brain Res 987:49–58

Richardson PJ, Brown SJ (1987) ATP release from affinity-purified rat cholinergic nerve terminals. J Neurochem 48:622–630

Roque FR, Soci UP, Angelis KD, Coelho MA, Furstenau CR, Vassallo DV, Irigoyen MC, Oliveira EM (2011) Moderate exercise training promotes adaptations in coronary blood flow and adenosine production in normotensive rats. Clinics (Sao Paulo) 66:2105–2111

Schrader J, Wahl M, Kuschinsky W, Kreutzberg GW (1980) Increase of adenosine content in cerebral cortex of the cat during bicuculline-induced seizure. Pflugers Arch 387:245–251

Shah GN, Ulmasov B, Waheed A, Becker T, Makani S, Svichar N, Chesler M, Sly WS (2005) Carbonic anhydrase IV and XIV knockout mice: roles of the respective carbonic anhydrases in buffering the extracellular space in brain. Proc Natl Acad Sci U S A 102:16771–16776

Sheldon C, Church J (2002) Intracellular pH response to anoxia in acutely dissociated adult rat hippocampal CA1 neurons. J Neurophysiol 87:2209–2224

Siqueira IR, Elsner VR, Rilho LS, Bahlis MG, Bertoldi K, Rozisky JR, Batasttini AM, Torres IL (2010) A neuroprotective exercise protocol reduces the adenine nucleotide hydrolysis in hippocampal synaptosomes and serum of rats. Brain Res 1316:173–180

Sosinsky GE, Boassa D, Dermietzel R, Duffy HS, Laird DW, MacVicar B, Naus CC, Penuela S, Scemes E, Spray DC, Thompson RJ, Zhao HB, Dahl G (2011) Pannexin channels are not gap junction hemichannels. Channels (Austin) 5:193–197

Sperlagh B, Sershen H, Lajtha A, Vizi ES (1998) Co-release of endogenous ATP and [3H]noradrenaline from rat hypothalamic slices: origin and modulation by alpha2-adrenoceptors. Neuroscience 82:511–520

Steiner JL, Murphy EA, McClellan JL, Carmichael MD, Davis JM (2011) Exercise training increases mitochondrial biogenesis in the brain. J Appl Physiol 111:1066–1071

Studer FE, Fedele DE, Marowsky A, Schwerdel C, Wernli K, Vogt K, Fritschy JM, Boison D (2006) Shift of adenosine kinase expression from neurons to astrocytes during postnatal development suggests dual functionality of the enzyme. Neuroscience 142:125–137

Sugiura Y, Taguchi R, Setou M (2011) Visualization of spatiotemporal energy dynamics of hippocampal neurons by mass spectrometry during a kainate-induced seizure. PLoS One 6: e17952

Supuran CT (2007) Carbonic anhydrases as drug targets—an overview. Curr Top Med Chem 7: 825–833

Supuran CT, Scozzafava A (2007) Carbonic anhydrases as targets for medicinal chemistry. Bioorg Med Chem 15:4336–4350

Swain JA (1983) Regulation of pH during hypothermia. J Thorac Cardiovasc Surg 85:147–149

Takahashi N, Sasaki T, Matsumoto W, Matsuki N, Ikegaya Y (2010) Circuit topology for synchronizing neurons in spontaneously active networks. Proc Natl Acad Sci U S A 107: 10244–10249

Thompson RJ, Jackson MF, Olah ME, Rungta RL, Hines DJ, Beazely MA, MacDonald JF, MacVicar BA (2008) Activation of pannexin-1 hemichannels augments aberrant bursting in the hippocampus. Science 322:1555–1559

Thorn JA, Jarvis SM (1996) Adenosine transporters. Gen Pharmacol 27:613–620

Tolner EA, Hochman DW, Hassinen P, Otahal J, Gaily E, Haglund MM, Kubova H, Schuchmann S, Vanhatalo S, Kaila K (2011) Five percent CO is a potent, fast-acting inhalation anticonvulsant. Epilepsia 52:104–114

Tonazzini I, Trincavelli ML, Storm-Mathisen J, Martini C, Bergersen LH (2007) Co-localization and functional cross-talk between A1 and P2Y1 purine receptors in rat hippocampus. Eur J Neurosci 26:890–902

Traynelis SF, Cull-Candy SG (1990) Proton inhibition of N-methyl-D-aspartate receptors in cerebellar neurons. Nature 345:347–350

Vogt KE, Gerharz S, Graham J, Canepari M (2011) High-resolution simultaneous voltage and Ca^{2+} imaging. J Physiol 589:489–494

Wall M, Dale N (2008) Activity-dependent release of adenosine: a critical re-evaluation of mechanism. Curr Neuropharmacol 6:329–337

Wang TF, Guidotti G (1998) Widespread expression of ecto-apyrase (CD39) in the central nervous system. Brain Res 790:318–322

Whittingham T, Warman E, Assaf H, Sick T, LaManna J (1989) Manipulating the intracellular environment of hippocampal slices; pH and high energy phosphates. J Neurosci Methods 28: 83–91

Winn HR, Rubio R, Berne RM (1979) Brain adenosine production in the rat during 60 seconds of ischemia. Circ Res 45:486–492

Winn HR, Welsh JE, Rubio R, Berne RM (1980) Changes in brain adenosine during bicuculline-induced seizures in rats. Effects of hypoxia and altered systemic blood pressure. Circ Res 47:868–877

Xiong Z, Saggau P, Stringer J (2000) Activity-dependent intracellular acidification correlates with the duration of seizure activity. J Neurosci 20:1290–1296

Yamada Y, Goto H, Ogasawara N (1980) Purification and properties of adenosine kinase from rat brain. Biochim Biophys Acta 616:199–207

Zhang D, Xiong W, Albensi BC, Parkinson FE (2011) Expression of human equilibrative nucleoside transporter 1 in mouse neurons regulates adenosine levels in physiological and hypoxic-ischemic conditions. J Neurochem 118:4–11

Zhu PJ, Krnjevic K (1994) Endogenous adenosine deaminase does not modulate synaptic transmission in rat hippocampal slices under normoxic or hypoxic conditions. Neuroscience 63:489–497

Ziemann AE, Schnizler MK, Albert GW, Severson MA, Howard MA III, Welsh MJ, Wemmie JA (2008) Seizure termination by acidosis depends on ASIC1a. Nat Neurosci 11:816–822

zur Nedden S, Hawley S, Pentland N, Hardie DG, Doney AS, Frenguelli BG (2011) Intracellular ATP influences synaptic plasticity in area CA1 of rat hippocampus via metabolism to adenosine and activity-dependent activation of adenosine A1 receptors. J Neurosci 31:6221–6234

Chapter 6
The Double-Edged Sword: Gaining Adenosine at the Expense of ATP. How to Balance the Books

Stephanie zur Nedden, Alexander S. Doney, and Bruno G. Frenguelli

Abstract Under physiological conditions the brain, via the *purine salvage pathway*, reuses the preformed purine bases hypoxanthine, derived from ATP degradation, and adenine (Ade), derived from polyamine synthesis, to restore its ATP pool. However, the massive degradation of ATP during ischemia, although providing valuable neuroprotective adenosine, results in the accumulation and loss of diffusible purine metabolites and thereby leads to a protracted reduction in the post-ischemic ATP pool size. In vivo, this may both limit the ability to deploy ATP-dependent reparative mechanisms and reduce the subsequent availability of adenosine, whilst in brain slices results in tissue with substantially lower levels of ATP than in vivo. In the present review, we describe the mechanisms by which brain tissue replenishes its ATP, how this can be improved with the clinically tolerated chemicals D-ribose and adenine, and the functional, and potential therapeutic, implications of doing so.

Keywords ATP • Adenosine • Purine salvage • Ischemia • Traumatic brain injury • Epilepsy • D-Ribose • Adenine • Rib/Ade

---

S. zur Nedden
School of Life Sciences, University of Warwick, Coventry CV4 7AL, UK

Molecular and Cellular Biophysics, Institute of Science and Technology Austria, Am Campus 1, Klosterneuburg A-3400, Austria

A.S. Doney
Department of Medicine and Therapeutics, Ninewells Hospital and Medical School, University of Dundee, Dundee DD1 9SY, UK

B.G. Frenguelli (✉)
School of Life Sciences, University of Warwick, Coventry CV4 7AL, UK
e-mail: b.g.frenguelli@warwick.ac.uk

S. Masino and D. Boison (eds.), *Adenosine: A Key Link between Metabolism and Brain Activity*, DOI 10.1007/978-1-4614-3903-5_6,

6.1 Introduction

The purine nucleoside adenosine plays a major role in the mammalian central nervous system ranging from the control of neurotransmitter release (Ribeiro and Sebastião 2010) to the regulation of cerebral blood flow (Kulik et al. 2008; Pelligrino et al. 2011), and from basal homeostatic functions, such as the regulation of sleep (Halassa 2011) to higher cognitive processes (Wei et al. 2011). Accordingly, the mechanisms governing the production and release of adenosine into the extracellular space, where it interacts with its four G-protein coupled receptors (A_1, A_{2A}, A_{2B}, A_3; (Fredholm et al. 2001)), have come under increasing scrutiny for a number of reasons, not least of which because of the recently appreciated importance of glial cells in purinergic signalling at neuronal synapses (Butt 2011; Dale and Frenguelli 2009; Fields 2011).

There is a general consensus that there are two main pathways for the production and appearance of adenosine in the extracellular space: First, adenosine can be generated via degradation of extracellular precursors, mainly adenine nucleotides, including cAMP. Second, adenosine can be generated from intracellular sources such as ATP and be released per se.

Extracellular adenosine concentrations in the in vivo brain are in the nanomolar range (90–120 nM), sufficiently high enough to activate A_1 and A_{2A} receptors (Dunwiddie and Masino 2001; Latini and Pedata 2001). These tonic levels reflect a balance between the release, metabolism and uptake of adenosine and its precursors. In this regard, a substantial component may arise from ATP since inhibition of ecto-ATPase results in a decrease in extracellular adenosine, as measured with adenosine sensors (Dulla et al. 2005; Frenguelli et al. 2007), and which may derive from astrocytes (Pascual et al. 2005). Adenosine may also arise from release and degradation of cAMP, by ecto-phosphodiesterase and 5′-nucleotidase, and this route may account for slow changes of extracellular adenosine levels (Brundege et al. 1997; Latini and Pedata 2001). The removal of extracellular adenosine is mediated by equilibrative nucleoside (ENT), as well as Na^+-dependent concentrative nucleoside (CNT), transporters. ENTs transport adenosine in both directions depending on the concentration gradient (Podgorska et al. 2005). Intracellular adenosine can be derived from *S*-adenosylhomocysteine (SAH) in a reversible reaction catalysed by SAH-hydrolase (Fig. 6.1, enzyme 7). This pathway is involved in transmethylation reactions and does not seem to contribute to adenosine release upon electrical stimulation or conditions that mimic ischemia (Latini et al. 1996).

The main focus of this review is on the degradation and restoration of intracellular ATP levels and the implications for adenosine production and release. During periods of high energy demand or metabolic stress, such as ischemia, ATP is rapidly degraded resulting in production and release of adenosine (Folbergrova et al. 1997; Frenguelli et al. 2007; Latini et al. 1996; Whittingham et al. 1984). Considering ATP as a primary source of intracellularly produced adenosine, any changes in the ATP pool size is likely to influence adenosine production and release. We therefore (1) discuss conditions where reduced ATP levels, such as after ischemia, may be the basis for reduced adenosine release observed after subsequent ischemic periods,

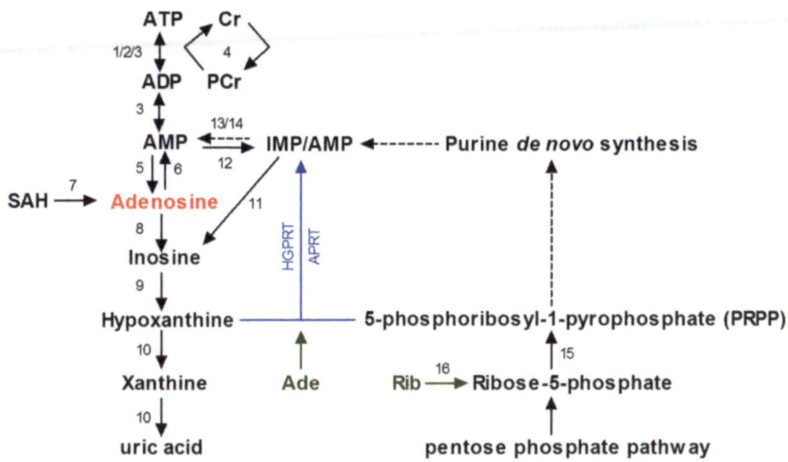

Fig. 6.1 Degradation/restoration of intracellular ATP levels: ATP is degraded to ADP by (*1*) ATPases and ATP is synthesised from ADP by (*2*) ATP synthase in the mitochondria as well as from phosphocreatine (PCr) in a reaction catalysed by creatine kinase (*4*) yielding creatine (Cr). Adenine nucleotides (ATP, ADP and AMP) are interconverted by (*3*) adenylate kinase. AMP is degraded to adenosine by (*5*) cytosolic 5′-nucleotidase (c-N-I). Adenosine can be dephosphorylated by (*6*) adenosine kinase or deaminated to inosine by (*8*) adenosine deaminase. Adenosine can also be derived from *S*-adenosylhomocysteine (SAH) in a reversible reaction catalysed by (*7*) SAH hydrolase. Inosine is degraded to the purine base hypoxanthine by (*9*) purine nucleoside phosphorylase and hypoxanthine can be either degraded to xanthine and uric acid by (*10*) xanthine oxidase or converted to IMP via the purine salvage pathway by hypoxanthine guanine phosphoribosyltransferase (HGPRT). Inosine can also be derived from IMP via dephosphorylation by (*11*) 5′-nucleotides (c-N-II). Likewise AMP can be converted to IMP by (*12*) AMP deaminase and IMP can be reconverted to AMP via the intermediate adenylosuccinate in two reactions catalysed by (*13*) adenylosuccinate synthetase and (*14*) adenylosuccinate lyase. In the purine salvage pathways (shown in blue) the free purine bases hypoxanthine or adenine (Ade) are combined with the activated sugar donor 5-phosphoribosyl-1-pyrophosphate (PRPP), derived from Ribose-5-phosphate in a reaction catalysed by (*15*) PRPP synthetase. This results in formation of the respective purine nucleotide IMP via HGPRT or AMP via adenine phosphoribosyl transferase (APRT). The pentose sugar D-Ribose (Rib) can bypass the pentose phosphate pathway as it is directly phosphorylated to Ribose-5-phosphate by (*16*) ribokinase

(2) summarise our own results showing that it is possible to increase tissue ATP with D-ribose and adenine and that this can result in enhanced activity-dependent adenosine release and (3) speculate as to the potential uses of D-ribose and adenine in neurological conditions associated with reductions in cellular ATP.

6.2 Degradation of Intracellular ATP

Intracellular ATP levels are consumed by ATPases (Fig. 6.1, enzyme 1) and restored by mitochondrial ATP-synthase (enzyme 2). Additionally ATP levels can be maintained by the activity of creatine kinase (enzyme 4), which transfers the phosphate group from phosphocreatine to ADP, yielding creatine and ATP. Creatine kinase can therefore buffer the local decline of ATP levels during increased metabolic demand. Creatine supplementation has been suggested as a cardio- and neuroprotective therapy, via actions extending beyond regulation of cellular ATP (Beal 2011; Perasso et al. 2011). However periods of metabolic stress, such as ischemia, result in a rapid degradation of ATP with a resulting increase in ADP levels (Gadalla et al. 2004; Phillis et al. 1996; Whittingham et al. 1984). The accumulated ADP is converted to AMP and ATP by adenylate kinase (enzyme 3), resulting in an increase in AMP. AMP can be deaminated to inosine monophosphate (IMP) by AMP deaminase (enzyme 12), or dephosphorylated to adenosine by 5′-nucleotidase (enzyme 5). There are two cytosolic 5′-nucleotidases (c-N-I and c-N-II) specific for purines as well as an ecto-5′-nucleotidase (e-N). c-N-I has a preference for AMP and c-N-II has been shown to hydrolyse GMP and IMP, resulting in guanosine and inosine formation, respectively (Hunsucker et al. 2005; Marques et al. 1998; Sala-Newby et al. 2000).

Adenosine can also be rephosphorylated to AMP by adenosine kinase (ADK, enzyme 6), which plays an important role in maintaining an inward directed adenosine gradient under physiological conditions (Boison 2006; Etherington et al. 2009). However, ADK is rapidly downregulated within 20 min of oxygen-glucose deprivation in vitro (Lynch et al. 1998), since it requires ATP as the phosphate donor and is inhibited by high intracellular adenosine concentrations (50 and 75 % inhibition at 5 and 50 μM adenosine, respectively) (Yamada et al. 1980), concentrations that are certainly achievable given the mM quantities of intracellular ATP. Furthermore, ADK is also subject to inhibition by ADP and AMP (Mimouni et al. 1994), the concentrations of which also rise during metabolic stress, whilst ADP may also induce aggregation of ADK (Sen et al. 2005).

Therefore, during metabolic stress, when the ability of ADK to convert intracellular adenosine to AMP is reduced, adenosine is either released to the extracellular space or irreversibly deaminated to inosine by adenosine deaminase (enzyme 8). Unlike adenosine, inosine cannot be rephosphorylated to IMP, due to the lack of a specific kinase (Mascia et al. 2000). Hence, inosine is either released or further degraded to the purine base hypoxanthine and the sugar moiety Ribose-1-phosphate by purine nucleoside phosphorylase (PNP, enzyme 9), whilst hypoxanthine can be reconverted to the purine nucleotide IMP via the purine salvage pathway, the main route for the brain to maintain its adenine nucleotide pool (Allsop and Watts 1980; Gerlach et al. 1971) (Fig. 6.1). Alternatively, hypoxanthine is released from cells or converted to xanthine and subsequently uric acid and hydrogen peroxide in two reactions catalysed by xanthine oxidase (enzyme 10). Unlike hypoxanthine, xanthine and uric acid cannot be reconverted to purine nucleotides and are therefore lost from the purine nucleotide pool. The conversion of IMP to AMP is catalysed by adenylosuccinate synthetase (enzyme 13) and adenylosuccinate lyase (enzyme 14) via the intermediate adenylosuccinate.

It should be noted that the concentration of purine metabolites differs between various brain regions (Kovács et al. 2010), due to a different expression/activity of purine metabolising enzymes, as well as nucleoside transporters. Indeed it has been shown that nucleoside degrading pathways are more active in glial cells than in neurons (Ceballos et al. 1994) and that the degradation of adenosine, released from cultured neurons during metabolic stress, to inosine and hypoxanthine is significantly higher if astrocytes are present (Zamzow et al. 2008).

From Fig. 6.1 it is obvious that AMP degradation to inosine can follow two distinct routes (via IMP or via adenosine). AMP deaminase (AMP→IMP; enzyme 12) and cytosolic c-N-II (IMP→inosine; enzyme 11) are stimulated by ATP, and c-N-II is inhibited by Pi (Chapman and Atkinson 1973; Ipata et al. 2011; Van den Berghe et al. 1992). Accordingly it has been shown that the AMP→IMP→inosine pathway is favoured in rat brain extracts at physiological ATP concentrations (5 mM) (Barsotti and Ipata 2004). In contrast, at low ATP concentrations (<2 mM ATP), as might be found during ischemia, ATP is mainly degraded via the adenosine pathway with little IMP formation (Barsotti and Ipata 2004; Ipata et al. 2011) since cytosolic 5′-nucleotidase c-N-I, which favours AMP as substrate, is stimulated by ADP not ATP (Sala-Newby et al. 2000; Yamazaki et al. 1991). Similarly, in primary neuronal cell cultures it has been shown that under physiological conditions the flux through both pathways is equal, whereas upon inhibition of mitochondrial respiration adenosine production is dominant (Brosh et al. 1996). Likewise, the activity of myocardial cytosolic 5′-nucleotidase activity is increased by decreasing ATP levels (Itoh et al. 1986), further showing that during metabolic stress ATP degradation is shifted towards adenosine formation.

These observations indicate that the metabolism of ATP and the production of adenosine are exquisitely geared to the prevailing metabolic status of the cell: physiological conditions hinder the production of excessive quantities of adenosine, which would suppress synaptic function, whereas crises in energy production favour the production and release of adenosine to mediate a retaliatory suppression of energy-consuming neuronal activity. In this regard, adenosine may be regarded as the extracellular counterpart to AMP-activated protein kinase (AMPK), which suppresses energy-consuming pathways and mobilises energy-producing pathways during metabolic stress (Hardie 2007).

6.3 Restoration of ATP

The two main pathways for restoring and maintaining a constant adenine nucleotide pool are the purine de novo synthesis and the purine salvage pathways. Whereas the purine de novo synthesis is an energy-dependent process where the purine base is built from low molecular weight precursors, the purine salvage pathway re-uses preformed purine bases to restore purine nucleotides.

The activated sugar donor for both pathways, 5-phosphoribosyl-1-pyrophosphate (PRPP) is derived from Ribose-5-phosphate, an intermediate of the pentose phosphate pathway (Fig. 6.1). Alternatively Ribose-1-phosphate, derived from PNP-mediated metabolism of inosine (Fig. 6.1, enzyme 9), can be isomerised to Ribose-5-phosphate by phosphopentomutase (Tozzi et al. 2006).

In the brain only 1–3 % of the consumed glucose is metabolised via the pentose phosphate pathway (Gaitonde and Evans 1982; Gaitonde et al. 1983) and the rate limiting step is the first reaction of the oxidative branch, catalysed by glucose-6-phosphate dehydrogenase, which converts glucose-6-phosphate to 6-phosphogluconate (Kauffman et al. 1969). Similarly, in heart tissue the activity of glucose-6-phosphate dehydrogenase has been shown to be rate limiting for purine de novo synthesis as well as purine salvage (Zimmer 1998). This leaves both brain and heart especially vulnerable to ATP-depleting conditions.

6.3.1 Purine De Novo Synthesis

Purine nucleotide de novo synthesis is an energy-dependent process, which results in formation of IMP from simple precursors upon expenditure of six molecules of ATP. This process involves ten steps and six enzymes (three monofunctional, two bifunctional and one trifunctional enzyme) all of which are located in the cytosol and can form clusters (the purinosome) in response to increased purine de novo synthesis (An et al. 2008). The availability of PRPP, various feedback inhibitions of purine de novo synthesis intermediates, as well as purine nucleotides are the main regulators of purine de novo synthesis (Watts 1983). The purine ring is formed by amino acids, CO_2 and derivatives of tetrahydrofolate, whilst the ribose phosphate moiety is derived from PRPP (An et al. 2008).

Purine de novo synthesis is dominant in the liver, which is responsible for the majority of circulating purine nucleosides and bases (Ipata et al. 2011), whereas it has been shown to be slow and weak in the brain (Allsop and Watts 1980; Gerlach et al. 1971).

6.3.2 Purine Salvage Pathway

In contrast to the energy-dependent reactions of purine de novo synthesis, preformed purine bases can be returned to the nucleotide pool by an energy saving pathway referred to as the purine salvage pathway. This is the predominant route for the brain to restore and maintain its adenine nucleotides (Allsop and Watts 1980; Barsotti and Ipata 2002; Gerlach et al. 1971; Mascia et al. 2000).

Hypoxanthine-guanosine-phosphoribosyltranserfase (HGPRT) and adenine phosphoribosyl-transferase (APRT) are the central enzymes of this pathway (Fig. 6.1). HGPRT catalyses the transfer of ribose-5-phosphate from PRPP to hypoxanthine or guanine yielding PPi and IMP or GMP, respectively (Murray 1971). Hypoxanthine and guanine are derived via PNP from inosine and guanosine, respectively, and the affinity of HGPRT for both is approximately equal (Manfredi and Holmes 1985). HGPRT is expressed in all tissues, and is very abundant in the brain, a further indicator of the importance of the salvage pathway in this organ

(Adams and Harkness 1976; Allsop and Watts 1980; Murray 1966, 1971). Purification of HGPRT from rat brain showed that most of the enzyme is present in the cytosol, although a small but significant fraction was found in synaptosomes (Gutensohn 1973). It is worth noting that an X-linked neurological disorder, Lesch–Nyhan syndrome, which is associated with a complete or severe deficiency of HGPRT activity, results in mental retardation, chorea and compulsive/aggressive self injury (Nyhan 2005).

APRT catalyses the PRPP-dependent phosphoribosylation of adenine to AMP and PPi. Like HGPRT, APRT is expressed in the brain, although its activity is lower compared to HGPRT (Allsop and Watts 1980). The purine base adenine is derived from methylthioadenosine phosphorylase, which catalyses the cleavage of methylthioadenosine, a by-product of polyamine synthesis, to adenine and methylthioribose-1-phosphate (Della Ragione et al. 1986; Ipata et al. 2011). Furthermore it has been shown that adenine can be derived from adenosine in primary rat neuronal cultures (Brosh et al. 1996), although it has been previously suggested that purine nucleoside phosphorylase does not react with adenosine (Murray 1971).

6.4 Modulation of Intracellular ATP Levels and the Effects on Adenosine Release

6.4.1 Incomplete Recovery of ATP After Ischemia May Contribute to the Reduced Adenosine Release During Subsequent Ischemic Periods

Since ATP degradation during metabolic stress, such as ischemia results in the accumulation and release of large quantities of adenosine, inosine and hypoxanthine (Kobayashi et al. 1998; Phillis et al. 1991) these substances can be washed out of the brain across the blood brain barrier into the systemic circulation via the activity of equilibrative nucleoside and nucleobase transporters (Ipata et al. 2011; Isakovic et al. 2008).

The release of purine metabolites during ischemia/hypoxia has not only been observed in in vitro (Dale et al. 2000; Frenguelli et al. 2003; Frenguelli et al. 2007; Latini et al. 1996) and in vivo animal models (Dux et al. 1990; Hillered et al. 1989; Kobayashi et al. 1998; Phillis et al. 1996, 1991; Valtysson et al. 1998) but also in humans. Adenosine, hypoxanthine, xanthine and uric acid were detected in blood and cerebrospinal fluid of stroke patients (Laghi Pasini et al. 2000; Stover et al. 1997). Adenosine was also detected in human plasma during moderate hypoxia (Saito et al. 1999) and in jugular venous blood of humans undergoing carotid endarterectomy with insufficient collateral blood supply of the brain (Weigand et al. 1999). These observations show that large quantities of purine metabolites are washed out from the brain and therefore unavailable for purine salvage pathway and subsequently lost from the adenine nucleotide pool.

Accordingly, in in vivo models of stroke, ATP recovery during reperfusion depends on the duration and severity of the ischemic period (Hermann et al. 2001; Ljunggren et al. 1974; Lust et al. 2002; Onodera et al. 1986; Sims and Zaidan 1995; Yatsu et al. 1975), can often undergo a secondary deterioration (Hata et al. 2000; Lust et al. 2002), and full recovery of ATP levels can take up to 24 h (Kleihues et al. 1975; Kleihues et al. 1974; Levy and Duffy 1977; Lipton 1999).

Additionally, repeated episodes of hypoxia and ischemia result in a reduced release of adenosine in vitro (Pearson et al. 2006; Pearson and Frenguelli 2004; Pearson et al. 2001) and in vivo (DiGeronimo et al. 1998; Dux et al. 1990; Valtysson et al. 1998). This suggests a depletable pool of adenosine (Pearson et al. 2006; Pearson and Frenguelli 2004; Pearson et al. 2001) that the brain cannot replenish at a sufficient rate to support non-depleting adenosine release, which can also be observed during prolonged ischemia in vivo (Matsumoto et al. 1992). Since adenosine is neuroprotective during ischemia (Cunha 2001; Dunwiddie and Masino 2001; Ribeiro 2005; Ribeiro et al. 2003) any reduction in adenosine release may leave the brain more susceptible to excitotoxic damage (Pearson et al. 2003). Indeed it has been shown that multiple sub-injurious ischemic insults at short intervals worsen brain damage in vivo (Kirino 2002). It is possible that this arises, at least in part, from the reduced availability of extracellular adenosine. This acute exacerbation is in contrast to the tolerance which develops over a longer time period after a sub-lethal ischemic insult, a phenomenon referred to as ischemic preconditioning, and which may involve adenosine release and the activation of adenosine receptors (Dirnagl and Meisel 2008).

Considering the brain's reliance on the purine salvage pathway, the loss of salvageable ATP degradation metabolites to the systemic circulation, as well as production of unsalvageable xanthine and uric acid by xanthine oxidase, might limit the potential recovery of post-ischemic total adenine nucleotide (TAN=ATP+ADP+AMP) and ATP levels (Onodera et al. 1986; Pearson et al. 2003). Since the purine de novo synthesis in the brain is not up-regulated after ischemia (Gerlach et al. 1971) a complete restoration of normal nucleotide levels may only be achieved through de novo synthesis at a relatively slow rate, which may last several hours or days (Gerlach et al. 1971; Kleihues et al. 1974; Ljunggren et al. 1974). This could explain the slow and incomplete recovery of ATP during short reperfusion periods, and, considering intracellular ATP as a primary source of adenosine, the reduced adenosine release after subsequent ischemic/hypoxic periods. Therefore, enhancing the resynthesis of ATP by manipulating the purine salvage pathway offers a promising strategy to improve post-ischemic ATP levels and potentially replenish the reduced adenosine release during subsequent ischemic periods. Indeed, in the next sections we will show that it is possible to increase basal ATP levels in brain slices upon addition of purine salvage metabolites and that this has measurable effects on the activity-dependent release of adenosine.

6.4.2 *Increasing Intracellular ATP Levels by Purine Salvage Metabolites: Brain Slices as a Model for Reduced ATP Levels in the Post-ischemic Brain*

Rodent hippocampal brain slices are a widely used in vitro model for studies into synaptic transmission and plasticity. The fact that high energy phosphate levels (ATP, phosphocreatine) of brain slices have traditionally been registered as lower than their in vivo values (Schurr and Rigor 1989; Thomas 1957; Whittingham et al. 1984) has not diminished their popularity as model systems for the study of CNS function due to the accessibility to neurones and glia, and the possibility of imaging, electrophysiological, biochemical and pharmacological experiments. Furthermore, although levels of high-energy phosphates are lower, in our recent studies (zur Nedden et al. 2009, 2011) the energy charge (an index of the relationship between the adenine nucleotides where EC = ([ATP] + ½[ADP])/([ATP] + [ADP] + [AMP])) was in the region of 0.93 or higher, which is in close agreement with estimates obtained in vivo. In addition, provided tissue is incubated close to physiological temperatures (~34 °C), the activity of AMPK, which is activated by metabolic stress, is very low (zur Nedden et al. 2011). These two lines of evidence suggest that given sufficient time and appropriate conditions after preparation, brain slices achieve a metabolic equilibrium not entirely dissimilar to that of the intact brain, despite lower absolute levels of adenine nucleotides.

Nonetheless, the reduction in tissue adenine nucleotides has similarities to that observed in the post-ischemic brain, and thus a model in which to study manipulations that might affect tissue ATP levels, and to establish what influence this may have on the release of adenosine.

We have recently shown that the reduced ATP levels of brain slices (typically 15 nmol/mg protein as opposed to 33 ± 4.7 nmol/mg protein in vivo (zur Nedden et al. 2011)) are primarily due to the loss of ATP precursor metabolites during the ischemic period during slice preparation, once the contribution of the protein content of the dead slice edges (~4 nmol/mg protein) is taken into account (zur Nedden et al. 2011).

Having established the basis of the ATP deficit, which, by extension is likely to occur in the post-ischemic brain in vivo, we used hippocampal brain slices to test the effect of the purine salvage metabolites D-Ribose (Rib, 1 mM) and adenine (Ade, 50 μM, see Fig. 6.1) on improving basal tissue ATP levels. Rib bypasses the rate limiting step of the pentose phosphate pathway as it can be directly phosphorylated to ribose-5-phosphate by ribokinase (Fig. 6.1; enzyme 16) and to PRPP by PRPP-synthase (Fig. 6.1; enzyme 15) (Zimmer 1996). PRPP and Ade are subsequently used by APRT to produce AMP (Barsotti and Ipata 2002).

The TAN content in slices incubated with Rib/Ade increased to 25.8 ± 0.7 nmol/mg protein after 3 hour incubation (compared to 19.1 ± 1.2 nmol/mg protein in slices incubated in standard artificial cerebrospinal fluid (aCSF); Fig. 6.2), with ATP

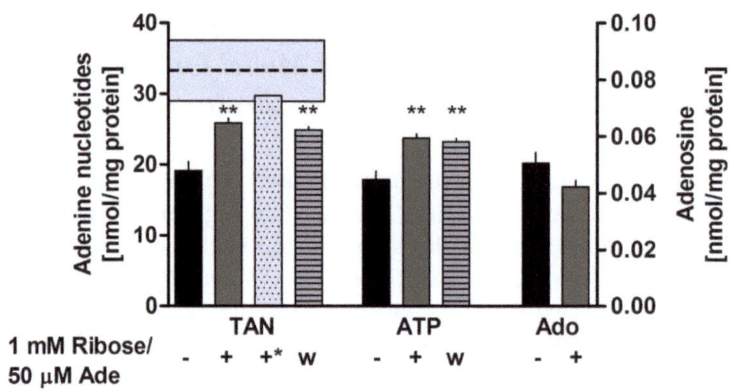

Fig. 6.2 The purine salvage metabolites D-Ribose (Rib) and adenine (Ade) elevate intracellular adenine nucleotides in hippocampal brain slices: Incubation of brain slices with 1 mM D-Ribose/50 μM Adenine for 3 h results in significant elevation of intracellular total adenine nucleotide (TAN=ATP+ADP+AMP) and ATP levels, without affecting adenosine (Ado) levels. TAN (*left y* axis) or ATP (*left y* axis) levels in slices after 3 h incubation in standard aCSF (–, *black bars*) or aCSF supplemented with 1 mM D-Ribose (Rib)/50 μM Ade (+, *grey bars*). Under these conditions, and with the correction for dead tissue on the edge of slices, slice TAN levels (+*, *dotted grey bar*) are close to those reported in vivo (*horizontal grey area* represents the arithmetic mean (*dotted line*)±SD). Elevated TAN levels are maintained when Rib/Ade are washed out (w) of the tissue (*dashed grey bars*) by transferring the slices to standard aCSF for 2 h. Despite increased adenine nucleotide levels Rib/Ade does not affect basal adenosine levels (right *y* axis). Figure adapted from (zur Nedden et al. 2011)

accounting for about 92 %. Furthermore, these elevated tissue TAN and ATP levels persisted when extracellular Rib/Ade was removed by transferring slices back to standard aCSF for 2 h (Fig. 6.2). Unlike ATP levels, the basal adenosine pool was not affected by Rib/Ade-supplementation (Fig. 6.2). This is in agreement with previous results, where addition of Ade increased adenine nucleotide but not adenosine levels in rat brain extracts (Barsotti and Ipata 2002).

When corrected for the influence of the protein content of the dead slice edges, basal TAN values of Rib/Ade-treated slices (~30 nmol/mg protein) are close to the values reported for in vivo tissue (33.6±4.7 nmol/mg protein), showing that the traditionally reported lower brain slice ATP levels are due to the loss of ATP precursors, and not a generally compromised energetic state of the tissue. Furthermore these results provide evidence that the post-ischemic brain has a strong capacity, via the purine salvage pathway, to restore its ATP pool.

6.4.2.1 Functional Implications of Improved Cellular ATP: Greater Activity-dependent Adenosine Release and Modulation of Synaptic Plasticity

The elevated tissue levels of ATP in Rib/Ade-treated slices had no discernable effect on basal synaptic transmission, paired-pulse facilitation or the handling/actions of extracellular adenosine in terms of activity of either adenosine A_1 receptors (A_1R) or ENTs (zur Nedden et al. 2011). Instead, long-term potentiation (LTP) induced by either tetanic stimulation (Fig. 6.3a) or theta burst stimulation (TBS, Fig. 6.3d, e) was reduced by Rib/Ade across a wide range of stimulation parameters, as seen in the amount of LTP 60 min after various stimulation protocols (Fig. 6.3c, f). Indeed, at weaker stimulation protocols (Fig. 6.3a, d, 1× tetanus and 0.5× TBS, respectively), persistent LTP was inhibited. This inhibition could be rescued by an A_1R antagonist (Fig. 6.3b), or by increasing the stimulation strength (1×, 2× or 3× TBS; Fig. 6.3e, f). These data suggest the increased activity-dependent release of adenosine in response to high-frequency stimulation in Rib/Ade-treated slices and the raising of the threshold for LTP induction.

This suggestion was confirmed by direct measurements of adenosine release using adenosine biosensors (Frenguelli et al. 2003) which revealed greater adenosine release in Rib/Ade-treated slices. Analysis of the mechanism of release revealed a predominant contribution of release via ENTs as release was inhibited by ~50 % by ENT inhibitors, but not by the ecto-ATPase inhibitor POM-1 (zur Nedden et al. 2011).

These data point to the release of adenosine upon electrical stimulation, which, via adenosine A_1Rs, inhibits the induction of LTP in Rib/Ade-treated slices. Importantly, this LTP deficit can be overcome by additional stimulation suggesting that the elevated levels of tissue ATP translate into raising the threshold for LTP induction. Interestingly, a ketogenic diet, which is believed to elevate intracellular levels of ATP (Masino and Geiger 2008), has recently been shown to result in reduced LTP in the dentate gyrus of freely moving rats (Koranda et al. 2011). It is tempting to speculate that this arises from increased release of adenosine in a manner analogous to that observed in Rib/Ade-treated hippocampal slices.

6.5 Therapeutic Implications of Rib/Ade

6.5.1 The Heart

The dramatic ability of Rib/Ade to normalise post-ischemic brain ATP levels (at least in these in vitro hippocampal slice models) is paralleled by a large body of work in the heart, which, like the brain, predominately utilises the purine salvage pathway to restore its levels of adenine nucleotides (Zimmer 1998). As in brain, brief periods of ischemia produces a rapid depletion in myocardial ATP levels

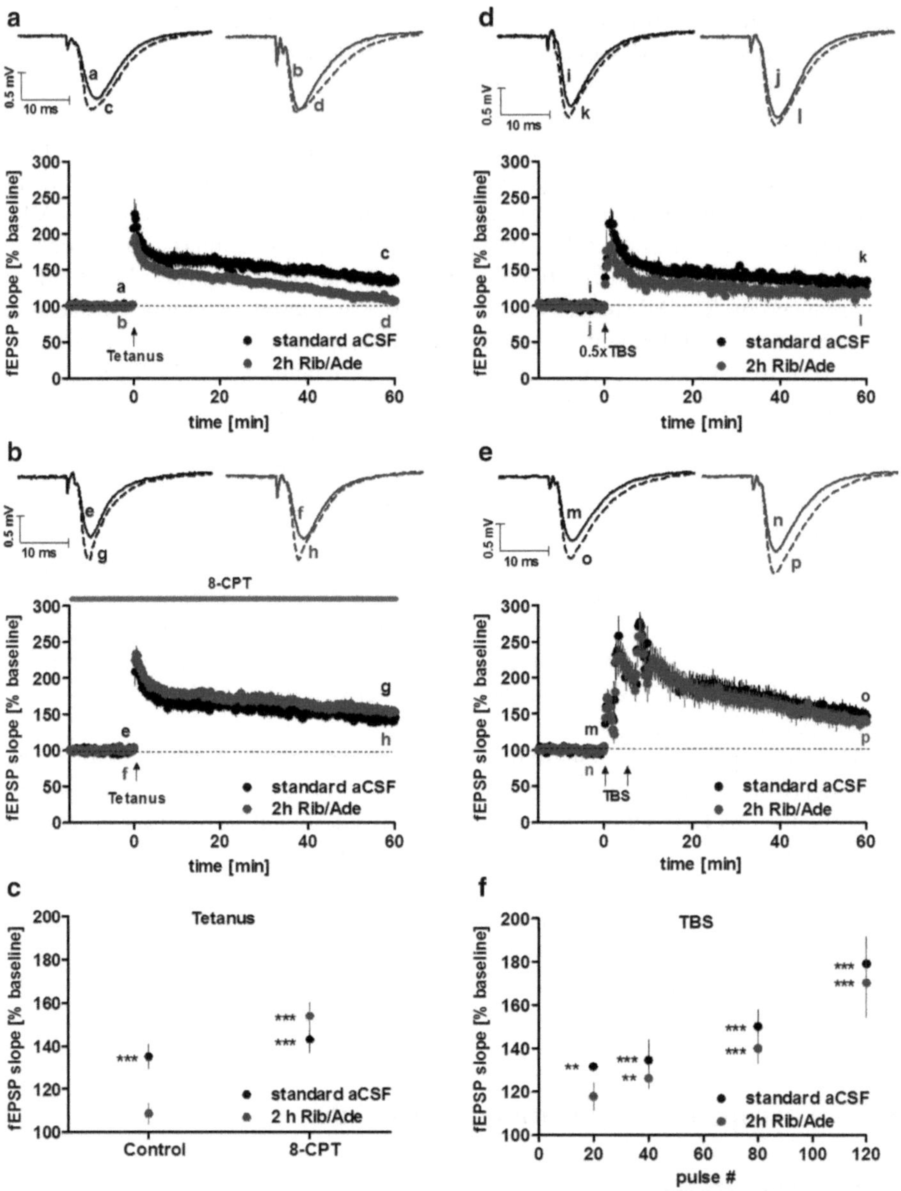

Fig. 6.3 Increasing tissue ATP levels by Rib/Ade-supplementation results in a raising of the threshold for LTP induction by enhanced adenosine release: (**a**) LTP after tetanic stimulation (1× 100 stimuli at 100 Hz) in slices incubated for 2 h in Rib/Ade-supplemented aCSF (*green dots*, $N=11$) and slices incubated in standard aCSF (*black dots*, $N=9$). (**b**) LTP after tetanic stimulation and in the presence of 8-CPT (8-cyclopentyltheophylline) in slices incubated for 2 h in Rib/Ade-supplemented aCSF (*green dots*, $N=6$) and slices incubated in standard aCSF (*black dots*, $N=5$). (**c**) The amount of LTP 55–60 min after tetanic stimulation is shown for control slices, incubated in standard aCSF (*black dots*) and slices which have been treated for 2 h with Rib/Ade (*green dots*). Under control conditions standard slices showed robust LTP (*** $p<0.001$ compared to baseline prior to LTP induction) whereas Rib/Ade-treated slices did not. Addition of the A_1 receptor antagonist 8-CPT (1 μM) rescued the decline in LTP in Rib/Ade-treated slices (*** $p<0.001$ compared

(Zimmer 1992). Likewise, the restoration of post-ischemic ATP levels is slow, due to the heart's dependence on the purine salvage pathway and the loss of salvageable purine metabolites to the systemic circulation (Zimmer 1992, 1998).

Addition of Rib increases myocardial PRPP levels and efficiently improves post-ischemic recovery of myocardial ATP in rat hearts by enhanced purine salvage as well as de novo synthesis (Mauser et al. 1985; Zimmer 1992, 1996, 1998; Zimmer and Gerlach 1978). Accordingly this metabolic approach has been utilised in different experimental in vivo models: for example during myocardial infarction in rats (Zimmer 1982; Zimmer et al. 1989) or after global myocardial ischemia in dogs (St Cyr et al. 1989). The improved post-ischemic ATP recovery has been linked to an enhanced functional recovery of the heart, as seen in improved cardiac contractility (Lamberts et al. 2007; Pasque and Wechsler 1984; Schneider et al. 2008).

Since Rib is known to be well tolerated in humans (Gross and Zollner 1991; Pauly et al. 2003; Salerno et al. 1999; Seifert et al. 2008), this metabolic approach has been successfully translated into clinical trials. In patients with coronary artery disease Rib improved cardiac function, exercise tolerance and general quality of life (Maccarter et al. 2008; Omran et al. 2003). Accordingly, Rib has been suggested as a supplement to improve cardiac energy metabolism (Pauly et al. 2003; Pauly and Pepine 2000; Shecterle et al. 2010). Rib has also been given to a patient with fibromyalgia, a disease which is correlated with reduced ATP levels in muscles, where it decreased certain symptoms such as muscle pain, weakness or sleeping disturbances (Gebhart and Jorgenson 2004).

6.5.2 The Brain

Considering our results, which show that brain slices have a strong capacity through the purine salvage pathway to restore their ATP levels, Rib/Ade might also be useful in improving the post-ischemic recovery of ATP levels after cerebral ischemia in

Fig. 6.3 (continued) to baseline prior to LTP induction). (**d**) LTP after weak theta-burst stimulation (0.5× TBS = 5 trains of 4 pulses at 100 Hz, 200 ms interval, 20 pulses in total), in slices incubated for 2 h in Rib/Ade-supplemented aCSF (*green dots*, N=4 from 3 animals) and slices incubated in standard aCSF (*black dots*, N=3 from 2 animals). (**e**) LTP after stronger TBS (2× TBS = 10 trains of 4 pulses at 100 Hz, 200 ms interval, repeated 5 min apart, 80 pulses in total; N=6 for both) in slices incubated for 2 h in Rib/Ade-supplemented aCSF (*green dots*, N=6) and slices incubated in standard aCSF (*black dots*, N=7). (**f**) The amount of LTP after various TBS protocols (0.5× TBS; 1× TBS = 10 trains of 4 pulses at 100 Hz, 200 ms interval, 40 pulses in total; 2× TBS; 3 TBS = 3× 1 TBS delivered 10 s apart, 120 pulses in total and) in control slices, incubated in standard aCSF (*black dots*) and slices treated for 2 h with Rib/Ade (*green dots*). Shown is a plot of pulse number versus magnitude of LTP 55–60 min (20, 40, 80 pulses) or 30 min (120 pulses) after induction. Note that throughout the pulse number range Rib/Ade-treated slices give rise to lower LTP, which falls below significant LTP induction threshold at 20 pulses. Figure adapted from (zur Nedden et al. 2011)

in vivo animal models, and potentially in stroke and traumatic brain injury patients. Indeed acetate therapy and the ketogenic diet, which both enhance ATP synthesis, have been shown to improve metabolism and neurological outcome after traumatic brain injury (Arun et al. 2010; Deng-Bryant et al. 2011), indicating that the approach of improving ATP recovery can exert neuroprotective or neurorestorative effects.

Cell membranes are highly permeable to Rib, by an order of magnitude greater than the permeability of its diastereomers arabinose, lyxose and xylose, as indicated by recent work using artificial fatty acid and phospholipid bilayers (Sacerdote and Szostak 2005; Wei and Pohorille 2009). Furthermore, it has been shown that Rib can enter the mammalian brain (Agnew and Crone 1967), cross the in vitro frog choroid plexus to the blood (Prather and Wright 1970) and that its uptake competes with that of glucose (Betz et al. 1975). Administration of Rib to a patient suffering from adenylosuccinate lyase deficiency reduced seizure frequency (Salerno et al. 1999), further suggesting that Rib is taken up by human brain. However, this effect on seizures was not replicated in another study (Jurecka et al. 2008), which could reflect differences in dosing and dietary regime, or the basis and severity of the seizures in these individuals. Ade is taken up by the brain by a transport mechanism shared with hypoxanthine (Betz 1985; Redzic et al. 2001; Spector 1989) and the K_m value of Ade transport across the blood brain barrier was estimated to be 0.027 mM (Cornford and Oldendorf 1975). These studies indicate that the blood brain barrier is likely to permit the bioavailability of these agents as metabolic substrates in the post-ischemic brain.

Like Rib, Ade is tolerated in humans. However, during prolonged treatment it may have to be administered with allopurinol, an inhibitor of xanthine oxidase commonly used for the treatment of gout (Simmonds 1986; Watts et al. 1974). Allopurinol prevents the conversion of Ade to the insoluble metabolite 2,8-dihydroxy-adenine, which can cause kidney stone formation. However, allopurinol may additionally have benefits for the post-ischemic brain as it would reduce the metabolism of salvageable hypoxanthine to non-salvageable xanthine, and thereby provide increased substrates for the purine salvage pathway (Phillis et al. 1995). Indeed pretreatment of brain tissue with allopurinol has been shown to be neuroprotective in various in vitro and in vivo models of cerebral ischemia (Lin and Phillis 1991, 1992; Marro et al. 2006).

In a wider context, the reduced tissue ATP levels observed in vivo after cerebral ischemia may, via reduced extracellular adenosine and reduced activation of the anticonvulsant A_1R (Boison and Stewart 2009; Dale and Frenguelli 2009), contribute to the development of acute seizures, which could promote entry into chronic post-ischemic epilepsy (Camilo and Goldstein 2004; Kadam et al. 2010). Indeed, the influence of intracellular ATP on extracellular adenosine and neuronal excitability has recently been described (Kawamura et al. 2010), whilst an increase in intracellular ATP with a concomitant increase in extracellular adenosine has been proposed of being the basis for reduced incidence of seizures during a ketogenic diet (Masino and Geiger 2008). Significantly, this proposal has recently been experimentally verified: a ketogenic diet did not alleviate seizures in mice lacking the adenosine A_1R (Masino et al. 2011). Accordingly, Rib/Ade (and allopurinol) may be of value in the post-ischemic brain in vivo and potentially in humans, since it is (a) safe, (b) clinically well-tolerated, (c) likely to improve the post-ischemic recovery of ATP and (d) thereby increase the pool of releasable neuroprotective adenosine.

6.6 Conclusions

The brain strikes a fine balance between the maintenance of ATP and the production of adenosine. This is tightly regulated by several intracellular enzymes capable of responding to the prevailing metabolic conditions to either favour ATP synthesis during physiological conditions, or to deploy adenosine when energy demand outstrips the means to produce it. This balance is vulnerable to pathological perturbations such as metabolic stress and ischemia where the ability to resynthesise ATP is compromised by the loss of salvageable purine metabolites, such as adenosine, inosine and hypoxanthine. Accordingly, re-establishing ATP homeostasis is slow and incomplete, and this in turn reduces the ability of the brain to activate restorative mechanisms and protect itself from acute secondary insults. However, hope may be at hand: post-ischemic brain tissue has the capacity to utilise exogenous substrates of the purine salvage pathway, which serve to both restore ATP levels and the releasable pool of adenosine. Simple and clinically tolerated manipulations such as D-ribose and adenine supplementation may provide the means to help the brain balance its energy expenditure with substrate income.

Acknowledgements We are grateful to Research into Ageing/Ageing UK and The Dunhill Trust for funding SzN's graduate studies, and to Prof Nicholas Dale for his valuable input.

References

Adams A, Harkness RA (1976) Developmental changes in purine phosphoribosyltransferases in human and rat tissues. Biochem J 160:565–576

Agnew WF, Crone C (1967) Permeability of brain capillaries to hexoses and pentoses in the rabbit. Acta Physiol Scan 70:168–175

Allsop J, Watts RWE (1980) Activities of amidophosphoribosyltransferase (EC 2.4.2.14) and the purine phosphoribosyltransferases (EC 2.4.2.7 and 2.4.2.8), and the phosphoribosylpyrophosphate content of rat central nervous system at different stages of development: Their possible relationship to the neurological dysfunction in the Lesch-Nyhan syndrome. J Neurol Sci 46:221–232

An S, Kumar R, Sheets ED, Benkovic SJ (2008) Reversible compartmentalization of de novo purine biosynthetic complexes in living cells. Science 320:103–106

Arun P, Ariyannur PS, Moffett JR, Xing G, Hamilton K, Grunberg NE, Ives JA, Namboodiri AMA (2010) Metabolic acetate therapy for the treatment of traumatic brain injury. J Neurotrauma 27:293–298

Barsotti C, Ipata PL (2002) Pathways for alpha-D-ribose utilization for nucleobase salvage and 5-fluorouracil activation in rat brain. Biochem Pharmacol 63:117–122

Barsotti C, Ipata PL (2004) Metabolic regulation of ATP breakdown and of adenosine production in rat brain extracts. Int J Biochem Cell Biol 36:2214–2225

Beal M (2011) Neuroprotective effects of creatine. Amino Acids 40:1305–1313

Betz AL (1985) Identification of hypoxanthine transport and xanthine oxidase activity in brain capillaries. J Neurochem 44:574–579

Betz AL, Drewes LR, Gilboe DD (1975) Inhibition of glucose transport into brain by phlorizin, phloretin and glucose analogues. Biochim Biophys Acta 406:505–515

Boison D (2006) Adenosine kinase, epilepsy and stroke: mechanisms and therapies. Trends Pharm Sci 27:652–658

Boison D, Stewart KA (2009) Therapeutic epilepsy research: from pharmacological rationale to focal adenosine augmentation. Biochem Pharmacol 78:1428–1437

Brosh S, Zoref-Shani E, Danziger E, Bromberg Y, Sperling O, Sidi Y (1996) Adenine nucleotide metabolism in primary rat neuronal cultures. Int J Biochem Cell Biol 28:319–328

Brundege JM, Diao L, Proctor WR, Dunwiddie TV (1997) The role of cyclic AMP as a precursor of extracellular adenosine in the rat hippocampus. Neuropharmacol 36:1201–1210

Butt AM (2011) ATP: a ubiquitous gliotransmitter integrating neuron-glial networks. Semin Cell Dev Biol 22:205–213

Camilo O, Goldstein LB (2004) Seizures and epilepsy after ischemic stroke. Stroke 35:1769–1775

Ceballos G, Tuttle JB, Rubio R (1994) Differential distribution of purine metabolizing enzymes between glia and neurons. J Neurochem 62:1144–1153

Chapman AG, Atkinson DE (1973) Stabilization of adenylate energy charge by the adenylate deaminase reaction. J Biol Chem 248:8309–8312

Cornford EM, Oldendorf WH (1975) Independent blood-brain barrier transport systems for nucleic acid precursors. Biochim Biophys Acta 394:211–219

Cunha RA (2001) Adenosine as a neuromodulator and as a homeostatic regulator in the nervous system: different roles, different sources and different receptors. Neurochem Int 38:107–125

Dale N, Frenguelli BG (2009) Release of adenosine and ATP during ischemia and epilepsy. Curr Neuropharmacol 7:160–179

Dale N, Pearson T, Frenguelli BG (2000) Direct measurement of adenosine release during hypoxia in the CA1 region of the rat hippocampal slice. J Physiol 526(Pt 1):143–155

Della Ragione F, CartenÃ-Farina M, Gragnaniello V, Schettino MI, Zappia V (1986) Purification and characterization of 5′-deoxy-5′-methylthioadenosine phosphorylase from human placenta. J Biol Chem 261:12324–12329

Deng-Bryant Y, Prins M, Hovda D, Harris N (2011) Ketogenic diet prevents alterations in brain metabolism in young but not adult rats after traumatic brain injury. J Neurotrauma 28:1813–1825

DiGeronimo RJ, Gegg CA, Zuckerman SL (1998) Adenosine depletion alters postictal hypoxic cerebral vasodilation in the newborn pig. Am J Physiol 274:H1495–H1501

Dirnagl U, Meisel A (2008) Endogenous neuroprotection: mitochondria as gateways to cerebral preconditioning? Neuropharmacol 55:334–344

Dulla CG, Dobelis P, Pearson T, Frenguelli BG, Staley KJ, Masino SA (2005) Adenosine and ATP link PCO2 to cortical excitability via pH. Neuron 48:1011–1023

Dunwiddie TV, Masino SA (2001) The role and regulation of adenosine in the central nervous system. Annu Rev Neurosci 24:31–55

Dux E, Fastbom J, Ungerstedt U, Rudolphi K, Fredholm BB (1990) Protective effect of adenosine and a novel xanthine derivative propentofylline on the cell damage after bilateral carotid occlusion in the gerbil hippocampus. Brain Res 516:248–256

Etherington LA, Patterson GE, Meechan L, Boison D, Irving AJ, Dale N, Frenguelli BG (2009) Astrocytic adenosine kinase regulates basal synaptic adenosine levels and seizure activity but not activity-dependent adenosine release in the hippocampus. Neuropharmacol 56:429–437

Fields RD (2011) Nonsynaptic and nonvesicular ATP release from neurons and relevance to neuron-glia signaling. Semin Cell Dev Biol 22:214–219

Folbergrova J, Li PA, Uchino H, Smith ML, Siesjo BK (1997) Changes in the bioenergetic state of rat hippocampus during 2.5 min of ischemia, and prevention of cell damage by cyclosporin A in hyperglycemic subjects. Exp Brain Res 114:44–50

Fredholm BB, Ijzerman AP, Jacobson KA, Klotz KN, Linden J (2001) International Union of Pharmacology. XXV. Nomenclature and classification of adenosine receptors. Pharmacol Rev 53:527–552

Frenguelli BG, Llaudet E, Dale N (2003) High-resolution real-time recording with microelectrode biosensors reveals novel aspects of adenosine release during hypoxia in rat hippocampal slices. J Neurochem 86:1506–1515

Frenguelli BG, Wigmore G, Llaudet E, Dale N (2007) Temporal and mechanistic dissociation of ATP and adenosine release during ischaemia in the mammalian hippocampus. J Neurochem 101:1400–1413

Gadalla AE, Pearson T, Currie AJ, Dale N, Hawley SA, Sheehan M, Hirst W, Michel AD, Randall A, Hardie DG et al (2004) AICA riboside both activates AMP-activated protein kinase and competes with adenosine for the nucleoside transporter in the CA1 region of the rat hippocampus. J Neurochem 88:1272–1282

Gaitonde MK, Evans GM (1982) The effect of inhibition of hexosemonophosphate shunt on the metabolism of glucose and function in rat brain in vivo. Neurochem Res 7:1163–1179

Gaitonde MK, Evison E, Evans GM (1983) The rate of utilization of glucose via hexosemonophosphate shunt in brain. J Neurochem 41:1253–1260

Gebhart B, Jorgenson JA (2004) Benefit of ribose in a patient with fibromyalgia. Pharmacotherapy 24:1646–1648

Gerlach E, Marko P, Zimmer HG, Pechan I, Trendelenburg C (1971) Different response of adenine nucleotide synthesis de novo in kidney and brain during aerobic recovery from anoxia and ischemia. Experientia 27:876–878

Gross M, Zollner N (1991) Serum levels of glucose, insulin, and C-peptide during long-term D-ribose administration in man. Klin Wochenschr 69:31–36

Gutensohn W (1973) Purification and characterization of a neural hypoxanthine-guanine-phosphoribosyltransferase (HGPRT). Adv Exp Med Biol 41:19–22

Halassa MM (2011) Thalamocortical dynamics of sleep: roles of purinergic neuromodulation. Semin Cell Dev Biol 22:245–251

Hardie DG (2007) Biochemistry. Balancing cellular energy. Science 315:1671–1672

Hata R, Maeda K, Hermann D, Mies G, Hossmann KA (2000) Evolution of brain infarction after transient focal cerebral ischemia in mice. J Cereb Blood Flow Metab 20:937–946

Hermann DM, Kilic E, Hata R, Hossmann KA, Mies G (2001) Relationship between metabolic dysfunctions, gene responses and delayed cell death after mild focal cerebral ischemia in mice. Neuroscience 104:947–955

Hillered L, Hallstrom A, Segersvard S, Persson L, Ungerstedt U (1989) Dynamics of extracellular metabolites in the striatum after middle cerebral artery occlusion in the rat monitored by intracerebral microdialysis. J Cereb Blood Flow Metab 9:607–616

Hunsucker SA, Mitchell BS, Spychala J (2005) The 5′-nucleotidases as regulators of nucleotide and drug metabolism. Pharmacol Therapeutics 107:1–30

Ipata PL, Camici M, Micheli V, Tozz MG (2011) Metabolic network of nucleosides in the brain. Curr Top Med Chem 11:909–922

Isakovic AJ, Dencic SM, Segal MB, Redzic ZB (2008) Transport of [14C]hypoxanthine by sheep choroid plexus epithelium as a monolayer in primary culture: Na^+-dependent and Na^+-independent uptake by the apical membrane and rapid intracellular metabolic conversion to nucleotides. Neurosci Lett 431:135–140

Itoh R, Oka J, Ozasa H (1986) Regulation of rat heart cytosol 5′-nucleotidase by adenylate energy charge. Biochem J 235:847–851

Jurecka A, Tylki-Szymanska A, Zikanova M, Krijt J, Kmoch S (2008) D-Ribose therapy in four Polish patients with adenylosuccinate lyase deficiency: absence of positive effect. J Inherit Metab Dis 31 (S2):S329–S332

Kadam SD, White AM, Staley KJ, Dudek FE (2010) Continuous electroencephalographic monitoring with radio-telemetry in a rat model of perinatal hypoxia-ischemia reveals progressive post-stroke epilepsy. J Neurosci 30:404–415

Kauffman FC, Brown JG, Passonneau JV, Lowry OH (1969) Effects of changes in brain metabolism on levels of pentose phosphate pathway intermediates. J Biol Chem 244:3647–3653
Kawamura M Jr, Ruskin DN, Masino SA (2010) Metabolic autocrine regulation of neurons involves cooperation among pannexin hemichannels, adenosine receptors, and KATP channels. J Neurosci 30:3886–3895
Kirino T (2002) Ischemic tolerance. J Cereb Blood Flow Metab 22:1283–1296
Kleihues P, Hossmann KA, Pegg AE, Kobayashi K, Zimmermann V (1975) Resuscitation of the monkey brain after one hour complete ischemia. III. Indications of metabolic recovery. Brain Res 95:61–73
Kleihues P, Kobayashi K, Hossmann KA (1974) Purine nucleotide metabolism in the cat brain after one hour of complete ischemia. J Neurochem 23:417–425
Kobayashi T, Yamada T, Okada Y (1998) The levels of adenosine and its metabolites in the guinea pig and rat brain during complete ischemia-in vivo study. Brain Res 787:211–219
Koranda JL, Ruskin DN, Masino SA, Blaise JH (2011) A ketogenic diet reduces long-term potentiation in the dentate gyrus of freely behaving rats. J Neurophysiol 106:662–666
Kovács Z, Dobolyi Á, Juhász G, Kékesi K (2010) Nucleoside map of the human central nervous system. Neurochem Res 35:452–464
Kulik T, Kusano Y, Aronhime S, Sandler AL, Winn HR (2008) Regulation of cerebral vasculature in normal and ischemic brain. Neuropharmacology 55:281–288
Laghi Pasini F, Guideri F, Picano E, Parenti G, Petersen C, Varga A, Di Perri T (2000) Increase in plasma adenosine during brain ischemia in man: a study during transient ischemic attacks, and stroke. Brain Res Bull 51:327–330
Lamberts RR, Caldenhoven E, Lansink M, Witte G, Vaessen RJ, St Cyr JA, Stienen GJM (2007) Preservation of diastolic function in monocrotaline-induced right ventricular hypertrophy in rats. Am J Physiol Heart Circ Physiol 293:H1869–H1876
Latini S, Corsi C, Pedata F, Pepeu G (1996) The source of brain adenosine outflow during ischemia and electrical stimulation. Neurochem Int 28:113–118
Latini S, Pedata F (2001) Adenosine in the central nervous system: release mechanisms and extracellular concentrations. J Neurochem 79:463–484
Levy DE, Duffy TE (1977) Cerebral energy metabolism during transient ischemia and recovery in the gerbil 1. J Neurochem 28:63–70
Lin Y, Phillis JW (1991) Oxypurinol reduces focal ischemic brain injury in the rat. Neurosci Lett 126:187–190
Lin Y, Phillis JW (1992) Deoxycoformycin and oxypurinol: protection against focal ischemic brain injury in the rat. Brain Res 571:272–280
Lipton P (1999) Ischemic cell death in brain neurons. Physiol Rev 79:1431–1568
Ljunggren B, Ratcheson RA, Siesjo BK (1974) Cerebral metabolic state following complete compression ischemia. Brain Res 73:291–307
Lust WD, Taylor C, Pundik S, Selman WR, Ratcheson RA (2002) Ischemic cell death: dynamics of delayed secondary energy failure during reperfusion following focal ischemia. Metab Brain Dis 17:113–121
Lynch JJ 3rd, Alexander KM, Jarvis MF, Kowaluk EA (1998) Inhibition of adenosine kinase during oxygen-glucose deprivation in rat cortical neuronal cultures. Neurosci Lett 252: 207–210
Maccarter D, Vijay N, Washam M, Shecterle L, Sierminski H, St Cyr J.A (2008) D-ribose aids advanced ischemic heart failure patients. Int J Cardiol 137:79–80
Manfredi JP, Holmes EW (1985) Purine salvage pathways in myocardium. Annu Rev Physiol 47:691–705
Marques AF, Teixeira NA, Gambaretto C, Sillero A, Sillero MA (1998) IMP-GMP 5′-nucleotidase from rat brain: activation by polyphosphates. J Neurochem 71:1241–1250
Marro PJ, Mishra OP, Delivoria-Papadopoulos M (2006) Effect of allopurinol on brain adenosine levels during hypoxia in newborn piglets. Brain Res 1073–1074:444–450
Mascia L, Cappiello M, Cherri S, Ipata PL (2000) In vitro recycling of alpha-D-ribose 1-phosphate for the salvage of purine bases. Biochim Biophys Acta 1474:70–74

Masino SA, Geiger JD (2008) Are purines mediators of the anticonvulsant/neuroprotective effects of ketogenic diets? Trends Neurosci 31:273–278

Masino SA, Li T, Theofilas P, Sandau US, Ruskin DN, Fredholm BB, Geiger JD, Aronica E, Boison D (2011) A ketogenic diet suppresses seizures in mice through adenosine A_1 receptors. J Clin Invest

Matsumoto K, Graf R, Rosner G, Shimada N, Heiss W-D (1992) Flow thresholds for extracellular purine catabolite elevation in cat focal ischemia. Brain Res 579:309–314

Mauser M, Hoffmeister HM, Nienaber C, Schaper W (1985) Influence of ribose, adenosine, and "AICAR" on the rate of myocardial adenosine triphosphate synthesis during reperfusion after coronary artery occlusion in the dog. Circ Res 56:220–230

Mimouni M, Bontemps F, Van den Berghe G (1994) Kinetic studies of rat liver adenosine kinase. Explanation of exchange reaction between adenosine and AMP. J Biol Chem 269:17820–17825

Murray AW (1966) Purine-phosphoribosyltransferase activities in rat and mouse tissues and in Ehrlich ascites-tumour cells. Biochem J 100:664–670

Murray AW (1971) The biological significance of purine salvage. Annu Rev Biochem 40:811–826

Nyhan WL (2005) Disorders of purine and pyrimidine metabolism. Mol Genet Metab 86:25–33

Omran H, Illien S, MacCarter D, St Cyr J, Luderitz B (2003) D-Ribose improves diastolic function and quality of life in congestive heart failure patients: a prospective feasibility study. Eur J Heart Fail 5:615–619

Onodera H, Iijima K, Kogure K (1986) Mononucleotide metabolism in the rat brain after transient ischemia. J Neurochem 46:1704–1710

Pascual O, Casper KB, Kubera C, Zhang J, Revilla-Sanchez R, Sul J-Y, Takano H, Moss SJ, McCarthy K, Haydon PG (2005) Astrocytic purinergic signaling coordinates synaptic networks. Science 310:113–116

Pasque MK, Wechsler AS (1984) Metabolic intervention to affect myocardial recovery following ischemia. Ann Surg 200:1–12

Pauly DF, Johnson C, St Cyr JA (2003) The benefits of ribose in cardiovascular disease. Med Hypotheses 60:149–151

Pauly DF, Pepine CJ (2000) D-Ribose as a supplement for cardiac energy metabolism. J Cardiovasc Pharmacol Ther 5:249–258

Pearson T, Currie AJ, Etherington LA, Gadalla AE, Damian K, Llaudet E, Dale N, Frenguelli BG (2003) Plasticity of purine release during cerebral ischemia: clinical implications? J Cell Mol Med 7:362–375

Pearson T, Damian K, Lynas RE, Frenguelli BG (2006) Sustained elevation of extracellular adenosine and activation of A_1 receptors underlie the post-ischaemic inhibition of neuronal function in rat hippocampus in vitro. J Neurochem 97:1357–1368

Pearson T, Frenguelli BG (2004) Adrenoceptor subtype-specific acceleration of the hypoxic depression of excitatory synaptic transmission in area CA1 of the rat hippocampus. Eur J Neurosci 20:1555–1565

Pearson T, Nuritova F, Caldwell D, Dale N, Frenguelli BG (2001) A depletable pool of adenosine in area CA1 of the rat hippocampus. J Neurosci 21:2298–2307

Pelligrino DA, Vetri F, Xu H-L (2011) Purinergic mechanisms in gliovascular coupling. Semin Cell Dev Biol 22:229–236

Perasso L, Spallarossa P, Gandolfo C, Ruggeri P, Balestrino M (2011) Therapeutic use of creatine in brain or heart ischemia: available data and future perspectives. Med Res Rev, (in press) doi: 10.1002/med.20255

Phillis JW, O'Regan MH, Estevez AY, Song D, VanderHeide SJ (1996) Cerebral energy metabolism during severe ischemia of varying duration and following reperfusion. J Neurochem 67:1525–1531

Phillis JW, Perkins LM, Smith-Barbour M, O'Regan MH (1995) Oxypurinol-enhanced postischemic recovery of the rat brain involves preservation of adenine nucleotides. J Neurochem 64:2177–2184

Phillis JW, Walter GA, Simpson RE (1991) Brain adenosine and transmitter amino acid release from the ischemic rat cerebral cortex: effects of the adenosine deaminase inhibitor deoxycoformycin. J Neurochem 56:644–650

Podgorska M, Kocbuch K, Pawelczyk T (2005) Recent advances in studies on biochemical and structural properties of equilibrative and concentrative nucleoside transporters. Acta Biochim Pol 52:749–758

Prather JW, Wright EM (1970) Molecular and kinetic parameters of sugar transport across the frog choroid plexus. J Membr Biol 2:150–172

Redzic ZB, Segal MB, Gasic JM, Markovic ID, Vojvodic VP, Isakovic A, Thomas SA, Rakic LM (2001) The characteristics of nucleobase transport and metabolism by the perfused sheep choroid plexus. Brain Res 888:66–74

Ribeiro JA (2005) What can adenosine neuromodulation do for neuroprotection? Curr Drug Targets CNS Neurol Disord 4:325–329

Ribeiro JA, Sebastião AM (2010) Modulation and metamodulation of synapses by adenosine. Acta Physiol 199:161–169

Ribeiro JA, Sebastiao AM, de Mendonca A (2003) Participation of adenosine receptors in neuroprotection. Drug News Perspect 16:80–86

Sacerdote MG, Szostak JW (2005) Semipermeable lipid bilayers exhibit diastereoselectivity favoring ribose. Proc Natl Acad Sci U S A 102:6004–6008

Saito H, Nishimura M, Shinano H, Makita H, Tsujino I, Shibuya E, Sato F, Miyamoto K, Kawakami Y (1999) Plasma concentration of adenosine during normoxia and moderate hypoxia in humans. Am J Respir Crit Care Med 159:1014–1018

Sala-Newby GB, Freeman NVE, Skladanowski AC, Newby AC (2000) Distinct Roles for Recombinant Cytosolic 5'-Nucleotidase-I and -II in AMP and IMP Catabolism in COS-7 and H9c2 Rat Myoblast Cell Lines. J Biol Chem 275:11666–11671

Salerno C, D'Eufemia P, Finocchiaro R, Celli M, Spalice A, Iannetti P, Crifo C, Giardini O (1999) Effect of D-ribose on purine synthesis and neurological symptoms in a patient with adenylosuccinase deficiency. Biochim Biophys Acta 1453:135–140

Schneider HJ, Rossner S, Pfeiffer D, Hagendorff A (2008) D-Ribose improves cardiac contractility and hemodynamics, and reduces expression of c-fos in the hippocampus during sustained slow ventricular tachycardia in rats. Int J Cardiol 125:49–56

Schurr A, Rigor BM (1989) Cerebral ischemia revisited: new insights as revealed using in vitro brain slice preparations. Experientia 45:684–695

Seifert J, Frelich A, Shecterle L, St Cyr J (2008) Assessment of Hematological and Biochemical parameters with extended D-Ribose ingestion. J Int Soc Sports Nutr 5:13

Sen B, Chakraborty A, Datta R, Bhattacharyya D, Datta AK (2005) Reversal of ADP-mediated aggregation of adenosine kinase by cyclophilin leads to its reactivation. Biochemistry 45:263–271

Shecterle LM, Terry KR, St Cyr JA (2010) The patented uses of D-ribose in cardiovascular diseases. Recent Pat Cardiovasc Drug Discov 5:138–142

Simmonds HA (1986) 2,8-Dihydroxyadenine lithiasis–epidemiology, pathogenesis and therapy. Verh Dtsch Ges Inn Med 92:503–508

Sims NR, Zaidan E (1995) Biochemical changes associated with selective neuronal death following short-term cerebral ischaemia. Int J Biochem Cell Biol 27:531–550

Spector R (1989) Micronutrient homeostasis in mammalian brain and cerebrospinal fluid. J Neurochem 53:1667–1674

St Cyr JA, Bianco RW, Schneider JR, Mahoney JR Jr, Tveter K, Einzig S, Foker JE (1989) Enhanced high energy phosphate recovery with ribose infusion after global myocardial ischemia in a canine model. J Surg Res 46:157–162

Stover JF, Lowitzsch K, Kempski OS (1997) Cerebrospinal fluid hypoxanthine, xanthine and uric acid levels may reflect glutamate-mediated excitotoxicity in different neurological diseases. Neurosci Lett 238:25–28

Thomas J (1957) The composition of isolated cerebral tissue; purines. Biochem J 66:655–658

Valtysson J, Persson L, Hillered L (1998) Extracellular ischaemia markers in repeated global ischaemia and secondary hypoxaemia monitored by microdialysis in rat brain. Acta Neurochir (Wien) 140:387–395

Van den Berghe G, Bontemps F, Vincent MF, Van den Bergh F (1992) The purine nucleotide cycle and its molecular defects. Prog Neurobiol 39:547–561

Watts RW, McKeran RO, Brown E, Andrews TM, Griffiths MI (1974) Clinical and biochemical studies on treatment of Lesch-Nyhan syndrome. Arch Dis Child 49:693–702

Watts RWE (1983) Some regulatory and integrative aspects of purine nucleotide biosynthesis and its control: an overview. Adv Enzyme Regul 21:33–51

Wei C, Pohorille A (2009) Permeation of membranes by ribose and its diastereomers. J Am Chem Soc 131:10237–10245

Wei CJ, Li W, Chen J-F (2011) Normal and abnormal functions of adenosine receptors in the central nervous system revealed by genetic knockout studies. Biochim Biophys Acta 1808:1358–1379

Weigand MA, Michel A, Eckstein HH, Martin E, Bardenheuer HJ (1999) Adenosine: a sensitive indicator of cerebral ischemia during carotid endarterectomy. Anesthesiology 91:414–421

Whittingham TS, Lust WD, Passonneau JV (1984) An in vitro model of ischemia: metabolic and electrical alterations in the hippocampal slice. J Neurosci 4:793–802

Yamada Y, Goto H, Ogasawara N (1980) Purification and properties of adenosine kinase from rat brain. Biochim Biophys Acta 616:199–207

Yamazaki Y, Truong VL, Lowenstein JM (1991) 5′-Nucleotidase I from rabbit heart. Biochemistry 30:1503–1509

Yatsu FM, Lee LW, Liao CL (1975) Energy metabolism during brain ischemia. Stability during reversible and irreversible damage. Stroke 6:678–683

Zamzow CR, Xiong W, Parkinson FE (2008) Adenosine produced by neurons is metabolized to hypoxanthine by astrocytes. J Neurosci Res 86:3447–3455

Zimmer HG (1982) Ribose enhances the isoproterenol-elicited positive inotropic effect in rats in vivo. J Mol Cell Cardiol 14:479–482

Zimmer HG (1992) The oxidative pentose phosphate pathway in the heart: regulation, physiological significance, and clinical implications. Basic Res Cardiol 87:303–316

Zimmer HG (1996) Regulation of and intervention into the oxidative pentose phosphate pathway and adenine nucleotide metabolism in the heart. Mol Cell Biochem 160–161:101–109

Zimmer HG (1998) Significance of the 5-phosphoribosyl-1-pyrophosphate pool for cardiac purine and pyrimidine nucleotide synthesis: studies with ribose, adenine, inosine, and orotic acid in rats. Cardiovasc Drugs Ther 12(Suppl 2):179–187

Zimmer HG, Gerlach E (1978) Stimulation of myocardial adenine nucleotide biosynthesis by pentoses and pentitols. Pflugers Arch 376:223–227

Zimmer HG, Martius PA, Marschner G (1989) Myocardial infarction in rats: effects of metabolic and pharmacologic interventions. Basic Res Cardiol 84:332–343

zur Nedden S, Eason R, Doney AS, Frenguelli BG (2009) An ion-pair reversed-phase HPLC method for determination of fresh tissue adenine nucleotides avoiding freeze-thaw degradation of ATP. Anal Biochem 388:108–114

zur Nedden S, Hawley S, Pentland N, Hardie DG, Doney AS, Frenguelli BG (2011) Intracellular ATP influences synaptic plasticity in area CA1 of rat hippocampus via metabolism to adenosine and activity-dependent activation of adenosine A_1 receptors. J Neurosci 31:6221–6234

Chapter 7
Downstream Pathways of Adenosine

Ana M. Sebastião, Sofia Cristóvão-Ferreira, and Joaquim A. Ribeiro

Abstract Adenosine belongs to the class of neuromodulators rather than neurotransmitters, since it is not stored in vesicles, nor released by exocytosis as a classical neurotransmitter. Moreover, it does not induce synaptic potentials but influences the release and the action of neurotransmitters. This mostly occurs through interactions with other G protein-coupled receptors as well as of receptors for neurotrophic factors, ion channels, ionotropic receptors, and neurotransmitter transporters. The actions of adenosine are operated by four different G protein-coupled membrane receptors (A_1, A_{2A}, A_{2B}, A_3), which activate several downstream signaling pathways, the main focus of the present review. Cross talk between adenosine receptors and receptors for neurotransmitters or other neuromodulators may result from interactions between common signaling cascades, as well as through receptor–receptor interactions, including receptor heteromerization. The key receptor in this synaptic interplay appears to be the A_{2A} receptor, whereas A_1 receptors mainly act as modulators of neurotransmitter release or by counteracting A_{2A} receptor-mediated actions. We herein review some of the most recent data on the regulation of adenosine availability, as well as on the consequences of adenosine actions in synapses and the corresponding downstream signaling pathways. Moreover, we discuss how activation of adenosine receptors and regulation of extracellular adenosine levels is operated by combined mechanisms. It is highlighted that modulation of neuronal activity by adenosine involves a diversity of enzymes, receptors and signaling cascades that act in a concerted way to fine tune the activity of neurons and glia, including astrocyte-to-neuron signaling.

A.M. Sebastião (✉) • S. Cristóvão-Ferreira • J.A. Ribeiro
Faculty of Medicine and Unit of Neurosciences, Institute of Pharmacology and Neurosciences, Institute of Molecular Medicine, University of Lisbon, Av. Professor Egas Moniz, Edifício Egas Moniz, Lisbon 1649-028, Portugal
e-mail: anaseb@fm.ul.pt

S. Masino and D. Boison (eds.), *Adenosine: A Key Link between Metabolism and Brain Activity*, DOI 10.1007/978-1-4614-3903-5_7,

Keywords Extracellular formation of adenosine • Nucleoside transporters • Intracellular adenosine clearance • Astrocyte control of extracellular adenosine levels • Signaling pathways operated by adenosine receptors • Neuromodulation • Fine-tuning of neuronal activity

7.1 Introduction

Adenosine, due to the way it operates in the nervous system, belongs to the class of neuromodulators rather than neurotransmitters (for a review on these concepts see, for instance, Ribeiro and Sebastião 2010). So, intracellular adenosine is not stored in vesicles, nor released by exocytosis as a classical neurotransmitter. Instead, adenosine is released to the extracellular space through equilibrative nucleoside transporters that function bidirectionally according to the gradient across the cell membrane. Adenosine is also formed in the extracellular space through degradation of released ATP. Once in the extracellular space, adenosine activates membrane located G protein-coupled receptors (GPCR) and through these receptors affects neuronal functioning at different levels, including changes in the ability to release or respond to neurotransmitters or even gliotransmitters, but so far, no neurotransmitter-like actions for adenosine have been identified.

Adenosine also behaves as a retaliatory metabolite, influencing and reflecting cell energy state, as well as metabolic demand and nutrient supply. More than 25 years have elapsed since it was first proposed that adenosine efficiently connects synaptic activity, energy expenditure, and nucleic acid metabolism by acting as a sensor of the bioenergetic state of the cell (Newby et al. 1985). Moreover, intracellular adenosine directly regulates transmethylation reactions, including DNA methylation (Boison et al. 2002), which can lead to long-lasting epigenetic modifications. Other than at the nerve tissue level, the influence of adenosine upon cerebral blood flow enables it to further act as an energy balancing metabolite. Thus, when metabolism is increased, the elevated ATP catabolism will produce higher amounts of adenosine, which through the activation of $A_{2A}R$, (Phillis 1989) will induce vasodilation, allowing an improvement of oxygen and nutrient delivery via the cerebral vasculature.

In the present work we discuss some of the most recent data on the regulation of adenosine availability and its effects through adenosine receptors and how their activation is regulated by extracellular adenosine levels. Downstream mechanisms of receptor activation and receptor cross talk are then reviewed on the basis of recently published data, highlighting the functional outcomes of the subtle ways adenosine fine-tunes neuronal activity.

7.2 Adenosine Formation

The main source of adenosine, in the central nervous system, is the dephosphorylation of 5′-AMP by 5′-nucleotidases (5′-NTs) (Meghji 1993). These enzymes dephosphorylate noncyclic nucleoside monophosphate to nucleosides and inorganic phosphate.

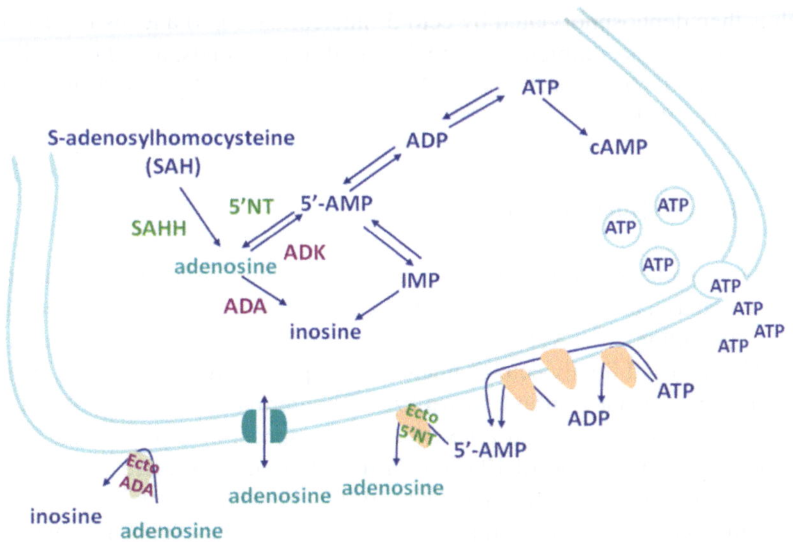

Fig. 7.1 Adenosine metabolism. Adenosine can be synthesized intra- and extracellularly. Inside the cell adenosine is formed from AMP metabolism through endo-5′-nucleotidase (5′NT) or by the transmethylation reaction catalyzed by SAHH, which converts SAH into adenosine and homocysteine. At the extracellular space, adenosine derives from the metabolism of ATP/ADP/AMP, being the last reaction catalyzed by the ectonucleotidase (Ecto 5′NT). The release of adenosine through equilibrative nucleoside transporters is an alternative source of adenosine. Regarding clearance of extracellular adenosine, in some cases it can be converted into inosine by ecto-adenosine deaminase (ecto-ADA) in the extracellular space, but in most cases adenosine is taken up by the equilibrative nucleoside transporter into cells where adenosine can be phosphorylated to AMP by adenosine kinase (ADK) or deaminated to inosine by adenosine deaminase (ADA)

So far, seven distinct nucleoside monophosphate phosphohydrolases or 5′-nucleotidases (EC 3.1.3.5 and EC 3.1.3.6) have been cloned. Five are localized in the cytosol; one is attached to the outer side of the plasma membrane and one in the mitochondrial matrix. Nucleotidases are responsible for both intracellular and extracellular synthesis of adenosine from the dephosphorylation of AMP. An alternative source of adenosine synthesis results from the hydrolysis of *S*-adenylhomocysteine (SAH), which is catalyzed by SAH hydrolase (SAHH) (Palmer and Abeles 1979) (see Fig. 7.1).

7.2.1 Extracellular Formation of Adenosine

Adenosine found in the extracellular space can be released via equilibrative nucleotide transporters or be synthesized locally via ATP catabolism, which involves several enzymes, including the enzymes of ectonucleoside triphosphate diphosphohydrolase (E-NTPDase) family, ectonucleotide pyrophosphatase/phosphodiesterase (E-NPP) family, ecto-5′-nucleotidase/CD73, and alkaline phosphatases (Yegutkin 2008). Through this cascade, adenine nucleotides are dephosphorylated into 5′-AMP, which is then dephosphorylated by ecto-5′-nucleotidase into adenosine. The entire catalytic pathway is complete in a few hundred milliseconds, and the rate-limiting step being the dephosphorylation of AMP into adenosine by ecto-5′-nucleotidase (Dunwiddie et al. 1997).

7.2.1.1 Ecto-5′-Nucleotidase

After the vesicular release of ATP (which is cosecreted with neurotransmitters or even released as a neurotransmitter), ATP is metabolized by a cascade of ectonucleotidases, including ecto-ATPase, ecto-ADPase (E-NTPDase family) and apyrase (E-NPP family) and finally ecto-5′-nucleotidase, producing adenosine (Ribeiro and Sebastião 1987; Richardson and Brown 1987; Zimmermann et al. 1986). Additionally cAMP can also be released into the extracellular space, by a probenecid-sensitive transporter (Rosenberg and Li 1995) in sufficient amounts to increase extracellular adenosine concentrations (Brundege et al. 1997; Dunwiddie et al. 1992).

In neuronal cells, ectonucleotidases are able to convert most adenine nucleotides (except cAMP) into adenosine in less than a second (Dunwiddie et al. 1997). Indeed, even stable ATP analogues can be converted into adenosine by ectonucleotidases (e.g., Cascalheira and Sebastião 1992; Cunha et al. 1998). Several ectonucleotidases, including alkaline phosphatase and nucleoside triphosphate diphosphohydrolase 2, are associated with subsets of progenitor cell populations in the mouse embryonic, postnatal, and adult neurogenic zones (Langer et al. 2007). Knockdown of tissue nonspecific alkaline phosphatase impairs neural stem cell proliferation and differentiation (Kermer et al. 2010), highlighting their relevance in neurogenesis, including adult neurogenesis (see Zimmermann 2011).

The ecto-5′-nucleotidase is a cell surface protein attached to the plasma membrane by a glycosyl phosphatidylinositol (GPI) anchor at its C terminal (Misumi et al. 1990). The hydrolysis of extracellular AMP is considered the main function of this enzyme, but 5′-nucleotidase is also involved in cell adhesion, as it also binds laminin and fibronectin (Mrhul et al. 1993; Olmo et al. 1992). Ecto-5′-nucleotidase acts also as a coreceptor in T cell activation (see Resta and Yamashita 1998).

Ecto-5′-nucleotidase is highly expressed in the brain, where it is mainly associated with glial cell membranes, namely, astrocytes, oligodendrocytes, and also microglia (Kreutzberg et al. 1978; Naidoo 1962; Schoen and Kreutzberg 1995). In fact, the predominant glial expression of ecto-5′-nucleotidase is related to an enhancement in the contribution of extracellular conversion of AMP into adenosine when astrocytes are cocultured with neurons (Zamzow et al. 2008a).

Regarding localization in neurons, some initial cytochemical studies associated ecto-5′-nucleotidase with the surface of migrating and immature nerve cells and with subsets of synapses during part of their regeneration period as well as during synapse remodeling and regeneration (Schoen et al. 1991, 1993). Later on, the ecto-5′-nucleotidase expression by mature neurons was also demonstrated in the cerebellum (Maienshein and Zimmermann 1996) and in hippocampal nerve terminals (Cunha et al. 2000). More recently, it has been demonstrated that ecto-5′-nucleotidase is expressed by nociceptive neurons in dorsal root ganglia and on terminals in substantia gelatinosa of spinal cord, where the conversion of AMP into adenosine promotes antinociception (Sowa et al. 2010).

Functionally, there is some evidence that ectonucleotidases are in close physical proximity with presynaptic adenosine receptors (Cunha et al. 1996; Dunwiddie and Masino 2001), so that recently formed adenosine becomes immediately available to the presynaptic receptors involved in modulation of neurotransmitter release. Topographical arrangement of membrane bound molecules involved in purinergic signaling may determine the type of receptor activated by adenosine, since there is evidence that adenosine formed from released ATP preferentially activates facilitatory receptors (Cunha et al. 1996).

7.2.2 *Intracellular Synthesis of Adenosine*

For the net extracellular adenosine levels, intracellular synthesis of adenosine is at least as important as adenosine formation from breakdown of extracellular ATP (Lloyd and Fredholm 1995). Intracellular synthesis of adenosine occurs mainly by AMP dephosphorylation, which is catalyzed by cytosolic nucleotidases. The presence of cytosolic 5′-nucleotidase in the brain was firstly demonstrated in 1982 (Montero and Fes 1982). Although differential expression of 5′-nucleotidase among different brain areas has not been established so far, its ubiquitous role in the intracellular synthesis of adenosine is well known. Additionally, the cytosolic nucleotidases participate in substrate cycles that regulate the cellular levels of ribo- and deoxyribonucleoside monophosphates, regulating the intracellular pools of ribo- and deoxyribonucleotides (Reichard 1988) which are crucial for DNA/RNA synthesis.

Furthermore, adenosine produced inside the cell contributes to restoring ATP levels by decreasing ATP utilization and increasing oxygen and nutrients supply via blood flow (Newby 1984). Thus, adenosine is commonly considered to be a retaliatory metabolite, since adenosine produced during cytosolic ATP degradation behaves as a metabolic stress sign promoting retaliatory effects against the stress-causing conditions (Newby 1984).

Another source of adenosine is the transmethylation pathway, where adenosine results from the hydrolysis of *S*-adenosylhomocysteine (SAH) catalyzed by SAH hydrolase (SAHH, EC 3.3.1.1), which also produces L-homocysteine (Palmer and Abeles 1979; Schrader et al. 1981). This enzyme was firstly described in 1959, in rat liver. SAHH catalyzes a reversible reaction, that preferentially evolves towards *S*-adenosylhomocysteine synthesis (de la Haba and Cantoni 1959). In the heart, this pathway provides one-third of the total cardiac adenosine at normoxic conditions but generates undetectable levels under hypoxic conditions (Deussen et al. 1989). SAHH expression is widespread in the brain, with higher expression levels present in cortex and cerebellum. Inside the cell, SAHH displays a nuclear expression where it is involved in transmethylation mechanisms. In detail, different methyltransferases convert *S*-adenosylmethionine (SAM) into SAH, which is then metabolized into adenosine and L-homocysteine by SAHH. So, adenosine is an obligatory end product of SAM-dependent transmethylation reactions and because of that is able to inhibit methylation reactions. To avoid this inhibition, adenosine is phosphorylated into AMP by a long isoform of adenosine kinase (ADK), which was described as a nuclear ADK (Cui et al. 2009).

Under normal conditions SAHH has low impact upon neuronal excitability (Pak et al. 1994), suggestive of a minor role in the control of cytoplasmic levels of adenosine in neurons.

7.3 Nucleoside Transporters

The relevance of adenosine uptake by nucleoside transporters in terminating adenosine effects was first supported by different studies showing that the nucleoside transporter blockade produces vasodilation, potentiates the ability of adenosine to decrease locomotor activity (Crawley et al. 1983), depresses neuronal activity (Motley and Collins 1983), increases nociceptive thresholds (Yarbrough and McGuffin-Clineschmidt 1981), and exerts anticonvulsive effects (Dragunow and Goddard 1984). Therefore, nucleoside transporter inhibitors exacerbate the effects mediated by adenosine.

Nucleoside transporters can be divided in two main classes: the equilibrative (Na^+-independent) nucleoside transporters (ENTs) and the concentrative (Na^+-dependent) nucleoside transporters (CNTs) (Baldwin et al. 1999). Six isoforms of CNTs (CNT1–CNT6) and four isoforms of ENTs (ENT1–ENT4) have been cloned, to date. Equilibrative transporters mediate nucleoside transport in both directions, depending on the nucleoside concentration gradient across the membrane. The four transporters are widely distributed, and all of them are able to transport adenosine but they have different abilities to transport other nucleosides (Baldwin et al. 2004). The transport mediated by concentrative transporters is independent of nucleoside gradient and is coupled to sodium gradient. As intracellular concentrations of adenosine are kept low due to its conversion into AMP, and as catabolism of released nucleotides constitutes an additional and transporter-independent source of extracellular adenosine, the extracellular concentrations of adenosine are usually higher than the intracellular ones. Therefore, the usual direction of equilibrative adenosine transport is uptake into cells, rather than release. Indeed, adenosine uptake in the brain occurs primarily by facilitated diffusion via equilibrative transporters, although some of it (10–20 %) can be mediated by concentrative transporters (Geiger and Fyda 1991; Parkinson et al. 1994).

Equilibrative nucleoside transporters are crucial to regulate the levels of extracellular adenosine, being the main entity responsible for removing adenosine from the extracellular space. They are thus responsible for restraining, both spatially and temporally, adenosinergic modulation. Due to the equilibrative nature of the adenosine transporters in neuronal cells, changes in the activity of enzymes involved in ADO metabolism will modify the transporters' activity. Accordingly, transporter inhibitors can either increase (Dunwiddie and Diao 1994; Phillis et al. 1989; Sanderson and Scholfield 1986) or decrease (Gu et al. 1995) extracellular adenosine levels, depending on the transmembrane adenosine gradient and consequently depending on transport direction, into or out of the cell. However, because the extracellular synthesis of adenosine from catabolism of nucleotides constitutes an alternative source of adenosine, which is not affected by transport blockade, the transporters inhibitors usually lead to an increase in the extracellular levels of the nucleoside.

The amount of adenosine released by nucleoside transporters is enhanced under some circumstances such as hypoxia or ischemia, when a massive increase in extracellular adenosine levels is observed, a process prevented by transporter blockade (Parkinson et al. 2002). At the synaptic level, however, the rise in extracellular levels of adenosine during hypoxia may increase rather than decrease upon inhibition of equilibrative nucleoside transporters (Frenguelli et al. 2007). Furthermore, recent evidence (Zhang et al. 2011) showed that neuronal nucleoside transporters contribute to the removal of extracellular adenosine from the synaptic space even during hypoxic or ischemic insults. Therefore, evidence now available allows suggesting that the control of extracellular adenosine levels may differ in different microdomains. As highlighted recently (Sebastião 2011), a deeper understanding of those microdomains as well as of the relative contribution of the different cell types (i.e., neurons vs. astrocytes) for the net production of adenosine is required to better predict the direction of the changes in adenosine levels after pharmacologic or genetic manipulation of adenosine transporters in pathological conditions.

Interestingly, adenosine release by nucleoside transporters is promoted by neurotransmitters. For example, glutamatergic agonists such as NMDA and kainate increase, in a dose-dependent manner, adenosine release (Carswell et al. 1997; Delaney et al. 1998). In fact, activation of NMDA receptors seems to promote release of adenosine itself instead of its precursor, ATP (Harvey and Lacey 1997; Manzoni et al. 1994). This may be part of a protective feedback loop since adenosine released through the transporters seems to preferentially activate adenosine A_1 receptors (A_1R; Cunha et al. 1996) and these are neuroprotective, namely, through inhibition of NMDA currents not only under normoxic (de Mendonça et al. 1995) but also under hypoxic (Sebastião et al. 2001) conditions.

7.4 Intracellular Adenosine Clearance

After being taken up through nucleoside transporters, adenosine is inactivated either by deamination through adenosine deaminase or by phosphorylation through adenosine kinase (Fig. 7.1). It is accepted that the pathway responsible for intracellular adenosine clearance is dependent on its concentration. As such, at low concentrations, adenosine is mainly inactivated by phosphorylation while at higher concentrations adenosine is predominantly deaminated by adenosine deaminase (Meghji and Newby 1990), in accordance with the affinity for adenosine and enzymatic capacity of those two enzymes.

7.4.1 Adenosine Kinase

ADK (EC 2.7.1.20) phosphorylates intracellular adenosine into AMP. Due to its low K_m for adenosine, it is the main enzyme responsible for intracellular adenosine catabolism, at least, for low adenosine concentration. ADK is therefore a key target whenever manipulation of the neuromodulatory actions of adenosine is desirable. By phosphorylating adenosine into AMP, ADK has a double role for maintaining a homeostatic energy flux: (1) a direct ability to influence the cellular energy pool (AMP, ADP, and ATP) and (2) an influence upon intra- and extracellular levels of the homeostatic regulator, adenosine. The relevance of ADK for the homeostatic control (Boison et al. 2011) is supported by several lines of evidence, namely, (1) the release of higher amounts of adenosine by ADK-deficient fibroblasts in cultures, when compared to that released by ADA-deficient fibroblasts (Huber et al. 2001), (2) the ability of ADK inhibition to depress neuronal activity in hippocampal slices, in a way sensitive to A_1R antagonists (Diógenes et al. 2004; Pak et al. 1994), and (3) the suppression of seizure activity caused by ADK inhibition in various models for epilepsy (Kowaluk and Jarvis 2000).

The immature brain is more vulnerable to seizure activity than the adult brain (Moshe 2000), an action probably related to the higher expression of ADK at early developmental stages (Studer et al. 2006). Interestingly, during maturation, there is a shift from neuronal to glial expression of ADK, suggestive of distinct functions of ADK and adenosine in immature and adult brain; thus, during neuronal development expression of ADK in neurons may provide a salvage pathway to utilize adenosine in nucleic acid synthesis, whereas in the mature brain predominant ADK expression in astrocytes contributes to maintenance of tonic adenosinergic inhibition in the central nervous system (Studer et al. 2006). Overexpression of adenosine kinase in epileptic hippocampus contributes to epileptogenesis (Gouder et al. 2004). Furthermore, there is a prominent upregulation of ADK in astrocytes after induction of status epilepticus (SE) in animals as well as in humans with temporal lobe epilepsy (Aronica et al. 2011). Selective ADK downregulation in astrocytes almost completely abolishes spontaneous recurrent seizures in epileptic mice (Theofilas et al. 2011). Thus, ADK emerges as a key link in astrocyte-to-neuron communication, and its dysregulation after intense neuronal activity may contribute to epileptogenesis. Permanent changes in ADK expression in astrocytes will be reflected in decreases in ambient adenosine, leading to a further enhancement of neuronal activity and in such a way being part of a positive feedback loop to promote epileptogenesis. Accordingly, focal adenosine augmentation therapeutic strategies, mainly based in local manipulation of ADK activity, have been proposed as a useful strategy to control pharmacoresistant seizures (Boison et al. 2011).

The regulation of ambient adenosine levels by ADK might also have a key role in the susceptibility of brain tissue to ischemic injury (Lynch et al. 1998; Pignataro et al. 2007; Shen et al. 2011). Indeed, pharmacological inhibition of ADK in animal models is also an effective strategy to protect from stroke (Boison 2006; Kowaluk and Jarvis 2000).

7.4.2 Adenosine Deaminase

Adenosine deaminase (ADA, EC 3.5.4.4) catalyzes the hydrolytic deamination of adenosine into inosine. It has been known for several years that inhibition of ADA causes adenosine-like effects such as sedation (Major et al. 1981; Radulovacki et al. 1983), reduction of the infarct area in the hippocampus and decrease in neuronal degeneration in animal models of global forebrain ischemia or focal ischemia (Lin and Phillis 1992; Phillis and O'Regan 1989). Although ADA is expressed by both neurons and astrocytes (Haun et al. 1996; Nagy et al. 1984) it seems that it is in glial cells that this enzyme assumes a major role in the control of adenosine levels. This role is more relevant during stress conditions (like trauma or ischemia), when adenosine levels rise and astrocytes became reactive, probably playing an important role in adenosine conversion to inosine (Zamzow et al. 2008b). Inosine by itself can have a protective effect in stroke models (Shen et al. 2005). However, neuroprotection conferred by inhibitors of ADA during hypoxia or ischemia (Lin and Phillis 1992) mostly results from potentiation of the stress-induced increase in intracellular adenosine, which leads to enhanced adenosine release through transport reversal (Phillis and O'Regan 1989).

Although the enzyme localization is mainly cytosolic, there is evidence of the existence of an ectoenzyme, bound to the extracellular side of the membrane (Franco et al. 1998). The A_1R may act as an anchoring protein for ecto-ADA, which through a nonenzymatic but allosteric interaction facilitates agonist and antagonist binding to A_1R (Ciruela et al. 1996; Gracia et al. 2008; Ruíz et al. 2000; Saura et al. 1998). Like A_1R, $A_{2B}R$ was also found to be anchored to ADA in lymphocytes and cultured cells. Similarly, binding of enzymatically active or inactive ADA to this receptor increases its affinity and signaling by a protein–protein interaction (Herrera et al. 2001).

7.5 Control of Extracellular Adenosine Levels by Astrocytes

In the brain, extracellular adenosine concentrations are normally kept in the range of 25–250 nM, therefore at concentrations that can tonically activate a substantial proportion of the high affinity A_1R and $A_{2A}R$ (Dunwiddie and Masino 2001). A major player in the steady-state levels of adenosine is ADK, which has high affinity for adenosine and is mostly expressed in astrocytes (see above and Boison et al. 2010). ADA also predominates in astrocytes. Equilibrative nucleoside transporters are also expressed in astrocytes. Therefore, under physiological conditions astrocytes probably function as a major sink for adenosine, since its uptake is driven by the intracellular activity of ADK. It is also likely that under conditions that prompt increases in intracellular adenosine, such as hypoxia or ischemia, astrocytes provide a major source of adenosine, which is released by reversal of transport direction, but direct evidence for a predominant astrocytic origin of extracellular adenosine during hypoxic/ischemic insults is still lacking.

The levels of extracellular adenosine can be regulated by both A_1R and A_2R activity since A_1R blockade increases extracellular levels of adenosine in cardiac fibroblasts (Andresen et al. 1999) and activation of A_2R promotes adenosine transport in chromaffin cells (Delicado et al. 1990). Furthermore, it has been shown that $A_{2A}R$ activation at nerve endings enhances the activity of nucleoside transporters, leading to a decrease in the availability of adenosine to activate A_1R under high frequency neuronal firing (Pinto-Duarte et al. 2005). Again, information is still lacking regarding the role of astrocytes in this process. In astrocytes, A_1R and $A_{2A}R$ form tetramers constituted by two A_1R and two $A_{2A}R$ molecules bound to $G_{i/0}$ and G_s proteins to regulate GABA transport in a deeply interactive and concerted way (Cristóvão-Ferreira et al. 2011). Whether these A_1R-$A_{2A}R$ tetramers control adenosine transporters is also unknown.

7.6 Purines and Intracellular Signaling

ATP acts upon different classes of membrane receptors, the ionotropic P2X and the metabotropic P2Y (for reviews see, for instance, Illes and Ribeiro 2004; Ralevic and Burnstock 1998). Adenosine operates through activation of four distinct metabotropic receptors: A_1R, $A_{2A}R$, $A_{2B}R$, and A_3R. All these receptors are expressed in the brain, where adenosine is involved in a variety of physiological and pathological processes, namely, regulation of sleep–arousal cycle, neuroprotection, epilepsy, pain, fine control of movement, fine-tuning of neurotransmission (see Boison 2006; Dunwiddie and Masino 2001; Ribeiro 2005; Sebastião and Ribeiro 2009a).

7.7 Adenosine Receptors and Signaling Pathways

All four adenosine receptors have been cloned. Being GPCRs, adenosine receptors are formed by a single peptide chain, with seven alpha-helical transmembrane domains, an intracellular C-terminal, and an N-terminal facing the extracellular space. The N-terminal usually contains one or more glycosylation sites. The C-terminal contains phosphorylation and palmitoylation sites, which are involved in regulation of receptor desensitization and internalization (Perez and Karnik 2005).

Adenosine receptors are widely distributed throughout the body. In the brain they can be found pre-, post-, or nonsynaptically, in neurons as well as in glia. Their expression is not homogenous in the central nervous system (Fig. 7.2). Higher A_1R expression levels are found in the cortex, hippocampus, cerebellum, thalamus, brain stem, and spinal cord (see Ribeiro et al. 2002 and references therein). Though at low density, A_1Rs are also present in basal ganglia, both on dopaminergic nigrostriatal and glutamatergic corticostriatal terminals. Adenosine $A_{2A}Rs$ are mostly expressed in the basal ganglia and olfactory bulb. However, it is

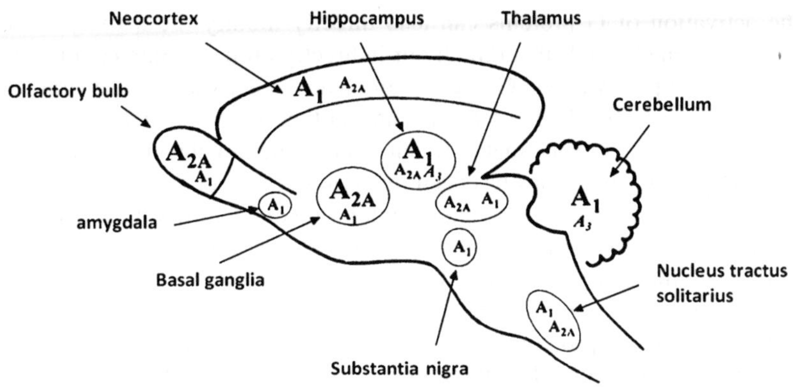

Fig. 7.2 Differential distribution of adenosine receptors (A_1, A_{2A}, A_3) among the brain (adapted from Ribeiro et al. 2002). Higher expression corresponds to *larger text size*

possible to find mRNA encoding A_{2A}Rs or the A_{2A}R protein in other brain regions where they are weakly expressed, namely, in the hippocampus and the cortex (see Sebastião and Ribeiro 1996). The A_{2B}Rs are mainly expressed in peripheral organs, being weakly expressed in the whole brain (Dixon et al. 1996). Finally, A_3Rs have an intermediate expression level in the human cerebellum, and as A_{2B}Rs, they display a low expression in the entire brain (see Fredholm et al. 2001). The A_1R and A_{2A}R are high affinity receptors, with adenosine K_d values of 70 and 150 nM, for A_1R and A_{2A}R respectively, which allow their tonic activation by basal adenosine levels. The A_{2B}R and A_3R are considered low affinity receptors, with adenosine affinity constant values around 5,100 and 6,500 nM, respectively (see Dunwiddie and Masino 2001).

Classically, A_1R and A_3R inhibit adenylate cyclase (AC) through coupling to $G_{i/0}$. A_{2A}R and A_{2B}R are coupled to G_s or G_{olf}, promoting AC activity. The A_{2B}R subtype is also coupled to $G_{q/11}$, through which it can activate phospholipase C (Ryzhov et al. 2006). The A_3R can also couple to $G_{q/11}$, also activating phospholipase C (Fredholm et al. 2001). The increase of cAMP mediated by AC leads to activation of cAMP dependent protein kinase (PKA), which then phosphorylates different targets such as ionotropic receptors or neurotransmitter transporters. On the other hand, activation of phospholipase C converts phosphatidylinositol 4,5-bisphosphate (PIP_2) into diacylglycerol (DAG) and inositol 1,4,5-trisphosphate (IP_3). Then DAG activates protein kinase C (PKC), which phosphorylates different substrates, while IP_3 triggers calcium release from intracellular stores. Then, elevation of cytosolic Ca^{2+} can stimulate a variety of signaling pathways, including a family of phosphatidyl serine-dependent serine/threonine-directed kinases collectively called protein kinase C (PKC), phospholipase A_2 (PLA_2), as well as Ca^{2+}-dependent K^+ channels, and nitric oxide synthase (NOS). IP_3 can also promote calcium influx from extracellular sources if Ca^{2+} intracellular stores are depleted due to previous activation of IP_3 receptors (see Ralevic and Burnstock 1998).

The activation of G proteins can also directly modify the activity of several enzymes and ion channels that directly or indirectly influence intracellular calcium levels. For example, A_1R, via $G_{i/0}$ activation, leads to activation of several types of K^+ channels and to blockade of N-, P-, and Q-type Ca^{2+} channels (see Fredholm et al. 2001). Direct evidence that A_1R can inhibit calcium channels in nerve terminals under hypoxic conditions has been reported (Coelho et al. 2002). A_3R are also able to modulate Ca^{2+} levels, through the inhibition of AC. Both $A_{2A}R$ and $A_{2B}R$ can also modify the levels of intracellular calcium (see Fredholm et al. 2001).

Aside from the involvement of AC/cAMP/PKA and PLC/IP_3-DAG/PKC, other transduction pathways are associated with adenosine receptor activation, namely, the mitogen-activated protein kinase (MAPKs) (Schulte and Fredholm 2003). The MAPK family is constituted by three main groups: extracellular regulated kinases (ERK) such as ERK1 and ERK2, stress-activated protein kinases (SAPK) such as p38, and jun-N-terminal kinase (JNK). These kinases are usually activated by receptors with tyrosine-kinase activity (Seger and Krebs 1995), but GPCR can also signal through them (Gutkind 1998; Liebmann 2001; Marinissen and Gutkind 2001). In fact, all adenosine receptors can affect the MAPK pathway. This was first shown in COS-7 cells transiently transfected with A_1R, leading to activation of ERK1/2 via $G_{i/0}$ (Faure et al. 1994). It was also early recognized that activation of $A_{2A}R$ can increase MAPK activity (Sexl et al. 1997). Interestingly, the signal pathway used by $A_{2A}R$ to activate MAPK can vary, depending on the cellular machinery available. Thus, in CHO cells, $A_{2A}R$-mediated ERK1/ERK2 activation involves G_s-AC-cAMP-PKA-MEK1, while in HEK-293 cells, MAPK activation by $A_{2A}R$ involves PKC but not PKA, even though cAMP levels are found to be enhanced by G_s activity (Seidel et al. 1999). $A_{2A}R$ can also inhibit ERK phosphorylation (Hirano et al. 1996). The activation of $A_{2B}R$ can trigger the three main branches of MAPK family (ERK1/2, p38, and JNK) (see Fredholm et al. 2001). Finally, the A_3R activate ERK1/2 in human fetal astrocytes (Neary et al. 1998). Also, the phosphorylation of ERK1/2 was also clearly demonstrated in CHO cells transfected with A_3R (see Schulte and Fredholm 2000).

To conclude, MAPK activation by adenosine receptors is quite similar to that prompted by other GPCR (Gutkind 1998; Luttrell et al. 1999; Sugden and Clerk 1998). Interestingly, ERK1/2 phosphorylation is promoted either by receptors coupled to G_s (A_{2A}/A_{2B}) or to $G_{i/0}$ (A_1/A_3) proteins. The MAPK-mediated effects of adenosine receptors mainly impact modulation of DNA synthesis, cellular differentiation, proliferation, and apoptosis (see Schulte and Fredholm 2003).

Adenosine receptor activity is regulated by its expression at the membrane level which results from a balance between endocytosis and exocytosis rates. An enhancement of the endocytosis rate, which restrains the intensity and duration of the signal, is often preceded by receptor phosphorylation and uncoupling from G proteins, a well-known process of receptor desensitization. Adenosine receptor subtypes desensitize differently. A_1R are slowly phosphorylated and internalized, a time frame of several hours being needed to complete the process. $A_{2A}R$ and $A_{2B}R$ desensitize in a faster way, with downregulation kinetics less than 1 h. The A_3R have the fastest desensitization profile, a process often occurring within minutes (Klaasse et al. 2008).

7.8 Implications for Modulation of Neuronal Function

The variety of downstream pathways operated by adenosine receptors highly supports the pluripotential of this nucleoside to interfere with a multiplicity of intracellular functions essential to regulate neuronal activity either directly or indirectly via interaction with several neurotransmitters and/or neuromodulators. These interactions can occur within the same cell, in some cases involving receptor heteromerization, or be a result of transcellular communication. Being a small and easily diffusible molecule, adenosine easily acts in a paracrine way, affecting cells away from the release point. Its role as a trans-synaptic modulator, involving neuron-to-astrocyte communication at tripartite synapses is now well accepted (Fields and Burnstock 2006; Hamilton and Attwell 2010; Perea et al. 2009).

Within the same cell there are many possibilities of cross talk between transduction pathways that have several kinases and other key molecules in common (see Fig. 7.3). GPCRs can interact at the G protein level, by sharing βγ-subunits or common α-subunits, affecting the activation kinetics of other GPCRs. This also applies to A_1R and $A_{2A}R$, and related mechanisms are most probably involved in the ability of adenosine receptors to interact with other receptors for neurotransmitters or neuromodulators (Sebastião and Ribeiro 2000).

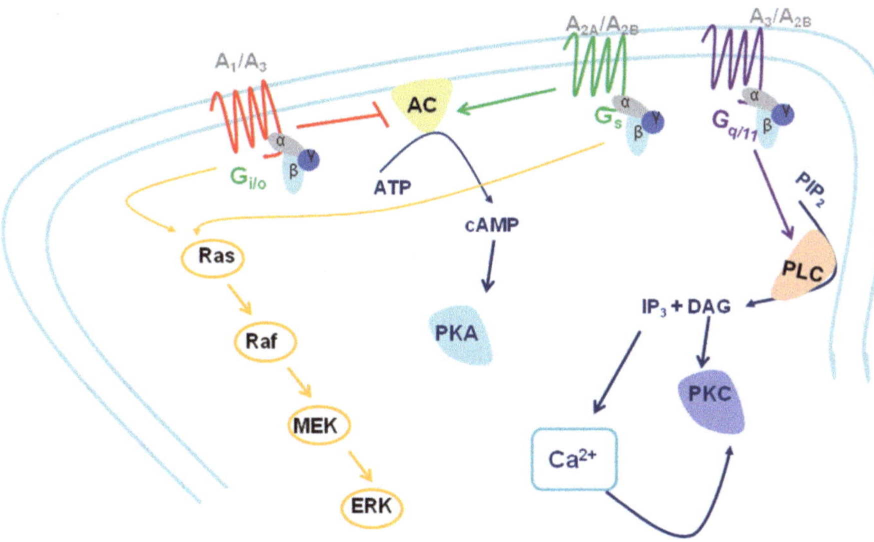

Fig. 7.3 Schematic representation of the different signaling pathways associated with adenosine receptors. Adenosine receptors are GPCRs. A_1R and A_3R couple to $G_{i/0}$, inhibiting AC, which will reduce cAMP levels and consequently decrease PKA activity. $A_{2A}R$ and $A_{2B}R$ are coupled to G_s, promoting AC activity and consequently PKA activity. A_3R and $A_{2B}R$ can also couple to $G_{q/11}$, enhancing PLC activity. PLC catalyzes PIP_2 into DAG and IP_3. DAG will directly activate PKC, while IP_3 will increase intracellular Ca^{2+} levels. Furthermore, all adenosine receptors can activate the MAPK pathway

A_1R and $A_{2A}R$ can form heteromeric complexes (Ciruela et al. 2006). As clearly shown in astrocytes, A_1R-$A_{2A}R$ heteromers appear as heteromers of homomers with a minimal structure consisting of an A_1R-A_1R-$A_{2A}R$-$A_{2A}R$ complex (Cristóvão-Ferreira et al. 2011). The heterotetramer makes it possible to accommodate the two different G proteins, and the A_1R-$A_{2A}R$ heteromer in astrocytes is clearly coupled to $G_{i/0}$ and G_s proteins (Cristóvão-Ferreira et al. 2011). Importantly, the blockade of a single partner in the $A_1R/A_{2A}R/G_{i/0}/G_s$ complex leads to adjustments in the functioning of the whole unit (Cristóvão-Ferreira et al. 2011).

Heteromerization between adenosine receptors and receptors of other neurotransmitters/neuromodulators also occurs, being the $A_{2A}R$-D_2R heteromer the first to be recognized (Hillion et al. 2002). Through this heteromer, adenosine restrains D_2R-mediated effects. The relevance of dopaminergic signaling and dysfunction in several pathologies turns $A_{2A}R$-D_2R heteromers into promising therapeutic targets (Altamura et al. 2005; Ferré et al. 1997). A close interaction between A_1R and D_1R was also described (Ginés et al. 2000). Once again, adenosine, through the activation of A_1R, inhibits D_1R-mediated effects. In detail, A_1R activation leads to uncoupling of D_1R from AC (Ginés et al. 2000), reinforcing the A_1R-induced inhibition of D_1R, which is mediated by $G_{i/0}$ activation (Ferré et al. 1994, 1998).

Adenosine $A_{2A}R$ can also heteromerize with cannabinoid CB_1R in the striatum (Carriba et al. 2007), and this has putative implications for pharmacotherapy drug addiction (Ferré et al. 2010). Psychomotor stimulation by $A_{2A}R$ antagonists also depends upon $A_{2A}R$-CB_1R cross talk (Lerner et al. 2010). A_1R also interact with CB_1R receptor-mediated actions in the hippocampus, a process with implications for CB_1R induced memory impairment, which is exacerbated by chronic caffeine consumption (Sousa et al. 2011).

The predominant neuromodulatory action of adenosine, inhibition of neurotransmitter release, is controlled by A_1R and relates to presynaptic inhibition of calcium responses (Fredholm et al. 1990; Fossier et al. 1999; Ribeiro 1978). However, $A_{2A}R$ can also modulate extracellular transmitter levels and they do so either by enhancing release and/or by influencing uptake. Indeed, $A_{2A}R$ activation in the hippocampus facilitates GABA release (Cunha and Ribeiro 2000) and GABA uptake into presynaptic terminals (Cristóvão-Ferreira et al. 2009) and astrocytes (Cristóvão-Ferreira et al. 2011). In the striatum, where $A_{2A}R$ expression is higher, the activation of these receptors leads to an inhibition of GABA uptake (Gonzalez et al. 2006). Glutamate release from hippocampal (Lopes et al. 2002) and cortical (Marchi et al. 2002) synaptosomes is also enhanced by $A_{2A}R$ activation. The same occurs with acetylcholine release from hippocampal nerve terminals (Cunha et al. 1995).

During excitotoxic conditions, such as hypoxia, the A_1R-mediated presynaptic inhibition of calcium responses (Coelho et al. 2002), together with inhibition of NMDA responses (Sebastião et al. 2001) confers protection against synaptic damage. In contrast, $A_{2A}R$ may facilitate ionotropic receptor activation, as it is the case of their ability to enhance AMPA receptor-mediated responses at the postsynaptic and extrasynaptic level, affecting the reserve of the $GluR_1$-containing AMPA receptors at the extrasynaptic pool, priming them for synaptic insertion and for reinforcement of synaptic strength (Dias et al. 2012). This action is sustained even after brief $A_{2A}R$ activation, involves cyclic AMP and PKA activation and leads to enhancement of long-term potentiation (LTP) in Schaffer collateral-CA1 synapses of the hippocampus (Dias et al. 2012). LTP facilitation by $A_{2A}R$ is also evident in aged animals (Costenla et al. 2011). Interestingly, $A_{2A}R$ blockade in vivo impairs conditional learning as well as potentiation of CA1 hippocampal potentials recorded concomitantly in freely moving animals (Fontinha et al. 2009). $A_{2A}R$ localized post-synaptically at synapses between mossy fibers and CA3 pyramidal cells are essential for a form of long-term potentiation (LTP) induced by short bursts of mossy fiber stimulation, which requires the activation of NMDA and metabotropic glutamate receptors ($mGluR_5$) to increase cytoplasmic Ca^{2+} levels (Rebola et al. 2008).

$A_{2A}R$ seem to be devoted to interacting with other metabotropic receptors, not only of the GPCR family (Sebastião and Ribeiro 2000) but also with tropomyosin-related kinase (Trk) receptors (Sebastião and Ribeiro 2009b). $A_{2A}R$ are able to transactivate TrkB receptors in the absence of the neurotrophin (Lee and Chao 2001). This transactivation requires long-term incubation with $A_{2A}R$ agonist and requires receptor internalization (Rajagopal et al. 2004). Furthermore, adenosine $A_{2A}R$ activation is also a crucial step for the functioning of neurotrophic receptors at synapses, through a mechanism most probably different from TrkB transactivation, and which involves translocation of TrkB molecules to lipid rafts (Assaife-Lopes et al. 2010). The $A_{2A}R$-mediated gating and/or facilitation of the actions of neurotrophins, has been shown for the facilitatory actions of brain derived neurotrophic factor (BDNF) on synaptic transmission (Diógenes et al. 2004; Tebano et al. 2008) and on plasticity (Fontinha et al. 2008) at the CA1 area of the hippocampus. The actions of BDNF are blocked by either $A_{2A}R$ blockade or inhibition of Trk phosphorylation, but a Trk phosphorylation inhibitor does not prevent $A_{2A}R$-mediated facilitation of synaptic transmission (Pousinha et al. 2006, indicating that $A_{2A}R$ operate upstream of TrkB activation. Synaptic actions of other neurotrophic factors, such as glial derived neurotrophic factor (GDNF) are also under influence of $A_{2A}R$ in the striatum (Gomes et al. 2006, 2009). Adenosine $A_{2A}R$ and BDNF TrkB receptors can coexist in the same nerve ending since the facilitatory action of $A_{2A}R$ upon TrkB-mediated BDNF action is also visible at the neuromuscular junction (Pousinha et al. 2006), a single nerve ending synapse model. Colocalization of $A_{2A}R$ and Ret, a component of the GDNF receptor complex, has also been shown in single axon terminals in the striatum (Gomes et al. 2009).

The ability of BDNF to facilitate synaptic transmission and synaptic plasticity is dependent on the age of the animals (Diógenes et al. 2007, 2011) and this may be related to the degree of activation of adenosine $A_{2A}R$ by endogenous adenosine at different ages. Thus, to trigger a BDNF facilitatory action at synapses of infant animals it is necessary to increase the extracellular levels of adenosine, either by inhibiting ADK or by a brief depolarization (Diógenes et al. 2004; Pousinha et al. 2006) or even by high frequency neuronal firing (Fontinha et al. 2008). These adenosine-triggered BDNF actions are lost by blocking adenosine $A_{2A}R$ with selective antagonists. In adult animals, BDNF per se can facilitate synaptic transmission, but this effect is also fully lost with blockade of adenosine $A_{2A}R$ (Diógenes et al. 2007) or in $A_{2A}R$ knockout mice (Tebano et al. 2008). Interestingly, the enhanced hippocampal synaptic plasticity in aged animals can be related not only to a higher influence of adenosine $A_{2A}R$ (Costenla et al. 2011) but also to an enhanced BDNF TrkB-mediated facilitatory tonus also dependent from cross talk with $A_{2A}R$ (Diógenes et al. 2011). Nicotinic alpha7 cholinergic currents in GABAergic hippocampal neurons are inhibited by BDNF, and this also requires coactivation of adenosine $A_{2A}R$ (Fernandes et al. 2008). Inhibition of GABA transporters (GAT) by BDNF at nerve terminals does not fully depend upon coactivation of $A_{2A}R$, since it is not abolished by $A_{2A}R$ blockade, though this inhibitory BDNF action can be exacerbated by $A_{2A}R$ coactivation (Vaz et al. 2008). Interestingly, in astrocytes BDNF facilitates GABA transport, and this facilitation is fully dependent upon $A_{2A}R$ activation (Vaz et al. 2011). Contrasting with $A_{2A}R$ which promote the actions of neurotrophic factors, A_1R inhibit neurite outgrowth of cultured dorsal root ganglion neurons, both in the absence and in the presence of NGF (Thevananther et al. 2001).

$A_{2A}R$, due to their ability to enhance excitotoxic phenomena, including glutamate release (Lopes et al. 2002; Marchi et al. 2002), are mostly regarded as promoters of neuronal death. However, in some cases, such as cultured retinal neurones, $A_{2A}R$ have been shown to protect neurones against glutamate-induced excitotoxicity (Ferreira and Paes-de-Carvalho 2001). Whether this is due to the ability of $A_{2A}R$ to facilitate actions of neurotrophic factors, as it has been shown to occur in relation to $A_{2A}R$-mediated neuroprotection of motor neurones (Wiese et al. 2007), is not yet known. The pathophysiological implications of the cross talk between $A_{2A}R$ and receptors for neurotrophic factors have been discussed in detail elsewhere (Sebastião and Ribeiro 2009c).

Adenosine receptor activation may also induce release of neurotrophic factors. Thus, the expression and/or release of NGF are enhanced by activation of $A_{2A}R$ in microglia (Heese et al. 1997) and by activation of A_1R in astrocytes (Ciccarelli et al. 1999). Adenosine $A_{2B}R$ in astrocytes are also able to enhance GDNF expression (Yamagata et al. 2007). In the whole hippocampus, $A_{2A}R$ are required for normal BDNF levels (Tebano et al. 2008). Interestingly, in a mouse model of Huntington's disease, $A_{2A}R$ are also required to keep striatal BDNF levels close to those obtained in wild-type mice (Potenza et al. 2007).

Finally, interactions among purinergic, growth factor, and cytokine signaling regulate neuronal and glial maturation as well as development. In neuronal-dependent glial maturation both ATP and adenosine purinoceptors are involved (see, for instance, Fields and Burnstock 2006). The extracellular adenosine levels during high frequency neuronal firing are sufficient to stimulate adenosine receptors in oligodendrocyte ancestor cells inhibiting their proliferation and stimulating their differentiation into myelinating oligodendrocytes (Stevens et al. 2002). In premyelinating Schwann cells, $A_{2A}R$ activate phosphorylation of extracellular signal-regulated kinases (ERKs), namely, ERK1/2, and inhibit Schwann cell proliferation without arresting differentiation (Stevens et al. 2004). Contrasting with $A_{2A}R$, which usually promote the actions of neurotrophic factors, adenosine A_1R inhibit neurite outgrowth of cultured dorsal root ganglion neurons, both in the absence and in the presence of NGF (Thevananther et al. 2001).

Besides influencing the activity of other neuromodulatory receptors, adenosine $A_{2A}R$ also affect the activity of equilibrative transporters, as in the case of adenosine transporters (Pinto-Duarte et al. 2005). Thus, activation of $A_{2A}R$ with the selective agonist CGS 21680 facilitates adenosine uptake and enhances release of adenosine, which points to a direct effect of $A_{2A}R$ on nucleoside transporters, rather than an indirect action resulting from a modification of the adenosine gradient of concentrations across the plasma membrane (i.e., a metabolic effect). Furthermore, high frequency neuronal firing activates $A_{2A}R$ and concomitantly enhances nucleoside transporters. The main consequence of this $A_{2A}R$-mediated enhancement of nucleoside transporters is a marked reduction of the tonic activation of inhibitory A_1R upon high-frequency firing. This action of $A_{2A}R$ on the activity of the adenosine transporters constitutes a clear demonstration that a neuromodulatory receptor is able to control the extracellular levels of its endogenous ligand and, hence, to influence its ability to control neurotransmitter release (Pinto-Duarte et al. 2005).

$A_{2A}Rs$ facilitate GAT-1 mediated GABA transport into nerve terminals, an effect that is mediated by AC activation, which restrains the inhibition of GAT-1 by PKC (Cristóvão-Ferreira et al. 2009). Transport facilitation by $A_{2A}Rs$ is due to an increase in the membrane expression of GAT-1 molecules, reflected in increased maximum transport velocity (Cristóvão-Ferreira et al. 2009). This $A_{2A}R$-mediated facilitation of GABA transport into nerve endings, if coupled to an increase in the release of GABA (Cunha and Ribeiro 2000), may contribute to faster neurotransmitter recycling, leading to an enhancement of phasic GABAergic signaling.

$A_{2A}Rs$ also facilitate GABA transport into astrocytes, by enhancing GAT-1 and GAT-3 mediated transport, an action under tight control of A_1R, due to formation of A_1R-$A_{2A}R$ heteromers (Cristóvão-Ferreira et al. 2011). While the A_1 protomer of the heteromer inhibits GABA transport, the A_{2A} protomer enhances it, the shift from inhibition to enhancement of GABA uptake probably occurring at low micromolar concentrations of extracellular adenosine (Cristóvão-Ferreira et al. 2011). These concentrations are easily attained at a tripartite synapse, where astrocytes and neurons release considerable amounts of ATP. The higher the release of ATP (as may occur at high neuronal firing rates in reciprocal neuron-to-astrocyte communication) the higher the expected concentration of extracellular adenosine. It is therefore likely that sustained neuronal firing promotes activation of the $A_{2A}R$ protomer of the A_1R-$A_{2A}R$ heteromer leading to facilitation of GABA uptake. Activation of GABA uptake by astrocytes will lead to a decrease in ambient GABA and a subsequent depression of tonic GABAergic inhibition resulting in enhanced excitatory tonus. Conversely, at submicromolar adenosine concentrations, there is a preferential activation of the A_1 protomer of the heteromer and so, GABA uptake by astrocytes would be inhibited. Consequently, tonic inhibition by GABA would be enhanced. Thus, through an adenosine action upon A_1R-$A_{2A}R$ heteromers, astrocytes might behave as dual amplifiers, facilitating excitation of intense astrocytic-to-neuronal signaling and increasing inhibition at low neuronal firing rates. This switch in neural activity requires a highly efficient control to avoid sudden state transitions, and this seems to be the main advantage of heteromerization of A_1R and $A_{2A}R$ in astrocytes. Indeed, overstimulation of just one of the receptor protomers leads to internalization of the whole functional unit (Cristóvão-Ferreira et al. 2011), therefore allowing a double brake in the system and avoiding an abrupt inhibitory signaling and a sudden switch from excitation to inhibition as a consequence of desensitization of only the excitatory protomer.

7.9 Concluding Remarks

Operating on multiple downstream signaling pathways, adenosine receptors influence the activity of other GPCRs as well as of receptors for neurotrophic factors, ion channels, ionotropic receptors, and neurotransmitter transporters. Modifications of extracellular adenosine levels, due to changes in its metabolic pathways, lead to alterations in the degree of activation of adenosine receptors, which will impact their ability to enhance or restrain the action of other neurotransmitters or neuromodulators. The cross talk between adenosine receptors and other membrane receptors results in part from intracellular cascade processes occurring between common transducing systems and through protein phosphorylation processes that involve PKC or PKA (Fig. 7.3). The key receptor in this synaptic interplay appears to be the $A_{2A}R$, whereas the A_1R can counteract $A_{2A}R$ mediated actions.

In summary, synaptic transmission is under tight control of endogenous extracellular adenosine, which through pre- and postsynaptic actions interplays with other synaptic molecules involved in neurotransmission as well as with membrane proteins (receptors, and transporters) essential for transmission to harmonically influence neuronal activity.

Acknowledgments The work in the authors' laboratory is supported by research grants from Faculdade de Medicina, Fundação para a Ciência e Tecnologia (FCT), Gulbenkian Foundation, and the European Union (COST B30).

References

Altamura AC, Bassetti R, Cattaneo E, Vismara S (2005) Some biological correlates of drug resistance in schizophrenia: a multidimensional approach. World J Biol Psychiatry 6(suppl 2): 23–30

Andresen BT, Gillespie DG, Mi Z, Dubey RK, Jackson EK (1999) Role of adenosine A(1) receptors in modulating extracellular adenosine levels. J Pharmacol Exp Ther 291:76–80

Aronica E, Zurolo E, Iyer A, de Groot M, Anink J, Carbonell C, van Vliet EA, Baayen JC, Boison D, Gorter JA (2011) Upregulation of adenosine kinase in astrocytes in experimental and human temporal lobe epilepsy. Epilepsia 52:1645–1655

Assaife-Lopes N, Sousa VC, Pereira DB, Ribeiro JA, Chao MV, Sebastião AM (2010) Activation of adenosine A_{2A} receptors induces TrkB translocation and increases BDNF-mediated phospho-TrkB localization in lipid rafts: implications for neuromodulation. J Neurosci 30:8468–8480

Baldwin SA, Mackey JR, Cass CE, Young JD (1999) Nucleoside transporters: molecular biology and implications for therapeutic development. Mol Med Today 5:216–224

Baldwin SA, Beal PR, Yao SY, King AE, Cass CE, Young JD (2004) The equilibrative nucleoside transporter family, SLC29. Pflugers Arch. Eur J Physiol 447:735–743

Boison D (2006) Adenosine kinase, epilepsy and stroke: mechanisms and therapies. Trends Pharmacol Sci 27:652–658

Boison D, Scheurer L, Zumsteg V, Rulicke T, Litynski P, Fowler B, Brandner S, Mohler H (2002) Neonatal hepatic steatosis by disruption of the adenosine kinase gene. Proc Natl Acad Sci U S A 99:6985–6990

Boison D, Chen JF, Fredholm BB (2010) Adenosine signaling and function in glial cells. Cell Death Differ 17:1071–1082

Boison D, Masino SA, Geiger JD (2011) Homeostatic bioenergetic network regulation—a novel concept to avoid pharmacoresistance in epilepsy. Expert Opin Drug Discov 6:713–724

Brundege JM, Diao L, Proctor WR, Dunwiddie TV (1997) The role of cyclic AMP as a precursor of extracellular adenosine in the rat hippocampus. Neuropharmacology 36:1201–1210

Carriba P, Ortiz O, Patkar K, Justinova Z, Stroik J, Themann A, Müller C, Woods AS, Hope BT, Ciruela F, Casadó V, Canela EI, Lluis C, Goldberg SR, Moratalla R, Franco R, Ferré S (2007) Striatal adenosine A2A and cannabinoid CB1 receptors form functional heteromeric complexes that mediate the motor effects of cannabinoids. Neuropsychopharmacology 32:2249–2259

Carswell HV, Graham DI, Stone TW (1997) Kainate-evoked release of adenosine from the hippocampus of the anaesthetised rat: possible involvement of free radicals. J Neurochem 68:240–247

Cascalheira JF, Sebastião AM (1992) Adenine nucleotide analogues, including gamma-phosphate-substituted analogues, are metabolised extracellularly in innervated frog sartorius muscle. Eur J Pharmacol 222:49–59

Ciccarelli R, Di Iorio P, Bruno V, Battaglia G, D'Alimonte I, D'Onofrio M, Nicoletti F, Caciagli F (1999) Activation of A(1) adenosine or mGlu3 metabotropic glutamate receptors enhances the release of nerve growth factor and S-100beta protein from cultured astrocytes. Glia 27:275–281

Ciruela F, Saura C, CanelaEI MJ, Lluis C, Franco R (1996) Adenosine deaminase affects ligand-induced signaling by interacting with cell surface adenosine receptors. FEBS Lett 380: 219–223

Ciruela F, Casadó V, Rodrigues RJ, Luján R, Burgueño J, Canals M, Borycz J, Rebola N, Goldberg SR, Mallol J, Cortés A, Canela EI, López-Giménez JF, Milligan G, Lluis C, Cunha RA, Ferré S, Franco R (2006) Presynaptic control of striatal glutamatergic neurotransmission by adenosine A_1-A_{2A} receptor heteromers. J Neurosci 26:2080–2087

Coelho JE, Faas GC, de Mendonça A, Ribeiro JA, Saggau P (2002) Effects of hypoxia on calcium signaling in excitatory presynaptic terminals of rat hippocampal slices. Society for Neuroscience, Orlando, Fl

Costenla AR, Diógenes MJ, Canas PM, Rodrigues RJ, Nogueira C, Maroco J, Agostinho PM, Ribeiro JA, Cunha RA, de Mendonça A (2011) Enhanced role of adenosine A(2A) receptors in the modulation of LTP in the rat hippocampus upon ageing. Eur J Neurosci 34:12–21

Crawley JN, Patel J, Marangos PJ (1983) Adenosine uptake inhibitors potentiate the sedative effects of adenosine. Neurosci Lett 36:169–174

Cristóvão-Ferreira S, Vaz SH, Ribeiro JA, Sebastião AM (2009) Adenosine A2A receptors enhance GABA transport into nerve terminals by restraining PKC inhibition of GAT-1. J Neurochem 10:336–347

Cristóvão-Ferreira S, Navarro G, Brugarolas M, Pérez-Capote K, Vaz SH, Fattorini G, Conti F, Lluis C, Ribeiro JA, McCormick PJ, Casadó V, Franco R, Sebastião AM (2011) Adenosine A1R-A2AR heteromers modulate GAT-1- and GAT-3-mediated GABA uptake by astrocytes. J Neurosci 31:15629–15639

Cui XA, Singh B, Park J, Gupta RS (2009) Subcellular localization of adenosine kinase in mammalian cells: the long isoform of AdK is localized in the nucleus. Biochem Biophys Res Commun 388:46–50

Cunha RA, Ribeiro JA (2000) Purinergic modulation of [$(^3)$H]GABA release from rat hippocampal nerve terminals. Neuropharmacology 39:1156–1167

Cunha RA, Johansson B, Fredholm BB, Ribeiro JA, Sebastião AM (1995) Adenosine A2A receptors stimulate acetylcholine release from nerve terminals of the rat hippocampus. Neurosci Lett 196:41–44
Cunha RA, Correia-de-Sá P, Sebastião AM, Ribeiro JA (1996) Preferential activation of excitatory adenosine receptors at rat hippocampal and neuromuscular synapses by adenosine formed from released adenine nucleotides. Br J Pharmacol 119:253–260
Cunha RA, Sebastião AM, Ribeiro JA (1998) Inhibition by ATP of hippocampal synaptic transmission requires localized extracellular catabolism by ecto-nucleotidases into adenosine and channeling to adenosine A1 receptors. J Neurosci 18:1987–1995
Cunha RA, Brendel P, Zimmermann H, Ribeiro JA (2000) Immunologically distinct isoforms of ecto-5′-nucleotidase in nerve terminals of different areas of rat hippocampus. J Neurochem 74:334–338
de la Haba G, Cantoni GL (1959) The enzymatic synthesis of S-adenosyl-L-homocysteine from adenosine and homocysteine. J Biol Chem 234:603–608
de Mendonça A, Sebastião AM, Ribeiro JA (1995) Inhibition of NMDA receptor-mediated currents in isolated rat hippocampal neurones by adenosine A1 receptor activation. Neuroreport 6:1097–1100
Delaney SM, Shepel PN, Geiger JD (1998) Levels of endogenous adenosine in rat striatum. I. Regulation by ionotropic glutamate receptors, nitric oxide and free radicals. J Pharmacol Exp Ther 285:561–567
Delicado EG, Rodrigues A, Sen RP, Sebastiao AM, Ribeiro JA, Miras-Portugal MT (1990) Effect of 5′-(N-ethylcarboxamido)adenosine on adenosine transport in cultured chromaffin cells. J Neurochem 54:1941–1946
Deussen A, Lloyd HG, Schrader J (1989) Contribution of S-adenosylhomocysteine to cardiac adenosine formation. J Mol Cell Cardiol 21:773–782
Dias RB, Ribeiro JA, Sebastião AM (2012) Enhancement of AMPA currents and GluR1 membrane expression through PKA-coupled adenosine A(2A) receptors. Hippocampus 22:276–291
Diógenes MJ, Fernandes CC, Sebastião AM, Ribeiro JA (2004) Activation of adenosine A2A receptor facilitates brain-derived neurotrophic factor modulation of synaptic transmission in hippocampal slices. J Neurosci 24:2905–2913
Diógenes MJ, Assaife-Lopes N, Pinto-Duarte A, Ribeiro JA, Sebastião AM (2007) Influence of age on BDNF modulation of hippocampal synaptic transmission: interplay with adenosine A2A receptors. Hippocampus 17:577–585
Diógenes MJ, Costenla AR, Lopes LV, Jerónimo-Santos A, Sousa VC, Fontinha BM, Ribeiro JA, Sebastião AM (2011) Enhancement of LTP in aged rats is dependent on endogenous BDNF. Neuropsychopharmacology 36:1823–1836
Dixon AK, Gubitz AK, Sirinathsinghji DJ, Richardson PJ, Freeman TC (1996) Tissue distribution of adenosine receptor mRNAs in the rat. Br J Pharmacol 118:1461–1468
Dragunow M, Goddard GV (1984) Adenosine modulation of amygdala kindling. Exp Neurol 84:654–665
Dunwiddie TV, Diao L (1994) Extracellular adenosine concentrations in hippocampal brain slices and the tonic inhibitory modulation of evoked excitatory responses. J Pharmacol Exp Ther 268:537–545
Dunwiddie TV, Masino SA (2001) The role and regulation of adenosine in the central nervous system. Annu Rev Neurosci 24:31–55
Dunwiddie TV, Taylor M, Heginbotham LR, Proctor WR (1992) Long-term increases in excitability in the CA1 region of rat hippocampus induced by beta-adrenergic stimulation: possible mediation by cAMP. J Neurosci 12:506–517
Dunwiddie TV, Diao L, Proctor WR (1997) Adenine nucleotides undergo rapid, quantitative conversion to adenosine in the extracellular space in rat hippocampus. J Neurosci 17:7673–7682
Faure M, Voyno-Yasenetskaya TA, Bourne HR (1994) cAMP and beta gamma subunits of heterotrimeric G proteins stimulate the mitogen-activated protein kinase pathway in COS-7 cells. J Biol Chem 269:7851–7854

Fernandes CC, Pinto-Duarte A, Ribeiro JA, Sebastião AM (2008) Postsynaptic action of brain-derived neurotrophic factor attenuates alpha7 nicotinic acetylcholine receptor-mediated responses in hippocampal interneurons. J Neurosci 28:5611–5618
Ferré S, Popoli P, Giménez-Llort L, Finnman U-B, Martínez E, Scotti de Carolis A, Fuxe K (1994) Postsynaptic antagonistic interaction between adenosine A1 and dopamine D1 receptors. Neuroreport 6:73–76
Ferré S, Fredholm BB, Morelli M, Popoli P, Fuxe K (1997) Adenosine-dopamine receptor-receptor interactions as an integrative mechanism in the basal ganglia. Trends Neurosci 20:482–487
Ferré S, Torvinen M, Antoniou K, Irenius E, Civelli O, Arenas E, Fredholm BB, Fuxe K (1998) Adenosine A1 receptor-mediated modulation of dopamine D1 receptors in stably cotransfected fibroblast cells. J Biol Chem 273:4718–4724
Ferré S, Lluís C, Justinova Z, Quiroz C, Orru M, Navarro G, Canela EI, Franco R, Goldberg SR (2010) Adenosine-cannabinoid receptor interactions. Implications for striatal function. Br J Pharmacol 160:443–453
Ferreira JM, Paes-de-Carvalho R (2001) Long-term activation of adenosine A(2a) receptors blocks glutamate excitotoxicity in cultures of avian retinal neurons. Brain Res 900:169–176
Fields RD, Burnstock G (2006) Purinergic signalling in neuron-glia interactions. Nat Rev Neurosci 7:423–436
Fontinha BM, Diógenes MJ, Ribeiro JA, Sebastião AM (2008) Enhancement of long-term potentiation by brain-derived neurotrophic factor requires adenosine A2A receptor activation by endogenous adenosine. Neuropharmacology 54:924–933
Fontinha BM, Delgado-García JM, Madroñal N, Ribeiro JA, Sebastião AM, Gruart A (2009) Adenosine A(2A) receptor modulation of hippocampal CA3-CA1 synapse plasticity during associative learning in behaving mice. Neuropsychopharmacology 34:1865–1874
Fossier P, Tauc L, Baux G (1999) Calcium transients and neurotransmitter release at an identified synapse. Trends Neurosci 22:161–166
Franco R, Valenzuela A, Lluis C, Blanco J (1998) Enzymatic and extraenzymatic role of ecto-adenosine deaminase in lymphocytes. Immunol Rev 161:27–42
Fredholm BB, Dunér-Engström M, Fastbom J, Hu PS, van der Ploeg I (1990) Role of G proteins, cyclic AMP, and ion channels in the inhibition of transmitter release by adenosine. Ann N Y Acad Sci 604:276–288
Fredholm BB, IJzerman AP, Jacobson KA, Klotz KN, Linden J (2001) International Union of Pharmacology. XXV. Nomenclature and classification of adenosine receptors. Pharmacol Rev 53:527–552
Frenguelli BG, Wigmore G, Llaudet E, Dale N (2007) Temporal and mechanistic dissociation of ATP and adenosine release during ischaemia in the mammalian hippocampus. J Neurochem 101:1400–1413
Geiger JD, Fyda DM (1991) Adenosine transport in nervous tissues. In: Stone TW (ed) Adenosine in the nervous system. Academic, London, pp 1–23
Ginés S, Hillion J, Torvinen M, Le Crom S, Casadó V, Canela EI, Rondin S, Lew JY, Watson S, Zolii M, Agnatii LF, Vernie P, Lluis C, Ferré S, Fuxe K, Franco R (2000) Dopamine D1 and adenosine A1 receptors form functionally interacting heteromeric complexes. Proc Natl Acad Sci U S A 97:8606–8611
Gomes CA, Vaz SH, Ribeiro JA, Sebastião AM (2006) Glial cell line-derived neurotrophic factor (GDNF) enhances dopamine release from striatal nerve endings in an adenosine A2A receptor-dependent manner. Brain Res 1113:129–136
Gomes CA, Simões PF, Canas PM, Quiroz C, Sebastião AM, Ferré S, Cunha RA, Ribeiro JA (2009) GDNF control of the glutamatergic cortico-striatal pathway requires tonic activation of adenosine A receptors. J Neurochem 108:1208–1219
Gonzalez B, Paz F, Florán L, Aceves J, Erlij D, Florán B (2006) Adenosine A2A receptor stimulation decreases GAT-1-mediated GABA uptake in the globus pallidus of the rat. Neuropharmacology 51:154–159

Gouder N, Scheurer L, Fritschy JM, Boison D (2004) Overexpression of adenosine kinase in epileptic hippocampus contributes to epileptogenesis. J Neurosci 24:692–701

Gracia E, Cortés A, Meana JJ, García-Sevilla J, Herhsfield MS, Canela EI, Mallol J, Lluís C, Franco R, Casadó V (2008) Human adenosine deaminase as an allosteric modulator of human A(1) adenosine receptor: abolishment of negative cooperativity for [H](R)-pia binding to the caudate nucleus. J Neurochem 107:161–170

Gu JG, Foga IO, Parkinson FE, Geiger JD (1995) Involvement of bidirectional adenosine transporters in the release of L-[^{3}H]adenosine from rat brain synaptosomal preparations. J Neurochem 64:2105–2107

Gutkind JS (1998) The pathways connecting G protein-coupled receptors to the nucleus through divergent mitogen-activated protein kinase cascades. J Biol Chem 273:1839–1842

Hamilton NB, Attwell D (2010) Do astrocytes really exocytose neurotransmitters? Nat Rev Neurosci 11:227–238

Harvey J, Lacey MG (1997) A postsynaptic interaction between dopamine D1 and NMDA receptors promotes presynaptic inhibition in the rat nucleus accumbens via adenosine release. J Neurosci 17:5271–5280

Haun SE, Segeleon JE, Trapp VL, Clotz MA, Horrocks LA (1996) Inosine mediates the protective effect of adenosine in rat astrocyte cultures subjected to combined glucose-oxygen deprivation. J Neurochem 67:2051–2059

Heese K, Fiebich BL, Bauer J, Otten U (1997) Nerve growth factor (NGF) expression in rat microglia is induced by adenosine A2a-receptors. Neurosci Lett 231:83–86

Herrera C, Casadó V, Ciruela F, Schofield P, Mallol J, Lluís C, Franco R (2001) Adenosine A2B receptors behave as an alternative anchoring protein for cell surface adenosine deaminase in lymphocytes and cultured cells. Mol Pharmacol 59:127–134

Hillion J, Canals M, Torvinen M, Casado V, Scott R, Terasmaa A, Hansson A, Watson S, Olah ME, Mallol J, Canela EI, Zoli M, Agnati LF, Ibanez CF, Lluis C, Franco R, Ferre S, Fuxe K (2002) Coaggregation, cointernalization and codesensitization of adenosine A2A eceptors and dopamine D2 receptors. J Biol Chem 277:18091–18097

Hirano D, Aoki Y, Ogasawara H, Kodama H, Waga I, Sakanaka C, Shimizu T, Nakamura M (1996) Functional coupling of adenosine A2a receptor to inhibition of the mitogen-activated protein kinase cascade in Chinese hamster ovary cells. Biochem J 316:81–86

Huber A, Padrun V, Déglon N, Aebischer P, Möhler H, Boison D (2001) Grafts of adenosine-releasing cells suppress seizures in kindling epilepsy. Proc Natl Acad Sci U S A 98:7611–7616

Illes P, Ribeiro JA (2004) Neuronal P2 receptors of the central nervous system. Curr Top Med Chem 4:831–838

Kermer V, Ritter M, Albuquerque B, Leib C, Stanke M, Zimmermann H (2010) Knockdown of tissue nonspecific alkaline phosphatase impairs neural stem cell proliferation and differentiation. Neurosci Lett 485:208–211

Klaasse EC, Ijzerman AP, de Grip WJ, Beukers MW (2008) Internalization and desensitization of adenosine receptors. Purinergic Signal 4:21–37

Kowaluk EA, Jarvis MF (2000) Therapeutic potential of adenosine kinase inhibitors. Expert Opin Investig Drugs 9:551–564

Kreutzberg GW, Barron KD, Schubert P (1978) Cytochemical localization of 5′-nucleotidasein glial plasma membranes. Brain Res 158:247–257

Langer D, Ikehara Y, Takebayashi H, Hawkes R, Zimmermann H (2007) The ectonucleotidases alkaline phosphatase and nucleoside triphosphate diphosphohydrolase 2 are associated with subsets of progenitor cell populations in the mouse embryonic, postnatal and adult neurogenic zones. Neuroscience 150:863–879

Lee FS, Chao MV (2001) Activation of Trk neurotrophin receptors in the absence of neurotrophins. Proc Natl Acad Sci U S A 98:3555–3560

Lerner TN, Horne EA, Stella N, Kreitzer AC (2010) Endocannabinoid signaling mediates psychomotor activation by adenosine A2A antagonists. J Neurosci 30:2160–2164

Liebmann C (2001) Regulation of MAP kinase activity by peptide receptor signaling pathway: paradigms of multiplicity. Cell Signal 13:777–785

Lin Y, Phillis JW (1992) Deoxycoformycin and oxypurinol: protection against focal ischemic brain injury in the rat. Brain Res 571:272–280
Lloyd HG, Fredholm BB (1995) Involvement of adenosine deaminase and adenosine kinase in regulating extracellular adenosine concentration in rat hippocampal slices. Neurochem Int 26:387–389
Lopes LV, Cunha RA, Kull B, Fredholm BB, Ribeiro JA (2002) Adenosine A(2A) receptor facilitation of hippocampal synaptic transmission is dependent on tonic A(1) receptor inhibition. Neuroscience 112:319–329
Luttrell LM, Daaka Y, Lefkowitz RJ (1999) Regulation of tyrosine kinase cascades by G-protein-coupled receptors. Curr Opin Cell Biol 11:177–183
Lynch JJ 3rd, Alexander KM, Jarvis MF, Kowaluk EA (1998) Inhibition of adenosine kinase during oxygen-glucose deprivation in rat cortical neuronal cultures. Neurosci Lett 252:207–210
Maienshein V, Zimmermann H (1996) Immunocytochemical localization of ecto-5′-nucleotidase in cultures of cerebellar granule cells. Neuroscience 70:429–438
Major PP, Agarwal RP, Kufe DW (1981) Clinical pharmacology of deoxycoformycin. Blood 58:91–96
Manzoni OJ, Manabe T, Nicoll RA (1994) Release of adenosine by activation of NMDA receptors in the hippocampus. Science 265:2098–2101
Marchi M, Raiteri L, Risso F, Vallarino A, Bonfanti A, Monopoli A, Ongini E, Raiteri M (2002) Effects of adenosine A1 and A2A receptor activation on the evoked release of glutamate from rat cerebrocortical synaptosomes. Br J Pharmacol 136:434–440
Marinissen MJ, Gutkind JS (2001) G-protein-coupled receptors and signaling networks: emerging paradigms. Trends Pharmacol Sci 22:368–376
Meghji P (1993) Storage, release, uptake and inactivation of purines. Drug Dev Res 28:214–219
Meghji P, Newby AC (1990) Sites of adenosine formation, action and inactivation in the brain. Neurochem Int 16:227–232
Misumi Y, Ogata S, Ohkubo K, Hirose S, Ikehara Y (1990) Primary structure of human placental 5′-nucleotidase and identification of the glycolipid anchor in the mature form. Eur J Biochem 191:563–569
Montero JM, Fes JB (1982) Purification and characterization of bovine brain 5′-nucleotidase. J Neurochem 39:982–989
Moshe SL (2000) Seizures early in life. Neurology 55(suppl 1):S15–S20
Motley SJ, Collins GG (1983) Endogenous adenosine inhibits excitatory transmission in the rat olfactory cortex slice. Neuropharmacology 22:1081–1086
Mrhul B, Aubery M, Mannherz HG, Codogno P (1993) Dual mechanism of laminin modulation of ecto-5′-nucleotidase activity. J Cell Biochem 52:266–274
Nagy JI, LaBella LA, Buss M, Daddona PE (1984) Immunohistochemistry of adenosine deaminase: implications for adenosine neurotransmission. Science 224:166–168
Naidoo D (1962) The activity of 5′-nucleotidase determined histochemically in the developing rat brain. J Histochem Cytochem 10:421–434
Neary JT, McCarthy M, Kang Y, Zuniga S (1998) Mitogenic signaling from P1 and P2 purinergic receptors to mitogen-activated protein kinase in human fetal astrocyte cultures. Neurosci Lett 242:159–162
Newby AC (1984) Adenosine and the concept of "retaliatory metabolites". Trends Biochem Sci 9:42–44
Newby AC, Worku Y, Holmquist CA (1985) Adenosine formation. Evidence for a direct biochemical link with energy metabolism. Adv Myocardiol 6:273–284
Olmo N, Turnay J, Risse G, Deutzmann R, vonder Mark K, Lizarbe A (1992) Modulation of 5′-nucleotidase activity in plasma membranes and intact cells by the extracellular matrix proteins laminin and fibronectin. Biochem J 282:181–188
Pak MA, Haas HL, Decking UK, Schrader J (1994) Inhibition of adenosine kinase increases endogenous adenosine and depresses neuronal activity in hippocampal slices. Neuropharmacology 33:1049–1053
Palmer JL, Abeles RH (1979) The mechanism of action of S-adenosylhomocysteinase. J Biol Chem 254:1217–1226

Parkinson FE, Rudolphi KA, Fredholm BB (1994) Propentofylline: a nucleoside transport inhibitor with neuroprotective effects in cerebral ischemia. Gen Pharmacol 25:1053–1058
Parkinson FE, Sinclair CJ, Othman T, Haughey NJ, Geiger JD (2002) Differences between rat primary cortical neurons and astrocytes in purine release evoked by ischemic conditions. Neuropharmacology 43:836–846
Perea G, Navarrete M, Araque A (2009) Tripartite synapses: astrocytes process and control synaptic information. Trends Neurosci 32:421–431
Perez DM, Karnik SS (2005) Multiple signaling states of G-protein-coupled receptors. Pharmacol Rev 57:147–161
Phillis JW (1989) Adenosine in the control of the cerebral circulation. Cerebrovasc Brain Metab Rev 1:26–54
Phillis JW, O'Regan MH (1989) Deoxycoformycin antagonizes ischemia-induced neuronal degeneration. Brain Res Bull 22:537–540
Phillis JW, O'Regan MH, Walter GA (1989) Effects of two nucleoside transport inhibitors, dipyridamole and soluflazine, on purine release from the rat cerebral cortex. Brain Res 481:309–316
Pignataro G, Simon RP, Boison D (2007) Transgenic overexpression of adenosine kinase aggravates cell death in ischemia. J Cereb Blood Flow Metab 27:1–5
Pinto-Duarte A, Coelho JE, Cunha RA, Ribeiro JA, Sebastião AM (2005) Adenosine A2A receptors control the extracellular levels of adenosine through modulation of nucleoside transporters activity in the rat hippocampus. J Neurochem 93:595–604
Potenza RL, Tebano MT, Martire A, Domenici MR, Pepponi R, Armida M, Pèzzola A, Minghetti L, Popoli P (2007) Adenosine A(2A) receptors modulate BDNF both in normal conditions and in experimental models of Huntington's disease. Purinergic Signal 3:333–338
Pousinha PA, Diogenes MJ, Ribeiro JA, Sebastião AM (2006) Triggering of BDNF facilitatory action on neuromuscular transmission by adenosine A2A receptors. Neurosci Lett 404:143–147
Radulovacki M, Virus RM, Djuricic-Nedelson M, Green RD (1983) Hypnotic effects of deoxycorformycin in rats. Brain Res 271:392–395
Rajagopal R, Chen ZY, Lee FS, Chao MV (2004) Transactivation of Trk neurotrophin receptors by G-protein-coupled receptor ligands occurs on intracellular membranes. J Neurosci 24: 6650–6658
Ralevic V, Burnstock G (1998) Receptors for purines and pyrimidines. Pharmacol Rev 50:413–492
Rebola N, Lujan R, Cunha RA, Mulle C (2008) Adenosine A2A receptors are essential for long-term potentiation of NMDA-EPSCs at hippocampal mossy fiber synapses. Neuron 57:121–134
Reichard P (1988) Interactions between deoxyribonucleotide and DNA synthesis. Annu Rev Biochem 57:349–374
Resta R, YamashitaY TLF (1998) Ecto-enzyme and signaling functions of lymphocyte CD73. Immunol Rev 161:95–109
Ribeiro JA (1978) ATP; related nucleotides and adenosine on neurotransmission. Life Sci 22:1373–1380
Ribeiro JA (2005) What can adenosine neuromodulation do for neuroprotection? Curr Drug Targets CNS Neurol Disord 4:325–329
Ribeiro JA, Sebastião AM (1987) On the role, inactivation and origin of endogenous adenosine at the frog neuromuscular junction. J Physiol 384:571–585
Ribeiro JA, Sebastião AM (2010) Modulation and metamodulation of synapses by adenosine. Acta Physiol 199:161–169
Ribeiro JA, Sebastião AM, de Mendonça A (2002) Adenosine receptors in the nervous system: pathophysiological implications. Prog Neurobiol 68:377–392
Richardson PJ, Brown SJ (1987) ATP release from affinity-purified cholinergic nerve terminals. J Neurochem 48:622–630
Rosenberg PA, Li Y (1995) Adenylyl cyclase activation underlies intracellular cyclic AMP accumulation, cyclic AMP transport, and extracellular adenosine accumulation evoked by beta-adrenergic receptor stimulation in mixed cultures of neurons and astrocytes derived from rat cerebral cortex. Brain Res 692:227–232

Ruíz MA, Escriche M, Lluís C, Franco R, Martín M, Andrés A, Ros M (2000) Adenosine A(1) receptor in cultured neurons from rat cerebral cortex: colocalization with adenosine deaminase. J Neurochem 75:656–664
Ryzhov S, Goldstein AE, Biaggioni I, Feoktistov I (2006) Cross-talk between G(s)- and G(q)-coupled pathways in regulation of interleukin-4 by A(2B) adenosine receptors in human mast cells. Mol Pharmacol 70:727–735
Sanderson G, Scholfield CN (1986) Effects of adenosine uptake blockers and adenosine on evoked potentials of guinea-pig olfactory cortex. Pflugers Arch 406:25–30
Saura CA, Mallol J, Canela EL, Lluis C, Franco R (1998) Adenosine deaminase and A1 adenosine receptors internalize together following agonist-induced receptor desensitization. J Biol Chem 273:17610–17617
Schoen SW, Kreutzberg GW (1995) Evidence that 5′-nucleotidaseis associated with malleable synapses. An enzyme cytochemical investigation of the olfactory bulb of adult rats. Neuroscience 65:37–50
Schoen SW, Graeber MB, Toth L, Kreutzberg GW (1991) Synaptic 5′-nucleotidase is transient and indicative of climbing fiber plasticity during the postnatal development of rat cerebellum. Brain Res Dev Brain Res 61:125–138
Schoen SW, Kreutzberg GW, Singer W (1993) Cytochemical redistribution of 5′-nucleotidase in the developing cat visual cortex. Eur J Neurosci 5:210–222
Schrader J, Schütz W, Bardenheuer H (1981) Role of S-adenosylhomocysteine hydrolase in adenosine metabolism in mammalian heart. Biochem J 196:65–70
Schulte G, Fredholm BB (2000) Human adenosine A(1), A(2A), A(2B), and A(3) receptors expressed in Chinese hamster ovary cells all mediate the phosphorylation of extracellular-regulated kinase 1/2. Mol Pharmacol 58:477–842
Schulte G, Fredholm BB (2003) Signaling from adenosine receptors to mitogen-activated protein kinases. Cell Signal 15:813–827
Sebastião AM (2011) Neuronal ENT1 takes up synaptic adenosine even under hypoxia/ischemia. J Neurochem 118:1–3
Sebastião AM, Ribeiro JA (1996) Adenosine A2 receptor-mediated excitatory actions on the nervous system. Prog Neurobiol 48:167–189
Sebastião AM, Ribeiro JA (2000) Fine-tuning neuromodulation by adenosine. Trends Pharmacol Sci 21:341–346
Sebastião AM, Ribeiro JA (2009a) Adenosine receptors and the central nervous system. Handb Exp Pharmacol 193:471–534
Sebastião AM, Ribeiro JA (2009b) Tuning and fine-tuning synapses with adenosine. Curr Neuropharmacol 7:180–194
Sebastião AM, Ribeiro JA (2009c) Triggering neurotrophic factor actions through adenosine A2A receptor activation: implications for neuroprotection. Br J Pharmacol 158:15–22
Sebastião AM, de Mendonca A, Moreira T, Ribeiro JA (2001) Activation of synaptic NMDA receptors by action potential-dependent release of transmitter during hypoxia impairs recovery of synaptic transmission on reoxygenation. J Neurosci 21:8564–8571
Seger R, Krebs EG (1995) The MAPK signaling cascade. FASEB J 9:726–735
Seidel MG, Klinger M, Freissmuth M, Höller C (1999) Activation of mitogen-activated protein kinase by the A(2A)-adenosine receptor via a rap1-dependent and via a p21(ras)-dependent pathway. J Biol Chem 274:25833–25841
Sexl V, Mancusi G, Höller C, Gloria-Maercker E, Schütz W, Freissmuth M (1997) Stimulation of the mitogen-activated protein kinase via the A2A-adenosine receptor in primary human endothelial cells. J Biol Chem 272:5792–5799
Shen H, Chen GJ, Harvey BK, Bickford PC, Wang Y (2005) Inosine reduces ischemic brain injury in rats. Stroke 36:654–659
Shen HY, Lusardi TA, Williams-Karnesky RL, Lan JQ, Poulsen DJ, Boison D (2011) Adenosine kinase determines the degree of brain injury after ischemic stroke in mice. J Cereb Blood Flow Metab 31:1648–1659

Sousa VC, Assaife-Lopes N, Ribeiro JA, Pratt JA, Brett RR, Sebastião AM (2011) Regulation of hippocampal cannabinoid CB1 receptor actions by adenosine A1 receptors and chronic caffeine administration: implications for the effects of Δ9-tetrahydrocannabinol on spatial memory. Neuropsychopharmacology 36:472–487

Sowa NA, Taylor-Blake B, Zylka MJ (2010) Ecto-5′-nucleotidase (CD73) inhibits nociception by hydrolyzing AMP to adenosine in nociceptive circuits. J Neurosci 30:2235–2244

Stevens B, Porta S, Haak LL, Gallo V, Fields RD (2002) Adenosine: a neuron-glial transmitter promoting myelination in the CNS in response to action potentials. Neuron 36:855–868

Stevens B, Ishibashi T, Chen JF, Fields RD (2004) Adenosine: an activity-dependent axonal signal regulating MAP kinase and proliferation in developing Schwann cells. Neuron Glia Biol 1:23–34

Studer FE, Fedele DE, Marowsky A, Schwerdel C, Wernli K, Vogt K, Fritschy JM, Boison D (2006) Shift of adenosine kinase expression from neurons to astrocytes during postnatal development suggests dual functionality of the enzyme. Neuroscience 142:125–137

Sugden PH, Clerk A (1998) Regulation of mitogen-activated protein kinase cascades in the heart. Adv Enzyme Regul 38:87–98

Tebano MT, Martire A, Potenza RL, Grò C, Pepponi R, Armida M, Domenici MR, Schwarzschild MA, Chen JF, Popoli P (2008) Adenosine A(2A) receptors are required for normal BDNF levels and BDNF-induced potentiation of synaptic transmission in the mouse hippocampus. J Neurochem 104:279–286

Theofilas P, Brar S, Stewart KA, Shen HY, Sandau US, Poulsen D, Boison D (2011) Adenosine kinase as a target for therapeutic antisense strategies in epilepsy. Epilepsia 52:589–601

Thevananther S, Rivera A, Rivkees SA (2001) A1 adenosine receptor activation inhibits neurite process formation by Rho kinase-mediated pathways. Neuroreport 12:3057–3063

Vaz SH, Cristóvão-Ferreira S, Ribeiro JA, Sebastião AM (2008) Brain-derived neurotrophic factor inhibits GABA uptake by the rat hippocampal nerve terminals. Brain Res 1219:19–25

Vaz SH, Jørgensen TN, Cristóvão-Ferreira S, Duflot S, Ribeiro JA, Gether U, Sebastião AM (2011) Brain-derived neurotrophic factor (BDNF) enhances GABA transport by modulating the trafficking of GABA transporter-1 (GAT-1) from the plasma membrane of rat cortical astrocytes. J Biol Chem 286:40464–40476

Wiese S, Jablonka S, Holtmann B, Orel N, Rajagopal R, Chao MV, Sendtner M (2007) Adenosine receptor A2A-R contributes to motoneuron survival by transactivating the tyrosine kinase receptor TrkB. Proc Natl Acad Sci U S A 104:17210–17215

Yamagata K, Hakata K, Maeda A, Mochizuki C, Matsufuji H, Chino M, Yamori Y (2007) Adenosine induces expression of glial cell line-derived neurotrophic factor (GDNF) in primary rat astrocytes. Neurosci Res 59:467–474

Yarbrough GG, McGuffin-Clineschmidt JC (1981) In vivo behavioral assessment of central nervous system purinergic receptors. Eur J Pharmacol 76:137–144

Yegutkin GG (2008) Nucleotide- and nucleoside-converting ectoenzymes: Important modulators of purinergic signalling cascade. Biochim Biophys Acta 1783:673–694

Zamzow CR, Xiong W, Parkinson FE (2008a) Astrocytes affect the profile of purines released from cultured cortical neurons. J Neurosci Res 86:2641–2649

Zamzow CR, Xiong W, Parkinson FE (2008b) Adenosine produced by neurons is metabolized to hypoxanthine by astrocytes. J Neurosci Res 86:3447–3455

Zhang D, Xiong W, Albensi BC, Parkinson FE (2011) Expression of human equilibrative nucleoside transporter 1 in mouse neurons regulates adenosine levels in physiological and hypoxic-ischemic conditions. J Neurochem 118:4–11

Zimmermann H (2011) Purinergic signaling in neural development. Semin Cell Dev Biol 22:194–204

Zimmermann H, Grondal EJM, Keller F (1986) Hydrolysis of ATP and formation of adenosine at the surface of cholinergic nerve endings. In: Kreutzberg GW, Reddington M, Zimmermann H (eds) Cellular biology of ectoenzymes. Springer, Berlin, pp 35–48

Chapter 8
Astrocytic ATP Release

Dustin J. Hines and Philip G. Haydon

Abstract For all its complexity, the brain can be broadly divided into two major types of cells, neurons and glial cells. Glia are defined as nonneuronal cells and function to maintain homeostasis, form myelin, and provide support and protection for neurons. Formerly thought to be passive support cells, we now know that astrocytes, the major subtype of glial cell in the brain, participate in signaling activities both with the vasculature and with neurons at the synapse. We have long known that astrocytes regulate neuronal signaling by controlling the ionic environment of the neuropil and controlling the supply of neurotransmitters to synapses, and evidence now exists for astrocytes directly signaling to neurons via regulated exocytic release, as well as signaling via gap junction mediated communication. The first indication of astrocytes actively signaling was via release of glutamate which acted to produce calcium signals in nearby neurons. Since this time astrocytes have also been shown to release D-serine, TNF-alpha, ANP, and ATP. ATP and its metabolites are well known as important signaling molecules in the nervous system, and astrocytes may be the most widespread source of ATP release in the nervous system. Astrocytes express the machinery necessary for exocytic and gap junction mediated release of ATP, as well as purinergic receptors and associated signaling molecules. In vivo studies using molecular and genetic tools have also demonstrated that astrocytic release of ATP and other signaling molecules has a major impact on synaptic transmission in multiple brain regions and under different physiological contexts. Via these actions, astrocytes have now been shown to mediate complex functions in the whole organism, such as respiration and homeostatic control of the sleep–wake cycle.

Keywords Glia • Synapse • Plasticity • Transmission • Adenosine • Adenosine triphosphate • Neurotransmitter

D.J. Hines • P.G. Haydon (✉)
Department of Neuroscience, Tufts University School of Medicine, Boston, MA 02111, USA
e-mail: philip.haydon@tufts.edu

S. Masino and D. Boison (eds.), *Adenosine: A Key Link between Metabolism and Brain Activity*, DOI 10.1007/978-1-4614-3903-5_8,

8.1 What Is an Astrocyte?

For all its complexity, the brain can be broadly divided into two major types of cells, neurons and glial cells. Glia are defined as nonneuronal cells and function to maintain homeostasis, form myelin, and provide support and protection for neurons. While glial cells have long been thought to serve as simple supportive cells for neurons, glial cells have more recently gained attention as active participants in many brain processes. However, this concept is not entirely new, in 1895 Santiago Ramón y Cajal proposed that astrocytes, the major subtype of glial cell in the brain, control sleep and waking states (Araque and Navarrete 2010). Cajal proposed that the processes of astrocytes act as electrical insulators that extend between neurons to facilitate sleep and retract to allow neuronal circuits to communicate thereby facilitating wakefulness (Garcia-Marin et al. 2007). A century of research has indeed provided support for parts of Cajal's original assertion. We now know that astrocytes do have both structural and functional links with neurons and are able to modulate complex behaviors including sleep and waking states (Halassa et al. 2009).

8.2 Properties of Astrocytes

Cajal was the first to systematically study astrocytes from a structural standpoint (Garcia-Marin et al. 2007), and until very recently, our view of astrocytic morphology has been based both on Cajal's metal impregnation methods, and on glial fibrillary acidic protein (GFAP) staining which both permit only static end point assessment of the actions of these cells. Advances in cellular labeling and imaging technologies have now demonstrated that astrocytic morphology is far more complicated than previously thought. By filling single astrocytes with fluorescent dyes, researchers in one study showed that GFAP staining reveals only 15 % of the true astrocytic volume and that astrocytes extend fine processes that occupy the surrounding neuropil (Bushong et al. 2002). Astrocytes occupy nonoverlapping spatial territories in which a single astrocyte contacts hundreds of neuronal processes and multiple neuronal cell bodies, and in addition also contact the microvasculature (Bushong et al. 2002).

Astrocytes are known to be intimately associated with the cerebral microvasculature, which they line with their endfeet (Simard et al. 2003). By being strategically positioned between neurons and blood vessels, astrocytes are thought to be mediators of neurovascular coupling, the process by which neuronal activity is coupled to cerebral blood flow. Astrocytes also play essential roles in brain energy homeostasis and metabolism, mediated by their close link with the microvasculature, and astrocytes are known to express transporters that mediate the uptake of glucose from cerebral microvessels.

Astrocytes interact with neurons at multiple spatial and temporal scales. By controlling the ionic and metabolic environment of the neuropil, astrocytes can dramatically impact neuronal activity. The processes of one astrocyte can contact tens of thousands of synapses (Fig. 8.1), and depending on the brain region being considered,

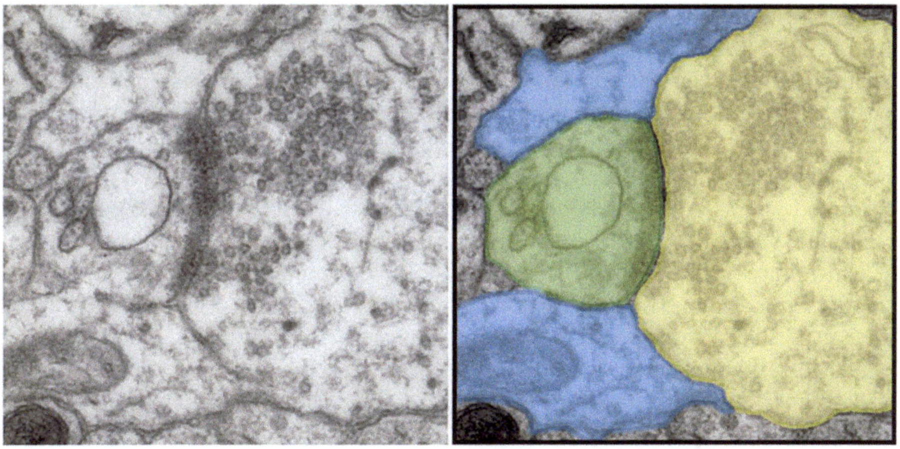

Fig. 8.1 Astrocytes and synapses have a close physical and functional relationship, and this close association has become known as the tripartite synapse. The electron micrograph on the left shows a hippocampal synapse with astrocyte end feet closely surrounding it. On the right, the colored overlay makes it apparent that the astrocytes (shown in *blue*) tightly surround the presynaptic (shown in *yellow*) and postsynaptic (shown in *green*) compartments, providing a tight seal around the sight of neurotransmission

more than 50 % of excitatory synapses are thought to be closely opposed to an astrocytic process (Ventura and Harris 1999). This close connection between the synapse and astrocytic process has been termed the tripartite synapse to recognize the structural and functional relationship between the astrocyte and the synaptic terminals (Araque et al. 1999). Evidence now supports the concept that astrocytes act as both "listening" and "talking" participants in the tripartite synapse via regulated signaling pathways (Haydon 2001).

8.3 How Do Astrocytes Signal?

Much attention has been focused recently on defining how astrocytes may also be sending signals in their complex relationships with neurons, as opposed to passively receiving the ongoing signaling activities of neurons. Although we have long known that astrocytes regulate neuronal signaling by controlling the ionic environment of the neuropil and controlling the supply of neurotransmitters to synapses, evidence now exists for astrocytes directly signaling to neurons via regulated release. Research has now shown that astrocytes signal via many chemical transmitters, including classical transmitters, peptides, chemokines, and cytokines, through a number of different release mechanisms. Astrocyte signaling through these various chemical compounds can have complex downstream effects on nearby neurons. For example, glutamate and D-serine released from astrocytes can boost NMDA receptor–mediated current, resulting in an excitatory feedback to neurons (Esser et al. 2007; Franken et al. 1999), whereas, the release of ATP from astrocytes leads to extracellular adenosine which then acts on adenosine A_1 receptors, ultimately resulting in inhibition of synaptic transmission (Haber and Murai 2006). There are four types of enzymes that are capable of hydrolyzing ATP to adenosine. The four family ectonucleotidases are ectonucleoside triphosphate diphosphohydrolases, ectonucleotide pyrophosphate/phosphodiesterases, alkaline phosphatases, and ecto-5′-nucleotidases.

Multiple release or signaling mechanisms have been proposed for astrocytes, all of which may operate under different physiological contexts or conditions. Studies have described connexin/gap junctions (Arcuino et al. 2002; Lazarowski et al. 2003; Stout et al. 2002); volume-regulated anion channels (Anderson et al. 2004; Queiroz et al. 1997); while other studies have focused on exocytotic release mechanism operating in astrocytes (Bal-Price et al. 2002; Coco et al. 2003; Pangrsic et al. 2007). ATP-binding cassette transporters, such as multidrug resistance P-glycoproteins (Abraham et al. 1993), or cystic fibrosis transmembrane conductance regulators (Schwiebert et al. 1995) have been demonstrated in CHO and airway epithelial cells respectively, and represent possible signaling mechanisms in astrocytes as well. For the purposes of this chapter we focus on two of the most well-characterized mechanisms, exocytosis and connexin/gap junction mediated release and signaling.

8.3.1 Exocytosis

Multiple cell types are known to release substances via the well characterized process of exocytosis. Regulated exocytosis is dependent on the docking and fusion of vesicles to the plasma membrane, which is orchestrated via the formation of the **s**oluble **NSF** (*N*-ethylmaleimide-sensitive fusion) protein **a**ttachment protein **re**ceptor (SNARE) complex. SNARE proteins are small, abundant and are primarily bound to the plasma membrane. Although they vary considerably in structure and size, all SNARE proteins share a segment in their cytosolic domain called a *SNARE motif* that consists of 60-70 amino acids capable of assembly into tight, four-helix bundles called *"trans"-SNARE complexes*. Astrocytes express the core machinery proteins involved in forming the SNARE complex, such as synaptobrevin II (Montana et al. 2004; Zhang et al. 2004b) and its homolog cellubrevin (Bezzi et al. 2004) and SNAP-23 (Montana et al. 2004; Zhang et al. 2004b). Astrocytes also express ancillary proteins to this complex, such as Munc 18 (Montana et al. 2004; Zhang et al. 2004b), complexin 2 (Montana et al. 2004; Zhang et al. 2004b), and synaptotagmin IV (Zhang et al. 2004a).

In addition to evidence from molecular biology, astrocytes have been shown to have vesicular structures. In cultured astrocytes, SNARE proteins have been shown to colocalize with a number of vesicular organelles, including small vesicles positive for vesicular glutamate transporters (VGlut 1–3)(Bezzi et al. 2004; Montana et al. 2004; Zhang et al. 2004a, 2004b), ATP-storing vesicles (Coco et al. 2003; Zhang et al. 2007), and D-serine- containing vesicles (Mothet et al. 2005a), suggesting the involvement of vesicular mechanisms in the release of these transmitters. The size of these vesicular organelles ranges between 30 and 100 nm in diameter in cultured astrocytes. Electron microscopic studies from tissue sections have demonstrated the existence of clear astrocytic vesicles, 30 nm in diameter, opposed to presynaptic terminals (Jourdain et al. 2007), which strongly supports the existence of a vesicular pathway of gliotransmission in the intact brain.

A significant advance in our understanding of astrocyte signals came in 1994 when Parpura et al. used calcium imaging techniques to demonstrate that cultured astrocytes release glutamate, which leads to elevation of calcium in nearby neurons (Parpura et al. 1994). Subsequent to this, several studies demonstrated that this process can also be observed in acute brain slices (Angulo et al. 2004; Fellin et al. 2004; Fiacco and McCarthy 2004; Kang et al. 2005; Pirttimaki et al. 2011; Porter and McCarthy 1996) and in vivo (Hirase et al. 2004). Since this time, astrocytes have been shown to release a number of chemical transmitters, including ATP (Darby et al. 2003; Koizumi et al. 2003; Newman 2003), D-serine (Henneberger et al. 2010; Panatier et al. 2006) (Mothet et al. 2005b), TNF-alpha (Santello et al. 2011), and ANP(Krzan et al. 2003), in a process that is now collectively termed gliotransmission.

8.3.2 *Gap Junctions*

One important feature of astrocytes in their role as metabolic regulators is their extensive intercellular coupling, mediated mainly by connexin-containing gap junctions (Theis et al. 2005). Gap junctions are specialized cell-to-cell contacts that provide exchange of the intracellular milieu and thereby communication of small molecules between cells. The estimated size is up to 1,200 Da which allows for the diffusion of glucose and other soluble factors among many astrocytes (Rouach et al. 2008).

A single gap junction consists of two hemichannels, or connexons, each of which is composed of six connexin proteins. Connexins consist of four transmembrane domains, connected by two extracellular loops and one intracellular loop, and have intracellular N- and C termini. In mammals, more than 20 connexin genes have been identified and connexins are most commonly named according to the molecular mass (KD) of the protein (Odermatt et al. 2003; Sohl and Willecke 2003). In the central nervous system, gap junctions occur between neurons, and glial cells including astrocytes, oligodendrocytes, and microglia, as well as between these different cell types. When one connexon is unopposed, referred to as a hemichannel, exchange between the intracellular and extracellular contents occurs. Under normal physiological conditions, gap junctions function in an open state, allowing free exchange of diffusible factors. In contrast, hemichannels are closed under normal physiological conditions and are opened to allow the passage of substances (Decrock et al. 2009a, b). The connexin hemichannel has been reported to be maintained in the closed state by extracellular divalent cations such as calcium and magnesium

(Ebihara et al., 2003). It is also well known that low extracellular calcium solution triggers connexin-mediated ATP release (Arcuino et al. 2002; Stout et al. 2002).

Of the connexin proteins identified connexin 43 (Cx43) is the most highly expressed and is enriched in astrocytes, activated microglia, the smooth muscle and endothelial cells of blood vessels, and developing neurons (Contreras et al. 2003; Contreras et al. 2002; Nagy et al. 1997; Rash et al. 2000; Zhu et al. 1991). In astrocytes, Cx43 containing gap junctions are abundant at end-foot processes along blood vessels, as well as in astrocyte processes that surround synaptic contacts (Danesh-Meyer and Green 2008; O'Carroll et al. 2008). Localized at these key locations, astrocyte gap junctions form a functional syncytium to maintain the brain environment (Cronin et al. 2008). Further, the diffusion of factors throughout astrocytic networks via gap junctions has been studied using fluorescent glucose analogs and has been shown to be a regulated process enhanced by neuronal activity (Rouach et al. 2008).

8.4 Astrocytes Express Channels and Receptors for ATP Signaling

In addition to possessing the proteins essential for both exocytic and gap junction mediated release, astrocytes also express a compliment of receptors and channels important for ATP signaling (Fig. 8.2). Astrocytes may be the most widespread source of ATP release in the CNS, and astroglial ATP and its metabolite adenosine activate purine receptors located on astrocytes, neurons, microglia, oligodendrocytes, and blood vessels. Purine receptors are broadly divided into categories according to activation by adenosine (P1) or ATP (P2). Evidence suggests that both groups of receptors are expressed in glial cells (James and Butt 2002). P2 receptors are further divided into ionotropic P2X and metabotropic P2Y receptors, both of which mediate intracellular calcium levels.

8.4.1 P1 Receptors

There are four subtypes of P1 receptors that respond to adenosine: A_1 and A_3 receptors inhibit cAMP via Gi/o, whereas A_{2A} and A_{2B} receptors stimulate cAMP via Gs. The expression of the different adenosine receptor subtypes has been described in astrocytes and microglia (Abbracchio et al. 2009; Ciccarelli et al. 2001; Stevens et al. 2002). Adenosine is well known to be an important neuromodulator, and adenosine signaling from astrocytic sources has been shown to modulate glial development, alter synaptic A_{2A} receptors, and mediate repulsive effects of ATP/adenosine on microglia, as well as participate in multiple pathological states (Butt 2006; Ciccarelli et al. 2001; Sperlagh and Illes 2007; Stevens et al. 2002).

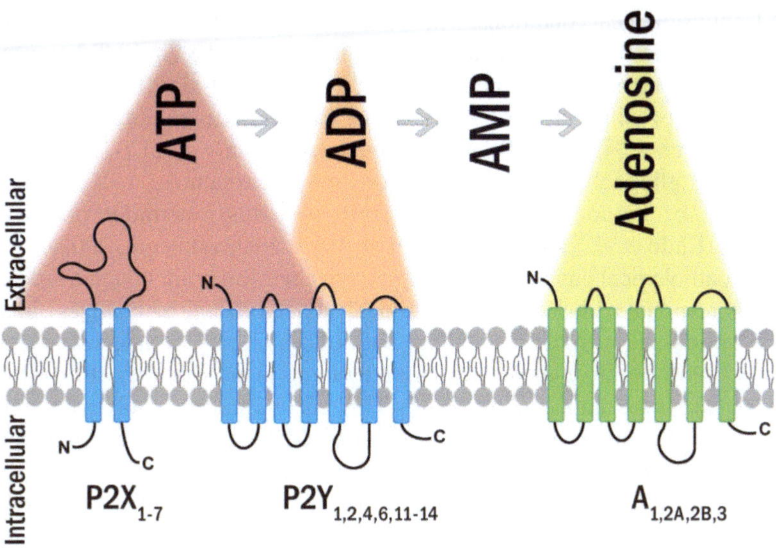

Fig. 8.2 Purinergic receptors are the site of action for ATP, and its metabolic products. ATP is converted to adenosine 5′-diphosphate (ADP), adenosine 5′-monophosphate (AMP), and adenosine by multiple classes of ecto-enzymes. The nucleotide/nucleoside receptors include P2X, P2Y, and adenosine categories of receptors. ATP acts on P2X ligand gated cation channels, and some P2Y receptors, which are G-protein-coupled, seven transmembrane domain receptors. Other P2Y receptors are preferentially activated by ADP, and other signaling molecules. Adenosine acts at four G-protein-coupled adenosine receptors

Physiologically, adenosine levels increase in situations when there is an imbalance between rates of energy utilization and rates of energy delivery. Under conditions of high neuronal activity and particularly during hypoxia or ischemia, markedly elevated levels of adenosine have been noted (Newby 1991). Adenosine has become known as a "retaliatory metabolite" due to its ability to adjust the energy supply and "retaliate" against the cellular environment (Newby et al. 1985; Sala-Newby et al. 1999), as such its potential role as an endogenous protective agent both in ischemia and following seizures has been repeatedly emphasized (Dunwiddie and Masino 2001; Rudolphi et al. 1992). However, these investigations focus primarily on the actions of adenosine on neurons, whereas glia are also important regulators and targets of adenosine signaling.

Adenosine can modulate astrocyte functions in multiple ways. Astrocytes are known to express all of the cloned and characterized subtypes of adenosine receptors, as well as other signaling systems required to mediate the effects of adenosine. The responses of astrocytes to varying environmental energy conditions depend on distinct signaling systems. For example, adenosine (and inosine)-mediated reduction of cell death in glucose deprived cultures of astrocytes relies upon intracellular formation of ribose-1-phosphate which in turn fuels intracellular production of ATP in astrocytes (Jurkowitz et al. 1998). In contrast, adenosine mediated increases in astrocyte glycogenesis are mediated by A_{2B} receptor signaling and induction of transcriptional changes in genes encoding energy metabolism proteins (Allaman et al. 2003).

Under physiological conditions, where adenosine concentrations are approximately 10 nM, adenosine signaling acts to inhibit glutamate uptake via GLT-1 transporters through an A_{2A} receptor mediated pathway (Nishizaki et al. 2002). At these concentrations, adenosine has also been shown to stimulate glutamate release from astrocytes via an A_{2A} receptor/PKA pathway, independently of GLT-1. Thus, adenosine action on glial cells rather than neurons might explain A_{2A} receptor mediated potentiation of hippocampal glutamatergic transmission under physiological conditions (Endo et al. 2007; Kikkawa et al. 1984; Nishizaki et al. 2002).

Under pathological conditions of traumatic or chemical brain injury, astrocytes are known to respond with both morphological and biochemical changes, including reactive gliosis and astrocytic swelling, both of which rely on adenosine signaling (Aschner 1998). Reactive gliosis involves alterations in the size of existing astrocytes, and an increase in the number of astrocytes. This reactive process involves up-regulation of enzymes and cytoskeletal proteins, including GFAP and vimentin, and increased levels of transforming growth factor β (TGF-β), nerve growth factor (NGF), basic-fibroblast growth factor (bFGF) and growth factor receptors, such as truncated froms of the neurotrophin receptor TrkB. Evidence suggests that these changes may be related to the elevated level of adenosine that follows brain injury. Following injection of the adenosine analog 5′-(N-cyclopropyl)-carboxamidoadenosine into rat cortex, increased numbers of GFAP-positive cells are observed (Hindley et al. 1994). Further, this paper also demonstrated that this effect could be blocked by coinfusion of the A_{2A} receptor antagonist 1,3-dipropyl-7-methylxanthine (DPMX), indicating a role for A_{2A} receptors in regulating astrogliosis via adenosine. In another study it was shown that the A_{2A} receptor antagonist SCH58261 reduced the formation of reactive astrocytes induced by bFGF treatment in culture (Brambilla et al. 2003). Adenosine has also been shown to induce hypertrophic changes in the size (stellation) of cultured astrocytes via activation of adenosine receptors (Abe and Saito 1998).

In addition to the involvement of A_2 receptors, A_3 receptors may also contribute to astrogliosis (Abbracchio et al. 2009), whereas A_1 receptors appear to inhibit astrocyte proliferation (Ciccarelli et al. 1994). Application of high levels of adenosine can induce cell death in cultured astrocytes, and has been shown to involve activation of the A_3 receptor and the internalization and metabolism of the excess adenosine (Di Iorio et al. 2002). The apoptosis inducing action of excessive adenosine has been interpreted as a mechanism to limit excessive astrocyte proliferation due to uncontrolled reactive gliosis [(Ciccarelli et al. 2001).

8.4.2 P2X Receptors

In mammals, seven P2X subunits ($P2X_{1-7}$) have been identified, with functional receptors being formed by the assembly of three subunits in either homomeric and heteromeric combinations (Evans 2010; Khakh and Burnstock 2009). P2X receptors are ligand-gated ion channels permeable to sodium, potassium and calcium, and most are activated at low concentrations of ATP, with EC50's of micro.

The properties of assembled receptors vary, and subtypes can be distinguished by their relative ion permeabilities, gating kinetics and sensitivity to ATP and a range of agonists and antagonists (Burnstock 2007). For example, $P2X_7$ receptors are only activated at high concentrations of ATP (0.1–1 mM), and are capable of pore formation, resulting in sustained influx of calcium (Burnstock 2007). With these properties, $P2X_7$ receptors are known to mediate glial pathological responses and are implicated in astrogliosis, microglia reactivity, and the loss of oligodendrocytes and myelin in ischemia and demyelination (Bianco et al. 2009; Domercq et al. 2010; Nobbio et al. 2009). Expression studies have detected $P2X_1$–$P2X_4$ and $P2X_7$ subunits in astrocytes (Domercq et al. 2010; Matute 2008). Some controversy exists as there is a lack of evidence from direct electrophysiological studies for astrocyte P2X receptors in the hippocampus (Jabs et al. 2007). However, ATP-induced currents with properties of $P2X_7$ receptors (Oliveira et al. 2011) and $P2X_{1/5}$ heteromers have been described in cortical astrocytes (Palygin et al. 2010).

8.4.3 P2Y Receptors

In mammals, eight subtypes of the P2Y receptor have been identified, and characterization studies have shown that they exhibit differential sensitivity to adenine nucleotides ATP/ADP ($P2Y_{1,11,12,13}$), the uracil nucleotides UTP/UDP ($P2Y_{4,6}$), both ($P2Y_2$), or UDP-glucose ($P2Y_{14}$). All P2Y receptors are G-protein-coupled and can lead to: activation of phospholipase C/inositol triphosphate and calcium-release from the endoplasmic reticulum via G-alpha(q/11) ($P2Y_1$, $P2Y_2$, $P2Y_4$, $P2Y_6$, and $P2Y_{11}$), stimulating adenylyl cyclase via G-alpha(s), or inhibiting adenylyl cyclase via Galpha(i/o) ($P2Y_{12}$, $P2Y_{13}$, and $P2Y_{14}$) (Burnstock 2007). P2Y receptors are broadly distributed in glia, although the specific subtypes expressed by astrocytes have not been completely defined (James and Butt 2002). Immunohistochemical confirmation of $P2Y_1$ expression has been demonstrated in astrocytes, oligodendrocytes and microglia (Franke et al. 2007; Gallagher and Salter 2003; Moran-Jimenez and Matute 2000). A role for $P2Y_1$ receptors has been identified in mediating ATP-evoked calcium signals coupled to inositol triphosphate and intracellular calcium-release (James and Butt 2002), and $P2Y_2$ receptors mediate different modes of calcium signaling in astrocytes (Fam et al. 2003). $P2Y_2$ and $P2Y_4$ receptors have been shown to be important for astrocyte calcium signals at the gliovascular interface (Simard et al. 2003).

8.5 ATP Signaling Through Exocytic Release

ATP is a major extracellular messenger important for coordinating the function of astrocytes as well as for the cross talk between them and other cell types (Lazarowski et al. 2003). The mechanisms of ATP release from astrocytes are not completely understood, but evidence is emerging to support the presence of an exocytotic

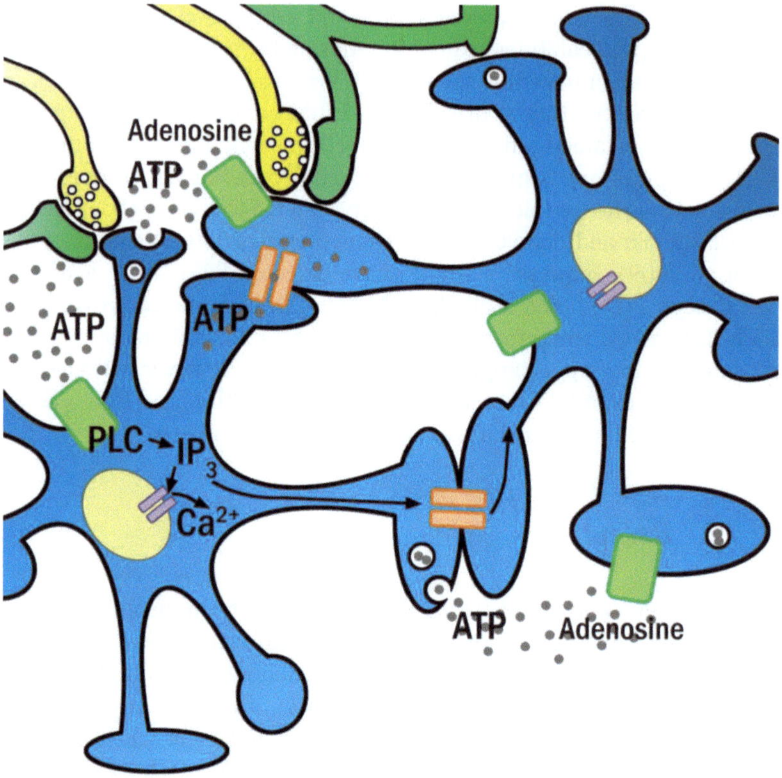

Fig. 8.3 ATP signaling via exoctyic release from astrocytes, and gap-junction mediated transmission of ATP through astrocyte networks. Astrocytes integrate neuronal and glial networks via multiple signaling mechanisms. Neuronal activity evokes a calcium signal in a neighboring astrocyte, generally through the release of glutamate acting on multiple receptor types expressed on astrocytes. The stimulated astrocyte releases ATP through multiple mechanisms, including $P2X_7$ receptors and vesicular release. Released ATP acts in both an autocrine fashion to amplify the initiating signal and in a paracrine fashion to propagate an intercellular wave

mechanism (Fig. 8.3) as discussed above. In particular, electrophysiological studies have shown calcium-dependent plasma membrane area changes in single astrocytes, directly reflective of calcium-regulated exocytosis. It has been suggested that ATP in astrocytes may be stored in vesicles along with glutamate or peptides (Bodin and Burnstock 2001). Using quinacrine, which binds ATP in peptidergic vesicles, it was demonstrated that ATP is in fact stored in secretory vesicles together with peptides, such as ANP within astrocytes (Pangrsic et al. 2007). These studies also showed that a significant number of the fluorescent quinacrine vesicles were exocytosed after ionomycin treatment, as an overall decrease in the total image fluorescence and in the number of quinacrine-stained vesicles was observed (Pangrsic et al.

2007) This result was further confirmed by total internal reflection fluorescence (TIRF) studies, showing a decrease in fluorescence of quinacrine-loaded vesicles upon stimulation with either ionomycin or glutamate. Using a sniffer cell method in heterologous cells other studies have also supported exocytic release of ATP from astrocytes (Pangrsic et al. 2007). Transfection of HEK-293T cells with a mutated form of the purinergic receptor $P2X_3$ (D266A) has been shown to reduce receptor desensitization while retaining receptor affinity in the low micromolar range (Fabbretti et al. 2004). Recordings from D266A expressing HEK-293T cells lying in close vicinity to astrocytes showed random, small, transient, inward currents (STICs) with kinetic properties suggestive of quantal origin (Pangrsic et al. 2007). Further, following glutamate stimulation of astrocytes the average frequency of STICs in expressing HEK-293T cells significantly increased, as would be expected for regulated exocytic signaling. Because HEK293T cells do not respond to glutamate directly, it is assumed that the increased frequency of recorded STICs resulted from vesicular release of ATP from astrocytes. STICs were reduced in both resting and glutamate-stimulated conditions in calcium-free solution, which is consistent with a role of calcium in SNARE mediated exocytosis. The release of ATP is also reduced by inhibitors of several anion channels (Anderson et al. 2004), ATP-binding cassette proteins or cystic fibrosis transmembrane conductance regulator (Schwiebert et al. 1995), gap junctions (Stout et al. 2002), and $P2X_7$ receptors, which suggests the involvement of multiple pathways in leading up to the release of ATP. In addition, data has also shown that the release of ATP is partly dependent on calcium and SNARE proteins (Pascual et al. 2005), and can be inhibited by vesicular ATPase inhibitors (Jeremic et al. 2001).

Although it had been known for quite some time that ATP hydrolysis leads to the accumulation of adenosine, the origin and mechanism of adenosine accumulation was not clearly revealed until the seminal study of Pascual et al. (2005). In this study a molecular genetics approach was used to perturb gliotransmission via conditional, astrocyte-specific expression of the cytoplasmic tail of synaptobrevin 2. Previous studies have shown that this portion of synaptobrevin 2 acts as a dominant-negative inhibitor of SNARE-dependent membrane fusion (dnSNARE). Functional studies performed at hippocampal Schaffer collateral-CA1 synapses in mice expressing dnSNARE revealed stronger synaptic transmission compared to wild-type littermates, or transgenic mice in which transgene expression was prevented by doxycycline in the diet Pascual et al. (2005). These studies went on to show that astrocytes released ATP through a mechanism of exocytosis, and that the astrocytic ATP and its metabolite adenosine tonically suppressed synaptic transmission. These results strongly suggest that astrocytes release ATP through a mechanism of exocytosis, which would be a key event for neuron–astrocyte communication. Recently, Sawada et al. (2008) have identified SLC17A9 (also known as VNUT), a novel member of an anion transporter family, as a vesicular nucleotide transporter that is essential for the storage of nucleotides within vesicles. Importantly, SLC17A9 is expressed in astrocytes, providing the mechanism by which ATP is transported into vesicles within astrocytes (Sawada et al. 2008).

8.6 ATP Signaling through Connexins

Evidence has also accumulated for a gap junction mediated system of ATP signaling in astrocytes (Fig. 8.3). The most highly characterized role for astrocyte ATP signaling through connexin gap junctions is in the propagation of calcium waves. Astrocytes respond to neuronal activity via intracellular calcium elevation, and astrocyte calcium elevations can be propagated to neighboring astrocytes in a wave-like manner both in cell culture (Cooper et al. 1990; Cornell-Bell et al. 1990; Nedergaard 1994) and in intact brain preparations (Dani et al. 1992; Newman and Zahs 1998; Schipke et al. 2002). Two principal models exist for the conduction of calcium waves through networks of astrocytes: 1. Calcium waves are mediated by diffusion of cytoplasmic inositol (1,4,5)-trisphosphate (IP3) through gap junctions. 2. Calcium waves are mediated by ATP released through hemichannels. Evidence exists for both mechanisms, and research has now clarified that both mechanisms of wave propagation can be observed in the same cells. A heterologous cell system, normally deficient of connexins, can be electrically coupled via transfection with eGFP-labeled connexins (Paemeleire 2002; Paemeleire et al. 2000). Using this system, it was shown that waves mediated by an extracellular signal could be blocked hydrolyzing ATP, antagonism of purine-receptors, and inhibited by perfusion of medium; while waves mediated by an intracellular signal were not blocked by ATP breakdown or purine-receptor antagonists (Paemeleire 2002; Paemeleire et al. 2000). During waves initiated by an extracellular signal, calcium levels first rise at the endoplasmic reticulum due to ATP-induced activation of a second messenger. It has also been demonstrated that intracellular Ins(1,4,5)P3 can induce release of ATP, meaning that ATP could act in a regenerative manner to propagate calcium waves (Braet et al. 2003a, b).

8.7 Functional Implications for Astrocytic Regulation of ATP

Despite the historical perspective that astrocytes are passive players in brain physiology, it is now widely accepted that these important cells play a functional role in complex behaviors. Release of ATP from astrocytes has now been shown to be important for modulation of multiple brain processes with implications for behavioral output. Three such examples that we discuss here involve astrocytic regulation of ATP and ATP signaling to impact synaptic transmission and the complex behaviors of respiration and sleep.

8.7.1 *Astrocytes and ATP at the Synapse*

For example, in hypothalamic slices, astrocytic release of ATP is both necessary and sufficient for noradrenaline-dependant synaptic potentiation (Gordon et al. 2005).

This mechanism has been shown to involve $\alpha 1$-adrenergic receptors expressed on astrocytes of the hypothalamus, and following adrenergic stimuli the astrocytes release ATP onto nearby magnocellular neurosecretory neurons. ATP in turn activates $P2X_7$ receptors on these neurons, causing the enhancement of AMPA receptor surface expression and increased amplitude of miniature excitatory postsynaptic current (mEPSC). Other experiments conducted in the retina have shown a contrasting suppression of neuronal activity resulting from astrocytic purinergic (Kurth-Nelson et al. 2009; Newman 2001, 2005). Using retinal whole mounts, stimulation of photoreceptors with light leads to glial calcium signaling (Newman 2004, 2005) and subsequent ATP release from Müller cells (Newman 2003; Newman and Zahs 1998). The resulting actions on ganglion cells of the retina are mediated by degradation of ATP to adenosine, which then acts on A_1 receptors to suppress neuronal activity. With similarity to the situation in the retina, suppressive actions of astrocyte-derived adenosine have been observed in the hippocampus (Pascual et al. 2005; Zhang et al. 2003). Hippocampal astrocytes release ATP, leading to adenosine production and A_1R-dependent presynaptic inhibition of synaptic transmission. Because both neurons and astrocytes express common receptors and use similar signaling pathways, it is difficult to use pharmacological manipulations to discern the specific role of astrocytes in the modulation of neuronal physiology. However, the role of astrocyte-derived adenosine has been confirmed through the use of glia-specific toxins (Zhang et al. 2003) and astrocyte-selective loading of the calcium chelator BAPTA (Serrano et al. 2006).

8.7.2 Astrocytes, ATP and Respiration

In a seminal study, Gourine et al. demonstrate that calcium rises elicit a depolarization of neurons in the primary locus for central respiratory chemosensitivity, which is evoked by vesicular release of ATP in neighboring astrocytes. The astrocytes in this case are stimulated to release ATP in response to the fall in extracellular pH. Specifically the authors induce local pH changes in chemosensitive brainstem areas via stereotaxic infusion of lower pH solutions while recording from the phrenic nerve to measure central respiratory drive. Local pH changes under these conditions were accompanied by an increase of calcium and the excitation of astrocytes. Calcium induced excitation of astrocytes in the ventral medulla was further confirmed in vitro using acutely prepared brainstem slices (Gourine et al. 2010). Using tetrodotoxin and muscimol to silence the chemosensitive neurons, the authors demonstrate the neuronal influence on pH-induced internal calcium levels. Using enzyme based biosensors, it is also shown that propagation of pH-evoked calcium increases in ventral medullary astrocytes is mediated by exocytic release of ATP, and that the ATP-hydrolyzing enzyme, apyrase, blocks the pH-induced astrocytic calcium responses (Gourine et al. 2010). Further, it has been demonstrated that carbon-dioxide dependent ATP release from brainstem astrocytes is independent of extracellular calcium and is likely mediated via the opening of a Cx26 hemichannel

(Huckstepp et al. 2010a, b). Huckstepp et al. went on to show that treatment with selective Cx26 blockers reduces hypercapnia-evoked ATP release and as such also block enhancement of breathing. Thus, brainstem astrocytes have the ability to sense changes in blood and brain CO_2 and pH directly, and in response to changes can control the activity of the respiratory neuronal networks leading to breathing regulation.

8.7.3 Astrocytes, ATP and Sleep

Astrocytic ATP signaling has also been linked to sleep regulation through a A_1 receptor mediated mechanisms (Basheer et al. 2004). The fundamental importance of sleep is well established, it is universal among multicellular animals (Greenspan et al. 2001; Tononi 2005), and all animals with a nervous system must sleep (Tononi and Cirelli 2006). Sleep is modulated by the circadian clock, which entrains multiple physiological processes to salient environmental cues, such as the light/dark cycle (Borbely 1982, 1987). Sleep is also controlled by a distinct homeostatic process, also known as Process S, which can be completely separated from circadian regulation.

Sleep is a highly ordered and highly complex state of brain activity. During sleep specific changes in brain activity occur (as measured by electroencephalography—EEG) (Franken et al. 1991a); in addition, organisms also exhibit an increased threshold to sensory stimulation (Tononi and Cirelli 2006). Based on combinations of these changes, mammalian sleep can be subdivided into rapid eye movement (REM) sleep and non-rapid-eye-movement (NREM) sleep. Slow rhythms (<5 Hz) dominate during NREM stages of sleep (Esser et al. 2007; Tononi and Cirelli 2006), and with increasing sleep pressure, the larger and more frequent these slower rhythms become in subsequent sleep (Franken and Dijk 2009; Franken et al. 2007; Franken et al. 1991b, 1995; Tobler and Franken 1993). This phenomenon is referred to as slow-wave activity and is thought to be an electrophysiological indication of sleep pressure.

Early investigations into the mechanisms regulating sleep investigated whether cerebrospinal fluid (CSF) extracts and cerebral venous blood from sleep-deprived animals could initiate sleep. Although no specific factors that induce sleep were identified from these early studies, gained the key understanding that factors regulating sleep were generated locally within the brain (Krueger and Obal 2003; Krueger et al. 1999). After many years of research, have now identified multiple factors that regulate the sleep wake cycle, one of which is adenosine.

Levels of adenosine are regulated throughout the sleep wake cycle, with adenosine progressively increasing during wakefulness, and subsiding during periods of sleep (Basheer et al. 2001; Porkka-Heiskanen et al. 2003; Stenberg et al. 2003). Adenosine antagonists (both A_1 and A_{2A} receptors) promote wakefulness (Basheer et al. 2001) and infusion of adenosine or adenosine agonists into the brain promotes sleep (Porkka-Heiskanen et al. 2000). It is also well known that caffeine can promote wakefulness (Huang et al. 2005). Thus, in addition to being a sleep factor, adenosine may also be a key mediator of the homeostatic sleep mechanism.

A role for astrocytic ATP release in sleep regulation was first indicated using genetic expression of the dominant-negative inhibitor of SNARE-dependent membrane fusion (dnSNARE) selectively in astrocytes. Expression of dnSNARE in astrocytes results in a blunting of NREM slow-wave activity and blocks the increase in total sleep time that typically follows a period of sleep deprivation (Halassa et al. 2009). This study demonstrates that both the electrophysiological and behavioral manifestations of sleep pressure rely on gliotransmission. The reduced SWA and reduced behavioral responses to sleep deprivation induced by dnSNARE expression in astrocytes were phenocopied by intraceberoventricular infusion of the A_1 receptor antagonist CPT into wild-type mice.

References

Abbracchio MP, Burnstock G, Verkhratsky A, Zimmermann H (2009) Purinergic signalling in the nervous system: an overview. Trends Neurosci 32:19–29

Abe K, Saito H (1998) Adenosine stimulates stellation of cultured rat cortical astrocytes. Brain Res 804:63–71

Abraham EH, Prat AG, Gerweck L, Seneveratne T, Arceci RJ, Kramer R, Guidotti G, Cantiello HF (1993) The multidrug resistance (mdr1) gene product functions as an ATP channel. Proc Natl Acad Sci U S A 90:312–316

Allaman I, Lengacher S, Magistretti PJ, Pellerin L (2003) A2B receptor activation promotes glycogen synthesis in astrocytes through modulation of gene expression. Am J Physiol Cell Physiol 284:C696–704

Anderson CM, Bergher JP, Swanson RA (2004) ATP-induced ATP release from astrocytes. J Neurochem 88:246–256

Angulo MC, Kozlov AS, Charpak S, Audinat E (2004) Glutamate released from glial cells synchronizes neuronal activity in the hippocampus. J Neurosci 24:6920–6927

Araque A, Navarrete M (2010) Glial cells in neuronal network function. Philos Trans R Soc Lond B Biol Sci 365:2375–2381

Araque A, Parpura V, Sanzgiri RP, Haydon PG (1999) Tripartite synapses: glia, the unacknowledged partner. Trends Neurosci 22:208–215

Arcuino G, Lin JH, Takano T, Liu C, Jiang L, Gao Q, Kang J, Nedergaard M (2002) Intercellular calcium signaling mediated by point-source burst release of ATP. Proc Natl Acad Sci U S A 99:9840–9845

Aschner M (1998) Astrocytic functions and physiological reactions to injury: the potential to induce and/or exacerbate neuronal dysfunction—a forum position paper. Neurotoxicology 19:7–17, discussion 37–18

Bal-Price A, Moneer Z, Brown GC (2002) Nitric oxide induces rapid, calcium-dependent release of vesicular glutamate and ATP from cultured rat astrocytes. Glia 40:312–323

Basheer R, Halldner L, Alanko L, McCarley RW, Fredholm BB, Porkka-Heiskanen T (2001) Opposite changes in adenosine A1 and A2A receptor mRNA in the rat following sleep deprivation. Neuroreport 12:1577–1580

Basheer R, Strecker RE, Thakkar MM, McCarley RW (2004) Adenosine and sleep-wake regulation. Prog Neurobiol 73:379–396

Bezzi P, Gundersen V, Galbete JL, Seifert G, Steinhauser C, Pilati E, Volterra A (2004) Astrocytes contain a vesicular compartment that is competent for regulated exocytosis of glutamate. Nat Neurosci 7:613–620

Bianco F, Colombo A, Saglietti L, Lecca D, Abbracchio MP, Matteoli M, Verderio C (2009) Different properties of P2X(7) receptor in hippocampal and cortical astrocytes. Purinergic Signal 5:233–240

Bodin P, Burnstock G (2001) Evidence that release of adenosine triphosphate from endothelial cells during increased shear stress is vesicular. J Cardiovasc Pharmacol 38:900–908
Borbely AA (1982) A two process model of sleep regulation. Hum Neurobiol 1:195–204
Borbely AA (1987) The S-deficiency hypothesis of depression and the two-process model of sleep regulation. Pharmacopsychiatry 20:23–29
Braet K, Aspeslagh S, Vandamme W, Willecke K, Martin PE, Evans WH, Leybaert L (2003a) Pharmacological sensitivity of ATP release triggered by photoliberation of inositol-1,4,5-trisphosphate and zero extracellular calcium in brain endothelial cells. J Cell Physiol 197:205–213
Braet K, Vandamme W, Martin PE, Evans WH, Leybaert L (2003b) Photoliberating inositol-1,4,5-trisphosphate triggers ATP release that is blocked by the connexin mimetic peptide gap 26. Cell Calcium 33:37–48
Brambilla R, Cottini L, Fumagalli M, Ceruti S, Abbracchio MP (2003) Blockade of A2A adenosine receptors prevents basic fibroblast growth factor-induced reactive astrogliosis in rat striatal primary astrocytes. Glia 43:190–194
Burnstock G (2007) Physiology and pathophysiology of purinergic neurotransmission. Physiol Rev 87:659–797
Bushong EA, Martone ME, Jones YZ, Ellisman MH (2002) Protoplasmic astrocytes in CA1 stratum radiatum occupy separate anatomical domains. J Neurosci 22:183–192
Butt AM (2006) Neurotransmitter-mediated calcium signalling in oligodendrocyte physiology and pathology. Glia 54:666–675
Ciccarelli R, Ballerini P, Sabatino G, Rathbone MP, D'Onofrio M, Caciagli F, Di Iorio P (2001) Involvement of astrocytes in purine-mediated reparative processes in the brain. Int J Dev Neurosci 19:395–414
Ciccarelli R, Di Iorio P, Ballerini P, Ambrosini G, Giuliani P, Tiboni GM, Caciagli F (1994) Effects of exogenous ATP and related analogues on the proliferation rate of dissociated primary cultures of rat astrocytes. J Neurosci Res 39:556–566
Coco S, Calegari F, Pravettoni E, Pozzi D, Taverna E, Rosa P, Matteoli M, Verderio C (2003) Storage and release of ATP from astrocytes in culture. J Biol Chem 278:1354–1362
Contreras JE, Saez JC, Bukauskas FF, Bennett MV (2003) Gating and regulation of connexin 43 (Cx43) hemichannels. Proc Natl Acad Sci U S A 100:11388–11393
Contreras JE, Sanchez HA, Eugenin EA, Speidel D, Theis M, Willecke K, Bukauskas FF, Bennett MV, Saez JC (2002) Metabolic inhibition induces opening of unapposed connexin 43 gap junction hemichannels and reduces gap junctional communication in cortical astrocytes in culture. Proc Natl Acad Sci U S A 99:495–500
Cooper MS, Cornell-Bell AH, Chernjavsky A, Dani JW, Smith SJ (1990) Tubulovesicular processes emerge from trans-Golgi cisternae, extend along microtubules, and interlink adjacent trans-golgi elements into a reticulum. Cell 61:135–145
Cornell-Bell AH, Finkbeiner SM, Cooper MS, Smith SJ (1990) Glutamate induces calcium waves in cultured astrocytes: long-range glial signaling. Science 247:470–473
Cronin M, Anderson PN, Cook JE, Green CR, Becker DL (2008) Blocking connexin43 expression reduces inflammation and improves functional recovery after spinal cord injury. Mol Cell Neurosci 39:152–160
Danesh-Meyer HV, Green CR (2008) Focus on molecules: connexin 43–mind the gap. Exp Eye Res 87:494–495
Dani JW, Chernjavsky A, Smith SJ (1992) Neuronal activity triggers calcium waves in hippocampal astrocyte networks. Neuron 8:429–440
Darby M, Kuzmiski JB, Panenka W, Feighan D, MacVicar BA (2003) ATP released from astrocytes during swelling activates chloride channels. J Neurophysiol 89:1870–1877
Decrock E, De Vuyst E, Vinken M, Van Moorhem M, Vranckx K, Wang N, Van Laeken L, De Bock M, D'Herde K, Lai CP et al (2009a) Connexin 43 hemichannels contribute to the propagation of apoptotic cell death in a rat C6 glioma cell model. Cell Death Differ 16:151–163
Decrock E, Vinken M, De Vuyst E, Krysko DV, D'Herde K, Vanhaecke T, Vandenabeele P, Rogiers V, Leybaert L (2009b) Connexin-related signaling in cell death: to live or let die? Cell Death Differ 16:524–536

Di Iorio P, Kleywegt S, Ciccarelli R, Traversa U, Andrew CM, Crocker CE, Werstiuk ES, Rathbone MP (2002) Mechanisms of apoptosis induced by purine nucleosides in astrocytes. Glia 38:179–190

Domercq M, Perez-Samartin A, Aparicio D, Alberdi E, Pampliega O, Matute C (2010) P2X7 receptors mediate ischemic damage to oligodendrocytes. Glia 58:730–740

Dunwiddie TV, Masino SA (2001) The role and regulation of adenosine in the central nervous system. Annu Rev Neurosci 24:31–55

Ebihara L, Liu X, Pal JD (2003) Effect of external magnesium and calcium on human connexin46 hemichannels. Biophys J 84:277–286

Endo Y, Obata T, Nomura M, Fukushima M, Yamada Y, Matsuda A, Sasaki T (2007) Development of an antitumor adenosine analog, 3''-ethynyladenosine. Nucleosides Nucleotides Nucleic Acids 26:691–694

Esser SK, Hill SL, Tononi G (2007) Sleep homeostasis and cortical synchronization: I. Modeling the effects of synaptic strength on sleep slow waves. Sleep 30:1617–1630

Evans RJ (2010) Structural interpretation of P2X receptor mutagenesis studies on drug action. Br J Pharmacol 161:961–971

Fabbretti E, Sokolova E, Masten L, D'Arco M, Fabbro A, Nistri A, Giniatullin R (2004) Identification of negative residues in the P2X3 ATP receptor ectodomain as structural determinants for desensitization and the Ca2+-sensing modulatory sites. J Biol Chem 279:53109–53115

Fam SR, Gallagher CJ, Kalia LV, Salter MW (2003) Differential frequency dependence of P2Y1- and P2Y2- mediated Ca 2+ signaling in astrocytes. J Neurosci 23:4437–4444

Fellin T, Pascual O, Gobbo S, Pozzan T, Haydon PG, Carmignoto G (2004) Neuronal synchrony mediated by astrocytic glutamate through activation of extrasynaptic NMDA receptors. Neuron 43:729–743

Fiacco TA, McCarthy KD (2004) Intracellular astrocyte calcium waves in situ increase the frequency of spontaneous AMPA receptor currents in CA1 pyramidal neurons. J Neurosci 24:722–732

Franke H, Schepper C, Illes P, Krugel U (2007) Involvement of P2X and P2Y receptors in microglial activation in vivo. Purinergic Signal 3:435–445

Franken P, Dijk DJ (2009) Circadian clock genes and sleep homeostasis. Eur J Neurosci 29:1820–1829

Franken P, Dijk DJ, Tobler I, Borbely AA (1991a) Sleep deprivation in rats: effects on EEG power spectra, vigilance states, and cortical temperature. Am J Physiol 261:R198–208

Franken P, Malafosse A, Tafti M (1999) Genetic determinants of sleep regulation in inbred mice. Sleep 22:155–169

Franken P, Thomason R, Heller HC, O'Hara BF (2007) A non-circadian role for clock-genes in sleep homeostasis: a strain comparison. BMC Neurosci 8:87

Franken P, Tobler I, Borbely AA (1991b) Sleep homeostasis in the rat: simulation of the time course of EEG slow-wave activity. Neurosci Lett 130:141–144

Franken P, Tobler I, Borbely AA (1995) Varying photoperiod in the laboratory rat: profound effect on 24-h sleep pattern but no effect on sleep homeostasis. Am J Physiol 269:R691–701

Gallagher CJ, Salter MW (2003) Differential properties of astrocyte calcium waves mediated by P2Y1 and P2Y2 receptors. J Neurosci 23:6728–6739

Garcia-Marin V, Garcia-Lopez P, Freire M (2007) Cajal's contributions to glia research. Trends Neurosci 30:479–487

Gordon GR, Baimoukhametova DV, Hewitt SA, Rajapaksha WR, Fisher TE, Bains JS (2005) Norepinephrine triggers release of glial ATP to increase postsynaptic efficacy. Nat Neurosci 8:1078–1086

Gourine AV, Kasymov V, Marina N, Tang F, Figueiredo MF, Lane S, Teschemacher AG, Spyer KM, Deisseroth K, Kasparov S (2010) Astrocytes control breathing through pH-dependent release of ATP. Science 329:571–575

Greenspan RJ, Tononi G, Cirelli C, Shaw PJ (2001) Sleep and the fruit fly. Trends Neurosci 24:142–145

Haber M, Murai KK (2006) Reshaping neuron-glial communication at hippocampal synapses. Neuron Glia Biol 2:59–66

Halassa MM, Florian C, Fellin T, Munoz JR, Lee SY, Abel T, Haydon PG, Frank MG (2009) Astrocytic modulation of sleep homeostasis and cognitive consequences of sleep loss. Neuron 61:213–219
Haydon PG (2001) GLIA: listening and talking to the synapse. Nat Rev Neurosci 2:185–193
Henneberger C, Papouin T, Oliet SH, Rusakov DA (2010) Long-term potentiation depends on release of D-serine from astrocytes. Nature 463:232–236
Hindley S, Herman MA, Rathbone MP (1994) Stimulation of reactive astrogliosis in vivo by extracellular adenosine diphosphate or an adenosine A2 receptor agonist. J Neurosci Res 38: 399–406
Hirase H, Qian L, Bartho P, Buzsaki G (2004) Calcium dynamics of cortical astrocytic networks in vivo. PLoS Biol 2:E96
Huang ZL, Qu WM, Eguchi N, Chen JF, Schwarzschild MA, Fredholm BB, Urade Y, Hayaishi O (2005) Adenosine A2A, but not A1, receptors mediate the arousal effect of caffeine. Nat Neurosci 8:858–859
Huckstepp RT, Eason R, Sachdev A, Dale N (2010a) CO2-dependent opening of connexin 26 and related beta connexins. J Physiol 588:3921–3931
Huckstepp RT, id Bihi R, Eason R, Spyer KM, Dicke N, Willecke K, Marina N, Gourine AV, Dale N (2010b) Connexin hemichannel-mediated CO2-dependent release of ATP in the medulla oblongata contributes to central respiratory chemosensitivity. J Physiol 588:3901–3920
Jabs R, Matthias K, Grote A, Grauer M, Seifert G, Steinhauser C (2007) Lack of P2X receptor mediated currents in astrocytes and GluR type glial cells of the hippocampal CA1 region. Glia 55:1648–1655
James G, Butt AM (2002) P2Y and P2X purinoceptor mediated Ca2+ signalling in glial cell pathology in the central nervous system. Eur J Pharmacol 447:247–260
Jeremic A, Jeftinija K, Stevanovic J, Glavaski A, Jeftinija S (2001) ATP stimulates calcium-dependent glutamate release from cultured astrocytes. J Neurochem 77:664–675
Jourdain P, Bergersen LH, Bhaukaurally K, Bezzi P, Santello M, Domercq M, Matute C, Tonello F, Gundersen V, Volterra A (2007) Glutamate exocytosis from astrocytes controls synaptic strength. Nat Neurosci 10:331–339
Jurkowitz MS, Litsky ML, Browning MJ, Hohl CM (1998) Adenosine, inosine, and guanosine protect glial cells during glucose deprivation and mitochondrial inhibition: correlation between protection and ATP preservation. J Neurochem 71:535–548
Kang N, Xu J, Xu Q, Nedergaard M, Kang J (2005) Astrocytic glutamate release-induced transient depolarization and epileptiform discharges in hippocampal CA1 pyramidal neurons. J Neurophysiol 94:4121–4130
Khakh BS, Burnstock G (2009) The double life of ATP. Sci Am 301(84–90):92
Kikkawa F, Nomura T, Hagino Y (1984) Adenosine 3′,5′-monophosphate dependent protein kinase of uterus in euthyroid and hypothyroid rats. Jpn J Pharmacol 36:161–167
Koizumi S, Fujishita K, Tsuda M, Shigemoto-Mogami Y, Inoue K (2003) Dynamic inhibition of excitatory synaptic transmission by astrocyte-derived ATP in hippocampal cultures. Proc Natl Acad Sci U S A 100:11023–11028
Krueger JM, Obal F Jr (2003) Sleep function. Front Biosci 8:d511–519
Krueger JM, Obal F Jr, Fang J (1999) Why we sleep: a theoretical view of sleep function. Sleep Med Rev 3:119–129
Krzan M, Stenovec M, Kreft M, Pangrsic T, Grilc S, Haydon PG, Zorec R (2003) Calcium-dependent exocytosis of atrial natriuretic peptide from astrocytes. J Neurosci 23:1580–1583
Kurth-Nelson ZL, Mishra A, Newman EA (2009) Spontaneous glial calcium waves in the retina develop over early adulthood. J Neurosci 29:11339–11346
Lazarowski ER, Boucher RC, Harden TK (2003) Mechanisms of release of nucleotides and integration of their action as P2X- and P2Y-receptor activating molecules. Mol Pharmacol 64:785–795
Matute C (2008) P2X7 receptors in oligodendrocytes: a novel target for neuroprotection. Mol Neurobiol 38:123–128

Montana V, Ni Y, Sunjara V, Hua X, Parpura V (2004) Vesicular glutamate transporter-dependent glutamate release from astrocytes. J Neurosci 24:2633–2642

Moran-Jimenez MJ, Matute C (2000) Immunohistochemical localization of the P2Y(1) purinergic receptor in neurons and glial cells of the central nervous system. Brain Res Mol Brain Res 78:50–58

Mothet JP, Pollegioni L, Ouanounou G, Martineau M, Fossier P, Baux G (2005a) Glutamate receptor activation triggers a calcium-dependent and SNARE protein-dependent release of the gliotransmitter D-serine. Proc Natl Acad Sci U S A 102:5606–5611

Mothet JP, Pollegioni L, Ouanounou G, Martineau M, Fossier P, Baux G (2005b) Glutamate receptor activation triggers a calcium-dependent and SNARE protein-dependent release of the gliotransmitter D-serine. Proc Natl Acad Sci U S A 102:5606–5611

Nagy JI, Ochalski PA, Li J, Hertzberg EL (1997) Evidence for the co-localization of another connexin with connexin-43 at astrocytic gap junctions in rat brain. Neuroscience 78:533–548

Nedergaard M (1994) Direct signaling from astrocytes to neurons in cultures of mammalian brain cells. Science 263:1768–1771

Newby AC (1991) Adenosine: origin and clinical roles. Adv Exp Med Biol 309A:265–270

Newby AC, Worku Y, Holmquist CA (1985) Adenosine formation. Evidence for a direct biochemical link with energy metabolism. Adv Myocardiol 6:273–284

Newman EA (2001) Calcium signaling in retinal glial cells and its effect on neuronal activity. Prog Brain Res 132:241–254

Newman EA (2003) Glial cell inhibition of neurons by release of ATP. J Neurosci 23:1659–1666

Newman EA (2004) Glial modulation of synaptic transmission in the retina. Glia 47:268–274

Newman EA (2005) Calcium increases in retinal glial cells evoked by light-induced neuronal activity. J Neurosci 25:5502–5510

Newman EA, Zahs KR (1998) Modulation of neuronal activity by glial cells in the retina. J Neurosci 18:4022–4028

Nishizaki T, Nagai K, Nomura T, Tada H, Kanno T, Tozaki H, Li XX, Kondoh T, Kodama N, Takahashi E et al (2002) A new neuromodulatory pathway with a glial contribution mediated via A(2a) adenosine receptors. Glia 39:133–147

Nobbio L, Sturla L, Fiorese F, Usai C, Basile G, Moreschi I, Benvenuto F, Zocchi E, De Flora A, Schenone A et al (2009) P2X7-mediated increased intracellular calcium causes functional derangement in Schwann cells from rats with CMT1A neuropathy. J Biol Chem 284:23146–23158

O'Carroll SJ, Alkadhi M, Nicholson LF, Green CR (2008) Connexin 43 mimetic peptides reduce swelling, astrogliosis, and neuronal cell death after spinal cord injury. Cell Commun Adhes 15:27–42

Odermatt B, Wellershaus K, Wallraff A, Seifert G, Degen J, Euwens C, Fuss B, Bussow H, Schilling K, Steinhauser C et al (2003) Connexin 47 (Cx47)-deficient mice with enhanced green fluorescent protein reporter gene reveal predominant oligodendrocytic expression of Cx47 and display vacuolized myelin in the CNS. J Neurosci 23:4549–4559

Oliveira JF, Riedel T, Leichsenring A, Heine C, Franke H, Krugel U, Norenberg W, Illes P (2011) Rodent cortical astroglia express in situ functional P2X7 receptors sensing pathologically high ATP concentrations. Cereb Cortex 21:806–820

Paemeleire K (2002) Calcium signaling in and between brain astrocytes and endothelial cells. Acta Neurol Belg 102:137–140

Paemeleire K, Martin PE, Coleman SL, Fogarty KE, Carrington WA, Leybaert L, Tuft RA, Evans WH, Sanderson MJ (2000) Intercellular calcium waves in HeLa cells expressing GFP-labeled connexin 43, 32, or 26. Mol Biol Cell 11:1815–1827

Palygin O, Lalo U, Verkhratsky A, Pankratov Y (2010) Ionotropic NMDA and P2X1/5 receptors mediate synaptically induced Ca2+ signalling in cortical astrocytes. Cell Calcium 48: 225–231

Panatier A, Theodosis DT, Mothet JP, Touquet B, Pollegioni L, Poulain DA, Oliet SH (2006) Glia-derived D-serine controls NMDA receptor activity and synaptic memory. Cell 125:775–784

Pangrsic T, Potokar M, Stenovec M, Kreft M, Fabbretti E, Nistri A, Pryazhnikov E, Khiroug L, Giniatullin R, Zorec R (2007) Exocytotic release of ATP from cultured astrocytes. J Biol Chem 282:28749–28758

Parpura V, Basarsky TA, Liu F, Jeftinija K, Jeftinija S, Haydon PG (1994) Glutamate-mediated astrocyte-neuron signalling. Nature 369:744–747

Pascual O, Casper KB, Kubera C, Zhang J, Revilla-Sanchez R, Sul JY, Takano H, Moss SJ, McCarthy K, Haydon PG (2005) Astrocytic purinergic signaling coordinates synaptic networks. Science 310:113–116

Pirttimaki TM, Hall SD, Parri HR (2011) Sustained neuronal activity generated by glial plasticity. J Neurosci 31:7637–7647

Porkka-Heiskanen T, Kalinchuk A, Alanko L, Urrila A, Stenberg D (2003) Adenosine, energy metabolism, and sleep. ScientificWorldJournal 3:790–798

Porkka-Heiskanen T, Strecker RE, McCarley RW (2000) Brain site-specificity of extracellular adenosine concentration changes during sleep deprivation and spontaneous sleep: an in vivo microdialysis study. Neuroscience 99:507–517

Porter JT, McCarthy KD (1996) Hippocampal astrocytes in situ respond to glutamate released from synaptic terminals. J Neurosci 16:5073–5081

Queiroz G, Gebicke-Haerter PJ, Schobert A, Starke K, von Kugelgen I (1997) Release of ATP from cultured rat astrocytes elicited by glutamate receptor activation. Neuroscience 78: 1203–1208

Rash JE, Staines WA, Yasumura T, Patel D, Furman CS, Stelmack GL, Nagy JI (2000) Immunogold evidence that neuronal gap junctions in adult rat brain and spinal cord contain connexin-36 but not connexin-32 or connexin-43. Proc Natl Acad Sci U S A 97:7573–7578

Rouach N, Koulakoff A, Abudara V, Willecke K, Giaume C (2008) Astroglial metabolic networks sustain hippocampal synaptic transmission. Science 322:1551–1555

Rudolphi KA, Schubert P, Parkinson FE, Fredholm BB (1992) Neuroprotective role of adenosine in cerebral ischaemia. Trends Pharmacol Sci 13:439–445

Sala-Newby GB, Skladanowski AC, Newby AC (1999) The mechanism of adenosine formation in cells. Cloning of cytosolic 5'-nucleotidase-I. J Biol Chem 274:17789–17793

Santello M, Bezzi P, Volterra A (2011) TNFalpha controls glutamatergic gliotransmission in the hippocampal dentate gyrus. Neuron 69:988–1001

Sawada K, Echigo N, Juge N, Miyaji T, Otsuka M, Omote H, Yamamoto A, Moriyama Y (2008) Identification of a vesicular nucleotide transporter. Proc Natl Acad Sci U S A 105:5683–5686

Schipke CG, Boucsein C, Ohlemeyer C, Kirchhoff F, Kettenmann H (2002) Astrocyte Ca2+ waves trigger responses in microglial cells in brain slices. Faseb J 16:255–257

Schwiebert EM, Egan ME, Hwang TH, Fulmer SB, Allen SS, Cutting GR, Guggino WB (1995) CFTR regulates outwardly rectifying chloride channels through an autocrine mechanism involving ATP. Cell 81:1063–1073

Serrano A, Haddjeri N, Lacaille JC, Robitaille R (2006) GABAergic network activation of glial cells underlies hippocampal heterosynaptic depression. J Neurosci 26:5370–5382

Simard M, Arcuino G, Takano T, Liu QS, Nedergaard M (2003) Signaling at the gliovascular interface. J Neurosci 23:9254–9262

Sohl G, Willecke K (2003) An update on connexin genes and their nomenclature in mouse and man. Cell Commun Adhes 10:173–180

Sperlagh B, Illes P (2007) Purinergic modulation of microglial cell activation. Purinergic Signal 3:117–127

Stenberg D, Litonius E, Halldner L, Johansson B, Fredholm BB, Porkka-Heiskanen T (2003) Sleep and its homeostatic regulation in mice lacking the adenosine A1 receptor. J Sleep Res 12:283–290

Stevens B, Porta S, Haak LL, Gallo V, Fields RD (2002) Adenosine: a neuron-glial transmitter promoting myelination in the CNS in response to action potentials. Neuron 36:855–868

Stout CE, Costantin JL, Naus CC, Charles AC (2002) Intercellular calcium signaling in astrocytes via ATP release through connexin hemichannels. J Biol Chem 277:10482–10488

Theis M, Sohl G, Eiberger J, Willecke K (2005) Emerging complexities in identity and function of glial connexins. Trends Neurosci 28:188–195

Tobler I, Franken P (1993) Sleep homeostasis in the guinea pig: similar response to sleep deprivation in the light and dark period. Neurosci Lett 164:105–108

Tononi G (2005) Regional sleep regulation. Sleep Med 6:575–576

Tononi G, Cirelli C (2006) Sleep function and synaptic homeostasis. Sleep Med Rev 10:49–62

Ventura R, Harris KM (1999) Three-dimensional relationships between hippocampal synapses and astrocytes. J Neurosci 19:6897–6906

Zhang JM, Wang HK, Ye CQ, Ge W, Chen Y, Jiang ZL, Wu CP, Poo MM, Duan S (2003) ATP released by astrocytes mediates glutamatergic activity-dependent heterosynaptic suppression. Neuron 40:971–982

Zhang Q, Fukuda M, Van Bockstaele E, Pascual O, Haydon PG (2004a) Synaptotagmin IV regulates glial glutamate release. Proc Natl Acad Sci U S A 101:9441–9446

Zhang Q, Pangrsic T, Kreft M, Krzan M, Li N, Sul JY, Halassa M, Van Bockstaele E, Zorec R, Haydon PG (2004b) Fusion-related release of glutamate from astrocytes. J Biol Chem 279:12724–12733

Zhang Z, Chen G, Zhou W, Song A, Xu T, Luo Q, Wang W, Gu XS, Duan S (2007) Regulated ATP release from astrocytes through lysosome exocytosis. Nat Cell Biol 9:945–953

Zhu D, Caveney S, Kidder GM, Naus CC (1991) Transfection of C6 glioma cells with connexin 43 cDNA: analysis of expression, intercellular coupling, and cell proliferation. Proc Natl Acad Sci U S A 88:1883–1887

Chapter 9
Role of Striatal A_{2A} Receptor Subpopulations in Neurological Disorders

Sergi Ferré, César Quiroz, Marco Orrú, Xavier Guitart, Seema Gulyani, Richard Allen, and Christopher J. Earley

Abstract A very significant density of adenosine A_{2A} receptors ($A_{2A}R$) is present in the striatum, where they are preferentially localized postsynaptically in enkephalinergic-GABAergic-medium spiny neurons (enkephalinergic MSN). In this localization, different subpopulations of $A_{2A}R$ with different functions exist. Their differential function seems to depend mostly on their ability to form heteromers with other G-protein-coupled receptors, such as dopamine D_2, cannabinoid CB_1 and glutamate $mGlu_5$ receptors. Furthermore, striatal $A_{2A}R$ are also localized presynaptically, in corticostriatal glutamatergic terminals that contact dynorphinergic-GABAergic-medium spiny neurons (dynorphinergic MSN). These presynaptic $A_{2A}R$ heteromerize with A_1 receptors and their activation facilitates glutamate release. Pharmacological tools are becoming available that allow the functional evaluation of some of these different subpopulations of $A_{2A}R$, which can therefore provide selective targets for drug development in different basal ganglia disorders. In fact, alterations in the function of different $A_{2A}R$ subpopulations have recently been observed in Parkinson's disease and in animal models of Huntington disease and Restless Legs Syndrome.

Keywords Adenosine A_{2A} receptor • Receptor heteromers • Striatum • Parkinson's disease • Huntington disease • Restless Legs Syndrome

S. Ferré (✉) • C. Quiroz • M. Orrú • X. Guitart
Intramural Research Program, Department of Health and Human Services, National Institute on Drug Abuse, National Institutes of Health, 251 Bayview Blvd, Baltimore, MD 21224, USA
e-mail: sferre@intra.nida.nih.gov

S. Gulyani • R. Allen • C.J. Earley
Department of Neurology, Johns Hopkins School of Medicine, Baltimore, MD, USA

S. Masino and D. Boison (eds.), *Adenosine: A Key Link between Metabolism and Brain Activity*, DOI 10.1007/978-1-4614-3903-5_9,

9.1 Striatal Subpopulations of $A_{2A}R$: Functional Aspects

9.1.1 *Postsynaptic Striatal Adenosine $A_{2A}R$*

Adenosine A_{2A} receptors ($A_{2A}R$) are more concentrated in the striatum than anywhere else in the brain (Rosin et al. 1998; Hettinger et al. 1998; Schiffmann et al. 2007; Quiroz et al. 2009). Striatal $A_{2A}R$ are preferentially localized postsynaptically in the soma and dendrites of enkephalinergic-GABAergic-medium spiny neurons (enkephalinergic MSN) (Rosin et al. 1998; Hettinger et al. 1998; Schiffmann et al. 2007; Quiroz et al. 2009). These neurons also show a high density of dopamine D_2 receptors (D_2R) with which $A_{2A}R$ establish two reciprocal antagonistic interactions (Ferré et al. 2008). In one type of interaction, which we can call *$A_{2A}R$–D_2R interaction*, stimulation of $A_{2A}R$ counteracts the D_2R-mediated inhibitory modulation of NMDA receptor (NMDAR)-mediated effects, which include modulation of Ca^{2+} influx, transition to the up-state and neuronal firing (Azdad et al. 2009; Higley and Sabatini 2010) (Fig. 9.1). This interaction depends on the ability of $A_{2A}R$ and D_2R to heteromerize and seems to be mostly responsible for the locomotor depressant and activating effects of $A_{2A}R$ agonists and antagonists, respectively (Ferré et al. 2007, 2008), which correlates with the results of behavioral experiments showing that $A_{2A}R$ activation or blockade decreases or increases, respectively, the motor effects elicited by D_2R activation (Ferré et al. 2008).

In the second type of interaction, which we can call *D_2R–$A_{2A}R$ interaction*, D_2R stimulation impedes $A_{2A}R$ to signal through adenylyl cyclase (AC) (Kull et al. 1999; Chen et al. 2001; Hillion et al. 2002; Håkansson et al. 2006) (Fig. 9.1). This D_2R–$A_{2A}R$ interaction takes place at the second messenger level, and stimulation of $G_{i/o}$-coupled D_2R counteracts the effects of $G_{s/olf}$-coupled $A_{2A}R$ stimulation (Ferré et al 2007, 2008). Due to a strong tonic effect of endogenous dopamine on striatal D_2R, this interaction keeps $A_{2A}R$ from signaling through AC. However, under conditions of dopamine depletion or with pharmacological D_2R blockade, $A_{2A}R$-mediated signaling through the cAMP-PKA cascade may be unleashed. Antagonism of D_2R is biochemically associated with a significant increase in the phosphorylation of PKA-dependent substrates, which increases gene expression and the activity of the enkephalinergic MSN, producing locomotor depression (reviewed in Ferré et al. 2008). This appears to be the main mechanism responsible for the locomotor depression induced by D_2R antagonists. Thus, the motor depressant and most biochemical effects induced by pharmacologic blockade of D_2R may be counteracted by pharmacological blockade of $A_{2A}R$ (Chen et al. 2001; Håkansson et al. 2006).

The two reciprocal antagonistic interactions, A_2R–D_2R and D_2R–$A_{2A}R$, take place simultaneously in the same cell, which suggests that they are most likely mediated by the existence of at least two different populations of postsynaptic striatal $A_{2A}R$ in the enkephalinergic MSN (Ferré et al. 2008). One population would be forming heteromers with D_2R and would determine that $A_{2A}R$ stimulation inhibits D_2R-mediated signaling ($A_{2A}R$–D_2R interaction), while another population would not be forming heteromers with D_2R and would determine that that D_2R stimulation

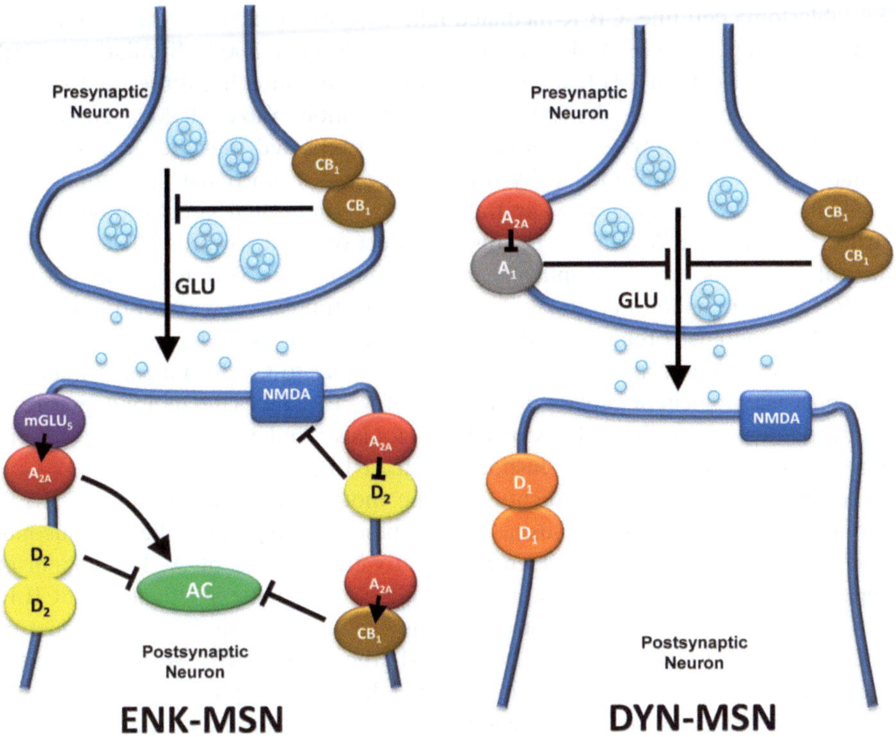

Fig. 9.1 Schematic representation of the different subpopulations of striatal $A_{2A}R$. Presynaptic $A_{2A}R$ are localized in glutamatergic terminals that contact the dynorphinergic MSN (DYN-MSN), where they form heteromers with A_1R. Postsynaptic $A_{2A}R$ are localized in the enkephalinergic MSN (ENK-MSN), where they form heteromers with D2R, CB1R, and mGlu5R. DYN-MSN preferentially express dopamine receptors of the D1 subtype. *AC* adenylyl cyclase, *GLU* glutamate

inhibits $A_{2A}R$-mediated signaling (D_2R–$A_{2A}R$ interaction). This second population of postsynaptic $A_{2A}R$ would either not form heteromers or would form heteromers with other receptors, such as glutamate $mGlu_5$ receptors ($mGlu_5R$) (Ferré et al. 2002) (Fig. 9.1). Importantly, heteromerization of $A_{2A}R$ with $mGlu_5R$ is associated with a synergistic effect upon $A_{2A}R$ and $mGlu_5R$ coactivation at the level of AC and MAPK, providing a physiological mechanism by which $A_{2A}R$ can overcome the D_2R-$A_{2A}R$ interaction (Ferré et al. 2002; Nishi et al. 2003). Costimulation of $A_{2A}R$ and $mGlu_5R$ in vivo, with the central administration of selective agonists, allowed $A_{2A}R$ to get rid of the inhibitory effect of the D_2R and signal through the cAMP-PKA cascade (Ferré et al. 2002). Since this $A_{2A}R$–D_2R–$mGlu_5R$ interaction could be demonstrated in animal models of Parkinson's disease (PD) (Popoli et al. 2001; Kachroo et al. 2005), it was postulated that coadministration of $A_{2A}R$ and $mGlu_5R$ antagonists could be useful as a therapeutic strategy in this disease (Popoli et al. 2001).

Still another subpopulation of postsynaptic $A_{2A}R$ would form heteromers with cannabinoid CB_1 receptors (CB_1R) (Carriba et al. 2007) (Fig. 9.1). In this heteromer, activation of $A_{2A}R$ is necessary to allow CB_1R-mediated signaling. Thus, in a human neuroblastoma cell line, CB_1R-mediated inhibition of AC activity was found to be completely dependent on $A_{2A}R$ coactivation (Carriba et al. 2007). Similarly, several biochemical effects of CB_1R agonists in primary striatal cell cultures and striatal slices have been shown to depend on $A_{2A}R$ coactivation (Yao et al. 2003; Andersson et al. 2005). Accordingly, Tebano et al. (2009) reported that the depression of synaptic transmission induced by a CB_1R agonist in corticostriatal slices was prevented by $A_{2A}R$ antagonists and also by the conditional genetic blockade of striatal postsynaptic $A_{2A}R$. The permissive effect of $A_{2A}R$ towards CB_1R function did not seem to occur presynaptically, as the ability of the CB_1R agonist to increase the R2–R1 ratio under a protocol of paired-pulse stimulation was not modified by an $A_{2A}R$ antagonist (Tebano et al. 2009). These results would predict that $A_{2A}R$ antagonists should produce behavioral effects similar to CB_1R antagonists, and in fact, pharmacological or genetic inactivation of $A_{2A}R$ reduces the motor depressant, cataleptic, and rewarding effects of CB_1R agonists (Andersson et al. 2005; Soria et al. 2004; Carriba et al. 2007; Justinova et al. 2010).

Although those studies indicate that the motor-depressant effects of CB_1R agonists might depend on adenosine A_{2A} receptor signaling, quite the opposite has also been suggested, that the locomotor activating effects of $A_{2A}R$ antagonists depend on CB_1R signaling. Thus, pharmacological or genetic inactivation of CB_1R was found to reduce the locomotor activation induced by an $A_{2A}R$ antagonist in mice habituated to the testing environment (Lerner et al. 2010). We have recently reevaluated these findings by studying in detail the effects of pharmacological interactions between $A_{2A}R$ antagonists and CB_1R antagonists on the locomotor activity in rats not habituated to the testing environment (Orrú et al. 2011a). We also found that a CB_1R antagonist significantly decreases the locomotor effects induced by an $A_{2A}R$ antagonist. However, the CB_1R antagonist produced a comparable decrease in locomotion in vehicle-treated animals and the statistical analysis indicated that the locomotor effects of $A_{2A}R$ and CB_1R antagonists were not interrelated (Orrú et al. 2011a). The use of habituated animals, which display very low locomotor activity in the testing environment, would have masked this depressant effect of CB_1R antagonist in the vehicle-treated animals in the study by Lerner et al. (2010).

In addition to the three subpopulations of postsynaptic striatal $A_{2A}R$ mentioned above, there is also experimental evidence for a potentially more complex picture, which includes the possibility of receptor heteromultimers. Thus, using a new biophysics-based technology, sequential resonance energy transfer, and bimolecular fluorescence complementation plus Bioluminescence Resonance Energy Transfer, evidence for $A_{2A}R$–CB_1R–D_2R and $A_{2A}R$–D_2R–$mGlu_5R$ heteromers in transfected cells has been recently obtained (Carriba et al. 2008; Cabello et al. 2009; Navarro et al. 2010). The stoichiometry of the different populations of postsynaptic striatal $A_{2A}R$ heteromers (and homomers) is not known, but taking into account the very high density of $A_{2A}R$ and D_2R in the enkephalinergic MSM, we postulate that $A_{2A}R$ and D_2R homomers and $A_{2A}R$–D_2R heteromers are the most common receptor populations, followed by combinations of those populations with CB_1R and $mGlu_5R$.

It is important to mention that there is evidence for the existence of $A_{2A}R$ receptors that also colocalized with D_2Rs, in the somatodendritic and nerve terminal regions of the cholinergic striatal interneurons and that their interactions modulate acetylcholine release (James and Richardson 1993; Jin et al. 1993; Preston et al. 2000; Tozzi et al. 2011). The study by Jin et al. (1993) showed evidence for an antagonistic $A_{2A}R$–$D_{2A}R$ interaction in the modulation of striatal acetylcholine release. Thus, $A_{2A}R$ stimulation counteracted the ability of D_2R activation to inhibit acetylcholine release. Similarly, a recent study showed that $A_{2A}R$ blockade potentiates D_2R-mediated modulation of acetylcholine release (Tozzi et al. 2011), again indicating the existence of an antagonistic $A_{2A}R$–D_2R interaction and, probably, $A_{2A}R$–$A_{2A}R$ homomers in striatal cholinergic interneurons.

9.1.2 *Presynaptic Striatal Adenosine A_{2A} Receptors*

Striatal $A_{2A}R$ are not only localized postsynaptically but also presynaptically, in glutamatergic terminals, where they heteromerize with A_1 receptors (A_1R) and where they perform a fine-tuned modulation of glutamate release (Ciruela et al. 2006; Quiroz et al. 2009) (Fig. 9.1). Thus, A_1R–$A_{2A}R$ heteromers seem to work as a concentration-dependent switch (Ferré et al. 2007), with adenosine acting primarily at A_1R at low concentrations, and at both A_1R and $A_{2A}R$ at higher concentrations. Activation of the A_1R in the A_1R–$A_{2A}R$ heteromer produces inhibition of glutamate release, while the additional activation of the $A_{2A}R$ produces the opposite effect (Ciruela et al. 2006). The mechanism by which $A_{2A}R$ stimulation shuts down the effects of A_1R stimulation could involve an allosteric interaction in the A_1R–$A_{2A}R$ heteromer. Thus, in membrane preparations from cotransfected mammalian cells and from rat striatum, stimulation of A_1R decreases the affinity of the $A_{2A}R$ for agonists (Ciruela et al. 2006). Another possibility would be a switch in G-protein coupling in the A_1R–$A_{2A}R$ heteromer, with activation of A_1R inducing a preferential coupling to G_i, while coactivation of the $A_{2A}R$ causing a preferential coupling to G_s in the heteromer, as it has been recently shown for the D_1-D_3 receptor heteromer (Fiorentini et al. 2008). Interestingly, presynaptic $A_{2A}R$ are preferentially localized in glutamatergic terminals of corticostriatal afferents to the dynorphinergic MSN (Quiroz et al. 2009), which preferentially express dopamine receptors of the D_1 subtype (Fig. 9.1). Apart from morphological evidence provided by immunohistochemical and electron microscopy experiments, patch-clamp experiments in identified enkephalinergic and dynorphinergic MSNs provided a functional demonstration of the segregation of striatal presynaptic $A_{2A}R$. Thus, an $A_{2A}R$ agonist and an $A_{2A}R$ antagonist significantly increased and decreased, respectively, the amplitude of excitatory postsynaptic currents induced by the intrastriatal stimulation of glutamatergic afferents measured in identified enkephalinergic, but not dynorphinergic MSNs. Mean-variance analysis indicated a presynaptic locus for the $A_{2A}R$-mediated modulation (Quiroz et al. 2009). Thus, there seems to be a selective $A_{2A}R$-mediated modulation of glutamate release to the dynorphinergic MSN.

The powerful modulatory role of presynaptic $A_{2A}R$ on striatal glutamate release was first demonstrated with in vivo microdialysis experiments by Popoli et al.

(1995), who showed that the striatal perfusion of an $A_{2A}R$ agonist produced a very pronounced increase in the basal striatal extracellular concentrations of glutamate. Also intrastriatal perfusion of an $A_{2A}R$ antagonist through a microdialysis probe could significantly counteract striatal glutamate release induced by cortical electrical stimulation in the orofacial premotor cortex (Quiroz et al. 2009). A striking unexpected finding was that the counteraction of glutamate release was also accompanied by a complete counteraction of the jaw movements induced by the cortical electrical stimulation, demonstrating the very important role of presynaptic $A_{2A}R$ in the control of corticostriatal glutamatergic neurotransmission. By combining cortical electrical stimulation and recording of EMG activity of the mastication muscles, a Power Correlation Coefficient (PCC) was established as a quantitative in vivo measure of corticostriatal neurotransmission (Quiroz et al. 2009). PCC was shown to be significantly and dose-dependently decreased by the systemic administration of an $A_{2A}R$ antagonist. PCC could therefore be used as a method to screen the presynaptic effect of $A_{2A}R$ antagonists (see below).

According to the widely accepted functional basal circuitry model (Obeso et al. 2002; DeLong and Wichmann 2007), blockade of postsynaptic striatal $A_{2A}R$ in the $A_{2A}R$-D_2R heteromer, localized in the enkephalinergic MSN should potentiate spontaneous or psychostimulant-induced motor activation. On the other hand, according to the same model, blockade of presynaptic striatal $A_{2A}R$ localized in the corticostriatal glutamatergic terminals that make synaptic contact with the dynorphinergic MSN should decrease motor activity. The clear locomotor-activating effects of systemically administered $A_{2A}R$ antagonists could be explained by the significantly higher density of postsynaptic *versus* presynaptic striatal $A_{2A}R$ and to a stronger influence of a tonic adenosine and $A_{2A}R$-mediated modulation of the enkephalinergic *versus* dynorphinergic MSNs under basal conditions. The results by Shen et al. (2008) about the differential effects of $A_{2A}R$ antagonists on psychostimulant-induced locomotor activation in conditional striatal postsynaptic $A_{2A}R$ KO mice *versus* forebrain-specific A2AR KO mice (deleting A2AR in the neurons of striatum as well as cerebral cortex and hippocampus potentiation *versus* counteraction, respectively) support this hypothesis. As previously suggested (Ferré et al. 2007), activation of presynaptic $A_{2A}R$ seems to be highly dependent on the level of adenosine generated upon corticostriatal glutamatergic input.

Striatal D_2Rs are also localized presynaptically, in dopaminergic and glutamatergic terminals (Higley and Sabatini 2010), giving the frame for the existence of interactions with $A_{2A}R$ at least in those terminals establishing contact with the dynorphinergic MSN. The experimental evidence suggest that there is also a presynaptic D_2R–$A_{2A}R$ interaction by which D_2R activation tonically inhibits the ability of endogenous adenosine to produce an $A_{2A}R$-mediated increase in the basal extracellular levels of glutamate. Thus, dopamine denervation significantly potentiates $A_{2A}R$ agonist-mediated stimulation of glutamate release (Tanganelli et al. 2004). This has the biochemical characteristics of an interaction between $A_{2A}R$ and D_2R at the AC level and not forming heteromers. Furthermore, results from Rodrigues et al. (2005) suggest the existence of $mGlu_5R$ in striatal (not known if cortical or thalamic in origin) glutamatergic terminals colocalized with $A_{2A}R$ and which facilitate glutamate release in a synergistic manner. The interplay between adenosine- and dopamine-mediated actions at the presynaptic level is therefore affected by the occurrence of $mGlu_5R$.

The presynaptic localization of CB_1R in striatal (not known if cortical or thalamic in origin) glutamatergic terminals is well established, and therefore they can be colocalized with $A_{2A}R$ in terminals establishing contact with the dynorphinergic MSN (Ferré et al. 2010). We have recently postulated that $A_{2A}R$ and CB_1R could form heteromers in striatal glutamatergic terminals, which could be involved in the reinforcing effects of cannabinoids (Ferré et al. 2001; Justinova et al. 2010). However, a recent study by Martire et al. (2011) indicates that cannabinoid–adenosine functional interactions result from an interaction at the second messenger level. In the frame of heteromerization $A_{2A}R$ activation should facilitate the Gi/o-mediated effect of CB_1R activation measured, as inhibition of glutamate release. Nevertheless, Martire et al. (2011), by recording extracellular field potentials in corticostriatal slices and superfused striatal nerve terminals, showed instead that $A_{2A}R$ activation prevents CB_1R-mediated inhibition of glutamate release. These results indicate that regulation of glutamate release by cannabinoids is not dependent on presynaptic $A_{2A}R$-CB_1R heteromers. In summary, a great amount of available data indicates that, presynaptically, $A_{2A}R$ form heteromers mostly with A_1R. In addition, there seems to be a population of $A_{2A}R$ not forming heteromers but establishing antagonistic interactions with D_2R and CB_1R and synergistic interactions with $mGlu_5R$.

9.2 Targeting Different Subpopulations of Striatal $A_{2A}R$

In a recent study, we reported that several $A_{2A}R$ antagonists that were previously thought to be pharmacologically similar present different striatal pre- and postsynaptic profiles (Orrú et al. 2011b). Six compounds, known to be selective $A_{2A}R$ antagonists, were first screened for their ability to block striatal pre- and postsynaptic $A_{2A}R$ with in vivo models. Locomotor activation was used to evaluate postsynaptic activity while PCC reduction was used to determine presynaptic activity (see above). The $A_{2A}R$ antagonists SCH-442416 and KW-6002 showed preferential pre- and postsynaptic profiles, respectively. Combining in vivo microdialysis with cortical electrical stimulation was used as an additional in vivo evaluation of presynaptic activity. Thus, SCH-442416 significantly counteracted striatal glutamate release induced by cortical stimulation at a dose that strongly decreased PCC but did not induce locomotor activation. On the other hand, according to its preferential postsynaptic profile, KW-6002 did not modify striatal glutamate release induced by cortical stimulation at a dose that produced a pronounced locomotor activation but did not decrease PCC.

An additional finding of that study was that at least part of these pharmacological differences between $A_{2A}R$ antagonists could be explained by the ability of pre- and postsynaptic $A_{2A}R$ to form different receptor heteromers, with A_1R and D_2R, respectively (Orrú et al. 2011b). Radioligand-binding experiments were performed in cells stably expressing $A_{2A}R$, $A_{2A}R–D_2R$ heteromers, or $A_1R–A_{2A}R$ heteromers to determine possible differences in the affinity of $A_{2A}R$ for $A_{2A}R$ antagonists in these different cell lines. The most dramatic difference was observed with SCH-442416. The affinity of $A_{2A}R$ for SCH-442416 in cells expressing $A_{2A}R–D_2R$ heteromers was markedly reduced (40 times higher B_{50} values in competitive-inhibition experiments with [3H]ZM-241385 in cells expressing $A_{2A}R–D_2R$ than $A_1R–A_{2A}R$ heteromers). On the other hand, KW-6002 showed the best relative affinity for $A_{2A}R–D_2R$ heteromers of all $A_{2A}R$ antagonists tested, which could at least partially explain its preferential postsynaptic profile. It must, however, be pointed out that to say that SCH-442416 is a selective presynaptic $A_{2A}R$ antagonist is an oversimplification. In fact, the in vitro data indicated that SCH-442416 binds equally well to the $A_{2A}R$ that do not form heteromers as to those that form $A_1R–A_{2A}R$ heteromers. Therefore, according to the previous description of the different subpopulations of striatal $A_{2A}R$, SCH-442416 should also be effective at counteracting D_2R antagonist-induced motor depression. In fact, at doses that do not produce locomotor activation, but that reduce PCC, SCH-442416 significantly counteracts the locomotor depression induced by the D_2R antagonist raclopride (Orrú et al. 2011a). On the other hand, KW-6002 produced the same locomotor activation with or without coadministration with raclopride, in agreement with its ability to block the three populations of $A_{2A}R$ studied so far in vitro, $A_{2A}R$, $A_{2A}–D_2R$, and $A_1R–A_{2A}R$. Importantly, KW-6002 also produced the same locomotor activation when coadministered with the $A_{2A}R$ agonist CGS-21680, while SCH-442416, at the same dose that counteracted the depressant effect of raclopride, did not significantly counteract the depressant effect of CGS-21680. These results, therefore agree with the hypothesis that the subpopulation of postsynaptic $A_{2A}R$ that form heteromers with D_2R, are mainly responsible for both the locomotor activation and depression induced by $A_{2A}R$ antagonists and agonists, respectively. In summary, SCH-442416 can be considered as a compound that at relatively low doses binds not only preferentially to presynaptic $A_{2A}R$ localized in corticostriatal glutamatergic terminals (Orrú et al. 2011b) but also to the subpopulation of postsynaptic $A_{2A}R$ not forming heteromers with D_2R (Orrú et al. 2011a). Interestingly, [^{11}C]SCH-442416 has been used in rats, monkeys, and humans as a PET radioligand and shown to nicely label striatal $A_{2A}R$ (Moresco et al. 2005; Schiffmann et al. 2007; Brooks et al. 2010). The low doses used in PET experiments indicate that [^{11}C]SCH-442416 is mostly labeling presynaptic $A_{2A}R$ and postsynaptic $A_{2A}R$ that do not form heteromers with D_2R. The use of [^{11}C]SCH-442416 and other less selective radioligand in combination with cold SCH-442416 could allow the identification of the different populations of $A_{2A}R$ in the human brain. The picture is still incomplete, and a further evaluation of the affinity of $A_{2A}R$ antagonists for $A_{2A}R–mGlu_5R$ and $A_{2A}R–CB_1$ heteromers (and of heterotrimers) is needed. Nevertheless, the information so far available can be very valuable to attempt the design of more efficient $A_{2A}R$ antagonists to be used in basal ganglia disorders.

Our studies support the notion that receptor heteromers may be used as selective targets for drug development (Ferré et al. 2009). Particularly significant are the very specific neuronal localization of receptor heteromers (even more specific than receptor subtypes themselves), and a differential ligand affinity of a receptor depending on its partner (or partners) in the receptor heteromer. Striatal A_{2A}R-containing heteromers become particularly interesting targets, eventually useful for a variety of neuropsychiatric disorders. Blocking postsynaptic A_{2A}Rs in the enkephalinergic MSN should be beneficial for PD because it would decrease the activity of the indirect striatal efferent pathway. On the one hand, one benefit would come from potentiating the effect of L-dopa or other dopamine receptor agonists on the D_2R-mediated signaling in the A_{2A}R–D_2R heteromer. On the other hand, blockade of A_{2A}R not forming heteromers with D_2R, but antagonistically interacting with D_2R at the AC level, should counteract the effects of the disinhibited A_{2A}R signaling (see below). However, blocking presynaptic A_{2A}R in glutamatergic terminals that are contacting dynorphinergic MSN (either forming or not heteromers with A_1R) should decrease glutamatergic transmission through the direct striatal efferent pathway, thus decreasing motor activity and, therefore, decreasing the antiparkinsonian efficacy of A_{2A}R antagonists. The most appropriate A_{2A}R antagonist to treat PD patients should have more affinity for postsynaptic than for presynaptic receptors. Additionally, a selective blockade of presynaptic A_{2A}Rs should be useful in dyskinetic disorders such as Huntington disease (HD) and could also be useful in obsessive-compulsive disorders and drug addiction. Effective treatment of L-dopa-induced dyskinesia using presynaptic A_{2A}R antagonists would also be a possibility to explore (see below).

9.3 Striatal A_{2A}R and Parkinson's Disease

Striatal adenosine A_{2A}R are becoming the focus of interest for the treatment of PD. Thus, a series of clinical studies indicate that an A_{2A}R antagonist (KW-6002) can improve the therapeutic index of L-dopa (Hauser et al. 2003; LeWitt et al. 2008; Mizuno et al. 2010; Factor et al. 2010; Fernandez et al. 2010). In 1992, we hypothesized that A_{2A}R antagonists associated with L-dopa or D_2R agonists could represent a new therapeutic approach for PD based on the A_{2A}R–D_2R interactions demonstrated in radioligand binding and behavioral experiments (Ferré et al. 1992). We further elaborated on that original hypothesis that A_{2A}R–D_2R heteromers could constitute a main target for antiparkinsonian drugs (Ferré et al. 2001). This would in fact be the most logical explanation when considering the different subpopulations of A_{2A}R and their interactions with D_2R. Thus, A_{2A}R blockade would enhance D_2R function in the A_{2A}R–D_2R heteromer and we already mentioned that those A_{2A}R seem to be responsible for the locomotor activating effects of A_{2A}R antagonists (A_{2A}R–D_2R interaction). However, the role of A_{2A}R not forming heteromers in Parkinson's disease cannot be underestimated.

As with D_2R blockade, dopamine depletion leads to the loss of the D_2R-mediated inhibition of the function of A_{2A}R that are not forming heteromers with D_2R. As mentioned before, unleashed A_{2A}R function leads, biochemically, to overexpression of PKA-dependent phosphorylation of several substrates and to the consequent increase in the expression of several genes, such as the gene for preproenkephalin, which is the precursor of enkephalin. Behaviorally, unleashed A_{2A}R function leads to motor depression. In the experimental animal both biochemical and behavioral effects can be counteracted by $A_{2A}R$ antagonists (reviewed in Ferré et al. 2008). It seems therefore that this subpopulation of $A_{2A}Rs$ plays an important physiopathological role in the symptomatology of PD. Furthermore, postmortem studies indicate that there is an overexpression of $A_{2A}Rs$ in the putamen of Parkinson's disease patients, which correlates with the severity of the disease and with the presence of L-dopa-induced dyskinesia (Calon et al. 2004; Varani et al. 2010). Similar results have been recently found in vivo using PET and [^{11}C]SCH442416 (Ramlackhansingh et al. 2011). In this study, patients with PD and L-dopa-induced dyskinesia showed an increase in $A_{2A}R$ availability in the putamen, which most probably represents an increase in the expression of the subpopulation of postsynaptic $A_{2A}Rs$ not forming heteromers with D_2Rs, in view of the very low affinity of this ligand for $A_{2A}R$–D_2R heteromers (see above). In fact, the experiments by Orrú et al. described before (Orrú et al. 2011b) were performed with transfected human $A_{2A}R$ cDNA.

In summary, the subpopulation of striatal postsynaptic $A_{2A}R$ not forming heteromers with D_2Rs seems to play an important pathophysiological role in at least advanced stages of PD. As mentioned by Calon et al. (2004), the increased $A_{2A}R$ expression could be involved in the increased preproenkephalin expression, which is well documented in MPTP-induced parkinsonian monkeys that develop L-dopa-induced dyskinesia (Morissette et al. 1997, Calon et al. 2000, Zeng et al. 2000). Those experiments show that the increase in enkephalin expression induced by dopamine depletion is not counteracted by dopamine agonists in the population of animals experiencing dyskinesia, suggesting a causal link between $A_{2A}Rs$, enkephalin and L-dopa-induced dyskinesia. In fact, Calon et al. (2002) have also demonstrated an increase in preproenkephalin expression in PD patients with L-dopa-induced dyskinesia. Unfortunately, although $A_{2A}R$ antagonist administration in MPTP-treated monkeys was reported to improve the antiparkinsonian and dyskinetic profile of L-dopa (Grondin et al. 1999; Kanda et al. 2000), clinical studies in PD patients have failed to demonstrate a significant dyskinetic effect of the $A_{2A}R$ antagonist KW-6002 (Hauser et al. 2003; LeWitt et al. 2008; Mizuno et al. 2010; Factor et al. 2010; Fernandez et al. 2010). More clinical studies with other $A_{2A}R$ antagonists are on the way, which might bring more significant results, maybe related to their preferential selectivity for specific $A_{2A}R$ subpopulations. It is possible that a compound like SCH-442416 might have a more significant effect on L-dopa-induced dyskinesia, not only because of its selectivity for the postsynaptic $A_{2A}R$ that are not forming heteromers with D_2R, but also because of its presynaptic effect of decreasing glutamatergic neurotransmission to the dynorphinergic MSN.

9.4 Striatal $A_{2A}R$ and Huntington Disease

Huntington disease (HD) is a neurodegenerative disorder characterized by progressive chorea, cognitive impairments and emotional disturbances. HD is caused by expansion of a CAG repeat within exon 1 of the huntingtin gene on chromosome 4.

The mechanism by which the mutant huntingtin may cause neurodegeneration is unknown and even controversial (Wellington et al. 2000; Yang et al. 2002, Ravikumar et al. 2004). The most vulnerable neurons in HD are the enkephalinergic MSN (Popoli et al. 2007). We have recently used the in vivo approach to study pre- and postsynaptic subpopulations of striatal A_{2A}Rs in different ages of a transgenic rat model of HD (Orrú et al. 2011c). This transgenic rat bears a human HD mutation with 51 CAG repeats and shows progressive neurological, neurochemical and neuropathological features similar to the slowly progressing disease (von Hörsten et al. 2003; Nguyen et al. 2006). According to previous studies (von Hörsten et al. 2003, Nguyen et al. 2006), at 10 months of age HD rats may start showing some cognitive decline signs and it is not until the age of 10 to 15 months that progressive motor impairment and striatal atrophy appear. Consequently and based on these studies, 3 and 6 months of age can be considered as presymptomatic stages, although some anxiety-related symptoms could be observed already at 2 months of age (Nguyen et al. 2006).

We found that both heterozygous and homozygous HD rats developed insensitivity to the locomotor activating effects of the A_{2A}R antagonist KW-6002 sometime between 3 and 6 months of age (Orrú et al. 2011c). This insensitivity was still observed in 12-month-old rats, and the effect was not observed in wild-type animals at any age. On the other hand, the ability of the A_{2A}R antagonist SCH-442416 to reduce PCC was preserved across at least 12 months of age and regardless the different genotype, with similar reductions in homozygous, heterozygous, and wild-type HD rats (Orrú et al. 2011c). These results indicate that the HD rats develop a significant and selective impairment of striatal postsynaptic A_{2A}R function, which agrees with the possibility of a selective impairment of the GABAergic striatopallidal neuronal function long before overt motor symptoms of HD appear (Popoli et al. 2007). In fact, A_{2A}R density has been found to be significantly decreased in early stages of HD (Glass et al. 2000). Interestingly, a genetic variation of the A_{2A}R has been shown to decrease the age of onset in HD (Dhaenens et al. 2009; Taherzadeh-Fard et al. 2010). Furthermore, an aberrant A_{2A}Rs signaling has been found in an in vitro model of the disease and in peripheral circulating cells of HD patients (Varani et al. 2001, 2003).

Previous results on a mouse model of HD (R6/2 mice) have been controversial, and both a downregulation and an upregulation of A_{2A}Rs have been reported during early presymptomatic stages (Cha et al. 1999; Tarditi et al. 2006). Cha et al. (1999) reported a significant decrease of the striatal density of A_{2A}Rs, while Tarditi et al. (2006) found a transient A_{2A}R upregulation associated with an increased biochemical response (cAMP accumulation) to A_{2A}R agonists. Our recent study showed that a striatal A_{2A}R dysfunction is also present in the transgenic HD rat model, which is believed to be a more appropriate animal model of HD than R6/2 mice (see von Hörsten et al. 2003). Our results fit better with those by Cha et al. (1999) and indicate that there is a permanent functional downregulation of postsynaptic striatal A_{2A}Rs. In a previous study, a significant reduction in the density of A_{2A}Rs was found in the striatum of 2-year old HD transgenic rats, but correlating with degeneration of the GABAergic striatopallidal neurons (Bauer et al. 2005). Additional biochemical studies in this model should provide a better understanding of a putative A_{2A}R dysfunction in the presymptomatic stages of the disease. Importantly, the fact that in our study presynaptic $A_{2A}R$ function seems not to be impaired (at least up to 12 months of age), still makes possible to think of this receptor as a target for the symptomatic treatment of the disease.

9.5 Striatal $A_{2A}R$, Iron Deficiency, and Restless Legs Syndrome

Restless Legs Syndrome (RLS) is a common sensory-motor disorder that affects about 7.5% of the population (Allen et al. 2005). The core feature of the syndrome is an overwhelming urge to move the legs that is triggered by rest and relieved with movement (Allen et al. 2003). Iron deficiency was first noted by Nordlander in the 1950s to be associated with symptom development and by treating the iron deficiency symptoms would improve or completely resolve (Nordlander 1954). Subsequent studies using cerebrospinal fluid analysis of iron proteins (Earley et al. 2000), MRI-determined brain iron levels (Earley et al. 2006) and autopsy studies (Connor et al. 2004) have all indicated the presence of low brain iron in RLS subjects, even in those with normal blood levels of iron. The most effective treatment for RLS has been the use of L-dopa, and based on this finding, the dopaminergic system has been implicated in RLS pathology along with iron deficiency (Allen 2004).

One of the most consistent and best characterized functional abnormalities associated to postweaning iron deficiency in the rodent is altered dopaminergic neurotransmission. In fact, the distribution of dopamine and iron in the brain are closely correlated, with particularly high concentrations of both in the adult striatum (Beard and Connor 2003). In the experimental animal, iron deficiency has been associated with a decrease in the striatal density of the dopamine transporter and dopamine D_1 and D_2 receptors and with an increase in striatal dopamine levels and nigral tyrosine hydroxylase (Youdim et al. 1983; Nelson et al. 1997; Erikson et al. 2000, 2001; Unger et al. 2008; Connor et al. 2009). As a confirmation of the predictive nature of the iron-deficiency rodent model as a tool for exploring the iron-dopamine relation in RLS, a recent autopsy study in RLS and control cases showed decreases in D_2 receptor density and increases in the putaminal and nigral tyrosine hydroxylase (Connor et al. 2009). However, the mechanisms linking brain iron deficiency and the alterations in the dopaminergic function still need to be determined.

In SH-SY5Y cells in culture, a step-wise decrease in iron induced by progressively increased concentrations of desferroxamine led to an inversely proportional increase in $A_{2A}R$ density (Gulyani et al. 2009). Dietary iron-deficiency in mice also led to a decrease in striatal iron and an increase in striatal $A_{2A}R$ density (Gulyani et al. 2009). Furthermore, when mice strains were selected on the basis of their natural variation in striatal iron concentration, strains with the lower striatal iron concentrations had significantly higher striatal $A_{2A}R$ density (Gulyani et al. 2009). More recently we have also found an increase in the density of striatal $A_{2A}R$ induced by brain iron deficiency in rats, which was accompanied by a functional upregulation of both pre- and postsynaptic $A_{2A}R$, as demonstrated by the increased efficacy of an $A_{2A}R$ antagonist to produce locomotor activation and its increased potency at decreasing PCC (Quiroz et al. 2010). Altogether, the results from cell lines, mice and rats subjected to iron deficiency strongly suggest the idea that $A_{2A}R$ upregulation is a general cellular response to iron-deficiency which can potentially develop in any cell that can express $A_{2A}R$. Exploratory activity was significantly decreased in iron-deficient rats, compared to controls (Quiroz et al. 2010), suggesting that a stronger tonic effect of endogenous adenosine on functionally upregulated postsynaptic $A_{2A}R$ is involved in the well-documented decrease in exploratory activity induced by iron deficiency (Weinberg et al. 1980; Beard 2003; Lozoff and Georgieff 2006). We therefore postulated that the same mechanism could be involved in the reported reduced activity of iron-deficient infants when exposed to unfamiliar environments (Angulo-Kinzler et al. 2002; Lozoff and Georgieff 2006).

The functional upregulation of striatal presynaptic $A_{2A}R$ induced by iron deficiency leads to both a stronger tonic facilitatory effect of endogenous adenosine on corticostriatal glutamatergic neurotransmission and to a reduction of corticostriatal filtering (Quiroz et al. 2009). It is therefore possible that functionally upregulated presynaptic $A_{2A}R$ are involved in the alterations that iron-deficient infants show on executive function tasks, which depend on prefrontal-striatal circuit function (Lozoff and Georgieff 2006). It is also possible that functional upregulation of striatal presynaptic $A_{2A}Rs$ could be involved in the alterations in corticostriatal processing that has been described as part of the pathophysiological mechanisms of RLS (Tergau et al. 1999). RLS symptomatology could in fact depend on a disruption of the normal balanced activity of the direct and indirect striatal efferent pathways, with an increased glutamatergic signaling in the dynorphynergic MSN of the direct pathway, which is due to the presynaptic upregulation of $A_{2A}R$, and a decreased dopaminergic signaling in the enkephalinergic MSN of the indirect pathway, which is due to the postsynaptic upregulation of $A_{2A}Rs$. The efficacy of L-dopa in RLS could be in part due to an amelioration of such imbalance, and the present results suggest that selective presynaptic $A_{2A}R$ antagonists could be beneficial in RLS.

9.6 Conclusion

It is becoming clear that different subpopulations of $A_{2A}R$ exist in the striatum, which can be differentiated by their ability to form heteromers with different GPCR, which endows $A_{2A}R$ with different biochemical properties, including different ligand recognition. In fact, pharmacological dissection of these subpopulations can be achieved by using different $A_{2A}R$ antagonists. These subpopulations of $A_{2A}R$ are strategically localized in different striatal neuronal elements in the striatum, playing altogether an important role in the modulation of striatal function. Furthermore, these subpopulations of $A_{2A}R$ are differentially altered in different animal models of PD, HD, and RLS (Table 9.1), which implies that they can be used as targets for these basal ganglia disorders.

Table 9.1 Putative differences in the expression of the different subpopulations of striatal $A_{2A}R$ in basal ganglia disorders (see text)

	Presynaptic $A_{2A}R$	Postsynaptic $A_{2A}R$	
	$A_{2A}R$–A_1R	$A_{2A}R$–D_2R	$A_{2A}R$
Parkinson's disease	No change	No change	Upregulation
Huntington disease	No change	Downregulation	not known
Restless Legs Syndrome	Upregulation	Upregulation	not known

Acknowledgments Work supported by NIDA IRP funds.

References

Allen RP, Picchietti D, Hening WA, Trenkwalder C, Walters AS, Montplaisir J (2003) Restless legs syndrome: diagnostic criteria, special considerations, and epidemiology. A report from the restless legs syndrome diagnosis and epidemiology workshop at the National Institutes of Health. Sleep Med 4:101–119

Allen R (2004) Dopamine and iron in the pathophysiology of restless legs syndrome (RLS). Sleep Med 5:385–391

Allen RP, Walter AS, Montplaisir J, Hening W, Myers A, Bell TJ, Ferini-Strambi L (2005) Restless legs syndrome prevalence and impact: REST general population study. Arch Intern Med 165:1286–1292

Andersson M, Usiello A, Borgkvist A, Pozzi L, Dominguez C, Fienberg AA, Svenningsson P, Fredholm BB, Borrelli E, Greengard P, Fisone G (2005) Cannabinoid action depends on phosphorylation of dopamine- and cAMP-regulated phosphoprotein of 32 kDa at the protein kinase A site in striatal projection neurons. J Neurosci 25:8432–8438

Angulo-Kinzler RM, Peirano P, Lin E, Algarin C, Garrido M, Lozoff B (2002) Twenty-four-hour motor activity in human infants with and without iron deficiency anemia. Early Hum Dev 70:85–101

Azdad K, Gall D, Woods AS, Ledent C, Ferré S, Schiffmann SN (2009) Dopamine D2 and adenosine A2A receptors regulate NMDA-excitation in accumbens neurons through A2A-D2 receptor heteromerization. Neuropsychopharmacology 34:972–986

Bauer A, Zilles K, Matusch A, Holzmann C, Riess O, von Hörsten S (2005) Regional and subtype selective changes of neurotransmitter receptor density in a rat transgenic for the Huntington's disease mutation. J Neurochem 94:639–650

Beard JL (2003) Iron deficiency alters brain development and functioning. J Nutr 133: 1468S–1472S

Beard JL, Connor JR (2003) Iron status and neural functioning. Annu Rev Nutr 23:41–58

Brooks DJ, Papapetropoulos S, Vandenhende F, Tomic D, He P, Coppell A, O'Neill G (2010) An open-label, positron emission tomography study to assess adenosine A2A brain receptor occupancy of vipadenant (BIIB014) at steady-state levels in healthy male volunteers. Clin Neuropharmacol 33:55–60

Cabello N, Gandia J, Bertarelli DC, Watanabe M, Lluis C, Franco R, Ferré S, Lujan R, Ciruela F (2009) Metabotropic glutamate type 5, dopamine D2 and adenosine A2A receptors form higher-order oligomers in living cells. J Neurochem 109:1497–1507

Calon F, Hadj Tahar A, Blanchet PJ, Morissette M, Grondin R, Goulet M, Doucet JP, Robertson GS, Nestler E, Di Paolo T, Bédard PJ (2000) Dopamine-receptor stimulation: biobehavioral and biochemical consequences. Trends Neurosci 23:S92–S100

Calon F, Birdi S, Rajput AH, Hornykiewicz O, Bédard PJ, Di Paolo T (2002) Increase of preproenkephalin mRNA levels in the putamen of Parkinson disease patients with levodopa-induced dyskinesias. J Neuropathol Exp Neurol 61:186–196

Calon F, Dridi M, Hornykiewicz O, Bédard PJ, Rajput AH, Di Paolo T (2004) Increased adenosine A2A receptors in the brain of Parkinson's disease patients with dyskinesias. Brain 127:1075–1084

Carriba P, Ortiz O, Patkar K, Justinova Z, Stroik J, Themann A, Müller C, Woods AS, Hope BT, Ciruela F, Casadó V, Canela EI, Lluis C, Goldberg SR, Moratalla R, Franco R, Ferré S (2007) Striatal adenosine A_{2A} and cannabinoid CB_1 receptors form functional heteromeric complexes that mediate the motor effects of cannabinoids. Neuropsychopharmacology 32:2249–2259

Carriba P, Navarro G, Ciruela F, Ferré S, Casado V, Agnati LF, Cortes A, Mallol J, Fuxe K, Canela EI, Lluis C, Franco R (2008) Detection of heteromerization of more than two proteins by sequential BRET-FRET. Nat Methods 5:727–733

Cha JH, Frey AS, Alsdorf SA, Kerner JA, Kosinski CM, Mangiarini L, Penney JB Jr, Davies SW, Bates GP, Young AB (1999) Altered neurotransmitter receptor expression in transgenic mouse models of Huntington's disease. Philos Trans R Soc Lond B Biol Sci 354:981–989

Chen JF, Moratalla R, Impagnatiello F, Grandy DK, Cuellar B, Rubinstein M, Beilstein MA, Hackett E, Fink JS, Low MJ, Ongini E, Schwarzschild MA (2001) The role of the D(2) dopamine receptor (D(2)R) in A(2A) adenosine receptor (A(2A)R)-mediated behavioral and cellular responses as revealed by A(2A) and D(2) receptor knockout mice. Proc Natl Acad Sci U S A 98:1970–1975

Ciruela F, Casado V, Rodrigues RJ, Luján R, Burgueno J, Canals M, Borycz J, Rebola N, Goldberg SR, Mallol J, Cortés A, Canela EI, López-Giménez JF, Milligan G, Lluis C, Cunha RA, Ferré S, Franco R (2006) Presynaptic control of striatal glutamatergic neurotransmission by adenosine A1-A2A receptor heteromers. J Neurosci 26:2080–2087

Connor JR, Wang XS, Patton SM, Menzies SL, Troncoso JC, Earley CJ, Allen RP (2004) Decreased transferrin receptor expression by neuromelanin cells in restless legs syndrome. Neurology 62:1563–1567

Connor JR, Wang XS, Allen RP, Beard JL, Wiesinger JA, Felt BT, Earley CJ (2009) Altered dopaminergic profile in the putamen and substantia nigra in Restless Leg Syndrome. Brain 132:2403–2412

Dhaenens CM, Burnouf S, Simonin C, Van Brussel E, Duhamel A, Defebvre L, Duru C, Vuillaume I, Cazeneuve C, Charles P, Maison P, Debruxelles S, Verny C, Gervais H, Azulay JP, Tranchant C, Bachoud-Levi AC, Dürr A, Buée L, Krystkowiak P, Sablonnière B, Blum D (2009) A genetic variation in the ADORA2A gene modifies age at onset in Huntington's disease. Neurobiol Dis 35:474–476

DeLong MR, Wichmann T (2007) Circuits and circuit disorders of the basal ganglia. Arch Neurol 64:20–24

Earley CJ, Connor JR, Beard JL, Malecki EA, Epstein DK, Allen RP (2000) Abnormalities in CSF concentrations of ferritin and transferrin in restless legs syndrome. Neurology 54:1698–1700

Earley CJ, Barker PB, Horska A, Allen RP (2006) MRI-determined regional brain iron concentrations in early- and late-onset restless legs syndrome. Sleep Med 7:459–461

Erikson KM, Jones BC, Beard JL (2000) Iron deficiency alters dopamine transporter functioning in rat striatum. J Nutr 130:2831–2837

Erikson KM, Jones BC, Hess EJ, Zhang Q, Beard JL (2001) Iron deficiency decreases dopamine D1 and D2 receptors in rat brain. Pharmacol Biochem Behav 69:409–418

Factor S, Mark MH, Watts R, Struck L, Mori A, Ballerini R, Sussman NM, Istradefylline 6002-US-007 Study Group (2010) A long-term study of istradefylline in subjects with fluctuating Parkinson's disease. Parkinsonism Relat Disord 16:423–426

Fernandez HH, Greeley DR, Zweig RM, Wojcieszek J, Mori A, Sussman NM, 6002-US-051 Study Group (2010) Istradefylline as monotherapy for Parkinson disease: results of the 6002-US-051 trial. Parkinsonism Relat Disord 16:16–20

Ferré S, Fuxe K, von Euler G, Johansson B, Fredholm BB (1992) Adenosine-dopamine interactions in the brain. Neuroscience 51:501–512

Ferré S, Popoli P, Gimenez-Llort L, Rimondini R, Müller CE, Stromberg I, Ogren SO, Fuxe K (2001) Adenosine/dopamine interaction: implications for the treatment of Parkinson's disease. Parkinsonism Relat Disord 7:235–241

Ferré S, Karcz-Kubicha M, Hope BT, Popoli P, Burgueno J, Casado V, Fuxe K, Lluis C, Goldberg SR, Franco R, Ciruela F (2002) Synergistic interaction between adenosine A_{2A} and glutamate $mGlu_5$ receptors: Implications for striatal neuronal function. 99:11940–11945

Ferré S, Ciruela F, Woods AS, Lluis C, Franco R (2007) Functional relevance of neurotransmitter receptor heteromers in the central nervous system. Trends Neurosci 30:440–446

Ferré S, Quiroz C, Woods AS, Cunha R, Popoli P, Ciruela F, Lluis C, Franco R, Azdad K, Schiffmann SN (2008) An update on adenosine A2A-dopamine D2 receptor interactions. Implications for the function of G protein-coupled receptors. Curr Pharm Des 14:1468–1474

Ferré S, Baler R, Bouvier M, Caron MG, Devi LA, Durroux T, Fuxe K, George SR, Javitch JA, Lohse MJ, Mackie K, Milligan G, Pfleger KDG, Pin J-P, Volkow N, Waldhoer M, Woods AS, Franco R (2009) Building a new conceptual framework for receptor heteromers. Nat Chem Biol 5:131–134

Ferré S, Lluis C, Justinova Z, Quiroz C, Orru M, Navarro G, Canela EI, Franco R, Goldberg SR (2010) Adenosine-cannabinoid receptor interactions. Implications for striatal function. Br J Pharmacol 160:443–453

Fiorentini C, Busi C, Gorruso E, Gotti C, Spano P, Missale C (2008) Reciprocal regulation of dopamine D1 and D3 receptor function and trafficking by heterodimerization. Mol Pharmacol 74:59–69

Glass M, Dragunow M, Faull RL (2000) The pattern of neurodegeneration in Huntington's disease: a comparative study of cannabinoid, dopamine, adenosine and GABA(A) receptor alterations in the human basal ganglia in Huntington's disease. Neuroscience 97:505–519

Grondin R, Bédard PJ, Hadj Tahar A, Grégoire L, Mori A, Kase H (1999) Antiparkinsonian effect of a new selective adenosine A2A receptor antagonist in MPTP-treated monkeys. Neurology 52:1673–1677

Gulyani S, Earley CJ, Camandola S, Maudsley S, Ferré S, Mughal MR, Martin B, Cheng A, Gleichmann M, Jones BC, Allen RP, Mattson MP (2009) Diminished iron concentrations increase adenosine A(2A) receptor levels in mouse striatum and cultured human neuroblastoma cells. Exp Neurol 215:236–242

Håkansson K, Galdi S, Hendrick J, Snyder G, Greengard P, Fisone G (2006) Regulation of phosphorylation of the GluR1 AMPA receptor by dopamine D2 receptors. J Neurochem 96:482–488

Hauser RA, Hubble JP, Truong DD, Istradefylline US-001 Study Group (2003) Randomized trial of the adenosine A(2A) receptor antagonist istradefylline in advanced PD. Neurology 61:297–303

Hettinger BD, Lee A, Linden J, Rosin DL (1998) Ultrastructural localization of adenosine A2A receptors suggests multiple cellular sites for modulation of GABAergic neurons in rat striatum. J Comp Neurol 431:331–346

Higley MJ, Sabatini BL (2010) Competitive regulation of synaptic Ca2+ influx by D2 dopamine and A2A adenosine receptors. Nat Neurosci 13:958–966

Hillion J, Canals M, Torvinen M, Casado V, Scott R, Terasmaa A, Hansson A, Watson S, Olah ME, Mallol J, Canela EI, Zoli M, Agnati LF, Ibanez CF, Lluis C, Franco R, Ferré S, Fuxe K (2002) Coaggregation, cointernalization and codesensitization of adenosine A_{2A} receptors and dopamine D_2 receptors. J Biol Chem 277:18091–18097

James S, Richardson PJ (1993) The subcellular distribution of [3H]-CGS 21680 binding sites in the rat striatum: copurification with cholinergic nerve terminals. Neurochem Int 23:115–122

Jin S, Johansson B, Fredholm BB (1993) Effects of adenosine A1 and A2 receptor activation on electrically evoked dopamine and acetylcholine release from rat striatal slices. J Pharmacol Exp Ther 267:801–808

Justinova Z, Ferré S, Redhi GH, Mascia P, Stroik J, Quarta D, Yasar S, Muller CE, Franco R, Goldberg SR (2010) Reinforcing and neurochemical effects of cannabinoid CB1 receptor agonists, but not cocaine, are altered by an adenosine A2A receptor antagonist. Addict Biol 16:405–415

Kachroo A, Orlando LR, Grandy DK, Chen JF, Young AB, Schwarzschild MA (2005) Interactions between metabotropic glutamate 5 and adenosine A2A receptors in normal and parkinsonian mice. J Neurosci 25:10414–10419

Kanda T, Jackson MJ, Smith LA, Pearce RK, Nakamura J, Kase H, Kuwana Y, Jenner P (2000) Combined use of the adenosine A(2A) antagonist KW-6002 with L-DOPA or with selective D1 or D2 dopamine agonists increases antiparkinsonian activity but not dyskinesia in MPTP-treated monkeys. Exp Neurol 162:321–327

Kull B, Ferré S, Arslan G, Svenningsson P, Fuxe K, Owman C, Fredholm BB (1999) Reciprocal interactions between adenosine A_{2A} and dopamine D_2 receptors in CHO cells co-transfected with the two receptors. Biochem Pharmacol 58:1035–1045

Lerner TN, Horne EA, Stella N, Kreitzer AC (2010) Endocannabinoid signaling mediates psychomotor activation by adenosine A2A antagonists. J Neurosci 30:2160–2164

LeWitt PA, Guttman M, Tetrud JW, Tuite PJ, Mori A, Chaikin P, Sussman NM, 6002-US-005 Study Group (2008) Adenosine A2A receptor antagonist istradefylline (KW-6002) reduces "off" time in Parkinson's disease: a double-blind, randomized, multicenter clinical trial (6002-US-005). Ann Neurol 63:295–302

Lozoff B, Georgieff MK (2006) Iron deficiency and brain development. Semin Pediatr Neurol 13:158–165

Martire A, Tebano MT, Chiodi V, Ferreira SG, Cunha RA, Köfalvi A, Popoli P (2011) Pre-synaptic adenosine A2A receptors control cannabinoid CB1 receptor-mediated inhibition of striatal glutamatergic neurotransmission. J Neurochem 116:273–280

Mizuno Y, Hasegawa K, Kondo T, Kuno S, Yamamoto M, Japanese Istradefylline Study Group (2010) Clinical efficacy of istradefylline (KW-6002) in Parkinson's disease: a randomized, controlled study. Mov Disord 25:1437–1443

Moresco RM, Todde S, Belloli S, Simonelli P, Panzacchi A, Rigamonti M, Galli-Kienle M, Fazio F (2005) In vivo imaging of adenosine A2A receptors in rat and primate brain using [11C] SCH442416. Eur J Nucl Med Mol Imaging 32:405–413

Morissette M, Goulet M, Soghomonian JJ, Blanchet PJ, Calon F, Bédard PJ, Di Paolo T (1997) Preproenkephalin mRNA expression in the caudate-putamen of MPTP monkeys after chronic treatment with the D2 agonist U91356A in continuous or intermittent mode of administration: comparison with L-DOPA therapy. Brain Res Mol Brain Res 49:55–62

Navarro G, Ferré S, Cordomi A, Moreno E, Mallol J, Casadó V, Cortés A, Hoffmann H, Ortiz J, Canela EI, Lluís C, Pardo L, Franco R, Woods AS (2010) Interactions between intracellular domains as key determinants of the quaternary structure and function of receptor heteromers. J Biol Chem 285:27346–27359

Nelson C, Erikson K, Piñero DJ, Beard JL (1997) In vivo dopamine metabolism is altered in iron-deficient anemic rats. J Nutr 127:2282–2288

Nguyen HP, Kobbe P, Rahne H, Wörpel T, Jäger B, Stephan M, Pabst R, Holzmann C, Riess O, Korr H, Kántor O, Petrasch-Parwez E, Wetzel R, Osmand A, von Hörsten S (2006) Behavioral abnormalities precede neuropathological markers in rats transgenic for Huntington's disease. Hum Mol Genet 15:3177–3194

Nishi A, Liu F, Matsuyama S, Hamada M, Higashi H, Nairn AC, Greengard P (2003) Metabotropic mGlu5 receptors regulate adenosine A2A receptor signaling. Proc Natl Acad Sci U S A 100:1322–1327

Nordlander NB (1954) Restless Legs. Brit J Phys Med 17:160–162

Obeso JA, Rodríguez-Oroz MC, Rodríguez M, Arbizu J, Giménez-Amaya JM (2002) The basal ganglia and disorders of movement: pathophysiological mechanisms. News Physiol Sci 17:51–55

Orrú M, Quiroz C, Guitart X, Ferré S (2011a) Pharmacological evidence for different populations of postsynaptic adenosine A2A receptors in the rat striatum. Neuropharmacology 61:967–974

Orrú M, Bakešová J, Brugarolas M, Quiroz C, Beaumont V, Goldberg SR, Lluís C, Cortés A, Franco R, Casadó V, Canela EI, Ferré S (2011b) Striatal Pre- and Postsynaptic Profile of Adenosine A(2A) Receptor Antagonists. PLoS One 6:e16088

Orrú M, Zanoveli JM, Quiroz C, Nguyen HP, Guitart X, Ferré S (2011c) Functional changes in postsynaptic adenosine A(2A) receptors during early stages of a rat model of Huntington disease. Exp Neurol 232:76–80

Popoli P, Betto P, Reggio R, Ricciarello G (1995) Adenosine A2A receptor stimulation enhances striatal extracellular glutamate levels in rats. Eur J Pharmacol 287:215–217

Popoli P, Pezzola A, Torvinen M, Reggio R, Pintor A, Scarchili L, Fuxe K, Ferré S (2001) The selective mGlu5 receptor agonist CHPG inhibits quinpirole-induced turning in 6-hydroxydopamine-lesioned rats and modulates the binding characteristics of dopamine D_2 receptors in the rat striatum: Interactions with adenosine A_{2A} receptors. Neuropsychopharmacology 25:505–513

Popoli P, Blum D, Martire A, Ledent C, Ceruti S, Abbracchio MP (2007) Functions, dysfunctions and possible therapeutic relevance of adenosine A2A receptors in Huntington's disease. Prog Neurobiol 81:331–348

Preston Z, Lee K, Widdowson L, Freeman TC, Dixon AK, Richardson PJ (2000) Adenosine receptor expression and function in rat striatal cholinergic interneurons. Br J Pharmacol 130: 886–890

Quiroz C, Lujan R, Uchigashima M, Simoes AP, Lerner TN, Borycz J, Kachroo A, Canas PM, Orru M, Schwarzschild MA, Rosin DL, Kreitzer AC, Cunha RA, Watanabe M, Ferré S (2009) Key modulatory role of presynaptic adenosine A_{2A} receptors in cortical neurotransmission to the striatal direct pathway. ScientificWorldJournal 9:1321–1344

Quiroz C, Pearson V, Gulyani S, Allen R, Earley C, Ferré S (2010) Up-regulation of striatal adenosine A_{2A} receptors with iron deficiency in rats. Effects on locomotion and cortico-striatal neurotransmission. Exp Neurol 224:292–298

Ramlackhansingh AF, Bose SK, Ahmed I, Turkheimer FE, Pavese N, Brooks DJ (2011) Adenosine 2A receptor availability in dyskinetic and nondyskinetic patients with Parkinson disease. Neurology 76:1811–1816

Ravikumar B, Vacher C, Berger Z (2004) Inhibition of mTOR induces autophagy and reduces toxicity of polyglutamine expansions in fly and mouse models of Huntington disease. Nat Genet 36:585–595

Rodrigues RJ, Alfaro TM, Rebola N, Oliveira CR, Cunha RA (2005) Co-localization and functional interaction between adenosine A(2A) and metabotropic group 5 receptors in glutamatergic nerve terminals of the rat striatum. J Neurochem 92:433–441

Rosin DL, Robeva A, Woodard RL, Guyenet PG, Linden J (1998) Immunohistochemical localization of adenosine A2A receptors in the rat central nervous system. J Comp Neurol 401:163–186

Schiffmann SN, Fisone G, Moresco R, Cunha R, Ferré S (2007) Adenosine A2A receptors and basal ganglia physiology. Prog Neurobiol 83:277–292

Shen HY, Coelho JE, Ohtsuka N, Canas PM, Day YJ, Huang QY, Rebola N, Yu L, Boison D, Cunha RA, Linden J, Tsien JZ, Chen JF (2008) A critical role of the adenosine A2A receptor in extrastriatal neurons in modulating psychomotor activity as revealed by opposite phenotypes of striatum and forebrain A2A receptor knock-outs. J Neurosci 28:2970–2975

Soria G, Castañé A, Ledent C, Parmentier M, Maldonado R, Valverde O (2004) Adenosine A2A receptors are involved in physical dependence and place conditioning induced by THC. Eur J Neurosci 20:2203–2213

Taherzadeh-Fard E, Saft C, Wieczorek S, Epplen JT, Arning L (2010) Age at onset in Huntington's disease: replication study on the associations of ADORA2A, HAP1 and OGG1. Neurogenetics 11:435–439

Tanganelli S, Sandager-Nielsen K, Ferraro L, Antonelli T, Kehr J, Franco R, Ferré S, Agnati LF, Fuxe K, Scheel-Krüger J (2004) Striatal plasticity at the network level. Focus on adenosine A2A and D2 interactions in models of Parkinson's disease. Parkinsonism Relat Disord 10:273–280

Tarditi A, Camurri A, Varani K, Borea PA, Woodman B, Bates G, Cattaneo E, Abbracchio MP (2006) Early and transient alteration of adenosine A_{2A} receptor signaling in a mouse model of Huntington disease. Neurobiol Dis 23:44–53

Tebano MT, Martire A, Chiodi V, Pepponi R, Ferrante A, Domenici MR, Frank C, Chen JF, Ledent C, Popoli P (2009) Adenosine A2A receptors enable the synaptic effects of cannabinoid CB1 receptors in the rodent striatum. J Neurochem 110:1921–1930

Tergau F, Wische S, Paulus W (1999) Motor system excitability in patients with Restless Legs Syndrome. Neurology 52:1060–1063

Tozzi A, de Iure A, Di Filippo M, Tantucci M, Costa C, Borsini F, Ghiglieri V, Giampà C, Fusco FR, Picconi B, Calabresi P (2011) The distinct role of medium spiny neurons and cholinergic interneurons in the D_2/A_2A receptor interaction in the striatum: implications for Parkinson's disease. J Neurosci 31:1850–1862

Unger EL, Wiesinger JA, Hao L, Beard JL (2008) Dopamine D2 receptor expression is altered by changes in cellular iron levels in PC12 cells and rat brain tissue. J Nutr 138:2487–2494

Varani K, Rigamonti D, Sipione S, Camurri A, Borea PA, Cattabeni F, Abbracchio MP, Cattaneo E (2001) Aberrant amplification of A_{2A} receptor signaling in striatal cells expressing mutant huntingtin. FASEB J 15:1245–1247

Varani K, Abbrachio MP, Cannella M, Cislaghi G, Giallonardo P, Mariotti C, Cattabriga E, Borea PA, Squitieri F, Cattaneo E (2003) Aberrant A_{2A} receptor function in peripheral blood cells in Huntington's disease. FASEB J 17:2148–2150

Varani K, Vincenzi F, Tosi A, Gessi S, Casetta I, Granieri G, Fazio P, Leung E, MacLennan S, Granieri E, Borea PA (2010) A2A adenosine receptor overexpression and functionality, as well as TNF-alpha levels, correlate with motor symptoms in Parkinson's disease. FASEB J 24:587–598

von Hörsten S, Schmitt I, Nguyen HP, Holzmann C, Schmidt T, Walther T, Bader M, Pabst R, Kobbe P, Krotova J, Stiller D, Kask A, Vaarmann A, Rathke-Hartlieb S, Schulz JB, Grasshoff U, Bauer I, Vieira-Saecker AM, Paul M, Jones L, Lindenberg KS, Landwehrmeyer B, Bauer A, Li XJ, Riess O (2003) Transgenic rat model of Huntington's disease. Hum Mol Genet 12:617–624

Weinberg J, Dallman PR, Levine S (1980) Iron deficiency during early development in the rat: behavioral and physiological consequences. Pharmacol Biochem Behav 12:493–502

Wellington CL, Leavitt BR, Hayden MR (2000) Huntington disease: new insights on the role of huntingtin cleavage. J Neural Transm Suppl 58:1–17

Yang W, Dunlap JR, Andrews RB, Wetzel R (2002) Aggregated polyglutamine peptides delivered to nuclei are toxic to mammalian cells. Hum Mol Genet 11:2905–2917

Yao L, Fan P, Jiang Z, Mailliard WS, Gordon AS, Diamond I (2003) Addicting drugs utilize a synergistic molecular mechanism in common requiring adenosine and Gi-beta gamma dimers. Proc Natl Acad Sci U S A 100:14379–14384

Youdim MB, Ben-Shachar D, Ashkenazi R, Yehuda S (1983) Brain iron and dopamine receptor function. Adv Biochem Psychopharmacol 37:309–321

Zeng BY, Pearce RK, MacKenzie GM, Jenner P (2000) Alterations in preproenkephalin and adenosine-2a receptor mRNA, but not preprotachykinin mRNA correlate with occurrence of dyskinesia in normal monkeys chronically treated with L-DOPA. Eur J Neurosci 12:1096–1104

Part III
Adenosine Function and Dysfunction in Brain Health and Disease

Chapter 10
Sleep and Adenosine: Human Studies

Tarja Porkka-Heiskanen

Abstract Adenosine, through A_1 receptors, acts as an inhibitory neuromodulator in the central nervous system. It is also directly linked to the energy metabolism of cells. It codes for energy depletion: adenosine levels increase as sign of problems in energy balance. Extracellular adenosine levels increase as response to prolonged wakefulness, starting locally in the basal forebrain area and later in the cortex. The increased extracellular adenosine concentrations increase signaling through A_1 receptors on the waking-active cells, inhibit their firing and thus promote sleep.

Human studies also indicate a connection between adenosine metabolism and sleep regulation. There are functional polymorphisms in adenosine transporters and adenosine deaminase, which will increase extracellular adenosine concentrations. Individuals with the adenosine deaminase variant have more and deeper slow wave sleep as well as more alpha activity in their waking EEG. Interestingly, their vigilance level and performance during waking are compromised.

Caffeine, a nonspecific adenosine receptor blocker, modulates effectively vigilance level and sleep in humans. Sensitivity to caffeine is modulated by an adenosine 2a receptor polymorphism: carriers of the different variants respond differently, both behaviorally and in EEG records, to caffeine. Prolonged wakefulness deteriorated performance more in caffeine-sensitive individuals, but their performance also benefitted more from caffeine intake.

Sleep problems predispose to depression, and as adenosine is one of the regulators of sleep, changes in adenosine metabolism can be expected also in depressed patients. Indeed, a suggestive finding is that a polymorphism in the adenosine transporter gene SLC29A3 was associated with depressive disorder. Interestingly, this association was found only in women but not in men.

T. Porkka-Heiskanen (✉)
Institute of Biomedicine, Biomedicum Helsinki,
University of Helsinki, Helsinki 00014, Finland
e-mail: porkka@cc.helsinki.fi

S. Masino and D. Boison (eds.), *Adenosine: A Key Link between Metabolism and Brain Activity*, DOI 10.1007/978-1-4614-3903-5_10,

In summary, both animal and human experiments evidence that low brain adenosine concentrations are associated with high vigilance levels and wakefulness, while increases in adenosine levels associate with increased sleep propensity and promotion of sleep.

Keywords Adenosine transporters • Depression • Microdialysis • Sleep • Sleep deprivation

10.1 Adenosine and Sleep: General Concept

10.1.1 What Is Sleep and How Can We Define It?

Sleep is more than absence of wakefulness: it is an active brain state with physiological function(s) and accurate regulation. Sleep as a behavioral state is featured by lack of consciousness, decreased reactivity for environmental signals, and characteristic sleep posture (in humans, lying) as well as closed eye lids. Sleep and wakefulness as behavioral states are mutually exclusive: a person is either sleeping or awake.

This is the classic view when the state of sleep is defined through electroencephalography (EEG), which is the gold standard to define sleep (see Box). Sleep initiation is characterized through the emergence of slower waves in the EEG recording. Based on EEG, the definition of sleep gets also more complicated. The definition is made by subjective evaluation of the wave forms and the appearance/disappearance of certain features in the recording that is obtained simultaneously from several electrodes attached to head.

Usually there is a short transition state from both wakefulness to sleep and also between sleep states where the definition of the state is not self-evident. After the transition, the definition of the vigilance state is regarded reliable and conclusive, including the entire, global brain. Recently, however, convincing evidence has shown that this is not entirely true: during behavioral sleep, waking EEG waveforms as well as neuronal activity can be recorded at local cortical loci (Nir et al. 2011; Nobili et al. 2011; Rector et al. 2009), providing evidence that separate parts of the cortex “sleep” and “wake” within the opposite behavioral state. This brings forth a need to define global sleep and local sleep separately. In this thinking global sleep would represent the classical, behaviorally evident state of sleep while local sleep could be defined as appearance of slow waves at local EEG recording sites and corresponding to local metabolic events. Sleep thus appears to have both local and global aspects, which most probably have at least partially different regulatory mechanisms. Adenosine may play a role in regulation of both of these aspects.

Measurement of Sleep Using EEG

Vigilance States

Waking (W) and REM sleep (REM) are characterized by a high-frequency but low-amplitude waveform, similarly. REM and W can be distinguished by muscle tone; muscle tone is high in W and very low in REM.

Non-REM (NREM) sleep is divided into stages N1–3, with deepening of sleep from stage 1 to N3. N3 is also referred to as *slow wave sleep*, or *slow-wave activity* (SWA).

Frequency Analysis

The EEG signal can be analyzed by dividing the signal into frequency bins, usually in range 0.1–25 Hz.

The low frequencies (0.1–4 Hz) are called *delta waves*, and they are typical for N3. Frequency from 4 to 8 Hz is called *theta frequency*, which is typical for REM sleep, but is expressed also during W. In rodents theta activity during W corresponds to exploratory behavior.

The higher frequencies are typical for W.

The signal from the frequency analysis can, through mathematical processing, be further developed to determine "power." Power can be expressed either as total power throughout all frequencies or as power in a specific frequency band, e.g., delta power. The delta power is used to describe the intensity of sleep.

10.2 Sleep Homeostasis

One of the key features of sleep is its homeostatic regulation: the duration and intensity of previous waking period is reflected as sleep duration and intensity during the subsequent sleep period. This means that in the brain there is a mechanism that can keep count of the duration (and intensity) of the waking period, i.e., keep that information and use it to modulate subsequent sleep. One suggestion as to the nature of this process is the concept of "sleep substance"—a molecule that would accumulate during waking and when in appropriate concentration, would force the brain to the state of sleep.

The oscillations, referred to as SWA, reflect sleep need and sleep intensity. The amount of SWA is homeostatically regulated. It is enhanced after sleep loss and declines during sleep. Animal studies suggested that sleep need is also genetically controlled (Cirelli 2002), yet the molecular level mechanisms remain unknown.

10.2.1 Regulation of Global Sleep

Waking is actively promoted by specific brain nuclei in the brainstem, the basal forebrain, and the hypothalamus. These "waking nuclei" secrete neurotransmitters—cholinergic, noradrenergic, serotonergic, histaminergic, and orexinergic—that reach cortex through long projections from the nuclei. Both the anatomy and the function of these nuclei are strikingly similar: They are most active during wakefulness and decrease their activity during slow wave sleep. During rapid eye movement (REM) sleep their activity decreases further, except for the cholinergic cells, which during REM sleep are activated to the same level as in wakefulness (Stenberg 2007). When the waking promoting nuclei are active, sleep cannot be initiated, meaning that one important aspect in regulation of sleep is the silencing of the activity of the waking nuclei.

Another important regulating system of the global state of sleep is the GABAergic sleep-active cells in the hypothalamus (Sherin et al. 1996). Their activity is higher during sleep than during wakefulness (Szymusiak et al. 1998), and their lesion decreases sleep (Lu et al. 2000).

Thus, adenosine at this level could promote sleep by either inhibiting the waking active cells through A_1 receptors (Basheer et al. 2004; Porkka-Heiskanen and Kalinchuk 2011) or activating sleep active cells, presumably through A_{2A} receptors, or disinhibiting the sleep active cells (Strecker et al. 2000).

10.2.2 Sleep Regulation at a Local Level

The local aspect of sleep has particularly been addressed in experiments where local brain areas have been stimulated during waking either by natural stimuli or instrumentally, forcing these areas to excessive neural activity as compared to neighboring brain areas or respective brain areas in the other brain half (Huber et al. 2006; Kattler et al. 1994). During subsequent sleep, more SWA has then been measured over the stimulated brain area. These experiments show that the production of slow waves during sleep is preceded by local neuronal activation during waking. In rodents the activation that induces more SWA is specific to high frequency theta range and is induced through activation of the NMDA receptors (Wigren et al. 2007).

10.2.2.1 Neuronal Activation and Energy Consumption

Neuronal activity consumes energy, which in neurons is provided either as ATP or lactate (Magistretti 2006). During wakefulness neuronal activity is at higher level that during sleep, and thus also energy consumption is increased, although local variations are large. Extracellular ATP levels increase during sleep (Dworak et al. 2010), possibly reflecting decreased consumption of energy substrates during sleep.

10.2.3 Function of Sleep

Present evidence suggests that the functions of sleep are related to synaptic down scaling (Liu et al. 2010) and memory and learning (Diekelmann and Born 2010; Stickgold and Walker 2007). It has been suggested that slow oscillations at delta frequency range may be a sign of synaptic plasticity that takes place during sleep (Tononi and Cirelli 2006). Studies on molecular mechanisms of sleep homeostasis have introduced adenosine as a key player in regulation of sleep recovery (see below). As adenosine signals for shortage of energy, we have proposed that its emergence in specific brain areas is a warning signals that, when reaching a certain level, will initiate sleep. It can be noted that these theories are not mutually exclusive—sleep can, and most probably does serve more than one physiological function.

10.3 Adenosine

Adenosine triphosphate (ATP) is the core molecule of energy metabolism through which the majority of energy is introduced for cell metabolism. The molecule consists of adenosine core linked to one to three phosphate groups, which represent the energy contents of the molecule. The maximum energy content is available when three phosphates are bound to adenosine, while adenosine without phosphates is void of energy. Thus, increased adenosine concentration in the cell (and extracellularly, through equilibrative transporters) represents a state of low energy availability and signals of bad energy balance.

Adenosine, working through A_1 receptors, is an inhibitory neurotransmitter. This route forms an ultrashort negative feedback loop: in an actively functioning cell, when energy level declines and adenosine concentration increases, there is increased signaling through A_1 receptors on the cell surface slowing down the activity of the cell. This simple regulatory mechanism is functional at local level and is thought to form the basis of homeostatic sleep regulation.

Animal experiments have shown that indeed, adenosine levels increase in the brain during prolonged wakefulness, particularly in the basal forebrain cholinergic area as well as in cortex (Porkka-Heiskanen et al. 1997, 2000). On the other hand, experimental induction of energy depletion using molecules that disrupt the mitochondrial electron transfer in the basal forebrain induces sleep and increase adenosine levels (Kalinchuk et al. 2003).

The results of these experiments can be interpreted to provide evidence for a role for energy depletion and adenosine in the regulation of sleep at both global and local levels, through inhibition of the waking active cholinergic cells in the basal forebrain as well as by contributing to development of sleep propensity at local level in the cortex.

Sleep in most, if not all, species appears to be influenced by adenosine metabolism. So far direct experimental evidence can be introduced from rodents, cats, and fruit flies (Kalinchuk et al. 2003; Porkka-Heiskanen et al. 1997; Shaw et al. 2000).

In humans many targets in adenosine's metabolism modify sleep and wakefulness—and perhaps most importantly, the adenosine receptor antagonist, caffeine.

10.3.1 Enzymes of Adenosine Metabolism

10.3.1.1 Adenosine Deaminase

Adenosine deaminase is an enzyme that metabolizes adenosine to inosine removing it from the energy circulation. Increases in adenosine deaminase activity are thus expected to decrease adenosine levels.

Animal studies have shown that although sleep deprivation does not affect the activity of adenosine deaminase (Mackiewicz et al. 2003), it has a significant circadian variation (Chagoya de Sanchez et al. 1993; Mackiewicz et al. 2003).

However, in humans the c.22 G>A polymorphism (rs73598374) in adenosine deaminase is a physiologically active variant, reducing the conversion of adenosine to inosine. Consequent increases in adenosine concentrations in brain are likely to induce increases in SWA during sleep. In a human experimental study it was found that an individual with the G>A genotype had higher amplitude and prevalence of EEG delta oscillations in N3 sleep in than an individual with the G>G genotype (Retey et al. 2005), consistent with previous results from animal work that higher adenosine levels would promote sleep (Porkka-Heiskanen et al. 1997). Later work confirmed that in individuals with the G>A polymorphism, EEG low oscillations had higher amplitude. Importantly, during waking these individuals showed higher alpha activity in EEG recordings, and they were more fatigued and sleepy, and their attention level was lower. Also, their saliva alpha-amylase concentration was elevated (Bachmann et al. 2012). These observations further support the view that adenosine levels are important regulators of sleep, sleep propensity, and waking behavior. A recent epidemiological study found that individuals carrying the G>A variant of the adenosine deaminase gene have higher sleep efficacy and more REM sleep compared to the G>G variant carrier, but this effect is evident only if they use caffeine (Mazzotti et al. 2011).

This observation would be in line with an earlier study where the responses of caffeine-sensitive and caffeine-insensitive subjects for sleep restriction were compared. Performance was more impaired in caffeine-sensitive individuals, but when both groups ingested caffeine during the sleep restriction, the caffeine-sensitive individuals' performance improved more (Retey et al. 2006). However, the caffeine sensitivity observed in this study was later associated with an adenosine 2A receptor polymorphism (Retey et al. 2007).

10.3.1.2 Adenosine Kinase

Adenosine kinase metabolizes adenosine to adenosine monophosphate, returning it to the energy circulation. Thus, the activity/increased concentration of this enzyme is also expected to reduce the concentration of adenosine.

No direct studies on the effect of adenosine kinase modulation on human sleep are available. Studies on animals show that modulation of the basal level of adenosine has effects on sleep. Also adenosine kinase shows clear diurnal variation (Alanko et al. 2003a; Chagoya de Sanchez et al. 1993), but no activity or increased expression during sleep deprivation (Alanko et al. 2003a; Mackiewicz et al. 2003). However, genetically engineered mice with overexpression of the cytoplasmic isoform of the adenosine kinase enzyme (but lacking the nuclear isoform) evidence a reduction in EEG low frequency power in all vigilance states. Also theta activity was reduced during REM sleep and waking. The mice spent also more time in waking. In addition, sleep homeostasis was affected as evidenced by more modest increase in slow wave activity during recovery sleep in the genetically engineered mice (Palchykova et al. 2010). Also these studies are consistent with the idea that lower adenosine concentrations promote an increase in wakefulness.

10.3.2 Adenosine Receptors

10.3.2.1 Caffeine

Caffeine, a nonselective adenosine receptor antagonist, is a powerful modulator of sleep also in humans. The common effects are prolonged sleep latency, reduced total sleep time, and decreased slow wave sleep. Subjective sleep quality is also worsened.

Caffeine ingested either in the morning or shortly before bedtime induces dose-dependent decrease in total sleep time, increase in sleep latency, and decrease in slow wave sleep (Karacan et al. 1976; Landolt et al. 1995b; Nicholson and Stone 1980), as well as decreased power in delta frequency range, but increased power in spindle frequency range (Landolt et al. 1995a). The caffeine-induced EEG changes in non-REM sleep were evident in both rested and sleep deprived subjects (Landolt et al. 1995b). Previous caffeine use also modulates the effects of caffeine on sleep: low consumers express more prolonged sleep latency, shortened total sleep time, and disturbances in different sleep phases than heavy users (Colton et al. 1968; Curatolo and Robertson 1983).

10.3.2.2 Adenosine A_{2A} Receptor

The inhibitory A_1 receptor has traditionally regarded the main target for caffeine: by preventing the action of A_1-mediated neural inhibition, cortical activity would be increased. Recent research has, however, provided strong evidence that the effects of caffeine, in fact, would be predominantly mediated through A_{2A} receptors (Huang et al. 2005). Studies in humans also give support to this view.

Humans have different sensitivity for the effect of caffeine: individuals who rate themselves as caffeine-sensitive also report reduced sleep quality as compared to those who report themselves as caffeine-insensitive (Retey et al. 2007). Indirect evidence suggests that this sensitivity is at least partially genetically predicted by a variation in the adenosine A_{2A} receptor gene: the c.1083T4C T>C polymorphism: the distribution of distinct c.1083T4C genotypes of the adenosine A_{2A} receptor gene differs between caffeine-sensitive and -insensitive individuals (Retey et al. 2007). EEG recordings provided objective differences in brain electrical activity between the variants: caffeine induced a lesser increase in EEG power density in the beta frequency in individuals carrying the T/T variant as compared to those carrying either the T/C or C/C variants (Retey et al. 2007).

Interestingly, caffeine sensitivity is reflected in the individual responses for prolonged wakefulness. In one experiment, sleep deprivation impaired performance (PVT) more in caffeine-sensitive than in caffeine-insensitive men. This difference was counteracted by caffeine, particularly in the caffeine-sensitive individuals (Retey et al. 2006).

10.3.3 Adenosine Transporters

Adenosine transporters regulate the balance between intra- and extracellular adenosine concentrations. Equilibrative transporters carry adenosine across the cell membrane according to the concentration gradient while concentrative transporters concentrate adenosine intracellularly. Only few studies have been performed on the relationships between sleep and adenosine transporters. One animal study indicated decreased adenosine transport by an equilibrative transporter (ENT1) in the basal forebrain during prolonged wakefulness, which may have contributed to the increased extracellular adenosine level in the basal forebrain during sleep deprivation (Alanko et al. 2003b). Another study found decreased expression of a concentrative transporter (CNT2) in the cortex after sleep deprivation (Guillen-Gomez et al. 2004).

10.3.3.1 Adenosine and Depression

Major depression is one of the most frequent psychiatric disorders with a remarkable component of heritability: 25–29 % for males and 42–49 % for female (Kendler et al. 2006; Orstavik et al. 2007). Sleep disturbances among these patients are frequent, including reduced amount of slow wave sleep, increased REM sleep amount and shortened REM sleep latency (Berger et al. 2003; Mendlewicz 2009). Also, early morning awakenings are characteristic of depression (Kiloh and Garside 1965). In a longitudinal study conducted on twins we showed that disturbed sleep predisposes to depression: subjective quality of sleep and life satisfaction (a commonly used indicator of depression) were assessed in both mono- and dizygotic twin pairs. Six years later the study was repeated. Those who reported problems with sleep at the earlier time point had elevated risk for depression measured at the later time point, but those who were depressed at the earlier time point did not have elevated risk for sleep problems at the later time point, suggesting a one-way causal relationship between sleep problems and elevated risk for depression (Paunio et al. 2009). As adenosine is one of the core regulators of sleep, we became interested in the possible role for adenosine in development of depression.

In a study conducted on 1,423 human individuals, polymorphisms in adenosine-related genes were associated with sleep disturbances and depression. Polymorphism rs12256138 in the adenosine transporter gene SLC29A3 was associated with depressive disorder. Interestingly, this association was found only in women but not in men (Gass et al. 2010). This may be one of the factors contributing to the fact that depressive disorder is more prevalent in women than in men.

SLC29A3 encodes equilibrative nucleoside transporter ENT3. The gene is expressed in multiple tissues, and the protein localizes intracellularly and may be involved in nucleoside transport from the lysosomes (Baldwin et al. 2005). As the function of this polymorphism presently remains to be determined, further speculation of the mechanism by which it could contribute to induction of depression remains futile.

10.3.4 Other Human Studies on Adenosine and Sleep

In vivo microdialysis has proved an excellent method to probe extracellular adenosine concentrations in behaving animals. These studies have established the role of increased extracellular adenosine concentrations in the basal forebrain in the regulation of sleep homeostasis. However, this invasive technique is rarely available in human studies. However, in some cases microdialysis probes have been implanted in human brain as part of diagnostic or curative treatments, and in some cases samples for adenosine measurements have also been collected.

One early study, conducted on trauma patients, showed that during jugular venous oxygen desaturation interstitial brain adenosine and lactate levels increased up to threefold, experimentally establishing a relationship between oxygen desaturation and adenosine concentrations in the brain and suggesting that adenosine is increased during periods of enhanced glycolytic metabolism (Bell et al. 2001).

In another study, deep electrodes with microdialysis probes were installed into different brain areas of epileptic patients (amygdala, hippocampus, cortex). Microdialysis samples were collected for adenosine analysis during sleep, waking, and prolonged wakefulness (40 h). No increases of extracellular adenosine were found in any of the probes during prolonged wakefulness. Interestingly, the epileptic brain activity did not affect adenosine levels in these patients. The authors concluded that adenosine increases during prolonged wakefulness are local, as shown previously in animal studies (Zeitzer et al. 2006).

10.4 Conclusions

Both animal and human studies consistently support the view that adenosine has an important role in modulation of sleep: low brain adenosine concentrations are associated with high vigilance levels and wakefulness, while increases in adenosine levels associated with increased sleep propensity and promotion of sleep. The nature of adenosine as a ubiquitous energy metabolite and neuromodulator makes it difficult to evaluate whether adenosine modulates sleep directly at cellular level affecting neurons in its immediate vicinity, or whether it modulates global sleep-regulating neuronal networks. Experimental evidence can be produced to support both views.

References

Alanko L, Heiskanen S, Stenberg D, Porkka-Heiskanen T (2003a) Adenosine kinase and 5′-nucleotidase activity after prolonged wakefulness in the cortex and the basal forebrain of rat. Neurochem Int 42:449–454

Alanko L, Stenberg D, Porkka-Heiskanen T (2003b) Nitrobenzylthioinosine (NBMPR) binding and nucleoside transporter ENT1 mRNA expression after prolonged wakefulness and recovery sleep in the cortex and basal forebrain of rat. J Sleep Res 12:299–304

Bachmann V, Klaus F, Bodenmann S, Schafer N, Brugger P, Huber S, Berger W, Landolt HP (2012) Functional ADA polymorphism increases sleep depth and reduces vigilant attention in humans. Cereb Cortex 22(4):962–970

Baldwin SA, Yao SY, Hyde RJ, Ng AM, Foppolo S, Barnes K, Ritzel MW, Cass CE, Young JD (2005) Functional characterization of novel human and mouse equilibrative nucleoside transporters (hENT3 and mENT3) located in intracellular membranes. J Biol Chem 280:15880–15887

Basheer R, Strecker RE, Thakkar MM, McCarley RW (2004) Adenosine and sleep-wake regulation. Prog Neurobiol 73:379–396

Bell MJ, Robertson CS, Kochanek PM, Goodman JC, Gopinath SP, Carcillo JA, Clark RS, Marion DW, Mi Z, Jackson EK (2001) Interstitial brain adenosine and xanthine increase during jugular venous oxygen desaturations in humans after traumatic brain injury. Crit Care Med 29:399–404

Berger M, van Calker D, Riemann D (2003) Sleep and manipulations of the sleep-wake rhythm in depression. Acta Psychiatr Scand Suppl (418): 83–91

Chagoya de Sanchez V, Hernandez Munoz R, Suarez J, Vidrio S, Yanez L, Diaz Munoz M (1993) Day-night variations of adenosine and its metabolizing enzymes in the brain cortex of the rat–possible physiological significance for the energetic homeostasis and the sleep-wake cycle. Brain Res 612:115–121

Cirelli C (2002) How sleep deprivation affects gene expression in the brain: a review of recent findings. J Appl Physiol 92:394–400

Colton T, Gosselin RE, Smith RP (1968) The tolerance of coffee drinkers to caffeine. Clin Pharmacol Ther 9:31–39

Curatolo PW, Robertson D (1983) The health consequences of caffeine. Ann Intern Med 98:641–653

Diekelmann S, Born J (2010) The memory function of sleep. Nat Rev Neurosci 11:114–126

Dworak M, McCarley RW, Kim T, Kalinchuk AV, Basheer R (2010) Sleep and brain energy levels: ATP changes during sleep. J Neurosci 30:9007–9016

Gass N, Ollila HM, Utge S, Partonen T, Kronholm E, Pirkola S, Suhonen J, Silander K, Porkka-Heiskanen T, Paunio T (2010) Contribution of adenosine related genes to the risk of depression with disturbed sleep. J Affect Disord 126:134–139

Guillen-Gomez E, Calbet M, Casado J, de Lecea L, Soriano E, Pastor-Anglada M, Burgaya F (2004) Distribution of CNT2 and ENT1 transcripts in rat brain: selective decrease of CNT2 mRNA in the cerebral cortex of sleep-deprived rats. J Neurochem 90:883–893

Huang ZL, Qu WM, Eguchi N, Chen JF, Schwarzschild MA, Fredholm BB, Urade Y, Hayaishi O (2005) Adenosine, but not A1, receptors mediate the arousal effect of caffeine. Nat Neurosci 8:858–859

Huber R, Ghilardi MF, Massimini M, Ferrarelli F, Riedner BA, Peterson MJ, Tononi G (2006) Arm immobilization causes cortical plastic changes and locally decreases sleep slow wave activity. Nat Neurosci 9:1169–1176

Kalinchuk AV, Urrila AS, Alanko L, Heiskanen S, Wigren HK, Suomela M, Stenberg D, Porkka-Heiskanen T (2003) Local energy depletion in the basal forebrain increases sleep. Eur J Neurosci 17:863–869

Karacan I, Thornby JI, Anch M, Booth GH, Williams RL, Salis PJ (1976) Dose-related sleep disturbances induced by coffee and caffeine. Clin Pharmacol Ther 20:682–689

Kattler H, Dijk DJ, Borbely AA (1994) Effect of unilateral somatosensory stimulation prior to sleep on the sleep EEG in humans. J Sleep Res 3:159–164

Kendler KS, Gatz M, Gardner CO, Pedersen NL (2006) A Swedish national twin study of lifetime major depression. Am J Psychiatry 163:109–114

Kiloh LG, Garside RF (1965) The independence of neurotic depression and endogenous depression. Int J Psychiatry 1:447–465

Landolt HP, Dijk DJ, Gaus SE, Borbely AA (1995a) Caffeine reduces low-frequency delta activity in the human sleep EEG. Neuropsychopharmacology 12:229–238

Landolt HP, Werth E, Borbely AA, Dijk DJ (1995b) Caffeine intake (200 mg) in the morning affects human sleep and EEG power spectra at night. Brain Res 675:67–74

Liu ZW, Faraguna U, Cirelli C, Tononi G, Gao XB (2010) Direct evidence for wake-related increases and sleep-related decreases in synaptic strength in rodent cortex. J Neurosci 30: 8671–8675

Lu J, Greco MA, Shiromani P, Saper CB (2000) Effect of lesions of the ventrolateral preoptic nucleus on NREM and REM sleep. J Neurosci 20:3830–3842

Mackiewicz M, Nikonova EV, Zimmerman JE, Galante RJ, Zhang L, Cater JR, Geiger JD, Pack AI (2003) Enzymes of adenosine metabolism in the brain: diurnal rhythm and the effect of sleep deprivation. J Neurochem 85:348–357

Magistretti PJ (2006) Neuron-glia metabolic coupling and plasticity. J Exp Biol 209:2304–2311

Mazzotti DR, Guindalini C, Pellegrino R, Barrueco KF, Santos-Silva R, Bittencourt LR, Tufik S (2011) Effects of the adenosine deaminase polymorphism and caffeine intake on sleep parameters in a large population sample. Sleep 34:399–402

Mendlewicz J (2009) Sleep disturbances: core symptoms of major depressive disorder rather than associated or comorbid disorders. World J Biol Psychiatry, 1–7

Nicholson AN, Stone BM (1980) Heterocyclic amphetamine derivatives and caffeine on sleep in man. Br J Clin Pharmacol 9:195–203

Nir Y, Staba RJ, Andrillon T, Vyazovskiy VV, Cirelli C, Fried I, Tononi G (2011) Regional slow waves and spindles in human sleep. Neuron 70:153–169

Nobili L, Ferrara M, Moroni F, De Gennaro L, Russo GL, Campus C, Cardinale F, De Carli F (2011) Dissociated wake-like and sleep-like electro-cortical activity during sleep. Neuroimage 58(2):612–619

Orstavik RE, Kendler KS, Czajkowski N, Tambs K, Reichborn-Kjennerud T (2007) Genetic and environmental contributions to depressive personality disorder in a population-based sample of Norwegian twins. J Affect Disord 99:181–189

Palchykova S, Winsky-Sommerer R, Shen HY, Boison D, Gerling A, Tobler I (2010) Manipulation of adenosine kinase affects sleep regulation in mice. J Neurosci 30:13157–13165

Paunio T, Korhonen T, Hublin C, Partinen M, Kivimaki M, Koskenvuo M, Kaprio J (2009) Longitudinal study on poor sleep and life dissatisfaction in a nationwide cohort of twins. Am J Epidemiol 169:206–213

Porkka-Heiskanen T, Kalinchuk AV (2011) Adenosine, energy metabolism and sleep homeostasis. Sleep Med Rev 15:123–135

Porkka-Heiskanen T, Strecker RE, McCarley RW (2000) Brain site-specificity of extracellular adenosine concentration changes during sleep deprivation and spontaneous sleep: an in vivo microdialysis study. Neuroscience 99:507–517

Porkka-Heiskanen T, Strecker RE, Thakkar M, Bjorkum AA, Greene RW, McCarley RW (1997) Adenosine: a mediator of the sleep-inducing effects of prolonged wakefulness. Science 276:1265–1268

Rector DM, Schei JL, Van Dongen HP, Belenky G, Krueger JM (2009) Physiological markers of local sleep. Eur J Neurosci 29:1771–1778

Retey JV, Adam M, Gottselig JM, Khatami R, Durr R, Achermann P, Landolt HP (2006) Adenosinergic mechanisms contribute to individual differences in sleep deprivation-induced changes in neurobehavioral function and brain rhythmic activity. J Neurosci 26:10472–10479

Retey JV, Adam M, Honegger E, Khatami R, Luhmann UF, Jung HH, Berger W, Landolt HP (2005) A functional genetic variation of adenosine deaminase affects the duration and intensity of deep sleep in humans. Proc Natl Acad Sci USA 102:15676–15681

Retey JV, Adam M, Khatami R, Luhmann UF, Jung HH, Berger W, Landolt HP (2007) A genetic variation in the adenosine nosine receptor gene (ADORA2A) contributes to individual sensitivity to caffeine effects on sleep. Clin Pharmacol Ther 81:692–698

Shaw PJ, Cirelli C, Greenspan RJ, Tononi G (2000) Correlates of sleep and waking in Drosophila melanogaster. Science 287:1834–1837

Sherin JE, Shiromani PJ, McCarley RW, Saper CB (1996) Activation of ventrolateral preoptic neurons during sleep. Science 271:216–219

Stenberg D (2007) Neuroanatomy and neurochemistry of sleep. Cell Mol Life Sci 64:1187–1204

Stickgold R, Walker MP (2007) Sleep-dependent memory consolidation and reconsolidation. Sleep Med 8:331–343

Strecker RE, Morairty S, Thakkar MM, Porkka-Heiskanen T, Basheer R, Dauphin LJ, Rainnie DG, Portas CM, Greene RW, McCarley RW (2000) Adenosinergic modulation of basal forebrain and preoptic/anterior hypothalamic neuronal activity in the control of behavioral state. Behav Brain Res 115:183–204

Szymusiak R, Alam N, Steininger TL, McGinty D (1998) Sleep-waking discharge patterns of ventrolateral preoptic/anterior hypothalamic neurons in rats. Brain Res 803:178–188

Tononi G, Cirelli C (2006) Sleep function and synaptic homeostasis. Sleep Med Rev 10:49–62

Wigren HK, Schepens M, Matto V, Stenberg D, Porkka-Heiskanen T (2007) Glutamatergic stimulation of the basal forebrain elevates extracellular adenosine and increases the subsequent sleep. Neuroscience 147:811–823

Zeitzer JM, Morales-Villagran A, Maidment NT, Behnke EJ, Ackerson LC, Lopez-Rodriguez F, Fried I, Engel J Jr, Wilson CL (2006) Extracellular adenosine in the human brain during sleep and sleep deprivation: an in vivo microdialysis study. Sleep 29:455–461

Chapter 11
Adenosine and Other Purinergic Products in Circadian Timing

Christine Muheim and Steven A. Brown

Abstract The circadian oscillator plays an important role in behavior and metabolic physiology. In turn, adenosine occupies a unique position as both a fundamental neuromodulator and a basic building block of cellular metabolism. Multiple connections exist between the two, both through the direct actions of adenosine and through the cellular signaling cascades regulating and regulated by its availability. Specifically, we show that the circadian clock is connected to adenosine and other purinergic products on three levels. At the level of circadian signaling, the adenosine-derived signaling molecule cAMP is itself a circadian clock component that indirectly induces transcription of many circadian genes, as well as influencing cell cycle timing. At the level of metabolism, AMP kinase, a cellular energy sensor dependent upon AMP, can phosphorylate multiple clock proteins. It phosphorylates cryptochromes and thereby enhances the activity of the inhibitory clock protein complex that contains them. The histone and clock protein deacetylase SIRT1 is also phosphorylated and upregulated by AMPK, leading to increased clock protein degradation and chromatin repression. SIRT1 activity is also regulated by NAD^+ cofactors, whose levels are themselves under both circadian and metabolic control. Finally, multiple adenosine receptor subtypes can control clock function. A3 receptors influence mammalian temperature control and therefore possibly the circadian oscillator. A1 receptor transcription can be induced indirectly via glucocorticoids which are under circadian control. In addition, A1 receptors modulate light responsiveness of the circadian clock. Taken together, this intricate regulatory web likely permits a complex dialogue between metabolism and diurnal behavior and physiology that allows organisms to exploit their circadian geophysical environment optimally.

C. Muheim • S.A. Brown (✉)
Institute of Pharmacology and Toxicology, University of Zurich,
Zurich, Switzerland
e-mail: steven.brown@pharma.uzh.ch

S. Masino and D. Boison (eds.), *Adenosine: A Key Link between Metabolism and Brain Activity*, DOI 10.1007/978-1-4614-3903-5_11,

Keywords Circadian clock • Adenosine • Sirtuin • Metabolism • NAD^+ • AMPK • Food entrainment • A1 receptor • Suprachiasmatic nucleus

In this chapter, we not only outline the existing connections between adenosine and the circadian oscillator defined by research of numerous different laboratories, but also show the vast potential for therapeutically relevant connections whose mechanisms remain to be elucidated. These connections are bidirectional: on the one hand, the circadian clock controls adenosine and its downstream products, which are also important signaling molecules; on the other, adenosine and related molecules directly regulate the circadian oscillator. The drama of this interrelationship plays at the level of both the single cell and the whole organism, in both brain and peripheral tissues. Below, after a general introduction to the circadian oscillator in which we briefly mention these connections, we discuss each of these links in detail, looking first at adenosine itself, and then at several of its products, with a focus on the phosphorylated nucleotides ATP and AMP.

Of course, no discussion of this topic would be complete without a consideration of the important role that adenosine also plays in sleep–wake behavior, the major behavioral output of the circadian clock. This complex and fascinating topic is the subject of a separate chapter.

11.1 Introduction to the Circadian Clock

Most life on earth is governed by 24 h light–dark changes. An internal clock has evolved in both unicellular and multicellular organisms in order to perceive diurnal environmental cues of light, temperature, and nutrients and to control behavioral and physiological output in order to anticipate these environmental changes. This "circadian" clock (from the Latin *circa diem*, "approximately a day"), is found in nearly all cells of multicellular organisms.

In complex organisms, the multitude of circadian oscillators in peripheral cells is controlled by a superior clock in the brain. Transplantation experiments in the late 1980s by Lehman et al. using Syrian hamsters revealed the suprachiasmatic nucleus (SCN) of the hypothalamus as the master circadian clock: animals with lesions in the SCN completely lost rhythmic daily behavior, and implantation of fetal SCN tissue was able to restore this circadian locomotor activity entirely (Lehman et al. 1987).

In order to be functionally useful, the circadian clock must be able to do more than just oscillate with a period of approximately 24 h. It should also be able to sense environmental cues and changes, and further to transmit this information on a behavioral and physiological level. Conceptually, the molecular circuitry responsible for each of these processes is separable. Thus, the circadian clock mechanism in general can be divided in three parts: an input part, where incoming stimuli are processed and passed on, a core part with a ticking molecular clock, and an output part, which transmits information to all target pathways of the circadian oscillator throughout the body. As we shall see, adenosine and its metabolites can influence all three of these parts independently. In turn, the levels of these compounds are influenced both directly and indirectly by the circadian oscillator.

11.2 How the Rising Sun Stimulates Signal Transmission in the Brain: The Morphology of Clock Input

In mammals, the most important and potent "zeitgeber" or timing cue from the environment, is light. After light is perceived in the retina by both conventional rods and cones and specialized retinal ganglion cells containing the alternative photopigment melanopsin (Berson et al. 2002; Drouyer et al. 2007), it is signaled glutamatergically directly via the retinohypothalamic tract (RHT) to the SCN (Castel et al. 1993). The SCN itself is a heterologous tissue consisting of at least two regions: a core and a shell. While the latter is thought to be primarily an oscillator, the former shows weak intrinsic oscillation but significant induction by RHT stimulation, and is believed to be crucial to clock input (Foley et al. 2011; Moore et al. 2002)

Brain region-specific lesion experiments in rats could show that in addition to the SCN at least two other nuclei in the brain are important for input processing: the intergeniculate leaflets (IGL) and the dorsale raphe nuclei (DRN). Although light signals are transmitted directly to the SCN by the RHT, all other nonlight signals are thought to proceed indirectly by first reaching the DRN (Moga and Moore 1997). The IGL seems to have a more general control function: light signals from the RHT and nonphotic input from the DRN are projected to the IGL before the information is passed on again to SCN neurons, giving rising evidence for an integrative function of the IGL for all environmental cue signaling. Nevertheless, its detailed function remains to be elucidated.

As discussed more completely below, one obvious way in which adenosine affects the circadian oscillator is via its actions as an inhibitory neurotransmitter that modulates the activities of these brain nuclei.

11.3 The Ticking Molecular Core Clock in the SCN

In mammals, the circadian clock mechanism itself depends upon a set of regulatory feedback loops containing both transcriptional and posttranslational components. The CLOCK/BMAL DNA binding complex enhances initially the transcription of Period genes (*Per1*, *Per2*, and *Per3*) and Cryptochromes (*Cry1* and *Cry2*). The PER and CRY proteins are translocated back to the nucleus and inhibit their own transcription by binding to the nuclear CLOCK/BMAL complex. Since both PER and CRY protein and RNA are unstable, the nuclear mRNA and protein level of both proteins drop as soon as this inhibition occurs. When their levels drop low enough, inhibition ceases and transcription of *PER* and *CRY* starts again (for review see Reppert and Weaver 2001).

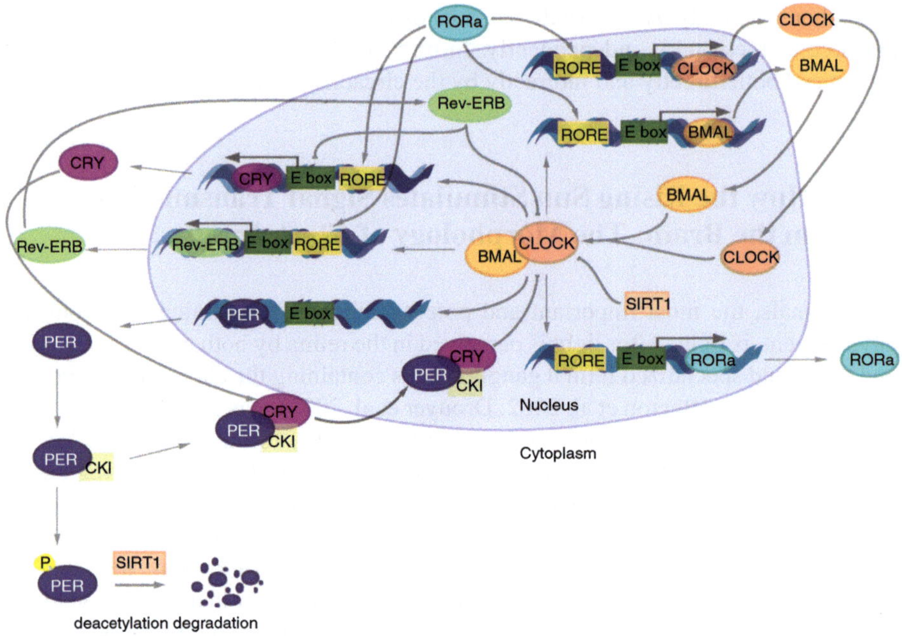

Fig. 11.1 Mechanistic overview of the circadian clock. The mammalian molecular clock responsible for diurnal behavior consists of several interlocked feedback loops of transcription and translation. A cycle begins with the enhanced transcription of Period (*Per1*, *Per2*, and *Per3*) and Cryptochrome (*Cry1* and *Cry2*) by binding of the CLOCK/BMAL heterodimer to DNA. The PER and CRY proteins are translated and then translocated back to the nucleus to inhibit their own transcription. Meanwhile, posttranslational modifications by casein kinase 1ε (CK1ε) and other kinases modify these proteins and target them for ubiquitination and proteosome-mediated degradation. As a result, the nuclear mRNA and protein level of both proteins drop to a level at which inhibition is reversed and transcription of PER and CRY starts again. A second interlocked feedback loop is formed by the action of REV-ERBα, the orphan nuclear receptor and repressor. Its transcription is enhanced by CLOCK/BMAL, but its protein product inhibits *Bmal1* transcription. Opposing it is RORα, whose transcription is also enhanced by CLOCK/BMAL, but whose activity activates *Bmal1* transcription. Metabolic feedback possibly occurs at the level of SIRT1, whose NAD^+-dependent histone deacetylation activity can associate with CLOCK proteins to deacetylate chromatin, BMAL, or PER. Adapted from Fu and Lee (2003)

Several mechanisms stabilize the oscillation of this core loop. For example, a second interlocked feedback loop is formed by the action of REV-ERBα, the orphan nuclear receptor and repressor, via inhibition of *Bmal1* transcription. Posttranslational modification of PER proteins, first to stabilize and then to destabilize them, likely also plays a key role via a number of kinases such as casein kinase 1ε (CK1ε) (Ripperger and Merrow 2011). A host of accessory factors involved in chromatin modification (WDR5, NONO, SFPQ, MLL) also participate (Brown et al. 2005; Duong et al. 2011; Katada and Sassone-Corsi 2010). Among these are the sirtuins, which are sensitive to cellular redox potential via their use of NAD^+ cofactors (Imai et al. 2000) (Fig. 11.1).

Recently, a separate independent circadian mechanism based upon redox modification of proteins has also been discovered. Its mechanism is as yet unknown. Because it is completely independent of either transcription or of known clock proteins, and in addition induces cyclic oscillations of a number of proteins important in buffering cellular redox capacity, it could prove important for many aspects of circadian metabolism (O'Neill et al. 2011). Hence, a second general way in which adenosine affects the circadian oscillator is via its effects upon cellular redox mechanisms. Since these mechanisms also affect adenosine levels, this pathway is bidirectional.

Although the SCN is the seat of the master circadian clock in the brain, the molecular clock mechanisms described above are in fact present in all of the cells of the body. First evidence for extra-SCN clocks came from experiments with retinal clocks identified in *X. laevis* (Besharse and Iuvone 1983) and in rabbits (Brandenburg et al. 1983). Light-induced entrainment of these oscillators could also be demonstrated (Tosini and Menaker 1996). These retinal clocks are independent of SCN rhythms and are important for proper photoreceptor metabolism (Besharse and Dunis 1983) and rod and cone shedding (Green and Besharse 2004). Adenosine plays a key role in the regulation of this process (Ribelayga and Mangel 2005).

11.4 Clock Output, from SCN to Behavior and Physiology

Subsequently, clocks were discovered in nearly all cells of the body, and therefore peripheral tissue like fibroblasts are often used as cellular models to monitor oscillator function (Cuninkova and Brown 2008). More directly, however, one way that the SCN controls circadian physiology is through the synchronization of these peripheral oscillators. Interestingly, the signals by which it does this are a mixture of direct and indirect ones: neural control of circadian behavior affects food intake and body temperature, both of which synchronize circadian clocks; SCN control of endocrine function provides hormonal synchronizing stimuli; and the sympathetic nervous system can also provide synchronizing cues (reviewed in Dibner et al. 2010; Hastings et al. 2007; Schibler et al. 2003).

In order to control behavior like rest–wake activity (Moore 2007) or memory consolidation (Eckel-Mahan and Storm 2009; Ruby et al. 2008) the SCN was shown to have several efferents to the diencephalon, into distinct nuclei in the hypothalamus like the supraventricular zone or the preoptic zone, as well as thalamic regions and the pineal gland (for review see Card 2000). Second, the SCN innervates several organs through the autonomic nervous system (reviewed in Bartness et al. 2001) by which it controls physiological aspects such as heart rate or liver activity (for review see Rüger and Scheer 2009; Shibata 2004). The adrenal cortex (for review see Dickmeis 2009) and glucocorticoid secretion (reviewed in Kalsbeek et al. 2012) is under direct SCN control via the hypothalamic pituitary adrenal gland axis and can furthermore be directly influenced by light (Ishida et al. 2005). Several other hormones like melatonin or luteinizing hormone are also under circadian control

(Lucas et al. 1999). Since adenosinergic modulation of neurotransmission occurs at most brain loci, it is plausible that adenosine might influence a number of these regulatory points.

The neural and hormonal pathways discussed above are directly responsible for a portion of circadian physiology. Another portion, however, is indirectly regulated by food intake. Even though it is tightly controlled, the circadian clock is still a flexible system. Light is known to be the strongest synchronizer, which means concretely that the clock can adjust to changes in seasonal light availability, but food also has a quite strong impact on circadian timing, especially upon peripheral oscillators. Normally, feeding time is controlled by behavior, which is SCN-governed in keeping with light. Thus, it is a potent timing cue from SCN to peripheral clocks. If feeding time is constrained, however, peripheral circadian physiology will adjust to match (Damiola et al. 2000; Stokkan et al. 2001) and in the absence of dominant light cues, behavior will follow. Feeding experiments with different diet-forms, high caloric, hypocaloric, or normocaloric, showed that the locomotor behavior adapted to the new Zeitgeber time (Challet 2010). It remains unclear what exactly in food in general has this strong shifting ability, and how this feedback reaches the clock. On the one hand, hormonal or neural signals in response to food probably play an essential role (Mistlberger 2011). On a cellular level, there is also rising evidence that factors from the sirtuin family link metabolism and therefore probably feeding to the circadian clock (for review see Asher and Schibler 2011; Bellet and Sassone-Corsi 2010), mainly by acting posttranslationally upon histones or clock proteins to modify transcription. All sirtuin families require nicotinamide adenine dinucleotide (NAD^+) as cofactor for their enzymatic activities of ADP-ribosylation and deacetylation (Imai et al. 2000). As mentioned earlier in the context of the core oscillator, the sirtuin SIRT1 interacts directly with clock proteins (Asher et al. 2008). The fact that SIRT1 is dependent on NAD^+ makes it a sensor of the metabolic state of the cell similar to ATP.

11.5 The Many Faces of Adenosine

Adenosine is well established as one of the principal inhibitory neuromodulators of the central nervous system. Numerous studies establish that it is critical in maintaining the delicate neurochemical balance that permits conscious thought, rest, and the maintenance of physiological equilibrium. Equally importantly, it is a precursor of intracellular purinergic signaling pathways and a basic building block of ATP, the energy currency of the cell (reviewed in Dunwiddie and Masino 2001). The functions of adenosine are therefore ubiquitous on both cellular and systemic levels. Interestingly, all of the aforementioned processes—energy regulation and metabolism, physiology, rest and activity, and memory consolidation—are regulated in mammals by the circadian oscillator via a complex bidirectional web of both intra- and inter-cellular signaling pathways. Not surprisingly, as we discuss below, adenosine figures prominently within these pathways.

Both the functions of adenosine and its sources in the CNS vary widely. Research in the past 30 years revealed many different functions and aspects of adenosine activity in the cell as well as in the extracellular matrix, be it in the form of a direct neuromodulator or as modifier of the activity of other receptors like D1 or NMDA (Manzoni et al. 1994; Popoli et al. 1996), to mention two examples. Due to its wide functionality, the metabolism of adenosine is a well-studied topic, and all precursors and breakdown products are known. Extracellular adenosine is dependent on two main processes, first on the degradation of extracellular AMP to adenosine by ecto-5′nucleotidase and second on the secretion of intracellular adenosine, which itself has two different origins. The intracellular adenosine level depends on the degradation of ATP to AMP and finally to adenosine via the cytoplasmic 5′nucleotidase as well as on the hydrolysis of S-adenosylhomocysteine (SAH). Interestingly the ATP-AMP dephosphorylation process is highly concentration-dependent and therefore changes in intracellular adenosine levels can reflect changes in metabolic rate. More specifically, the formation of adenosine from AMP occurs as the concentration of ATP changes in a physiological range, most likely due to the activity of a specific subtype of cytoplasmic 5′nucleotidase (for review see Latini and Pedata 2001) (Fig. 11.2). However, as ATP is not the only source of adenosine, the relationship between adenosine and ATP concentration is most probably not a linear one.

On the contrary, to reduce levels of adenosine two major reactions can occur, either via phosphorylation of adenosine by adenosine kinase (ADK) resulting in raising concentrations of 5′AMP, or via the generation of inosine and later on inosine monophosphate (IMP) by an irreversible reaction of adenosine with adenosine deaminase (ADA). Both reactions are necessary tools to stabilize intra-and extracellular pools of adenosine (for review see Dunwiddie and Masino 2001).

11.6 From Bottom to Top: Adenosine as Regulator of the Circadian Clock

11.6.1 Adenosine Receptors

The first adenosine agonist experiments more than 15 years ago were carried out with hamsters, where application of the selective adenosine agonist N6-cyclohexyladenosine (CHA) revealed a regulative aspect of adenosine for the clock (Watanabe et al. 1996). In particular, phase advances induced by changed light settings were diminished after CHA application to the SCN (Elliott et al. 2001). Administration of the adenosine antagonist caffeine did not affect the circadian clock, though it did diminish sleep and stimulate arousal (Antle et al. 2001).

Surprisingly, subsequent experiments in mice showed that the observed changes in phase-shifting were regulated by A1 adenosine receptors (Sigworth and Rea 2003). It was already known that relatively few adenosine A1 receptor mRNAs are present in the hypothalamus or SCN, so it remained unclear how A1 receptor

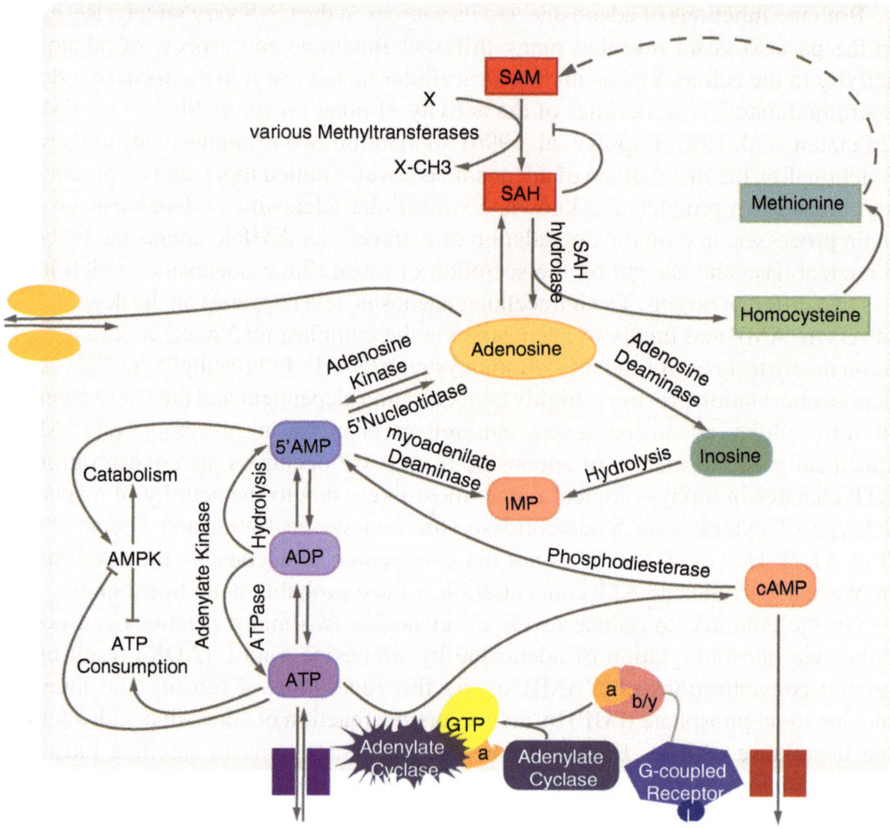

Fig. 11.2 Adenosine metabolism. Adenosine is formed by the sequential breakdown of ATP into first ADP and then 5′AMP via the enzymatic activity of a cytosolic 5′nucleotidase. To a lesser extent, S-adenosylhomocystein (SAH) derived from S-adenosylmethionine (SAM) also serves as a source of cellular adenosine. Adenosine kinase (ADK) and adenosine deaminase (ADA) instead promote the breakdown reaction of adenosine. ADK activity leads to the production of 5′AMP by adenosine phosphorylation. ADA converts adenosine to inosine in an irreversible reaction. Subsequent hydrolyis of inosine results in inosine monophosphate (IMP), an intermediate of purine metabolism. In addition to sequential breakdown to form adenosine, ATP is also a substrate for adenylate cyclase, whose activation by G-coupled receptors promotes the formation of the signaling molecule cAMP, which can itself be transformed into 5′AMP. The "energy metabolites" of the cell (ATP, ADP, and AMP) can bind to AMP kinase (AMPK) to regulate its activity. AMPK is itself a sensor of the cellular energy status and can promote ATP formation

signaling acts on SCN clock activity. Work by Hallworth and coworkers was able to show that A1 receptors located in the retinohypothalamic tract play an important role in attenuation of the release of glutamate and therefore regulate the transmission of light information to the SCN (Hallworth et al. 2002).

The A2a and A2b receptors are both expressed in the brain—A2a rather ubiquitously, and A2b in specific regions—and both are also expressed in somatic tissue (for details see review Dunwiddie and Masino 2001). For both of them, agonists show no effects upon the circadian clock and there is no further indication that they might play a role in controlling the transmission of input cues to or within the SCN (Fredholm et al. 2000; Hallworth et al. 2002; Latini and Pedata 2001). The A3 receptor was also tested with several different agonists but light-induced phase advances were not observed (Elliott et al. 2001).

11.6.2 Adenosine Monophosphate

The first metabolic product of adenosine by adenosine kinase (ADK) is AMP, and in subsequent enzymatic steps ADP and ATP are produced. Together with adenosine deaminase, ADK is the principal regulator of intracellular adenosine availability. AMP levels themselves, or more specifically the cellular ratio of AMP–ATP, is highly controlled. AMP protein kinase (AMPK) is a direct sensor for changed AMP–ATP ratios, or better, energy levels in the cell in general (reviewed in Hardie 2007). Due to the fact that the adenylate kinase enzyme keeps the catalysis of $2ADP \rightarrow 1ATP + 1AMP$ near to equilibrium, the ATP–AMP ratio varies approximately as the square of the ATP–ADP ratios, which makes the AMP–ATP ratio a reliable index for metabolic processes in the cell (Kahn et al. 2005). Moreover, AMP has the ability to activate AMPK through three different mechanisms—allosteric activation, phosphorylation, and inhibition of dephosphorylation—AMPK is therefore a very sensitive sensor for changes in cellular energy status (reviewed in Hardie 2007). In its turn, through phosphorylation of a diversity of targets, AMPK can organize metabolic pathways depending upon hormonal inputs like leptin or adiponectin, as well as coordinate the response to stress-induced depletion of ATP (Kemp et al. 1999).

What does this have to do with the regulation of circadian clocks? In fact, one of the targets of AMPK is the circadian clock! In 2009 Lamia and coauthors demonstrated nicely by using the agonist aminoimidazole carboxamide ribonucleotide (AICA), that AMPK phosphorylates the circadian repressor protein CRY1, thereby stabilizing its binding to PER2 while decreasing its degradation. As a result, the transcription normally induced by the CLOCK/BMAL complex is suppressed, and the "trigger" for a new clock cycle is inactivated (Lamia et al. 2009). In a related story 2 years before, another group showed that AMPK phosphorylates CK1ε, which itself controls the degradation of PER2 (Um et al. 2007). The same group published data this year about tissue- and isoform-specific activity of AMPK and concluded that AMPK controls expression of peripheral clock genes (Um et al. 2011). Beside this direct interaction of AMPK on the circadian clock, it has been observed in muscle tissue that AMPK also triggers the activation of SIRT1 (Cantó et al. 2009, 2010). SIRT1, as discussed previously, deacetylates PER2 in order to degrade PER2 protein and create the environment for a new cycle (Asher et al. 2008).

11.6.3 Adenosine Triphosphate

Recent research shows that adenosine in the form of ATP also seems to be an important factor to link metabolism to clock function. In the simplest existing circadian oscillator, found in the cyanobacterium *Synechococcus elongates*, the protein products of the *KaiABC* locus, together with ATP, are themselves sufficient to generate a 24 h rhythm, even in vitro. One of those proteins, KaiC, has two ATP binding sites (Ishiura et al. 1998) and was later on found to display ATPase activity under the control of KaiA (Pattanayek et al. 2009). In fact, the rate of ATP hydrolysis is crucial to set clock speed (Rust et al. 2011).

Even if the KaiABC clock is not conserved directly in mammalian cells, ATP hydrolysis equally provides timing and directionality to circadian processes in mammals. For example, diurnal modulation of a Na^+/K^+-ATPase in rat neurons directly influences firing rate (Wang and Huang 2004). Highest ATPase enzyme activity was found during the subjective day, when calcium concentrations were also increased. In fact, the membrane potential and firing of rat SCN neurons is highly dependent on the actual activity status of the Na^+/K^+-ATPase (Wang and Huang 2006). Similar regulation of firing could explain the spontaneous circadian electrical activity of SCN neurons.

Although the ATPase of the previous paragraph was governed by circadian mechanisms, ATP itself is mostly regulated by metabolic function and energy consumption. Briefly, ATP is generated or (more accurately) recycled in several catabolic subprocesses during glycolysis or in the respiration chain reaction in mitochondria. In the same process in which ATP is synthesized, NAD^+ is oxygenated and vice versa. In the context of circadian clocks, it has been suggested that food/energy uptake modulates ATP and NAD^+ levels (Ishikawa and Shimanzu 1976; Mistlberger 2011) thereby regulating clock function. The effects of this energy input on clock output are discussed in the next paragraph. Apart from these metabolic considerations, however, ATP may directly affect clock function via P2X7 purinergic receptors, a class of ligand-gated ion channel whose ATP-dependent activation results in subsequent upregulation of *Per1* clock gene expression in some tissues (Nakazato et al. 2011).

11.7 Metabolism and Clock Output: NAD^+ as a Linker

The ability of food to entrain the circadian clock has itself hinted repeatedly at a long-suspected connection between metabolism and circadian function (reviewed in Asher and Schibler 2011). Nevertheless, the molecular currency of this link is not clear. Beside ATP, nicotinamide adenine dinucleotide (NAD^+) is a second potent linker between energy metabolism and the circadian clock. As stated above, recent data has shown that the histone acetylase SIRT1 is able to counterbalance the activity of CLOCK on a chromatin level (Nakahata et al. 2009), and also to modulate PER protein function (Asher et al. 2008). In turn, SIRT1 activity is directly modulated by intracellular levels of NAD^+. (As explained below, because NAD^+ salvage is itself under circadian control via the CLOCK/BMAL-controlled transcription of nicotinamide phosphoribosyltransferase (NAMPT) (Nakahata et al. 2009; Revollo et al. 2004), this pathway is bidirectional.)

11.8 From Top to Bottom: Diurnal Regulation of Adenosine and Adenosinergic Factors

As discussed above, adenosine levels in the retinohypothalamic tract are critical to regulating light-induced phase-shifting in the SCN. In studying these adenosine-dependent currents, however, Ribelayga found in a rabbit retinal model system additionally a strong direct dependency upon light and darkness (Ribelayga and Mangel 2005). Thus, adenosine levels in the extracellular matrix of the rabbit retina are highly sensitive to light, and this sensitivity is more pronounced during the subjective night than day (Brandenburg et al. 1983). The authors therefore suggested a threshold light intensity for the modulation of adenosine as a function of the circadian clock. Concretely, this double regulation means that extracellular baseline levels of adenosine are circadian, with high levels at the subjective night and lower levels during the day. The underlying mechanism is still not fully understood, but it appears that not only less adenosine is transported to the intracellular compartment but also the intracellular purinergic fraction is increased. Both factors lead to an extracellular accumulation of adenosine in the observed circadian pattern (Ribelayga and Mangel 2005). The physiological effects of this regulation upon neural activity remain unclear. Of course, adenosine is also an important regulator of sleep–wake behavior—the most important behavioral output of the circadian system—but this topic is treated in a separate chapter.

Metabolic products of adenosine are also regulated in circadian fashion, most notably ATP, in many examined tissues. For example, Womac et al. reported circadian ATP fluctuations in brain, with highest levels at night. The authors suggest that one consequence of this oscillation would be a circadian modulation of intracellular signaling between astrocytes and neurons in the brain (Womac et al. 2009). In this context, ATP serves as an auto- as well as a paracrine factor, which is thought to regulate intracellular calcium waves, the interconnection between astrocytes and neurons, and even brain metabolism in general (Bernardinelli et al. 2004; Haydon 2001; Womac et al. 2009).

More generally still, metabolomic analysis of cultured human or mouse cells (R. Dallmann, personal communication) shows circadian levels of ATP as well as most redox molecules. This suggests a global circadian regulation of mitochondrial function, which in turn affects ATP and adenosine levels.

11.9 Outlook: The Extending Front of Circadian Adenosine Regulation

11.9.1 Adenosine and cAMP Signaling Pathways in the Central Nervous System

So far, we have considered the circadian effects of adenosine and the metabolic products whose production is directly dependent upon it—AMP and ATP. However, these products are themselves the source of signaling molecules crucial to circadian function. ATP is not only a source of adenosine, but also cAMP, which is a cellular signaling molecule for both the circadian oscillator and other processes. For example, a study focusing on melatonin synthesis in the pineal gland revealed a special role for cAMP in controlling the circadian expression of genes (Kim et al. 2005). Melatonin synthesis requires (S)-Adenosylmethionine (SAM) activity, which in turn is dependent on methionine adenosyltransferase II (METII), an enzyme with increased activity during the night. It is clear that the circadian expression of METII is directly under the control of cAMP signaling pathways, though various mechanisms have been proposed (Chik et al. 2007; Ho et al. 2007; Kim et al. 2005). A study some years later was even able to show that the adrenergic/cAMP signaling pathway controls more than 600 genes in the pineal gland and drives their nocturnal expression (Bailey et al. 2009). Another study published comparable results, showing that a homeobox gene with strong circadian expression in the pineal gland itself serves as a transcription factor in the adult pineal gland. Interestingly, the authors as well postulated adrenergic/cAMP signaling mechanisms as mediators of the circadian oscillator (Rath et al. 2009).

A completely different aspect of clock-controlled cAMP is its direct interference in glucogenesis. The impact of the circadian pacemaker on liver glucogenesis is generally accepted, but Zhang et al. recently showed that CRY1, a core clock protein, is able to block cAMP increase and as a consequence pCREB-induced gene expression. CRY1 inhibition interferes upstream of cAMP synthesis by repression of the Gs alpha subunit from activated G protein-coupled receptors and therefore inhibits the activity of the adenylate cyclase (Zhang et al. 2010).

At the same time, cAMP is an important regulator of circadian function. Indeed, inhibitors of cAMP signaling disrupt circadian oscillations at a cellular level and cAMP levels themselves oscillate in circadian fashion. Thus, purinergic signaling in the form of cAMP forms an auxiliary feedback loop within the circadian clock itself (O'Neill et al. 2008). Similarly, Amelio et al. discovered in 2007 a NONO-TORC2 complex which activates CREB targets dependent on cAMP levels (Amelio et al. 2007). NONO itself plays a crucial part of the internal circadian clock (Brown et al. 2005) and might therefore be a clock output transmitter to transcription via cAMP.

11.10 Feedback Loops from Adenosine to the Clock and Back

In the case just described, cAMP is both an effector of circadian output, and itself a regulator of clock function. Such a feedback loop forms a mechanism by which pathways controlled by the clock can themselves influence clock timing, and has become a common theme in the circadian regulation of physiology. It is possible that two evolutionary benefits are obtained by this strategy. First, by coupling multiple feedback loops of cellular signaling, additional resilience is achieved. Very simply, two interlocked feedback loops are more robust to perturbation than just one. In this way, at a cellular level the timing of the circadian oscillator even withstands cell division (Nagoshi et al. 2004). Second, through clock control of a given pathway, an organism can anticipate an environmental change and induce signaling cascades to respond appropriately. By linking some output of this change back to clock timing, repeatedly mistimed inductions can change the phase of the clock to better anticipate these changes. In this way, for example, the circadian clock regulates feeding behavior, but repeatedly mistimed feeding resets the phase of the clock (Damiola et al. 2000; Stokkan et al. 2001).

For all of the cases above, it is likely that adenosine and the purinergic signaling that it governs form parts of such clock-controlled feedback loops. In the case of cAMP, which is one of the most important signaling metabolites in the CNS and peripheral tissues, the case is clear. On the one hand, it shows circadian expression, with highest concentrations in the subjective night. The effects of this are wide-ranging, as we discuss above. Not only is it responsible for different types of indirect circadian signaling (An et al. 2011; Atkinson et al. 2011), but it also activates transcription at *Per* clock gene promoters themselves and is itself therefore necessary for robust clock oscillations (Hastings et al. 2007).

Adenosine itself occupies a similar central position. For example, in retinal cells, its levels vary directly with light (Ribelayga and Mangel 2005). At the same time, however, its abundance affects the ability of the circadian oscillator to be phase-shifted by light (Sigworth and Rea 2003). The elegant result is that the more light there is in the environment, the less the clock is systematically affected by it! Adding additional complexity, an alternative promoter of the A1 receptor itself is induced by glucocorticoids themselves (Ren and Stiles 1999). The physiological implications of this further feedback are as yet unclear.

A final and more complex example is furnished by AMP and ATP in metabolism. Both extracellular and intracellular ATP levels vary in circadian fashion, both in brain regions including the SCN and in retina (Burkeen et al. 2011; Womac et al. 2009), and in cultured astrocytes (Marpegan et al. 2011). Similarly, there exists a global circadian regulation of redox molecules like NAD^+ (Nakahata et al. 2009). Extracellular ATP acts as a signaling molecule for glia, and is thought to play a role in the regulation of brain metabolism (Womac et al. 2009). Meanwhile, as discussed earlier, the AMP–ATP ratio is a cellular "thermometer" for energy levels and regulates mitochondrial function. Lastly, AMP (via AMPK) and NAD^+ (via SIRT1)

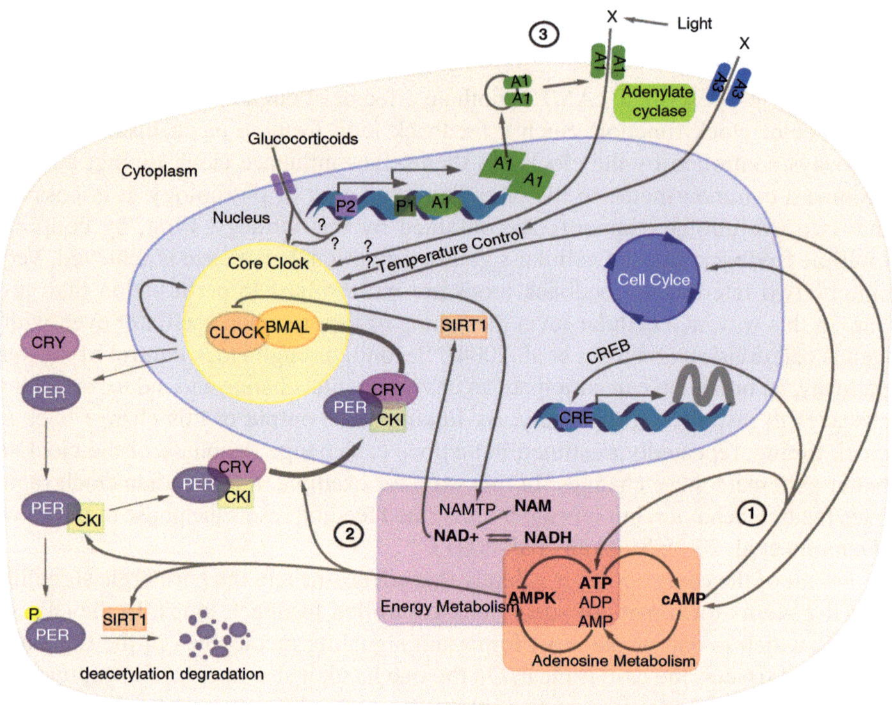

Fig. 11.3 Adenosine and the circadian clock: a model. In this review, we propose that the circadian clock and adenosine are connected on three levels. (1) Circadian signaling pathways. cAMP is itself a circadian signaling molecule and induces transcription of several circadian genes, as well as influencing cell cycle timing. (2) Metabolic clock feedback. AMP kinase, a cellular energy sensor, phosphorylates CRY and thereby enhances the activity of the PER/CRY inhibitory clock protein complex. Moreover, SIRT1 deacetylase activity is upregulated by AMPK, leading to increased PER degradation and chromatin repression. SIRT1 activity is also regulated by NAD+ cofactors, whose levels are themselves under both circadian and metabolic control. (3) Adenosine receptor control of clock function. A3 receptor subtypes influence mammalian temperature control and therefore possibly the circadian oscillator. A1 receptor subtype transcription can be induced indirectly via glucocorticoids which are under circadian control. In addition, A1 receptors modulate light responsiveness of the circadian clock

directly regulate circadian function (Asher et al. 2008; Lamia et al. 2009; Nakahata et al. 2009; Um et al. 2011). Circadian entrainment to feeding is additionally regulated via NAD^+ through the activity of poly-ADP ribose polymerase 1 (PARP1) (Asher et al. 2010), and AMPK itself regulates SIRT1 (Cantó et al. 2009, 2010). These interconnections form in effect another feedback loop (Fig. 11.3). Since ATP signaling also regulates intracellular Ca^{2+} waves, this loop might have more than just metabolic implications: it could also be an important factor for synaptic plasticity and long-term potentiation, regulated in parallel by the homeostatic process of sleep itself (Tononi and Cirelli 2006).

11.11 The Tip of the Iceberg

As discussed extensively elsewhere in this book, adenosine receptors exist in many brain tissues that are also enervated by the SCN, including the hypothalamus, thalamus, or the amygdale. Thus, in the brain alone, the potential exists for many more links between the circadian clock and adenosine that await discovery. For example, the adenosine A3 receptor plays an important role in thermoregulation in mammals (Yang et al. 2010). In fact, however, mammalian body temperature varies slightly in circadian fashion, probably via rest–activity regulation, and these fluctuations are sufficient to entrain and phase-shift peripheral circadian oscillators (Brown et al. 2002). Since circadian body temperature is one of the cues by which the SCN entrains peripheral circadian oscillators, adenosine might play an important unexplored role in this process.

Peripheral tissues also express a variety of adenosine receptors, like A2a, A2b or A3. Some tissues, like muscle, are able to link adenosine metabolism to PER2 degradation via SIRT1 activation (Cantó et al. 2009). In cells throughout the body, the molecular basis of circadian clocks is mostly the same. Thus, where it exists, adrenergic input/output could influence many diverse aspects of circadian physiology.

A variety of pathophysiologies have been directly linked to circadian dysfunction, and in parallel to purinergic signaling dysfunction. It is possible that these correlations are marks of the interplay of circadian clocks with adenosine. Circadian disruption from night shiftwork, for example, has been linked to metabolic syndrome (Knutsson et al. 1999; Pietroiusti et al. 2010). Metabolic disorders have also been tied to adenosinergic dysfunction (Figler et al. 2011; Westermeier et al. 2011). Additionally, many psychiatric disorders are linked to a circadian context—for example bipolar disorders (Mansour et al. 2005) or schizophrenia (Lamont et al. 2010; Moons et al. 2011). Similarly, some depressive disorders have been proposed to arise as consequences of manipulations in the adenosine neuromodulatory system (reviewed in Gomes et al. 2011). Of course, experimental evidence is still missing that directly links the circadian clock to purinergic signaling in any major disease, but the plethora of molecular connections between these pathways makes it likely that such connections exist.

To conclude, research in the past 40 years has discovered more and more connections between adenosine, its downstream products, and the circadian oscillator, in the brain as well as in peripheral tissues. The mechanism of this interaction seems to vary from tissue to tissue, and many interesting connections remain to be found.

References

Amelio AL, Miraglia LJ, Conkright JJ, Mercer BA, Batalov S, Cavett V, Orth AP, Busby J, Hogenesch JB, Conkright MD (2007) A coactivator trap identifies NONO (p54nrb) as a component of the cAMP-signaling pathway. Proc Natl Acad Sci USA 104:20314–20319

An S, Irwin RP, Allen CN, Tsai C, Herzog ED (2011) Vasoactive intestinal polypeptide requires parallel changes in adenylate cyclase and phospholipase C to entrain circadian rhythms to a predictable phase. J Neurophysiol 105:2289–2296

Antle MC, Steen NM, Mistlberger RE (2001) Adenosine and caffeine modulate circadian rhythms in the Syrian hamster. Neuroreport 12:2901–2905

Asher G, Gatfield D, Stratmann M, Reinke H, Dibner C, Kreppel F, Mostoslavsky R, Alt FW, Schibler U (2008) SIRT1 regulates circadian clock gene expression through PER2 deacetylation. Cell 134:317–328

Asher G, Reinke H, Altmeyer M, Gutierrez-Arcelus M, Hottiger MO, Schibler U (2010) Poly(ADP-ribose) polymerase 1 participates in the phase entrainment of circadian clocks to feeding. Cell 142:943–953

Asher G, Schibler U (2011) Crosstalk between components of circadian and metabolic cycles in mammals. Cell Metab 13:125–137

Atkinson SE, Maywood ES, Chesham JE, Wozny C, Colwell CS, Hastings MH, Williams SR (2011) Cyclic AMP signaling control of action potential firing rate and molecular circadian pacemaking in the suprachiasmatic nucleus. J Biol Rhythms 26:210–220

Bailey MJ, Coon SL, Carter DA, Humphries A, Kim JS, Shi Q, Gaildrat P, Morin F, Ganguly S, Hogenesch JB et al (2009) Night/day changes in pineal expression of >600 genes. J Biol Chem 284:7606–7622

Bartness TJ, Song CK, Demas GE (2001) SCN efferents to peripheral tissues: implications for biological rhythms. J Biol Rhythms 16:196–204

Bellet MM, Sassone-Corsi P (2010) Mammalian circadian clock and metabolism—the epigenetic link. J Cell Sci 123:3837–3848

Bernardinelli Y, Magistretti PJ, Chatton J-Y (2004) Astrocytes generate Na^+-mediated metabolic waves. Proc Natl Acad Sci USA 101:14937–14942

Berson DM, Dunn FA, Takao M (2002) Phototransduction by retinal ganglion cells that set the circadian clock. Science 295:1070–1073

Besharse J, Dunis D (1983) Methoxyindoles and photoreceptor metabolism: activation of rod shedding. Science 219:1341–1343

Besharse JC, Iuvone PM (1983) Circadian clock in Xenopus eye controlling retinal serotonin N-acetyltransferase. Nature 305:133–135

Brandenburg J, Bobbert AC, Eggelmeyer F (1983) Circadian changes in the response of the rabbit's retina to flashes. Behav Brain Res 7:113–123

Brown SA, Ripperger J, Kadener S, Fleury-Olela F, Vilbois F, Rosbash M, Schibler U (2005) PERIOD1-associated proteins modulate the negative limb of the mammalian circadian oscillator. Science 308:693–696

Brown SA, Zumbrunn G, Fleury-Olela F, Preitner N, Schibler U (2002) Rhythms of mammalian body temperature can sustain peripheral circadian clocks. Curr Biol 12:1574–1583

Burkeen JF, Womac AD, Earnest DJ, Zoran MJ (2011) Mitochondrial calcium signaling mediates rhythmic extracellular ATP accumulation in suprachiasmatic nucleus astrocytes. J Neurosci 31:8432–8440

Cantó C, Gerhart-Hines Z, Feige JN, Lagouge M, Noriega L, Milne JC, Elliott PJ, Puigserver P, Auwerx J (2009) AMPK regulates energy expenditure by modulating NAD^+ metabolism and SIRT1 activity. Nature 458:1056–1060

Cantó C, Jiang LQ, Deshmukh AS, Mataki C, Coste A, Lagouge M, Zierath JR, Auwerx J (2010) Interdependence of AMPK and SIRT1 for metabolic adaptation to fasting and exercise in skeletal muscle. Cell Metab 11:213–219

Card JP (2000) Pseudorabies virus and the functional architecture of the circadian timing system. J Biol Rhythms 15:453–461

Castel M, Belenky M, Cohen S, Ottersen OP, Storm-Mathisen J (1993) Glutamate-like immunoreactivity in retinal terminals of the mouse suprachiasmatic nucleus. Eur J Neurosci 5:368–381

Challet E (2010) Interactions between light, mealtime and calorie restriction to control daily timing in mammals. J Comp Physiol B 180:631–644

Chik CL, Arnason TG, Dukewich WG, Price DM, Ranger A, Ho AK (2007) Histone H3 phosphorylation in the rat pineal gland: adrenergic regulation and diurnal variation. Endocrinology 148:1465–1472

Cuninkova L, Brown SA (2008) Peripheral circadian oscillators. Ann N Y Acad Sci 1129:358–370
Damiola F, Le Minh N, Preitner N, Kornmann B, Fleury-Olela F, Schibler U (2000) Restricted feeding uncouples circadian oscillators in peripheral tissues from the central pacemaker in the suprachiasmatic nucleus. Genes Dev 14:2950–2961
Dibner C, Schibler U, Albrecht U (2010) The mammalian circadian timing system: organization and coordination of central and peripheral clocks. Annu Rev Physiol 72:517–549
Dickmeis T (2009) Glucocorticoids and the circadian clock. J Endocrinol 200:3–22
Drouyer E, Rieux C, Hut RA, Cooper HM (2007) Responses of suprachiasmatic nucleus neurons to light and dark adaptation: relative contributions of melanopsin and rod cone inputs. J Neurosci 27:9623–9631
Dunwiddie TV, Masino SA (2001) The role and regulation of adenosine in the central nervous system. Annu Rev Neurosci 24:31–55
Duong HA, Robles MS, Knutti D, Weitz CJ (2011) A molecular mechanism for circadian clock negative feedback. Science 332:1436–1439
Eckel-Mahan KL, Storm DR (2009) Circadian rhythms and memory: not so simple as cogs and gears. EMBO Rep 10:584–591
Elliott KJ, Todd Weber E, Rea MA (2001) Adenosine A1 receptors regulate the response of the hamster circadian clock to light. Eur J Pharmacol 414:45–53
Figler RA, Wang G, Srinivasan S, Jung DY, Zhang Z, Pankow JS, Ravid K, Fredholm B, Hedrick CC, Rich SS et al (2011) Links between insulin resistance, adenosine A2B receptors, and inflammatory markers in mice and humans. Diabetes 60:669–679
Foley NC, Tong TY, Foley D, LeSauter J, Welsh DK, Silver R (2011) Characterization of orderly spatiotemporal patterns of clock gene activation in mammalian suprachiasmatic nucleus. Eur J Neurosci 33:1851–1865
Fredholm BB, Arslan G, Halldner L, Kull B, Schulte G, Wasserman W (2000) Structure and function of adenosine receptors and their genes. Naunyn Schmiedebergs Arch Pharmacol 362:364–374
Fu L, Lee CC (2003) The circadian clock: pacemaker and tumour suppressor. Nat Rev Cancer 3:350–361
Gomes CV, Kaster MP, Tomé AR, Agostinho PM, Cunha RA (2011) Adenosine receptors and brain diseases: neuroprotection and neurodegeneration. Biochim Biophys Acta 1808:1380
Green CB, Besharse JC (2004) Retinal circadian clocks and control of retinal physiology. J Biol Rhythms 19:91–102
Hallworth R, Cato M, Colbert C, Rea MA (2002) Presynaptic adenosine A1 receptors regulate retinohypothalamic neurotransmission in the hamster suprachiasmatic nucleus. J Neurobiol 52:230–240
Hardie DG (2007) AMP-activated/SNF1 protein kinases: conserved guardians of cellular energy. Nat Rev Mol Cell Biol 8:774–785
Hastings M, O'Neill JS, Maywood ES (2007) Circadian clocks: regulators of endocrine and metabolic rhythms. J Endocrinol 195:187–198
Haydon PG (2001) Glia: listening and talking to the synapse. Nat Rev Neurosci 2:185–193
Ho AK, Price DM, Dukewich WG, Steinberg N, Arnason TG, Chik CL (2007) Acetylation of histone H3 and adrenergic-regulated gene transcription in rat pinealocytes. Endocrinology 148:4592–4600
Imai SI, Armstrong CM, Kaeberlein M, Guarente L (2000) Transcriptional silencing and longevity protein Sir2 is an NAD-dependent histone deacetylase. Nature 403:795–800
Ishida A, Mutoh T, Ueyama T, Bando H, Masubuchi S, Nakahara D, Tsujimoto G, Okamura H (2005) Light activates the adrenal gland: timing of gene expression and glucocorticoid release. Cell Metab 2:297–307
Ishikawa K, Shimanzu T (1976) Daily rhythm of glycogen synthetase and phosphorylase activities in rat liver: influence of food and light. Life Sci 19:1873–1878
Ishiura M, Kutsuna S, Aoki S, Iwasaki H, Andersson CR, Tanabe A, Golden SS, Johnson CH, Kondo T (1998) Expression of a gene cluster kaiABC as a circadian feedback process in cyanobacteria. Science 281:1519–1523

Kahn BB, Alquier T, Carling D, Hardie DG (2005) AMP-activated protein kinase: ancient energy gauge provides clues to modern understanding of metabolism. Cell Metab 1:15–25
Kalsbeek A, van der Spek R, Lei J, Endert E, Buijs RM, Fliers E (2012) Circadian rhythms in the hypothalamo-pituitary-adrenal (HPA) axis. Mol Cell Endocrinol 349(1):20–29
Katada S, Sassone-Corsi P (2010) The histone methyltransferase MLL1 permits the oscillation of circadian gene expression. Nat Struct Mol Biol 17:1414–1421
Kemp BE, Mitchelhill KI, Stapleton D, Michell BJ, Chen Z-P, Witters LA (1999) Dealing with energy demand: the AMP-activated protein kinase. Trends Biochem Sci 24:22–25
Kim J-S, Coon SL, Blackshaw S, Cepko CL, Møller M, Mukda S, Zhao W-Q, Charlton CG, Klein DC (2005) Methionine adenosyltransferase:adrenergic-cAMP mechanism regulates a daily rhythm in pineal expression. J Biol Chem 280:677–684
Knutsson A, Hallquist J, Reuterwall C, Theorell T, Akerstedt T (1999) Shiftwork and myocardial infarction: a case-control study. Occup Environ Med 56:46–50
Lamia KA, Sachdeva UM, DiTacchio L, Williams EC, Alvarez JG, Egan DF, Vasquez DS, Juguilon H, Panda S, Shaw RJ et al (2009) AMPK regulates the circadian clock by cryptochrome phosphorylation and degradation. Science 326:437–440
Lamont E, Coutu D, Cermakian N, Boivin D (2010) Circadian rhythms and clock genes in psychotic disorders. Isr J Psychiatry Relat Sci 47:27–35
Latini S, Pedata F (2001) Adenosine in the central nervous system: release mechanisms and extracellular concentrations. J Neurochem 79:463–484
Lehman MN, Silver R, Gladstone WR, Kahn RM, Gibson M, Bittman EL (1987) Circadian rhythmicity restored by neural transplant. Immunocytochemical characterization of the graft and its integration with the host brain. J Neurosci 7:1626–1638
Lucas RJ, Stirland JA, Darrow JM, Menaker M, Loudon ASI (1999) Free running circadian rhythms of melatonin, luteinizing hormone, and cortisol in Syrian hamsters bearing the circadian tau mutation. Endocrinology 140:758–764
Mansour HA, Monk TH, Nimgaonkar VL (2005) Circadian genes and bipolar disorder. Ann Med 37:196–205
Manzoni OJ, Manabe T, Nicoll RA (1994) Release of adenosine by activation of NMDA receptors in the hippocampus. Science 265:2098–2101
Marpegan L, Swanstrom AE, Chung K, Simon T, Haydon PG, Khan SK, Liu AC, Herzog ED, Beaulé C (2011) Circadian regulation of ATP release in astrocytes. J Neurosci 31:8342–8350
Mistlberger RE (2011) Neurobiology of food anticipatory circadian rhythms. Physiol Behav 104(4):535–545
Moga MM, Moore RY (1997) Organization of neural inputs to the suprachiasmatic nucleus in the rat. J Comp Neurol 389:508–534
Moons T, Claes S, Martens GJM, Peuskens J, Van Loo KMJ, Van Schijndel JE, De Hert M, van Winkel R (2011) Clock genes and body composition in patients with schizophrenia under treatment with antipsychotic drugs. Schizophr Res 125:187–193
Moore R, Speh J, Leak R (2002) Suprachiasmatic nucleus organization. Cell Tissue Res 309:89–98
Moore RY (2007) Suprachiasmatic nucleus in sleep-wake regulation. Sleep Med 8:27–33
Nagoshi E, Saini C, Bauer C, Laroche T, Naef F, Schibler U (2004) Circadian gene expression in individual fibroblasts: cell-autonomous and self-sustained oscillators pass time to daughter cells. Cell 119:693–705
Nakahata Y, Sahar S, Astarita G, Kaluzova M, Sassone-Corsi P (2009) Circadian control of the NAD+ salvage pathway by CLOCK-SIRT1. Science 324:654–657
Nakazato R, Takarada T, Yamamoto T, Hotta S, Hinoi E, Yoneda Y (2011) Selective upregulation of Per1 mRNA expression by ATP through activation of P2X7 purinergic receptors expressed in microglial cells. J Pharmacol Sciences 116, 350–361
O'Neill JS, Maywood ES, Chesham JE, Takahashi JS, Hastings MH (2008) cAMP-dependent signaling as a core component of the mammalian circadian pacemaker. Science 320:949–953
O'Neill JS, van Ooijen G, Dixon LE, Troein C, Corellou F, Bouget F-Y, Reddy AB, Millar AJ (2011) Circadian rhythms persist without transcription in a eukaryote. Nature 469:554–558

Pattanayek R, Mori T, Xu Y, Pattanayek S, Johnson CH, Egli M (2009) Structures of KaiC circadian clock mutant proteins: a new phosphorylation site at T426 and mechanisms of Kinase, ATPase and phosphatase. PLoS One 4:e7529

Pietroiusti A, Neri A, Somma G, Coppeta L, Iavicoli I, Bergamaschi A, Magrini A (2010) Incidence of metabolic syndrome among night-shift healthcare workers. Occup Environ Med 67:54–57

Popoli P, Ferré S, Pezzola A, Reggio R, Scotti de Carolis A, Fuxe K (1996) Stimulation of adenosine A1 receptors prevents the EEG arousal due to dopamine D1 receptor activation in rabbits. Eur J Pharmacol 305:123–126

Rath MF, Bailey MJ, Kim J-S, Ho AK, Gaildrat P, Coon SL, Møller M, Klein DC (2009) Developmental and diurnal dynamics of Pax4 expression in the mammalian pineal gland: nocturnal down-regulation is mediated by adrenergic-cyclic adenosine 3′,5′-monophosphate signaling. Endocrinology 150:803–811

Ren H, Stiles GL (1999) Dexamethasone stimulates human A1 adenosine receptor (A1AR) gene expression through multiple regulatory sites in promoter B. Mol Pharmacol 55:309–316

Reppert SM, Weaver DR (2001) Molecular analysis of mammalina circadian rhythms. Annu Rev Physiol 63:647–676

Revollo JR, Grimm AA, Imai SI (2004) The NAD biosynthesis pathway mediated by nicotinamide phosphoribosyltransferase regulates Sir2 activity in mammalian cells. J Biol Chem 279:50754–50763

Ribelayga C, Mangel SC (2005) A circadian clock and light/dark adaptation differentially regulate adenosine in the mammalian retina. J Neurosci 25:215–222

Ripperger JA, Merrow M (2011) Perfect timing: epigenetic regulation of the circadian clock. FEBS Lett 585:1406–1411

Ruby NF, Hwang CE, Wessells C, Fernandez F, Zhang P, Sapolsky R, Heller HC (2008) Hippocampal-dependent learning requires a functional circadian system. Proc Natl Acad Sci USA 105:15593–15598

Rüger M, Scheer F (2009) Effects of circadian disruption on the cardiometabolic system. Rev Endocr Metab Disord 10:245–260

Rust MJ, Golden SS, O'Shea EK (2011) Light-driven changes in energy metabolism directly entrain the cyanobacterial circadian oscillator. Science 331:220–223

Schibler U, Ripperger J, Brown SA (2003) Peripheral circadian oscillators in mammals: time and food. J Biol Rhythms 18:250–260

Shibata S (2004) Neural regulation of the hepatic circadian rhythm. Anat Rec A Discov Mol Cell Evol Biol 280A:901–909

Sigworth LA, Rea MA (2003) Adenosine A1 receptors regulate the response of the mouse circadian clock to light. Brain Res 960:246–251

Stokkan K-A, Yamazaki S, Tei H, Sakaki Y, Menaker M (2001) Entrainment of the circadian clock in the liver by feeding. Science 291:490–493

Tononi G, Cirelli C (2006) Sleep function and synaptic homeostasis. Sleep Med Rev 10:49–62

Tosini G, Menaker M (1996) Circadian rhythms in cultured mammalian retina. Science 272:419–421

Um J-H, Pendergast JS, Springer DA, Foretz M, Viollet B, Brown A, Kim MK, Yamazaki S, Chung JH (2011) AMPK regulates circadian rhythms in a tissue- and isoform-specific manner. PLoS One 6:e18450

Um JH, Yang S, Yamazaki S, Kang H, Viollet B, Foretz M, Chung JH (2007) Activation of 5′-AMP-activated kinase with diabetes drug metformin induces casein kinase Iε (CKIε)-dependent degradation of clock protein mPer2. J Biol Chem 282:20794–20798

Wang H-Y, Huang R-C (2004) Diurnal modulation of the Na+/K+-ATPase and spontaneous firing in the rat retinorecipient clock neurons. J Neurophysiol 92:2295–2301

Wang Y-C, Huang R-C (2006) Effects of sodium pump activity on spontaneous firing in neurons of the rat suprachiasmatic nucleus. J Neurophysiol 96:109–118

Watanabe A, Moriya T, Nisikawa Y, Araki T, Hamada T, Shibata S, Watanabe S (1996) Adenosine A1-receptor agonist attenuates the light-induced phase shifts and fos expression in vivo and optic nerve stimulation-evoked field potentials in the suprachiasmatic nucleus in vitro. Brain Res 740:329–336

Westermeier F, Salomón C, González M, Puebla C, Guzmán-Gutiérrez E, Cifuentes F, Leiva A, Casanello P, Sobrevia L (2011) Insulin restores gestational diabetes mellitus-reduced adenosine transport involving differential expression of insulin receptor isoforms in human umbilical vein endothelium. Diabetes 60:1677–1687

Womac AD, Burkeen JF, Neuendorff N, Earnest DJ, Zoran MJ (2009) Circadian rhythms of extracellular ATP accumulation in suprachiasmatic nucleus cells and cultured astrocytes. Eur J Neurosci 30:869–876

Yang JN, Wang Y, Garcia-Roves PM, Björnholm M, Fredholm BB (2010) Adenosine A3 receptors regulate heart rate, motor activity and body temperature. Acta Physiol 199:221–230

Zhang EE, Liu Y, Dentin R, Pongsawakul PY, Liu AC, Hirota T, Nusinow DA, Sun X, Landais S, Kodama Y et al (2010) Cryptochrome mediates circadian regulation of cAMP signaling and hepatic gluconeogenesis. Nat Med 16:1152–1156

Chapter 12
Adenosine in the Immune System

György Haskó, Balázs Koscsó, and Balázs Csóka

Abstract Adenosine is an endogenous purine nucleoside, which is produced during metabolic stress, including inflammatory insults and tissue injury, and its extracellular level robustly increases in response to tissue damage. Extracellular adenosine orchestrates immune/inflammatory responses by activating adenosine receptors on the surface of innate and adaptive immune cells. The four adenosine receptors, A_1, A_{2A}, A_{2B}, and A_3, are G protein-coupled membrane receptors and can be expressed by every immune cell. This book chapter summarizes our knowledge about the immune modulatory effect of adenosine on immune/inflammatory responses by focusing on its effects on various immune cell populations.

Keywords Asthma • Atherosclerosis • Caffeine • Cancer • Metabolic syndrome • Theophylline • Vascular

12.1 Introduction

The purine nucleoside adenosine is an endogenous modulatory molecule, and its physiologic role has been intensively examined for decades because of the plethora of its actions on different organ functions. Under physiological conditions, adenosine

G. Haskó (✉)
Department of Surgery, New Jersey Medical School,
University of Medicine and Dentistry of New Jersey,
185 South Orange Ave, University Heights, Newark, NJ 07103, USA

Department of Medical Chemistry, Medical and Health Science Center,
University of Debrecen, Egyetem Tér 1, Debrecen 4032, Hungary
e-mail: haskoge@umdnj.edu

B. Koscsó • B. Csóka
Department of Surgery, New Jersey Medical School,
University of Medicine and Dentistry of New Jersey,
185 South Orange Ave, University Heights, Newark, NJ 07103, USA

S. Masino and D. Boison (eds.), *Adenosine: A Key Link between Metabolism and Brain Activity*, DOI 10.1007/978-1-4614-3903-5_12,

is constantly present in the tissues at low concentration, mostly as a by-product of ATP degradation. However, in response to cellular stress and damage, its extracellular level robustly increases (up to 200-fold), and elevated levels of adenosine are found in conditions of ischemia, hypoxia, inflammation, trauma, and sepsis (Hasko and Cronstein 2004; Hasko et al. 2002, 2004; Latini et al. 1999; Linden 2001; Martin et al. 2000; Sottofattori et al. 2001; Sperlagh et al. 2000). The bioavailability of adenosine depends on its production, release, cellular uptake, and enzymatic degradation, the processes which are strictly controlled. Under pathophysiological conditions, such as hypoxia and ischemia, intracellular dephosphorylation of ATP to adenosine is increased because of intensive energy demand. Adenosine, when reaching a critical intracellular concentration, is carried into the extracellular space via nucleoside transporters. The other dominant pathway of extracellular adenosine accumulation becomes activated during metabolic stress, when ATP and ADP are released from intracellular pools to the surrounding tissues, and extracellular adenosine is formed from ATP and ADP by a cascade of ectonucleotidases, including CD39 (nucleoside triphosphate diphosphohydrolase, [NTPDase]) and CD73 (5′-ectonucleotidase, [Ecto5′NTase]) (Eltzschig et al. 2004; Linden 2001; Thompson et al. 2004; Zimmermann 2000). These metabolic enzymes are abundant on both leukocytes and endothelial cells. Adenosine-converting enzymes such as adenosine deaminase and adenosine kinase also regulate the availability of the adenosine (Fredholm et al. 2001b; Pastor-Anglada et al. 2001). After being formed, adenosine diffuses to the cell membrane of adjacent cells where it binds its specific G-protein-coupled receptors, termed adenosine receptors (Fredholm et al. 2001b, 2011; Ralevic and Burnstock 1998). There are four different subtypes of adenosine receptor, A_1, A_{2A}, A_{2B}, and A_3 (Fredholm et al. 2001a, 2000), and adenosine receptors are expressed on a wide range of immune cell types. A remarkable body of evidence has confirmed the role of adenosine as an endogenous immune modulator acting through its receptors (Hasko et al. 2008). Consequently, adenosine and adenosine receptors recently have become eminent therapeutic targets in various inflammatory diseases. In this chapter, we provide a general overview about the role of extracellular adenosine in immunity and inflammation by focusing on its effects on various immune cell populations.

12.2 Effect of Adenosine on the Innate Immune System

The innate immune system comprises the first line of defense against infectious agents. Adenosine is recognized for controlling numerous innate immune functions, and adenosine receptors are abundant on the surface of the various innate immune cell types. In this section, we focus on the numerous roles of adenosine in modulating the functions of the prominent cell types of the innate immune system, including neutrophils, monocytes/macrophages, dendritic cells (DC)s, mast cells, and natural killer (NK) cells.

12.2.1 Adenosine and Neutrophils

During the acute phase of inflammation, which can be a result of infection or injury, neutrophils are recruited to the site of infection or injury, respectively, within minutes. These immune cells are essential players in the destruction of pathogens and are important for an appropriate immune response. Stringent regulation of neutrophil function is critical, and adenosine is a potent modulator of virtually every neutrophil function (Cronstein 1994; Fredholm 1997). Neutrophils express all four adenosine receptors (Bouma et al. 1997; Gessi et al. 2002; Martini et al. 1991; Salmon et al. 1993), and their expression pattern changes during neutrophil activation (Fortin et al. 2006). Importantly, neutrophils release ATP and adenosine during their endothelial transmigration, and endothelial cells express CD39 and CD73, which sequentially generate adenosine from ATP and ADP (Eltzschig et al. 2004). Neutrophil migration is facilitated by the adenosine generated, and A_3 receptors that have been found abundantly at the leading edge of neutrophils appear to mediate the effect of adenosine in promoting migration (Chen et al. 2006).

In addition to A_3 receptors, adenosine binds to A_{2A} receptors, and activation of A_{2A} receptors on neutrophils reduces collateral damage caused by neutrophil activation, and the protective mechanisms include inhibition of the generation of reactive free radicals, production of cytokines and leukotriene B4, and expression of adhesion molecules (Cronstein et al. 1985; Krump et al. 1997). In 1983 an influential study by Cronstein et al. (1983) revealed that adenosine inhibited the generation of superoxide anion by neutrophils. The adenosine A_{2A} receptor turned out to be responsible for this effect (Cronstein et al. 1985). Additionally, contributions of the A_{2B} and A_3 receptors to downregulating the release of microbicidal compounds have also been described (Bouma et al. 1997; van der Hoeven et al. 2011; Visser et al. 2000). Neutrophils produce a variety of inflammatory cytokines and chemokines and neutrophil A_{2A} receptors are able to decrease the synthesis of a range of these factors including tumor necrosis factor (TNF)-α, CCL3, CCL4, CXCL2, and CCL20 (McColl et al. 2006; Thiel and Chouker 1995).

The effect of adenosine on neutrophil adhesion is controversial. Adenosine at low concentration enhances adhesion but the effect may be due to endothelial A_1 receptors. On the contrary, higher levels of adenosine, mainly acting via A_{2A} receptors, inhibit neutrophil recruitment to the endothelium probably by decreasing neutrophil integrin expression that mediates vascular cell adhesion molecule (VCAM)-1-dependent adhesion (Sullivan et al. 2004). A_{2B} receptors on neutrophils also have an impact on the adhesion of neutrophil granulocytes to endothelial cells, as it has been documented that A_{2B} receptor activation diminishes production of vascular-endothelial growth factor (VEGF), and reduces the ability of neutrophil leukocytes to transmigrate through endothelial membranes (Wakai et al. 2001). Furthermore, the involvement of endothelial A_{2B} receptors in regulating neutrophil transmigration is also well documented. Namely, A_{2B} receptors on endothelial cells inhibit leukocyte adhesion and are important in regulating barrier function to leukocyte migration (Yang et al. 2006).

12.2.2 Adenosine and Monocytes/Macrophages

Macrophages are a heterogeneous population of mononuclear phagocytes that exist in various tissues as resident macrophages or can be found in the circulation as monocytes. After their recruitment into infected or damaged tissues, monocytes become macrophages through a variety of differentiation processes. The functions of activated macrophages are diverse. These cells are professional phagocytes and during Fcγ receptor-mediated phagocytosis macrophages ingest, degrade and present peptides derived from pathogens. Macrophages are a major source of various pro- and anti-inflammatory mediators and are master regulators of immune and inflammatory responses. Moreover, macrophages have pivotal roles in the resolution of inflammation and restriction of excessive tissue injury (Gilroy et al. 2004) and there is growing evidence that adenosine can interfere with these responses (Csoka et al. 2007).

All four types of adenosine receptors are expressed on both monocytes (Merrill et al. 1997; Munro et al. 1998; Nguyen et al. 1999; Suzuki et al. 2000) and macrophages (Eppell et al. 1989; McWhinney et al. 1996; Salmon et al. 1993). The cell surface pattern of adenosine receptors changes during maturation, and these changes can determine the effect of adenosine on this cell population. One classic example of this effect is the regulation of Fcγ receptor-mediated phagocytosis by adenosine; as in undifferentiated monocytes the dominant stimulatory influence of A_1 receptors is replaced by A_2 receptor-mediated inhibitory effect on phagocytosis, as reviewed by Hasko et al. (Csoka et al. 2007). Macrophages produce reactive nitrogen and oxygen species during phagocytosis and adenosine potently diminishes the secretion of these molecules. The first evidence of the inhibitory role of adenosine on respiratory burst by fMLP-activated human monocytes was demonstrated by Leonard et al. (1987). Subsequent reports indicated the inhibitory role of A_3 receptors in respiratory burst by human monocytes/macrophages (Nguyen et al. 1999; Thiele et al. 2004). Nitric oxide (NO) is another product produced during the respiratory burst of macrophages and inducible NO synthetase (iNOS) is the main producing enzyme of NO. Our group has demonstrated that A_1 and A_{2A} receptor agonists diminished NO generation (Hasko et al. 1996). Subsequently, A_{2B} receptors have also been shown to decrease iNOS expression (Xaus et al. 1999). Conversely, other studies demonstrated the increasing effect of adenosine or various adenosine analogues on NO production by macrophages (Hon et al. 1997; Min et al. 2000), highlighting that further studies are warranted in deciphering the involvement of the different adenosine receptors in regulating NO release.

The involvement of adenosine in regulating of monocyte/macrophage cytokine production has drawn major interest, because these macrophage-derived inflammatory mediators are critical orchestrators of inflammatory responses. Pioneering studies demonstrated employing adenosine receptor analogues, that adenosine acting probably via A_{2A} or A_{2B} receptors diminishes TNF-α release by human monocytes (Bouma et al. 1994; Le Vraux et al. 1993). We and others found that adenosine receptors suppressed TNF-α production by lipopolysaccharide

(LPS)-stimulated mouse macrophages (Hasko et al. 1996; Martin et al. 2006); however, the receptor subtypes were poorly characterized. Recent studies utilizing adenosine receptor knock-out (KO) animals have helped to solve the question of which adenosine receptors were involved in decreasing TNF-α production. Our results with A_{2A}R KO mice confirmed the contribution of A_{2A} receptors in the inhibition of TNF-α production by LPS-challenged macrophages (Hasko et al. 2000), and later A_{2A} receptor-independent—A_{2B} receptor-mediated—effect of adenosine was also confirmed (Chen et al. 2009; Kreckler et al. 2006).

Fewer studies have been performed to assess the role of adenosine receptors in the regulation of interleukin (IL)-12 production by monocytes/macrophages. Our laboratory documented that adenosine acting via both A_{2A} receptors and independently of them is able to suppress IL-12 release by LPS-induced murine macrophages (Hasko et al. 2000). More recently, the contribution of A_3 receptor activation in downregulating IL-12 production has been documented in human monocytes (la Sala et al. 2005).

In addition to suppressing proinflammatory cytokine production by monocytes/macrophages, adenosine augments the production of anti-inflammatory cytokine IL-10. In human monocytes stimulated by LPS, TNF-α or H_2O_2, adenosine upregulated IL-10 production (Le Moine et al. 1996). The enhancing effect of adenosine on IL-10 release in LPS-stimulated mouse macrophages was confirmed in our laboratory in the same year (Hasko et al. 1996). Subsequent studies employing selective adenosine receptor agonists and antagonists documented the predominant role of A_{2A} and A_{2B} receptors in mediating the stimulatory effect of adenosine on LPS-elicited IL-10 production by human monocytes (Link et al. 2000) or by murine RAW 264.7 macrophages (Nemeth et al. 2005). Additionally, Koscso and colleagues have documented the stimulatory role of A_{2B} receptor activation on LPS- or peptidoglycan-activated IL-10 production by murine microglial cells (Koscso et al. 2012). The augmenting effect of adenosine receptor activation on whole bacteria-induced IL-10 production has been investigated in detail. Khoa and coworkers using specific adenosine receptor agonists found that A_{2A} receptor specific agonist 2-*p*-(2-carboxyethyl)phenethyl-amino-5′-*N*-ethyl-carboxamido-adenosine (CGS21680) enhanced *Staphylococcus aureus*-induced IL-10 production by human monocytes (Khoa et al. 2001). Recently, our group has confirmed the dominant role of A_{2A} receptor activation in enhancing bacteria-stimulated IL-10 production by murine macrophages, as CGS21680 augmented heat-killed *Escherichia coli*-elicited IL-10 production, and the stimulatory effect of adenosine was abolished in A_{2A}R KO peritoneal macrophages (Csoka et al. 2007).

Unlike classical macrophage activation, alternative macrophage activation takes place in a Th2 cytokine -IL-4 and IL-13-rich environment, and imparts immunomodulatory and anti-inflammatory properties on macrophages (Van Ginderachter et al. 2006) with enhanced expression of specific genes, including *arginase-1*, *Ym1*, and the tissue inhibitor of metalloproteinase *(TIMP)-1*. Alternative activated macrophages (aaMφ)s possess high tissue repairing capacity and promote the resolution of inflammation by supporting angiogenesis and removing damaged tissue debris. The term "aaMφ" sometimes is more widely used and includes various anti-inflammatory macrophage phenotypes induced by as diverse stimuli as immune complexes, IL-10, glucocorticoids, and apoptotic cells. As we detailed above, the role of adenosine receptors in regulating classical macrophage activation has been studied extensively, while the role of adenosine receptors in governing alternative macrophage activation still is less well understood. We have recently observed that adenosine stimulates alternative macrophage activation, as it upregulated IL-4 and IL-13-induced arginase-1 and TIMP-1 expression via A_{2A} and A_{2B} adenosine receptor activation (Csoka et al. 2012).

12.2.3 Adenosine and Dendritic Cells

DCs are professional antigen-presenting immune cells with a specialized function to trigger the activation of naïve T lymphocytes and thereby initiate adaptive immune responses. These heterogeneous cells reside in the peripheral tissues or in the bloodstream in an immature state. In response to infection, DCs are able to sense the invaders and to take up antigens, which processes are then followed by the maturation and migration of these cells to the lymph nodes in order to activate T cells.

Adenosine has been shown to regulate both immature and mature DC functions and the effect of adenosine depends on the maturation level of DCs. The pattern of adenosine receptor expression on DCs changes during their development. Immature human myeloid DCs (mDC)s express dominantly A_1 and A_3 receptors, while mature mDCs predominantly express A_{2A} receptors (Fossetta et al. 2003; Panther et al. 2001). Similarly, human immature plasmacytoid DCs (pDC)s expressed only functional A_1 receptors, and upon their maturation, the level of A_1 receptor was decreased and the expression of A_{2A} receptor was robustly increased (Schnurr et al. 2004). In harmony with these expressional changes throughout DC maturation, the effect of adenosine on these cells is bidirectional. Adenosine enhanced chemotaxis of immature human mDCs and pDCs via their A_1 receptors (Panther et al. 2001; Schnurr et al. 2004). Besides the promoting effect on DC chemotaxis, adenosine increased intracellular calcium levels (Fossetta et al. 2003) and antigen uptake by immature DCs (Panther et al. 2003). Altogether, it seems that stimulation of A_1 and A_3 adenosine receptors on immature DC population promotes their migration to the site of infection or tissue injury. On the other hand, mature human DCs expressing dominantly A_{2A} receptors and adenosine acting on this receptor subtype acts as a negative regulator of DC functions. Adenosine A_{2A} receptor activation on mature DCs changes their cytokine expression, shifting their profile toward an anti-inflammatory type, with reduced IL-12, IL-6, IFN-α, and TNF-α release and enhanced IL-10 production (Panther et al. 2003, 2001; Schnurr et al. 2004). Additionally, adenosine also affects the chemokine profile of mature DCs; it decreases the production of type 1 lymphocyte chemokine CXCL10 and augments the level of type 2 lymphocyte chemokine CCL17, thereby skewing DCs toward a T helper (Th)2 phenotype (Panther et al. 2003).

Recently, A_{2B} adenosine receptors have emerged as a dominant regulator of DCs. Novitskiy and colleagues have demonstrated that A_{2B} receptor activation shifts the differentiation of DCs to a unique cell population that shows both DC and macrophage characteristics (Novitskiy et al. 2008). The ability of this cell population to induce T-cell proliferation was significantly decreased, but they expressed high levels of VEGF and IL-8, suggesting increased angiogenic activity of these cells. Moreover, this cell population induced by adenosine also showed an anti-inflammatory cytokine profile with elevated IL-10 and transforming growth factor (TGF)-β secretion. In another set of studies, the nonselective adenosine receptor agonist adenosine-5′*N*-ethylcarboxamide (NECA) was found to decrease IL-12 and TNF-α release by LPS-stimulated DCs (Wilson et al. 2009). NECA was able to augment the stimulatory effect of LPS on IL-10 production; moreover, NECA-treated DCs expressed lower level MHC II and CD86, and the effect of NECA disappeared in A_{2B}R KO DCs (Wilson et al. 2009).

In recent years, a prominent role of A_{2B} receptors in governing DC function in hypoxic conditions has also become well documented. Matrix metalloproteinase (MMP)-9 is important during DC migration across extracellular matrix and Zhao et al. (2008) showed that hypoxia diminished the production of MMP-9 by human monocyte-derived DCs, and since this effect was abolished by the A_{2B} receptor antagonist *N*-(4-Cyanophenyl)-2-[4-(2,3,6,7-tetrahydro-2,6-dioxo-1,3-dipropyl-1*H*-purin-8-yl)phenoxy]-acetamide (MRS1754), it was suggested that A_{2B} receptors mediated the effect of hypoxia. Moreover, the same group found that MRS1754 potently increased the production IL-12p70 and TNF-α by hypoxic mDCs (Yang et al. 2010).

Finally, very recently, a study from Ernst and coworkers demonstrated that A_{2B} receptor activation on DC shifted the differentiation of naïve Th cells toward the Th17 phenotype in a coculture system where naïve T cells were cultured together with DCs and were activated by CD3 engagement (Wilson et al. 2011). Mechanistically, NECA robustly increased the production of IL-6 by DCs, and this increased IL-6 forced naïve T cell differentiation toward the Th17 phenotype (Wilson et al. 2011).

12.2.4 Adenosine and Mast Cells

Mast cells are tissue resident innate immune cells that contain many granules, which are rich in histamine and heparin. Although best known for their role in allergy and anaphylaxis, mast cells play an important tissue protective role as well, being intimately involved in wound healing and defense against pathogens. During activation, mast cells rapidly release the contents of their granules into the surrounding tissue. There is a large body of evidence that mast cell mediators, such as histamine and inflammatory cytokines are key components in the airway constricting effect of adenosine that occurs in chronic airway inflammation. All four types of adenosine receptors are present on mast cells, but it is still not clear which adenosine receptor is responsible for the increased mast cell activation in response to adenosine (Polosa and Holgate 2006). Adenosine by activating A_1 and A_{2A} adenosine receptors on mast cells decreases the release of histamine and other inflammatory mediators (Spicuzza et al. 2003; Yip et al. 2011). However, adenosine enhances degranulation and inflammatory mediator production of human mast cells by acting via A_{2B} and A_3 receptors (Feoktistov and Biaggioni 1995; Feoktistov et al. 2003; Ramkumar et al. 1993; Ryzhov et al. 2004, 2008). Recently, Ryzhov and colleagues (Ryzhov et al. 2004) have shown that A_{2B} receptor stimulation on human mast cells induces the production of IL-4, which is a well-known inducer of chronic airway inflammation. Additionally, they have documented that adenosine through A_{2B} receptor activation increases the secretion of IL-13, another key asthma-promoting cytokine by murine mast cells (Ryzhov et al. 2008). These observations together suggest a strong connection between mast cell adenosine A_{2B} receptor activation and the pathogenesis of asthma.

12.2.5 Adenosine and NK Cells

NK cells have common progenitor cells with T cells, but unlike progenitor T cells that acquire appropriate T cell characteristics in the thymus, NK cells develop in the bone marrow. NK cells are a type of cytotoxic lymphocyte that constitutes a major component of the innate immune system. NK cells play a major role in the rejection of tumors and cells infected by viruses. They destroy malfunctioning cells by releasing small cytoplasmic granules of proteins including perforin and granzyme that cause the target cell to die by apoptosis. These cells are a prominent source of various cytokines and chemokines, such as TNF-α, IFN-γ, macrophage inflammatory protein (MIP)-1α, MIP-1β, and regulated on activation T cells expressed (RANTES); that are released by NK cells in response to pathogens, such as viral or parasitic infection.

Adenosine A_1, A_{2A}, and A_{2B} receptor expression has been detected on murine NK cells (Raskovalova et al. 2005) and the possible presence of functional A_3 receptor was suggested based on observations that selective A_3 receptor agonists were efficacious in modulating NK cell functions (Harish et al. 2003; Jeffe et al. 2009). Upon stimulation with cytokines, such as IL-2 and IL-12, NK cells develop into lymphokine-activated killer (LAK) cells, and adenosine is able to modify the activity of LAK cells. A_{2A} adenosine receptor agonists decrease the cytotoxic activity of these cells (Lokshin et al. 2006; Miller et al. 1999; Raskovalova et al. 2006; Williams et al. 1997), while A_1 and A_3 receptor agonists have promoting effects on NK and LAK cell functions (Harish et al. 2003). Both the stable adenosine analogue 2-chloro-adenosine and adenosine were able to suppress the killing of lung carcinoma cells by mouse LAK cells dominantly through activation of A_{2A} receptors (Raskovalova et al. 2005). Furthermore, a marginal involvement of A_{2B} receptors was also suggested in the adenosine-regulated killing activity, but the role of A_1 and A_3 receptors were excluded, because adenosine and 2-chloro-adenosine-treated

LAK cells generated from A_1 and A_3 KO mice showed similar killing activity when compared to LAK cell obtained from WT littermates (Raskovalova et al. 2005). Conversely, oral administration of the A_3 receptor agonist 2-chloro-*N*6-(3-iodobenzyl)-adenosine-5′-*N*-methyluronamide (Cl-IB-MECA) enhanced the cytotoxic activity of NK cells against B16-F10 melanoma cells in mice (Harish et al. 2003). Moreover, Cl-IB-MECA augmented the systemic level of IL-12 and suppressed the growth of B16-F10 melanoma cells. Additionally, another A_3 receptor agonist, N^6-(3-Iodobenzyl) adenosine-5′-*N*-methyluronamide (IB-MECA) synergistically increased IFN-α-induced IFN-γ production by human NK cells in vitro (Jeffe et al. 2009). In addition to the stimulatory effect of A_3 receptor agonists, adenosine predominantly via A_{2A} receptor activation has an overall suppressive effect on NK cell activity, indicating that extracellular adenosine can also be a potent inhibitor of tumor cell killing initiated by NK cells.

12.3 Effect of Adenosine on the Adaptive Immune System

The adaptive immune system is composed of highly specialized cells that eliminate or prevent pathogenic infection or cancer. The adaptive immune response has an ability to recognize and remember specific pathogens, and to mount stronger attacks each time the pathogen is encountered. The characteristic cells of the adaptive immune system are B and T lymphocytes. B cells and T cells are derived from the same multipotent hematopoietic stem cells and are morphologically indistinguishable from one another until after they are activated. B cells play a major role in the humoral immune response, whereas T cells are intimately involved in cell-mediated immune responses. In an adult organism, the peripheral lymphoid organs contain a mixture of B and T cells in at least three stages of differentiation. Naïve cells are mature cells that have left the bone marrow or thymus and entered the lymphatic system but have not yet met with antigen. Effector cells are antigen-activated mature cells that are involved in defeating pathogens; and memory cells are survivors of past infections with preserved "memory." The members of the adaptive immune system express adenosine receptors and myriad lines of evidence have shown that adenosine, via mainly A_{2A} adenosine receptor occupation, regulates the function of these cells.

12.3.1 Adenosine and T Lymphocytes

T lymphocytes are composed of a heterogeneous population of cells, including $CD4^+$ (Th), $CD8^+$ cytotoxic T lymphocytes (CTL)s, memory T cells, regulatory (Treg) cells, Natural Killer T (NKT) cells, and γδT cells. Professional antigen presenting cells, such as dendritic cells present processed antigens to naïve T lymphocytes. In the presence of appropriate costimulatory signals, T cells differentiate into various subtypes directed by a specific cytokine/chemokine milieu. CD4$^+$ T cells can differentiate into divergent classes of Th cells, including Th1, Th2, Th17, or Treg cells. Th1 cells produce IFN-γ, a potent inducer of macrophage activation that promotes cell-mediated immunity and inflammation. Th2 cells produce IL-4 and IL-13, cytokines that induce humoral immunity against extracellular pathogens and which are able to skew macrophage activation toward an alternatively activated state that is involved in resolving chronic inflammatory conditions. T cells express A_{2A}, A_{2B}, and A_3 receptors, but the level of A_1 receptor on these cells is very low (Huang et al. 1997; Koshiba et al. 1997, 1999; Mirabet et al. 1999). Ample evidence has been presented in the literature that A_{2A} receptors are the dominant receptors influencing CD4$^+$ cell functions. Studies employing A_{2A}R KO animals have shown a major inhibitory effect of A_{2A} receptor activation on IL-2 production and proliferation of naïve CD4$^+$ cells after T cell receptor engagement (Naganuma et al. 2006; Sevigny et al. 2007). A_{2A} receptor activation also suppresses the synthesis of both Th1 and Th2 cytokines by naïve CD4$^+$ (Lappas et al. 2005; Naganuma et al. 2006), as well as polarized Th1 and Th2 lymphocytes (Csoka et al. 2008). These recent studies have questioned a previous hypothesis that lymphocyte A_{2A} receptors skew immune cell responses in the direction of Th2 dominance. A_{2A} receptors downregulate expression of the positive costimulatory protein CD40L and augment expression of the negative costimulatory molecules cytotoxic T-lymphocyte antigen (CTLA)-4 and programmed cell death (PD)-1 (Sevigny et al. 2007). TCR reactivation of previously activated mature T cells by an antigen or mitogen causes activation-induced cell death, which is mediated by the induction of the expression of Fas ligand (FasL) and its interaction with Fas. Recent results from our laboratory have demonstrated that A_{2A} receptor activation suppresses the activation-induced cell death of peripheral T cells via inhibiting the extrinsic apoptotic pathway and Fas/FasL mRNA accumulation (Himer et al. 2010).

It is now well accepted that in addition to Th1 and Th2 cells, Th17 cells represent functionally unique cells that develop from a common naive precursor, and recent evidence suggests a role of adenosine in regulating the differentiation and function of these cell populations. Zarek and colleagues demonstrated that A_{2A} receptor activation led to a marked inhibition of IL-2 production, which then limited the generation of Th1 and Th17 cells in vivo (Zarek et al. 2008). In contrast, as alluded to earlier in the text, a subsequent study revealed that adenosine preferentially promoted Th17 polarization with a characteristic cytokine and transcription factor signature, including IL-17, IL-22, IL-23R, and retinoic acid receptor (RAR)-related orphan receptor (ROR)γ (Wilson et al. 2011). In this case, DC A_{2B} receptors were necessary to upregulate IL-6 production by DCs, and this IL-6, in turn, promoted the Th17 polarization of naïve T cells.

CTLs destroy virally infected cells and tumor cells, and are also implicated in transplant rejection. Adenosine generally inhibits a wide variety of CTL responses, including their adhesion to tumor cells. Adenosine diminished IL-2 and TNF-α release by murine type 1 (Tc1) and type 2 (Tc2) CTLs through A_{2A} receptors, but the production of IFN-γ or IL-4 and IL-5 were not affected by A_{2A} receptor engagement (Erdmann et al. 2005). By inhibiting IL-2 production by T cells in the tumor environment, adenosine blocks the clonal expansion of activated CTLs and this phenomenon together with the decreased TNF-α levels results in an anergic, anti-inflammatory milieu at the surrounding cancerous tissue. In agreement with these results, pharmacological A_{2A} receptor blockade or siRNA-mediated A_{2A} receptor downregulation increased the ability of CTLs to fight against cancerous cells (Ohta et al. 2006). Based on recent studies, A_{2A} receptor activation actually allows the expansion of CTLs, but it prohibits their activation thereby preventing collateral damage of the surrounding tissue. This protective effect occurs by a strong inhibition of the cytotoxic and cytokine producing activity of T cells (Ohta et al. 2009). In another compelling study, Clayton and colleagues found that cancer cells release exosomes that express CD39 and CD73, and these exosomes effectively suppress T cell cytotoxicity by their adenosine generating activity (Clayton et al. 2011).

Treg cells can produce adenosine and thereby regulate immune cell function. The first evidence that adenosine might be involved in Treg functions was derived from an in vivo experimental colitis study, where Tregs failed to control colitis by effector T cells that lacked the A_{2A} receptor, suggesting that adenosine contributed to the anti-inflammatory effect conferred by Tregs (Naganuma et al. 2006). In subsequent studies, murine Tregs were shown to express CD73 and CD39 and that Tregs can regulate inflammation in an adenosine-dependent manner (Borsellino et al. 2007; Deaglio et al. 2007). Furthermore, Tregs from CD73 deficient mice were unsuccessful in preventing gastritis when compared with Tregs derived from WT animals (Alam et al. 2009). Tregs express A_{2A} receptors and A_{2A} receptor activation augments the level of (forkhead box P) FOXP3 transcription factor, which is a key Treg-inducing factor, in Tregs (Zarek et al. 2008). Additionally, FOXP3 upregulates CD39 expression thereby creating a positive feedback circle for amplifying adenosine signaling by Tregs (Borsellino et al. 2007). Altogether, these data strongly support the concept that adenosine is a crucial mediator of Treg responses.

Invariant NKT (iNKT) cells are specialized lymphocytes that bridge the innate and adaptive immune systems. These cells express an invariant type of T cell receptor (TCR) (Vα/β TCR) alongside the NK cell marker NK1.1. iNKT cells, unlike conventional T cells, sense pathogen- or injured tissue-derived glycolipid antigens that are presented on the MHCI-related molecule CD1d. In response to antigenic activation, iNKT cells rapidly become activated and produce various cytokines. iNKT cells express all four types of adenosine receptors, but the level of A_{2A} receptors is highest (Nowak et al. 2010). A_{2A} receptor activation abolishes α-galactoceramid-elicited IFN-γ release by iNKT cells, but the production of IL-10, IL-4, and TGFβ is upregulated (Nowak et al. 2010). These initial results suggest that A_{2A} receptors on iNKT cells could become a therapeutic target, as activation of these cells is involved in the progression of many diseases, including arthritis, type 1 diabetes, and atherosclerosis (Yamamura et al. 2007).

12.3.2 Adenosine and B Lymphocytes

Mouse B cells possess abundant A_{2A} receptor mRNA, but very low levels of A_{2B} and A_3 receptor mRNA, and do not express A_1 mRNA (Lukashev et al. 2003) In contrast, the A_{2A} receptor is not detectable at the protein level on peripheral human B lymphocytes (Koshiba et al. 1999), and there is no evidence for the expression of the other three types of adenosine receptors on human B cells. Adenosine was shown to suppress the proliferation and antibody production of mitogen-activated murine B cells three decades ago (Seegmiller et al. 1977). Subsequent studies found that adenosine affects B cell functions through inhibiting antigen- or TNF-α-induced IκBα phosphorylation/degradation and consequently the activation of the NF-κB system (Majumdar and Aggarwal 2003; Minguet et al. 2005). Finally, adenosine has been shown to protect human B cells from B cell receptor-induced apoptosis mostly via A_{2A} and A_{2B} receptors and by downregulating caspase-3 activation (Sakata et al. 2000).

12.4 Conclusion

Inflammatory responses can be categorized into three overlapping phases, including (1) initiation of inflammation, (2) modulation of ongoing immune responses, and (3) downregulation of immune responses and tissue repair. Adenosine acting on the different immune cell populations fine-tunes various pro- and anti-inflammatory responses mainly in the second and third stages. These immunomodulatory effects of adenosine protect the tissue from collateral damage and shift the activation of various immune cells toward the resolution of inflammation. These protective effects of adenosine are important for the preservation of tissue homeostasis. During the second phase of inflammatory processes, adenosine, which is produced through enzymatic metabolism of ATP by surrounding tissue and immune cells, efficiently downregulates excessive neutrophil responses, inhibits lymphocyte effector and helper functions, and contributes to the skewing of macrophage activation toward an alternatively activated profile with high tissue repairing capacity. These suppressive effects of adenosine are generally attributable to A_{2A} and A_{2B} receptor activation on innate and adaptive immune cells. In the final phase of inflammation, adenosine activates A_{2A} and A_{2B} receptors and enhances the tissue restoring capacity of macrophages by promoting angiogenesis and removing damaged tissue debris (Macedo et al. 2007). It should be noted that this protective action of adenosine is operative mostly under acute inflammatory conditions. In contrast, in certain chronic pathological conditions, including asthma and chronic obstructive pulmonary disease, adenosine acting via A_{2B} receptors can aggravate organ dysfunction by promoting mast cell activation, fibrosis, and skewing T cell responses (Polosa and Blackburn 2009; Sun et al. 2006; Zaynagetdinov et al. 2010; Zhou et al. 2009). Moreover, in certain scenarios, such as sepsis when excessive immunosuppression could be harmful for the host in fighting against invading pathogens, the activation of the adenosine system through A_{2A} receptor engagement can be deleterious (Belikoff et al. 2011; Nemeth et al. 2006).

Advances during the last few decades in deciphering the regulatory role of adenosine receptors in different diseases have raised the potential for new pharmacological therapies to certain pathophysiological conditions, such as asthma, arthritis, sepsis, and inflammatory bowel disease. However, the broad tissue/organ distribution of adenosine receptors may limit the applicability of targeting adenosine receptors directly with agonists or antagonists in the treatment of immune-related illness. As an emerging alternative strategy, compounds that modify the function of adenosine metabolizing enzymes and action of adenosine transporter systems at the site of injury may turn out to be more useful to treat immune system related disorders.

Acknowledgment This work was supported by National Institutes of Health Grant R01GM66189, LOG#09065004 (Contract W81XWH-10-1-1015) grant from DOD, and the Hungarian Research Fund (CK 78275).

References

Alam MS, Kurtz CC, Rowlett RM, Reuter BK, Wiznerowicz E, Das S, Linden J, Crowe SE, Ernst PB (2009) CD73 is expressed by human regulatory T helper cells and suppresses proinflammatory cytokine production and Helicobacter felis-induced gastritis in mice. J Infect Dis 199:494–504

Belikoff B, Hatfield S, Sitkovsky M, Remick DG (2011) Adenosine negative feedback on A2A adenosine receptors mediates hyporesponsiveness in chronically septic mice. Shock 35: 382–387

Borsellino G, Kleinewietfeld M, Di Mitri D, Sternjak A, Diamantini A, Giometto R, Hopner S, Centonze D, Bernardi G, Dell'Acqua ML et al (2007) Expression of ectonucleotidase CD39 by Foxp3+ Treg cells: hydrolysis of extracellular ATP and immune suppression. Blood 110:1225–1232

Bouma MG, Stad RK, van den Wildenberg FA, Buurman WA (1994) Differential regulatory effects of adenosine on cytokine release by activated human monocytes. J Immunol 153:4159–4168

Bouma MG, Jeunhomme TM, Boyle DL, Dentener MA, Voitenok NN, van den Wildenberg FA, Buurman WA (1997) Adenosine inhibits neutrophil degranulation in activated human whole blood: involvement of adenosine A2 and A3 receptors. J Immunol 158:5400–5408

Chen Y, Corriden R, Inoue Y, Yip L, Hashiguchi N, Zinkernagel A, Nizet V, Insel PA, Junger WG (2006) ATP release guides neutrophil chemotaxis via P2Y2 and A3 receptors. Science 314:1792–1795

Chen H, Yang D, Carroll SH, Eltzschig HK, Ravid K (2009) Activation of the macrophage A2b adenosine receptor regulates tumor necrosis factor-alpha levels following vascular injury. Exp Hematol 37:533–538

Clayton A, Al-Taei S, Webber J, Mason MD, Tabi Z (2011) Cancer exosomes express CD39 and CD73, which suppress T cells through adenosine production. J Immunol 187:676–683

Cronstein BN (1994) Adenosine, an endogenous anti-inflammatory agent. J Appl Physiol 76:5–13

Cronstein BN, Kramer SB, Weissmann G, Hirschhorn R (1983) Adenosine: a physiological modulator of superoxide anion generation by human neutrophils. J Exp Med 158:1160–1177

Cronstein BN, Rosenstein ED, Kramer SB, Weissmann G, Hirschhorn R (1985) Adenosine; a physiologic modulator of superoxide anion generation by human neutrophils. Adenosine acts via an A2 receptor on human neutrophils. J Immunol 135:1366–1371

Csoka B, Nemeth ZH, Virag L, Gergely P, Leibovich SJ, Pacher P, Sun CX, Blackburn MR, Vizi ES, Deitch EA et al (2007) A2A adenosine receptors and C/EBPbeta are crucially required for IL-10 production by macrophages exposed to Escherichia coli. Blood 110:2685–2695

Csoka B, Himer L, Selmeczy Z, Vizi ES, Pacher P, Ledent C, Deitch EA, Spolarics Z, Nemeth ZH, Hasko G (2008) Adenosine A2A receptor activation inhibits T helper 1 and T helper 2 cell development and effector function. FASEB J 22:3491–3499

Csoka B, Selmeczy Z, Koscso B, Nemeth ZH, Pacher P, Murray PJ, Kepka-Lenhart D, Morris SM Jr, Gause WC, Leibovich SJ et al (2012) Adenosine promotes alternative macrophage activation via A2A and A2B receptors. FASEB J 26:376–386

Deaglio S, Dwyer KM, Gao W, Friedman D, Usheva A, Erat A, Chen JF, Enjyoji K, Linden J, Oukka M et al (2007) Adenosine generation catalyzed by CD39 and CD73 expressed on regulatory T cells mediates immune suppression. J Exp Med 204:1257–1265

Eltzschig HK, Thompson LF, Karhausen J, Cotta RJ, Ibla JC, Robson SC, Colgan SP (2004) Endogenous adenosine produced during hypoxia attenuates neutrophil accumulation: coordination by extracellular nucleotide metabolism. Blood 104:3986–3992

Eppell BA, Newell AM, Brown EJ (1989) Adenosine receptors are expressed during differentiation of monocytes to macrophages in vitro. Implications for regulation of phagocytosis. J Immunol 143:4141–4145

Erdmann AA, Gao ZG, Jung U, Foley J, Borenstein T, Jacobson KA, Fowler DH (2005) Activation of Th1 and Tc1 cell adenosine A2A receptors directly inhibits IL-2 secretion in vitro and IL-2-driven expansion in vivo. Blood 105:4707–4714

Feoktistov I, Biaggioni I (1995) Adenosine A2b receptors evoke interleukin-8 secretion in human mast cells. An enprofylline-sensitive mechanism with implications for asthma. J Clin Invest 96:1979–1986

Feoktistov I, Ryzhov S, Goldstein AE, Biaggioni I (2003) Mast cell-mediated stimulation of angiogenesis: cooperative interaction between A2B and A3 adenosine receptors. Circ Res 92:485–492

Fortin A, Harbour D, Fernandes M, Borgeat P, Bourgoin S (2006) Differential expression of adenosine receptors in human neutrophils: up-regulation by specific Th1 cytokines and lipopolysaccharide. J Leukoc Biol 79:574–585

Fossetta J, Jackson J, Deno G, Fan X, Du XK, Bober L, Soude-Bermejo A, de Bouteiller O, Caux C, Lunn C et al (2003) Pharmacological analysis of calcium responses mediated by the human A3 adenosine receptor in monocyte-derived dendritic cells and recombinant cells. Mol Pharmacol 63:342–350

Fredholm BB (1997) Purines and neutrophil leukocytes. Gen Pharmacol 28:345–350

Fredholm BB, Arslan G, Halldner L, Kull B, Schulte G, Wasserman W (2000) Structure and function of adenosine receptors and their genes. Naunyn Schmiedebergs Arch Pharmacol 362:364–374

Fredholm BB, IJzerman AP, Jacobson KA, Klotz KN, Linden J (2001) International Union of Pharmacology. XXV. Nomenclature and classification of adenosine receptors. Pharmacol Rev 53:527–552

Fredholm BB, IJzerman AP, Jacobson KA, Linden J, Muller CE (2011) International Union of Basic and Clinical Pharmacology. LXXXI. Nomenclature and classification of adenosine receptors—an update. Pharmacol Rev 63:1–34

Gessi S, Varani K, Merighi S, Cattabriga E, Iannotta V, Leung E, Baraldi PG, Borea PA (2002) A(3) adenosine receptors in human neutrophils and promyelocytic HL60 cells: a pharmacological and biochemical study. Mol Pharmacol 61:415–424

Gilroy DW, Lawrence T, Perretti M, Rossi AG (2004) Inflammatory resolution: new opportunities for drug discovery. Nat Rev Drug Discov 3:401–416

Harish A, Hohana G, Fishman P, Arnon O, Bar-Yehuda S (2003) A3 adenosine receptor agonist potentiates natural killer cell activity. Int J Oncol 23:1245–1249

Hasko G, Cronstein BN (2004) Adenosine: an endogenous regulator of innate immunity. Trends Immunol 25:33–39

Hasko G, Szabo C, Nemeth ZH, Kvetan V, Pastores SM, Vizi ES (1996) Adenosine receptor agonists differentially regulate IL-10, TNF-alpha, and nitric oxide production in RAW 264.7 macrophages and in endotoxemic mice. J Immunol 157:4634–4640

Hasko G, Kuhel DG, Chen JF, Schwarzschild MA, Deitch EA, Mabley JG, Marton A, Szabo C (2000) Adenosine inhibits IL-12 and TNF-[alpha] production via adenosine A2a receptor-dependent and independent mechanisms. FASEB J 14:2065–2074

Hasko G, Deitch EA, Szabo C, Nemeth ZH, Vizi ES (2002) Adenosine: a potential mediator of immunosuppression in multiple organ failure. Curr Opin Pharmacol 2:440–444

Hasko G, Sitkovsky MV, Szabo C (2004) Immunomodulatory and neuroprotective effects of inosine. Trends Pharmacol Sci 25:152–157

Hasko G, Linden J, Cronstein B, Pacher P (2008) Adenosine receptors: therapeutic aspects for inflammatory and immune diseases. Nat Rev Drug Discov 7:759–770

Himer L, Csoka B, Selmeczy Z, Koscso B, Pocza T, Pacher P, Nemeth ZH, Deitch EA, Vizi ES, Cronstein BN et al (2010) Adenosine A2A receptor activation protects CD4+ T lymphocytes against activation-induced cell death. FASEB J 24:2631–2640

Hon WM, Moochhala S, Khoo HE (1997) Adenosine and its receptor agonists potentiate nitric oxide synthase expression induced by lipopolysaccharide in RAW 264.7 murine macrophages. Life Sci 60:1327–1335

Huang S, Apasov S, Koshiba M, Sitkovsky M (1997) Role of A2a extracellular adenosine receptor-mediated signaling in adenosine-mediated inhibition of T-cell activation and expansion. Blood 90:1600–1610

Jeffe F, Stegmann KA, Broelsch F, Manns MP, Cornberg M, Wedemeyer H (2009) Adenosine and IFN-{alpha} synergistically increase IFN-gamma production of human NK cells. J Leukoc Biol 85:452–461

Khoa ND, Montesinos MC, Reiss AB, Delano D, Awadallah N, Cronstein BN (2001) Inflammatory cytokines regulate function and expression of adenosine A(2A) receptors in human monocytic THP-1 cells. J Immunol 167:4026–4032

Koscso B, Csoka B, Selmeczy Z, Himer L, Pacher P, Virag L, Hasko G (2012) Adenosine augments IL-10 production by microglial cells through an A2B adenosine receptor-mediated process. J Immunol 188:445–453

Koshiba M, Kojima H, Huang S, Apasov S, Sitkovsky MV (1997) Memory of extracellular adenosine A2A purinergic receptor-mediated signaling in murine T cells. J Biol Chem 272:25881–25889

Koshiba M, Rosin DL, Hayashi N, Linden J, Sitkovsky MV (1999) Patterns of A2A extracellular adenosine receptor expression in different functional subsets of human peripheral T cells. Flow cytometry studies with anti-A2A receptor monoclonal antibodies. Mol Pharmacol 55:614–624

Kreckler LM, Wan TC, Ge ZD, Auchampach JA (2006) Adenosine inhibits tumor necrosis factor-alpha release from mouse peritoneal macrophages via A2A and A2B but not the A3 adenosine receptor. J Pharmacol Exp Ther 317:172–180

Krump E, Picard S, Mancini J, Borgeat P (1997) Suppression of leukotriene B4 biosynthesis by endogenous adenosine in ligand-activated human neutrophils. J Exp Med 186:1401–1406

la Sala A, Gadina M, Kelsall BL (2005) G(i)-protein-dependent inhibition of IL-12 production is mediated by activation of the phosphatidylinositol 3-kinase-protein 3 kinase B/Akt pathway and JNK. J Immunol 175:2994–2999

Lappas CM, Rieger JM, Linden J (2005) A2A adenosine receptor induction inhibits IFN-gamma production in murine CD4+ T cells. J Immunol 174:1073–1080

Latini S, Bordoni F, Pedata F, Corradetti R (1999) Extracellular adenosine concentrations during in vitro ischaemia in rat hippocampal slices. Br J Pharmacol 127:729–739

Le Moine O, Stordeur P, Schandene L, Marchant A, de Groote D, Goldman M, Deviere J (1996) Adenosine enhances IL-10 secretion by human monocytes. J Immunol 156:4408–4414

Le Vraux V, Chen YL, Masson I, De Sousa M, Giroud JP, Florentin I, Chauvelot-Moachon L (1993) Inhibition of human monocyte TNF production by adenosine receptor agonists. Life Sci 52:1917–1924

Leonard EJ, Shenai A, Skeel A (1987) Dynamics of chemotactic peptide-induced superoxide generation by human monocytes. Inflammation 11:229–240

Linden J (2001) Molecular approach to adenosine receptors: receptor-mediated mechanisms of tissue protection. Annu Rev Pharmacol Toxicol 41:775–787

Link AA, Kino T, Worth JA, McGuire JL, Crane ML, Chrousos GP, Wilder RL, Elenkov IJ (2000) Ligand-activation of the adenosine A2a receptors inhibits IL-12 production by human monocytes. J Immunol 164:436–442

Lokshin A, Raskovalova T, Huang X, Zacharia LC, Jackson EK, Gorelik E (2006) Adenosine-mediated inhibition of the cytotoxic activity and cytokine production by activated natural killer cells. Cancer Res 66:7758–7765

Lukashev DE, Smith PT, Caldwell CC, Ohta A, Apasov SG, Sitkovsky MV (2003) Analysis of A2a receptor-deficient mice reveals no significant compensatory increases in the expression of A2b, A1, and A3 adenosine receptors in lymphoid organs. Biochem Pharmacol 65:2081–2090

Macedo L, Pinhal-Enfield G, Alshits V, Elson G, Cronstein BN, Leibovich SJ (2007) Wound healing is impaired in MyD88-deficient mice: a role for MyD88 in the regulation of wound healing by adenosine A2A receptors. Am J Pathol 171:1774–1788

Majumdar S, Aggarwal BB (2003) Adenosine suppresses activation of nuclear factor-kappaB selectively induced by tumor necrosis factor in different cell types. Oncogene 22:1206–1218

Martin C, Leone M, Viviand X, Ayem ML, Guieu R (2000) High adenosine plasma concentration as a prognostic index for outcome in patients with septic shock. Crit Care Med 28:3198–3202

Martin L, Pingle SC, Hallam DM, Rybak LP, Ramkumar V (2006) Activation of the adenosine A3 receptor in RAW 264.7 cells inhibits lipopolysaccharide-stimulated tumor necrosis factor-alpha release by reducing calcium-dependent activation of nuclear factor-kappaB and extracellular signal-regulated kinase 1/2. J Pharmacol Exp Ther 316:71–78

Martini C, Di Sacco S, Tacchi P, Bazzichi L, Soletti A, Bondi F, Ciompi ML, Lucacchini A (1991) A2 adenosine receptors in neutrophils from healthy volunteers and patients with rheumatic disease. Adv Exp Med Biol 309A:459–462

McColl SR, St-Onge M, Dussault AA, Laflamme C, Bouchard L, Boulanger J, Pouliot M (2006) Immunomodulatory impact of the A2A adenosine receptor on the profile of chemokines produced by neutrophils. FASEB J 20:187–189

McWhinney CD, Dudley MW, Bowlin TL, Peet NP, Schook L, Bradshaw M, De M, Borcherding DR, Edwards CK (1996) Activation of adenosine A(3) receptors on macrophages inhibits tumor necrosis factor-alpha. Eur J Pharmacol 310:209–216

Merrill JT, Shen C, Schreibman D, Coffey D, Zakharenko O, Fisher R, Lahita RG, Salmon J, Cronstein BN (1997) Adenosine A1 receptor promotion of multinucleated giant cell formation by human monocytes: a mechanism for methotrexate-induced nodulosis in rheumatoid arthritis. Arthritis Rheum 40:1308–1315

Miller JS, Cervenka T, Lund J, Okazaki IJ, Moss J (1999) Purine metabolites suppress proliferation of human NK cells through a lineage-specific purine receptor. J Immunol 162:7376–7382

Min HW, Moochhala S, Eng KH (2000) Adenosine and its receptor agonists regulate nitric oxide production and RAW 264.7 macrophages via both receptor binding and its downstream metabolites-inosine. Life Sci 66:1781–1793

Minguet S, Huber M, Rosenkranz L, Schamel WW, Reth M, Brummer T (2005) Adenosine and cAMP are potent inhibitors of the NF-kappa B pathway downstream of immunoreceptors. Eur J Immunol 35:31–41

Mirabet M, Herrera C, Cordero OJ, Mallol J, Lluis C, Franco R (1999) Expression of A2B adenosine receptors in human lymphocytes: their role in T cell activation. J Cell Sci 112(Pt 4):491–502

Munro R, Ressner R, Stone M, Gessner G, Jarvis MF, Saltzman A (1998) Differential expression of adenosine A(2A) and A(2B) receptor subtypes on myeloid U937 and THP-1 cells: adenosine A(2B) receptor activation selectively stimulates cAMP formation and inhibition of TNF alpha release in THP-1 cells. Drug Dev Res 44:41–47

Naganuma M, Wiznerowicz EB, Lappas CM, Linden J, Worthington MT, Ernst PB (2006) Cutting edge: critical role for A2A adenosine receptors in the T cell-mediated regulation of colitis. J Immunol 177:2765–2769

Nemeth ZH, Lutz CS, Csoka B, Deitch EA, Leibovich SJ, Gause WC, Tone M, Pacher P, Vizi ES, Hasko G (2005) Adenosine augments IL-10 production by macrophages through an A2B receptor-mediated posttranscriptional mechanism. J Immunol 175:8260–8270

Nemeth ZH, Csoka B, Wilmanski J, Xu D, Lu Q, Ledent C, Deitch EA, Pacher P, Spolarics Z, Hasko G (2006) Adenosine A2A receptor inactivation increases survival in polymicrobial sepsis. J Immunol 176:5616–5626

Nguyen P, Broussas M, Cornillet-Lefebvre P, Potron G (1999) Inhibition of fMLP-triggered respiratory burst of human monocytes by adenosine: involvement of A3 adenosine receptor. J Leukoc Biol 66:495–501

Novitskiy SV, Ryzhov S, Zaynagetdinov R, Goldstein AE, Huang Y, Tikhomirov OY, Blackburn MR, Biaggioni I, Carbone DP, Feoktistov I et al (2008) Adenosine receptors in regulation of dendritic cell differentiation and function. Blood 112:1822–1831

Nowak M, Lynch L, Yue S, Ohta A, Sitkovsky M, Balk SP, Exley MA (2010) The A2aR adenosine receptor controls cytokine production in iNKT cells. Eur J Immunol 40:682–687

Ohta A, Gorelik E, Prasad SJ, Ronchese F, Lukashev D, Wong MK, Huang X, Caldwell S, Liu K, Smith P et al (2006) A2A adenosine receptor protects tumors from antitumor T cells. Proc Natl Acad Sci U S A 103:13132–13137

Ohta A, Madasu M, Kini R, Subramanian M, Goel N, Sitkovsky M (2009) A2A adenosine receptor may allow expansion of T cells lacking effector functions in extracellular adenosine-rich microenvironments. J Immunol 183:5487–5493

Panther E, Idzko M, Herouy Y, Rheinen H, Gebicke-Haerter PJ, Mrowietz U, Dichmann S, Norgauer J (2001) Expression and function of adenosine receptors in human dendritic cells. FASEB J 15:1963–1970

Panther E, Corinti S, Idzko M, Herouy Y, Napp M, la Sala A, Girolomoni G, Norgauer J (2003) Adenosine affects expression of membrane molecules, cytokine and chemokine release, and the T-cell stimulatory capacity of human dendritic cells. Blood 101:3985–3990

Pastor-Anglada M, Casado FJ, Valdes R, Mata J, Garcia-Manteiga J, Molina M (2001) Complex regulation of nucleoside transporter expression in epithelial and immune system cells. Mol Membr Biol 18:81–85

Polosa R, Blackburn MR (2009) Adenosine receptors as targets for therapeutic intervention in asthma and chronic obstructive pulmonary disease. Trends Pharmacol Sci 30:528–535

Polosa R, Holgate ST (2006) Adenosine receptors as promising therapeutic targets for drug development in chronic airway inflammation. Curr Drug Targets 7:699–706

Ralevic V, Burnstock G (1998) Receptors for purines and pyrimidines. Pharmacol Rev 50: 413–492

Ramkumar V, Stiles GL, Beaven MA, Ali H (1993) The A3 adenosine receptor is the unique adenosine receptor which facilitates release of allergic mediators in mast cells. J Biol Chem 268: 16887–16890

Raskovalova T, Huang X, Sitkovsky M, Zacharia LC, Jackson EK, Gorelik E (2005) Gs protein-coupled adenosine receptor signaling and lytic function of activated NK cells. J Immunol 175:4383–4391

Raskovalova T, Lokshin A, Huang X, Jackson EK, Gorelik E (2006) Adenosine-mediated inhibition of cytotoxic activity and cytokine production by IL-2/NKp46-activated NK cells: involvement of protein kinase A isozyme I (PKA I). Immunol Res 36:91–99

Ryzhov S, Goldstein AE, Matafonov A, Zeng D, Biaggioni I, Feoktistov I (2004) Adenosine-activated mast cells induce IgE synthesis by B lymphocytes: an A2B-mediated process involving Th2 cytokines IL-4 and IL-13 with implications for asthma. J Immunol 172:7726–7733

Ryzhov S, Zaynagetdinov R, Goldstein AE, Novitskiy SV, Dikov MM, Blackburn MR, Biaggioni I, Feoktistov I (2008) Effect of A2B adenosine receptor gene ablation on proinflammatory adenosine signaling in mast cells. J Immunol 180:7212–7220

Sakata N, Kawasome H, Terada N, Johnson GL, Gelfand EW (2000) CD40 and adenosine A2 receptor agonist-cyclic adenosine monophosphate rescue B-cell antigen receptor-induced apoptosis through independent pathways and converge to prevent caspase activation. J Allergy Clin Immunol 105:522–531

Salmon JE, Brogle N, Brownlie C, Edberg JC, Kimberly RP, Chen BX, Erlanger BF (1993) Human mononuclear phagocytes express adenosine A1 receptors. A novel mechanism for differential regulation of Fc gamma receptor function. J Immunol 151:2775–2785

Schnurr M, Toy T, Shin A, Hartmann G, Rothenfusser S, Soellner J, Davis ID, Cebon J, Maraskovsky E (2004) Role of adenosine receptors in regulating chemotaxis and cytokine production of plasmacytoid dendritic cells. Blood 103:1391–1397

Seegmiller JE, Watanabe T, Schreier MH (1977) The effect of adenosine on lymphoid cell proliferation and antibody formation. Ciba Found Symp 48:249–276

Sevigny CP, Li L, Awad AS, Huang L, McDuffie M, Linden J, Lobo PI, Okusa MD (2007) Activation of adenosine 2A receptors attenuates allograft rejection and alloantigen recognition. J Immunol 178:4240–4249

Sottofattori E, Anzaldi M, Ottonello L (2001) HPLC determination of adenosine in human synovial fluid. J Pharm Biomed Anal 24:1143–1146

Sperlagh B, Doda M, Baranyi M, Hasko G (2000) Ischemic-like condition releases norepinephrine and purines from different sources in superfused rat spleen strips. J Neuroimmunol 111: 45–54

Spicuzza L, Bonfiglio C, Polosa R (2003) Research applications and implications of adenosine in diseased airways. Trends Pharmacol Sci 24:409–413

Sullivan GW, Lee DD, Ross WG, DiVietro JA, Lappas CM, Lawrence MB, Linden J (2004) Activation of A2A adenosine receptors inhibits expression of alpha 4/beta 1 integrin (very late antigen-4) on stimulated human neutrophils. J Leukoc Biol 75:127–134

Sun CX, Zhong H, Mohsenin A, Morschl E, Chunn JL, Molina JG, Belardinelli L, Zeng D, Blackburn MR (2006) Role of A2B adenosine receptor signaling in adenosine-dependent pulmonary inflammation and injury. J Clin Invest 116:2173–2182

Suzuki T, Hashimoto S, Toyoda N, Nagai S, Yamazaki N, Dong HY, Sakai J, Yamashita T, Nukiwa T, Matsushima K (2000) Comprehensive gene expression profile of LPS-stimulated human monocytes by SAGE. Blood 96:2584–2591

Thiel M, Chouker A (1995) Acting via A2 receptors, adenosine inhibits the production of tumor necrosis factor-alpha of endotoxin-stimulated human polymorphonuclear leukocytes. J Lab Clin Med 126:275–282

Thiele A, Kronstein R, Wetzel A, Gerth A, Nieber K, Hauschildt S (2004) Regulation of adenosine receptor subtypes during cultivation of human monocytes: role of receptors in preventing lipopolysaccharide-triggered respiratory burst. Infect Immun 72:1349–1357

Thompson LF, Eltzschig HK, Ibla JC, Van De Wiele CJ, Resta R, Morote-Garcia JC, Colgan SP (2004) Crucial role for ecto-5′-nucleotidase (CD73) in vascular leakage during hypoxia. J Exp Med 200:1395–1405

van der Hoeven D, Wan TC, Gizewski ET, Kreckler LM, Maas JE, Van Orman J, Ravid K, Auchampach JA (2011) A role for the low-affinity A2B adenosine receptor in regulating superoxide generation by murine neutrophils. J Pharmacol Exp Ther 338(3):1004–1012

Van Ginderachter JA, Movahedi K, Hassanzadeh Ghassabeh G, Meerschaut S, Beschin A, Raes G, De Baetselier P (2006) Classical and alternative activation of mononuclear phagocytes: picking the best of both worlds for tumor promotion. Immunobiology 211:487–501

Visser SS, Theron AJ, Ramafi G, Ker JA, Anderson R (2000) Apparent involvement of the A(2A) subtype adenosine receptor in the anti-inflammatory interactions of CGS 21680, cyclopentyladenosine, and IB-MECA with human neutrophils. Biochem Pharmacol 60:993–999

Wakai A, Wang JH, Winter DC, Street JT, O'Sullivan RG, Redmond HP (2001) Adenosine inhibits neutrophil vascular endothelial growth factor release and transendothelial migration via A2B receptor activation. Shock 15:297–301

Williams BA, Manzer A, Blay J, Hoskin DW (1997) Adenosine acts through a novel extracellular receptor to inhibit granule exocytosis by natural killer cells. Biochem Biophys Res Commun 231:264–269

Wilson JM, Ross WG, Agbai ON, Frazier R, Figler RA, Rieger J, Linden J, Ernst PB (2009) The A2B adenosine receptor impairs the maturation and immunogenicity of dendritic cells. J Immunol 182:4616–4623

Wilson JM, Kurtz CC, Black SG, Ross WG, Alam MS, Linden J, Ernst PB (2011) The A2B adenosine receptor promotes Th17 differentiation via stimulation of dendritic cell IL-6. J Immunol 186:6746–6752

Xaus J, Mirabet M, Lloberas J, Soler C, Lluis C, Franco R, Celada A (1999) IFN-gamma up-regulates the A2B adenosine receptor expression in macrophages: a mechanism of macrophage deactivation. J Immunol 162:3607–3614

Yamamura T, Sakuishi K, Illes Z, Miyake S (2007) Understanding the behavior of invariant NKT cells in autoimmune diseases. J Neuroimmunol 191:8–15

Yang D, Zhang Y, Nguyen HG, Koupenova M, Chauhan AK, Makitalo M, Jones MR, St Hilaire C, Seldin DC, Toselli P et al (2006) The A2B adenosine receptor protects against inflammation and excessive vascular adhesion. J Clin Invest 116:1913–1923

Yang M, Ma C, Liu S, Shao Q, Gao W, Song B, Sun J, Xie Q, Zhang Y, Feng A et al (2010) HIF-dependent induction of adenosine receptor A2b skews human dendritic cells to a Th2-stimulating phenotype under hypoxia. Immunol Cell Biol 88:165–171

Yip KH, Lau HY, Wise H (2011) Reciprocal modulation of anti-IgE induced histamine release from human mast cells by A(1) and A(2B) adenosine receptors. Br J Pharmacol 164:807–819

Zarek PE, Huang CT, Lutz ER, Kowalski J, Horton MR, Linden J, Drake CG, Powell JD (2008) A2A receptor signaling promotes peripheral tolerance by inducing T-cell anergy and the generation of adaptive regulatory T cells. Blood 111:251–259

Zaynagetdinov R, Ryzhov S, Goldstein AE, Yin H, Novitskiy SV, Goleniewska K, Polosukhin VV, Newcomb DC, Mitchell D, Morschl E et al (2010) Attenuation of chronic pulmonary inflammation in A2B adenosine receptor knockout mice. Am J Respir Cell Mol Biol 42:564–571

Zhao P, Li XG, Yang M, Shao Q, Wang D, Liu S, Song H, Song B, Zhang Y, Qu X (2008) Hypoxia suppresses the production of MMP-9 by human monocyte-derived dendritic cells and requires activation of adenosine receptor A2b via cAMP/PKA signaling pathway. Mol Immunol 45:2187–2195

Zhou Y, Mohsenin A, Morschl E, Young HW, Molina JG, Ma W, Sun CX, Martinez-Valdez H, Blackburn MR (2009) Enhanced airway inflammation and remodeling in adenosine deaminase-deficient mice lacking the A2B adenosine receptor. J Immunol 182:8037–8046

Zimmermann H (2000) Extracellular metabolism of ATP and other nucleotides. Naunyn Schmiedebergs Arch Pharmacol 362:299–309

Chapter 13
The Bioenergetic Network of Adenosine in Hibernation, Sleep, and Thermoregulation

Kelly L. Drew and Tulasi R. Jinka

Abstract Adenosine is a homeostatic bioenergetic network regulator that plays a fundamental role in energy homeostasis through biochemical, bioenergetic, and receptor dependent processes. Hibernation, torpor, and sleep are integral to energy homeostasis. Here we review evidence that adenosine receptor dependent signaling as well as biochemical and bioenergetic influences of adenosine are essential to all three of these processes placing adenosine at the core of mammalian energy homeostasis. Central A_1 adenosine receptor (A_1R) dependent signaling is necessary for onset of hibernation and fasting-induced torpor in ground squirrels, hamsters, and mice. Activation of A_1R within the central nervous system is sufficient to induce hibernation. A seasonally mediated change in sensitivity to central A_1R stimulation is necessary for A_1R agonist-induced hibernation in ground squirrels and may underlie the distinction between sleep and hibernation. One function of sleep is to restore brain energy homeostasis, while the primary function of hibernation and torpor is to restore or protect somatic energy homeostasis. Where in the brain A_1R agonists act to induce torpor and how central A_1R dependent signaling reduces metabolic rate to 1–2 % of resting metabolic rate in hibernating animals is a topic for further research. Understanding mechanisms of energy homeostasis may have implications for treatment of stroke, cardiac arrest, and other conditions where delivery of blood fails to meet demand.

Keywords Purinergic signaling • AMPK • ATP • Torpor • Ground squirrel

K.L. Drew (✉) • T.R. Jinka
Alaska Basic Neuroscience Program, Department of Chemistry and Biochemistry, Institute of Arctic Biology, University of Alaska Fairbanks, 902 N. Koyukuk Drive STE 311, Fairbanks, AK 99775-7000, USA
e-mail: kdrew@alaska.edu

S. Masino and D. Boison (eds.), *Adenosine: A Key Link between Metabolism and Brain Activity*, DOI 10.1007/978-1-4614-3903-5_13,

13.1 The Bioenergetic Network of Adenosine in Hibernation, Sleep, and Thermoregulation

Adenosine has been termed a homeostatic bioenergetic network regulator because it plays a fundamental role in energy homeostasis (Boison et al. 2011). Adenosine regulates energy homeostasis through biochemical, bioenergetic, and receptor dependent processes (Fig. 13.1). Hibernation, torpor and sleep are phenomena integral to systemic energy homeostasis (Berger and Phillips 1995). Electrophysiological and thermoregulatory continuities between slow wave sleep (SWS), hibernation and daily torpor argue for common regulatory mechanisms (Florant et al. 1978). Here we review evidence that adenosine receptor dependent signaling as well as biochemical and bioenergetic influences of adenosine are essential to all three of these processes placing adenosine at the core of mammalian energy homeostasis.

13.1.1 Energy Homeostasis in Hibernation

Hibernation conserves energy through an adaptive transformation in physiology and behavior that allows survival during seasonal deficiencies in food supply that are often associated with extreme environmental challenges. Hibernation consists of prolonged torpor bouts during which animals decrease metabolic rate (MR) to as low as 1–2 % of basal metabolic rate (BMR) (Buck and Barnes 2000; Geiser 1988). This decrease in metabolic demand allows animals to survive without food for 5–8 months despite ambient temperatures that can fall to −20 °C (Barnes 1989; Sheriff et al. 2010). During prolonged torpor in hibernating ground squirrels core body temperature (T_b) decreases and is regulated to within 1–2 °C above ambient temperature (T_a) (Buck and Barnes 2000).

Prolonged torpor is interrupted by brief (12–24 h) periods of euthermic T_b and MR known as interbout arousals. These events occur when T_a and thus torpid T_b falls below 30 °C, (Barnes 1989; Dausmann et al. 2004). Arousal from torpid T_b requires significant energy expenditure (Karpovich et al. 2009; Toien et al. 2001) suggesting that interbout arousals are essential for successful hibernation. Interbout arousals are associated with the accumulation of metabolic waste and the need for glucose (Galster and Morrison 1970, 1975; Serkova et al. 2007) and serve multiple functions such as restoration of synaptic spines (Magarinos et al. 2006; Popov and Bocharova 1992; von der Ohe et al. 2007), gluconeogenesis (Galster and Morrison 1975), and biomolecular synthesis (Epperson et al. 2011; French 1985).

Torpor bouts last from a few days in species such as the Syrian hamster (*Mesocricetus auratus*) to 2–3 weeks in larger rodents such as the Arctic ground squirrel (AGS; *Urocitellus parryii*) (Carey et al. 2003; Dausmann et al. 2004; Geiser 2004; Heldmaier et al. 2004). Black bears (*Ursus americanus*), the largest hibernators for which metabolic data are available, display many of the same attributes of hibernation as smaller mammalian hibernators including a decrease in metabolic rate, reduced heart rate, and long periods of aphagia and adipsia (Toien et al. 2011).

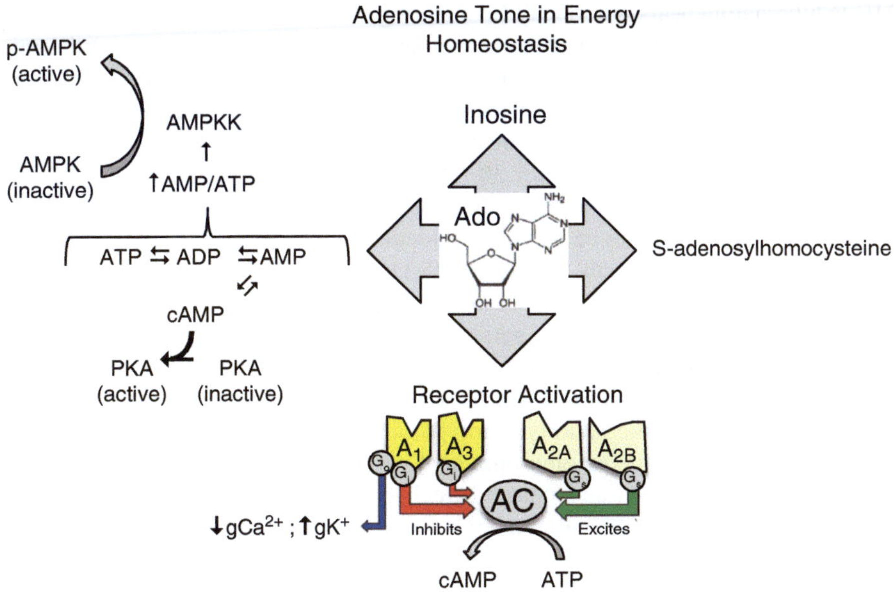

Fig. 13.1 Adenosine is a purine nucleoside, comprising a purine base, adenine, and a ribose sugar moiety. Adenosine is the precursor for and metabolite of 5′AMP (AMP), 3′5′-monophosphate (cyclic AMP; cAMP), and ATP and is thus central to energy charge as well as a host of signaling pathways involved in energy homeostasis. One pathway involves activation of AMP-activated protein kinase (AMPK) via several mechanisms including allosteric activation of AMPK kinase (AMPKK), an enzyme that phosphorylates AMPK. Many of these activating effects are antagonized by ATP making the system highly responsive to changes in cellular energy charge. AMPK regulates energy balance at both the cellular and whole-animal level. Activated AMPK generally stimulates catabolic pathways (glycolysis, fatty acid oxidation, and mitochondrial biogenesis) and inhibits anabolic pathways (gluconeogenesis, and glycogen, fatty acid, and protein synthesis). AMPK also enhances feeding and nutrient uptake. AMP is a precursor to cAMP which activates protein kinase A (PKA) and other signaling molecules that influence multiple aspects of energy homeostasis. Examples include leptin resistance (Fukuda et al. 2011) and triglyceride hydrolysis (Zimmermann et al. 2004). In addition, adenosine influences cAMP levels directly through activation of four types of G-protein coupled receptors (A_1, A_{2A}, A_{2B}, A_3). Adenosine receptors distributed throughout the body including the central and peripheral nervous systems influence multiple processes including immune function, lipolysis, rate and force of cardiac contraction, lung mast cells and neuronal excitation (Fredholm et al. 2011; Wilson 2009). Within the brain, A_1R and A_{2A}R are expressed in high density. Activation of A_1R hyperpolarizes neurons and decreases presynaptic neurotransmitter release by increasing gK^+ and decreasing gCa^{2+} (Latini and Pedata 2001). Adenosine may also influence energy homeostasis through gene expression via S-adenosylhomocysteine-mediated inhibition of transmethylation reactions (Garcia et al. 2011). Inosine is the product of adenosine deamination. Although its role in energy homeostasis in mammals is not well understood, inosine serves as a sink for adenine nucleotides under fermentative conditions and thus plays an essential role in energy homeostasis in yeast (Walther et al. 2010)

MR in hibernating bears is about 47 % of resting metabolic rate and remains low for weeks at the end of the hibernation season even after T_b returns to normothermic levels. Lower levels of thermal conductance, due in part to larger body size, prevent T_b in hibernating bears from decreasing to below about 30 °C. Hibernating bears do not show spontaneous, episodic arousals like smaller hibernators consistent with evidence that T_b needs to fall below 30 °C to produce a pattern of periodic arousals that are often considered to be an attribute of true hibernation.

The hibernation phenotype is typically identified by a series of torpor bouts that repeat at predictable intervals (Barnes 1989; Drew et al. 2007; Russell et al. 2010). In some hibernating species, termed obligate hibernators, the hibernation season runs on a circannual rhythm that persists in constant photoperiods and with food provided ad libitum and at high ambient temperatures (Lee and Zucker 1991; Pengelley et al. 1976). Ground squirrels are obligate hibernators and therefore display aspects of the hibernation phenotype according to an endogenous circannual cycle that is affected little by environmental variables. By contrast, other hibernating species such as the Syrian hamster are called facultative hibernators. Facultative hibernators express the hibernation phenotype in response to changing environmental variables such as decreasing day length, limited food or water availability, and decreases in ambient temperature (Hoffman et al. 1968). In both obligate and facultative hibernators a single torpor bout consists of three phases: onset, maintenance, and arousal. Evidence suggests that each phase is regulated by distinct mechanisms (Tamura et al. 2005). Other small mammals display daily torpor that is expressed during acute periods of food or water deprivation. Daily torpor differs from hibernation in limited duration, by less extreme decreases in MR and T_b, and by more immediate expression in response to environmental challenge (Geiser 2004). Torpor onset in all cases, however, involves pronounced decreases in oxygen consumption that precedes a decline in T_b. Here we review evidence that A_1R signaling may be a common mediator of torpor in hibernation and daily torpor.

Torpor onset during obligate and facultative hibernation as well as fasting-induced torpor involves an initial decrease in MR that is followed by a more gradual decrease in T_b. Proposed mechanisms for suppressing MR fall into three general categories. One proposed mechanism includes inhibition of thermogenesis within the CNS and a subsequent decrease in the brain temperature that stimulates thermogenesis (Heller et al. 1977). Much evidence supports this hypothesis and is discussed below. A second mechanism involves thermodynamic effects of cooling on metabolic rate described by the Arrhenius equation. In biological systems temperature effects are often referred to as Q_{10} effects where reaction rates generally double or triple with every 10 °C increase in temperature (Atkins and De Paula 2006; Schmidt-Nielsen 1997). A third set of mechanisms involves "active" suppression of cellular processes such as ion channel arrest (Hochachka 1986) or inhibition of cellular respiration (Muleme et al. 2006). In mammals evidence is largely lacking for ion channel arrest or inhibition of cellular respiration (Staples and Brown 2008). However, in nonmammalian "hibernation" such as winter dormancy of freshwater turtles, adenosine mediated downregulation of ion channels contributes to energy homeostasis (Pek and Lutz 1997).

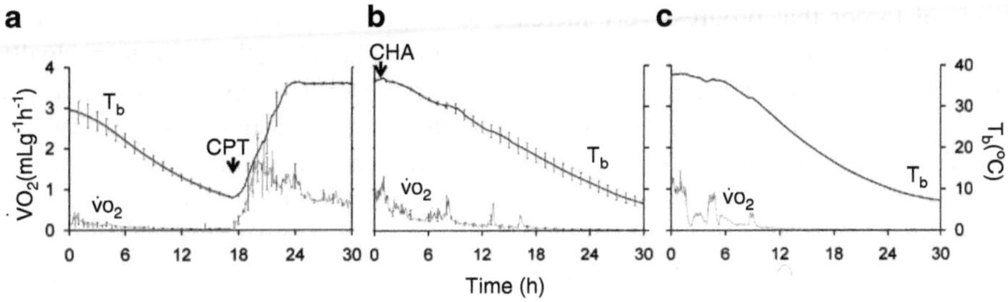

Fig. 13.2 Central A_1R stimulation is necessary and sufficient to induce torpor in AGS. Spontaneous torpor (**a**) is reversed by cyclopentyltheophylline (CPT; 3 nmol, icv), an A_1R antagonist (n=6). N^6-cyclohexyladenosine (CHA), an A_1R agonist, (0.5 nmol, icv) delivered during the mid-hibernation season induces torpor (n=6) (**b**). The same dose of CHA delivered to the same AGS during the summer season fails to induce torpor (*not shown*). CHA-induced torpor resembles the temporal profile characteristic of spontaneous torpor in representative AGS where the rate of oxygen consumption ($\dot{V}O_2$) decreases prior to a decrease in T_b (**c**). Redrawn from Jinka et al. (2011)

13.1.1.1 Adenosine Signaling in Hibernation and Torpor

Recent evidence points towards A_1R dependent signaling within the CNS as a means to inhibit thermogenesis and induce hibernation. Using the seasonal aspect of hibernation in arctic ground squirrels (AGS) as a clue we found that A_1R activation within the CNS meets all of the necessary requirements for an endogenous mediator of torpor. However, A_1R activation is effective only in the winter season (Jinka et al. 2011). The A_1R antagonist cyclopentyltheophylline (CPT), but not the $A_{2A}R$ antagonist (MSX-3), delivered into the lateral ventricle (icv) reverses spontaneous torpor (Fig. 13.2a). The A_1R agonist, N^6-cyclohexyladenosine (CHA), delivered icv to AGS during the summer season produces a moderate (~4 °C) decrease in T_b and a moderate (~30 %) decrease in the rate of O_2 consumption ($\dot{V}O_2$) with a maximal decrease in T_b and $\dot{V}O_2$ at about 1 h (Jinka et al. 2011). When administered to the same animals later in the year, after these animals had begun to show spontaneous torpor but were not yet at the midpoint of the hibernation season, CHA produced a summer-like response in four of the six animals and induced a torpor-like response in the remaining two animals. When administered in the middle of the hibernation season the same dose of CHA (0.5 nmol, icv) induced torpor in all six animals tested (Fig. 13.2b). Importantly the T_b and $\dot{V}O_2$ decline after CHA in winter followed the same temporal profile as during onset of spontaneous torpor (Fig. 13.2c) and contrasted with a rapid, short-lived hypothermic-like CHA-induced decrease in T_b and $\dot{V}O_2$ noted in summer. Vehicle, the A_3R agonist Cl-IB MECA and the somnogenic pentobarbital failed to induce torpor regardless of season (Jinka et al. 2011). Thus, A_1R dependent signaling within the CNS is both necessary and sufficient to induce and sustain the onset of torpor in AGS, an obligate hibernator.

The progressive increase in the sensitivity of AGS to A_1R mediated signaling described above parallels the seasonal transition into hibernation. In obligate hibernators the hibernation season is characterized by the display of short (3–7 day long) bouts of torpor that progressively increase to longer (2–3 week long) bouts. Long torpor bouts continue to occur through midwinter until bout length begins to gradually shorten in the spring. Ground squirrels are more resistant to arousal early in a torpor bout than late in the bout and resistance to arousal is greatest in midwinter (Twente and Twente 1968). When ground squirrels are housed at warm ambient temperatures such as 18–21 °C torpor bouts occur less regularly and torpor bout length is less consistent than when animals are housed at cooler ambient temperatures (Drew et al., unpublished observations). Nonetheless, torpor does occur at warm ambient temperatures (Russell et al. 2010). Under these housing conditions seasonal changes in sleep and T_b can be observed with less interference from spontaneous bouts of torpor and seasonal changes in sensitivity to adenosine signaling remains evident. Golden mantled ground squirrels sleep more during the hibernation season (Walker et al. 1980). In AGS, resting, euthermic T_b is lower during the hibernation season than during the summer season. A trough in resting T_b in AGS predicts spontaneous torpor and a larger CHA-induced decrease in resting T_b (Olson et al., submitted). In addition, the Eastern chipmunk (*Tamias striatus*), another obligate hibernator, shows a seasonal change in the thermoregulatory response to hypoxia (Levesque and Tattersall 2009).

In facultative hibernators, such as the Syrian hamster (*Mesocricetus auratus*), A_1R dependent signaling is also necessary for torpor onset. The A_1R antagonist CPT reverses spontaneous torpor when administered during torpor onset in Syrian hamsters (Shiomi and Tamura 2000). Interestingly, CPT has no effect when delivered icv later in the torpor bout demonstrating that onset, maintenance, and interbout arousal are regulated by distinct mechanisms. Whether A_1R activation is sufficient to induce torpor in facultative hibernators is less clear. This uncertainty stems from a lack of data on the effects of A_1R agonists in hamsters showing spontaneous torpor vs. those not showing spontaneous torpor. Uncertainty is exacerbated by relying solely on T_b to identify torpor when the characteristic temporal profile in decline of T_b and MR is a more definitive marker of torpor. With these caveats in mind, evidence suggests that A_1R activation is sufficient to induce torpor in Syrian hamsters. CHA induces a torpor-like decline in T_b in Syrian hamsters and the rate of fall in T_b after CHA is similar to the rate of fall in T_b during spontaneous torpor (Shiomi and Tamura 2000; Tamura et al. 2005). The question remains then as to whether an increase in sensitivity to A_1R signaling is necessary for hibernation in hamsters and whether environmental cues such as shortened day length and food restriction increase sensitivity to central A_1R signaling in these animals.

Daily torpor also depends on central A_1R activation despite several differences between daily torpor and hibernation. Daily torpor is induced by fasting in many strains of mice and lasts for less than 48 h. During daily torpor T_b falls from around 37 °C to between 15 and 25 °C, while during hibernation T_b follows ambient temperature down to near 0 °C (Geiser 2004). MR during daily torpor in the house mouse (*Mus musculus*), weighing 45 g, decreases from a BMR of about 1.5 $mLg^{-1}\ h^{-1}$ to a torpid MR (TMR) of 0.3 $mLg^{-1}\ h^{-1}$, while T_b decreases from 37 to 19 °C. By contrast MR during hibernation in a Western Jumping Mouse (*Zapus princeps*) weighing 33 g decreases from a BMR of 1.5 $mLg^{-1}\ h^{-1}$ to a TMR of 0.04 $mLg^{-1}\ h^{-1}$ while T_b decreases from 37 °C to 5 °C (Geiser 2004). Some of the differences in torpid metabolic rates may be due to colder torpid T_bs. Mice do not tolerate ambient temperatures below about 15 °C. Consequently study of daily torpor is usually at ambient temperatures of 20–21 °C while study of hibernation is usually at ambient temperatures of 2–5 °C. C57BL mice display fasting induced torpor within 6–7 h when food is withheld during the dark period when mice eat. MR falls prior to a fall in temperature during onset of fasting-induced torpor and, as with hibernation, it is this slow rate of cooling that distinguishes torpor from hypothermia.

Like torpor in obligate and facultative hibernation, inhibition of A_1R within the CNS reverses fasting-induced torpor. Intracerebroventricular delivery of the hydrophilic adenosine receptor antagonist 8-(p-sulfophenyl)theophylline (8-SPT) reverses fasting-induced torpor in C57BL mice. Systemic delivery (sc) of this same drug that does not cross the blood–brain barrier does not reverse torpor (Swoap and Lliff 2011). These results suggest that central adenosine receptor activation is necessary for fasting induced torpor. Similar to the effects of fasting, AMP, ADP, or ATP decrease T_b to about 25 °C when delivered ip to fed mice. This effect was originally attributed to activation of AMPK (Zhang et al. 2006) but later found to depend on an adenosine receptor dependent mechanism (Swoap et al. 2007; Zhang et al. 2006). Adenosine-induced suppression of T_b to torpor-like levels differs from torpor because the rate of cooling induced by AMP, ADP or ATP is greater than fasting-induced torpor (Swoap et al. 2007). It remains to be determined if a short-term fast sensitizes adenosine receptor dependent signaling or if an increase in sensitivity is necessary for torpor or distinguishes torpor from A_1R-induced hypothermia.

Adenosine via A_1R signaling within the CNS is essential for all three types of torpor (obligate, facultative, and fasting-induced), since in all cases torpor is reversed by central application of an A_1R antagonist. It does not necessarily follow, however, that A_1R activation is a common means to induce torpor. A_1R agonists mimic the decline in T_b characteristic of torpor but in some cases it is unclear if animals enter torpor or simply cool to torpor-like T_b. In general, hypothermia is distinguished from torpor by a faster rate of cooling that precedes or parallels a decrease in MR. Although not tested in all three types of torpor a change in A_1R signaling within the CNS during preparation for hibernation or torpor may alter the thermoregulatory response to central A_1R agonists and transform a hypothermic response into torpor. In addition, other mechanisms downstream to A_1R signaling may modulate the rate of cooling. Vasoconstriction regulates heterogeneous warming during arousal and is reversed by phentolamine, an α-adrenergic antagonist (Osborne et al. 2005). Onset of daily torpor requires intact noradrenergic signaling to decrease circulating leptin. Sympathetic activation may also modulate the rate of cooling during torpor onset as it does during rewarming (Braulke and Heldmaier 2010; Swoap et al. 2006; Swoap and Weinshenker 2008).

H_2S-induced "suspended animation" could also involve an A_1R dependent mechanism. Exposing mice to H_2S, an inhibitor of cellular respiration, decreases MR (measured as CO_2 output) to about 30 % of pretreatment values, similar to daily torpor. Gradually decreasing ambient temperature during H_2S exposure causes T_b to decrease gradually so that the subsequent temporal profile of metabolic suppression and T_b mimics torpor (Blackstone et al. 2005). Evidence discussed above begs the question, does H_2S-induced chemical hypoxia suppress metabolism via an A_1R dependent mechanism? Hypoxia and ischemia increase adenosine release (Dunwiddie and Masino 2001). Hypoxia suppresses MR (termed the hypoxic metabolic response)

(Tattersall and Milsom 2003) and T_b (termed hypoxia-induced anapyrexia). Hypoxia-induced decrease in T_b depends on central adenosine receptors (Barros and Branco 2000; Steiner and Branco 2002), and chemical hypoxia induced by cyanide suppresses neuronal activity via adenosine (Zhu and Krnjevic 1997). Although not studied directly, A_1R activation when combined with a controlled decrease in T_a, remains a viable mechanism for H_2S-induced suspended animation.

13.1.1.2 Central Site of Adenosine Signaling in Hibernation and Torpor

It is unknown where in the brain adenosine acts to induce torpor. A_1R are ubiquitous throughout the CNS and periphery. In brain, A_1R density is highest in hippocampus (dentate gyrus and CA3), cerebral cortex (layers III and V), throughout the thalamus and habenula and in basket cells of the cerebellum. A_1R immunoreactivity is also evident, albeit less dense, in basal forebrain and many hypothalamic nuclei including the magnocellular preoptic area, supraoptic nucleus, premammillary nucleus, lateral mammillary nucleus, and the lateral hypothalamic area (Rivkees et al. 1995). In the Columbian ground squirrel whole brain adenosine levels are lower in brains of torpid hibernators than in animals during the nonhibernating season. By contrast, density of A_1R is higher in many brain regions including the solitary and vagus motor nuclei during torpor (Lee et al. 1993).

Brain regions involved in A_1R mediated torpor onset must be identified before changes in adenosine levels or A_1R receptor function within these regions can be studied. Sleep and thermoregulatory regions within the CNS are likely sites of action of adenosine during torpor onset. The ventrolateral medial preoptic area, a nucleus involved in thermoregulation, is active during onset of torpor (Bratincsak et al. 2007) although A_1R density is not increased in the medial preoptic nucleus of torpid hibernators (Lee et al. 1993). Figure 13.3 illustrates areas in the brain where

Fig. 13.3 (continued) neurons of the perifornical-lateral hypothalamus (PF-LH), and histaminergic (HA) neurons of the tuberomammillary nucleus (TMN). *Negative signs* indicate a hyperpolarizing influence. Cholinergic BF neurons and hypocretin/orexin neurons originating in the PF-LH provide excitatory, wake-promoting influences on the CNS indicated by the color green and a (+)CNS. One means by which A_1R agonists promote sleep is via direct inhibition of these wake-active neuronal populations. The POAH GABAergic neurons promote sleep by inhibiting the wake-active hypocretin/orexin neurons in the PF-LH as well as wake-active histaminergic neurons in the TMN. This inhibitory influence is indicated by the color *red*. To promote sleep via the POAH, an A_1R agonist must stimulate these POAH GABAergic neurons that then inhibit wake-promoting effects of the PF-LH and TMN. A stimulatory influence is achieved by inhibiting an unknown intermediate inhibitory pathway indicated by (*question mark*). Because these pathways are known to contribute to sleep drive they are hypothesized to be involved in A_1R-mediated torpor. The POAH is also of interest because it plays a role in thermoregulation. Within the POAH, the ventrolateral medial preoptic area, a nucleus involved in thermoregulation, is active during onset of torpor. (**c**) A single bout of torpor in hibernating mammals includes three processes: onset, maintenance, and arousal that are regulated by distinct mechanisms. Torpor is induced by a central A_1R dependent mechanism. An unknown mechanism drives TMR below BMR and maintains TMR until a signal associated with accumulation of metabolic wastes and depletion of glucose induces arousal

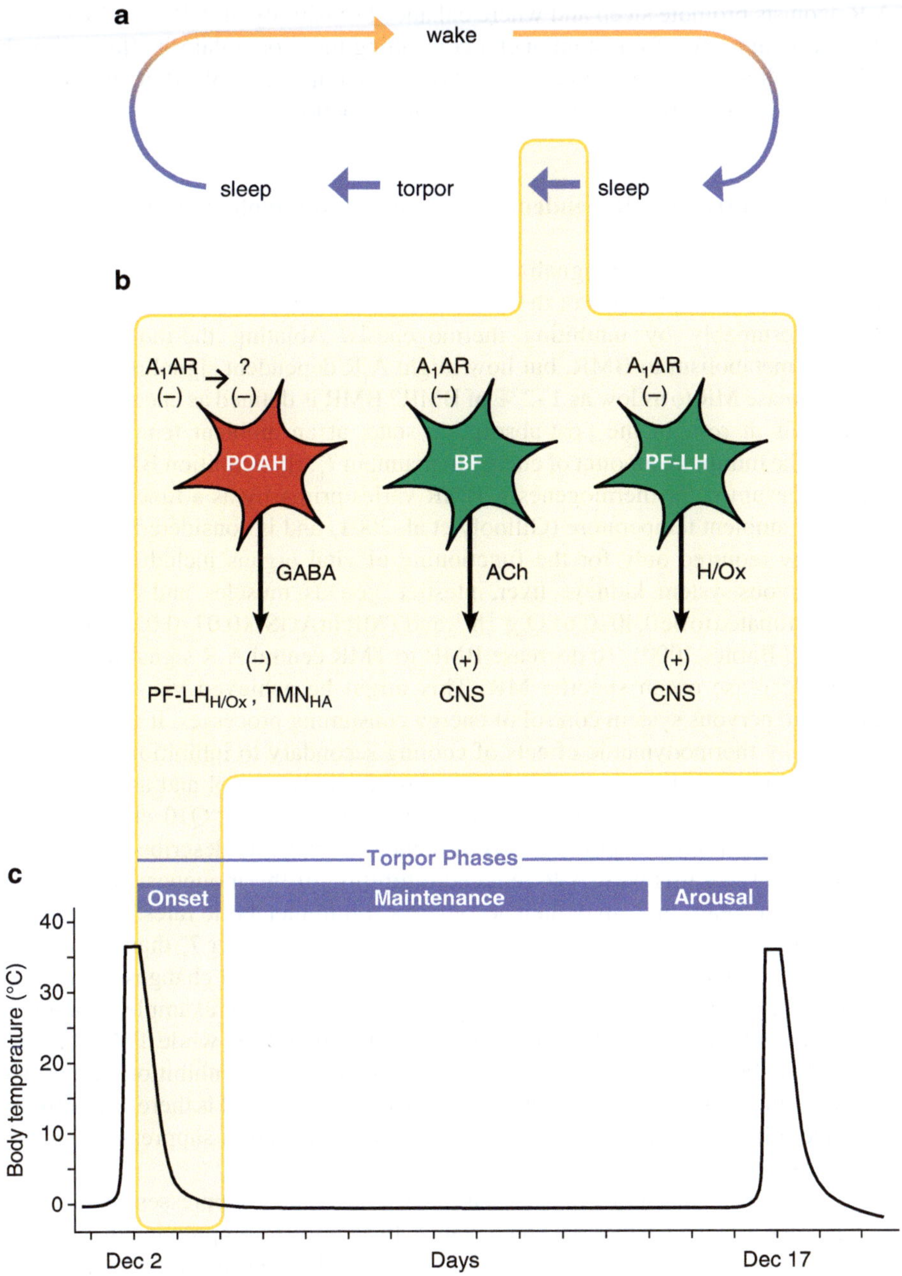

Fig. 13.3 (**a**) Cycle of sleep, torpor, and wake expressed during a single torpor bout that includes torpor onset, maintenance, and interbout arousal as shown in (**c**). (**b**) Where adenosine acts in the brain to induce torpor is unknown. However, because characteristics of the EEG during onset of torpor are an extension of characteristics of NREM sleep, torpor onset is hypothesized to result from increased sensitivity of A_1R dependent signaling within neural pathways known to promote sleep. These sleep-promoting pathways include GABAergic neurons of the preoptic anterior hypothalamus (POAH), cholinergic (ACh) neurons of the basal forebrain (BF), hypocretin/orexin (H/Ox)

A_1R agonists promote sleep and where enhanced sensitivity of A_1R signaling could play a role in torpor. Brainstem nuclei controlling thermoregulatory effectors such as brown adipose tissue (Cano et al. 2003) and other nuclei involved in thermoregulation (Nakamura 2011) are also potential sites of action.

13.1.1.3 Central A_1R Dependent Signaling and Metabolic Suppression

How does A_1R dependent signaling within the CNS decrease MR? Activation of A_1R within the preoptic area of the hypothalamus decreases T_b (Barros and Branco 2000) presumably by inhibiting thermogenesis. Ablating thermogenesis will decrease metabolism to BMR, but how might A_1R dependent signaling within the CNS decrease MR to as low as 1–2 % of BMR? BMR is defined as energy expended by animals at rest, in the post-absorptive state, at an ambient temperature that requires the minimum amount of energy to maintain T_b. By definition BMR excludes energy consumed for thermogenesis. BMR varies primarily as a function of body mass and ambient temperature (Gillooly et al. 2001) and is considered to represent the energy required only for the functioning of vital organs including the heart, lungs, nervous system, kidneys, liver, intestine, gonads, muscles, and skin. BMR in AGS is estimated to be 0.40–0.61 O_2g^{-1} h^{-1}, and TMR in AGS is 0.01–0.02 mLO_2g^{-1} h^{-1} (Buck and Barnes 2000). To decrease BMR to TMR central A_1R signaling must, in theory, decrease organ specific MR. This might be achieved via influence on autonomic nervous system control of energy consuming processes. It could also be achieved by thermodynamic effects of cooling secondary to inhibition of thermogenesis. Inhibiting thermogenesis would cause animals to cool and as T_b declines thermodynamic effects on rates of chemical reactions (or a "Q10 effect") could decrease oxygen consumption to torpid metabolic rates, as described previously (Geiser 2004). In this way, A_1R-induced inhibition of thermogenesis may lower T_b and the decrease in T_b may then account for torpid metabolic rates. However, if torpid MR depends on T_b then TMR would be higher at warmer T_b than at lower T_b and this is not the case. Indeed in AGS changes in T_b parallel changes in ambient temperature (T_a) between 0 and 20 °C while TMR does not. For example, at ambient temperatures between 4 and 12 °C, T_b increases by about 8 °C while TMR remains constant at about 0.02 mLO_2g^{-1} h^{-1} (Buck and Barnes 2000). Inhibition of thermogenesis with a subsequent temperature dependent decline in MR is therefore unlikely to be the primary mechanism by which central A_1R activation suppresses metabolism during onset of torpor.

An alternative means by which central A_1R activation suppresses metabolism may involve a change in thermal conductance. In addition to suppressing thermogenesis, Jinka et al. (2011) show that A_1R-induced cooling during onset of torpor is strikingly slower than A_1R-induced cooling in summer when animals do not enter torpor. Slower rate of cooling is consistent with a decrease in thermal conductance which is lower for torpid animals than for euthermic animals (Snyder and Nestler 1990). Can the relationship between T_a, T_b and TMR reported by Buck and Barnes (2000) be explained by a decrease in thermal conductance rather than reduced

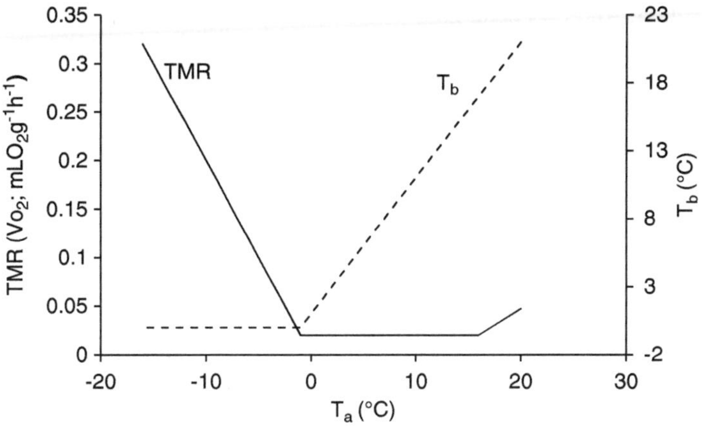

Fig. 13.4 Torpid T_b and metabolic rate in AGS were measured at T_a between 20 °C and −16 °C (Buck and Barnes 2000). Torpid T_b was found to decline in parallel with T_a down to about 0 °C after which it remained at approximately 0 °C. Torpid metabolic rate, by contrast, was found to fall within a range of minimal values between 0 and 16 °C showing that torpid metabolic rate does not depend on T_b. Here MR is calculated as a function of thermal conductance [MR = $C(T_b - T_a)$] estimated from data illustrated in Buck and Barnes (2000). The calculated minimum for C of 0.02 $mLO_2 g^{-1} h^{-1} °C^{-1}$ was used to calculate TMR at T_a less than 16 °C. When C is assumed to be minimal at T_a's below 16 °C the data generated approximates the affect of T_a on T_b and TMR reported by Buck and Barnes (2000)

metabolic demand at the systemic cellular level? If torpid AGS are assumed to reach and sustain minimal thermal conductance at and below an ambient temperature of 16 °C, TMR estimated from the results illustrated in Buck and Barnes (2000) approximates MR calculated as a function of thermal conductance (C) and the difference between T_a and T_b; [MR = $C(T_b - T_a)$ (Snyder and Nestler 1990)] (Fig. 13.4). Central A_1R activation could influence thermal conductance via postural adjustments, vasoconstriction, piloerection, and other heat conserving behavior and physiology. Findings in a small (25 g) marsupial, however, do not support the hypothesis that a decrease in thermal conductance accounts for TMR. Exposing torpid stripe-faced dunnarts (*Sminthopsis macroura*) to He-O_2 increases thermal conductance but decreases or has no effect on TMR (Geiser et al. 1996). More research is required to determine how central A_1R stimulation regulates suppression of systemic energy consumption beyond what can be explained by a decrease in thermogenesis.

Adenosine clearly plays a fundamental role in energy homeostasis in hibernation and torpor and as illustrated in Fig. 13.1 adenosine is poised to contribute to other aspects of hibernation via biochemical pathways. Activation of AMPK stimulates feeding and prevents torpor in marmots (Florant et al. 2010; Lim et al. 2010). Quantities of stored lipid that may be under influence by AMPK and/or peripheral effects of adenosine influence when hibernation begins (Pulawa and Florant 2000). Hibernation consists of a suite of physiological adaptations associated with energy homeostasis in addition to metabolic suppression and cold body temperature (Drew et al. 2001). Hibernating animals switch metabolic fuel use from carbohydrates to fat

(Andrews 2004; Andrews et al. 2009), dramatically suppress immune responsiveness (Bouma et al. 2010, 2011), suppress cellular division (Kruman 1992; Popov et al. 2011) and inhibit protein synthesis (Frerichs et al. 1998). Although this chapter addresses the central effects of adenosine on torpor in hibernating animals other aspects of hibernation may also be governed by adenosine. Adenosine has wide ranging influence over processes involved in energy homeostasis (Fredholm et al. 2011) and is clearly poised to regulate many aspects of the hibernation phenotype.

13.1.2 Energy Homeostasis and Sleep

Prolonged torpor in hibernating animals is in several aspects an extension of sleep suggesting that adenosine may share a regulatory role in sleep, hibernation and torpor (Berger and Phillips 1995). Slow wave sleep precedes entry into torpor and immediately follows arousal (Kilduff et al. 1993; Pastukhov 1997; Strijkstra and Daan 1997; Walker et al. 1977). During slow wave sleep, the hypothalamic temperature threshold for initiating metabolic heat production is decreased similar to altered thermoregulation and metabolism seen in torpor (Glotzbach and Heller 1976). Moreover, like torpor, sleep involves adenosine receptor-dependent signaling.

13.1.2.1 Adenosine Signaling in Sleep

Extracellular adenosine promotes homeostatic sleep and may originate from an imbalance in brain energy homeostasis. Sleep deprivation leads to an increase in homeostatic sleep drive and once sleep ensues, a rebound increase in EEG delta power, a measure of NREM sleep. Adenosine induces sleep (Benington et al. 1995; Porkka-Heiskanen and Kalinchuk 2011; Scharf et al. 2008) and may restore brain energy homeostasis (Benington and Heller 1995; Porkka-Heiskanen and Kalinchuk 2011). Benington and Heller (1995) rationalized that only adenosine mediated increase in gK^+ could explain the increased delta power characteristic of rebound sleep and predicted that sleep deprivation would be associated with an increase in extracellular adenosine. Experimental evidence supports this theory. The A_1R agonist N^6-cyclopentyladenosine administered systemically or icv mimics the EEG effects of sleep deprivation (Benington et al. 1995). Moreover increasing extracellular adenosine with a transport inhibitor induces sleep (Porkka-Heiskanen et al. 1997). Extracellular adenosine progressively increases in basal forebrain and cortex during sleep deprivation and in these and other brain regions extracellular adenosine levels decline during sleep (Porkka-Heiskanen et al. 2000). The increase in adenosine is associated with local energy deficits as indicated by increases in levels of pyruvate and lactate and increased phosphorylation of AMPK (Porkka-Heiskanen and Kalinchuk 2011). Adenosine contributes to homeostatic sleep by suppressing cortical arousal through inhibition or disinhibition of cell groups that promote arousal or sleep. In this way, adenosine that originates from an imbalance in brain energy homeostasis may restore homeostasis through biochemical and hyperpolarizing influences.

Although adenosine accumulates with sleep need higher levels of adenosine do not persist throughout NREM sleep (Porkka-Heiskanen et al. 2000) as predicted by Benington and Heller (1995). However, viewing adenosine as a bioenergetic network regulator illustrates how adenosine may restore brain energy homeostasis through influence that extends beyond receptor dependent signaling. Activated AMPK generally stimulates catabolic pathways (glycolysis, fatty acid oxidation and mitochondrial biogenesis) and inhibits anabolic pathways (gluconeogenesis and glycogen, fatty acid, and protein synthesis) (Hardie 2008; Lim et al. 2010). AMPK is activated during sleep deprivation (Chikahisa et al. 2009) and inactivated during sleep (Dworak et al. 2010). During sleep ATP increases within silent, wake-active neurons (Dworak et al. 2010). In this way, sleep restores energy balance in these wake-active neurons through biochemical and bioenergetic influences of adenosine. Thus, while receptor dependent signaling may contribute to homeostatic sleep drive, the biochemical influence of adenosine mediated through activation of AMPK restores ATP and biosynthetic processes during sleep. ATP is an important source of adenosine (zur Nedden et al. 2011) and is released from astrocytes to give rise to adenosine during sleep deprivation (Florian et al. 2011). In a reciprocal fashion, infusion of adenosine into basal forebrain increases tissue levels of ATP that may then facilitate anabolic processes during sleep (Dworak et al. 2010). Likewise, sleep expressed prior to and after arousal from torpor may restore biosynthetic processes.

Understanding the biochemical influence of adenosine in sleep may guide understanding of the biochemical influence of adenosine in hibernation and torpor with the caveat that although sleep precedes and follows torpor the function of sleep differs from the function of hibernation. Sleep lowers MR and T_b and leads to some somatic energy savings principally due to a decrease in MR during NREM sleep (Heller 2005; Katayose et al. 2009). However, comparison between sleep and quiet rest suggests that sleep plays a greater role in restoration of brain energy and synaptic homeostasis than in somatic energy homeostasis (Benington and Heller 1995; Bushey et al. 2011). As such, through affects on the CNS, sleep deprivation disrupts cognitive function and CNS coordination of energy balance (St-Onge et al. 2011) which may influence somatic energy homeostasis and lead to metabolic syndrome (Beccuti and Pannain 2011). By contrast, hibernation and torpor are means of somatic energy conservation that through a direct influence on MR prevent starvation and dehydration when food and water availability is limited. Although torpid animals do not eat or drink and heart rate, blood pressure, and respiratory rates decrease to 1–2 % of resting levels, torpid animals show no sign of energy deprivation (Lust et al. 1989). Indeed, O_2 saturation and other indices of cellular stress are less in the brains of torpid AGS than in the brains of summer euthermic or interbout euthermic AGS (Ma et al. 2005).

Adenosine receptor dependent signaling contributes to homeostatic sleep through a global inhibition of cortical arousal (Benington et al. 1995) and through focal affects on sleep–wake neurons. By contrast to torpor, effects of adenosine on homeostatic sleep drive are mediated through both A_1R and $A_{2A}R$ dependent processes (Huang et al. 2005, 2007; Oishi et al. 2008; Porkka-Heiskanen et al. 1997). A_1R inhibit wake-active cholinergic and noncholinergic neurons in basal forebrain (Arrigoni et al. 2006; Porkka-Heiskanen et al. 1997), histaminergic neurons in the tuberomammillary nucleus (Oishi et al. 2008), and hypocretin (orexin) neurons in the lateral hypothalamus (Liu and Gao 2007). Just dorsal to the cholinergic zone of the basal forebrain, in the $A_{2A}R$ rich ventral striatum, $A_{2A}R$ agonists promote sleep by disinhibition of sleep active neurons within the ventrolateral preoptic area (VLPO) of the hypothalamus (POAH) (Huang et al. 2007; Satoh et al. 2006; Sherin et al. 1996; Strecker et al. 2000). In addition, both A_1R and $A_{2A}R$ activation in the POAH promote sleep. Postsynaptic activation of $A_{2A}R$ in the VLPO excites sleep active neurons (Gallopin et al. 2005) and A_1R activation within the POAH removes inhibition of sleep-active neurons (Ticho and Radulovacki 1991).

13.1.2.2 Distinctions Between Hibernation and Sleep

Involvement of $A_{2A}R$ and differences in the regulated rate of cooling distinguishes sleep from torpor. Onset of sleep is normally coupled to an increase in the rate of heat loss, and a subsequent fall in body temperature as well as a decrease in MR (Heller 2005). By contrast, hibernating animals decrease the rate of heat loss. This decrease in the rate of heat loss during onset of torpor is evident from differences in the rate of cooling in AGS induced by CHA in summer and winter (Jinka et al. 2011). Differences in the rate of heat loss are consistent with vasodilatation during onset of sleep and vasoconstriction during hibernation (Hampton et al. 2010; Heller 2005; Lyman and O'Brien 1960). Nonetheless, both sleep and hibernation involve a change rather than a loss in thermoregulation. The capacity to thermoregulate distinguishes hibernation and sleep from hypothermia.

13.1.3 Adenosine and Thermoregulation

Adenosine receptor dependent signaling contributes to thermoregulation and maintains core body temperature (T_b) in part via A_1R activation in the POAH (Barros et al. 2006; Fredholm et al. 2011). During both NREM sleep and torpor (Glotzbach and Heller 1984; Heller et al. 1977) thermoregulatory mechanisms are intact but body temperature is regulated at lower levels than during wakefulness or euthermy. Altered CNS control of thermoregulatory processes during torpor (Heller et al. 1977) mirror similar processes that occur during sleep (Walker et al. 1977, 1980) except, as mentioned above, these processes slow the rate of cooling during torpor onset and enhance the rate of cooling during onset of sleep. Given the role of adenosine in sleep and hibernation differences in adenosine modulation of thermoregulation inherent to these two processes may underlie fundamental differences in the function of sleep and torpor. Adenosine also regulates metabolism and body temperature via receptor dependent and biochemical branches of the bioenergetic network outside of the CNS (Asakura 2004; Bauwens et al. 2011; Fredholm et al. 2011; Mulligan et al. 2007).

13.1.4 Therapeutic Significance of Adenosine Mediated Torpor

The capacity to translate reduced metabolic demand displayed during hibernation as well as daily torpor to humans could have therapeutic applications for patients suffering from disruptions in blood flow and oxygen delivery (Drew et al. 2001). We have begun to explore the effects of dietary restriction on sensitivity to A_1R agonists as a means to suppress metabolic demand in nonhibernating species. Dietary restriction lowers T_b and enhances longevity presumably through a decrease in oxidative metabolism (Contestabile 2009; Ungvari et al. 2008). Dietary restriction-induced modification of thermoregulation is associated with changes in components of the purinergic neuromodulatory system. Dietary restriction imposed by every other day feeding sensitizes rats to the cooling and metabolic depressant effects of CHA and is associated with increased surface expression of A_1R in the hypothalamus (Jinka et al. 2010). These data suggest that dietary restriction sensitizes A_1R and in this way contributes to the decline in T_b and MR during dietary restriction. It is interesting to speculate that ketogenic diets may have similar effects; however, ketogenic diets are not known to affect T_b in the same way as dietary restriction.

In summary, adenosine plays a fundamental, regulatory role in mammalian hibernation, sleep, and thermoregulation all of which are integral to systemic energy homeostasis. Regulatory influence of adenosine in these processes has been described best in terms of central A_1R dependent signaling. However, adenosine also influences energy homeostasis through biochemical and bioenergetic signaling pathways within and outside of the CNS.

Acknowledgments This work was supported by US Army Research Office Grant W911NF-05-1-0280, US Army Medical Research and Materiel Command Grant 05178001, and National Institute of Neurological Disorders and Stroke Grants NS041069-06 and R15NS070779.

References

Andrews MT (2004) Genes controlling the metabolic switch in hibernating mammals. Biochem Soc Trans 32:1021–1024

Andrews MT, Russeth KP, Drewes LR, Henry PG (2009) Adaptive mechanisms regulate preferred utilization of ketones in the heart and brain of a hibernating mammal during arousal from torpor. Am J Physiol Regul Integr Comp Physiol 296:R383–R393

Arrigoni E, Chamberlin NL, Saper CB, McCarley RW (2006) Adenosine inhibits basal forebrain cholinergic and noncholinergic neurons in vitro. Neuroscience 140:403–413

Asakura H (2004) Fetal and neonatal thermoregulation. J Nippon Med Sch 71:360–370

Atkins PW, De Paula J (2006) Atkins' physical chemistry, 8th edn. W.H. Freeman, New York

Barnes BM (1989) Freeze avoidance in a mammal: body temperatures below 0 degree C in an Arctic hibernator. Science 244:1593–1595

Barros RC, Branco LG (2000) Role of central adenosine in the respiratory and thermoregulatory responses to hypoxia. Neuroreport 11:193–197

Barros RC, Branco LG, Carnio EC (2006) Respiratory and body temperature modulation by adenosine A1 receptors in the anteroventral preoptic region during normoxia and hypoxia. Respir Physiol Neurobiol 153:115–125

Bauwens JD, Schmuck EG, Lindholm CR, Ertel RL, Mulligan JD, Hovis I, Viollet B, Saupe KW (2011) Cold tolerance, cold-induced hyperphagia and non-shivering thermogenesis are normal in {alpha}1 AMPK-/- mice. Am J Physiol Regul Integr Comp Physiol 301(2): R473–R483

Beccuti G, Pannain S (2011) Sleep and obesity. Curr Opin Clin Nutr Metab Care 14:402–412

Benington JH, Heller HC (1995) Restoration of brain energy metabolism as the function of sleep. Prog Neurobiol 45:347–360

Benington JH, Kodali SK, Heller HC (1995) Stimulation of A1 adenosine receptors mimics the electroencephalographic effects of sleep deprivation. Brain Res 692:79–85

Berger RJ, Phillips NH (1995) Energy conservation and sleep. Behav Brain Res 69:65–73

Blackstone E, Morrison M, Roth MB (2005) H2S induces a suspended animation-like state in mice. Science 308:518

Boison D, Masino SA, Geiger JD (2011) Homeostatic bioenergetic network regulation: a novel concept to avoid pharmacoresistance in epilepsy. Expert Opin Drug Discov 6:1–12

Bouma HR, Carey HV, Kroese FG (2010) Hibernation: the immune system at rest? J Leukoc Biol 88:619–624

Bouma HR, Kroese FG, Kok JW, Talaei F, Boerema AS, Herwig A, Draghiciu O, van Buiten A, Epema AH, van Dam A et al (2011) Low body temperature governs the decline of circulating lymphocytes during hibernation through sphingosine-1-phosphate. Proc Natl Acad Sci USA 108:2052–2057

Bratincsak A, McMullen D, Miyake S, Toth ZE, Hallenbeck JM, Palkovits M (2007) Spatial and temporal activation of brain regions in hibernation: c-fos expression during the hibernation bout in thirteen-lined ground squirrel. J Comp Neurol 505:443–458

Braulke LJ, Heldmaier G (2010) Torpor and ultradian rhythms require an intact signalling of the sympathetic nervous system. Cryobiology 60:198–203

Buck CL, Barnes BM (2000) Effects of ambient temperature on metabolic rate, respiratory quotient, and torpor in an arctic hibernator. Am J Physiol Regul Integr Comp Physiol 279:R255–R262

Bushey D, Tononi G, Cirelli C (2011) Sleep and synaptic homeostasis: structural evidence in Drosophila. Science 332:1576–1581

Cano G, Passerin AM, Schiltz JC, Card JP, Morrison SF, Sved AF (2003) Anatomical substrates for the central control of sympathetic outflow to interscapular adipose tissue during cold exposure. J Comp Neurol 460:303–326

Carey HV, Andrews MT, Martin SL (2003) Mammalian hibernation: cellular and molecular responses to depressed metabolism and low temperature. Physiol Rev 83:1153–1181

Chikahisa S, Fujiki N, Kitaoka K, Shimizu N, Sei H (2009) Central AMPK contributes to sleep homeostasis in mice. Neuropharmacology 57:369–374

Contestabile A (2009) Benefits of caloric restriction on brain aging and related pathological States: understanding mechanisms to devise novel therapies. Curr Med Chem 16:350–361

Dausmann KH, Glos J, Ganzhorn JU, Heldmaier G (2004) Physiology: hibernation in a tropical primate. Nature 429:825–826

Drew KL, Buck CL, Barnes BM, Christian SL, Rasley BT, Harris MB (2007) Central nervous system regulation of mammalian hibernation: implications for metabolic suppression and ischemia tolerance. J Neurochem 102:1713–1726

Drew KL, Rice ME, Kuhn TB, Smith MA (2001) Neuroprotective adaptations in hibernation: therapeutic implications for ischemia-reperfusion, traumatic brain injury and neurodegenerative diseases. Free Radic Biol Med 31:563–573

Dunwiddie TV, Masino SA (2001) The role and regulation of adenosine in the central nervous system. Annu Rev Neurosci 24:31–55

Dworak M, McCarley RW, Kim T, Kalinchuk AV, Basheer R (2010) Sleep and brain energy levels: ATP changes during sleep. J Neurosci 30:9007–9016

Epperson LE, Karimpour-Fard A, Hunter LE, Martin SL (2011) Metabolic cycles in a circannual hibernator. Physiol Genomics 43(13):799–807

Florant GL, Fenn AM, Healy JE, Wilkerson GK, Handa RJ (2010) To eat or not to eat: the effect of AICAR on food intake regulation in yellow-bellied marmots (Marmota flaviventris). J Exp Biol 213:2031–2037

Florant GL, Turner BM, Heller HC (1978) Temperature regulation during wakefulness, sleep, and hibernation in marmots. Am J Physiol 235:R82–R88

Florian C, Vecsey CG, Halassa MM, Haydon PG, Abel T (2011) Astrocyte-derived adenosine and A1 receptor activity contribute to sleep loss-induced deficits in hippocampal synaptic plasticity and memory in mice. J Neurosci 31:6956–6962

Fredholm BB, Johansson S, Wang YQ (2011) Adenosine and the regulation of metabolism and body temperature. Adv Pharmacol 61:77–94

French AR (1985) Allometries of the durations of torpid and euthermic intervals during mammalian hibernation: a test of the theory of metabolic control of the timing of changes in body temperature. J Comp Physiol B 156:13–19

Frerichs KU, Smith CB, Brenner M, DeGracia DJ, Krause GS, Marrone L, Dever TE, Hallenbeck JM (1998) Suppression of protein synthesis in brain during hibernation involves inhibition of protein initiation and elongation. Proc Natl Acad Sci USA 95:14511–14516

Fukuda M, Williams KW, Gautron L, Elmquist JK (2011) Induction of leptin resistance by activation of cAMP-Epac signaling. Cell Metab 13:331–339

Gallopin T, Luppi PH, Cauli B, Urade Y, Rossier J, Hayaishi O, Lambolez B, Fort P (2005) The endogenous somnogen adenosine excites a subset of sleep-promoting neurons via A2A receptors in the ventrolateral preoptic nucleus. Neuroscience 134:1377–1390

Galster W, Morrison PR (1975) Gluconeogenesis in arctic ground squirrels between periods of hibernation. Am J Physiol 228:325–330

Galster WA, Morrison P (1970) Cyclic changes in carbohydrate concentrations during hibernation in the arctic ground squirrel. Am J Physiol 218:1228–1232

Garcia MM, Gueant-Rodriguez RM, Pooya S, Brachet P, Alberto JM, Jeannesson E, Maskali F, Gueguen N, Marie PY, Lacolley P et al (2011) Methyl donor deficiency induces cardiomyopathy through altered methylation/acetylation of PGC-1alpha by PRMT1 and SIRT1. J Pathol 225(3):324–335

Geiser F (1988) Reduction of metabolism during hibernation and daily torpor in mammals and birds: temperature effect or physiological inhibition? J Comp Physiol B 158:25–37

Geiser F (2004) Metabolic rate and body temperature reduction during hibernation and daily torpor. Annu Rev Physiol 66:239–274

Geiser F, Song X, Körtner G (1996) The effect of He-O_2 exposure on metabolic rate, thermoregulation and thermal conductance during normothermia and daily torpor. J Comp Physiol B 166:190–196

Gillooly JF, Brown JH, West GB, Savage VM, Charnov EL (2001) Effects of size and temperature on metabolic rate. Science 293:2248–2251

Glotzbach SF, Heller HC (1976) Central nervous regulation of body temperature during sleep. Science 194:537–539

Glotzbach SF, Heller HC (1984) Changes in the thermal characteristics of hypothalamic neurons during sleep and wakefulness. Brain Res 309:17–26

Hampton M, Nelson BT, Andrews MT (2010) Circulation and metabolic rates in a natural hibernator: an integrative physiological model. Am J Physiol Regul Integr Comp Physiol 299:R1478–R1488

Hardie DG (2008) AMPK: a key regulator of energy balance in the single cell and the whole organism. Int J Obes (Lond) 32 Suppl 4: S7–S12

Heldmaier G, Ortmann S, Elvert R (2004) Natural hypometabolism during hibernation and daily torpor in mammals. Respir Physiol Neurobiol 141:317–329

Heller H (2005) Temperature, thermoregulation and sleep. In: Kryger M, Roth T, Dement W (eds) Principles and practice of sleep medicine. Elsevier Saunders, Philadelphia, pp 292–304

Heller HC, Colliver GW, Beard J (1977) Thermoregulation during entrance into hibernation. Pflugers Arch 369:55–59

Hochachka PW (1986) Defense strategies against hypoxia and hypothermia. Science 231:234–241

Hoffman RA, Robinson PF, Magalhaes H (eds) (1968) The golden hamster; its biology and use in medical research. The Iowa State University Press, Ames Iowa, USA

Huang ZL, Qu WM, Eguchi N, Chen JF, Schwarzschild MA, Fredholm BB, Urade Y, Hayaishi O (2005) Adenosine A2A, but not A1, receptors mediate the arousal effect of caffeine. Nat Neurosci 8:858–859

Huang ZL, Urade Y, Hayaishi O (2007) Prostaglandins and adenosine in the regulation of sleep and wakefulness. Curr Opin Pharmacol 7:33–38

Jinka TJ, Toien O, Drew KL (2011) Season primes the brain in an arctic hibernator to facilitate entrance into torpor mediated by adenosine A_1 receptors. J Neurosci 31(30):10752–10758

Jinka TR, Carlson ZA, Moore JT, Drew KL (2010) Altered thermoregulation via sensitization of A1 adenosine receptors in dietary restricted rats. Psychopharmacology (Berl) 209(3):217–224

Karpovich SA, Toien O, Buck CL, Barnes BM (2009) Energetics of arousal episodes in hibernating arctic ground squirrels. J Comp Physiol B 179:691–700

Katayose Y, Tasaki M, Ogata H, Nakata Y, Tokuyama K, Satoh M (2009) Metabolic rate and fuel utilization during sleep assessed by whole-body indirect calorimetry. Metabolism 58:920–926

Kilduff TS, Krilowicz B, Milsom WK, Trachsel L, Wang LC (1993) Sleep and mammalian hibernation: homologous adaptations and homologous processes? Sleep 16:372–386

Kruman II (1992) Comparative analysis of cell replacement in hibernators. Comp Biochem Physiol A 101:11–18

Latini S, Pedata F (2001) Adenosine in the central nervous system: release mechanisms and extracellular concentrations. J Neurochem 79:463–484

Lee TF, Nurnberger F, Wang LCH (eds) (1993) Possible involvement of endogenous adenosine in hibernation. Westview, Boulder

Lee TM, Zucker I (1991) Suprachiasmatic nucleus and photic entrainment of circannual rhythms in ground squirrels. J Biol Rhythms 6:315–330

Levesque DL, Tattersall GJ (2009) Seasonal changes in thermoregulatory responses to hypoxia in the Eastern chipmunk (Tamias striatus). J Exp Biol 212:1801–1810

Lim CT, Kola B, Korbonits M (2010) AMPK as a mediator of hormonal signalling. J Mol Endocrinol 44:87–97

Liu ZW, Gao XB (2007) Adenosine inhibits activity of hypocretin/orexin neurons by the A1 receptor in the lateral hypothalamus: a possible sleep-promoting effect. J Neurophysiol 97:837–848

Lust WD, Wheaton AB, Feussner G, Passonneau J (1989) Metabolism in the hamster brain during hibernation and arousal. Brain Res 489:12–20

Lyman C, O'Brien R (1960) Circulatory changes in the thirteen-lined ground squirrel during the hibernation cycle. In Mammalian Hibernation Bull Mus Comp Zool, 353–372

Ma YL, Zhu X, Rivera PM, Toien O, Barnes BM, LaManna JC, Smith MA, Drew KL (2005) Absence of cellular stress in brain after hypoxia induced by arousal from hibernation in Arctic ground squirrels. Am J Physiol Regul Integr Comp Physiol 289:R1297–R1306

Magarinos AM, McEwen BS, Saboureau M, Pevet P (2006) Rapid and reversible changes in intrahippocampal connectivity during the course of hibernation in European hamsters. Proc Natl Acad Sci USA 103:18775–18780

Muleme HM, Walpole AC, Staples JF (2006) Mitochondrial metabolism in hibernation: metabolic suppression, temperature effects, and substrate preferences. Physiol Biochem Zool 79:474–483

Mulligan JD, Gonzalez AA, Stewart AM, Carey HV, Saupe KW (2007) Upregulation of AMPK during cold exposure occurs via distinct mechanisms in brown and white adipose tissue of the mouse. J Physiol 580:677–684

Nakamura K (2011) Central circuitries for body temperature regulation and fever. Am J Physiol Regul Integr Comp Physiol 301:R1207–R1228

Oishi Y, Huang ZL, Fredholm BB, Urade Y, Hayaishi O (2008) Adenosine in the tuberomammillary nucleus inhibits the histaminergic system via A1 receptors and promotes non-rapid eye movement sleep. Proc Natl Acad Sci USA 105:19992–19997

Osborne PG, Sato J, Shuke N, Hashimoto M (2005) Sympathetic alpha-adrenergic regulation of blood flow and volume in hamsters arousing from hibernation. Am J Physiol Regul Integr Comp Physiol 289:R554–R562

Pastukhov YF (1997) REM sleep as a criterion of temperature comfort and temperature homeostasis "well-being" in euthermic and hibernating mammals. Ann N Y Acad Sci 813:71–72
Pek M, Lutz PL (1997) Role for adenosine in channel arrest in the anoxic turtle brain. J Exp Biol 200:1913–1917
Pengelley ET, Asmundson SJ, Barnes B, Aloia RC (1976) Relationship of light intensity and photoperiod to circannual rhythmicity in the hibernating ground squirrel, Citellus lateralis. Comp Biochem Physiol A 53:273–277
Popov VI, Bocharova LS (1992) Hibernation-induced structural changes in synaptic contacts between mossy fibres and hippocampal pyramidal neurons. Neuroscience 48:53–62
Popov VI, Kraev IV, Ignat'ev DA, Stewart MG (2011) Suspension of mitotic activity in dentate gyrus of the hibernating ground squirrel. Neural Plast 2011:867525
Porkka-Heiskanen T, Kalinchuk AV (2011) Adenosine, energy metabolism and sleep homeostasis. Sleep Med Rev 15:123–135
Porkka-Heiskanen T, Strecker RE, McCarley RW (2000) Brain site-specificity of extracellular adenosine concentration changes during sleep deprivation and spontaneous sleep: an in vivo microdialysis study. Neuroscience 99:507–517
Porkka-Heiskanen T, Strecker RE, Thakkar M, Bjorkum AA, Greene RW, McCarley RW (1997) Adenosine: a mediator of the sleep-inducing effects of prolonged wakefulness. Science 276:1265–1268
Pulawa LK, Florant GL (2000) The effects of caloric restriction on the body composition and hibernation of the golden-mantled ground squirrel (*Spermophilus lateralis*). Physiol Biochem Zool 73:538–546
Rivkees SA, Price SL, Zhou FC (1995) Immunohistochemical detection of A1 adenosine receptors in rat brain with emphasis on localization in the hippocampal formation, cerebral cortex, cerebellum, and basal ganglia. Brain Res 677:193–203
Russell RL, O'Neill PH, Epperson LE, Martin SL (2010) Extensive use of torpor in 13-lined ground squirrels in the fall prior to cold exposure. J Comp Physiol B 180:1165–1172
Satoh S, Matsumura H, Kanbayashi T, Yoshida Y, Urakami T, Nakajima T, Kimura N, Nishino S, Yoneda H (2006) Expression pattern of FOS in orexin neurons during sleep induced by an adenosine A2A receptor agonist. Behav Brain Res 170:277–286
Scharf MT, Naidoo N, Zimmerman JE, Pack AI (2008) The energy hypothesis of sleep revisited. Prog Neurobiol 86:264–280
Schmidt-Nielsen K (1997) Animal physiology: adaptation and environment, 5th edn. Cambridge University Press, Cambridge
Serkova NJ, Rose JC, Epperson LE, Carey HV, Martin SL (2007) Quantitative analysis of liver metabolites in three stages of the circannual hibernation cycle in 13-lined ground squirrels by NMR. Physiol Genomics 31:15–24
Sheriff MJ, Kenagy GJ, Richter M, Lee T, Toien O, Kohl F, Buck CL, Barnes BM (2010) Phenological variation in annual timing of hibernation and breeding in nearby populations of Arctic ground squirrels. Proc Biol Sci 278(1716):2369–2375
Sherin JE, Shiromani PJ, McCarley RW, Saper CB (1996) Activation of ventrolateral preoptic neurons during sleep. Science 271:216–219
Shiomi H, Tamura Y (2000) Pharmacological aspects of mammalian hibernation: central thermoregulation factors in hibernation cycle. Nippon Yakurigaku Zasshi (Folia Pharmacol Jpn) 116:304–312
Snyder GK, Nestler JR (1990) Relationships between body temperature, thermal conductance, Q10 and energy metabolism during daily torpor and hibernation in rodents. J Comp Physiol B 159:667–675
St-Onge MP, Roberts AL, Chen J, Kelleman M, O'Keeffe M, Roychoudhury A, Jones PJ (2011) Short sleep duration increases energy intakes but does not change energy expenditure in normal-weight individuals. Am J Clin Nutr 94:410–416
Staples JF, Brown JC (2008) Mitochondrial metabolism in hibernation and daily torpor: a review. J Comp Physiol B 178:811–827

Steiner AA, Branco LG (2002) Hypoxia-induced anapyrexia: implications and putative mediators. Annu Rev Physiol 64:263–288

Strecker RE, Morairty S, Thakkar MM, Porkka-Heiskanen T, Basheer R, Dauphin LJ, Rainnie DG, Portas CM, Greene RW, McCarley RW (2000) Adenosinergic modulation of basal forebrain and preoptic/anterior hypothalamic neuronal activity in the control of behavioral state. Behav Brain Res 115:183–204

Strijkstra AM, Daan S (1997) Sleep during arousal episodes as a function of prior torpor duration in hibernating European ground squirrels. J Sleep Res 6:36–43

Swoap S, Lliff B (2011) AMP vs. adenosine as a mediator of fasting-induced torpor. Paper presented at metabolic responses to extreme conditions, keystone symposia on molecular and cellular biology, Big Sky, Montana

Swoap SJ, Gutilla MJ, Liles LC, Smith RO, Weinshenker D (2006) The full expression of fasting-induced torpor requires beta 3-adrenergic receptor signaling. J Neurosci 26:241–245

Swoap SJ, Rathvon M, Gutilla M (2007) AMP does not induce torpor. Am J Physiol Regul Integr Comp Physiol 293:R468–R473

Swoap SJ, Weinshenker D (2008) Norepinephrine controls both torpor initiation and emergence via distinct mechanisms in the mouse. PLoS One 3:e4038

Tamura Y, Shintani M, Nakamura A, Monden M, Shiomi H (2005) Phase-specific central regulatory systems of hibernation in Syrian hamsters. Brain Res 1045:88–96

Tattersall GJ, Milsom WK (2003) Transient peripheral warming accompanies the hypoxic metabolic response in the golden-mantled ground squirrel. J Exp Biol 206:33–42

Ticho SR, Radulovacki M (1991) Role of adenosine in sleep and temperature regulation in the preoptic area of rats. Pharmacol Biochem Behav 40:33–40

Toien O, Blake J, Edgar DM, Grahn DA, Heller HC, Barnes BM (2011) Hibernation in black bears: independence of metabolic suppression from body temperature. Science 331:906–909

Toien O, Drew KL, Chao ML, Rice ME (2001) Ascorbate dynamics and oxygen consumption during arousal from hibernation in Arctic ground squirrels. Am J Physiol Regul Integr Comp Physiol 281:R572–R583

Twente JW, Twente JA (1968) Progressive irritability of hibernating Citellus lateralis. Comp Biochem Physiol 25:467–474

Ungvari Z, Parrado-Fernandez C, Csiszar A, de Cabo R (2008) Mechanisms underlying caloric restriction and lifespan regulation: implications for vascular aging. Circ Res 102:519–528

von der Ohe CG, Garner CC, Darian-Smith C, Heller HC (2007) Synaptic protein dynamics in hibernation. J Neurosci 27:84–92

Walker JM, Glotzbach SF, Berger RJ, Heller HC (1977) Sleep and hibernation in ground squirrels (Citellus spp): electrophysiological observations. Am J Physiol 233:R213–R221

Walker JM, Haskell EH, Berger RJ, Heller CH (1980) Hibernation and circannual rhythms of sleep. Physiol Zool 53:8–11

Walther T, Novo M, Rossger K, Letisse F, Loret MO, Portais JC, Francois JM (2010) Control of ATP homeostasis during the respiro-fermentative transition in yeast. Mol Syst Biol 6:344

Wilson CN (2009) Adenosine receptors in health and disease. Springer, New York

Zhang J, Kaasik K, Blackburn MR, Lee CC (2006) Constant darkness is a circadian metabolic signal in mammals. Nature 439:340–343

Zhu PJ, Krnjevic K (1997) Adenosine release mediates cyanide-induced suppression of CA1 neuronal activity. J Neurosci 17:2355–2364

Zimmermann R, Strauss JG, Haemmerle G, Schoiswohl G, Birner-Gruenberger R, Riederer M, Lass A, Neuberger G, Eisenhaber F, Hermetter A et al (2004) Fat mobilization in adipose tissue is promoted by adipose triglyceride lipase. Science 306:1383–1386

zur Nedden S, Hawley S, Pentland N, Hardie DG, Doney AS, Frenguelli BG (2011) Intracellular ATP influences synaptic plasticity in area CA1 of rat hippocampus via metabolism to adenosine and activity-dependent activation of adenosine A1 receptors. J Neurosci 31:6221–6234

Chapter 14
Adenosine and Stroke

Felicita Pedata, Anna Maria Pugliese, Francesca Corti, and Alessia Melani

Abstract The concentration of adenosine in the brain increases dramatically during ischemia, and adenosine has long been known to act predominantly as a neuroprotectant during ischemia. In agreement, adenosine infusion into the ischemic striatum significantly ameliorates neurological outcome and reduces the infarct volume after transient focal cerebral ischemia. Despite the neuroprotective effect of adenosine through A_1 receptors during ischemia, the use of selective A_1 agonists is hampered by unwanted peripheral effects such as sedation, bradycardia, and hypotension. An alternative therapeutic approach may consist of using agents that elevate the local concentration of adenosine at areas of injury-induced adenosine release, thus minimizing undesirable peripheral and central effects mediated by A_1 receptors. Adenosine-potentiating agents at injury sites may act by (1) inhibiting its metabolism by adenosine kinase or adenosine deaminase, (2) preventing its transport through equilibrative membrane transporters (ENTs), (3) enhancing hydrolysis of extracellular ATP by use of nucleoside triphosphate diphosphohydrolases (NTPDases) and ecto-5′-nucleotidase (e5′-NTs). The role of A_{2A} receptors also is important for neuroprotection during ischemia. Evidence suggests that low doses of A_{2A} antagonists (that do not modify hemodynamic parameters) provide protection via centrally mediated control of precocious excessive excitotoxicity, while A_{2A} agonists provide protection by controlling massive cell infiltration in the hours after ischemia. Information suggests that A_{2A} receptors are a potentially attractive therapeutic target in ischemia. In perspective of using adenosine A_{2A} active drugs to protect against brain ischemia, attention should be given to the dose and administration time after injury. Although much is still to be learned about the function of the A_3 receptor, at the present time it can be speculated that A_3 receptor selective ligands may be useful in the treatment of ischemic conditions involving inflammation.

F. Pedata (✉) • A.M. Pugliese • F. Corti • A. Melani
Department of Preclinical and Clinical Pharmacology,
University of Florence, Viale Pieraccini 6, Florence 50139, Italy
e-mail: felicita.pedata@unifi.it

S. Masino and D. Boison (eds.), *Adenosine: A Key Link between Metabolism and Brain Activity*, DOI 10.1007/978-1-4614-3903-5_14,

Keywords Adenosine • Adenosine receptors • Excitotoxicity • Infiltration • Neuroprotection • Stroke

14.1 Introduction

Ischemic stroke is the third leading cause of death in major industrialized countries, with a mortality rate of around 30 %, and the major cause of long-lasting disabilities. Ischemic stroke results from a transient or permanent reduction in cerebral blood flow which is, in most cases, caused by the occlusion of a major brain artery, either by an embolus or by local thrombosis. There still is no good therapy for this pathology. Neuroprotective drugs that have been developed and have shown therapeutic potential in animal stroke trials have failed to be efficacious during clinical trials (De Keyser et al. 1999). Because the extracellular adenosine concentration increases dramatically during ischemia (Hagberg et al. 1987; Dux et al. 1990; Matsumoto et al. 1992; Sciotti et al. 1992; Phillis et al. 1994a, 1996; Melani et al. 1999, 2003; Latini and Pedata 2001), numerous authors have indicated adenosine and its receptors as a target for therapeutic implementation in the treatment of stroke (see Fig. 14.1).

Adenosine is a naturally occurring nucleoside that is distributed ubiquitously throughout the body as a metabolic intermediary. Extracellular adenosine acts through multiple G-protein coupled receptors (adenosine receptor subtypes A_1R, $A_{2A}R$, $A_{2B}R$, and A_3R) to exert a variety of physiological effects (Fredholm et al. 2001). During ischemia, extracellular adenosine levels increase and adenosine has long been known to act predominantly as a neuroprotectant endogenous agent (Ongini et al. 1997; Cunha 2001; Ribeiro et al. 2002; Schwarzschild et al. 2002; Fredholm et al. 2003; Pedata et al. 2007). Adenosine infusion into the ischemic striatum has been shown to significantly ameliorate neurological outcome and reduce infarct volume after transient focal cerebral ischemia (Kitagawa et al. 2002).

In this chapter, we summarize recent developments that have contributed to current understanding of how adenosine receptors modulate tissue damage in brain ischemia models. Adenosine receptors are expressed at significant levels in neurons and glial cells and in peripheral inflammatory cells (such as lymphocytes and neutrophils) (Fiebich et al. 1996; Peterfreund et al. 1996; Brodie et al. 1998; Svenningsson et al. 1999; Hettinger et al. 2001; Yu et al. 2004a). The wide distribution is consistent with the multifaceted neurochemical and molecular effects of adenosine receptor activation and suggests that the role of adenosine in ischemia is the consequence of interplay among different receptor activation in neuronal, glial and inflammatory cells, which may vary depending on the time-related development of the pathological condition.

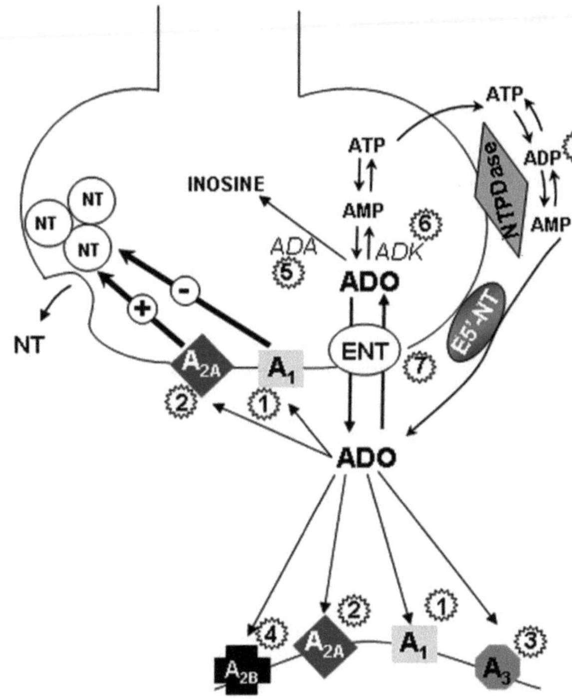

Fig. 14.1 Adenosine receptors and enzymes involved in adenosine metabolism. Putative protective drugs in ischemia include: (1) *A_1R agonists*; (2) *$A_{2A}R$ antagonists or agonists*; (4) *A_3R agonists or antagonists*; (5) *ADA inhibitors*; (6) *ADK inhibitors*; (7) ETN *inhibitors*; (8) *NTPDases*; *ADA* adenosine deaminase, ADK adenosine kinase, ADO adenosine, ADP adenosine diphosphate, AMP adenosine monophosphate, ATP adenosine triphosphate, E5′-NT ecto-5′-nucleotidase, ENT equilibrative nucleoside transporter, NT neurotransmitter, NTPDase nucleoside triphosphate diphosphohydrolase

14.2 Roles of A_1 Receptors

14.2.1 A_1 Receptors Are Protective

The increase in extracellular adenosine during ischemia may be an endogenous neuroprotective response (Latini and Pedata 2001). One of the prime adaptive mechanisms in response to hypoxia–ischemia is the cellular activation of A_1R which inhibits excitatory synaptic transmission as demonstrated in vitro (Fowler 1989, 1990; Gribkoff et al. 1990; Latini et al. 1999a; Dale et al. 2000; Sebastião et al. 2000) and in vivo (Gervitz et al. 2001, 2003; Fowler et al. 2003).

Protective effects are greatly attributed to A_1R activation due to reduced Ca^{2+} influx, thus lowering presynaptic release of excitatory neurotransmitters (Corradetti et al. 1984; Dunwiddie 1984; Zetterstrom and Fillenz 1990; Pedata et al. 1993; Andiné 1993; Kitagawa et al. 2002) and in particular glutamate which exerts an excitotoxic effect during ischemia mainly by overstimulation of NMDA receptors (Choi 1990). In addition, by directly increasing the K^+ and Cl^- ion conductances, adenosine stabilizes the neuronal membrane potential, thus reducing neuronal excitability (Choi 1990). Consequent reductions in cellular metabolism and energy consumption (Greene and Haas 1991) and moderate lowering of the body/brain temperature (Gourine et al. 2004) protect against ischemia. A_1R agonists have been shown to attenuate ischemic or excitotoxic neuronal damage in both in vitro and in vivo models of cerebral ischemia. In in vitro hypoxia/ischemia models, it was demonstrated that both adenosine and selective A_1R agonists reduce neuronal damage following hypoxia and/or oxygen-glucose deprivation (OGD) in primary cortical or hippocampal cell cultures (Goldberg et al. 1988; Daval and Nicolas 1994; Logan and Sweeney 1997) and brain slices (Mori et al. 1992; Newman et al. 1998). A_1R agonists decrease reactive oxygen species (ROS) production and increase survival in anoxia and anoxia/reoxygenation, while A_1R blockade increases ROS release and cell death in primary neuronal cultures (Milton et al. 2007). Marcoli and coworkers (2003) demonstrated that the selective A_1R antagonist, 1,3-dipropyl-8-cyclopentyladenosine (DPCPX) (Weyler et al. 2006), increases OGD-evoked aspartate and glutamate efflux in rat cerebrocortical slices. In accordance, the A_1R -mediated OGD-induced depression of synaptic transmission is reversed by administration of selective A_1R antagonists to rat hippocampal slices (Pedata et al. 1993; Latini et al. 1999a; Tanaka et al. 2001). A_1R antagonists reduce inhibition of synaptic transmission, impair the recovery of synaptic potentials (Sebastião et al. 2001), and shorten the onset of anoxic depolarization (AD) induced by hypoxia in the CA1 region of hippocampal slices (Lee and Lowenkopf 1993). A depression of excitatory synaptic transmission such as that brought about by adenosine A_1Rs during hypoxia/ischemia is neuroprotective and crucial for the functional recovery of hippocampal circuits upon reoxygenation (Johansson et al. 2001; Sebastião et al. 2001; Arrigoni et al. 2005). Furthermore, activation of A_1R, by endogenous adenosine, transiently enhances inhibitory synaptic transmission in CA1 pyramidal slices from rats exposed to transient forebrain ischemia (Liang et al. 2009). Therefore, it is likely that A_1R activation can protect CA1 pyramidal neurons and delay the process of neuronal death after cerebral ischemia. Studies using A_1R knockout (KO) mice support the neuroprotective role of A_1R receptor stimulation. Hippocampal slices from A_1R KO mice showed a markedly reduced and delayed protective response to hypoxia compared to slices from wild-type mice (Johansson et al. 2001). In astrocytes prepared from A_1R KO mice more pronounced hypoxic cytotoxicity was observed (Bjorklund et al. 2008). However, Olsson et al. (2004) reported that hippocampal slices from A_1R KO mice subjected to hypoxia show a small, though not statistically significant decrease in damage to neurons in the CA1 region.

Although the mechanisms that underlie A_1R-mediated effects are unclear, it is likely that desensitization of A_1R after ischemia might be critical in interfering with maintaining the neuroprotective efficiency of adenosine. In rat hippocampal slices, hypoxia leads to a rapid (<90 min) desensitization of A_1R -mediated inhibition of synaptic transmission that is likely due to an internalization of A_1R in nerve terminals (Coelho et al. 2006), a process that may result in hyperexcitability and increased brain damage. One recent interesting observation is that DPCPX reversed cannabidiol (CBD)-induced neuroprotection after hypoxic/ischemic insult in forebrain slices (Castillo et al. 2010), demonstrating that A_1R are involved in cannabinoid protection during ischemia.

In in vivo animal models of global cerebral ischemia, it has been demonstrated that local administration of an adenosine analogue, 2-chloroadenosine (CADO), and of a nonselective A_1R agonist, N6-(L-2-phenylisopropyl) adenosine (L-PIA), attenuates neuronal loss in the CA1 region of the rat hippocampus (Evans et al. 1987; Domenici et al. 1996) and that acute systemic or intracerebroventricular injection of the A_1R agonists cyclohexyladenosine (CHA) and R-phenylisopropyladenosine (R-PIA) improves neurological deficits (Von Lubitz and Marangos 1990; Heron et al. 1994; Zhou et al. 1994), protects the CA1 region of the hippocampus (Von Lubitz et al. 1988), and prevents the reduction of A_1R (Daval et al. 1989) in rats or gerbils. Similarly, acute administration of the A_1R agonists N6-cyclopentyladenosine (CPA) and 2-chloro-N(6)-cyclopentyladenosine (CCPA) reduces mortality and loss of neurons after global forebrain ischemia in the gerbil (Von Lubitz et al. 1994a). Administration of the A_1R agonist adenosine amine congener (ADAC), which induces fewer undesirable effects, after global ischemia in the gerbil, increased survival, protected neuronal morphology, induced preservation of microtubule associated protein 2 (MAP-2) and of spatial memory and learning ability (von Lubitz et al. 1996; Phillis and Goshgarian 2001). In accordance, A_1R antagonists given acutely exacerbate the changes induced by global ischemia. A nonspecific A_1R antagonist, theophylline, increased mortality (Jarrott and Domer 1980) and ischemic cell damage (Rudolphi et al. 1987, 1989) after global ischemia in gerbils. Similar detrimental effects were also observed after acute administration of the selective A_1R antagonists DPCPX and 8-cyclopentyl-1,3-dimethylxanthine (CPT) (Boissard et al. 1992; Von Lubitz et al. 1994a; Phillis 1995). Interestingly, a reverse effect appeared after chronic administration of adenosine receptor antagonists. Both caffeine and DPCPX given chronically weeks before an ischemic insult reduced the neuronal injury assessed by magnetic resonance imaging and histopathological examination in rats and gerbils (Rudolphi et al. 1989; Sutherland et al. 1991; Von Lubitz et al. 1994a). It has been suggested that the beneficial effects seen after chronic administration of adenosine antagonists may be due to upregulation of A_1R (Jacobson et al. 1996). In agreement and unlike acute treatment, chronic administration of A_1R agonists worsened survival and increased neuronal loss (Jacobson et al. 1996), a phenomenon thought to depend on A_1R desensitization.

Plastic changes in A_1Rs are critical to understand the effects of A_1R agonists/antagonists but also to understand whether adenosine maintains its neuroprotective efficiency after ischemia. In this regard, several studies have shown that short periods of focal or global ischemia produced a long-lasting decrease in the density of A_1Rs (Nagasawa et al. 1994; Onodera et al. 1987). On the contrary, an increase in the mRNA level of A_1R was reported in the cerebral cortex after global ischemia in rats and normalization under reperfusion (Lai et al. 2005). Relevantly, it was reported that A_1R KO mice, when exposed to global ischemia, do not show increased neuronal damage in the CA1 region of the hippocampus, in the cortex, or in the striatum compared to wild-type controls (Olsson et al. 2004). These discrepancies may reflect development of compensatory mechanisms after genetic deletion (Wei et al. 2011).

Unlike in adult rodents, pharmacologic A_1R stimulation in immature brain exacerbates rather than attenuates ischemic injury (Turner et al. 2002). It should be remembered that in the formation of the central nervous system, A_1R activation potently inhibits the development of axons and can lead to leukomalacia (Rivkees et al. 2001). Leukomalacia is the most common ischemic brain injury in premature infants and it is characterized by the death or damage and softening of the white matter near the lateral ventricules. Consistent with this result, genetic deletion of A_1R was neuroprotective in immature mice under hypoxic conditions under which adenosine ectracellular levels are increased (Turner et al. 2003).

Recent studies show that sublethal anoxic/ischemic insults may "precondition" and protect the brain from subsequent insults and adenosine, by stimulating A_1Rs, plays a crucial role in this phenomenon. In fact, an A_1R agonist, CADO, markedly enhanced (Perez-Pinzon et al. 1996) and A_1R antagonists completely prevented (Perez-Pinzon et al. 1996; Pugliese et al. 2003; Lange-Asschenfeldt et al. 2004) the protective effect of ischemic preconditioning in rat hippocampal slices. In accordance with in vitro data, the selective A_1R antagonist, DPCPX, attenuated the neuroprotective effect of ischemic preconditioning in models of global (Heurteaux et al. 1995) and focal (Nakamura et al. 2002; Yoshida et al. 2004) cerebral ischemia. A list of selectivity of different ligands of adenosine receptors cited in the text is reported in Table 14.1.

Although data demonstrate a neuroprotective effect of adenosine through A_1Rs during ischemia, the use of selective A_1R agonists is hampered by undesirable peripheral effects such as sedation, bradycardia, and hypotension (White et al. 1996; Fredholm et al. 2005).

14.2.2 Elevation of Local Adenosine Levels in Injured Areas

An alternative therapeutic strategy may include use of agents that enhance local concentration of adenosine at areas of injury-induced adenosine release, thus minimizing unwanted peripheral and central effects mediated by A_1Rs (von Lubitz et al. 1996). Adenosine-potentiating agents at injury sites may act by (1) inhibiting its metabolism by adenosine kinase (ADK) or adenosine deaminase (ADA), (2) preventing its transport through equilibrative membrane transporters (ENTs), (3) enhancing hydrolysis of extracellular ATP by use of nucleoside triphosphate diphosphohydrolases (NTPDases). Both inhibitors of adenosine deaminase or kinase or agents preventing its transport have been shown to offer protection against ischemic neuronal damage in different in vivo ischemia models.

Adenosine levels are regulated mainly by ADK, an enzyme that is responsible for removal of adenosine via phosphorylation to AMP (Boison 2006), both in basal conditions and during energy depletion due to ischemia (Lloyd and Fredholm 1995).

Table 14.1 Identity and selectivity of different ligands of adenosine receptors (ARs) cited in the text

AR subtype	Compound	Ki value for ARs (nM)				References
Agonists		A_1[a]	A_{2A}[a]	A_{2B}[a]	A_3[a]	
A_1	CADO	6,7	63	N.D.	1,890	Daly et al. (1992), Pugliese et al. (2007b)
	CCPA	0,6(0,83[b])	10,3(2,270[b])	18,800[b]	113(38[b])	Gao et al. (2003), Jacobson and Gao (2006)
	CHA	0,9	514	N.D.	167	Daly et al. (1992), Van Galen et al. (1994)
	CPA	0.45(2,3[b])	462(794[b])	18,600[b]	240(72[b])	Gao et al. (2003), Jacobson and Gao (2006)
	R-PIA	1,2	124	ND	158	Daly et al. (1986), Van Galen et al. (1994)
A_{2A}	CGS21680	2,600(289[b])	15(27[b])	>10,000[b]	584(67[b])	Jacobson and Gao (2006), Van Galen et al. (1994)
A_3	IB-MECA	51[b]	2,900[b]	11,000[b]	1,8[b]	Jacobson and Gao (2006)
	Cl-IB-MECA	820(220[b])	470(5,360[b])	>100,000[b]	0,33(1,4[b])	Gao et al. (2003), Jacobson and Gao (2006)
Antagonists						
A_1	DPCPX	3,9[b]	129[b]	56[b]	3,980[b]	Jacobson and Gao (2006)
A_{2A}	CSC	28,000	54	N.D.	N.D.	Jacobson and Gao (2006)
	SCH58261	128(725[b])	2,3(5,0[b])	1,110[b]	>10,000(1,200[b])	Jacobson and Gao (2006), Moro et al. (2006)
	ZM241385	2,040(774[b])	0,3(1,6[b])	75[b]	>10,000(743[b])	Jacobson and Gao (2006), Moro et al. (2006)
A_{2B}	MRS1706	157[b]	112[b]	1,39[b]	230[b]	Xi et al. (2009)
A_3	MRS1523	15,600	2,050	N.D.	18,9(113[b])	Jacobson and Gao (2006), Moro et al. (2006)
	LJ1251	2490[b]	341[b]	N.D.	3,89(4,16[b])	Moro et al. (2006), Pugliese et al. (2007b)

[a]Binding experiments at rat A_1, A_{2A}, A_{2B} and A_3 ARs, unless noted
[b]Binding experiments at recombinant human ARs

The ADK inhibitors, 5′-deoxyiodotubercidin (Miller et al. 1996; Jiang et al. 1997) and GP683 (Tatlisumak et al. 1998), significantly reduced cerebral ischemic damage after middle cerebral artery occlusion (MCAo). Pignataro and coworkers (2008) have recently demonstrated that focal cerebral ischemia induces acute transient down regulation of ADK and increased extracellular adenosine levels within hours after MCAo (an innate self-repair neuroprotective mechanism). Transgenic mice with increased ADK expression in the striatum and reduced ADK expression in the cortex have an increased striatal infarct volume and almost complete protection in cortex after MCAo (Shen et al. 2011). The regional changes in ADK expression correlate positively with adenosine levels, suggesting that the degree of acute brain injury directly depends on ADK expression levels and on the resulting tone in adenosine (Shen et al. 2011). These findings confirm that adenosine-dependent acute neuroprotection depends on local rather than systemic responses (Shen et al. 2011).

ADA is another enzyme that regulates adenosine levels in the CNS (Phillis et al. 1979). The activity of ADA is unevenly distributed in the brain (Geiger and Nagy 1986; Patel and Tudball 1986; Schrader et al. 1987) and becomes important in regulating adenosine levels when they are increased by energy depletion (Lloyd and Fredholm 1995) and complete ischemia (Kobayashi et al. 1998). Numerous papers have demonstrated that the treatment with ADA inhibitor, deoxycoformycin, reduces brain injury in different rat models of cerebral ischemia (Phillis and O'Regan 1989; Phillis et al. 1991; Lin and Phillis 1992; Gidday et al. 1995).

Adenosine is a hydrophilic molecule and the transport across cell membranes is mediated by nucleoside transporters: bidirectional equilibrative transporters (ENTs), driven by chemical gradients, and unilateral concentrative transporters (CNTs), driven by sodium electrochemical gradients (Parkinson et al. 2011; King et al. 2006). In the brain these transporters have the potential to regulate synaptic levels of neuroactive purines such as adenosine and thereby modulate physiological and pathophysiological responses (King et al. 2006). Several studies show that different adenosine uptake inhibitors, propentofylline (DeLeo et al. 1988; Andiné et al. 1990; Dux et al. 1990; Park and Rudolphi 1994; Kano et al. 1994; Gidday et al. 1995; Matsumoto et al. 1996; Kawahara et al. 1998; Johnson et al. 1998; Turcáni and Turèáni 2001), dipyridamole (Seif-el-Nasr and Khattab 2002; Parkinson et al. 2005), indomethacin (Phillis et al. 1994b), and nitrobenzylmercaptopurine ribonucleoside (NBMPR) (Parkinson et al. 2005) reduce brain ischemic injury in rat cerebral ischemia models. The neuroprotective effect of adenosine uptake inhibitors is attributed to increased extracellular adenosine levels and thereby to stimulation of A_1Rs. It was recently demonstrated that transgenic mice overexpressing neuronal human ENT1 show a reduced hypoxia- and ischemia-induced increase in extracellular adenosine levels (Zhang et al. 2011). The higher uptake led to decreased A_1R signaling and decreased adenosine neuroprotective effect (Zhang et al. 2011).

It is well known that, during ischemia, mitochondria consume ATP reducing it by about 30 % 15 min after focal ischemia induction (Latini et al. 1995). Under these conditions the equilibrium is toward AMP that, due to the lack of oxygen and glucose, cannot be reconverted to ATP by nucleotide kinase enzyme (Latini and Pedata 2001). ATP and ADP levels go below the threshold that inhibits the enzyme 5′-nucleotidase and intracellular formation of adenosine from AMP occurs as demonstrated in vitro

(Latini et al. 1995) and in vivo (Madrigal et al. 2003). Under these conditions adenosine is formed and may escape cells by ENTs. However, during ischemia adenosine might be produced in the synaptic cleft also from ATP hydrolysis via the concerted action of two enzymes: NTPDase and e5′-NT. In vitro evidence suggests that released ATP does not substantially contribute to the adenosine concentration in the extracellular milieu following an ischemic-like event in hippocampal slices (Frenguelli et al. 2007). On the contrary, in a rat model of focal cerebral ischemia in vivo, Melani and coworkers (2012) demonstrated that, in the first minutes after cerebral ischemia, the extracellular adenosine concentration mainly derives from extracellular ATP hydrolysis and thereafter as adenosine per se from inside cells. NTPDases have a dual function in modulating purinergic neurotransmission: (1) they rapidly interrupt ATP-mediated signaling by degrading extracellular nucleotides and (2) they give rise to extracellular adenosine, which activates purinergic P1 receptors. During ischemia, both NTPDases and e5′-NT are upregulated in the brain (Braun et al. 1998; Chitolina Schetinger et al. 1998; Frassetto et al. 2000). Braun and coworkers (1997) observed upregulation of e5′-NT on astrocytes and microglial cells in the infarcted brain. In agreement, Parkinson et al. (2005) observed that the e5′-NT inhibitor, alpha,beta-methylene ADP, inhibited the increase in extracellular adenosine concentration evoked by ischemic-like conditions from primary cultures of rat forebrain astrocytes, but not from cultured neurons. In rat stroke models, treatment with SolCD39, a soluble form of human NTPDase1, reduced cerebral infarct, improved neurological deficit, and restored postischemic cerebral perfusion (Pinsky et al. 2002; Belayev et al. 2003; Marcus et al. 2003).

Conversion of ATP to adenosine by ectonucleotidases represents an important mechanism regulating extracellular ATP and adenosine levels during ischemia. Degradation of ATP in ischemia might be important as a mechanism of removal of any cytotoxic effect mediated by ATP (Melani et al. 2006a; Burnstock 2007). NTPDase and e5′-NT upregulation after ischemia might be another innate self-repair neuroprotective mechanism involving purinergic signaling (Braun et al. 1998; Chitolina Schetinger et al. 1998; Frassetto et al. 2000). In its turn, the extracellular catabolism of adenine nucleotides is associated with adenosine formation. The understanding of the contribution of extracellular ATP to adenosine during ischemia is important because such information might constitute the basis for devising a correct putative purinergic strategy aimed at protection from ischemic damage.

14.3 Roles of A_{2A} Receptors

14.3.1 A_{2A} Receptor Antagonists Are Protective Against Ischemic Brain Injury

More recently, the role of A_{2A}Rs in ischemic neuroprotection has been studied. Gao and Phillis (1994) demonstrated for the first time that the nonselective A_{2A}R antagonist, 9-chloro-2-(2-furanyl)-[1,2,4] triazolo[1,5-c]quinazolin-5-amine (CGS15943), reduces cerebral ischemic injury in the gerbil following global forebrain ischemia. Thereafter many reports have confirmed the neuroprotective role of A_{2A}R antagonists in different models of ischemia. The selective A_{2A}R antagonist 8-(3-chlorostyryl) caffeine (CSC), as well as the less selective antagonists CGS15943 and 4-amino [1,2,4] triazolo [4,3a] quinoxalines (CP66713), protect against hippocampal cell injury during global forebrain ischemia in gerbils (Phillis 1995; Von Lubitz et al. 1995). Similarly, the selective A_{2A}R antagonist 7-(2-phenylethyl)-5-amino-2-(2-furyl)-pyrazolo-[4,3-e]-1,2,4,triazolo[1,5-c]pyrimidine (SCH58261) reduces ischemic brain damage in a rat neonatal model of hypoxia/ischemia (Bona et al. 1997) and in an adult rat model of focal cerebral ischemia (Monopoli et al. 1998; Melani et al. 2003). The same antagonist, administered subchronically, was protective against both brain damage and neurological deficit (Pedata et al. 2005; Melani et al. 2006b) and disorganization of myelin (Melani et al. 2009) in the adult rat model of focal cerebral ischemia. The selective A_{2A}R antagonist, 4-(2-[7-amino-2-(2-furyl) [1,2,4]triazolo[2,3-a][1,3,5]triazin-5-yl-amino]ethyl)phenol (ZM241385), reduced hippocampal injury and improved performance in the Morris water maze after four-vessel occlusion in the rat (Higashi et al. 2002). Studies in A_{2A}R KO mice provided support to the noxious role played by A_{2A}Rs during ischemia induced by focal cerebral ischemia (Chen et al. 1999).

The selective A_{2A}R antagonist SCH58261 is a significant protective agent in the MCAo model at a low dose that does not have cardiovascular effects. This low dose does not affect motor behavior in naive animals but decreases contralateral turning after unilateral focal ischemia induced by the monofilament technique (Melani et al. 2003, 2006b). At a higher dose, in the range that is effective in different models of Parkinson's Disease, the same drug significantly increases motility and rearing in the rat (Svenningsson et al. 1997).

14.3.1.1 A_{2A} Receptors Modulate Glutamate Extracellular Concentrations

Major protective effects of A_{2A}R antagonists in stroke are attributed to reduced glutamate outflow (Marcoli et al. 2003; Pedata et al. 2001, 2003, 2005, 2007). Adenosine, by A_{2A}R stimulation, promotes glutamate release under normoxic and ischemic conditions (O'Regan et al. 1992; Simpson et al. 1992; Popoli et al. 1995; Corsi et al. 1999, 2000). In rat hippocampal slices, the selective A_{2A}R agonist CGS 21680 clearly reduces the depression of synaptic activity brought about by OGD (Latini et al. 1999b). Such an effect is attributable to stimulation of glutamate outflow that counteracts depression brought about by A_1Rs. In agreement, selective A_{2A}R antagonists ZM241385 and SCH58261 protect from CA1 death and delay the appearance of AD, a phenomenon strictly related to cell damage and death (Somjen 2001; Gáspárová et al. 2008; Pugliese et al. 2009). This suggests that after acute ischemia A_{2A}Rs regulate excitatory transmission. The increased outflow of excitatory amino acids soon after ischemia, presumably mediated by A_{2A}Rs, is thought to prime an excitotoxic cascade which ultimately results in cell death (Bruce-Keller 1999). In vivo, the low dose of SCH58261 that protects against tissue damage induced by MCAo or quinolinic acid excitotoxicity, also reduces glutamate outflow (Popoli et al. 2002; Melani et al. 2003). This result is consistent with the observation that $A_{2A}R$ KO mice are protected from an excess of striatal glutamate outflow and damage induced by transient MCAo (Chen et al. 1999; Gui et al. 2009). In agreement, in the traumatic brain injury model, protective effects are dictated by local glutamate concentrations; antagonists are protective 15 min after trauma when CSF extracellular glutamate rises (Dai et al. 2010). The $A_{2A}R$ antagonist ZM241385 injected directly intra hippocampus is protective against excitotoxicity induced by kainate (Jones et al. 1998a) and by the combinations of quinolinic acid and IL-1beta (Stone and Behan 2007). These data suggest that the neuroprotective properties of $A_{2A}R$ antagonists largely reside in control of central excitotoxicity.

$A_{2A}Rs$ are highly concentrated in the striatum where most receptors are present on GABA-enkephalin neurons (Schiffmann et al. 1991) but are also located presynaptically on glutamatergic terminals (Hettinger et al. 2001; Rosin et al. 2003) where they can directly regulate glutamate outflow (Marcoli et al. 2003, 2004). A definite overexpression of $A_{2A}Rs$ was found in neurons in the striatum and cortex 24 h after focal ischemia (Trincavelli et al. 2008a). $A_{2A}Rs$ located on glial cells mediate inhibition of glutamate uptake transporter GLT-1 and stimulate glutamate outflow (Nishizaki et al. 2002; Pintor et al. 2004). However, 24 h after focal ischemia, $A_{2A}Rs$ are not overepressed on astrocytes (Trincavelli et al. 2008a). Evidence indicates that protective effects of low doses of $A_{2A}R$ antagonists early after brain ischemia are largely due to reduced glutamate outflow from neurons.

A further protective effect of $A_{2A}R$ antagonism may be attributed to the capability of $A_{2A}R$ antagonists to increase GABA outflow during ischemia. The major part of excitatory glutamatergic innervation is modulated by inhibitory GABA-releasing interneurons. Potentiation of GABAergic synaptic transmission has neuroprotective effects in several experimental models of cerebral ischemia (Schwartz-Bloom and Sah 2001). GABA is strongly increased in the cortex and striatum during ischemia (O'Regan et al. 1992; Melani et al. 1999) and evidence shows that selective $A_{2A}R$ stimulation decreases ischemia-evoked GABA outflow (O'Regan et al. 1992; Saransaari and Oja 2005).

14.3.1.2 A_{2A} Receptor Antagonists Protect from Mitogen-Activated Protein Kinase Activation

All members of the mitogen-activated protein kinase (MAPK) family are activated up to 24 h after ischemia (Irving et al. 2000; Wu et al. 2000). Whereas p38 and ERK1/2 are activated in neurons and in microglia (Irving et al. 2000; Takagi et al. 2000; Piao et al. 2003; Melani et al. 2009) JNK is activated in neurons and mainly in oligodendrocytes (Melani et al. 2009). Twenty-four hours after MCAo, $A_{2A}Rs$ are overexpressed on microglia (Trincavelli et al. 2008a). Subchronic administration of the $A_{2A}R$ antagonist SCH58261 reduced phospho-p38 in microglia (Melani et al. 2006b) and phospho-JNK in neurons and oligodendrocytes, whereas it did not affect ERK1/2 activation (Melani et al. 2009). It is known that soon after early excitotoxic phenomena, microglial cells initiate a rapid change in their phenotype that is referred to as microglial cell activation (Bruce-Keller 1999). These cells start to proliferate and migrate toward the site of damage (Stoll et al. 1998; Marks et al. 2001) and by producing cytotoxic substances and cytokines may contribute to the inflammatory response that follows ischemic insult, further exacerbating brain damage (Dirnagl et al. 1999). An inhibitor of p38 activation has direct neuroprotective effects in hippocampal brain slices after OGD (Barone et al. 2001). Therefore, a control of p38 activation by $A_{2A}R$ antagonist might account for protection against secondary damage induced by ischemia. Subchronic administration of the $A_{2A}R$ antagonist SCH58261 also reduces phospho-JNK in oligodendrocytes (Melani et al. 2009). Interestingly activation of JNK has been described in oligodendrocytes in multiple sclerosis lesions where oligodendrocytes are major targets of the disease (Bonetti et al. 1999). Activation of JNK is involved in oligodendrocyte death (Howe et al. 2004; Jurewicz et al. 2006). A specific peptide inhibitor of JNK protects against cell death induced by OGD in vitro (Hirt et al. 2004) and by MCAo in vivo (Borsello et al. 2003; Hirt et al. 2004). JNK2/3 KO mice are protected from damage following cerebral ischemia (Kuan et al. 2003; Gelderblom et al. 2004). Therefore, we must assume that JNK activation in oligodendrocytes and neurons represents a noxious event after ischemia that can damage both adult and precursor oligodendrocytes. Phospho-JNK is in fact expressed mainly by oligodendrocyte precursor cells (OPC: stained by NG2 antibodies) (Melani et al. 2011). Therefore, by reducing JNK activation, the $A_{2A}R$ antagonist not only may protect from demyelination but also might stimulate OPC differentiation to mature cells. Such a possibility is suggested by the observation that the $A_{2A}R$ antagonist reduces Olig2 which is a transcription factor expressed mostly by OPC whereas mature oligodendrocytes are characterized by lower levels of Olig2 (Kitada and Rowitch 2006).

We should consider that since SCH58261 reduces glutamate outflow in the initial hours after ischemia (Melani et al. 2003), reduced MAPK activation might be secondary to an overall reduction in the excitotoxic cascade that in turn primes JNK activation (Kawasaki et al. 1997). Oligodendroglia are extremely sensitive to glutamate receptor overactivation and ensuing oxidative stress (Matute et al. 1997, 2002; McDonald et al. 1998) as well as to cytokines and adenosine (Back 2006). However, we cannot exclude a direct effect of the $A_{2A}R$ antagonist on oligodendrocytes. $A_{2A}Rs$ have been identified also in oligodendrocytes (Stevens et al. 2002). No data are present regarding a direct action of $A_{2A}Rs$ in regulating MAPK activation in microglia or oligodendrocytes. We have recently observed that in OPC in culture, selective stimulation of $A_{2A}Rs$ inhibit K^+ currents that have characteristics of K^+ "delayed rectifier" channels (unpublished observations). Interestingly, inhibition of these channels inhibits proliferation and differentiation of OPC to mature oligodendrocytes, preventing myelin deposition (Shrager and Novakovic 1995; Attali et al. 1997).

Twenty-four hours after permanent MCAo, the $A_{2A}R$ antagonist also reduces gene c-fos expression in glial cells (Petroni et al. 2002). Products of the Fos family are players in inducing inflammatory gene expression in glial cells (Lewis and Manning 1999). Data indicate that regulation of proteins involved in transcriptional or posttranslational mechanisms plays an important role in the neuroprotective effect of $A_{2A}R$ antagonism in ischemia.

14.3.2 A_{2A} Receptor Agonists Are Protective Against Ischemic Brain Injury

Several studies show that A_{2A}R agonists systemically administered are protective in the global ischemia model in the gerbil (Von Lubitz et al. 1995; Sheardown and Knutsen 1996) and A_{2A}R KO neonatal mice show aggravated hypoxic/ischemic injury in comparison to wild-type mice (Adén et al. 2003). Recently, in immature brain forebrain slices, it was demonstrated that cannabinoids induce robust neuroprotection through both CB(2) receptors and A_{2A}R (Castillo et al. 2010).

While many data support that A_{2A}R antagonists protect against central excitotoxicity, the protective effect of A_{2A}R agonists appears unrelated to glutamate. In the traumatic brain injury model, A_{2A}R agonists and antagonists were administered systemically and their effect was related to extracellular concentration of glutamate (Dai et al. 2010). An A_{2A}R antagonist administered repeatedly i.p. in the 12 h after trauma was protective 15 min after trauma when CSF extracellular glutamate rose; conversely, an A_{2A}R agonist administered repeatedly i.p. in the 12 h after trauma was protective 3 h after trauma when CSF glutamate concentrations were down (Dai et al. 2010).

Evidence indicates that the protective mechanisms of A_{2A}R agonists are attributable in part to peripheral effects. Jones and coworkers (1998a) showed that peripheral administration of the A_{2A}R agonist, CGS21680, protects the hippocampus against kainate-induced excitotoxicity. However, the direct injection of CGS21680 into the hippocampus failed to afford protection (Jones et al. 1998b). Similar results were obtained after spinal cord trauma where the A_{2A}R agonist CGS21680 protected from damage when injected systemically but not when centrally injected in the injured spinal cord (Paterniti et al. 2011). Li et al. (2006) demonstrated that the protective effect of A_{2A}R agonists systemically administered after spinal trauma against motor deficits is lost in mice lacking A_{2A}Rs on bone marrow-derived cells (BMDCs), but is restored in A_{2A}R KO mice reconstituted with A_{2A}Rs on BMDCs. This suggests that BMDCs are targets of A_{2A}R agonist protective effects after spinal cord injury (SCI).

A_{2A}Rs located on BMDCs reduce production of cell adhesion factors, reduce platelet aggregation and neutrophil activation, thereby exerting antithrombotic, antioxidant, and antiinflammatory effects (Sitkovsky et al. 2004). Most studies have reported that selective activation of A_{2A}Rs inhibits proinflammatory responses directly in BMDCs, including platelets, monocytes, some mast cells, neutrophils and T cells (Mirabet et al. 1999; Lappas et al. 2006; Naganuma et al. 2006; Haskó et al. 2008). Consistent with its antiinflammatory and immunosuppressive role, the protective effects of A_{2A}R stimulation have been observed in various models of autoimmune disease, such as rheumatoid arthritis (Odashima et al. 2005), colitis (Choukèr et al. 2008; Genovese et al. 2010), and hepatitis (Howe et al. 2004). Protection due to A_{2A}R agonists systemically administered after brain ischemia is most probably exerted at peripheral BMDCs resulting ultimately in reduced leukocyte infiltration and inflammatory cascade at the central level.

The notion that A_{2A}Rs on BMDCs are the target of the protective effects of A_{2A}R agonists should be reconciled with the fact that inactivation of A_{2A}Rs on BMDCs attenuates ischemic brain injury, inhibits inflammatory cytokine production, and increases the expression of antiinflammatory cytokines 22 h after focal ischemia. This neuroprotection however did not relate to infiltration of circulating BMDCs into brain infarct areas since >90 % of the microglial cells present in the brain infarct area, originated from local brain microglia (Yu et al. 2004b).

Besides effects related to a control of excitotoxicity or inflammation, we must remember that adenosine is also a potent vasodilator implicated in cerebral blood flow regulation, acting on A_{2A}R on endothelial cells (Phillis 2004). It has been suggested that adenosine has an autoregulatory role on blood flow during hypotension through not only A_{2A}R but also A_{2B}R. The A_{2A}R antagonist (A_{2A}>A_{2B}:10–85>1), ZM241385, systemically administered, attenuates autoregulation in A_{2A}R KO mice, and treatment with dipyridamole (which increases extracellular concentrations of adenosine) improved autoregulation in A_{2A}R KO mice (Kusano et al. 2010).

In A_{2A}R KO mice, severe demyelination-related damage, proliferation of glia, and increased levels of proinflammatory cytokines were observed, and working memory was seriously affected from 7 to 30 days after chronic cerebral hypoperfusion induced by the two vessel occlusion (2VO) (Duan et al. 2009). In the model of 2VO, infusion of dipyridamole, improved spatial working memory (Melani et al. 2010). An important autoregulatory role of adenosine through A_{2A}Rs is also suggested during hypotension: autoregulation is significantly impaired in A_{2A}R KO mice and the A_{2A}R antagonist ZM241385, in a dose-related manner, attenuated autoregulation in wild-type mice (Kusano et al. 2010).

14.3.2.1 A_{2A} Receptor and Neuroinflammation

In view of the evidence that A_{2A}R agonists protect through peripheral effects acting on BMDCs, we might consider that protection by A_{2A}R agonists after ischemia is due to control of infiltration and ensuing neuroinflammation. An early phenomenon after ischemia is a massive increase in extracellular glutamate that can prime an excitotoxic cascade which in turn activates immune cells and produces mediators of inflammation (Dirnagl et al. 1999). After transient (1 h) focal ischemia induced by MCAo, definite microglial activation is present after 12 h; thereafter macrophages infiltrate (Gelderblom et al. 2009) (Fig. 14.2). Activated immune cells probably exert an important role in promoting neutrophil accumulation by producing proinflammatory cytokines that upregulate cell adhesion molecules (Stoll et al. 1998; Huang et al. 2006). Infiltration is maximal at 3 days when the majority of immune cells are neutrophils and to a lesser extent, lymphocytes (Gelderblom et al. 2009). Changes in blood–brain barrier (BBB) permeability are responsible for cell infiltration. The nature of BBB permeability is dependent on the duration of ischemia, the degree of reperfusion, and the animal stroke model. After transient MCAo, initial BBB permeability is associated with reperfusion, followed by a biphasic increase at 5 and 72 h (Kuroiwa et al. 1985; Sandoval and Witt 2008). Studies in the human brain after

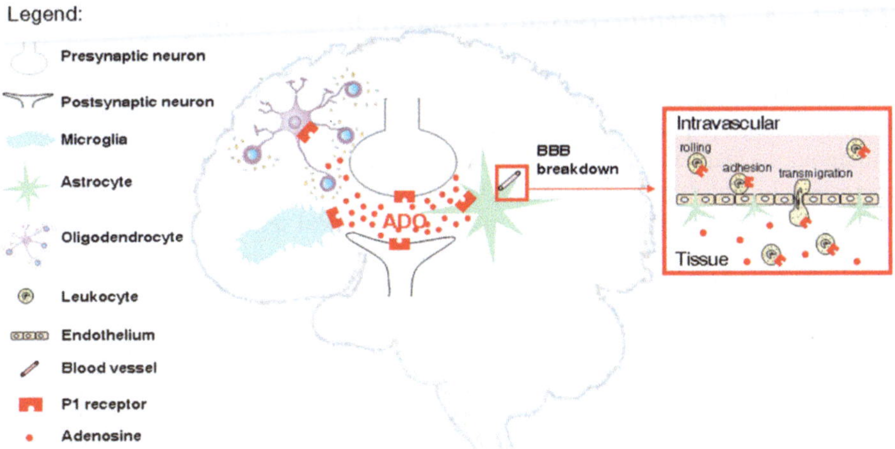

Fig. 14.2 Schematic drawing of events during brain ischemia. Immediately after ischemia, adenosine levels increase reaching extracellular concentrations able to stimulate all adenosine receptor subtypes. Adenosine receptors are present on neurons, astrocytes, microglia and oligodendrocytes. After ischemia, blood–brain barrier (BBB) breakdown allows massive infiltration of leukocytes into brain ischemic areas. All adenosine receptors subtypes are present on leukocytes. Infiltrating leukocytes, together with resident activated microglial cells, are responsible for the formation of inflammation products. Neuroinflammation is responsible for secondary ischemic damage

ischemic stroke confirm that neutrophils intensively accumulate in the regions of cerebral infarction (Akopov et al. 1996; Dirnagl et al. 1999). Neutrophil accumulation correlates with the severity of brain tissue damage and poor neurological outcome (Akopov et al. 1996; Dirnagl et al. 1999). Neuroinflammation is now recognized as a predominant mechanism of secondary progression of brain injury after ischemia.

Interestingly, neuroinflammation results in inhibition of adult neurogenesis (Ekdahl et al. 2003). Very little is known about the effects of adenosine on adult neurogenesis, but there is evidence about its involvement: enzymes involved in extracellular ATP metabolism such as alkaline phosphate (Langer et al. 2007) and e5′-NT (Lie et al. 1999) are expressed in the neurogenetic areas (olfactory bulb and subventricular zone); functional adenosine receptors are expressed in neurosphere cells (cultured neuronal stem cell aggregates) (Migita et al. 2008); and caffeine depresses adult hippocampal neurogenesis (Han et al. 2007; Wentz and Magavi 2009). Because A_{2A}Rs are involved in control of neuroinflammation they might have a potential for influencing neurogenesis (Ekdahl et al. 2009; Gomes et al. 2011) and reparative neurogenetic processes after ischemia.

A_{2A}R agonists have undergone clinical trials as antinflammatory agents (Fredholm et al. 2011). A problem is that they show hypotensive effects in relation to dose. There are still major problems in this field, including side effects, low brain penetration (for the targeting of CNS diseases), short half-life, or lack of in vivo effects. The design and development of new A_{2A}R ligands is an area of intense research activity (Cristalli et al. 2009).

14.3.3 A_{2A} Receptor-Based Protection in Ischemia?

Because the activation of resident immune cells and the spatial–temporal infiltration of immune cell populations vary according to ischemic model, the time-course of inflammation markers needs to be considered for any approach using antiinflammatory treatment in patients with acute ischemic stroke. A careful characterization of immune cell infiltration is therefore essential in clinical settings of new therapies in stroke. Thus, stimulation or antagonism of A_{2A}Rs might be a protective strategy secondary to the time-related development of phenomena typical of trauma and ischemia. Altogether, evidence suggests that low doses of A_{2A}R antagonists (that do not modify hemodynamic parameters) in brain ischemia provide protection centrally by moderating excitotoxicity, whereas A_{2A}R agonists provide protection by controlling massive infiltration in the hours after SCI. When considering use of A_{2A}R active drugs to protect against brain ischemia, attention should be given to the dose and administration time after injury. A novel therapeutic strategy could involve early treatment by an antagonist in what is called the therapeutic window of 4–6 h followed by A_{2A}R agonist treatment for the control of later secondary injury. These observations highlight that a therapeutic strategy in different cerebrovascular pathologies should be carefully evaluated in terms of time.

14.4 A_{2B} Receptors

Because of an apparent paucity of A_{2B}R selective agonists and antagonists (see Müller and Jacobson 2011) there have been few relatively studies indicating a putative role for A_{2B}R during brain ischemic damage. In the stratum radiatum of CA1 hippocampal slices, the number and immunostaining density of immunoreactive cells for A_{2B}Rs after ischemic preconditioning were increased (Zhou et al. 2004). Some evidence suggests that A_{2B}Rs are involved in the inflammatory response to ischemia. For example, in human astroglial cells, N-(4-acetylphenyl)-2-[4-(2,3,6,7-tetrahydro-2,6-dioxo-1,3-dipropyl-1H-purin-8-yl) henoxy]acetamide (MRS1706), a selective A_{2B}R antagonist, completely prevents elongation of astrocytic processes, which is classified as a morphological hallmark of in vivo reactive astrogliosis, induced by selective stimulation of A_{2B}R (Trincavelli et al. 2004). Short-term TNF-alpha treatment that simulates the acute phase of cerebral damage, which is characterized by both cytokine and adenosine high release, induces A_{2B}R desensitization in human astroglial cells (Trincavelli et al. 2008b). The results may indicate that A_{2B}R functional impairment is a cell defense mechanism to counteract the A_{2B}R-mediated effects during the acute phase of brain damage, underlying A_{2B}Rs as a target for modulation of early inflammatory responses (Trincavelli et al. 2008b).

14.5 Roles of A_3 Receptors

A_3R mRNA was detected in the rat and mouse brain by RT-PCR (Dixon et al. 1996) and radioligand binding (Jacobson et al. 1993) at relatively low levels in comparison to A_1R and A_{2A}R. This receptor is widespread in the rat and mouse brain (see Gessi et al. 2008). Compared to A_1R and A_{2A}R, A_3Rs have less affinity for adenosine (10–30 nM versus 1 μM). However, because extracellular adenosine concentrations increase dramatically during ischemia (see Latini and Pedata 2001), evidence now indicates that all three adenosine receptors are potential targets for therapeutic treatment of stroke.

The studies currently in the literature concerning the role of A_3R in the pathophysiology of cerebral ischemia are rather contradictory and have been a matter of intense debate (Jacobson 1998; Jacobson et al. 1999; Abbracchio and Cattabeni 1999; von Lubitz 1999; von Lubitz et al. 1999; Baraldi et al. 2000; Borea et al. 2009; Pedata et al. 2010). An early in vivo study in the model of global forebrain ischemia in the gerbil showed that a selective agonist of A_3R, IB-MECA, acutely administered 15 min prior to global ischemia, impaired postischemic blood flow, increased mortality and exacerbated the loss of hippocampal neurons (von Lubitz et al. 1994b). IB-MECA administration 20 min prior to transient middle cerebral ischemia also resulted in a significant increase in infarct size (von Lubitz et al. 2001). In agreement with a possible noxious role of A_3R in ischemia, it has been shown that several different selective antagonists of A_3R consistently abolish or delay the occurrence of AD and significantly prevent the irreversible failure of excitatory neurotransmission caused by a severe (7 min) ischemic episode in rat hippocampal slices (Pugliese et al. 2006, 2007a; Colotta et al. 2007, 2008, 2009). Moreover, in a model of in vitro ischemic preconditioning, the selective A_3R antagonist, 5-propyl-2-ethyl-4-propyl-3-(ethylsulfanylcarbonyl)-6-phenylpyridine-5-carboxylate (MRS1523), facilitated the full recovery of CA1 hippocampal neurotransmission after a severe (7 min) irreversible OGD period (Pugliese et al. 2003) when applied before and during OGD. The block of A_3R, by removing A_3R-mediated impairment of the feedback inhibition of glutamate release exerted by specific metabotropic glutamate receptor subtypes (Macek et al. 1998), may reduce the participation of glutamate in triggering AD.

Contrary to the above information, Hentschel and colleagues (2003) demonstrated that under hypoxic conditions (5 min 95 % N2–5 % CO_2) in vitro, selective activation of A_3R by a brief (5 min) application of IB-MECA, inhibits excitatory neurotransmission on cortical neurons. These data indicate that A_3R, similar to A_1R, may sustain inhibition of synaptic activity during hypoxia, a phenomenon that is considered neuroprotective.

Conflicting results on the excitatory or inhibitory role of A_3R on synaptic activity in hypoxia/ischemia may be reconciled by data reported by Pugliese et al. (2007a). Although the selective A_3R antagonists, MRS1523 and LJ1251, applied before a severe (7 min) OGD stimulate a recovery of neurotransmission, application of the antagonist before a brief (2 min) OGD reduces the depression of fEPSP in the CA1 hippocampal area. The latter finding indicates an inhibitory role of A_3Rs on synaptic transmission during brief OGD. These results suggest A_3Rs may have a synergistic role with A_1Rs in decreasing synaptic transmission and exert a protective effect in ischemia. Prolonged ischemic conditions could play a pivotal role in switching the effects of A_3R stimulation from A_1R-like inhibition to potentiation of an excitotoxic glutamate effect. Rat cortical neurons exposed to hypoxia in vitro show an increase in activation of protein kinase C (PKC) after selective A_3R stimulation (Nieber and Hentschel 2006). If OGD is applied long enough to be considered severe, PKC activation induced by A_3R could account for an increase in intracellular calcium, which may participate in increasing tissue excitability and thus lead to irreversible synaptic failure. We may speculate that A_3R-mediated effects would become particularly deleterious during ischemia, when high extracellular levels of adenosine are reached and detrimental effects of A_3R activation may be due, at least in part, to sustained stimulation. We must take into account that during ischemic conditions in hippocampal slices, adenosine progressively increases, reaching concentrations up to 30 μM after 5 min OGD (Latini et al. 1998; Pearson et al. 2006).

A protective role of A_3Rs in ischemia is supported by the observation that IB-MECA acutely administered 20 min after transient (30 min) focal cerebral ischemia decreases the infarct volume (von Lubitz et al. 2001) and mice deleted for A_3R showed more pronounced hippocampal pyramidal neuron damage following repeated episodes of moderate hypoxia (Fedorova et al. 2003) and an increase in cerebral infarction after transient ligation of the MCA (Chen et al. 2006).

Plastic changes in A_3Rs following ischemia might be relevant to the effects of A_3R agonists/antagonists in ischemia. Both human and rat A_3Rs are desensitized within a few minutes after agonist exposure (Palmer et al. 1995; Trincavelli et al. 2000, 2002; Baines et al. 2011) and A_3Rs are very sensitive to prolonged stress in vitro (von Arnim et al. 2000, 2006).

In an in vitro OGD model in hippocampal slices, it was found that a long application (before and during OGD) of Cl-IB-MECA and of new selective A_3R agonists (Volpini et al. 2002, 2007) have effects similar to those of antagonists upon the OGD-induced depression of synaptic transmission and on the appearance of AD after the severe (7 min) OGD period (Pugliese et al. 2007a). These effects may be attributed to desensitization of A_3Rs. A stimulation as massive as that reached in the presence of endogenous adenosine plus exogenous A_3R agonists might induce substantial A_3R plastic adjustments such as desensitization. These in vitro results agree with the observation that chronic administration (10-day preischemic) of IB-MECA improves postischemic cerebral flow circulation, survival and neuron preservation in the global forebrain ischemia model in the gerbil (von Lubitz et al. 1994b, 1999) and that repeated intravenous administration of Cl-IB MECA before MCAo increases locomotor activity and decreases cerebral infarction (Chen et al. 2006).

Altogether these data suggest that the outcome of A_3R stimulation on synaptic transmission during hypoxic/ischemic phenomena depends on the severity and duration of the ischemic episode. Initially, massive excitotoxicity may be controlled by A_3Rs but later, an ensuing cascade of neuroinflammatory events could be potentiated by prolonged A_3R stimulation. Moreover, A_3R desensitization may account for the different effects of acute versus chronic agonist treatments reported in different studies.

14.5.1 A_3 Receptor and Neuroinflammation

The evidence summarized in the previous section suggests that A_3R may control ischemic brain injury by controlling excitotoxicity. Although excitotoxicity is invoked in the pathophysiology of most neurodegenerative central diseases, the putative involvement of A_3Rs in modulation of brain injury remains unclear. However, recent evidence suggests that central A_3Rs affect not only neurons but also glia, and control important intracellular signaling pathways involved in neuroinflammation (Daré et al. 2007; Gessi et al. 2008; Luo et al. 2010). A_3R mRNA has been identified by Northern blot analysis in mouse astrocytes (Zhao et al. 1999) and A_3R mRNA is expressed in microglia (Fiebich et al. 1996) and in oligodendrocytes (Stevens et al. 2002).

Early evidence indicated that A_3Rs on astrocytes mediate both protection and cell death, depending on A_3R agonist concentration (Abbracchio et al. 1997; Yao et al. 1997; Jacobson et al. 1999; Di Iorio et al. 2002). Recently, in human D384 astrocytoma cells, Cl-IB-MECA at relatively low concentration (0.8 μM), reduced ATP depletion and apoptosis caused by hypoxic conditions (Bjorklund et al. 2008). Furthermore, primary astrocytes prepared from A_3R KO mice were more affected by hypoxia than those prepared from WT mice (Bjorklund et al. 2008). In the in vivo model of global forebrain ischemia in the gerbil, chronic preischemic administration of IB-MECA results in enhancement of the expression of glial fibrillary acidic protein (GFAP) and depression of nitric oxide synthase in ischemic brain tissue (von Lubitz 1999), whereas in the ischemia model of MCAo (transient, 30 min occlusions), IB-MECA administered after ischemia proved to decrease the intensity of reactive gliosis involving microglia and astrocytes as evaluated 7 days after ischemia (von Lubitz et al. 2001). Injury-induced plasticity in A_3R may be relevant in modulation of intracellular pathways and cell safety (von Arnim et al. 2000, 2006).

The state of the art of the role of A_3Rs in inflammatory responses appears conflicting also because exposure of blood peripheral cell lines to selective A_3R agonists results in both anti- and proinflammatory effects (see Fishman and Bar-Yehuda 2003; Gessi et al. 2008; Borea et al. 2009; Luo et al. 2010). Although much is still to be learned about the function of the A_3R, at the present time it can be speculated that A_3R selective ligands might be helpful in the treatment of ischemic conditions in which inflammation is a feature. At present, oral active A_3R agonists are being tested in inflammatory diseases (Haskó et al. 2008; Fredholm et al. 2011).

14.5.2 A_3 Receptor-Based Protection in Ischemia?

Results raise the question of the time-related utility of A_3R antagonists/agonists for treatment of ischemia. It may be speculated that after ischemia, prolonged treatment with A_3R agonists first protects by reducing glutamate-mediated excitotoxicity. This supports a depression of neuronal activity and saves cellular energy. After ischemia, desensitization of A_3Rs may then avoid late onset of any deleterious A_3R influences.

A protective effect of drugs acting on A_3R receptors is still controversial and needs to be examined further (see Jacobson and Gao 2006). Further studies aimed at verifying the effect of agonists versus antagonists at different times after in vivo ischemia will help clarify the utility of this class of drugs.

14.6 Conclusions and Perspectives

Under neurodegenerative conditions involving ischemia, excitotoxicity is a first phenomenon. Thereafter, the interplay of resident glial cells with infiltrating peripheral BMDCs produces neuroinflammation. Presently, the understanding of postischemic inflammation and the functions of regulatory immune cells in ischemia is very limited. Understanding the dynamics of postischemic inflammation, i.e., the definition of temporal and spatial activation and infiltration of immune cell populations, is a prerequisite for therapeutic intervention in this fragile system to prevent harmful side effects. The state of art of a putative adenosinergic strategy as protective in ischemia foresees that a first approach could be that of minimizing excitotoxicity. This can be achieved by increasing adenosine at the site of injury with the aim of obtaining A_1R-mediated protection and of avoiding undesirable peripheral and central effects of A_1R agonist administration. $A_{2A}R$ antagonists at doses that do not modify heart frequency and blood pressure are efficacious at the central level in decreasing early excitotoxicity and ensuing tissue damage. $A_{2A}R$ agonists appear efficacious in control of later secondary injury due to massive cell infiltration in terms of hours and days after ischemia. A possible protective effect of adenosine in ischemia through activation of A_3R is still controversial. At present, orally administered A_3R agonists are being tested in patients with inflammatory diseases. This leaves $A_{2A}Rs$ as a potentially attractive target for therapeutic exploitation.

References

Abbracchio MP, Cattabeni F (1999) Brain adenosine receptors as targets for therapeutic intervention in neurodegenerative diseases. Ann N Y Acad Sci 890:79–92

Abbracchio MP, Ceruti S, Brambilla R, Franceschi C, Malorni W, Jacobson KA, von Lubitz DK, Cattabeni F (1997) Modulation of apoptosis by adenosine in the central nervous system: a possible role for the A3 receptor. Pathophysiological significance and therapeutic implications for neurodegenerative disorders. Ann N Y Acad Sci 825:11–22

Adén U, Halldner L, Lagercrantz H, Dalmau I, Ledent C, Fredholm BB (2003) Aggravated brain damage after hypoxic ischemia in immature adenosine A2A knockout mice. Stroke 34: 739–744

Akopov SE, Simonian NA, Grigorian GS (1996) Dynamics of polymorphonuclear leukocyte accumulation in acute cerebral infarction and their correlation with brain tissue damage. Stroke 27:1739–1743

Andiné P (1993) Involvement of adenosine in ischemic and postischemic calcium regulation. Mol Chem Neuropathol 18:35–49

Andiné P, Rudolphi KA, Fredholm BB, Hagberg H (1990) Effect of propentofylline (HWA 285) on extracellular purines and excitatory amino acids in CA1 of rat hippocampus during transient ischaemia. Br J Pharmacol 100:814–818
Arrigoni E, Crocker AJ, Saper CB, Greene RW, Scammell TE (2005) Deletion of presynaptic adenosine A1 receptors impairs the recovery of synaptic transmission after hypoxia. Neuroscience 132:575–580
Attali B, Wang N, Kolot A, Sobko A, Cherepanov V, Soliven B (1997) Characterization of delayed rectifier Kv channels in oligodendrocytes and progenitor cells. J Neurosci 17:8234–8245
Back SA (2006) Perinatal white matter injury: the changing spectrum of pathology and emerging insights into pathogenetic mechanisms. Ment Retard Dev Disabil Res Rev 12:129–140
Baines AE, Corrêa SA, Irving AJ, Frenguelli BG (2011) Differential trafficking of adenosine receptors in hippocampal neurons monitored using GFP- and super-ecliptic pHluorin-tagged receptors. Neuropharmacology 61:1–11
Baraldi PG, Cacciari B, Romagnoli R, Merighi S, Varani K, Borea PA, Spalluto G (2000) A(3) adenosine receptor ligands: history and perspectives. Med Res Rev 20:103–128
Barone FC, Irving EA, Ray AM, Lee JC, Kassis S, Kumar S, Badger AM, Legos JJ, Erhardt JA, Ohlstein EH, Hunter AJ, Harrison DC, Philpott K, Smith BR, Adams JL, Parsons AA (2001) Inhibition of p38 mitogen-activated protein kinase provides neuroprotection in cerebral focal ischemia. Med Res Rev 21:129–145
Belayev L, Khoutorova L, Deisher TA, Belayev A, Busto R, Zhang Y, Zhao W, Ginsberg MD (2003) Neuroprotective effect of SolCD39, a novel platelet aggregation inhibitor, on transient middle cerebral artery occlusion in rats. Stroke 34:758–763
Bjorklund O, Shang M, Tonazzini I, Dare E, Fredholm BB (2008) Adenosine A1 and A3 receptors protect astrocytes from hypoxic damage. Eur J Pharmacol 596:6–13
Boison D (2006) Adenosine kinase, epilepsy and stroke: mechanisms and therapies. Trends Pharmacol Sci 27:652–658
Boissard CG, Lindner MD, Gribkoff VK (1992) Hypoxia produces cell death in the rat hippocampus in the presence of an A1 adenosine receptor antagonist: an anatomical and behavioral study. Neuroscience 48:807–812
Bona E, Aden U, Gilland E, Fredholm BB, Hagberg H (1997) Neonatal cerebral hypoxia-ischemia: the effect of adenosine receptor antagonists. Neuropharmacology 36:1327–1338
Bonetti B, Stegagno C, Cannella B, Rizzato N, Moretto G, Raine CS (1999) Activation of NF-kappaB and c-jun transcription factors in multiple sclerosis lesions. Implications for oligodendrocyte pathology. Am J Pathol 155:1433–1438
Borea PA, Gessi S, Bar-Yehuda S, Fishman P (2009) A3 adenosine receptor: pharmacology and role in disease. Handb Exp Pharmacol 193:297–327
Borsello T, Clarke PG, Hirt L, Vercelli A, Repici M, Schorderet DF, Bogousslavsky J, Bonny C (2003) A peptide inhibitor of c-Jun N-terminal kinase protects against excitotoxicity and cerebral ischemia. Nat Med 9:1180–1186
Braun N, Lenz C, Gillardon F, Zimmermann M, Zimmermann H (1997) Focal cerebral ischemia enhances glial expression of ecto-5′-nucleotidase. Brain Res 766:213–226
Braun N, Zhu Y, Kriegistein J, Culmsee C, Zimmermann H (1998) Upregulation of the enzyme chain hydrolyzing extracellular ATP after transient forebrain ischemia in the rat. J Neurosci 18:4891–4900
Brodie C, Blumberg PM, Jacobson KA (1998) Activation of the A2A adenosine receptor inhibits nitric oxide production in glial cells. FEBS Lett 429:139–142
Bruce-Keller AJ (1999) Microglial-neuronal interactions in synaptic damage and recovery. J Neurosci Res 58:191–201
Burnstock G (2007) Physiology and pathophysiology of purinergic neurotransmission. Physiol Rev 87:659–797
Castillo A, Tolón MR, Fernández-Ruiz J, Romero J, Martinez-Orgado J (2010) The neuroprotective effect of cannabidiol in an in vitro model of newborn hypoxic-ischemic brain damage in mice is mediated by CB(2) and adenosine receptors. Neurobiol Dis 37:434–440

Chen JF, Huang Z, Ma J, Zhu J, Moratalla R, Standaert D, Moskowitz MA, Fink JS, Schwarzschild, MA (1999) A(2A) adenosine receptor deficiency attenuates brain injury induced by transient focal ischemia in mice. J Neurosci 19:9192–9200

Chen GJ, Harvey BK, Shen H, Chou J, Victor A, Wang Y (2006) Activation of adenosine A3 receptors reduces ischemic brain injury in rodents. J Neurosci Res 84:1848–1855

Chitolina Schetinger MR, Bonan CD, Schierholt RC, Webber A, Arteni N, Emanuelli T, Dias RD, Freitas Sarkis JJ, Netto CA (1998) Nucleotide hydrolysis in rats submitted to global cerebral ischemia: a possible link between preconditioning and adenosine production. J Stroke Cerebrovasc Dis 7:281–286

Choi DW (1990) Possible mechanisms limiting N-methyl-D-aspartate receptor overactivation and the therapeutic efficacy of N-methyl-D-aspartate antagonists. Stroke 21:III20–III22

Choukèr A, Thiel M, Lukashev D, Ward JM, Kaufmann I, Apasov S, Sitkovsky MV, Ohta A (2008) Critical role of hypoxia and A2A adenosine receptors in liver tissue-protecting physiological anti-inflammatory pathway. Mol Med 14:116–123

Coelho JE, Rebola N, Fragata I, Ribeiro JA, de Mendonça A, Cunha RA (2006) Hypoxia-induced desensitization and internalization of adenosine A1 receptors in the rat hippocampus. Neuroscience 138:1195–1203

Colotta V, Catarzi D, Varano F, Capelli F, Lenzi O, Filacchioni G, Martini C, Trincavelli L, Ciampi O, Pugliese AM, Pedata F, Schiesaro A, Morizzo E, Moro S (2007) New 2-arylpyrazolo[3,4-c] quinoline derivatives as potent and selective human A3 adenosine receptor antagonists. Synthesis, pharmacological evaluation, and ligand-receptor modeling studies. J Med Chem 50:4061–4074

Colotta V, Catarzi D, Varano F, Lenzi O, Filacchioni G, Martini C, Trincavelli L, Ciampi O, Traini C, Pugliese AM, Pedata F, Morizzo E, Moro S (2008) Synthesis, ligand-receptor modeling studies and pharmacological evaluation of novel 4-modified-2-aryl-1,2,4-triazolo[4,3-a]quinoxalin-1-one derivatives as potent and selective human A3 adenosine receptor antagonists. Bioorg Med Chem 16:6086–6102

Colotta V, Lenzi O, Catarzi D, Varano F, Filacchioni G, Martini C, Trincavelli L, Ciampi O, Pugliese AM, Traini C, Pedata F, Morizzo E, Moro S (2009) Pyrido[2,3-e]-1,2,4-triazolo[4,3-a] pyrazin-1-one as a new scaffold to develop potent and selective human A3 adenosine receptor antagonists. Synthesis, pharmacological evaluation, and ligand-receptor modeling studies. J Med Chem 52:2407–2419

Corradetti R, Lo CG, Moroni F, Passani MB, Pepeu G (1984) Adenosine decreases aspartate and glutamate release from rat hippocampal slices. Eur J Pharmacol 104:19–26

Corsi C, Melani A, Bianchi L, Pepeu G, Pedata F (1999) Striatal A_{2A} adenosine receptors differentially regulate spontaneous and K+-evoked glutamate release in vivo in young and aged rats. Neuroreport 10:687–691

Corsi C, Melani A, Bianchi L, Pedata F (2000) Striatal A_{2A} adenosine receptor antagonism differentially modifies striatal glutamate outflow in vivo in young and aged rats. Neuroreport 11:2591–2595

Cristalli G, Müller CE, Volpini R (2009) Recent developments in adenosine A2A receptor ligands. Handb Exp Pharmacol 193:59–98

Cunha RA (2001) Adenosine as a neuromodulator and as a homeostatic regulator in the nervous system: different roles, different sources and different receptors. Neurochem Int 38: 107–125

Dai SS, Zhou YG, Li W, An JH, Li P, Yang N, Chen XY, Xiong RP, Liu P, Zhao Y, Shen HY, Zhu PF, Chen JF (2010) Local glutamate level dictates adenosine A2A receptor regulation of neuroinflammation and traumatic brain injury. J Neurosci 30:5802–5810

Dale N, Pearson T, Fringuelli BG (2000) Direct measurement of adenosine release during hypoxia in the CA1 region of the rat hippocampal slice. J Physiol 526 Pt 1:143–155

Daly JW, Padgett W, Thompson RD, Kusachi S, Bugni WJ, Olsson RA (1986) Structure-activity relationships for N6-substituted adenosines at a brain A1-adenosine receptor with a comparison to an A2-adenosine receptor regulating coronary blood flow. Biochem Pharmacol 35: 2467–2481

Daly JW, Padgett WL (1992) Agonist activity of 2- and 5′-substituted adenosine analogs and their N6-cycloalkyl derivatives at A1- and A2-adenosine receptors coupled to adenylate cyclase. Biochem Pharmacol 43:1089–1093

Daré E, Schulte G, Karovic O, Hammarberg C, Fredholm BB (2007) Modulation of glial cell functions by adenosine receptors. Physiol Behav 92:15–20

Daval JL, Nicolas F (1994) Opposite effects of cyclohexyladenosine and theophylline on hypoxic damage in cultured neurons. Neurosci Lett 175:114–116

Daval JL, Von Lubitz DK, Deckert J, Redmond DJ, Marangos PJ (1989) Protective effect of cyclohexyladenosine on adenosine A1-receptors, guanine nucleotide and forskolin binding sites following transient brain ischemia: a quantitative autoradiographic study. Brain Res 491:212–226

De Keyser J, Sulter G, Luiten PG (1999) Clinical trials with neuroprotective drugs in acute ischaemic stroke: are we doing the right thing? Trends Neurosci 22:535–540

DeLeo J, Schubert P, Kreutzberg GW (1988) Protection against ischemic brain damage using propentofylline in gerbils. Stroke 19:1535–1539

Di Iorio P, Kleywegt S, Ciccarelli R, Traversa U, Andrew CM, Crocker CE, Werstiuk ES, Rathbone MP (2002) Mechanisms of apoptosis induced by purine nucleosides in astrocytes. Glia 38: 179–190

Dirnagl U, Iadecola C, Moskowitz MA (1999) Pathobiology of ischaemic stroke: an integrated view. Trends Neurosci 22:391–397

Dixon AK, Gubitz AK, Sirinathsinghji DJ, Richardson PJ, Freeman TC (1996) Tissue distribution of adenosine receptor mRNAs in the rat. Br J Pharmacol 118:1461–1468

Domenici MR, de Carolis AS, Sagratella S (1996) Block by N6-L-phenylisopropyl-adenosine of the electrophysiological and morphological correlates of hippocampal ischaemic injury in the gerbil. Br J Pharmacol 118:1551–1557

Duan W, Gui L, Zhou Z, Liu Y, Tian H, Chen JF, Zheng J (2009) Adenosine A2A receptor deficiency exacerbates white matter lesions and cognitive deficits induced by chronic cerebral hypoperfusion in mice. J Neurol Sci 285:39–45

Dunwiddie TV (1984) Interactions between the effects of adenosine and calcium on synaptic responses in rat hippocampus in vitro. J Physiol 350:545–559

Dux E, Fastbom J, Ungerstedt U, Rudolphi K, Fredholm BB (1990) Protective effect of adenosine and a novel xanthine derivative propentofylline on the cell damage after bilateral carotid occlusion in the gerbil hippocampus. Brain Res 516:248–256

Ekdahl CT, Claasen JH, Bonde S, Kokaia Z, Lindvall O (2003) Inflammation is detrimental for neurogenesis in adult brain. Proc Natl Acad Sci USA 100:13632–13637

Ekdahl CT, Kokaia Z, Lindvall O (2009) Brain inflammation and adult neurogenesis: the dual role of microglia. Neuroscience 158:1021–1029

Evans MC, Swan JH, Meldrum BS (1987) An adenosine analogue, 2-chloroadenosine, protects against long term development of ischaemic cell loss in the rat hippocampus. Neurosci Lett 83:287–292

Fedorova IM, Jacobson MA, Basile A, Jacobson KA (2003) Behavioral characterization of mice lacking the A3 adenosine receptor: sensitivity to hypoxic neurodegeneration. Cell Mol Neurobiol 23:431–447

Fiebich BL, Biber K, Lieb K, van Calker D, Berger M, Bauer J, Gebicke-Haerter PJ (1996) Cyclooxygenase-2 expression in rat microglia is induced by adenosine A2a-receptors. Glia 18:152–160

Fishman P, Bar-Yehuda S (2003) Pharmacology and therapeutic applications of A3 receptor subtype. Curr Top Med Chem 3:463–469

Fowler JC (1989) Adenosine antagonists delay hypoxia-induced depression of neuronal activity in hippocampal brain slice. Brain Res 490:378–384

Fowler JC (1990) Adenosine antagonists alter the synaptic response to in vitro ischemia in the rat hippocampus. Brain Res 509:331–334

Fowler JC, Gervitz LM, Hamilton ME, Walker JA (2003) Systemic hypoxia and the depression of synaptic transmission in rat hippocampus after carotid artery occlusion. J Physiol 550: 961–972

Frassetto SS, Schetinger MR, Schierholt R, Webber A, Bonan CD, Wyse AT, Dias RD, Netto CA, Sarkis JJ (2000) Brain ischemia alters platelet ATP diphosphohydrolase and 5′-nucleotidase activities in naive and preconditioned rats. Braz J Med Biol Res 33:1369–1377

Fredholm BB, Jacobson KA, Klotz KN, Linden J (2001) International union of pharmacology. XXV. Nomenclature and classification of adenosine receptors. Pharmacol Rev 53:527–552

Fredholm BB, Cunha RA, Svenningsson P (2003) Pharmacology of adenosine A(2A) receptors and therapeutic applications. Curr Top Med Chem 3:413–426

Fredholm BB, Chen JF, Masino SA, Vaugeois JM (2005) Actions of adenosine at its receptors in the CNS: insights from knockouts and drugs. Annu Rev Pharmacol Toxicol 45:385–412

Fredholm BB, Ijzerman AP, Jacobson KA, Linden J, Müller CE (2011) International Union of Basic and Clinical Pharmacology. LXXXI. Nomenclature and classification of adenosine receptors—an update. Pharmacol Rev 63:1–34

Frenguelli BG, Wigmore G, Llaudet E et al (2007) Temporal and mechanistic dissociation of ATP and adenosine release during ischaemia in the mammalian hippocampus. J Neurochem 101:1400–1413

Gao Y, Phillis JW (1994) CGS 15943, an adenosine A2 receptor antagonist, reduces cerebral ischemic injury in the Mongolian gerbil. Life Sci 55:61–65

Gao ZG, Blaustein JB, Gross AS, Melman N, Jacobson KA (2003) N6-Substituted adenosine derivatives: selectivity, efficacy, and species differences at A3 adenosine receptors. Biochem Pharmacol 65(10): 1675–1684

Gáspárová Z, Jariabka P, Stolc S (2008) Effect of the pyridoindole SMe1EC2 and compounds affecting A(1) and A(2A) adenosine receptors in rat hippocampus under ischemia in vitro. Pharmacol Rep 60:353–360

Geiger JD, Nagy JI (1986) Distribution of adenosine deaminase activity in rat brain and spinal cord. J Neurosci 6:2707–2714

Gelderblom M, Eminel S, Herdegen T, Waetzig V (2004) c-Jun N-terminal kinases (JNKs) and the cytoskeleton—functions beyond neurodegeneration. Int J Dev Neurosci 22:559–564

Gelderblom M, Leypoldt F, Steinbach K, Behrens D, Choe CU, Siler DA, Arumugam TV, Orthey E, Gerloff C, Tolosa E, Magnus T (2009) Temporal and spatial dynamics of cerebral immune cell accumulation in stroke. Stroke 40:1849–1857

Genovese T, Melani A, Esposito E, Paterniti I, Mazzon E, Di Paola R, Bramanti P, Linden J, Pedata F, Cuzzocrea S (2010) Selective adenosine A(2a) receptor agonists reduce the apoptosis in an experimental model of spinal cord trauma. J Biol Regul Homeost Agents 24:73–86

Gervitz LM, Lutherer LO, Davies DG, Pirch JH, Fowler JC (2001) Adenosine induces initial hypoxic-ischemic depression of synaptic transmission in the rat hippocampus in vivo. Am J Physiol Regul Integr Comp Physiol 280:R639–R645

Gervitz LM, Davies DG, Omidvar K, Fowler JC (2003) The effect of acute hypoxemia and hypotension on adenosine-mediated depression of evoked hippocampal synaptic transmission. Exp Neurol 182:507–517

Gessi S, Merighi S, Varani K, Leung E, Mac Lennan S, Borea PA (2008) The A3 adenosine receptor: an enigmatic player in cell biology. Pharmacol Ther 117(1):123–140

Gidday JM, Fitzgibbons JC, Shah AR, Kraujalis MJ, Park TS (1995) Reduction in cerebral ischemic injury in the newborn rat by potentiation of endogenous adenosine. Pediatr Res 38:306–311

Goldberg MP, Monyer H, Weiss JH, Choi DW (1988) Adenosine reduces cortical neuronal injury induced by oxygen or glucose deprivation in vitro. Neurosci Lett 89:323–327

Gomes CV, Kaster MP, Tomé AR, Agostinho PM, Cunha RA (2011) Adenosine receptors and brain diseases: neuroprotection and neurodegeneration. Biochim Biophys Acta. 1808(5): 1380–1399

Gourine AV, Dale N, Gourine VN, Spyer KM (2004) Fever in systemic inflammation: roles of purines. Front Biosci 9:1011–1022

Greene RW, Haas HL (1991) The electrophysiology of adenosine in the mammalian central nervous system. Prog Neurobiol 36:329–341

Gribkoff VK, Bauman LA, VanderMaelen CP (1990) The adenosine antagonist 8-cyclopentyltheophylline reduces the depression of hippocampal neuronal responses during hypoxia. Brain Res 512:353–357

Gui L, Duan W, Tian H, Li C, Zhu J, Chen JF, Zheng J (2009) Adenosine A2A receptor deficiency reduces striatal glutamate outflow and attenuates brain injury induced by transient focal cerebral ischemia in mice. Brain Res 1297:185–193

Hagberg H, Andersson P, Lacarewicz J, Jacobson I, Butcher S, Sandberg M (1987) Extracellular adenosine, inosine, hypoxanthine, and xanthine in relation to tissue nucleotides and purines in rat striatum during transient ischemia. J Neurochem 49:227–231

Han ME, Park KH, Baek SY, Kim BS, Kim JB, Kim HJ, Oh SO (2007) Inhibitory effects of caffeine on hippocampal neurogenesis and function. Biochem Biophys Res Commun 356:976–980

Haskó G, Linden J, Cronstein B, Pacher P (2008) Adenosine receptors: therapeutic aspects for inflammatory and immune diseases. Nat Rev Drug Discov 7:759–770

Hentschel S, Lewerenz A, Nieber K (2003) Activation of A(3) receptors by endogenous adenosine inhibits synaptic transmission during hypoxia in rat cortical neurons. Restor Neurol Neurosci 21:55–63

Heron A, Lekieffre D, Le Peillet E, Lasbennes F, Seylaz J, Plotkine M, Boulu RG (1994) Effects of an A1 adenosine receptor agonist on the neurochemical, behavioral and histological consequences of ischemia. Brain Res 641:217–224

Hettinger BD, Lee A, Linden J, Rosin DL (2001) Ultrastructural localization of adenosine A2A receptors suggests multiple cellular sites for modulation of GABAergic neurons in rat striatum. J Comp Neurol 431:331–346

Heurteaux C, Lauritzen I, Widmann C, Lazdunski M (1995) Essential role of adenosine, adenosine A1 receptors, and ATP-sensitive K+channels in cerebral ischemic preconditioning. Proc Natl Acad Sci USA 92:4666–4670

Higashi H, Meno JR, Marwaha AS, Winn HR (2002) Hippocampal injury and neurobehavioral deficits following hyperglycemic cerebral ischemia: effect of theophylline and ZM 241385. J Neurosurg 96:117–126

Hirt L, Badaut J, Thevenet J, Granziera C, Regli L, Maurer F, Bonny C, Bogousslavsky J (2004) D-JNKI1, a cell-penetrating c-Jun-N-terminal kinase inhibitor, protects against cell death in severe cerebral ischemia. Stroke 35:1738–1743

Howe CL, Bieber AJ, Warrington AE, Pease LR, Rodriguez M (2004) Antiapoptotic signaling by a remyelination-promoting human antimyelin antibody. Neurobiol Dis 15:120–131

Huang J, Upadhyay UM, Tamargo RJ (2006) Inflammation in stroke and focal cerebral ischemia. Surg Neurol 66:232–245

Irving EA, Barone FC, Reith AD, Hadingham SJ, Parsons AA (2000) Differential activation of MAPK/ERK and p38/SAPK in neurones and glia following focal cerebral ischaemia in the rat. Brain Res Mol Brain Res 77:65–75

Jacobson KA (1998) Adenosine A3 receptors: novel ligands and paradoxical effects. Trends Pharmacol Sci 19:184–191

Jacobson KA, Gao ZG (2006) Adenosine receptors as therapeutic targets. Nature Rev 5:247–264

Jacobson KA, Nikodijević O, Shi D, Gallo-Rodriguez C, Olah ME, Stiles GL, Daly JW (1993) A role for central A3-adenosine receptors. Mediation of behavioral depressant effects. FEBS Lett 336:57–60

Jacobson KA, Von Lubitz DK, Daly JW, Fredholm BB (1996) Adenosine receptor ligands: differences with acute versus chronic treatment. Trends Pharmacol Sci 17:108–113

Jacobson KA, Hoffmann C, Cattabeni F, Abbracchio MP (1999) Adenosine-induced cell death: evidence for receptor-mediated signalling. Apoptosis 4:197–211

Jarrott DM, Domer FR (1980) A gerbil model of cerebral ischemia suitable for drug evaluation. Stroke 11:203–209

Jiang N, Kowaluk EA, Lee CH, Mazdiyasni H, Chopp M (1997) Adenosine kinase inhibition protects brain against transient focal ischemia in rats. Eur J Pharmacol 320:131–137

Johansson B, Halldner L, Dunwiddie TV, Masino SA, Poelchen W, Giménez-Llort L, Escorihuela RM, Fernández-Teruel A, Wiesenfeld-Hallin Z, Xu XJ, Hårdemark A, Betsholtz C, Herlenius E, Fredholm BB (2001) Hyperalgesia, anxiety, and decreased hypoxic neuroprotection in mice lacking the adenosine A1 receptor. Proc Natl Acad Sci USA 98:9407–9412

Johnson MP, McCarty DR, Chmielewski PA (1998) Temporal dependent neuroprotection with propentofylline (HWA 285) in a temporary focal ischemia model. Eur J Pharmacol 346:151–157
Jones PA, Smith RA, Stone TW (1998a) Protection against kainate-induced excitotoxicity by adenosine A_{2A} receptor agonists and antagonists. Neuroscience 85:229–237
Jones PA, Smith RA, Stone TW (1998b) Protection against hippocampal kainate excitotoxicity by intracerebral administration of an adenosine A_{2A} receptor antagonist. Brain Res 800:328–335
Jurewicz A, Matysiak M, Andrzejak S, Selmaj K (2006) TRAIL-induced death of human adult oligodendrocytes is mediated by JNK pathway. Glia 53:158–166
Kano T, Katayama Y, Kawamata T, Hirota H, Tsubokawa T (1994) Propentofylline administered by microdialysis attenuates ischemia-induced hippocampal damage but not excitatory amino acid release in gerbils. Brain Res 641:149–154
Kawahara N, Ide T, Saito N, Kawai K, Kirino T (1998) Propentofylline potentiates induced ischemic tolerance in gerbil hippocampal neurons via adenosine receptor. J Cereb Blood Flow Metab 18:472–475
Kawasaki H, Morooka T, Shimohama S, Kimura J, Hirano T, Gotoh Y, Nishida E (1997) Activation and involvement of p38 mitogen-activated protein kinase in glutamate-induced apoptosis in rat cerebellar granule cells. J Biol Chem 272:18518–18521
King AE, Ackley MA, Cass CE et al (2006) Nucleoside transporters: from scavengers to novel therapeutic targets. Trends Pharmacol Sci 27:416–425
Kitada M, Rowitch DH (2006) Transcription factor co-expression patterns indicate heterogeneity of oligodendroglial subpopulations in adult spinal cord. Glia 54:35–46
Kitagawa H, Mori A, Shimada J, Mitsumoto Y, Kikuchi T (2002) Intracerebral adenosine infusion improves neurological outcome after transient focal ischemia in rats. Neurol Res 24:317–323
Kobayashi T, Yamada T, Okada Y (1998) The levels of adenosine and its metabolites in the guinea pig and rat brain during complete ischemia-in vivo study. Brain Res 787:211–219
Kuan CY, Whitmarsh AJ, Yang DD, Liao G, Schloemer AJ, Dong C, Bao J, Banasiak KJ, Haddad GG, Flavell RA, Davis RJ, Rakic P (2003) A critical role of neural-specific JNK3 for ischemic apoptosis. Proc Natl Acad Sci USA 100:15184–15189
Kuroiwa T, Ting P, Martinez H, Klatzo I (1985) The biphasic opening of the blood-brain barrier to proteins following temporary middle cerebral artery occlusion. Acta Neuropathol 68:122–129
Kusano Y, Echeverry G, Miekisiak G, Kulik TB, Aronhime SN, Chen JF, Winn HR (2010) Role of adenosine A2 receptors in regulation of cerebral blood flow during induced hypotension. J Cereb Blood Flow Metab 30:808–815
Lai D-M, Tu Y-K, Liu I-M, Cheng J-T (2005) Increase of adenosine A_1 receptor gene expression in cerebral ischemia of Wistar rats. Neurosci Lett 387:59–61
Lange-Asschenfeldt C, Raval AP, Dave KR, Mochly-Rosen D, Sick TJ, Perez-Pinzon MA (2004) Epsilon protein kinase C mediated ischemic tolerance requires activation of the extracellular regulated kinase pathway in the organotypic hippocampal slice. J Cereb Blood Flow Metab 24:636–645
Langer D, Ikehara Y, Takebayashi H, Hawkes R, Zimmermann H (2007) The ectonucleotidases alkaline phosphatase and nucleoside triphosphate diphosphohydrolase 2 are associated with subsets of progenitor cell populations in the mouse embryonic, postnatal and adult neurogenic zones. Neuroscience 150:863–879
Lappas CM, Day YJ, Marshall MA, Engelhard VH, Linden J (2006) Adenosine A2A receptor activation reduces hepatic ischemia reperfusion injury by inhibiting CD1d-dependent NKT cell activation. J Exp Med 203:2639–2648
Latini S, Pedata F (2001) Adenosine in the central nervous system: release mechanisms and extracellular concentrations. J Neurochem 79:463–484
Latini S, Corsi C, Pedata F et al (1995) The source of brain adenosine outflow during ischemia and electrical stimulation. Neurochem Int 27:239–244
Latini S, Bordoni F, Corradetti R, Pepeu G, Pedata F (1998) Temporal correlation between adenosine outflow and synaptic potential inhibition in rat hippocampal slices during ischemia-like conditions. Brain Res 794:325–328

Latini S, Bordoni F, Pedata F, Corradetti R (1999a) Extracellular adenosine concentrations during in vitro ischaemia in rat hippocampal slices. Br J Pharmacol 127:729–739
Latini S, Bordoni F, Corradetti R, Pepeu G, Pedata F (1999b) Effect of A2A adenosine receptor stimulation and antagonism on synaptic depression induced by in vitro ischaemia in rat hippocampal slices. Br J Pharmacol 128:1035–1044
Lee KS, Lowenkopf T (1993) Endogenous adenosine delays the onset of hypoxic depolarization in the rat hippocampus in vitro via an action at A1 receptors. Brain Res 609:313–315
Lewis AJ, Manning AM (1999) New targets for anti-inflammatory drugs. Curr Opin Chem Biol 3:489–494
Li Y, Oskouian RJ, Day YJ, Rieger JM, Liu L, Kern JA, Linden J (2006) Mouse spinal cord compression injury is reduced by either activation of the adenosine A2A receptor on bone marrow-derived cells or deletion of the A2A receptor on non-bone marrow-derived cells. Neuroscience 141:2029–2039
Liang R, Pang ZP, Deng P, Xu ZC (2009) Transient enhancement of inhibitory synaptic transmission in hippocampal CA1 pyramidal neurons after cerebral ischemia. Neuroscience 160: 412–418
Lie AA, Blümcke I, Beck H, Wiestler OD, Elger CE, Schoen SW (1999) 5′-Nucleotidase activity indicates sites of synaptic plasticity and reactive synaptogenesis in the human brain. J Neuropathol Exp Neurol 58:451–458
Lin Y, Phillis JW (1992) Deoxycoformycin and oxypurinol: protection against focal ischemic brain injury in the rat. Brain Res 571:272–280
Lloyd HG, Fredholm BB (1995) Involvement of adenosine deaminase and adenosine kinase in regulating extracellular adenosine concentration in rat hippocampal slices. Neurochem Int 26:387–395
Logan M, Sweeney MI (1997) Adenosine A1 receptor activation preferentially protects cultured cerebellar neurons versus astrocytes against hypoxia-induced death. Mol Chem Neuropathol 31:119–133
Luo C, Yi B, Tao G, Li M, Chen Z, Tang W, Zhang JH, Feng H (2010) Adenosine A3 receptor agonist reduces early brain injury in subarachnoid haemorrhage. Neuroreport 21:892–896
Macek TA, Schaffhauser H, Conn PJ (1998) Protein kinase C and A3 adenosine receptor activation inhibit presynaptic metabotropic glutamate receptor (mGluR) function and uncouple mGluRs from GTP-binding proteins. J Neurosci 18:6138–6146
Madrigal JL, Caso JR, de Cristóbal J et al (2003) Effect of subacute and chronic immobilisation stress on the outcome of permanent focal cerebral ischaemia in rats. Brain Res 979:137–145
Marcoli M, Raiteri L, Bonfanti A, Monopoli A, Ongini E, Raiteri M, Maura G (2003) Sensitivity to selective adenosine A1 and A2A receptor antagonists of the release of glutamate induced by ischemia in rat cerebrocortical slices. Neuropharmacology 45:201–210
Marcoli M, Bonfanti A, Roccatagliata P, Chiaramonte G, Ongini E, Raiteri M, Maura G (2004) Glutamate efflux from human cerebrocortical slices during ischemia: vesicular-like mode of glutamate release and sensitivity to A(2A) adenosine receptor blockade. Neuropharmacology 47:884–891
Marcus AJ, Broekman MJ, Drosopoulos JH, Islam N, Pinsky DJ, Sesti C, Levi R (2003) Metabolic control of excessive extracellular nucleotide accumulation by CD39/ecto-nucleotidase-1: implications for ischemic vascular diseases. J Pharmacol Exp Ther 305:9–16
Marks L, Carswell HV, Peters EE, Graham DI, Patterson J, Dominiczak AF, Macrae IM (2001) Characterization of the microglial response to cerebral ischemia in the stroke-prone spontaneously hypertensive rat. Hypertension 38:116–122
Matsumoto K, Graf R, Rosner G, Shimada N, Heiss WD (1992) Flow thresholds for extracellular purine catabolite elevation in cat focal ischemia. Brain Res 579:309–314
Matsumoto K, Sakaki T, Kohmura E, Hayakawa T, Yamada K (1996) Amelioration of ischemic brain damage by the preischemic administration of propentofylline (HWA285) in a rat focal ischemia. Brain Res 723:228–230
Matute C, Sanchez-Gomez MV, Martinez-Millan L, Miledi R (1997) Glutamate receptor-mediated toxicity in optic nerve oligodendrocytes. Proc Natl Acad Sci USA 94:8830–8835

Matute C, Alberdi E, Ibarretxe G, Sanchez-Gomez MV (2002) Excitotoxicity in glial cells. Eur J Pharmacol 447:239–246

McDonald JW, Althomsons SP, Hyrc KL, Choi DW, Goldberg MP (1998) Oligodendrocytes from forebrain are highly vulnerable to AMPA/kainate receptor-mediated excitotoxicity. Nat Med 4:291–297

Melani A, Pantoni L, Corsi C, Bianchi L, Monopoli A, Bertorelli R, Pepeu G, Pedata F (1999) Striatal outflow of adenosine, excitatory amino acids, γ-aminobutyric acid, and taurine in awake freely moving rats after middle cerebral artery occlusion. Correlations with neurological deficit and histopatological damage. Stroke 30:2448–2455

Melani A, Pantoni L, Bordoni F, Gianfriddo M, Bianchi L, Vannucchi MG, Bertorelli R, Monopoli A, Pedata F (2003) The selective A2A receptor antagonist SCH 58261 reduces striatal transmitter outflow, turning behavior and ischemic brain damage induced by permanent focal ischemia in the rat. Brain Res 959:243–250

Melani A, Amadio S, Gianfriddo M, Vannucchi MG, Volontè C, Bernardi G, Pedata F (2006a) Sancesario G. P2X7 receptor modulation on microglial cells and reduction of brain infarct caused by middle cerebral artery occlusion in rat. J Cereb Blood Flow Metab 26:974–982

Melani A, Gianfriddo M, Vannucchi MG, Cipriani S, Baraldi PG, Giovannini MG, Pedata F (2006b) The selective A(2A) receptor antagonist SCH 58261 protects from neurological deficit, brain damage and activation of p38 MAPK in rat focal cerebral ischemia. Brain Res 1073–1074:470–480

Melani A, Cipriani S, Vannucchi MG, Nosi D, Donati C, Bruni P, Giovannini MG, Pedata F (2009) Selective adenosine A_{2A} receptor antagonism reduces JNK activation in oligodendrocytes after cerebral ischaemia. Brain 132:1480–1495

Melani A, Cipriani S, Corti F, Pedata F (2010) Effect of intravenous administration of dipyridamole in a rat model of chronic cerebral ischemia. Ann N Y Acad Sci 1207:89–96

Melani A, Corti F, Vannucchi MG, Nosi D, Giovannini MG, Pedata F (2011) Role of adenosine A2A receptors on modulation of oligodendroglia in cerebral ischemia. Shock 36(Suppl 2):24–25

Melani A, Corti F, Stephan H, Müller CE, Donati C, Bruni P, Vannucchi MG, Pedata F (2012) Ecto-ATPase inhibition: ATP and adenosine release under physiological and ischemic *in vivo* conditions in the rat striatum. Exp Neurol 233:193–204

Migita H, Kominami K, Higashida M, Maruyama R, Tuchida N, McDonald F, Shimada F, Sakurada K (2008) Activation of adenosine A1 receptor-induced neural stem cell proliferation via MEK/ERK and Akt signaling pathways. J Neurosci Res 86:2820–2828

Miller LP, Jelovich LA, Yao L, DaRe J, Ugarkar B, Foster AC (1996) Pre- and peristroke treatment with the adenosine kinase inhibitor, 5′-deoxyiodotubercidin, significantly reduces infarct volume after temporary occlusion of the middle cerebral artery in rats. Neurosci Lett 220:73–76

Milton SL, Nayak G, Kesaraju S, Kara L, Prentice HM (2007) Suppression of reactive oxygen species production enhances neuronal survival in vitro and in vivo in the anoxia-tolerant turtle Trachemys scripta. J Neurochem 101:993–1001

Mirabet M, Herrera C, Cordero OJ, Mallol J, Lluis C, Franco R (1999) Expression of A2B adenosine receptors in human lymphocytes: their role in T cell activation. J Cell Sci 112:491–502

Monopoli A, Lozza G, Forlani A, Mattavelli A, Ongini E (1998) Blockade of A2A adenosine receptors by SCH 58261 results in neuroprotective effects in cerebral ischaemia in rats. Neuroreport 9:3955–3959

Mori M, Nishizaki T, Okada Y (1992) Protective effect of adenosine on the anoxic damage of hippocampal slice. Neuroscience 46:301–307

Moro S, Gao ZG, Jacobson KA, Spalluto G (2006) Progress in the pursuit of therapeutic adenosine receptor antagonists. Med Res Rev 26: 131–159

Müller CE, Jacobson KA (2011) Recent developments in adenosine receptor ligands and their potential as novel drugs. Biochim Biophys Acta 1808:1290–1308

Naganuma M, Wiznerowicz EB, Lappas CM, Linden J, Worthington MT, Ernst PB (2006) Cutting edge: critical role for A2A adenosine receptors in the T cell-mediated regulation of colitis. J Immunol 177:2765–2769

Nagasawa H, Araki T, Kogure K (1994) Alteration of adenosine A1 receptor binding in the post-ischaemic rat brain. Neuroreport 5:1453–1456
Nakamura M, Nakakimura K, Matsumoto M, Sakabe T (2002) Rapid tolerance to focal cerebral ischemia in rats is attenuated by adenosine A1 receptor antagonist. J Cereb Blood Flow Metab 22:161–170
Newman GC, Hospod FE, Trowbridge SD, Motwani S, Liu Y (1998) Restoring adenine nucleotides in a brain slice model of cerebral reperfusion. J Cereb Blood Flow Metab 18:675–685
Nieber K, Hentschel S (2006) Signalling pathways of the adenosine A_3 receptors in rat cortical neurons. In: Proceedings of the 8th international symposium on adenosine and adenine nucleotides, Ferrara, Italy, May 24–28
Nishizaki T, Nagai K, Nomura T, Tada H, Kanno T, Tozaki H, Li XX, Kondoh T, Kodama N, Takahashi E, Sakai N, Tanaka K, Saito N (2002) A new neuromodulatory pathway with a glial contribution mediated via A_{2A} adenosine receptors. Glia 39:133–147
Odashima M, Bamias G, Rivera-Nieves J, Linden J, Nast CC, Moskaluk CA, Marini M, Sugawara K, Kozaiwa K, Otaka M, Watanabe S, Cominelli F (2005) Activation of A2A adenosine receptor attenuates intestinal inflammation in animal models of inflammatory bowel disease. Gastroenterology 129:26–33
Olsson T, Cronberg T, Rytter A, Asztély F, Fredholm BB, Smith ML, Wieloch T (2004) Deletion of the adenosine A1 receptor gene does not alter neuronal damage following ischaemia in vivo or in vitro. Eur J Neurosci 20:1197–1204
Ongini E, Adami M, Ferri C, Bertorelli R (1997) Adenosine A2A receptors and neuroprotection. Ann N Y Acad Sci 825:30–48
Onodera H, Sato G, Kogure K (1987) Quantitative autoradiographic analysis of muscarinic cholinergic and adenosine A1 binding sites after transient forebrain ischemia in the gerbil. Brain Res 415:309–322
O'Regan MH, Simpson RE, Perkins LM, Phillis JW (1992) The selective A_2 adenosine receptor agonist CGS 21680 enhances excitatory transmitter amino acid release from the ischemic rat cerebral cortex. Neurosci Lett 138:169–172
Palmer TM, Benovic JL, Stiles GL (1995) Agonist-dependent phosphorylation and desensitization of the rat A3 adenosine receptor. Evidence for a G-protein-coupled receptor kinase-mediated mechanism. J Biol Chem 270:29607–29613
Park CK, Rudolphi KA (1994) Antiischemic effects of propentofylline (HWA 285) against focal cerebral infarction in rats. Neurosci Lett 178:235–238
Parkinson FE, Xiong W, Zamzow CR (2005) Astrocytes and neurons: different roles in regulating adenosine levels. Neurol Res 27:153–160
Parkinson FE, Damaraju VL, Graham K, Yao SY, Baldwin SA, Cass CE, Young JD (2011) Molecular biology of nucleoside transporters and their distributions and functions in the brain. Curr Top Med Chem 11:948–972
Patel BT, Tudball N (1986) Localization of S-adenosylhomocysteine hydrolase and adenosine deaminase immunoreactivities in rat brain. Brain Res 370:250–264
Paterniti I, Melani A, Cipriani S, Corti F, Mello T, Mazzon E, Esposito E, Bramanti P, Cuzzocrea S, Pedata F (2011) Selective adenosine A2A receptor agonists and antagonists protect against spinal cord injury through peripheral and central effects. J Neuroinflammation 8:31
Pearson T, Damian K, Lynas RE, Frenguelli BG (2006) Sustained elevation of extracellular adenosine and activation of A1 receptors underlie the post-ischaemic inhibition of neuronal function in rat hippocampus in vitro. J Neurochem 97:1357–1368
Pedata F, Latini S, Pugliese AM, Pepeu G (1993) Investigations into the adenosine outflow from hippocampal slices evoked by ischemia-like conditions. J Neurochem 61:284–289
Pedata F, Corsi C, Melani A, Bordoni F, Latini S (2001) Adenosine extracellular brain concentrations and role of A_{2A} receptors in ischemia. Ann N Y Acad Sci 939:74–84
Pedata F, Pugliese AM, Melani A, Gianfriddo M (2003) A_{2A} receptors in neuroprotection of dopaminergic neurons. Neurology 61:S49–S50
Pedata F, Gianfriddo M, Turchi D, Melani A (2005) The protective effect of adenosine A_{2A} receptor antagonism in cerebral ischemia. Neurol Res 27:169–174

Pedata F, Pugliese AM, Coppi E, Popoli P, Morelli M, Schwarzschild MA, Melani A (2007) Adenosine in the central nervous system: effects on neurotransmission and neuroprotection. Immunol Endocr Metab Agents Med Chem 4:304–321

Pedata F, Pugliese AM, Sebastiao AM, Ribeiro JA (2010) Adenosine A3 receptor signaling in the central nervous system. In: Borea PA (ed) A3 adenosine receptor. Springer Science+Business Media B.V. 2010, pp 165–188

Perez-Pinzon MA, Mumford PL, Rosenthal M, Sick TJ (1996) Anoxic preconditioning in hippocampal slices: role of adenosine. Neuroscience 75:687–694

Peterfreund RA, MacCollin M, Gusella J, Fink JS (1996) Characterization and expression of the human A2a adenosine receptor gene. J Neurochem 66:362–368

Petroni A, Papini N, Blasevich M, Galli C (2002) Blockade of A(2A) adenosine receptors leads to c-fos inhibition in a rat model of brain ischemia. Pharmacol Res 45:125–128

Phillis JW (1995) The effects of selective A1 and A2a adenosine receptor antagonists on cerebral ischemic injury in the gerbil. Brain Res 705:79–84

Phillis JW (2004) Adenosine and adenine nucleotides as regulators of cerebral blood flow: roles of acidosis, cell swelling, and KATP channels. Crit Rev Neurobiol 16:237–270

Phillis JW, Goshgarian HG (2001) Adenosine and neurotrauma: therapeutic perspectives. Neurol Res 23:183–189

Phillis JW, O'Regan MH (1989) Deoxycoformycin antagonizes ischemia-induced neuronal degeneration. Brain Res Bull 22:537–540

Phillis JW, Edstrom JP, Kostopoulos GK, Kirkpatrick JR (1979) Effects of adenosine and adenine nucleotides on synaptic transmission in the cerebral cortex. Can J Physiol Pharmacol 57:1289–1312

Phillis JW, Walter GA, Simpson RE (1991) Brain adenosine and transmitter amino acid release from the ischemic rat cerebral cortex: effects of the adenosine deaminase inhibitor deoxycoformycin. J Neurochem 56:644–650

Phillis JW, Smith-Barbour M, O'Regan MH, Perkins LM (1994a) Amino acid and purine release in rat brain following temporary middle cerebral artery occlusion. Neurochem Res 19:1125–1130

Phillis JW, Smith-Barbour M, Perkins LM, O'Regan MH (1994b) Indomethacin modulates ischemia-evoked release of glutamate and adenosine from the rat cerebral cortex. Brain Res 652:353–356

Phillis JW, Smith-Barbour M, O'Regan MH (1996) Changes in extracellular amino acid neurotransmitters and purines during and following ischemias of different durations in the rat cerebral cortex. Neurochem Int 29:115–120

Piao CS, Kim JB, Han PL, Lee JK (2003) Administration of the p38 MAPK inhibitor SB203580 affords brain protection with a wide therapeutic window against focal ischemic insult. J Neurosci Res 73:537–544

Pignataro G, Maysami S, Studer FE, Wilz A, Simon RP, Boison D (2008) Downregulation of hippocampal adenosine kinase after focal ischemia as potential endogenous neuroprotective mechanism. J Cereb Blood Flow Metab 28:17–23

Pinsky DJ, Broekman MJ, Peschon JJ, Stocking KL, Fujita T, Ramasamy R, Connolly ES Jr, Huang J, Kiss S, Zhang Y, Choudhri TF, McTaggart RA, Liao H, Drosopoulos JH, Price VL, Marcus AJ, Maliszewski CR (2002) Elucidation of the thromboregulatory role of CD39/ectoapyrase in the ischemic brain. J Clin Invest 109:1031–1040

Pintor A, Galluzzo M, Grieco R, Pezzola A, Reggio R, Popoli P (2004) Adenosine A_{2A} receptor antagonists prevent the increase in striatal glutamate levels induced by glutamate uptake inhibitors. J Neurochem 89:152–156

Popoli P, Betto P, Reggio R, Ricciarello G (1995) Adenosine A_{2A} receptor stimulation enhances striatal extracellular glutamate levels in rats. Eur J Pharmacol 287:215–217

Popoli P, Pintor A, Domenici MR, Frank C, Tebano MT, Pezzola A, Scarchilli L, Quarta D, Reggio R, Malchiodi-Albedi F, Falchi M, Massotti M (2002) Blockade of striatal adenosine A_{2A} receptor reduces, through a presynaptic mechanism, quinolinic acid-induced excitotoxicity: possible relevance to neuroprotective interventions in neurodegenerative diseases of the striatum. J Neurosci 22:1967–1975

Pugliese AM, Latini S, Corradetti R, Pedata F (2003) Brief, repeated, oxygen-glucose deprivation episodes protect neurotransmission from a longer ischemic episode in the in vitro hippocampus: role of adenosine receptors. Br J Pharmacol 140:305–314

Pugliese AM, Coppi E, Spalluto G, Corradetti R, Pedata F (2006) A3 adenosine receptor antagonists delay irreversible synaptic failure caused by oxygen and glucose deprivation in the rat CA1 hippocampus in vitro. Br J Pharmacol 147:524–532

Pugliese AM, Coppi E, Volpini R, Cristalli G, Corradetti R, Jeong LS, Jacobson KA, Pedata F (2007a) Role of adenosine A3 receptors on CA1 hippocampal neurotransmission during oxygen-glucose deprivation episodes of different duration. Biochem Pharmacol 74:768–779

Pugliese AM, Coppi E, Volpini R, Cristalli G, Corradetti R, Jeong LS, Jacobson KA, Pedata F (2007b) Role of adenosine A3 receptors on CA1 hippocampal neurotransmission during oxygen-glucose deprivation episodes of different duration. Biochem Pharmacol 74(5): 768–779

Pugliese AM, Traini C, Cipriani S, Gianfriddo M, Mello T, Giovannini MG, Galli A, Pedata F (2009) The adenosine A2A receptor antagonist ZM241385 enhances neuronal survival after oxygen-glucose deprivation in rat CA1 hippocampal slices. Br J Pharmacol 157:818–830

Ribeiro JA, Sebastiao AM, de Mendonca A (2002) Adenosine receptors in the nervous system: pathophysiological implications. Prog Neurobiol 68:377–392

Rivkees SA, Zhao Z, Porter G, Turner C (2001) Influences of adenosine on the fetus and newborn. Mol Genet Metab 74:160–171

Rosin DL, Hettinger BD, Lee A, Linden J (2003) Anatomy of adenosine A2A receptors in brain: morphological substrates for integration of striatal function. Neurology 61:S10–S11

Rudolphi KA, Keil M, Hinze HJ (1987) Effect of theophylline on ischemically induced hippocampal damage in Mongolian gerbils: a behavioral and histopathological study. J Cereb Blood Flow Metab 7:74–81

Rudolphi KA, Keil M, Fastbom J, Fredholm BB (1989) Ischaemic damage in gerbil hippocampus is reduced following upregulation of adenosine (A1) receptors by caffeine treatment. Neurosci Lett 103:275–280

Sandoval KE, Witt KA (2008) Blood-brain barrier tight junction permeability and ischemic stroke. Neurobiol Dis 32:200–219

Saransaari P, Oja SS (2005) GABA release modified by adenosine receptors in mouse hippocampal slices under normal and ischemic conditions. Neurochem Res 30:467–473

Schiffmann SN, Jacobs O, Vanderhaeghen JJ (1991) Striatal restricted adenosine A2 receptor (RDC8) is expressed by enkephalin but not by substance P neurons: an in situ hybridization histochemistry study. J Neurochem 57:1062–1067

Schrader WP, West CA, Strominger NL (1987) Localization of adenosine deaminase and adenosine deaminase complexing protein in rabbit brain. J Histochem Cytochem 35:443–451

Schwartz-Bloom RD, Sah R (2001) Gamma-Aminobutyric acid(A) neurotransmission and cerebral ischemia. J Neurochem 77:353–371

Schwarzschild MA, Chen JF, Ascherio A (2002) Caffeinated clues and the promise of adenosine A(2A) antagonists in PD. Neurology 58:1154–1160

Sciotti VM, Roche FM, Grabb MC, Van Wylen DG (1992) Adenosine receptor blockade augments interstitial fluid levels of excitatory amino acids during cerebral ischemia. J Cereb Blood Flow Metab 12:646–655

Sebastião AM, Cunha RA, de Mendonca A, Ribeiro JA (2000) Modification of adenosine modulation of synaptic transmission in the hippocampus of aged rats. Br J Pharmacol 131:1629–1634

Sebastião AM, de Mendonca A, Moreira T, Ribeiro JA (2001) Activation of synaptic NMDA receptors by action potential-dependent release of transmitter during hypoxia impairs recovery of synaptic transmission on reoxygenation. J Neurosci 21:8564–8571

Seif-el-Nasr M, Khattab M (2002) Influence of inhibition of adenosine uptake on the gamma-aminobutyric acid level of the ischemic rat brain. Arzneimittelforschung 52:353–357

Sheardown MJ, Knutsen LJS (1996) Unexpected neuroprotection observed with the adenosine A_{2A} receptor agonist CGS21680. Drug Dev Res 39:108–114

Shen H-Y, Lusardi TA, Williams-Karnesky RL, Lan J-Q, Poulsen DJ, Boison D (2011) Adenosine kinase determines the degree of brain injury after ischemic stroke in mice. J Cereb Blood Flow Metab 31(7):1648–59

Shrager P, Novakovic SD (1995) Control of myelination, axonal growth, and synapse formation in spinal cord explants by ion channels and electrical activity. Brain Res Dev Brain Res 88:68–78

Simpson RE, O'Regan MH, Perkins LM, Phillis JW (1992) Excitatory transmitter amino acid release from the ischemic rat cerebral cortex: effects of adenosine receptor agonists and antagonists. J Neurochem 58:1683–1690

Sitkovsky MV, Lukashev D, Apasov S, Kojima H, Koshiba M, Caldwell C, Ohta A, Thiel M (2004) Physiological control of immune response and inflammatory tissue damage by hypoxia-inducible factors and adenosine A2A receptors. Ann Rev Immunol 22:657–682

Somjen GG (2001) Mechanisms of spreading depression and hypoxic spreading depression-like depolarization. Physiol Rev 81:1065–1096

Stevens B, Porta S, Haak LL, Gallo V, Fields RD (2002) Adenosine: a neuron-glial transmitter promoting myelination in the CNS in response to action potentials. Neuron 36:855–868

Stoll G, Jander S, Schroeter M (1998) Inflammation and glial responses in ischemic brain lesions. Prog Neurobiol 56:149–171

Stone TW, Behan WM (2007) Interleukin-1beta but not tumor necrosis factor-alpha potentiates neuronal damage by quinolinic acid: protection by an adenosine A2A receptor antagonist. J Neurosci Res 85:1077–1085

Sutherland GR, Peeling J, Lesiuk HJ, Brownstone RM, Rydzy M, Saunders JK, Geiger JD (1991) The effects of caffeine on ischemic neuronal injury as determined by magnetic resonance imaging and histopathology. Neuroscience 42:171–182

Svenningsson P, Nomikos GG, Ongini E, Fredholm BB (1997) Antagonism of adenosine A2A receptors underlies the behavioural activating effect of caffeine and is associated with reduced expression of messenger RNA for NGFI-A and NGFI-B in caudate-putamen and nucleus accumbens. Neuroscience 79:753–764

Svenningsson P, Le Moine C, Fisone G, Fredholm BB (1999) Distribution, biochemistry and function of striatal adenosine A2A receptors. Prog Neurobiol 59:355–396

Takagi Y, Nozaki K, Sugino T, Hattori I, Hashimoto N (2000) Phosphorylation of c-Jun NH(2)-terminal kinase and p38 mitogen-activated protein kinase after transient forebrain ischemia in mice. Neurosci Lett 294:117–120

Tanaka E, Yasumoto S, Hattori G, Niiyama S, Matsuyama S, Higashi H (2001) Mechanisms underlying the depression of evoked fast EPSCs following in vitro ischemia in rat hippocampal CA1 neurons. J Neurophysiol 86:1095–1103

Tatlisumak T, Takano K, Carano RA, Miller LP, Foster AC, Fisher M (1998) Delayed treatment with an adenosine kinase inhibitor, GP683, attenuates infarct size in rats with temporary middle cerebral artery occlusion. Stroke 29:1952–1958

Trincavelli ML, Tuscano D, Cecchetti P, Falleni A, Benzi L, Klotz KN, Gremigni V, Cattabeni F, Lucacchini A, Martini C (2000) Agonist-induced internalization and recycling of the human A(3) adenosine receptors: role in receptor desensitization and resensitization. J Neurochem 75:1493–1501

Trincavelli ML, Tuscano D, Marroni M, Falleni A, Gremigni V, Ceruti S, Abbracchio MP, Jacobson KA, Cattabeni F, Martini C (2002) A3 adenosine receptors in human astrocytoma cells: agonist-mediated desensitization, internalization, and down-regulation. Mol Pharmacol 62: 1373–1384

Trincavelli ML, Marroni M, Tuscano D, Ceruti S, Mazzola A, Mitro N, Abbracchio MP, Martini C (2004) Regulation of A2B adenosine receptor functioning by tumour necrosis factor a in human astroglial cells. J Neurochem 91:1180–1190

Trincavelli ML, Melani A, Guidi S, Cipriani S, Pedata F, Martini C (2008a) Regulation of A2A adenosine receptor expression and functioning following permanent focal ischemia in rat brain. J Neurochem 104:479–490

Trincavelli ML, Tonazzini I, Montali M, Abbracchio MP, Martini C (2008b) Short-term TNF-Alpha treatment induced A2B adenosine receptor desensitization in human astroglial cells. J Cell Biochem 104:150–161
Turcáni P, Turèáni M (2001) Effect of propentofylline on cerebral blood flow in a gerbil focal cerebral ischemia. J Neurol Sci 183:57–60
Turner CP, Yan H, Schwartz M, Othman T, Rivkees SA (2002) A1 adenosine receptor activation induces ventriculomegaly and white matter loss. Neuroreport 13:1199–1204
Turner CP, Seli M, Ment L, Stewart W, Yan H, Johansson B, Fredholm BB, Blackburn M, Rivkees SA (2003) A1 adenosine receptors mediate hypoxia-induced ventriculomegaly. Proc Natl Acad Sci USA 100:11718–11722
Van Galen PJ, van Bergen AH, Gallo-Rodriguez C, Melman N, Olah ME, IJzerman AP, Stiles GL, Jacobson KA (1994) A binding site model and structure-activity relationships for the rat A3 adenosine receptor. Mol Pharmacol 45:1101–1111
Volpini R, Costanzi S, Lambertucci C, Taffi S, Vittori S, Klotz KN, Cristalli G (2002) N(6)-alkyl-2-alkynyl derivatives of adenosine as potent and selective agonists at the human adenosine A(3) receptor and a starting point for searching A(2B) ligands. J Med Chem 45:3271–3279
Volpini R, Dal Ben D, Lambertucci C, Taffi S, Vittori S, Klotz KN, Cristalli G (2007) N6-methoxy-2-alkynyladenosine derivatives as highly potent and selective ligands at the human A3 adenosine receptor. J Med Chem 50:1222–1230
von Arnim CA, Timmler M, Ludolph AC, Riepe MW (2000) Adenosine receptor up-regulation: initiated upon preconditioning but not upheld. Neuroreport 11:1223–1226
von Arnim CA, Spoelgen R, Peltan ID, Deng M, Courchesne S, Koker M, Matsui T, Kowa H, Lichtenthaler SF, Irizarry MC, Hyman BT (2006) GGA1 acts as a spatial switch altering amyloid precursor protein trafficking and processing. J Neurosci 26:9913–9922
von Lubitz DK (1999) Adenosine and cerebral ischemia: therapeutic future or death of a brave concept? Eur J Pharmacol 371:85–102
von Lubitz DK, Marangos PJ (1990) Cerebral ischemia in gerbils: postischemic administration of cyclohexyl adenosine and 8-sulfophenyl-theophylline. J Mol Neurosci 2:53–59
von Lubitz DK, Dambrosia JM, Kempski O, Redmond DJ (1988) Cyclohexyl adenosine protects against neuronal death following ischemia in the CA1 region of gerbil hippocampus. Stroke 19:1133–1139
von Lubitz DK, Lin RC, Melman N, Ji XD, Carter MF, Jacobson KA (1994a) Chronic administration of selective adenosine A1 receptor agonist or antagonist in cerebral ischemia. Eur J Pharmacol 256:161–167
von Lubitz DK, Lin RC, Popik P, Carter MF, Jacobson KA (1994b) Adenosine A3 receptor stimulation and cerebral ischemia. Eur J Pharmacol 263:59–67
von Lubitz DK, Lin RC, Jacobson KA (1995) Cerebral ischemia in gerbils: effects of acute and chronic treatment with adenosine A2A receptor agonist and antagonist. Eur J Pharmacol 287:295–302
von Lubitz DK, Lin RC, Paul IA, Beenhakker M, Boyd M, Bischofberger N, Jacobson KA (1996) Postischemic administration of adenosine amine congener (ADAC): analysis of recovery in gerbils. Eur J Pharmacol 1996:171–179
von Lubitz DK, Ye W, McClellan J, Lin RC (1999) Stimulation of adenosine A3 receptors in cerebral ischemia. Neuronal death, recovery, or both? Ann N Y Acad Sci 890:93–106
von Lubitz DK, Simpson KL, Lin RC (2001) Right thing at a wrong time? Adenosine A3 receptors and cerebroprotection in stroke. Ann N Y Acad Sci 939:85–96
Wei CJ, Li W, Chen JF (2011) Normal and abnormal functions of adenosine receptors in the central nervous system revealed by genetic knockout studies. Biochim Biophys Acta 1808:1358–1379
Wentz CT, Magavi SS (2009) Caffeine alters proliferation of neuronal precursors in the adult hippocampus. Neuropharmacology 56:994–1000
Weyler S, Fülle F, Diekmann M, Schumacher B, Hinz S, Klotz KN, Müller CE (2006) Improving potency, selectivity, and water solubility of adenosine A1 receptor antagonists: xanthines modified at position 3 and related pyrimido[1,2,3-cd]purinediones. Chem Med Chem 1(8):891–902

White PJ, Rose'Meyer RB, Hope W (1996) Functional characterization of adenosine receptors in the nucleus tractus solitarius mediating hypotensive responses in the rat. Br J Pharmacol 117:305–308

Wu DC, Ye W, Che XM, Yang GY (2000) Activation of mitogen-activated protein kinases after permanent cerebral artery occlusion in mouse brain. J Cereb Blood Flow Metab 20: 1320–1330

Xi J, McIntosh R, Shen X, Lee S, Chanoit G, Criswell H, Zvara DA, Xu Z (2009) Adenosine A2A and A2B receptors work in concert to induce a strong protection against reperfusion injury in rat hearts. J Mol Cell Cardiol 47(5):684–690

Yao Y, Sei Y, Abbracchio MP, Jiang JL, Kim YC, Jacobson KA (1997) Adenosine A3 receptor agonists protect HL-60 and U-937 cells from apoptosis induced by A3 antagonists. Biochem Biophys Res Commun 232:317–322

Yoshida M, Nakakimura K, Cui YJ, Matsumoto M, Sakabe T (2004) Adenosine A(1) receptor antagonist and mitochondrial ATP-sensitive potassium channel blocker attenuate the tolerance to focal cerebral ischemia in rats. J Cereb Blood Flow Metab 24:771–779

Yu L, Frith MC, Suzuki Y, Peterfreund RA, Gearan T, Sugano S, Schwarzschild MA, Weng Z, Fink JS, Chen JF (2004a) Characterization of genomic organization of the adenosine A(2A) receptor gene by molecular and bioinformatics analyses. Brain Res 1000:156–173

Yu L, Huang Z, Mariani J, Wang Y, Moskowitz M, Chen JF (2004b) Selective inactivation or reconstitution of adenosine A2A receptors in bone marrow cells reveals their significant contribution to the development of ischemic brain injury. Nat Med 10:1081–1087

Zetterstrom T, Fillenz M (1990) Adenosine agonists can both inhibit and enhance in vivo striatal dopamine release. Eur J Pharmacol 180:137–143

Zhang D, Xiong W, Albensi BC, Parkinson FE (2011) Expression of human equilibrative nucleoside transporter 1 in mouse neurons regulates adenosine levels in physiological and hypoxic-ischemic conditions. J Neurochem 118:4–11

Zhao Z, Francis C, Ravid K (1999) Characterization of the mouse A3 adenosine receptor gene: exon/intron organization and promoter activity. Genomics 57:152–155

Zhou JG, Meno JR, Hsu SS, Winn HR (1994) Effects of theophylline and cyclohexyladenosine on brain injury following normo- and hyperglycemic ischemia: a histopathologic study in the rat. J Cereb Blood Flow Metab 14:166–173

Zhou AM, Li WB, Li QJ, Liu HQ, Feng RF, Zhao HG (2004) A short cerebral ischemic preconditioning up-regulates adenosine receptors in the hippocampal CA1 region of rats. Neurosci Res 48:397–404

Chapter 15
The Many Roles of Adenosine in Traumatic Brain Injury

Patrick M. Kochanek, Jonathan D. Verrier, Amy K. Wagner, and Edwin K. Jackson

Abstract Many secondary mechanisms regulate the evolution of damage and repair after traumatic brain injury (TBI). Bench-to-bedside studies from our team have shown that adenosine plays a key role in several secondary injury pathways in the brain, namely, excitotoxicity, neuroinflammation, and cerebral blood flow (CBF) dysregulation. Studies of adenosine A_1-receptor (A_1R) knockout (KO) mice strongly support a key role for adenosine-dependent effects at this receptor in attenuating posttraumatic excitotoxicty. A_1R KO mice develop lethal status epilepticus after experimental TBI. Evidence also supports a role of the A_1R in human TBI since A_1R gene variants are associated with posttraumatic seizures in patients with TBI. A_1R also modulates the neuroinflammatory response to TBI. A_1R KO mice exhibit enhanced microglial proliferation early after TBI. Adenosine acting at A_{2A} and/or A_{2B} receptors also modulates posttraumatic CBF. Local injection of either the nonselective AR agonist 2-chloroadenosine or the $A_{2A}R$ agonist CGS21680 markedly increases CBF in naïve or brain-injured rats. However, whether the effects of adenosine on CBF after TBI are beneficial or detrimental is controversial. Mitigation of ischemia by adenosine produced from breakdown of ATP after TBI has been suggested, and consistent with this hypothesis, we reported increases in brain interstitial adenosine levels in patients with severe TBI during episodes of ischemia. However, in studies of CBF in patients after severe TBI, we also noted an association between cerebrospinal fluid (CSF) adenosine levels and uncoupling of CBF and metabolism (cerebral

P.M. Kochanek, M.D., M.C.C.M. (✉)
Department of Critical Care Medicine, University of Pittsburgh School of Medicine, Pittsburgh, PA 15260, USA
e-mail: kochanekpm@ccm.upmc.edu

J.D. Verrier, Ph.D. • E.K. Jackson, Ph.D.
Department of Pharmacology and Chemical Biology, University of Pittsburgh School of Medicine, Pittsburgh, PA 15217, USA

A.K. Wagner, M.D.
Department of Physical Medicine & Rehabilitation, University of Pittsburgh School of Medicine, Pittsburgh, PA 15213, USA

S. Masino and D. Boison (eds.), *Adenosine: A Key Link between Metabolism and Brain Activity*, DOI 10.1007/978-1-4614-3903-5_15,

metabolic rate for oxygen [$CMRO_2$]). Uncoupling of CBF and $CMRO_2$ is associated with intracranial hypertension. Thus, adenosine may represent a two-edged sword after TBI. Additional evidence supporting a protective role of adenosine in TBI comes from clinical studies showing a powerful association between CSF caffeine levels and favorable outcome after TBI—given the interplay between caffeine and adenosine in neuroprotection. Finally, our work also suggests that the newly discovered 2,3-cyclic AMP pathway represents an important component of the adenosine response to TBI.

Keywords Head injury • Neuroinflammation • Excitotoxicity • Microglia • Seizure • Controlled cortical impact • Cerebral blood flow • Caffeine

15.1 Introduction

It has become clear that traumatic brain injury (TBI) is an enormous public health problem worldwide. In the United States, data from the Centers for Disease Control and Prevention indicate that greater than 1.7 million people suffer a TBI annually, 275,000 are hospitalized, and over 50,000 die (http://www.CDC.gov, 2011). TBI has also taken on special importance in the setting of combat casualty care within the US Department of Defense. For example, the RAND Corporation estimates that more than 400,000 US soldiers have suffered TBI in Operation Iraqi Freedom and Operation Enduring Freedom (http://www.rand.org, 2011)—largely from blast injury. The long-term consequences of repeated exposure to even a mild TBI, such as that encountered in sport concussion, may be linked to a broad spectrum of neurodegenerative diseases, like dementia, Parkinson's disease, and amyotrophic lateral sclerosis, among others (DeKosky et al. 2010; McKee et al. 2010). Currently, patients with severe TBI are treated with guideline-based therapies; however, the available treatments are limited largely to agents targeting control of raised intracranial pressure. Development of novel neuroprotective therapies is badly needed. Finally, there is no current therapy for cases of mild TBI—which represent the most common presentation of this disease. It is now well recognized that repeated mild TBI in the setting of concussion could be very detrimental, and there is considerable effort ongoing to develop new diagnostic and therapeutic modalities for this specific condition.

15.2 The Evolution of Secondary Damage and Repair After TBI

TBI is a highly complex disease. Indeed, some have questioned whether or not it represents a single disease entity (Saatman et al. 2008). For example, severe TBI can present with a variety of injury patterns such as contusion, epidural, subdural, subarachnoid, or parenchymal hemorrhage, diffuse swelling, diffuse axonal injury,

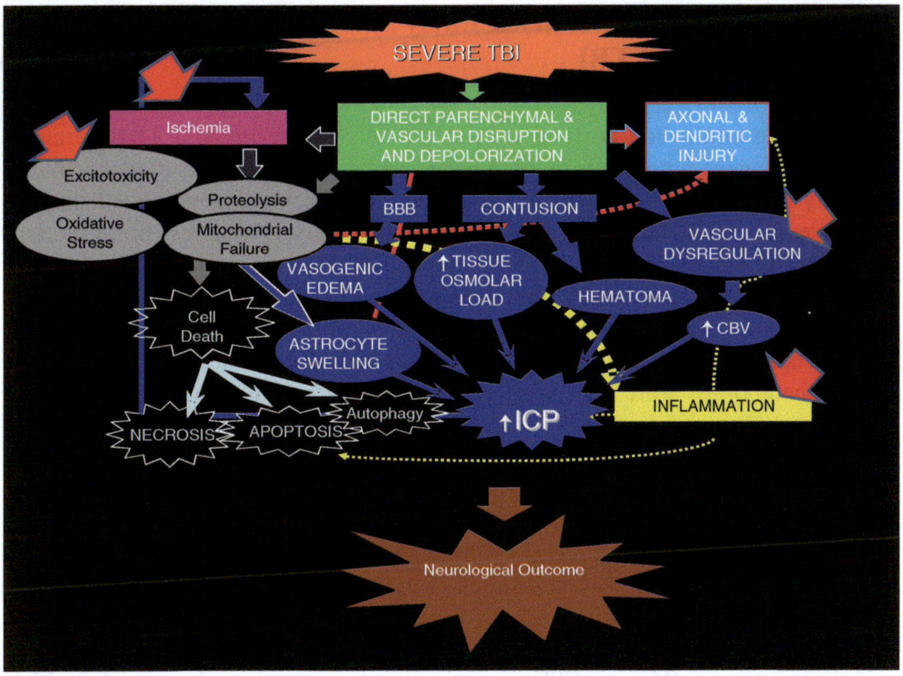

Fig. 15.1 Schematic of the putative mechanisms of secondary injury and repair after severe traumatic brain injury. Ischemia, neuronal death, brain swelling, axonal injury, and neuroinflammation are believed to play important roles after the primary injury that disrupts the parenchyma. Adenosine has effects on multiple aspects of this cascade. The most important are identified by the red arrows and include inhibition of excitotoxicity via effects at A_1 receptors, attenuation of ischemia and possible vascular dysregulation via effects at A_{2A} and A_{2B} receptors, and complex effects on neuroinflammation via effects at both A_1 and A_{2A} receptors. Please see text for additional details. *TBI* traumatic brain injury, *ICP* intracranial pressure, *CBV* cerebral blood volume, *BBB* blood–brain barrier

or combinations of these presentations. Similarly, within each of these presentations, myriad mechanisms are involved in the secondary events that follow the primary injury. And these secondary mechanisms can be detrimental, beneficial, or both. Emerging data suggest that adenosine influences secondary injury and repair after TBI. Figure 15.1 provides an overview of secondary injury and repair pathways after severe TBI that will serve as the basis for a discussion of the role of adenosine in this condition (Kochanek et al. 2000). In general, secondary processes after TBI can be categorized into those effecting excitation, neuronal death cascades, cerebral blood flow (CBF) dysregulation (resulting in brain swelling and intracranial hypertension), axonal injury, inflammation, and regeneration. Our group has been studying the role of adenosine in TBI for the past decade using a bench-to-bedside approach, and we discuss the findings of these and studies by other laboratories that show substantial production of adenosine early after TBI and suggest important roles of adenosine in modulating three important secondary injury pathways, namely, excitotoxicity, neuroinflammation, and CBF dysregulation.

15.3 Adenosine Is Increased Early After Experimental and Clinical TBI

We used the controlled cortical impact model of experimental TBI in rats to monitor changes in the levels of adenosine early after injury. This model provides a highly reproducible cerebral contusion along with more generalized damage (Dixon et al. 1991; Hall et al. 2005). Using microdialysis to assess brain interstitial levels of adenosine in the cerebral cortex, we reported marked (over 60-fold) increases in adenosine in the initial hour after controlled cortical impact, with peak levels observed in the initial 20 min (Bell et al. 1998). Increases in adenosine were directly correlated with injury severity. Marked increases in levels of other purine metabolites were also seen, and a possible role for energy failure with resultant ATP breakdown was suggested as a potential source of adenosine. In three separate studies, we also examined adenosine levels in comatose patients with severe TBI (Clark et al. 1997; Bell et al. 2001; Robertson et al. 2001). In two studies, we measured cerebrospinal fluid (CSF) levels of adenosine, and in a third study we assessed brain interstitial levels of adenosine using intracerebral microdialysis in these patients during episodes of metabolic crisis. Using CSF analysis, we demonstrated increases in levels of adenosine in both children and adults. In these studies, adenosine levels were directly associated with severity of injury and poor outcome. Of note, CSF adenosine levels were approximately 4–8 nM in controls, on average approximately twofold higher after injury, and as high as 100–800 nM in some patients (Clark et al. 1997). In addition, CSF adenosine levels were increased for days after injury.

In contrast, in rats after controlled cortical impact, brain interstitial levels of adenosine were much higher than CSF, specifically over 10,000 nM after severe injury, but these increases were much more transient—lasting less than 1 h. It is likely that local metabolism of adenosine by adenosine deaminase and adenosine kinase in the interstitial space tightly regulates adenosine levels, while adenosine that enters the CSF compartment may be less vulnerable to breakdown. Consistent with this hypothesis, we studied brain interstitial adenosine concentration in six patients using microdialysis (Bell et al. 2001). These patients demonstrated episodes of metabolic crisis defined by periods of jugular venous oxygen desaturation to levels < 50 %—a level that is known to be associated with brain ischemia and increased mortality rate in clinical TBI (Gopinath et al. 1994). During these episodes, brain interstitial adenosine levels increased from baseline values of ~200 nM to levels as high as 1,000 nM. Adenosine has often been suggested to represent a retaliatory metabolite that is an endogenous neuroprotectant (Miller and Hsu 1992). Thus, the association between high levels of adenosine and both severity of injury and poor outcome has generally been attributed to a putative endogenous neuroprotectant response to TBI. Specific studies of adenosine effects on various secondary injury mechanisms in TBI are discussed later in this chapter.

15.4 The Role of Adenosine in Modulating Excitotoxicity and Seizures After TBI

The most dramatic effect of manipulating the adenosine pathway in experimental TBI was observed when A_1-receptor (A_1R) KO mice were subjected to controlled cortical impact TBI (Kochanek et al. 2006). In this commonly used model of experimental TBI, overt seizures are observed in ~33 % of wild-type mice, but are very transient—lasting only seconds in duration, and status epilepticus has not been observed. Post-injury mortality rates are near zero at injury levels routinely used in this model in both rats and mice. In contrast, 80 % of A_1R KO mice demonstrated seizure activity and 50 % exhibited lethal sustained status epilepticus (Fig. 15.2). Although both γ-amino butyric acid and adenosine play important roles in maintaining inhibitory tone in brain, it has been suggested that adenosine is the principal agent responsible for seizure arrest—terminating brief episodes of seizure activity and inhibiting progression to status epilepticus (Franklin et al. 1989; Avsar and Empson 2004). Our data in the A_1R KO mice are consistent with this concept. In addition, in the aforementioned clinical study of CSF levels of adenosine in children (Robertson et al. 2001), spikes in adenosine were temporally associated with increases in CSF levels of the excitotoxic neurotransmitter glutamate. Further supporting the role of adenosine as an important endogenous anti-excitotoxic and anticonvulsant

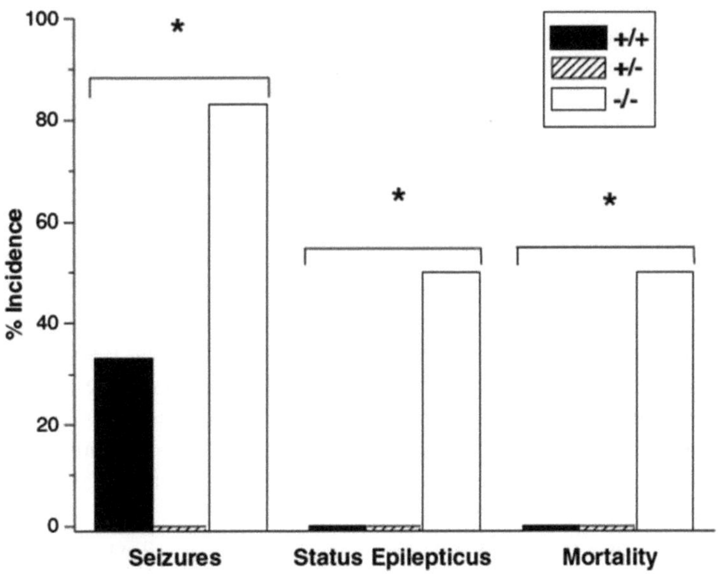

Fig. 15.2 Incidence of seizures, status epilepticus, and mortality in male adenosine A_1 receptor +/+, +/−, and −/− mice in the initial 2 h after traumatic brain injury induced by controlled cortical impact. Adenosine A_1 receptor −/− mice showed an increased incidence of seizures, status epilepticus, and mortality after injury versus both +/+ and +/− genotypes, *$P<0.05$. Reprinted from Kochanek et al. (2006) with permission

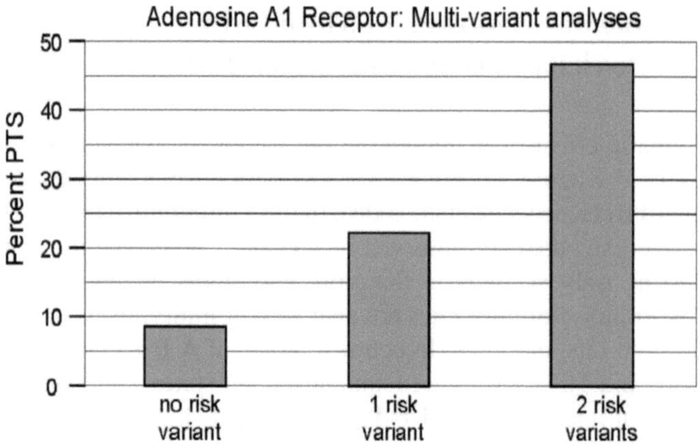

Fig. 15.3 Multiple tagging single-nucleotide polymorphisms within the adenosine A_1 receptor gene are associated with posttraumatic seizures in severe traumatic brain injury. Late seizure rates (seizures occurring in the first week after traumatic brain injury) increase substantially as the number of risk variants increases from zero to two. Please see text for details

molecule after TBI is a recent work by Wagner et al. (2010). In that study five single-nucleotide polymorphisms were examined in over 200 patients with severe TBI. A_1R polymorphisms were associated with an increased incidence of both early- and delayed-onset posttraumatic seizures, and multiple risk genotype analysis showed that subjects with two high-risk genotypes had a 46.7 % chance of developing late posttraumatic seizures (Fig. 15.3). The early posttraumatic seizures could be linked to excitotoxicity and the late posttraumatic seizures could be related to interactions between adenosine and glial response to injury, as discussed below. Recent work by Vespa et al. (2007) has demonstrated a remarkably high incidence of early posttraumatic subclinical seizure activity in humans after severe TBI assessed with continuous electroencephalographic monitoring. This suggests that augmentation of adenosine might have therapeutic potential. Similarly, our data are consistent with the work of Gouder et al. (2004) and Fedele et al. (2005) who reported that over-expression of adenosine kinase by astrocytes in the glial scar after CNS injury reduces adenosine levels and aggravates seizures. Because of the low incidence of late posttraumatic seizures (as low as 3 % in some series) and the potential long-term side effects of anticonvulsants, the routine long-term use of anticonvulsants in patients with TBI is not recommended (Brain Trauma Foundation et al. 2007). However, given our findings, it is possible that A_1R-genotype-guided anticonvulsant therapy could be useful in patients with TBI.

In addition to beneficial effects on excitoxicity and seizures after TBI, A_1R receptor manipulation after experimental TBI suggests a neuroprotective role of this receptor. Local injection of the A_1R antagonist DPCPX into the hippocampus after controlled cortical impact in rats exacerbated both motor deficits and hippocampal neuronal death, while treatment with the selective A_1R agonist CCPA attenuated hippocampal neuronal death (Varma et al. 2002). It is unclear what the contribution of effects on excitotoxicity and seizures was to the observed effects on behavior and neuropathology. Nevertheless, taken together, these studies suggest that augmentation of A_1R signaling may represent a therapeutic target after TBI. What remains unclear is how to best augment the effects of A_1R early after injury without inducing adverse cardiovascular effects or potential deleterious consequences of sustained A_1R activation on plasticity (Lu et al. 2010). A related issue is what duration of A_1R augmentation might be optimal to maximize benefit without causing the mentioned undesired effects.

15.5 Adenosine and Neuroinflammation After TBI

Adenosine, in addition to acting as a neuromodulator, has long been known to serve as an immunomodulator, and in particular regulates neuroinflammation after various central nervous system insults (Yu et al. 2004; Chen and Pedata 2008; Dai and Zhou 2011). Recently, the A_1R has been shown to have an important role in blunting the endogenous microglial response in several neuroinflammatory conditions. For example, Tsutsui et al. (2004) reported that A_1R KO mice have a markedly enhanced pro-inflammatory response in experimental allergic encephalitis—including exacerbation of demyelination and marked microglial proliferation. Similarly, enhanced microglial proliferation was reported in A_1R KO mice in a model of experimental glioblastoma (Synowitz et al. 2006). Recently, we reported that A_1R KO mice subjected to experimental TBI from controlled cortical impact exhibited a global exacerbation of microglial proliferation at 7 days after the insult as assessed with Iba-1 immunohistochemistry, a microglial marker (Haselkorn et al. 2010). Corroborating this in vivo finding, we also reported that treatment of immortalized microglial cells (BV-2) with the selective A_1R agonist CCPA or selective A_1R antagonist DPCPX significantly decreased or increased proliferation, respectively, as assessed by thymidine incorporation (Haselkorn et al. 2010). Although a contribution of activated microglia to secondary injury has been suggested in many studies across a variety of models of experimental brain injury, some reports suggest that microglia influence the regenerative response after brain injury (Lalancette-Hébert et al. 2007). It is, thus, essential that the effects on behavioral and neuropathological outcomes be carefully examined after any intervention targeting neuroinflammation in TBI—including manipulation of adenosine effects. Finally, it is noteworthy that in BV-2 cells, A_{2A}, A_{2B}, and A_3 selective AR antagonists all increased microglial proliferation. In addition, mRNA expression of A_{2B} receptors was greatest in this system, suggesting a possible important and unexplored role of the A_{2B} receptor on the endogenous microglial response to TBI or possibly other central nervous system diseases.

In contrast to the potential anti-inflammatory role of A_1R activation after experimental TBI, Li et al. (2009) suggested a pro-inflammatory role of $A_{2A}R$ activation after experimental TBI, reporting that genetic inactivation of $A_{2A}R$ was associated with reduced neuroinflammation in mice. However, a complex pattern was observed, with exacerbation of increases in TNFα and IL-1β at 12 h after injury but inhibition of these two cytokines at 24 h after injury in $A_{2A}R$ KO versus wild-type controls. In a subsequent study by this group, inactivation of $A_{2A}R$ in both bone marrow-derived and non-bone marrow-derived cells attenuated inflammatory cytokines and secondary damage after experimental TBI in mice—likely suggesting roles of both circulating macrophages that enter the injured brain and endogenous microglia or other resident cell types in brain (Dai et al. 2010).

Robust neutrophil accumulation has been reported in numerous studies early after experimental TBI (Clark et al. 1994) and also macrophage accumulation in injured brain has been demonstrated using MPIO labeling, which specifically identifies infiltration of circulating macrophages (Foley et al. 2009). Macrophage accumulation in brain has also been reported in the initial 72 h after severe TBI in humans (Sinz et al. 1998). Thus, the impact of circulating and endogenous inflammatory processes could be important in influencing the secondary injury response after TBI and may be influenced by the adenosine pathway. In any case, it appears that the effects of adenosine on the neuroinflammatory response to TBI are complex, dependent on the receptor that is activated, and also dependent on the temporal profile studied.

Given the ability to manipulate adenosine receptors pharmacologically, further study of the role of adenosine in the neuroinflammatory response is needed. The effect of adenosine on neuroinflammation after TBI could have substantial importance given the fact that TBI is now being linked to many neuroinflammatory and neurodegenerative diseases such as amytrophic lateral sclerosis, Parkinson's disease, and dementia (DeKosky et al. 2010; McKee et al. 2010). Similarly, microglial activation and proliferation appear to be synergistically produced in the setting of repetitive mild TBI—at injury levels where a single exposure induces no injury or neuroinflammatory response. In addition, several laboratories have reported remarkable chronic microglial activation after experimental TBI suggesting the possibility of a sustained therapeutic target for adenosine-related pharmacological approaches. How genetic differences with regard to adenosine receptors, as shown for A_1R, impact on these long-term neuroinflammatory effects deserves additional investigation. Given the aforementioned link between genetic polymorphisms related to the A_1R and posttraumatic seizures, it is likely that similar genetic predispositions play a role in adenosine's regulation of neuroinflammation after brain injury.

15.6 Adenosine and Cerebral Blood Flow Dysregulation After TBI

Early after TBI, CBF is reduced. Given that the early posttraumatic period is complicated by an excitotoxic response, the combination of low flow and increased metabolic demands has been suggested to contribute to secondary damage (reviewed in Kochanek et al. 2000). Early low CBF has been correlated with poor outcome after TBI in humans (Bouma et al. 1992). However, after several hours, a state of metabolic depression ensues and CBF may, in some patients, increase to levels that represent either relative or absolute hyperemia (Hovda et al. 1995). Increases in CBF during the delayed post-injury period could contribute to an increase in cerebral blood volume and potentially complicate intracranial hypertension. Thus either low or high CBF could be problematic in patients with severe TBI.

Traditionally, adenosine has been touted as an important retaliatory metabolite that defends the brain against ischemia—with adenosine production from ATP breakdown during states of low CBF mediating this response (Miller and Hsu 1992; Laghi Pasini et al. 2000). Effects at both A_{2A} and A_{2B} receptors have been suggested in mediating the vasodilatory effects of adenosine in the cerebral circulation (Ngai et al. 2001; Kusano et al. 2010). We carried out studies in both experimental and clinical TBI that support this hypothesis. In the controlled cortical impact model in rats, we observed marked CBF reductions early after injury in the injured hemisphere (Hendrich et al. 1999). These reductions could be readily attenuated by local injection of either the nonselective AR agonist 2-chloroadenosine or the A_{2A} selective agonist CGS 21680 (Kochanek et al. 2005). Intra-hippocampal injection of CGS 21680 at doses of only 6 nmol produced marked and sustained global increases in CBF (Fig. 15.4). However, neither agent altered cerebral glucose utilization after injury—suggesting improvements in flow without pathologically increasing metabolic demands.

In the aforementioned study in patients undergoing microdialysis to assess brain interstitial levels of adenosine during periods of ischemia defined as a jugular venous desaturation to <50 %, we demonstrated robust and immediate increases in adenosine (Bell et al. 2001). The jugular venous desaturation episodes were associated with events such as systemic hypotension or intracranial hypertension in these patients. A single episode of jugular venous desaturation to this level was previously demonstrated to be associated with a doubling of mortality rate in patients with severe TBI—supporting the highly deleterious role of secondary ischemic insults in these patients (Gopinath et al. 1994). Although we were not able to assess simultaneously CBF at the bedside in these critically ill patients with severe TBI, our data support the aforementioned retaliatory role of adenosine that has been demonstrated in numerous experimental studies of cerebral ischemia (Miller and Hsu 1992). Therapeutically, given the highly deleterious consequences of early posttraumatic ischemia after injury, augmentation of adenosine early after severe TBI deserves further study in experimental models of TBI. This approach might have greater potential impact in certain forms of severe TBI such as cerebral contusion, where CBF decrements are greatest.

In contrast to the potential beneficial effects of adenosine on CBF early after TBI, delayed increases in adenosine may contribute to CBF dysregulation after TBI and mediate luxury perfusion and possibly exacerbate intracranial hypertension by increasing cerebral blood volume. In this regard, Clark et al. (1997) studied 13 adults with severe TBI using radioactive Xenon 133 to assess CBF, jugular venous bulb catheters to assess cerebral metabolic rate for oxygen ($CMRO_2$), and cerebrospinal fluid analysis of adenosine concentration. CBF was reduced between 6 and 18 h after injury. But thereafter, CBF and adenosine levels increased, and adenosine levels were associated with uncoupling of CBF and $CMRO_2$. Thus, adenosine appears to have potent effects on CBF after TBI.

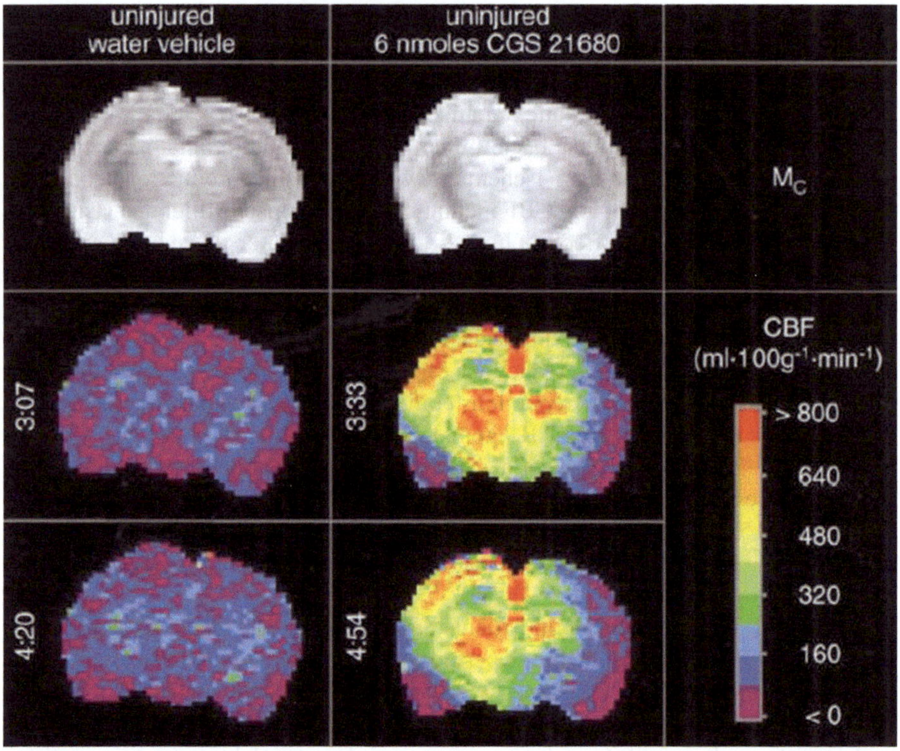

Fig. 15.4 Cerebral blood flow maps (as assessed by arterial spin-labeled magnetic resonance imaging) demonstrating a dramatic increase in flow between 3 and 5 h after intra-hippocampal injection of the adenosine A_{2A} receptor agonist GCS21680. Left three vertical panels show the anatomical image (*top panel*) and serial cerebral blood flow maps after injection of vehicle. Right three vertical panels show the anatomical image (*top panel*) and serial cerebral blood flow maps after injection of 6 nanomoles of GCS21680. Reprinted from Kochanek et al. (2005) with permission

15.7 Caffeine and Outcome in Severe TBI

Caffeine is the most widely consumed psychoactive drug and a weak adenosine receptor antagonist. We explored its role in clinical TBI by serially measuring caffeine levels in CSF of 30 patients with severe TBI (Sachse et al. 2008). Caffeine was detected in 24 of the 30 patients, and marked increases in CSF levels of caffeine were observed in some patients—as high as 2,000 ng/mL (Fig. 15.5). Surprisingly, CSF caffeine levels on admission were highly associated with favorable neurological outcome at 6 months. One possibility is that chronic caffeine consumption leads to A_1R up-regulation and ultimately enhanced neuroprotection and anti-excitotoxic effects from the large amount of adenosine that is produced early after TBI. Recently Li et al. (2008) studied the effect of caffeine administration to mice after experimental TBI and reported that chronic but not acute caffeine treatment was neuroprotective. Specifically, chronic

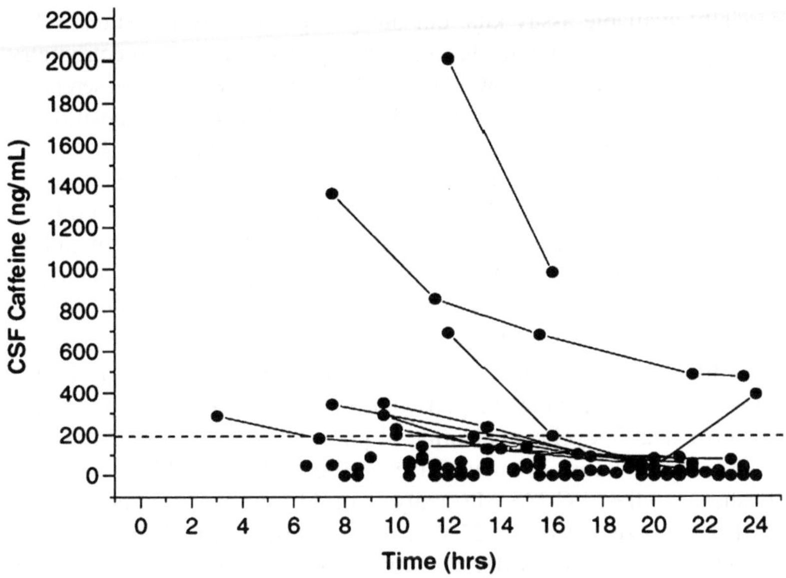

Fig. 15.5 Time course of cerebrospinal fluid caffeine levels in patients after severe traumatic brain injury. Marked increases were seen in many patients and caffeine level was directly correlated with favorable outcome. The *dotted line* represents a putative caffeine level of 194 ng/mL (1 micromolar) that would be expected to produce clinically significant effects. Reprinted from Sachse et al. (2008) with permission

caffeine administration was correlated with up-regulation of A_1R and attenuated glutamate release, corroborating our clinical findings. A reduction in inflammatory cytokine production in the mice treated with chronic caffeine was also observed, suggesting a possible link to modulation of the microglial response by A_1R that was previously described in this chapter. Related to our findings, Saharan and Nantwi (2006) reported that administration of theophylline produced enhanced affinity and an apparent unmasking of a second binding site in the profile of A_1R in the cervical spinal cord. Further study is needed, but the possible utility of caffeine as a neuroprotectant in TBI deserves additional exploration. This approach could have special relevance in the setting of blast TBI in combat casualty care or sports concussion where pharmacological pretreatment in high-risk settings is a potential option.

15.8 The Possible Role of the 2′,3′-cAMP-Adenosine Pathway in the Adenosine Response to TBI

The term "cAMP" is used universally to refer to 3′,5′-cyclic adenosine monophosphate (3′,5′-cAMP), the famous "second messenger" discovered by Dr. Earl Sutherland. The vast majority of investigators measure 3′,5′-cAMP using various commercially available assay kits, but our group routinely measures 3′,5′-cAMP using high-performance liquid chromatography-tandem mass spectrometry (LC-MS/MS). Due to the analytical power and flexibility of LC-MS/MS, we serendipitously discovered while measuring 3′,5′-cAMP that tissues also produce a positional isomer of 3′,5′-cAMP, namely, 2′,3′-cyclic adenosine monophosphate (2′,3′-cAMP) (Ren et al. 2009). Follow-up experiments showed that 2′,3′-cAMP arises from degradation of mRNA (Jackson et al. 2009) and that tissues (Jackson et al. 2009) and cells (Jackson et al. 2010, 2011a, b, c) metabolize 2′,3′-cAMP to 2′-AMP and 3′-AMP, with further conversion of 2′-AMP and 3′-AMP to adenosine (a mechanism we call the 2′,3′-cAMP-adenosine pathway). Because tissue injury increases mRNA breakdown (Akahane et al. 2001; Catts et al. 2005; Chevyreva et al. 2008; Del Prete et al. 2002), we hypothesize that the 2′,3′-cAMP-adenosine pathway may be activated during tissue trauma. This may be particularly important in the brain because of the abundance of a 2′,3′-cyclic nucleotide-3′-phosphodiesterase (CNPase) (Sprinkle 1989), an enzyme know to metabolize 2′,3′-cAMP to 2′-AMP. In preliminary (unpublished) studies, LC-MS/MS analysis of 44 samples of cerebral spinal fluid from TBI patients revealed that 2′,3′-cAMP is present in CSF from patients with TBI. In addition to 2′,3′-cAMP, the levels of 2′-AMP, adenosine, and inosine (metabolite of adenosine) were also measured in these same samples. Importantly, there was a large and significant correlation between 2′,3′-cAMP and 2′-AMP, 2′-AMP and adenosine, and 2′-AMP and inosine in CSF from TBI patients. Moreover, in mouse primary microglia and astrocytes in cell culture, we observed that 2′,3′-cAMP is readily metabolized to 2′-AMP and 3′-AMP and that 2′-AMP and 3′-AMP are converted to adenosine (Verrier et al. 2011). The existence of the 2′,3′-cAMP pathway in both microglia and astrocytes potentially reveals a new mechanism of adenosine production in the brain. It is tempting to hypothesize that localized AMP metabolism to adenosine may be a previously unappreciated role for infiltrating and proliferating glial cells post injury. These discoveries suggest a need to explore the role of the 2′,3′-cAMP-adenosine pathway in TBI pathophysiology.

15.9 Conclusions

TBI has emerged as a leading public health problem, as we have become aware of the acute and long-term consequences of the range of insults, from mild repetitive injury to severe. Evidence has been presented demonstrating that adenosine levels markedly increase early after injury and sustained increases are seen in some patients in CSF. Adenosine effects at A_1R confer potent anti-excitotoxic, anticonvulsant effects, and some degree of neuroprotection in experimental BI. Moreover, genetic studies suggest a clinical correlation of these effects both in the acute and delayed period. Adenosine also serves as an immunomodulator in brain after TBI with anti-inflammatory actions mediated by A_1R and pro-inflammatory effects mediated by $A_{2A}Rs$—although A_{2B} and A_3 receptors may also play a role. Modulation of the adenosine pathway can dramatically influence CBF after experimental TBI, likely via

A_{2A} and A_{2B} receptor activation, but the effects of adenosine on CBF in clinical TBI are complex—and could mediate both a beneficial response to ischemia and delayed uncoupling of flow and metabolism. The source of adenosine after TBI is being reevaluated in light of recognition of the emerging role of the 2′,3′-cAMP-adenosine pathway, and related to this pathway, novel therapeutic targets may be identified, such as 2′-AMP. Several factors have been viewed as potential limitations to the development of drugs enhancing effects of the endogenous neuroprotectant adenosine after brain injury such as the fact that adenosine levels are already markedly elevated after injury and that systemic toxicity can be seen with parenteral administration of adenosine-augmenting therapies. However, given the expanding importance of TBI and the increasing recognition of the influence of adenosine and related metabolites on the secondary injury process such as the CNPase pathway and posttraumatic seizures, the potential for adenosine-related therapies deserves additional exploration.

Acknowledgement We thank the National Institutes of Neurological Disorders and Stroke NS070003 (PMK, EKJ) and the American Heart Association (JDV) for support.

References

Akahane M, Ono H, Ohgushi H, Tamai S (2001) Viability of ischemia/reperfused muscles in rat: a new evaluation method by RNA degradation. J Orthop Res 19:559–564

Avsar E, Empson RM (2004) Adenosine acting via A1 receptors, controls the transition to status epilepticus-like behaviour in an in vitro model of epilepsy. Neuropharmacology 47:427–437

Bell MJ, Kochanek PM, Carcillo JA, Mi Z, Schiding JK, Wisniewski SR, Clark RS, Dixon CE, Marion DW, Jackson E (1998) Interstitial adenosine, inosine, and hypoxanthine are increased after experimental traumatic brain injury in the rat. J Neurotrauma 15:163–170

Bell MJ, Robertson CS, Kochanek PM, Goodman JC, Gopinath SP, Carcillo JA, Clark RS, Marion DW, Mi Z, Jackson EK (2001) Interstitial brain adenosine and xanthine increase during jugular venous oxygen desaturations in humans after traumatic brain injury. Crit Care Med 29:399–404

Bouma GJ, Muizelaar JP, Stringer WA, Choi SC, Fatouros P, Young HF (1992) Ultra-early evaluation of regional cerebral blood flow in severely head-injured patients using xenon-enhanced computerized tomography. J Neurosurg 77:360–368

Brain Trauma Foundation; American Association of Neurological Surgeons; Congress of Neurological Surgeons; Joint Section on Neurotrauma and Critical Care, AANS/CNS, Bratton SL, Chestnut RM, Ghajar J, McConnell Hammond FF, Harris OA, Hartl R, Manley GT, Nemecek A, Newell DW, Rosenthal G, Schouten J, Shutter L, Timmons SD, Ullman JS, Videtta W, Wilberger JE, Wright DW (2007) Guidelines for the management of severe traumatic brain injury. XIII. Antiseizure prophylaxis. J Neurotrauma 24 (suppl 1): S83–S86

Catts VS, Catts SV, Fernandez HR, Taylor JM, Coulson EJ, Lutze-Mann LH (2005) A microarray study of post-mortem mRNA degradation in mouse brain tissue. Brain Res 138:164–177

Centers for Disease Control and Prevention. http://www.cdc.gov/TraumaticBrainInjury/statistics.html. Accessed 26 June 2011

Chen JF, Pedata F (2008) Modulation of ischemic brain injury and neuroinflammation by adenosine A2A receptors. Curr Pharm Des 14:1490–1499

Chevyreva I, Faull RLM, Green CR, Nicholson LFB (2008) Assessing RNA quality in postmortem human brain tissue. Exp Mol Pathol 84:71–77

Clark RS, Carcillo JA, Kochanek PM, Obrist WD, Jackson EK, Mi Z, Wisneiwski SR, Bell MJ, Marion DW (1997) Cerebrospinal fluid adenosine concentration and uncoupling of cerebral

blood flow and oxidative metabolism after severe head injury in humans. Neurosurgery 41:1284–1292, discussion 1292–1293
Clark RS, Schiding JK, Kaczorowski SL, Marion DW, Kochanek PM (1994) Neutrophil accumulation after traumatic brain injury in rats: comparison of weight drop and controlled cortical impact models. J Neurotrauma 11:499–506
Dai SS, Li W, An JH, Wang H, Yang N, Chen XY, Zhao Y, Li P, Liu P, Chen JF, Zhou YG (2010) Adenosine A2A receptors in both bone marrow cells and non-bone marrow cells contribute to traumatic brain injury. J Neurochem 113:1536–1544
Dai SS, Zhou YG (2011) Adenosine 2A receptor: a crucial neuromodulator with bidirectional effect in neuroinflammation and brain injury. Rev Neurosci 22:231–239
DeKosky ST, Ikonomovic MD, Gandy S (2010) Traumatic brain injury—football, warfare, and long-term effects. N Engl J Med 363:1293–1296
Del Prete MJ, Robles MS, Guao A, Martinez-A C, Izquierdo M, Garcia-Sanz JA (2002) Degradation of cellular mRNA is a general early apoptosis-induced event. FASEB J 16:2003–2005
Dixon CE, Clifton GL, Lighthall JW, Yaghmai AA, Hayes RL (1991) A controlled cortical impact model of traumatic brain injury in the rat. J Neurosci Methods 39:253–262
Fedele DE, Gouder N, Güttinger M, Gabernet L, Scheurer L, Rülicke T, Crestani F, Boison D (2005) Astrogliosis in epilepsy leads to overexpression of adenosine kinase, resulting in seizure aggravation. Brain 128:2383–2395
Foley LM, Hitchens TK, Ho C, Janesko-Feldman KL, Melick JA, Bayir H, Kochanek PM (2009) Magnetic resonance imaging assessment of macrophage accumulation in mouse brain after experimental traumatic brain injury. J Neurotrauma 26:1509–1519
Franklin PH, Zhang G, Tripp ED, Murray TF (1989) Adenosine A1 receptor activation mediates suppression of (-) bicuculline methiodide-induced seizures in rat prepiriform cortex. J Pharmacol Exp Ther 251:1229–1236
Gopinath SP, Robertson CS, Contant CF, Hayes C, Feldman Z, Narayan RK, Grossman RG (1994) Jugular venous desaturation and outcome after head injury. J Neurol Neurosurg Psychiatry 57:717–723
Gouder N, Scheurer L, Fritschy JM, Boison D (2004) Overexpression of adenosine kinase in epileptic hippocampus contributes to epileptogenesis. J Neurosci 24:692–701
Hall ED, Sullivan PG, Gibson TR, Pavel KM, Thompson BM, Scheff SW (2005) Spatial and temporal characteristics of neurodegeneration after controlled cortical impact in mice: more than a focal brain injury. J Neurotrauma 22:252–265
Haselkorn ML, Shellington DK, Jackson EK, Vagni VA, Janesko-Feldman K, Dubey RK, Gillespie DG, Cheng D, Bell MJ, Jenkins LW, Homanics GE, Schnermann J, Kochanek PM (2010) Adenosine A1 receptor activation as a brake on the microglial response after experimental traumatic brain injury in mice. J Neurotrauma 27:901–910
Hendrich KS, Kochanek PM, Williams DS, Schiding JK, Marion DW, Ho C (1999) Early perfusion after controlled cortical impact in rats: quantification by arterial spin-labeled MRI and the influence of spin-lattice relaxation time heterogeneity. Magn Reson Med 42:673–681
Hovda DA, Lee SM, Smith ML, Von Stuck S, Bergsneider M, Kelly D, Shalmon E, Martin N, Caron M, Mazziotta J et al (1995) The neurochemical and metabolic cascade following brain injury: moving from animal models to man. J Neurotrauma 12:903–906
Jackson EK, Gillespie DG, Dubey RK (2011a) 2′-AMP and 3′-AMP inhibit proliferation of preglomerular vascular smooth muscle cells and glomerular mesangial cells via A_{2B} receptors. J Pharmacol Exp Ther 337:444–450
Jackson EK, Ren J, Cheng D, Mi Z (2011b) Extracellular cAMP-adenosine pathways in the mouse kidney. Am J Physiol Renal Physiol 301(3):F565–F573
Jackson EK, Ren J, Gillespie DG (2011c) 2′,3′-cAMP, 3′-AMP and 2′-AMP inhibit human aortic and coronary vascular smooth muscle cell proliferation via A_{2B} receptors. Am J Physiol Heart Circ Physiol 301(2):H391–H401
Jackson EK, Ren J, Gillespie DG, Dubey RK (2010) Extracellular 2′,3′-cyclic adenosine 5′-monophosphate is a potent inhibitor of preglomerular vascular smooth muscle cell and mesangial cell growth. Hypertension 56:151–158

Jackson EK, Ren J, Mi Z (2009) Extracellular 2′,3′-cAMP is a source of adenosine. J Biol Chem 284:33097–33106

Kochanek PM, Clark RS, Ruppel RA, Adelson PD, Bell MJ, Whalen MJ, Robertson CL, Satchell MA, Seidberg NA, Marion DW, Jenkins LW (2000) Biochemical, cellular, and molecular mechanisms in the evolution of secondary damage after severe traumatic brain injury in infants and children: lessons learned from the bedside. Pediatr Crit Care Med 1:4–19

Kochanek PM, Hendrich KS, Jackson EK, Wisniewski SR, Melick JA, Shore PM, Janesko KL, Zacharia L, Ho C (2005) Characterization of the effects of adenosine receptor agonists on cerebral blood flow in uninjured and traumatically injured rat brain using continuous arterial spin-labeled magnetic resonance imaging. J Cereb Blood Flow Metab 25:1596–1612

Kochanek PM, Vagni VA, Janesko KL, Washington CB, Crumrine PK, Garman RH, Jenkins LW, Clark RS, Homanics GE, Dixon CE, Schnermann J, Jackson EK (2006) Adenosine A1 receptor knockout mice develop lethal status epilepticus after experimental traumatic brain injury. J Cereb Blood Flow Metab 26:565–575

Kusano Y, Echeverry G, Miekisiak G, Kulik TB, Aronhime SN, Chen JF, Winn HR (2010) Role of adenosine A2 receptors in regulation of cerebral blood flow during induced hypotension. J Cereb Blood Flow Metab 30:808–815

Laghi Pasini F, Guideri F, Picano E, Parenti G, Petersen C, Varga A, Di Perri T (2000) Increase in plasma adenosine during brain ischemia in man: a study during transient ischemic attacks, and stroke. Brain Res Bull 51:327–330

Lalancette-Hébert M, Gowing G, Simard A, Weng YC, Kriz J (2007) Selective ablation of proliferating microglial cells exacerbates ischemic injury in the brain. J Neurosci 27:2596–2605

Li W, Dai S, An J, Li P, Chen X, Xiong R, Liu P, Wang H, Zhao Y, Zhu M, Liu X, Zhu P, Chen JF, Zhou Y (2008) Chronic but not acute treatment with caffeine attenuates traumatic brain injury in the mouse cortical impact model. Neuroscience 151:1198–1207

Li W, Dai S, An J, Xiong R, Li P, Chen X, Zhao Y, Liu P, Wang H, Zhu P, Chen J, Zhou Y (2009) Genetic inactivation of adenosine A2A receptors attenuates acute traumatic brain injury in the mouse cortical impact model. Exp Neurol 215:69–76

Lu G, Zhou QX, Kang S, Li QL, Zhao LC, Chen JD, Sun JF, Cao J, Wang YJ, Chen J, Chen XY, Zhong DF, Chi ZQ, Xu L, Liu JG (2010) Chronic morphine treatment impaired hippocampal long-term potentiation and spatial memory via accumulation of extracellular adenosine acting on adenosine A1 receptors. J Neurosci 30:5058–5070

McKee AC, Gavett BE, Stern RA, Nowinski CJ, Cantu RC, Kowall NW, Perl DP, Hedley-Whyte ET, Price B, Sullivan C, Morin P, Lee HS, Kubilus CA, Daneshvar DH, Wulff M, Budson AE (2010) TDP-43 proteinopathy and motor neuron disease in chronic traumatic encephalopathy. J Neuropathol Exp Neurol 69:918–929

Miller LP, Hsu C (1992) Therapeutic potential for adenosine receptor activation in ischemic brain injury. J Neurotrauma 9 (Suppl 2): S563–S577

Ngai AC, Coyne EF, Meno JR, West GA, Winn HR (2001) Receptor subtypes mediating adenosine-induced dilation of cerebral arterioles. Am J Physiol Heart Circ Physiol 280: H2329–H2335

RAND invisible wounds of war study. http://www.rand.org/multi/military/veterans.html. Accessed 26 June 2011

Ren J, Mi Z, Stewart NA, Jackson EK (2009) Identification and quantification of 2′,3′-cAMP release by the kidney. J Pharmacol Exp Ther 328:855–865

Robertson CL, Bell MJ, Kochanek PM, Adelson PD, Ruppel RA, Carcillo JA, Wisniewski SR, Mi Z, Janesko KL, Clark RS, Marion DW, Graham SH, Jackson EK (2001) Increased adenosine in cerebrospinal fluid after severe traumatic brain injury in infants and children: association with severity of injury and excitotoxicity. Crit Care Med 29:2287–2293

Saatman KE, Duhaime AC, Bullock R, Maas AI, Valadka A, Manley GT, Workshop Scientific Team and Advisory Panel Members (2008) Classification of traumatic brain injury for targeted therapies. J Neurotrauma 25:719–738

Sachse KT, Jackson EK, Wisniewski SR, Gillespie DG, Puccio AM, Clark RS, Dixon CE, Kochanek PM (2008) Increases in cerebrospinal fluid caffeine concentration are associated with favorable outcome after severe traumatic brain injury in humans. J Cereb Blood Flow Metab 28:395–401

Saharan RS, Nantwi KD (2006) Changes in the biochemical profiles of mid-cervically located adenosine A1 receptors after repeated theophylline administration in adult rats. J Spinal Cord Med 29:520–526

Sinz EH, Kochanek PM, Heyes MP, Wisniewski SR, Bell MJ, Clark RS, DeKosky ST, Blight AR, Marion DW (1998) Quinolinic acid is increased in CSF and associated with mortality after traumatic brain injury in humans. J Cereb Blood Flow Metab 18:610–615

Sprinkle TJ (1989) 2′,3′-cyclic nucleotide 3′-phosphodiesterase, an oligodendrocyte-Schwann cell and myelin-associated enzyme of the nervous system. Crit Rev Neurobiol 4:235–301

Synowitz M, Glass R, Färber K, Markovic D, Kronenberg G, Herrmann K, Schnermann J, Nolte C, van Rooijen N, Kiwit J, Kettenmann H (2006) A1 adenosine receptors in microglia control glioblastoma-host interaction. Cancer Res 66:8550–8557

Tsutsui S, Schnermann J, Noorbakhsh F, Henry S, Yong VW, Winston BW, Warren K, Power C (2004) A1 adenosine receptor upregulation and activation attenuates neuroinflammation and demyelination in a model of multiple sclerosis. J Neurosci 24:1521–1529

Varma MR, Dixon CE, Jackson EK, Peters GW, Melick JA, Griffith RP, Vagni VA, Clark RS, Jenkins LW, Kochanek PM (2002) Administration of adenosine receptor agonists or antagonists after controlled cortical impact in mice: effects on function and histopathology. Brain Res 951:191–201

Verrier JD, Exo JL, Jackson TC, Ren J, Gillespie DG, Dubey RK, Kochanek PM, Jackson EK (2011) Robust expression of the 2′,3′-cAMP-adenosine pathway, but not the 3′,5′-cAMP-adenosine pathway, in microglia. J Neurochem 118:979–987

Vespa PM, Miller C, McArthur D, Eliseo M, Etchepare M, Hirt D, Glenn TC, Martin N, Hovda D (2007) Nonconvulsive electrographic seizures after traumatic brain injury result in a delayed, prolonged increase in intracranial pressure and metabolic crisis. Crit Care Med 35:2830–2836

Wagner AK, Miller MA, Scanlon J, Ren D, Kochanek PM, Conley YP (2010) Adenosine A1 receptor gene variants associated with post-traumatic seizures after severe TBI. Epilepsy Res 90:259–272

Yu L, Huang Z, Mariani J, Wang Y, Moskowitz M, Chen JF (2004) Selective inactivation or reconstitution of adenosine A2A receptors in bone marrow cells reveals their significant contribution to the development of ischemic brain injury. Nat Med 10:1081–1087

Chapter 16
Therapeutic Perspectives of Adenosine Receptor Compounds in Functional Restitution After Spinal Cord Injury

Kwaku D. Nantwi

Abstract Adenosine is a nucleoside that occurs naturally in mammalian tissues. It is a neuromodulator that is involved in a variety of physiological processes. It is generally accepted that the effects of adenosine are mediated through four G-protein-coupled receptors classified as A_1, A_{2A}, A_{2B}, and A_3 Rs. There is evidence that following traumatic injury in the brain and/or spinal cord excessive amounts of adenosine are released, and that this neuromodulator is involved in inflammatory processes and subsequent secondary effects of the injury. There is also evidence that excessive release of adenosine contributes to progressive neuronal injury and eventual loss of function. In several experimental animal models of spinal cord injury (SCI) pharmacologic manipulation of adenosine receptors following trauma can be beneficial in restoration of function.

The purpose of this review is to summarize the roles of adenosine receptors in two rodent models of SCI. In the first model, an upper cervical spinal cord hemisection paralyzes the hemidiaphragm ipsilateral to the injury; a latent respiratory motor pathway can be activated to restore respiratory function after injury. In the model, restoration of respiratory activity following systemic administration of theophylline, a nonspecific adenosine receptor antagonist, can be demonstrated. In the second model, involving thoracic contusion (T5–T8), neuro-protective effects can be induced by activation of specific adenosine receptor subtypes. Finally, potential therapeutic perspectives of adenosine compounds in SCI are discussed.

Keywords Respiration • Upper cervical spinal cord injury • Hemi-paralysis • Theophylline • Contusion • Adenosine antagonists/agonists

K.D. Nantwi (✉)
Department of Anatomy and Cell Biology, Wayne State University
School of Medicine, 540 E. Canfield, Detroit, MI 48202, USA
e-mail: knantwi@med.wayne.edu

S. Masino and D. Boison (eds.), *Adenosine: A Key Link between Metabolism and Brain Activity*, DOI 10.1007/978-1-4614-3903-5_16,

16.1 Introduction

According to the national spinal cord database, the major cause of mortality and morbidity following spinal cord injury (SCI) is respiratory failure. The primary focus of current therapy for patients with severe respiratory deficits after upper cervical spinal cord injury is mechanical ventilatory support (National Spinal Cord Injury Statistical Center 2008). Invariably, such patients feel a loss of independence; in addition, long-term use of mechanical support can often lead to severe infections and pneumonia (Winslow and Rozovsky 2003). There is (therefore) a need for alternative approaches to alleviate respiratory deficits. In order to understand the respiratory system and explore ways to improve respiratory dysfunction after injury, several animal models have been employed in basic research. In this review, we focus on the rat C2 hemisection model (C2HS) that has been extensively employed to examine respiratory dysfunction post SCI. In addition, recent advances in a mouse contusion model of SCI (T5–T8) are highlighted. In both models, the therapeutic potential of adenosine compounds is addressed.

In the C2HS model, an upper cervical (C2) hemisection interrupts the major descending bulbospinal respiratory motor pathways and effectively paralyzes the hemidiaphragm ipsilateral to the hemisection. However, a latent respiratory motor pathway can be activated to restore function to the paralyzed hemidiaphragm (Goshgarian 2009). The latent respiratory motor pathway is referred to as the "crossed phrenic pathway" (CPP). Elucidation of the physiological basis for the CPP led to the conclusion that the CPP could be attenuated by artificial respiration and enhanced by conditions that increased central respiratory drive (Lewis and Brookhart 1951). The conclusion of Lewis and Brookhart that the amount of CPP activity was directly proportional to the intensity of central respiratory drive and independent of the state of conduction in the contralateral phrenic nerve led to studies that investigated the use of pharmacologic agents that enhance central respiratory drive and activate the CPP without the need to transect the functionally intact contralateral phrenic nerve.

16.1.1 Pharmacologic Activation of the CPP

Pharmacologic manipulation to activate the CPP preserves the integrity of the contralateral phrenic nerve and simultaneously induces functional recovery in the previously paralyzed ipsilateral hemidiaphragm (Nantwi et al. 1996; Goshgarian 2009). With this approach the animal is better off functionally, since both halves of the diaphragm are functional, i.e., function in the ipsilateral hemidiaphragm is induced and function in the contralateral hemidiaphragm is initially left intact. The clinical relevance is clear. The entire diaphragm becomes bilaterally functional and, as a consequence, the animal is better able to cope with respiratory distress.

Drugs that stimulate respiration and excite neuronal activity have been extensively studied in functional recovery in the C2HS model. Adenosine-like compounds are an example of such a class of drugs.

Theophylline (1,3-dimethylxanthine) is a respiratory stimulant used to treat apnea in infants, asthmatic patients, and patients with "chronic obstructive pulmonary disease" (COPD) (Newman et al. 1994). It enhances respiratory frequency in healthy adult humans (Gorini et al. 1994), humans with a compromised respiratory function (Chevrolet et al. 1983), as well as human infants (Bona et al. 1995). Pharmacologically, it is a nonspecific adenosine receptor antagonist (blocks adenosine A_1R and $A_{2A}Rs$ with comparable affinity), phosphodiesterase (PDE) inhibitor, and bronchodilatory agent.

16.1.2 Acute Theophylline Administration

In a seminal study on the use of theophylline to restore function in the C2HS model, respiratory effects of the drug in three groups of rats were investigated (Nantwi et al. 1996). In the first group, the CPP was induced with varying doses (2.5–15 mg/kg) of theophylline under standardized electrophysiologic recording conditions; the stroke volume of the small animal ventilator was set at 60 breaths/min, tidal volume was 2.5 ml, end tidal CO_2 was set at 25 mmHg, and rats were paralyzed with pancuronium bromide (0.5 mg/kg). In the second group, activity in the ipsilateral and contralateral phrenic nerves was monitored continuously before and after systemic administration of theophylline in spontaneously breathing rats. In the third group, respiratory activity in both the ipsilateral phrenic nerve and the ipsilateral hemidiaphragm was assessed under spontaneous breathing conditions (Fig. 16.1). In this study, Nantwi and colleagues (1996) demonstrated that functional recovery of respiratory activity could be attained without the need to transect the contralateral phrenic nerve to enhance central respiratory drive to activate the CPP. This was a significant advancement in SCI research, since never before had such a level of motor recovery been achieved in a mammalian model. Previous studies had demonstrated that in rats subjected to phrenicotomy, systemic administration of aminophylline, a more soluble form of theophylline, improved ventilation (Nachazel and Palecek 1990). The findings confirmed conclusions by other investigators that theophylline/aminophylline stimulates respiration via a centrally mediated mechanism. More importantly, the study by Nantwi and colleagues (1996) was the first study to describe restoration of respiratory function after SCI. At a higher dose, 30 mg/kg, theophylline depressed respiration. Thus, theophylline evokes excitation at 15 mg/kg and depression at 30 mg/kg, a finding consistent with the demonstration that theophylline induces a biphasic response in cats and in clinical studies (Eldridge et al. 1983, 1985; Woodcock et al. 1983).

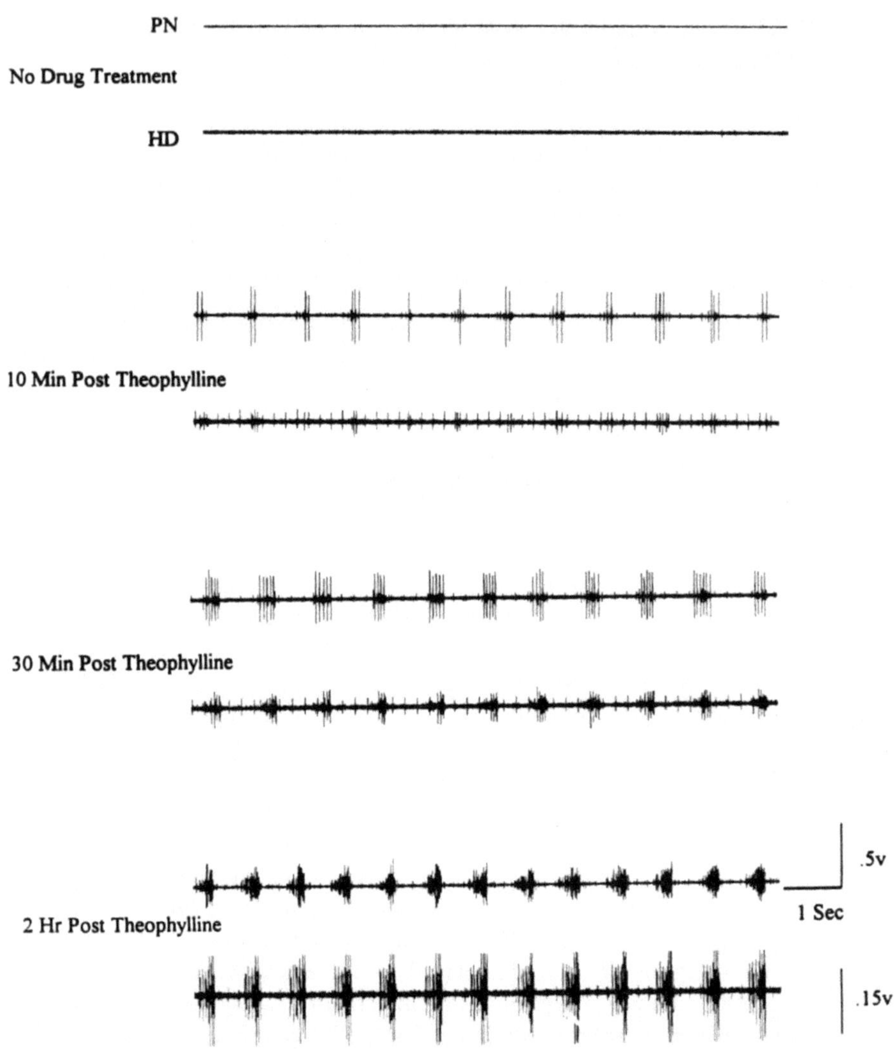

Fig. 16.1 A representative neurogram from a hemisected and spontaneously breathing animal, showing respiratory-related activity in the phrenic nerve (PN) and hemidiaphragm (HD) ipsilateral to hemisection. In the upper two traces, the total absence of activity in both the nerve and hemidiaphragm is indicative of a functionally complete hemisection. 10 min after administration of theophylline 15 mg/kg (third and fourth traces), activity in the left nerve and left hemidiaphragm becomes evident. The induced respiratory-related activity in the nerve and hemidiaphragm becomes progressively enhanced; and 30 min after the drug (fifth and sixth traces), induced activity is more distinct and 2 h after drug administration (seventh and eighth traces), the induced respiratory-related activity in both the nerve and hemidiaphragm is very similar (synchronous) to normal activity (reprinted with permission from Nantwi et al. 1996). Note the similarity in timing and duration of activity in both the nerve and hemidiaphragm

16.1.3 Chronic Theophylline Administration

Since acute administration of theophylline induced functional recovery in respiration, it was important to ascertain whether chronic drug administration would maintain recovery or induce tolerance/inverse effect similar to that shown after chronic caffeine administration (Chou et al. 1985). Briefly, theophylline was injected (i.p. 20 mg/kg 3× daily) for 3, 7, 14, or 30 days in C2HS rats. At the conclusion of the respective injection periods, respiratory activity was assessed (Nantwi and Goshgarian 1998a). At the termination of recording respiratory activity, approximately 1 ml of blood was withdrawn via the femoral vein for serum analysis of theophylline. Serum levels of drug were subsequently correlated with recovery. Theophylline serum levels (mean ± SEM) of rats treated for 3 and 7 days amounted to 2.075 ± 0.54 μg/ml and 18.20 ± 12.50 μg/ml, respectively. The levels were significantly lower than those detected after administration of pro-convulsive doses of theophylline (Chu 1981). It was concluded that the amount of recovered function correlated positively with serum levels at 3 and 7 days and that chronic theophylline administration did not induce tolerance. Theophylline thus induced and maintained function when administered chronically in C2HS rats.

Earlier work by Eldridge and coworkers (1983, 1985) suggested that theophylline stimulates respiration via adenosine receptor antagonism; however, a mechanism of action had not been established or explored in theophylline-induced recovery in the C2HS model. In addition, theophylline is a bronchodilator and has been shown to increase diaphragm contractility in vitro (Murciano et al. 1987). In their study on theophylline-induced functional recovery, the authors suggested that the drug may be exerting its effects via adenosine receptor antagonism. This proposal was based on the following: (1) adenosine modulates respiratory activity (McQueen and Ribeiro 1986; Maxwell et al. 1986; Fuller et al. 1987), (2) theophylline reverses adenosine-mediated respiratory depression (Eldridge et al. 1985), and (3) adenosine A_1 receptor is located on phrenic motoneurons, putative sites of the CPP, and modulates respiratory activity in vitro (Dong and Feldman 1995).

16.2 Mechanism of Action

16.2.1 Adenosine A_1 Receptor Antagonism in Theophylline-Induced Recovery

In a series of experiments with theophylline and an adenosine A_1 receptor agonist, N^6 (L-2-phenylisopropyl) adenosine (L-PIA), it was confirmed that adenosine A_1 receptors are involved in theophylline-induced functional recovery (Nantwi and Goshgarian 1998b). This conclusion extended the seminal work of Eldridge and coworkers (1983, 1985) in identifying adenosine receptor antagonism as a mechanism of action of theophylline. However, theophylline is a nonspecific adenosine receptor antagonist as well as a bronchodilator and a PDE inhibitor. To further identify mechanism(s) of actions, a variety of alkylxanthines with diverse pharmacologic profiles were tested. Theophylline and 8-Phenyltheophylline, 3-propylxanthine (enprofylline), and 8(*p*-sulfophenyl)theophylline (8-PST) were tested for their relative effects on functional recovery. Enprofylline is a methylxanthine, like theophylline, but a more potent bronchodilator than theophylline (Fredholm, 1995) and, unlike theophylline, is not an adenosine antagonist (Persson et al. 1986). 8-Phenyltheophylline is a more potent adenosine receptor antagonist than theophylline and 8-PST is a peripherally specific adenosine receptor antagonist (Evoniuk et al. 1986; Nikodijevic et al. 1991).

The magnitude of recovered activity was assessed by (1) expressing recovery as a function of pre-drug activity in the right phrenic nerve or (2) expressing recovery as a function of activity in the homo-lateral nerve of non-injured age-matched control animals. It has been demonstrated that either approach results in similar outcome conclusions (Hadley et al. 1999). Expressing recovery as a function of the contralateral activity has also been employed by other investigators (Vinit et al. 2007). It must be stressed, however, that there is currently no consensus on how to accurately quantify recovered activity after C2HS. A simple approach of using the raw or unprocessed respiratory signal of motor output has been used (Fuller et al. 2006; Golder and Mitchell 2005); however in an effort to offset variability inherent with this approach, the raw signal can be "normalized" to a maximum during a challenge (Fuller et al. 2005), or a minimum during baseline levels (Fuller et al. 2008; Golder et al. 2008). More recent studies on respiratory dysfunction after SCI have employed whole body plethysmography (WBP) to describe ventilator/breathing parameters post injury (Choi et al. 2005; Golder et al. 2008). It is more likely that a combination of approaches will yield the most meaningful information in assessing the magnitude of recovered function after injury.

With either approach, Nantwi and Goshgarian (1998b) showed that 8-phenyltheophylline was more potent than theophylline in restoring respiratory function; however, enprofylline was ineffective. At a dose similar to that employed by Thomas and colleagues (1994) to monitor involvement of adenosine in systemic hypoxia, 8-SPT did not induce recovery. It was concluded that (1) central (spinal and brainstem) adenosine receptor antagonism underlies theophylline-induced functional recovery, (2) effects of theophylline and 8-phenyltheophylline are unrelated to bronchodilatory actions, (3) peripheral adenosine receptor antagonism may not be involved, and (4) effects of theophylline and 8-phenyltheophylline in inducing functional recovery after acute administration are unrelated to PDE inhibition since both drugs have minimal PDE inhibitory properties (Bergstrand 1980; however, see Kajana and Goshgarian 2008). Perhaps the most relevant finding was that the two alkylxanthines that induce functional recovery are both nonspecific adenosine receptor antagonists. This principal finding led to a series of experiments to establish the relative contribution(s) of adenosine A_1 and A_2 receptor antagonism in functional recovery.

Adenosine is formed inside cells or on cell surfaces by the action of membrane-bound endonucleotides, and acts via receptor-mediated mechanisms. Adenosine receptors are classified into four subtypes (A_1, A_{2A}, A_{2B}, and A_3). All four receptor subtypes are coupled to G-proteins. The A_1 and A_3 receptors are coupled to an inhibitory G-protein (G_i) while the A_{2A} and A_{2B} receptors are coupled to a stimulatory G-protein (G_s). The four receptor subtypes have been cloned and pharmacologically characterized (Fredholm et al. 2005; Latini and Pedata 2001). Adenosine A_1, A_{2A}, and A_{2B} receptor subtypes have a higher affinity for adenosine and mediate the physiologic actions of endogenous adenosine (Fredholm et al. 2005). The expression and distribution of adenosine receptors are wide (Table 16.1). Upon activation, adenosine receptors may exert their distinct effects by acting with different G-proteins that in turn activate enzyme systems (adenyl cyclase) to alter intracellular levels of cyclic AMP. In the C2HS model, contribution of specific adenosine receptor subtypes to respiratory recovery has been addressed.

16.2.2 Adenosine A_1R- and A_2R-Specific Compounds in Functional Recovery

Previous studies from several investigators had established that adenosine itself modulates respiratory activity through activation of central and peripherally mediated mechanisms (Fuller et al. 1987; Reid et al. 1991). It is known that the A_1 receptor has a higher affinity for the endogenous ligand than the A_2 receptor (Latini et al. 1996). The A_2 receptor is subdivided into the A_{2A}, which is localized primarily in the hippocampus, and the low-affinity A_{2B} receptor, localized throughout the CNS. In a series of studies, Nantwi and Goshgarian (2002) assessed the actions of specific adenosine receptor agonists and antagonists alone and in combination. Selective adenosine A_1R compounds chosen were the specific A_1R antagonist, 1,3-dipropyl-8-cyclopentylxanthine (DPCPX), and the A_1R agonist, N^6-cyclohexyladenosine (CHA). Selective adenosine A_2R compounds chosen were the specific A_2R antagonist, 3,7-dimethyl-1-proparglyxanthine and the specific A_2R agonist (2-*p*-(2-carboxyethyl) phenethylamino-5-*N*-ethylcarboxyamido adenosine hydrochloride(CGS 21680).

It was demonstrated that 24 h after C2HS, systemic administration (i.v.) of DPCPX (0.05–0.2 mg/kg) induced recovery of respiration while DMPX (0.5–2 mg/kg) was ineffective (Nantwi and Goshgarian 2002). The effect of DPCPX confirmed and extended previous findings that adenosine A_1 receptor antagonism mediates functional recovery induced with theophylline (Nantwi and Goshgarian 1998b, 2001). DPCPX at 0.1 mg/kg administered i.v. induced a greater degree of recovered activity than that elicited by the optimal theophylline or 8-phenyltheophylline dose. Prior administration of the A_1R agonist CHA (0.1 mg/kg) blunted DPCPX-induced recovery. Administration of CHA after DPCPX resulted in attenuation and eventual blockade of recovered activity.

Systemic administration of CGS 21680 (0.5–2 mg/kg; i.v.) enhanced tonic activity which was blocked by DMPX. Functional recovery, however, was not elicited in the majority (6/8) of C2HS rats 24 h following injury. In age-matched control non-injured rats, CGS21680 induced respiratory excitation. It was concluded from these

Table 16.1 Distribution of adenosine receptors in mammalian systems

A_1 receptor	A_{2A} receptor	A_{2B} receptor	A_3 receptor	References
High expression				
Brain (cortex, cerebellum, hippocampus); dorsal horn of spinal cord, pons, medulla, eye, adrenal gland, atria	Spleen, thymus, leucocytes (lymphocytes and granulocytes), blood platelets, GABAergic neurons (in caudate putamen, nucleus accumbens, tuberculum olfactorium), olfactory bulb	Caecum, colon, bladder	Testes (rat) Mast cells (rat)	Fredholm et al. (2000), Gaytan et al. (2006)
Intermediate expression				
Other brain regions, skeletal muscle, liver, kidney adipose tissue, salivary glands, esophagus, colon, antrum, testis	Heart, lung, blood vessels	Lung, blood vessels, eye, median eminence, mast cells	Cerebellum (human) Hippocampus (human) Lung, spleen (sheep) Pineal	Fredholm et al. (2000), Fozard et al. (2003)

experiments that activation of A_2 receptors induces changes in respiratory activity, characterized by enhanced amplitude of respiratory bursts in the phrenic nerves and decreases in inspiratory burst duration and inter burst interval (Nantwi and Goshgarian 2002). These findings were intriguing since the kinetics of CGS 21680 may preclude its access into the CNS (Hutchinson et al. 1990; Nikodijevic et al. 1991). The kinetics of the drug, coupled with evidence that peripherally localized adenosine receptor subtypes modulate respiratory activity, led to studies that investigated putative roles of peripheral adenosine receptors in respiratory function (described below).

A recent study by Golder and colleagues (2008) showed that activation of A_{2A} receptors via spinal administration of CGS 21680 elicits long-lasting facilitation of phrenic motor activity, a form of plasticity. Perhaps more importantly, the studies demonstrated that systemic activation of A_{2A} receptors stimulates ventilation (monitored with WBP) with a magnitude and time course similar to phrenic motor facilitation. The authors suggested that while spinal mechanisms are involved in mediating long-term facilitation, putative involvement of adenosine receptors in the carotid body may also be involved (Golder et al. 2008 Sect. 3.3). In the C2HS model, interactions between A_1 and A_2 receptors led to the conclusion that activation of A_2 receptors was necessary to subserve A_1 receptor-mediated effects but may not be necessary to maintain the recovery (Nantwi and Goshgarian 2002). An important caveat to the findings with the use of selective adenosine compounds in combination is that this approach reinforces the suggestion by Ribeiro (1999) that adenosine-related medicines that combine A_1 receptor blockade with A_2 receptor activation may be clinically beneficial.

16.2.3 Adenosine Receptors in the Carotid Bodies

Adenosine receptors involved in the modulation of respiration are located in the carotid bodies of several mammalian species including man (Monteiro and Ribeiro 1987; Watt and Routledge 1985). Adenosine A_2 receptors mediate respiratory excitation while depression is mediated by A_1 receptors (Monteiro and Ribeiro 1987; Carley and Radulovacki 1999).

Previous studies have shown that adenosine A_1 receptor activation can reduce centrally mediated respiratory disturbances that occur during sleep (Carley and Radulovacki 1999; Monti et al. 1996). To directly assess putative involvement of peripherally located adenosine receptors in recovery of respiratory-related activity in C2HS, Bae and colleagues (2005) tested the hypothesis that adenosine A_1 receptors localized in the carotid bodies can be specifically activated to modulate recovered respiratory activity. The investigators proposed that in the clinical application of theophylline to improve respiratory function post SCI, enhanced respiratory rate (theophylline-induced) would lead to diaphragm muscle fatigue. In order to preserve the potential therapeutic benefit of theophylline, and concurrently minimize any potential of respiratory muscle fatigue, the authors designed a series of experiments using specific adenosine receptor compounds to differentially target adenosine receptors localized at peripheral sites (carotid bodies) of action.

Adenosine A_1 receptors were targeted with a specific agonist in C2HS rats with intact or excised carotid bodies. The A_1R agonist, N^6-p-sulfophenyladenosine (*p*-SPA), which does not cross the blood–brain barrier and is classified as peripherally specific was employed (Gleeson and Zwillich 1992; Bae et al. 2005). Total carotid body excision was conducted according to the method of Olson and colleagues (1988). Using immunohistochemistry, Bae and colleagues demonstrated positive immunoreactivity for adenosine receptors in the carotid body, confirming that adenosine A_1 and A_2 receptors are located in the carotid bodies (Bae et al. 2005; Gauda 2000). Bae and colleagues (2005) demonstrated that in C2HS rats with the carotid body intact (H-CBI), the discharge rates of theophylline-induced respiratory-related recovery were significantly reduced to near-basal discharge levels by concurrent administration of *p*-SPA. More clinically relevant was the observation that the recovered function was preserved. In C2HS rats with carotid bodies excised (H-CBE), administration of *p*-SPA (0.25 mg/kg, i.v.) did not attenuate theophylline-induced respiratory frequency (Bae et al. 2005).

The authors concluded that (1) carotid body excision by itself does not adversely affect the expression of theophylline-induced functional recovery, (2) adenosine A_1 and A_2 receptors are located in the carotid bodies, (3) specific activation of peripherally located adenosine A_1 receptors can modulate theophylline-induced enhanced frequency back to normal discharge rates, and (4) a novel approach of selective activation of peripheral adenosine A_1 receptors, combined with blockade of central adenosine A_1 receptors, can be employed to attain the desired pharmacologic action of recovered function and maintain basal rates to near-normal levels.

16.2.4 Peripheral Adenosine A_2 Receptors and Functional Recovery

The study of Bae and colleagues (2005) confirmed and extended previous studies that adenosine A_1 and A_2 receptors are localized in the carotid bodies (Fuller et al. 1987; McQueen and Ribeiro 1986; Gauda 2000). In light of the discovery that the adenosine A_1 receptor antagonist, DPCPX, elicited robust recovery, we posited that concurrent administration of the A_2R agonist, CGS 21680, and A_1R antagonist, DPCPX, may evoke an even greater amount of recovery than that had been realized with the A_1R antagonist alone. The initial study of Bae and colleagues (2005) was extended to assess involvement of peripheral A_2 receptors in A_1 receptor-mediated recovery. Briefly, H-CBI and H-CBE rats were tested following systemic administration of DPCPX alone, or a combination of DPCPX and CGS 21680. In cases of concurrent drug administration, the effects were assessed based on the order of drug administration, i.e., whether CGS 21680 was administered before or after DPCPX. The investigators showed that concurrent administration of CGS 21680 and DPCPX in H-CBI rats elicited more recovery than DPCPX alone, and concluded that activation of A_2

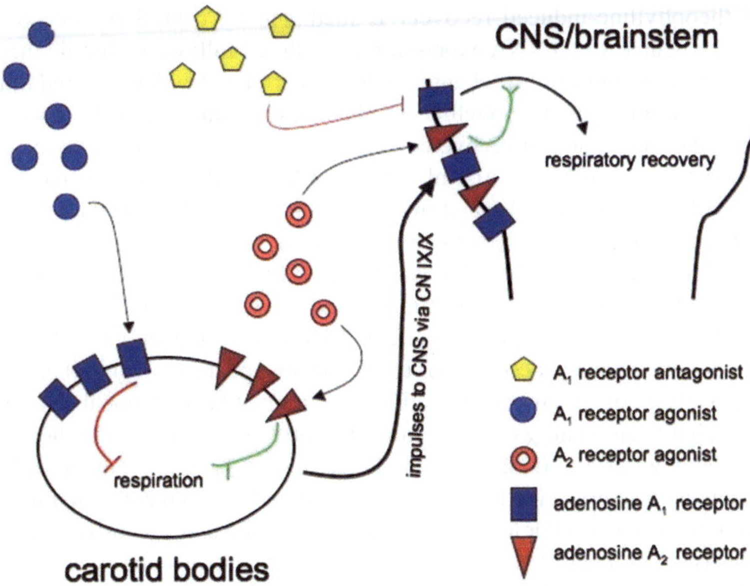

Fig. 16.2 Schematic illustration of the interaction of peripheral and central adenosine receptors in respiratory recovery. Activation of peripherally located A_{2A} receptors in the carotid bodies excites respiratory activity which is transmitted to the CNS via the IX/X cranial nerves. On the other hand, activation of peripheral adenosine A_1 receptors inhibits respiration and can dampen induced recovery. Central blockade of adenosine A_1 receptors prevents the inhibitory effect of A_1 receptor activation on respiration and the result is disinhibition or excitation. Simultaneous administration of the adenosine A_1 receptor antagonist DPCPX and peripherally acting A_{2A} agonist CGS21680 can therefore elicit enhanced recovery via an apparently additive effect through disinhibition combined with a direct excitation

receptors in the carotid bodies can enhance the magnitude of DPCPX-induced recovery (James and Nantwi 2006). This study once again reinforces the therapeutic potential for adenosine-related strategies that combine A_1 receptor blockade with A_2 receptor activation. Putative interactions of peripherally located adenosine A_2 receptor activation and central adenosine A_1 receptor antagonism are shown in Fig. 16.2.

16.3 Adenosine A_1R mRNA Expression

Thus far, studies of adenosine receptor involvement in functional recovery in the C2HS model have implicated blockade of central A_1 receptors and activation of peripheral A_2 receptors. Basura and colleagues (2002) demonstrated a positive co-localization of adenosine A_1 receptor mRNA and immunoreactivity with phrenic motoneurons from cervical segments C3–C6 putative sites of the CPP (Basura et al. 2002; Goshgarian 2009).

Since theophylline-induced recovery is mediated by central A_1 receptors, we determined whether A_1R mRNA expression was altered following chronic theophylline administration, and whether changes in levels correlated with recovered function. We showed that following theophylline administration (orally 20 mg/kg, 3× daily) to C2HS rats, the magnitude of recovered activity correlates positively (over a theophylline serum range of 1.2–1.9 μg/ml) for 3, 7, 12, or 30 days. Perhaps most importantly, when C2HS rats were weaned from the drug (for 7, 12, or 30 days) after only 3 days of administration, the beneficial effects (i.e., recovered respiratory activity) persisted (Nantwi et al. 2003b). This was the first finding in which a drug-induced effect persisted when the initial drug stimulus had been withdrawn and is an example of drug-induced plasticity or preomic effect. This observation suggests that chronic administration may not be necessary to sustain the therapeutic benefit of theophylline. However, analysis of in situ hybridization and immunohistochemical data did *not* reveal any significant changes in A_1R mRNA expression after chronic theophylline treatment (Nantwi et al. 2003a, b). This latter finding is in agreement with other investigators who assessed the long-term effects of caffeine, a methylxanthine with a pharmacologic profile similar to theophylline, and showed that adenosine A_1R mRNA expression is unchanged after chronic caffeine treatment in mice (Georgiev et al. 1993; Johansson et al. 1993; however, see Sousa et al. 2011). The demonstration that there was no apparent correlation in levels of A_1R mRNA expression and recovery after theophylline led to a series of studies aimed at assessing changes in adenosine receptor levels and/or affinity after theophylline administration.

16.3.1 Biochemical Binding Assays After Chronic Theophylline

Saharan and Nantwi (2006) demonstrated that after 3 days of oral theophylline administration, the profile of adenosine A_1 receptors located in the phrenic motoneurons (C3–C6) is altered. In naïve untreated rats, the specific adenosine A_1 receptor antagonist [^{3}H]-DPCPX detected a single binding site; however, after chronic administration of theophylline, [^{3}H]-DPCPX detected two binding sites: one site is identical to the single site detected in naïve non-injured rats and a second site is unmasked (Fig. 16.3). The second site may be due to binding with the A_2 receptors, since DPCPX has been shown to have affinity for the A_2 site at higher concentrations (Klotz 2000; McAdoo et al. 2000). Alternatively, the unmasked site may represent a conformational change in the A_1 receptor.

The clinical relevance of changes in adenosine A_1 receptor profile reported by Saharan and Nantwi (2006) may support the findings of Varani and colleagues (2006) that altered adenosine receptor numbers and/or affinity may be involved in respiratory dysfunction. The findings of Saharan and Nantwi (2006) may offer insights into the potential use of adenosine compounds in respiratory deficits. If the unmasked site is characterized as an A_2 receptor site, then it would suggest that activation of A_2 receptors may be an important therapeutic target for improving respiratory impairment after injury.

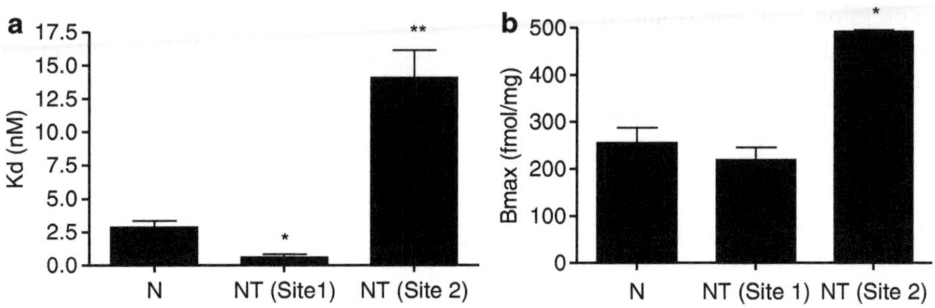

Fig. 16.3 (**a**) The histogram was generated from saturation binding studies for A_1 receptors. Displacement of 1nM [^{3}H]-DPCPX binding, to rat spinal cord (C3–C5) homogenate from naïve rats (N) and theophylline-treated (NT) rats by 2.7 pM to 27 mM theophylline. Values represent the mean specific binding as a percentage increase above basal binding (± SEM) of at least three experiments, each performed in triplicate. All values are corrected for nonspecific binding defined with 100 μM adenosine. (**b**) Histogram showing the respective Bmax values in N and NT animals. Bmax for N and NT animals appears to be the same; however, a second binding site with a greater Bmax is unmasked and persisted for 12 days after drug weaning (reprinted with permission from Saharan and Nantwi 2006)

16.3.2 Adenosine A_1 and A_2 Receptor mRNA/Protein Levels After Chronic Theophylline

In our previous studies on the putative involvement of adenosine A_1R mRNA in functional recovery, we concluded that theophylline-induced recovery may not be due to alterations at the message level but changes in A_2 receptor proteins profile (Nantwi et al. 2003a, b; Saharan and Nantwi 2006). Petrov and colleagues (2007) assessed expression of adenosine A_1R and A_2R proteins in C3–C6 spinal cord segments containing phrenic motoneurons in non-injured, C2HS rats, C2HS rats treated with theophylline for 3 days, and C2HS rats treated with theophylline for 3 days followed by 12 drug-free days.

Briefly, Petrov and colleagues (2007) showed that 24 h following C2HS, adenosine A_1 and A_2 receptors were differentially altered, i.e., A_1 receptors were decreased significantly while the A_2 receptors were unchanged. However, after chronic administration of theophylline, A_1 receptors increased significantly. Adenosine A_2 receptor protein levels were not changed by C2HS. Theophylline administration significantly decreased basal levels, a decrease that persisted after theophylline withdrawal for 12 days. It was concluded that theophylline mitigates the effects of C2HS by attenuating the injury-induced decrease in A_1R protein levels and that A_2R protein levels are unaltered by C2HS but decrease after theophylline withdrawal (Petrov et al. 2007). The significance of the findings from this study is that the decrease in A_1 receptors may dampen the inhibitory actions of endogenous adenosine and could contribute to the hyperexcitability of motoneurons which may underlie spasticity after SCI. In addition, it also suggests that altered levels of adenosine A_2 receptor expression may contribute to the apparent persistent actions of theophylline in functional recovery after drug withdrawal. It further suggests that if A_2 receptors are down regulated after theophylline cessation, then activation of the same receptors may be a tool to further improve recovery.

16.3.3 Clinical Use of Theophylline in SCI

Two case studies demonstrated that theophylline administration increased respiratory motor output after SCI (Ferguson et al. 1999; Bascom et al. 2005). The drug enhanced central respiratory drive and inspiratory muscle force in one patient with C5–C7 injury and a history of respiratory insufficiency (Ferguson et al. 1999). In another case study, a patient with a C5–C6 injury was weaned from ventilator support after aminophylline (a more soluble form of theophylline) treatment (Bascom et al. 2005). This particular case was rather promising since the patient had developed respiratory insufficiency after prolonged bouts of pneumonia and septicemia which had to be cleared first. In a third case study, it was demonstrated that small doses of theophylline were successfully used in three SCI patients with persistent bradycardia (Schulz-Stubner 2005). In one of the three patients, the dose of theophylline used increased respiratory rate and improved minute ventilation (Schulz-Stubner 2005). After 6 weeks, treatment with theophylline was discontinued with no further episodes of bradycardia. However, the apparent promise shown in the case studies mentioned above has not been documented in a larger patient population. In a double-blind placebo-controlled crossover study in ten patients with chronic tetraplegia, theophylline administration did not significantly improve functional outcome in the entire patient pool (Tzelepis et al. 2006). Future studies will need to thoroughly examine clinical effects of theophylline in selected SCI patients with respiratory deficits, in particular insufficiency. It is anticipated that a more rigorous screening of patients, with a particular focus on duration of injury and other underlying conditions, can provide reliable results.

16.4 Adenosine A_2 Receptors in Recovery: Compression Model

Adenosine A_2 receptor activation has been implicated in functional recovery in other motor systems. Functional recovery using SCI compression is one area of intense investigation to identify molecules important in two phases of injury: the initial phase and the chronic phase. The initial phase is characterized by destruction of various elements of tissues, while the chronic phase is characterized by further tissue damage and lesion expansion. It is generally accepted that inflammatory and immune responses are intimately involved in both phases of injury (Genovese et al. 2009). The goal of this line of investigation is identification of molecules that are overexpressed soon after injury and target therapies to mitigate responses that define the secondary damage.

Adenosine A_{2A} receptor agonists and antagonists have been demonstrated to be neuroprotective against ischemia reperfusion and locomotor dysfunction (Reece et al. 2008; Cassada et al. 2002; Li et al. 2006; Genovese et al. 2009).

In a rat compression model of SCI, compression at T5–T8 spinal segments induces locomotor dysfunction (Rivlin and Tator 1978). Using this model, Genovese and colleagues (2009) showed that selective activation of A_2 receptors with agonists after injury reduces cell death by limiting tissue damage, via anti-inflammatory and anti-apoptotic actions, to improve locomotor function. The study provides a strategy to harness the early molecular and cellular signals critical in secondary damage in SCI. Interestingly, Li and colleagues (2006) demonstrated that systemic administration of A_{2A}R antagonists offers protection against locomotor dysfunction. Current evidence suggests that both A_{2A}R agonists and antagonists may mitigate locomotor dysfunction.

Paterniti and colleagues (2011) suggested that the main mechanism of action of agonists or antagonists may be central or peripheral. In their study, the investigators demonstrated that neuroprotection mediated via adenosinergic mechanisms involves (1) central modulation by the A_{2A}R antagonists, through attenuating/blocking glutamate excitotoxicity (Genevese et al. 2010), and (2) peripheral modulation mediated by A_{2A}R agonists. With both mechanisms, the net effect is neuroprotection and improved locomotor activity. Based on studies from our laboratory as well as others, it is likely that the peripheral site of action is the carotid bodies (Bae et al. 2005; Gauda 2000). The findings from these studies emphasize that selective targeting of A_2 receptors offers a therapeutic strategy that can be utilized to enhance function after SCI.

16.5 Conclusion

We have provided a summary of theophylline-mediated actions in functional recovery in the C2HS model of SCI. The mechanism of action of theophylline has been shown to be via adenosine receptors. Although blockade of the A_1 receptor subtype has been emphasized, we have also underscored the role of A_2 receptors, with a particular emphasis on the contribution of activation of A_2 receptors localized in the carotid body to improved respiratory activity. The basic research that has led to the clinical application of theophylline in SCI patients has been emphasized. It has been shown that while theophylline induces recovery in acute C2HS, it is ineffective after chronic C2HS (Nantwi and Goshgarian 2005). Clinical case studies have shown that in chronic SCI patients, theophylline is beneficial (Ferguson et al. 1999; Bascom et al. 2005; Schulz-Stubner 2005). As far as this author is aware, theophylline has not been used in acute clinical SCI cases so far. This author suggests that early treatment with theophylline may provide a more positive response. Based on advances in studies on adenosine agonists (Golder et al. 2008; Genovese et al. 2009, 2010; Paterniti et al. 2011) use of such compounds offers a therapeutic approach with promise.

In an attempt to identify molecular targets that may underlie the apparent persistent theophylline-induced effects, we recently conducted a series of molecular studies that targeted neurotrophic factors known to be involved in respiratory plasticity. Thus far, initial results from ongoing studies in our laboratory have identified neurotrophic and survival factors in drug-induced plasticity in C2HS. It is anticipated that this will pave the way to elucidation of putative molecular signals/ pathways that sustain drug-induced plasticity; identification of such targets may suggest therapeutic strategies to harness the inherent plasticity within the respiratory system to improve respiratory function after upper cervical SCI.

Acknowledgements This work was supported by National Institute of Child and Human Development Grants 35766 and 31550. The author expresses sincere appreciation for the contributions of Drs. Alilain, Bae, Basura, James, Kreipke, Petrov, and Ms. Saharan. The contributions of Dr. Singh and Ms. Moore-Langston in ongoing studies targeting molecular/survival factors are acknowledged. A special appreciation to Dr. Harry Goshgarian is acknowledged.

References

Bae H, Nantwi KD, Goshgarian HG (2005) Recovery of respiratory function following C2 hemi and carotid body denervation in adult rats: influence of peripheral adenosine receptors. Exp Neurol 191:94–103

Bascom AT, Lattin CD, Aboussouan LS, Goshgarian HG (2005) Effect of acute aminophylline administration on diaphragm function in high cervical tetraplegia: a case report. Chest 127(2):658–661

Basura GJ, Nantwi KD, Goshgarian HG (2002) Theophylline-induced respiratory recovery following cervical spinal cord hemisection is augmented by serotonin 2 receptor stimulation. Brain Res 956:1–13

Bergstrand H (1980) Phosphodiesterase inhibition and theophylline. Eur J Respir Dis 61(109):37–44

Bona E, Aden U, Fredholm BB, Hagberg H (1995) The effect of long term caffeine treatment on hypoxic-ischemic brain damage in the neonate. Pediatr Res 38(2):312–318

Carley DW, Radulovacki M (1999) Role of peripheral adenosine A1 receptors in regulation of sleep apneas in rats. Exp Neurol 159:545–550

Cassada DC, Tribble GC, Young JS, Gangemi JJ, Gohari AR, Butler PD, Rieger JM, Kron IL, Linden J, Kern JA (2002) Adenosine A2A analogue improves neurologic outcome after spinal cord trauma in rabbits. J Trauma 53(2):225–229

Chevrolet J-C, Reverdin A, Suter PM, Tschopp JM, Junod A (1983) Ventilatory dysfunction resulting from bilateral anterolateral high cervical cordotomy. Chest 84(1):112–115

Choi H, Liao W-L, Newton KM, Onario RC, Knig AM, Desilets FC, Woodward EJ, Eichler ME, Frontera WR, Sabharwal S, Teng YD (2005) Respiratory abnormalities resulting from midcervical spinal cord injury and their reversal by serotonin 1A agonists in conscious rats. J Neurosci 25(18):4550–4559

Chou TD, Khan S, Forde J, Hirsh KR (1985) Caffiene tolerance: electrophysiological and neurochemical evidence. Life Sci 36:2347–2358

Chu N-S (1981) Caffeine- and aminophylline-induced seizures. Epilepsia 22:85–94

Dong X-W, Feldman JL (1995) Modulation of inspiratory drive to phrenic motoneurons by presynaptic adenosine A1 receptors. J Neurosci 15:3458–3467

Eldridge FL, Millhorn DE, Killey JP (1985) Antagonism by theophylline of respiratory inhibition induced by adenosine. J Appl Physiol 59(5):1428–1433

Eldridge FL, Millhorn DE, Waldrop TG, Killey JP (1983) Mechanism of respiratory effects of methylxanthines. Respir Physiol 53:239–261
Evoniuk G, Von Borstel RW, Wurtman RJ (1986) Antagonism of the cardiovascular effects of adenosine by caffeine or 8-(-p-Sulfophenyl)theophylline. J Pharmacol Exp Ther 240(20):428–432
Ferguson GT, Narendra K, Lattin CD, Goshgarian HG (1999) Clinical effects of theophylline therapy on inspiratory muscle in tetraplegia. Neurorehab Neural Repair 13(3):191–197
Fozard JR, Baur F, Wobler C (2003) Antagonist pharmacology of adenosine A2B receptors from rat, guinea pig, and dog. Eur J Pharmacol 475:79–84
Fredholm BB (1995) Purinoceptors in the nervous system. Pharmacol Toxicol. 76(4):228–239
Fredholm BB, Arslan G, Halldner L, Kull B, Schulte G, Wasserman W (2000) Structure and function of adenosine receptors and their genes. Naunyn Schmiedeberges Arch Pharmacol 362(4–5):364–374
Fredholm BB, Chen J-F, Masino SA, Vaugeois J-M (2005) Actions of adenosine at its receptors in the CNS: insights from knockouts and drugs. Ann Rev Pharmacol Toxicol 45:385–412
Fuller RW, Maxwell DL, Conradson T-BG, Dixon CMS, Barnes PJ (1987) Circulatory and respiratory effects of infused adenosine in conscious man. Br J Clin Pharmacol 24:309–317
Fuller DD, Golder FJ, Olson Jr. EB, Mitchell GS (2005) Recovery of phrenic activity and ventilation after cervical spinal cord hemisection in rats. J Appl. Physiol 100:800–806
Fuller DD, Golder FJ, Olson EB Jr, Mitchell GS (2006) Recovery of phrenic activity and ventilation after cervical spinal hemisection in rats. J Appl Physiol 100(3):800–806
Fuller DD, Doperalski NJ, Dougherty BJ, Sandhu MS, Bolser DC, Reier PJ (2008) Modest spontaneous recovery of ventilation following chronic high cervical hemisection in rats. Exp Neurol 211(1):97–106
Gauda E (2000) Expression and localization of A2A- and A1-adenosine receptor genes in the carotid body and petrosal ganglia. A2A and A1-adenosine receptor m RNA in the carotid body. Adv Exp Med Biol 475:549–558
Gaytan SP, Saadani-Makki F, Bodineau L, Frugiere A, Larnicol N, Pasaro R (2006) Effect of postnatal exposure to caffeine on the pattern of adenosine A1 receptor distribution in the respiration-related nuclei of the rat brainstem. Auton Neurosci 126–127:339–346
Genovese T, Melani A, Esposito E, Mazzon E, Di Paolo R, Bramanti P, Pedata F, Cuzzocrea S (2009) The selective adenosine A2A receptor agonist CGS-21680 reduces JNK MAPK activation in oligodendrocytes in injured spinal cord. Shock 32:578–585
Genovese T, Melani A, Esposito E, Paterniti I, Mazzon E, Di Paolo R, Bramanti P, Linden J, Pedata F, Cuzzocrea S (2010) Selective adenosine A2A receptor agonists reduce the apoptosis in an experimental model of spinal cord trauma. J Biol Regul Homeost Agents 24(1):73–86
Georgiev V, Johansson B, Fredholm BB (1993) Long term caffeine treatment leads to a decreased susceptibility to NMDA-induced clonic seizures in mice without changes in adenosine A1 receptor number. Brain Res 612:271–277
Gleeson K, Zwillich CW (1992) Adenosine stimulation, ventilation and arousal from sleep. Am Rev Respir Dis 145:453–457
Golder FJ, Ranganathan L, Satriomoto I, Hoffman M, Lovett-Barr MR, Walters JJ, Baker-Herman TL, Mitchell GS (2008) Spinal adenosine $A2_A$ receptor activation elicits long-lasting phrenic motor facilitation. J Neurosci 28(9):2033–2042
Golder FJ, Mitchell GS (2005) Spinal synaptic enhancement with acute intermittent hypoxia improves respiratory function after chronic cervical spinal cord injury. J Neurosci 25: 2925–2932
Gorini M, Duranti R, Misuri G, Valenza T, Spinelli A, Goti P, Gigliotti F, Scano G (1994) Aminophylline and respiratory muscle interaction in normal humans. Am J Physiol 149(5):1227–1234
Goshgarian HG (2009) The crossed phrenic phenomenon and recovery following spinal cord injury. Respir Physiol Neurobiol 169:85–93
Hadley SD, Walker PD, Goshgarian HG (1999) Effects of serotonin synthesis inhibitor p-CPA on expression of the crossed phrenic phenomenon 4 h following C2 spinal cord hemisection. Exp Neurol 160(2):479–488

Hutchinson AJ, Williams M, De Jesus R, Yokoyama R, Oei HH, Ghai GR, Webb RL, Zoganas HC, Stone GA, Jarvis MF (1990) 2-(Arylakylamino) adenosine-5-uronamides. A new class of selective adenosine A2 receptor ligands. J Med Chem 33:1919–1924

James E, Nantwi KD (2006) Involvement of peripheral adenosine A2 receptors in adenosine A1-receptor mediated recovery of respiratory motor function following upper cervical spinal cord hemisection. J Spinal Cord Med 29(1):57–66

Johansson B, Ahlberg S, van der Ploeg I, Brene S, Lindefors N, Peterson H, Fredholm BB (1993) Effects of long term caffeine treatment on A1 and A2 adenosine receptor binding and on mRNA levels in rat brain. Naunyn Schmiedelbergs Arch Pharmacol 347:407–414

Kajana S, Goshgarian HG (2008) Administration of phosphodiesterase inhibitors and an adenosine A1 receptor antagonist induces phrenic nerve recovery in high cervical spinal cord injured rats. Exp Neurol 210(2):671–80

Klotz KN (2000) Adenosine receptors and their ligands. Naunyn Schmiedelbergs Arch Pharmacol 362(4):382–91

Latini S, Pedata F (2001) Adenosine in the central nervous system: release mechanisms and extracellular concentrations. J Neurochem 79:463–484

Latini S, Pazzagli M, Pepeu G, Pedata F (1996) A2 adenosine receptors. Their presence and neuromodulatory role in the central nervous system. Gen Pharmacol 27:925–933

Lewis JL, Brookhart JM (1951) Significance of the crossed phrenic phenomenon. J Neurophysiol 166:241–254

Li Y, Oskuian RJ, Day YJ, Rieger JM, Liu L, Kern JA, Linden J (2006) Mouse spinal cord compression injury model is reduced either by activation of A2A receptor on bone marrow-derived cells or on non-bone marrow-derived cells. Neuroscience 141:2029–2039

Maxwell DL, Fuller RW, Dolop KB, Dixon CMS, Hughes JMB (1986) Effects of adenosine on ventilatory responses to hypoxia and hypercapnia in humans. J Appl Physiol 1:1762–1766

McAdoo DJ, Robak G, Xu G-Y, Hughes MG (2000) Adenosine release upon spinal cord injury. Brain Res 854:152–157

McQueen DS, Ribeiro JA (1986) Pharmacological characterization of the receptor involved in chemoexcitation induced by adenosine. Br J Clin Pharmacol 88:615–620

Monteiro EC, Ribeiro JA (1987) Ventilatory effects of adenosine mediated by carotid body chemoreceptors in the rat. Nauyn-Schmiedeberg Arch Pharmacol 335:143–148

Monti D, Carley DW, Radulovacki M (1996) p-SPA, a peripheral adenosine A1 analog, reduces sleep apnea in rats. Pharmacol Biochem Behav 53(2):341–345

Murciano D, Aubier M, Vires N, Mal H, Pariente R (1987) Effects of theophylline and enprofylline on diaphragmatic contractility. J Appl Physiol 63(1):51–57

Nachazel J, Palecek F (1990) Aminophylline enhances ventilation in phrenicotomized rats. Eur Respir J 3:311–317

Nantwi KD, El-Bohy AA, Goshgarian HG (1996) Actions of systematic theophylline on hemidiaphragm recovery in rats following cervical spinal cord hemisection. Exp Neurol 140:53–59

Nantwi KD, Goshgarian HG (1998a) Effects of chronic systematic theophylline injections on recovery of hemidiaphragmatic function after cervical spinal cord injury in adult rats. Brain Res 789:126–129

Nantwi KD, Goshgarian HG (1998b) Theophylline-induced recovery in a hemidiaphragm paralyzed by hemisection in rats. Contribution of adenosine receptors. Neuropharmacology 37:113–121

Nantwi KD, Goshgarian HG (2001) Alkylxanthine-induced recovery of respiratory function following cervical spinal cord injury in adult rats. Exp Neurol 168(1):123–134

Nantwi KD, Basura JG, Goshgarian HG (2003a) Adenosine A1 mRNA expression and the effects of systemic theophylline on respiratory function four months after C2 hemisection. J Spinal Cord Med 26(4):364–371

Nantwi KD, Basura JG, Goshgarian HG (2003b) Effects of long-term theophylline exposure on recovery of respiratory function and expression of adenosine A1 mRNA in cervical spinal cord hemisected adult rats. Exp Neurol 182:232–239

Nantwi KD, Goshgarian HG (2002) Actions of specific adenosine receptor A1 and A2 compounds in recovery of phrenic motor output following upper cervical spinal cord injury in adult rats. Clin Exp Pharmacol Physiol 29(10):915–923

Nantwi KD, Goshgarian HG (2005) Adenosinergic mechanisms underlying recovery of diaphragm motor function following cervical spinal cord injury: potential therapeutic implications. Neurol Res 27:195–205

National Spinal Cord Injury Statistical Center (2008) Spinal cord injury: facts and figures at a glance. UAB, Database, 1–2

Newman D, Tamir J, Speedy L, Newman JP, Ben-Dov I (1994) Physiological and neuropsychological effects of theophylline in chronic obstructive pulmonary disease. Isr J Med Sci 30: 811–817

Nikodijevic O, Sarges R, Daly JW, Jacobson KA (1991) Behavioral effects of A1- and A2-selective adenosine agonists and antagonists: evidence for synergism and antagonism. J Pharmacol Exp Ther 259(1):286–294

Olson EB Jr, Vidruk EH, Dempsey JA (1988) Carotid body excision significantly changes ventilatory control in awake rats. Am J Physiol 64(2):666–671

Paterniti I, Melani A, Cipriani S, Corti F, Mello T, Mazzon E, Esposito E, Bramanti P, Cuzzocrea S (2011) Selective adenosine A2A receptor agonists and antagonists protect against spinal cord injury through peripheral and central effects. J Neuroinflammation. doi:10.1186/1742

Persson CGA, Anderson KE, Kjellin G (1986) Effects of enprofylline and theophylline may show the role of adenosine. Life Sci 38(12):1057–72

Petrov T, Alilain WJ, Kreipke C, Nantwi KD (2007) Differential expression of adenosine A1 and A2A receptors after upper cervical (C2) spinal cord hemisection in adult rats. J Spinal Cord Med 30(4):331–7

Reece TB, Tribble CG, Okonkwo DO, Davis JD, Maxey TS, Gazoni LM, Linden J, Kron IL, Kern JA (2008) Early adenosine receptor activation ameliorates spinal cord reperfusion injury. J Cardiovasc Med 9(4):363–367

Reid PG, Watt AH, Penny WJ, Newby AC, Smith AP, Routledge PA (1991) Plasma adenosine concentrations during adenosine-induced respiratory stimulation in man. Eur J Clin Pharmacol 40:175–180

Ribeiro JA (1999) Adenosine A2A receptor interactions with receptors for other neurotransmitters and neuromodulators. Eur J Pharmacol 375(1–3):101–113

Rivlin AS, Tator CH (1978) Effect of duration of acute spinal cord compression in a new acute cord injury model in the rat. Surg Neurol 10:38–43

Saharan RS, Nantwi KD (2006) Changes in the biochemical profiles of mid-cervically located adenosine A1 receptors after long term theophylline administration. J Spinal Cord Med 29(5):520–526

Schulz-Stubner S (2005) The use of small-dose theophylline for the treatment of bradycardia in patients with spinal cord injury. Anesth Analg 101(6):1809–1811

Sousa VC, Assaifa-Lopes N, Ribeiro JA, Pratt JA, Brett RR, Sebastiao AM (2011) Regulation of hippocampal cannabinoid CB1 receptor actions by adenosine A1 receptors and chronic caffeine administration: implications for the effects of Δ^9-tetrahydrocannabinol on spatial memory. Neuropsychopharmacology 36:472–487

Thomas T, Elnazir BK, Marshall JM (1994) Differentiation of the peripherally mediated from the centrally mediated influences of adenosine in the rat during systemic hypoxia. Exp Physiol 79:809–822

Tzelepis GE, Bascom AT, Safwan Badr M, Goshgarian HG (2006) Effects of theophylline on pulmonary function in patients with traumatic tetraplegia. J Spinal Cord Med 29(3):227–233

Varani K, Caramori G, Vincenzi F, Adcock I, Casolari P, Leung E, Maclennan S, Gessi S, Morello S, Barnes PJ, Ito K, Chung KF, Cavallesco G, Azzena G, Papi A, Borea PA (2006) Alteration of adenosine receptors in patients with obstructive pulmonary disease. Am J Respir Crit Care Med 173:398–406

Vinit S, Stamegna JC, Boulenguez P, Gauthier P, Kastner A (2007) Restorative respiratory pathways after partial cervical spinal cord injury: role of ipsilateral phrenic afferents. Eur J Neurosci 12:3551–3560

Watt AH, Routledge PA (1985) Adenosine stimulates respiration in man. Br J Pharmacol 20:503–506

Winslow C, Rozovsky J (2003) Effect of spinal cord injury on the respiratory system. Arch Phys Med Rehabil 82:803–814

Woodcock AA, Johnson MA, Geddes DM (1983) Theophylline prescribing, serum concentrations and toxicity. Lancet 2:610–613

Chapter 17
Adenosine and Pain

Jana Sawynok

Abstract Adenosine A_1 receptors (A_1Rs) have been shown to be involved in antinociception in preclinical models for several decades. Thus, systemic, peripheral, spinal, and supraspinal administration of A_1R agonists universally produces antinociception in nociceptive, inflammatory, and neuropathic pain models. The clinical potential for adenosine (given intravenously), or A_1R ligands (given systemically or spinally) to produce analgesia in humans was supported by earlier trials, but more recent, larger controlled trials have generally not demonstrated analgesic activity for postoperative pain. Adenosine A_{2A}Rs have more complex effects on pain, generating pronociceptive effects peripherally and spinally, but antinociceptive effects supraspinally. The presence of A_{2A}Rs on astrocytes and microglia within the spinal cord may be particularly important for their pronociceptive actions in states of nerve injury. There is also a report that ultralow doses of A_{2A}R agonists produce long lasting antinociception in such states. Manipulation of endogenous levels of adenosine by inhibiting adenosine kinase represented a promising novel approach, but development in this area is no longer active. Recent data have demonstrated that specific ectonucleotidases are localized on sensory afferent neurons, and that spinal delivery of recombinant forms of these enzymes produce long-lasting antinociceptive actions; this led to the suggestion that manipulating ectonucleotidases may represent a potential new approach for development. Additional observations have implicated tissue release of nucleotides and adenosine in acupuncture analgesia, and shown analgesia results from peripheral actions at adenosine A_1Rs. Finally, other recent observations indicate that caffeine, which inhibits both A_1- and A_{2A}Rs with high affinity, blocks antinociception in preclinical studies by several drugs currently used to treat pain in humans. As caffeine is widely consumed, it will be important to attend to caffeine intake in future trial design with

J. Sawynok (✉)
Department of Pharmacology, Dalhousie University,
Halifax, Nova Scotia B3H 4R2, Canada
e-mail: Jana.Sawynok@dal.ca

S. Masino and D. Boison (eds.), *Adenosine: A Key Link between Metabolism and Brain Activity*, DOI 10.1007/978-1-4614-3903-5_17,

respect to evaluating novel therapies that use these receptor systems, some existing analgesics, as well as acupuncture analgesia.

Keywords Adenosine • Antinociception • Analgesia • Adenosine kinase inhibitors • Ectonucleotidases • Acupuncture • Caffeine • Spinal cord

17.1 Introduction

Adenosine is a ubiquitous endogenous cell signaling and regulatory agent that acts at four extracellular G-protein coupled receptors (Rs) –A_1R, $A_{2A}R$, $A_{2B}R$, and A_3R; it has similar in vitro affinities for A_1- and $A_{2A}Rs$ and a lower affinity for A_{2B}- and A_3Rs (Dunwiddie and Masino 2001; Fredholm et al. 2001a). The functions of adenosine are elaborated in several contexts—physiological regulation (fine-tuning of systems, restoration of homeostasis), nonphysiological regulation (role in perturbed states, attempts to restore homeostasis), and a role in dysfunctional or disease states; endogenous tissue levels of adenosine, relative affinity for receptor subtypes, and interactions with other systems (receptors, transducers) orchestrate the actions of adenosine in these various states. Nociceptive signaling and pain involve sensory afferent input which brings information from the periphery into the central nervous system (CNS) via the dorsal spinal cord or brainstem, transmission of such information via ascending systems to supraspinal regions, descending modulation of such signaling, and cognitive interpretation. Adenosine receptors are positioned at many steps and on many cell types within these general pathways, and can regulate signaling in nociceptive (physiological), inflammatory, and neuropathic pain states.

In preclinical models, the antinociceptive properties of A_1R ligands, given systemically and spinally, were elaborated during the 1980s; in the 1990s, additional sites of action in the periphery were identified and efficacy in neuropathic pain states was demonstrated when nerve injury models were developed. This body of information led to exploration of the potential for adenosine and A_1R ligands to represent novel analgesics, and the clinical potential was explored by administering adenosine itself or selective A_1R ligands. $A_{2A}Rs$ exhibit pain facilitatory actions in several models, and while $A_{2A}Rs$ have not been explored as analgesic targets, recent data with gene-deletion animals and selective $A_{2A}R$ antagonists, and some intriguing observations with ultralow doses of $A_{2A}R$ agonists, suggest that this receptor may be worthy of investigation in this arena. A further approach has been to increase endogenous levels of adenosine by inhibiting metabolism via adenosine kinase, or by increasing generation by manipulating ectonucleotidase activity, and recent developments with this latter approach have provided an interesting perspective. Finally, adenosine appears to contribute to antinociception by several pharmacological agents, as well as by certain nonpharmacological approaches. The present review considers recent developments with respect to (a) receptor-mediated regulation of pain, (b) the potential for manipulating endogenous adenosine as a therapeutic approach, and (c) the involvement of endogenous adenosine in pain relieving actions of several modalities currently used for treating pain.

17.2 Adenosine A_1 Receptors and Pain

There is an extensive preclinical literature indicating that adenosine ligands, via activation of A_1Rs, produce pain relieving effects in nociceptive, inflammatory, and neuropathic pain models (Sawynok 1998; Dickenson et al. 2000). Studies using A_1R knockout (A_1R-/-) mice reveal there is enhanced pain signaling following inflammation and nerve injury (Johansson et al. 2001; Wu et al. 2005; Zylka et al. 2008), and this indicates that deficiencies in endogenous adenosine signaling can contribute to regulating nociception in these states. A_1Rs are expressed throughout the neuraxis at many sites relevant to pain transmission and modulation (Riberio et al. 2003). A_1Rs are expressed on small- and medium-sized sensory neurons, and colocalize with markers for several classes of C-fiber sensory neurons (TRPV1, substance P and IB_4)—few receptors are on large diameter sensory neurons (Lima et al. 2010). Within the dorsal spinal cord, A_1Rs are concentrated in the substantia gelatinosa (Reppert et al. 1991; Deuchars et al. 2001; Ackley et al. 2003; Schulte et al. 2003) and are prominent on intrinsic neurons (Choca et al. 1988; Schulte et al. 2003); they are also observed in the intermediolateral cell column and on motor neurons (Deuchars et al. 2001). Within supraspinal structures, A_1Rs are present in areas known to be involved in pain modulation (e.g., periaqueductal gray, brainstem) (Maione et al. 2007). Mechanisms involved in pain modulation by A_1Rs at peripheral sites, within the spinal cord, and at supraspinal sites are summarized in Fig. 17.1.

Spinal and peripheral sites have been examined the most extensively, but supraspinal sites and discrete brain regions are now explicitly implicated in adenosine actions (Bagley et al. 1999; Tanase et al. 2002; Ramos-Zepeda et al. 2004; Maione et al. 2007). The effects of systemically administered agents are often attributed to spinal actions at A_1Rs—this site of action is well characterized in terms of efficacy in many pain states and with respect to mechanisms of action, and there is direct evidence for this. Thus, when an adenosine kinase inhibitor (A-134974) (presumed to act via A_1Rs) was examined for its site of action, intrathecal (i.t.) spinal delivery exhibited a lower ED_{50} compared to intracerebroventricular (i.c.v., supraspinal) or intraplantar (i.pl., peripheral) delivery against inflammatory hyperalgesia (McGaraughty et al. 2001) and against allodynia following nerve injury (Zhu et al. 2001). Furthermore, i.t. administration of theophylline (nonselective adenosine receptor antagonist) antagonized the action of the systemically administered agent in both tests, while i.c.v. administration was not effective in the inflammatory hyperalgesia model. For other analgesics (e.g., morphine, acetaminophen), site-site synergy is known to occur (Kolesnikov et al. 1996; Raffa et al. 2000) but this phenomenon has not been examined directly for A_1R agonists.

The clinical potential for A_1Rs to represent a novel approach to analgesia has been explored by administering adenosine via intravenous (i.v.) infusions or via spinal administration. The earlier clinical literature investigated analgesic actions of i.v. and i.t. adenosine on nociceptive threshold tests in healthy volunteers, in ischemic pain models, in skin inflammation or sensitization models, and in chronic neuropathic pain, and reported antinociception in several instances—this earlier literature

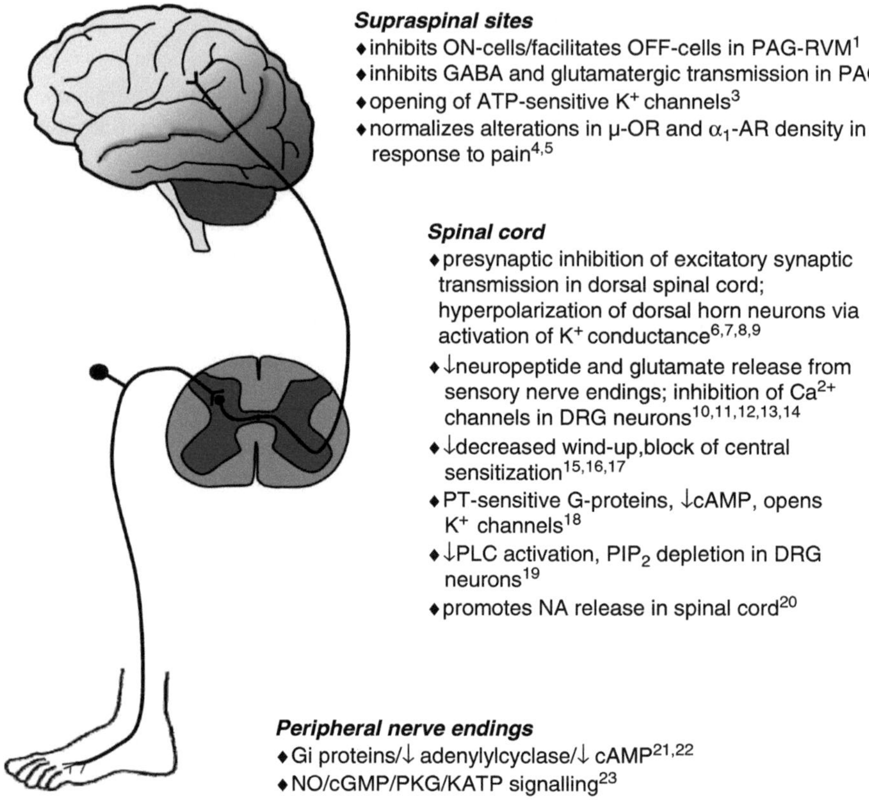

Fig. 17.1 Mechanisms involved in antinociception by adenosine A_1 receptors at various levels of the pain signaling neuraxis. *References*: [1]Maione et al. 2007; [2]Bagley et al. 1999; [3]Ocaña and Bayens 1994; [4]Borghi et al. 2002; [5]Nelapa et al. 2010; [6]Li and Perl 1994; [7]Patel et al. 2001; [8]Tian et al. 2010; [9]Lao et al. 2004; [10]Santicioli et al. 1993; [11]Haas and Selbach 2000; [12]Carruthers et al. 2001; [13]Mauborgne et al. 2002; [14]Li and Eisenach 2005; [15]Reeve and Dickenson 1995; [16]Curros-Criado and Herrero 2005; [17]Gong et al. 2010; [18]Zahn et al. 2007; [19]Sowa et al. 2010a; [20]Bantel et al. 2003; [21]Taiwo and Levine 1990; [22]Khasar et al. 1995; [23]Lima et al. 2010. *Abbreviations*: *α_1-AR* α_1-Adrenergic receptor, *cAMP* Cyclic adenosine 3′,5′-monophosphate, *cGMP* Cyclic 3′,5′-guanosine monophosphate, *DRG* Dorsal root ganglion, *GABA* Gamma-aminobutyric acid, *NA* Noradrenaline, *NO* Nitric oxide, *μ-OR* μ-Opioid receptor, *PAG* Periaqueductal gray, *PIP_2* Phosphoinositidylinositol 4,5-bisphosphate, *PKG* Protein kinase G, *PLC* Phospholipase C, *PT* Pertussis toxin, *RVM* Rostroventral medulla

has been systematically reviewed (Hayashida et al. 2005). There is also a clinical literature on effects of i.v. adenosine in a perioperative setting reporting stable hemodynamics, anesthetic sparing effects (reduced requirement for anesthetic during surgery), and improvement in postoperative pain (lower pain scores, less opioid consumption) (Hayashida et al. 2005; Gan and Habib 2007).

The most recent published clinical trials using i.t. and i.v. adenosine, and i.v. delivery of a selective A_1R agonist in postoperative pain settings are presented in

Table 17.1 Recent clinical trials examining analgesic actions of adenosine and related ligands

Method	Indication	Outcome	References
i.t. Adenosine 1,000 μg 30 min preanesthesia or following surgery	Postsurgical pain; hysterectomy ($N=90$ subjects)	No analgesia with early or late applications compared to no drug	Sharma et al. (2006)
i.v. GR79236X[a] 4 or10 μg kg^{-1} for 15 min	Third molar extraction ($N=79$ subjects)	No analgesia compared to saline; trial included active comparator, diclofenac, which exhibited analgesia	Sneyd et al. (2007)
i.v. Adenosine 25,50,100, 200 μg kg^{-1} min^{-1}	Gynecological surgery ($N=166$ subjects)	No difference in opioid Consumption in any group	Habib et al. (2008)
i.v. ATP160 μg kg^{-1} min^{-1}	Orofacial surgery ($N=30$ subjects)	Opioid sparing effects lasting 72 h	Handa et al. (2009)

[a]See the end of text for abbreviations of adenosine ligand names

Table 17.1. Sharma et al. (2006) observed no analgesia with i.t. adenosine given 30 min prior to anesthesia, or following completion of hysterectomy surgery. Sneyd et al. (2007) observed no analgesia with GR79236X, a selective A_1R agonist, following third molar extraction using a dose that had modestly inhibited experimental pain. Habib et al. (2008) tested four doses of i.v. adenosine (25, 50, 100, 200 μg kg^{-1} min^{-1}) for gynecological surgery; infusions lasted from the time of incision to the end of surgery, and no analgesia was observed in any group. In considering differences from previous reports, Habib et al. suggested that methodological issues in earlier studies may have contributed to positive outcomes. Furthermore, they noted that in two major trials where analgesia *was* reported, this was in comparison with infusions of remifentanil—the analgesia reported with adenosine in those trials could be apparent and due to the acute opioid tolerance and hyperalgesia which is known to occur with remifentanil. Collectively, these more recent, larger, and controlled trials question earlier positive reports of analgesic properties of adenosine in humans. A recent survey of unpublished clinical trials that target adenosine receptors for pain (via www.clinicaltrials.gov) reveals no completed studies that demonstrate the usefulness of this approach (Zylka 2011).

A further set of observations relevant to the analgesic potential of adenosine in humans occurs with i.v. infusions of ATP. Such infusions produce anesthetic sparing effects during surgery, and analgesia in chronic pain subjects (Hayashida et al. 2005). ATP is converted rapidly to adenosine by ectonucleotidases, and its analgesic actions are attributed to actions on A_1Rs (Hayashida et al. 2005). Handa et al. (2009) recently have reported that i.v. ATP (160 μg kg^{-1} min^{-1}), given immediately after the loss of consciousness until the end of surgery, produced a significant opioid-sparing effect (less morphine consumption) for 72 h following orofacial surgery. One difference between Handa et al. (2009) and the Habib et al. (2008) study noted above was the time of initiation of treatment infusions (loss of consciousness in former, time of incision in latter), and this could contribute to different outcomes. The potential for preemptive effects of adenosine given at different times in relation to surgery will need to be examined in preclinical models to better understand such observations.

Neuropathic pain is a condition that has received particular attention with respect to potential analgesic actions of adenosine. In humans, several studies report antinociception with i.v. infusions of adenosine (50 μg kg^{-1} min^{-1} for 45–60 min) for neuropathic pain (Hayashida et al. 2005). ATP has also been given i.v. (160 μg kg^{-1} min^{-1} for 60 min) for neuropathic pain in an open-label study of intractable orofacial pain (Fukuda et al. 2007). Ten of 16 patients with neuropathic pain, but 0/8 with nonneuropathic pain, responded with reductions in pain lasting 10–26 h. Efficacy was again attributed to A_1R activation following metabolic conversion to adenosine. Preclinical studies indicate actions of A_1Rs on several aspects of central sensitization (Reeve and Dickenson 1995; Suzuki et al. 2000; Curros-Criado and Herrero 2005; Gong et al. 2010), providing plausibility for this indication. However, in a recently developed and novel conditioned place preference model for examining affective actions of drugs on spontaneous pain following nerve injury, spinal administration of adenosine has been without effect (King et al. 2009). Whether neuropathic pain in humans is amenable to adenosine-based mechanisms remains to be determined.

17.3 Adenosine A_{2A} Receptors and Pain

There is a considerable body of evidence that suggests the role of $A_{2A}Rs$ in pain signaling is to facilitate pain (pronociception), and this includes the following observations. (1) Peripheral administration of $A_{2A}R$ agonists produces pronociceptive actions when administered locally to the hindpaw of rodents in several models (Taiwo and Levine 1990; Karlsten et al. 1992; Doak and Sawynok 1995; Li et al. 2010). (2) Systemic administration of a selective $A_{2A}R$ antagonist (SCH58261) produces antinociception in several preclinical models (Bastia et al. 2002; Godfrey et al. 2006; Hussey et al. 2007). (3) Gene deletion of $A_{2A}Rs$ produces hypoalgesia (reduced pain thresholds and responses) in nociceptive (Ledent et al. 1997), inflammatory (Hussey et al. 2007; Li et al. 2010), and neuropathic pain models (Bura et al. 2008). There are also reports of antinociception with CGS21680, a selective $A_{2A}R$ agonist, given i.t. (Lee and Yaksh 1996; Poon and Sawynok 1998). However, CGS21680 binds to adenosine A_1Rs at higher doses (Fredholm et al. 2001a), and this may contribute to activity with methods that deliver high local concentrations of drug. Ideally, in order to clearly implicate $A_{2A}Rs$ in a pharmacological process, selective receptor antagonists or gene deletion animals need to be used.

$A_{2A}Rs$ may influence pain signaling at several sites. In sensory afferents, mRNA for $A_{2A}Rs$ has been identified in dorsal root ganglia (DRGs) (Kaelin-Lang et al. 1998; Li et al. 2010); however, it has been difficult to demonstrate the presence of

A_{2A}Rs in DRGs using immunohistochemistry (Li et al. 2010). In the dorsal spinal cord, the presence of A_{2A}Rs has been reported in rats (Choca et al. 1988; Brooke et al. 2004), but not in mice (Bailey et al. 2002). At supraspinal sites, A_{2A}Rs are predominantly localized in the striatum which does not play a major role in pain transmission; however, other regions (somatosensory cortex, amygdala, thalamus, hypothalamus) which are involved in pain integration, do express lower levels of A_{2A}Rs (Ferré et al. 2007). A_{2A}Rs on glial cells regulate release of inflammatory mediators that may contribute to pain and central sensitization (Boison et al. 2010). A_{2A}R expression in microglia and astrocytes is usually low under physiological conditions, but expression is induced following brain insult and inflammatory signals (Boison et al. 2010). In neuropathic pain conditions, glia are activated and functionally involved in central sensitization (DeLeo et al. 2006; Scholz and Woolf 2007). Such a role for A_{2A}Rs is consistent with recent data that indicates suppression of microglial and astrocytic activation in response to nerve injury in the spinal cord in A_{2A}R -/- mice, as well as with attenuation of mechanical allodynia and thermal hyperalgesia following nerve injury (Bura et al. 2008).

There are recent intriguing observations that suggest A_{2A}R mechanisms can lead to antinociception at spinal sites. Thus, Loram et al. (2009) reported that a single i.t. injection of the selective A_{2A}R agonists ATL313 and CGS21680 10–14 days following surgery (chronic constriction injury in rats) produced a long-lasting reversal of mechanical allodynia and thermal hyperalgesia for 4 weeks. The effect of ATL313 was reversed by coadministration of ZM241385 (selective A_{2A}R antagonist), but no reversal was seen 1 week later. In contrast, delayed effects were reversed with i.t. administration of antibody to the anti-inflammatory cytokine interleukin-10 (IL-10). Finally, spinal administration of ATL313 reduced activation of microglia and astrocytes in the spinal cord following nerve injury. Such observations suggest a protracted pain relieving effect resulting from an initial activation of A_{2A}R, but sustained in the longer-term by release of IL-10. It is important to note two features of this study. (1) Drugs were administered via an acutely inserted lumbar entry intrathecal cannula, but many other studies used rats with chronic cervical entry intrathecal cannulas; it is now appreciated that implantation of the latter cannulas leads to activation of spinal microglia and astrocytes (DeLeo et al. 1997) and this could influence the spinal pharmacology of A_{2A}R agonists (Bura et al. 2008). (2) The dose of CGS21680 that led to the long-lasting antihyperalgesic effect was 1 pmol, which is considerably lower than those used in other studies (1–40 nmol) (Lee and Yaksh 1996; Poon and Sawynok 1998). Opioids produce ultralow dose sensitizing effects that contrast with their analgesic effects at more conventional doses (Crain and Shen 2000), and it is possible that A_{2A}Rs also exhibit dual effects—only in this case the effect of an ultralow dose is to produce *antinociception*. This possibility will require direct exploration.

Another recent report indicates that a novel method of activation of A_{2A}Rs leads to antinociception. Thus, i.c.v. administration of an antibody to A_{2A}Rs in mice, which leads to in vitro activation of A_{2A}Rs, produces antinociception in several nociceptive tests (thermal threshold)—this is completely blocked by a selective A_{2A}R antagonist (ZM241385) (By et al. 2011). In accounting for the mechanisms involved, the authors considered supraspinal effects of A_{2A}Rs on motor systems because of the i.c.v. method of delivery and the prominence of such receptors in brain regions controlling motor function and areas involved in pain integration (Ferré et al. 2007). The effect was partially blocked by naloxone, and a role for opioid receptors, perhaps in the PAG, was considered (By et al. 2011). An earlier report had shown that i.c.v administration of the A_{2A}R ligand CGS21680 produced antinociception in mice, and implicated calcium-activated K^+ channels in such actions (Regaya et al. 2004). However, the selectivity of CGS21680 actions was not verified using additional approaches.

17.4 Adenosine A_{2B} and A_3 Receptors and Pain

Adenosine has a higher affinity for human A_1, A_{2A}, and A_3 receptors than for A_{2B} receptors—it exhibits physiological regulatory actions at the former receptors, while the latter may be involved in pathophysiological actions (Fredholm et al. 2001a). A_3Rs exhibit complex effects on inflammation and immune responses, and exhibit both anti- and proinflammatory effects, as well as immunosuppressive and immunostimulant effects (Gessi et al. 2008). A_{2B}Rs promote inflammatory responses, and may contribute to inflammatory and immune system conditions (Haskó et al. 2009). Effects of these receptors on pain are likely to be secondary to their effects on these responses.

Some studies have directly examined the influence of A_3Rs and A_{2B}Rs on nociception. The local peripheral administration of A_3R agonists to the rodent hindpaw results in enhanced nociception and edema, and this involves release of histamine, serotonin and neuropeptides (Reeves et al. 1997; Sawynok et al. 1997). Local peripheral administration of antagonists selective for both A_{2B}Rs and A_3Rs inhibits pain behaviors and paw edema resulting from an inflammatory stimulus (Bilkei-Gorzo et al. 2008). Furthermore, systemic administration of selective A_{2B}R antagonists also produces antinociceptive actions in a nociceptive threshold test (Abo-Salem et al. 2004). A_3R-/- mice exhibit reduced hyperalgesia and edema following an inflammatory stimulus (Wu et al. 2002). Mice lacking A_{2B}Rs have now been developed (Yang et al. 2006), but these have not been characterized for effects on nociception. Collectively, these observations support peripheral pronociceptive and proinflammatory actions for these two receptor subtypes.

17.5 Recruitment of Endogenous Adenosine by Altering Metabolism

Extracellular adenosine, which is then available to act on cell surface receptors, can be released from many cell types either as adenosine itself, or as nucleotides (e.g., ATP, cyclic AMP) which are subsequently converted to adenosine following metabolism by ectonucleotidases—such release is enhanced by conditions of cell stress such as hypoxia and by tissue injury (Dunwiddie and Masino 2001; Fredholm et al. 2001a). Efflux from cells can occur following vesicular release, or by mechanisms involving transporters or anion channels which can vary in importance according to cell type. Factors regulating endogenous generation and metabolism of adenosine, specifically in relation to pain, have recently been reviewed (Zylka 2011). Adenosine in the extracellular space can (a) interact with cell surface adenosine receptors, (b) be metabolized by adenosine deaminase extracellularly (to inosine), (c) be removed from the extracellular space by equilibrative transporters, or (d) be metabolized by adenosine kinase (to AMP) inside the cell.

Inhibitors of adenosine metabolism, and particularly of adenosine kinase, have received attention as potential therapeutic agents. Such agents exhibit a wider therapeutic index (separation of doses producing analgesic effects compared to cardiovascular and motor effects) than do directly acting adenosine receptor agonists, and were considered as a potential novel therapeutic strategy for several conditions (McGaraughty et al. 2005). The antinociceptive profile of the adenosine kinase inhibitor A-134974 was characterized widely in a range of pain models (nociceptive, inflammatory, neuropathic) and with respect to sites of action (spinal vs. supraspinal and peripheral) (McGaraughty et al. 2001; Zhu et al. 2001). Other inhibitors (ABT-702, ABT-501) also were characterized for their profile of activity in several pain models and for their relative potential adverse effects (McGaraughty et al. 2005). While such results appeared promising, clinical development of this class of agents was halted because of a preliminary report of CNS hemorrhage in preclinical models (Erion et al. 2000; McGaraughty et al. 2005). Whether this side effect was due to elevated levels of adenosine or was unrelated to adenosine is unknown.

Inhibitors of adenosine deaminase, in contrast to adenosine kinase inhibitors, did not exhibit antinociception when given alone (Keil and DeLander 1994; Goldman et al. 2010). However, when combined with exogenous adenosine (Keil and DeLander 1994), or procedures that increase endogenous levels of adenosine (Goldman et al. 2010), the adenosine deaminase inhibitor deoxycoformycin produced a marked enhancement of antinociception. Furthermore, in biochemical studies, deoxycoformycin produced a marked enhancement in recovery of adenosine from the spinal cord (Golembiowska et al. 1996) and in the periphery (Goldman et al. 2010). It appears that, provided there is a stimulus for enhancing extracellular adenosine availability, inhibition of adenosine deaminase can impact that event. Curiously, inosine, which is the end product of adenosine deamination, itself leads to antinociception when given systemically, spinally, and supraspinally, and this activity is seen in nociceptive, inflammatory and neuropathic pain tests (Nascimento et al. 2010). A_1Rs are involved in systemic inosine actions, as antinociception is blocked by selective inhibitors of A_1Rs (DPCPX, 8-phenyltheophylline) (Nascimento et al. 2010). The selective A_{2A}R antagonist (ZM241385) also inhibited the action of inosine (Nascimento et al. 2010). Inosine has limited direct affinity for A_1Rs and does not bind to A_{2A}Rs (Fredholm et al. 2001b), so the nature of the adenosine receptor involvement in these inosine actions will require elaboration.

A recent body of work has suggested that altering endogenous adenosine levels specifically within the vicinity of sensory afferent neurons by manipulating ecto-nucleotidase activity may be a promising area for exploration in relation to pain. Sensory neurons have been known to release adenosine and ATP for some time (Sawynok and Liu 2003), and release of both entities is altered by nerve injury (Liu et al. 2002; Matsuka et al. 2008). Recently, prostatic acid phosphatase (PAP), which is now known to convert AMP to adenosine, has been shown to be expressed in peptidergic and nonpeptidergic sensory neurons, and to be localized in axon terminals in the substantia gelatinosa of the spinal cord (Taylor-Blake and Zylka 2010). I.t. administration of PAP leads to long-lasting (several days) antinociception in inflammatory and neuropathic pain models—this involves A_1Rs, as it was inhibited by a selective A_1R antagonist (transiently, on the day of injection of the antagonist) and abolished in A_1R-/- mice (Zylka et al. 2008; Sowa et al. 2009). Ecto-5′-nucleotidase (NT5E) colocalizes with PAP in peptidergic and nonpeptidergic sensory neurons, and also is present on axon terminals in the substantia gelatinosa, as well as on nerve endings in the epidermis (Zylka et al. 2008; Sowa et al. 2010b). Both PAP-/- and NT5E-/- mice exhibit reduced hydrolysis of AMP and hypersensitivity in several pain models, indicating that these ectonucleotidases are involved in generating adenosine that leads to tonic regulation of sensory transmission (Zylka et al. 2008; Sowa et al. 2010b). Tonic regulation of excitatory transmission within the spinal cord by A_1Rs has recently also been demonstrated using electrophysiological approaches (Tian et al. 2010). As with PAP, i.t. administration of recombinant NT5E leads to long-lasting (several days) antinociception in several preclinical models, and this involves A_1Rs, as it is no longer observed in A_1R-/- mice (Sowa et al. 2010c). The durability of the sensory effect was notable, and it has been suggested that applications of recombinant nucleotidases might represent a novel class of analgesic agents with therapeutic applications and advantages compared to other agents (Sowa et al. 2009). The selective localization of ectonucleotidases in proximity to sensory afferents and key areas of transmission in the dorsal spinal cord might contribute to the unique analgesic efficacy of ATP infusions (Sect. 17.2).

17.6 Adenosine and Nonpharmacological Modalities for Pain

Acupuncture has been used for thousands of years and is an integral part of Traditional Chinese Medicine. Since the 1990s, with the formation of the National Centre for Complementary and Alternative Medicine and publication of National Institutes of Health consensus conference proceedings (NIH Consensus Conference 1998), acupuncture has received increasing attention. Several biological mediators for electroacupuncture analgesia (e.g., endorphins, monoamines, peptides, hormones) have been identified (Zhao 2008), but those involved in the more traditional needle insertion and mechanical rotation methods of acupuncture are less well understood. Recently, in an elegant series of experiments, adenosine has been implicated as a mediator of acupuncture analgesia (Goldman et al. 2010). This report demonstrated the following: (1) increased tissue release of adenosine, AMP, ADP, and ATP at stimulated acupoints, as determined by tissue microdialysis; (2) local injection of a selective A_1R agonist (CCPA) into the acupoint reduced inflammatory pain and sensory hypersensitivity following nerve injury; (3) absence of effects of local CCPA and acupuncture in A_1R-/- mice; (4) marked augmentation of acupuncture analgesia by deoxycoformycin, an inhibitor of adenosine deaminase. Needle insertion and rotation alters collagen fiber alignment and fibroblast conformation (Langevin et al. 2006), and such changes may lead to ATP release from several cell types within the tissue close to the needle insertion. ATP has recently been shown to be released from sensory nerves in response to activity within the nerve (Fields and Ni 2010), and acupoints in proximity to nerve tracts that result in activity could lead to ATP release. While A_1Rs are known to be present at several steps along pain signaling pathways (Sect. 17.2), adenosine has a short tissue half-life and its actions are most likely to be mediated peripherally close to the site of its endogenous generation. The ability of selective A_1R agonists administered locally into the acupuncture site to produce antinociception (Goldman et al. 2010) provides strong supportive evidence for such a local mechanism.

There are practical implications to the involvement of adenosine and A_1Rs in acupuncture analgesia. Thus, caffeine, which is widely consumed by 80–90% of adults for its stimulant actions (Frary et al. 2005), blocks A_1Rs and $A_{2A}Rs$ with a high affinity (Fredholm et al. 1999) and could potentially interfere with the clinical efficacy of acupuncture (Zylka 2010). Prior administration of caffeine has been reported to interfere with analgesia resulting from transcutaneous electrical nerve stimulation (TENS) in an experimental trial in humans (Marchand et al. 1995). While refraining from caffeine intake prior to an acupuncture session could certainly be considered within trial design, the analgesic efficacy of acupuncture is often determined over extended intervals—e.g., at 2, 4, and 6 months following 1–2 weekly treatments over a course of 6 weeks—and the potential influence of caffeine may need to be determined pragmatically. This could involve monitoring caffeine intake and determining if there are different outcomes in low and high caffeine intake groups (Sect. 17.7).

17.7 Involvement of Adenosine in the Action of Drugs Used to Manage Pain

In recent years, adenosine has been implicated in antinociception produced by several pharmacological agents that are currently used to treat pain. This is based on the ability of low doses of caffeine (10 mg/kg) and of a selective A_1R antagonist (DPCPX) to inhibit antinociception by such agents in several preclinical models of pain. Inhibition of antinociception by caffeine has been observed with amitriptyline (Esser and Sawynok 2000; Sawynok et al. 2008), carbamazepine and oxcarbazepine (Tomić et al. 2004; Sawynok et al. 2010), allopurinol (Schmidt et al. 2009), and even acetaminophen (Godfrey et al. 2006). Caffeine is widely used as an adjuvant analgesic in combination with acetaminophen (Palmer et al. 2010), and both pharmacokinetic and pharmacodynamic mechanisms are implicated in augmentation of analgesia (Granados-Soto and Castaneda-Hernández 1999; Renner et al. 2007). In view of this, the ability of low doses of caffeine, which are relevant to dietary intake levels, to inhibit antinociception by acetaminophen is both curious and worthy of attention. Given the widespread prevalence of caffeine use in society, it is important that the potential influence of caffeine intake on the clinical effects of such agents is addressed. The most practical design for such clinical studies may be to monitor caffeine intake and consider outcomes in relation to this intake—this approach has been used to determine whether caffeine interferes with the anti-inflammatory action of methotrexate, another agent where endogenous adenosine is implicated in its biological actions (Nesher et al. 2003; Benito-Garcia et al. 2006).

17.8 Conclusions and Future Directions

Despite considerable promise for adenosine-based mechanisms to offer a novel target for pain management, more recent larger and well-controlled clinical studies have provided limited results in terms of analgesia with adenosine and A_1R ligands (Table 1; Zylka 2011). The lack of translation of preclinical observations into the clinical domain is not unique to adenosine, and reasons potentially involved in lack of translation have been considered (Rice et al. 2008). Despite this, there is an emerging potential for prodrugs to be able to recruit adenosine-based mechanisms for analgesia. This approach is supported by clinical observations using i.v. infusions of ATP in several pain conditions (Hayashida et al. 2005; Fukuda et al. 2007; Handa et al. 2009), and preclinical observations which indicate that spinal administration of recombinant forms of ectonucleotidases leads to long-lasting antinociception (Zylka et al. 2008; Sowa et al. 2009, 2010c). Unique forms of localized delivery, such as injections into acupoints, have also been suggested. Thus, injections of A_1R agonists into acupuncture points in preclinical models results in antinociception (Goldman et al. 2010), and it has been proposed that this could also be done with ectonucleotidases (Zylka 2011). This would be a variation on injections of local anesthetics (or steroids, or botulinum toxin) into trigger points, which is an approach currently used in pain management. Finally, the intriguing long-lasting analgesic effect of ultralow doses of $A_{2A}R$ agonists (Loram et al. 2009) needs further exploration, both to replicate the observation and to understand possible mechanisms involved. The possibility of developing novel agents based on adenosine systems to manage pain appears to be entering a new phase of elaboration, one in which the complexity of adenosine receptor signaling and pain is understood more completely. This elaboration will need to use models specifically relevant to particular clinical conditions (Rice et al. 2008), and consider the clinical context more closely (e.g., perisurgical administration where opioids are routinely used; drug effects in the presence and absence of opioids need to be evaluated). Finally, there will be a need to attend to caffeine consumption, as chronic caffeine intake could compromise clinical effects using such approaches.

Abbreviations for adenosine ligands referred to in text and tables:

ATL313 4-(3-(6-amino-9-(5-cyclopropylcarbamoyl-13,4-dihydrotetrahydrofuran-2-yl)-9*H*-purin-2-yl)prop-2-ynyl)piperidine-1-carboxylic acid methyl ester, *CCPA* 2-chloro-*N*(6)-cyclopentyladenosine, *CGS21680* 2-*p*-(2-carboxyethyl)phenethylamino-5′-*N*-ethylcarboxamido adenosine HCl, *DPCPX* 1,3-dipropyl-8-cyclopentylxanthine, *GR79236X* *N*-[(1*S*,*trans*)-2-hydroxycyclopentyl]adenosine, *SCH58261* 5-amino-7-(2-phenylethyl)-2-(2-furyl)pyrazolo[4,3-*e*]-1,2,4-triazolo[1,5-*c*]pyrimidine, *ZM241385* 4-(2-[7-amino-2-(2-furyl)(1,2,4)triazolo(2,3-a)(1,3,5)triazin-5-ylamino]ethyl)phenol.

References

Abo-Salem OM, Hayallah AM, Bilkei-Gorzo A, Filipek B, Zimmer A, Müller CE (2004) Antinociceptive effects of novel A_{2B} adenosine receptor antagonists. J Pharmacol Exp Ther 308:358–366

Ackley MA, Governo RJM, Cass CE, Young JD, Baldwin SA, King AE (2003) Control of glutamatergic neurotransmission in the rat spinal dorsal horn by nucleoside transporter ENT1. J Physiol 548:507–517

Bagley EE, Vaughan CW, Christie MJ (1999) Inhibition by adenosine receptor agonists of synaptic transmission in rat periaqueductal grey neurons. J Physiol 516:219–225

Bailey A, Matthes H, Kieffer B, Slowe S, Hourani SMO, Kitchen I (2002) Quantitative autoradiography of adenosine receptors and NBTI-sensitive adenosine transporters in the brains and spinal cords of mice deficient in the μ-opioid receptor gene. Brain Res 943:68–79

Bantel C, Li X, Eisehach JC (2003) Intraspinal adenosine induces spinal cord norepinephrine release in spinal nerve-ligated rats but not in normal or sham controls. Anesthesiology 98:1461–1466

Bastia E, Varani K, Monopoli A, Bertorelli R (2002) Effects of A_1 and A_{2A} adenosine receptor ligands in mouse acute models of pain. Neurosci Lett 328:241–244

Benito-Garcia E, Heller JE, Chibnik LB, Maher NE, Matthews HM, Bilics JA, Weinblatt ME, Shadick NA (2006) Dietary caffeine intake does not affect methotrexate efficacy in patients with rheumatoid arthritis. J Rheumatol 33:1275–1281

Bilkei-Gorzo A, Abo-Salem OM, Hayallah AM, Michel K, Müller CE, Zimmer A (2008) Adenosine receptor subtype-selective antagonists in inflammation and hyperalgesia. Naunyn-Schmiedeberg's Arch Pharmacol 377:65–76

Boison D, Chen JF, Fredholm BB (2010) Adenosine signalling and function in glial cells. Cell Death Differ 17:1071–1082

Borghi V, Przewlocka B, Labuz D, Maj M, Ilona O, Pavone F (2002) Formalin-induced pain and μ-opioid receptor density in brain and spinal cord are modulated by A_1 and A_{2A} adenosine agonists in mice. Brain Res 956:339–348

Brooke RE, Deuchars J, Deuchars SA (2004) Input-specific modulation of neurotransmitter release in the lateral horn of the spinal cord via adenosine receptors. J Neurosci 24:127–137

Bura SA, Nadal X, Ledent C, Maldonado R, Valverde O (2008) A_{2A} adenosine receptor regulates glia proliferation and pain after peripheral nerve injury. Pain 140:95–103

By Y, Condo J, Durand-Gorde JM, Lejeune PJ, Mallet B, Guieu R, Ruf G (2011) Intracerebroventricular injection of an agonist-like monoclonal antibody to adenosine A_{2A} receptor has antinociceptive effects in mice. J Neuroimmunol 230:178–182

Carruthers AM, Sellers LA, Jenkins DW, Jarvie EM, Feniuk W, Humphrey PPA (2001) Adenosine A_1 receptor-mediated inhibition of protein kinase A-induced calcitonin gene-related peptide release from rat trigeminal neurons. Mol Pharmacol 59:1533–1541

Choca JI, Green RD, Proudfit HK (1988) Adenosine A_1 and A_2 receptors of the substantia gelatinosa are located predominantly on intrinsic neurons: an autoradiography study. J Pharmacol Exp Ther 247:757–764

Crain SM, Shen K-F (2000) Antagonists of excitatory opioid receptor functions enhance morphine's analgesic potency and attenuate opioid tolerance/dependence liability. Pain 84:121–131

Curros-Criado MM, Herrero JF (2005) The antinociceptive effects of the systemic adenosine A_1 receptor agonist CPA in the absence and in the presence of spinal cord sensitization. Pharmacol Biochem Behav 82:721–726

DeLeo JA, Colburn RW, Rickman AJ, Yeager MP (1997) Intrathecal catheterization alone induces neuroimmune activation in the rat. Eur J Pain 1:115–122

DeLeo JA, Tawfik VL, LaCroix-Fralish ML (2006) The tetrapartite synapse: path to CNS sensitization and chronic pain. Pain 122:17–21

Deuchars SA, Brooke RE, Deuchars J (2001) Adenosine A_1 receptors reduce release from excitatory but not inhibitory synaptic inputs onto lateral horn neurons. J Neurosci 21:6308–6320

Dickenson A, Suzuki R, Reeve AJ (2000) Adenosine as a potential analgesic target in inflammatory and neuropathic pains. CNS Drugs 13:77–85

Doak GJ, Sawynok J (1995) Complex role of peripheral adenosine in the genesis of the response to subcutaneous formalin in the rat. Eur J Pharmacol 281:311–318

Dunwiddie TV, Masino SA (2001) The role and regulation of adenosine in the central nervous system. Annu Rev Neurosci 24:31–55

Erion MD, Wiesner JB, Rosengren S, Ugarkar B, Boyer SH, Tsuchiya M, Nakane M, Pettersen BA, Nagahisa A (2000) Therapeutic potential of adenosine kinase inhibitors as analgesic agents. Drug Dev Res 50:22, S14-06

Esser MJ, Sawynok J (2000) Caffeine blockade of the thermal antihyperalgesic effect of acute amitriptyline in a rat model of neuropathic pain. Eur J Pharmacol 399:131–139

Ferré S, Diamond I, Goldberg SR, Yao L, Hourani SMO, Huang ZL, Urade Y, Kitchen I (2007) Adenosine A_{2A} receptors in ventral striatum, hypothalamus and nociceptive circuitry. Implications for drug addiction, sleep and pain. Prog Neurobiol 83:332–347

Fields RD, Ni Y (2010) Nonsynaptic communication through ATP release from volume-activated anion channels in axons. Sci Signal 3(142):ra73

Frary CD, Johnson RK, Wang MQ (2005) Food sources and intake of caffeine in the diets of persons in the United States. J Am Diet Assoc 105:110–113

Fredholm BB, Bättig K, Holmén J, Nehlig A, Zvaratau EE (1999) Actions of caffeine in the brain with special reference to factors that contribute to its widespread use. Pharmacol Rev 51:83–133

Fredholm BB, Ijzerman AP, Jacobson KA, Klotz KN, Linden J (2001a) International Union of Pharmacology. XXV. Nomenclature and classification of adenosine receptors. Pharmacol Rev 53:527–552

Fredholm BB, Irenius E, Kull B, Schulte G (2001b) Comparison of the potency of adenosine as an agonist at human adenosine receptors expressed in Chinese hamster ovary cells. Biochem Pharmacol 61:443–448

Fukuda KI, Hayashida M, Fukunaga A, Kasahara M, Koukita Y, Ichinohe T, Kaneko Y (2007) Pain-relieving effects of intravenous ATP in chronic intractable orofacial pain: an open-label study. J Anesth 21:24–30

Gan TJ, Habib AS (2007) Adenosine as a non-opioid analgesic in the perioperative setting. Anesth Analg 105:487–494

Gessi S, Merighi S, Vareni K, Leung E, MacLennan S, Borea PA (2008) The A_3 adenosine receptor: an enigmatic player in cell biology. Pharmacol Ther 117:123–140

Godfrey L, Yan L, Clarke GD, Ledent C, Kitchen I, Hourani SMO (2006) Modulation of paracetamol antinociception by caffeine and by selective adenosine A_2 receptor antagonists in mice. Eur J Pharmacol 531:80–86

Goldman N, Chen M, Fujita T, Wu Q, Peng W, Liu W, Jensen TK, Pai Y, Wang F, Han X, Chen JF, Schnermann J, Takano T, Bekar L, Tieu K, Nedergaard M (2010) Adenosine A_1 receptors mediate local antinociceptive effects of acupuncture. Nat Neurosci 13:883–889

Golembiowska K, White TD, Sawynok J (1996) Adenosine kinase inhibitors augment release of adenosine from spinal cord slices. Eur J Pharmacol 307:157–162

Gong Q-J, Li Y-Y, Xin W-J, Wei X-H, Wang J, Liu Y, Liu C-C, Li Y-Y, Liu X-G (2010) Differential effects of adenosine A_1 receptor on pain-related behavior in normal and nerve-injured rats. Brain Res 1361:23–30

Granados-Soto V, Castaneda-Hernández G (1999) A review of the pharmacokinetic and pharmacodynamic factors in the potentiation of the antinociceptive effect of nonsteroidal anti-inflammatory drugs by caffeine. J Pharmacol Toxicol 42:67–72

Haas HL, Selbach O (2000) Functions of neuronal adenosine receptors. Naunyn-Schmiedeberg's Arch Pharmacol 362:375–381

Habib AS, Minkowitz H, Osborn T, Ogunnaike B, Candiotti K, Viscusi E, Gu J, Creed MR, Gan TJ (2008) Phase 2, double-blind, placebo-controlled, dose-response trial of intravenous adenosine for perioperative analgesia. Anesthesiology 109:1085–1091

Handa T, Fukuda KI, Hayashida M, Koukita Y, Ichinohe T, Kaneko Y (2009) Effects of intravenous adenosine 5'-triphosphate on intraoperative hemodynamics and postoperative pain in patients undergoing major orofacial surgery: a double-blind placebo-controlled study. J Anesth 23:315–322

Haskó G, Csóka B, Németh ZH, Vizi ES, Pacher P (2009) A_{2B} adenosine receptors in immunity and inflammation. Trends Immunol 30:263–270

Hayashida M, Fukuda KI, Fukunaga A (2005) Clinical application of adenosine and ATP for pain control. J Anesth 19:225–235

Hussey MJ, Clarke GD, Ledent C, Hourani SMO, Kitchen I (2007) Reduced response to the formalin test and lowered spinal NMDA glutamate receptor binding in adenosine A_{2A} receptor knockout mice. Pain 129:287–294

Johansson B, Halldner L, Dunwiddie TV, Masino SA, Poelchen W, Giménez-Llort L, Escorihuela RM, Fernández-Teruel A, Wiesenfeld-Hallin Z, Xu X-J, Hårdemark A, Betsholtz C, Herlenius E, Fredholm BB (2001) Hyperalgesia, anxiety, and decreased hypoxic neuroprotection in mice lacking the adenosine A_1 receptor. Proc Natl Acad Sci 98:9407–9412

Kaelin-Lang A, Lauterburg T, Burgunder JM (1998) Expression of adenosine A_{2A} receptor gene in rat dorsal root and autonomic ganglia. Neurosci Lett 246:21–24

Karlsten R, Gordh T, Post C (1992) Local antinociceptive and hyperalgesic effects in the formalin test after peripheral administration of adenosine analogues in mice. Pharmacol Toxicol 70:434–438

Keil GJ, DeLander GE (1994) Adenosine kinase and adenosine deaminase inhibition modulate spinal adenosine- and opioid agonist-induced antinociception in mice. Eur J Pharmacol 271:37–46

Khasar SG, Wang JF, Taiwo YO, Heller PH, Green PG, Levine JD (1995) Mu-opioid agonist enhancement of prostaglandin-induced hyperalgesia in the rat: a G-protein βγ subunit-mediated effect? Neuroscience 67:189–195

King T, Vera-Portocarrero L, Gutierrez T, Vanderah TW, Dussor G, Lai J, Fields HL, Porreca F (2009) Unmasking the tonic-aversive state in neuropathic pain. Nat Neurosci 12:1364–1366

Kolesnikov YA, Jain S, Wilson R, Pasternak G (1996) Peripheral morphine analgesia: synergy with central sites and a target of morphine tolerance. J Pharmacol Exp Ther 279:502–506

Langevin HM, Bouffard NA, Badger GJ, Churchill DL, Howe AK (2006) Subcutaneous tissue fibroblast cytoskeletal remodeling induced by acupuncture: evidence for a mechanotransduction-based mechanism. J Cell Physiol 207:767–774

Lao L-J, Kawasaki Y, Yang K, Fujita T, Kumamoto E (2004) Modulation by adenosine of Aδ and C primary-afferent glutamatergic transmission in adult rat substantia gelatinosa neurons. Neuroscience 125:221–231

Ledent C, Vaugeois JM, Schiffmann SN, Pedrazzini T, El Yacoubi M, Vanderhaeghen JJ, Costentin J, Heath JK, Vassart G, Parmentier M (1997) Aggressiveness, hypoalgesia and high blood pressure in mice lacking the adenosine A_{2A} receptor. Nature 388:674–678

Lee Y-W, Yaksh TL (1996) Pharmacology of the spinal adenosine receptor which mediates the antiallodynic action of intrathecal adenosine agonists. J Pharmacol Exp Ther 277:1642–1648

Li X, Eisenach JC (2005) Adenosine reduces glutamate release in rat spinal synaptosomes. Anesthesiology 103:1060–1065

Li J, Perl ER (1994) Adenosine inhibition of synaptic transmission in the substantia gelatinosa. J Neurophysiol 72:1611–1621

Li L, Hao JX, Fredholm BB, Schulte G, Wiesenfeld-Hallin Z, Xu XJ (2010) Peripheral adenosine A_{2A} receptors are involved in carrageenan-induced mechanical hyperalgesia in mice. Neuroscience 170:923–928

Lima FO, Souza GR, Verri WA, Parada CA, Ferreira SH, Cunha FQ, Cunha TM (2010) Direct blockade of inflammatory hypernociception by peripheral A_1 adenosine receptors: involvement of the NO/cGMP/PKG/KATP signaling pathway. Pain 151:506–515

Liu XJ, White TD, Sawynok J (2002) Enhanced release of adenosine in rat hindpaw following spinal nerve ligation: involvement of capsaicin-sensitive sensory afferents. Neuroscience 114:379–387

Loram LC, Harrison JA, Sloane EM, Hutchison MR, Sholar P, Taylor FR, Berkelhammer D, Coats BD, Poole S, Milligan ED, Maier SF, Riegers J, Watkins LR (2009) Enduring reversal of neuropathic pain by a single intrathecal injection of adenosine 2A receptor agonists: a novel therapy for neuropathic pain. J Neurosci 29:14015–14025

Maione S, de Novellis V, Cappellacci L, Palazzo E, Vita D, Luongo L, Stella L, Franchetti P, Marabese I, Rossi F, Grifantini M (2007) The antinociceptive effect of 2-chloro-2'-*C*-methyl-N^6-cyclopentyladenosine (2'-Me-CCPA), a highly selective adenosine A_1 receptor agonist, in the rat. Pain 131:281–292

Marchand S, Li J, Charest J (1995) Effects of caffeine on analgesia from transcutaneous electrical nerve stimulation. N Eng J Med 333:325–326

Matsuka Y, Ono T, Iwase H, Mitrirattanakul S, Omote KS, Cho T, Lam YYN, Synder B, Spigelman I (2008) Altered ATP release and metabolism in dorsal root ganglia of neuropathic rats. Mol Pain 4:66

Mauborgne A, Poliénor H, Hamon M, Cesselin F, Bourgoin S (2002) Adenosine receptor-mediated control of in vitro release of pain-related neuropeptides from the rat spinal cord. Eur J Pharmacol 441:47–55

McGaraughty S, Chu K, Wismer CT, Mikusa J, Zhu CZ, Cowart M, Kowaluk EA, Jarvis MF (2001) Effects of A-134974, a novel ADO kinase inhibitor, on carrageenan-induced inflammatory hyperalgesia and locomotor activity in rats: evaluation of the sites of action. J Pharmacol Exp Ther 296:501–509

McGaraughty S, Cowart M, Jarvis MF, Berman RF (2005) Anticonvulsant and antinociceptive actions of novel adenosine kinase inhibitors. Curr Top Med Chem 5:43–58

Nascimento FP, Figueredo SM, Marcon R, Martins DF, Macedo SJ, Lima DAN, Almeida RC, Ostroski RM, Rodrigues ALS, Santos ARS (2010) Inosine reduces pain-related behavior in mice: involvement of adenosine A_1 and A_{2A} receptor subtypes and protein kinase C pathways. J Pharmacol Exp Ther 334:590–598

Nelapa I, Vetulani J, Borghi V, Kowalska M, Przewlocka B, Roman AD, Pavone F (2010) Changes induced by formalin pain in central α_1-receptor density are modulated by adenosine receptor agonists. J Neural Transm 117:549–558

Nesher G, Mates M, Zevin S (2003) Effect of caffeine consumption on efficacy of methotrexate in rheumatoid arthritis. Arthritis Rheum 48:571–572

NIH Consensus Conference (1998) Acupuncture. J Am Med Assoc 280:1518–1524

Ocaña M, Bayens JM (1994) Role of ATP-sensitive K^+-channels in antinociception induced by R-PIA, an adenosine A_1 receptor agonist. Naunyn Schmiedeberg's Arch Pharmacol 350:57–62

Palmer H, Graham G, Williams K, Day R (2010) A risk-benefit assessment of paracetamol (acetaminophen) combined with caffeine. Pain Med 11:951–965

Patel MK, Pinnock RD, Lee K (2001) Adenosine exerts multiple effects in dorsal horn neurones of the adult rat spinal cord. Brain Res 920:19–26

Poon A, Sawynok J (1998) Antinociception by adenosine analogs and inhibitors of adenosine metabolism in an inflammatory thermal hyperalgesia model in the rat. Pain 74:235–245

Raffa RB, Stone DJJR, Tallarida RJ (2000) Discovery of "self-synergistic" spinal/supraspinal antinociception produced by acetaminophen (paracetamol). J Pharmacol Exp Ther 295:291–294
Ramos-Zepeda G, Schröder W, Rosenow S, Herrero JF (2004) Spinal vs. supraspinal antinociceptive activity of the adenosine A_1 receptor agonist cyclopentyl-adenosine in rats with inflammation. Eur J Pharmacol 499:247–256
Reeve AJ, Dickenson AH (1995) The roles of spinal adenosine receptors in the control of acute and more persistent nociceptive responses of dorsal horn neurons in the anaesthetized rat. Br J Pharmacol 116:2221–2228
Reeves JJ, Jones CA, Sheehan MJ, Cardey CJ, Whelan CJ (1997) Adenosine A_3 receptors promote degranulation of rat mast cells both in vitro and in vivo. Inflamm Res 46:180–184
Regaya I, Pham T, Andreotti N, Sauze N, Carrega L, Martin-Eauclaire MF, Jouirou B, Peragut JC, Vacher H, Rochat H, Devaux C, Sabatier JM, Guieu R (2004) Small conductance calcium-activated K^+ channels, SkCa, but not voltage-gated K^+ (Kv) channels, are implicated in the antinociception induced by CGS21680, a A_{2A} adenosine receptor agonist. Life Sci 76:367–377
Renner B, Clarke G, Grattan T, Beisel A, Mueller C, Werner U, Kobal G, Brune K (2007) Caffeine accelerates absorption and enhances the analgesic effect of acetaminophen. J Clin Pharmacol 47:715–726
Reppert SM, Weaver DR, Stehle JH, Rivkees SA (1991) Molecular cloning and characterization of a rat A_1-adenosine receptor that is widely expressed in brain and spinal cord. Mol Endocrinol 5:1037–1048
Riberio JA, Sebastiao AM, Mendoca A (2003) Adenosine receptors in the nervous system: pathophysiological implications. Prog Neurobiol 68:377–392
Rice ASC, Cimino-Brown D, Eisenach JC, Kontinen VK, Lacroix-Fralish ML, Machin I (2008) Animal models and the prediction of efficacy in clinical trials of analgesic drugs: a critical appraisal and call for uniform reporting standards. Pain 139:243–247
Santicioli P, Del Bianco E, Maggi CA (1993) Adenosine A_1 receptors mediate the presynaptic inhibition of calcitonin gene-related peptide release by adenosine in the rat spinal cord. Eur J Pharmacol 231:139–142
Sawynok J (1998) Adenosine receptor activation and nociception. Eur J Pharmacol 347:1–11
Sawynok J, Liu XJ (2003) Adenosine in the spinal cord and periphery: release and regulation of pain. Prog Neurobiol 69:313–340
Sawynok J, Zarrindast MR, Reid AR, Doak GJ (1997) Adenosine A_3 receptor activation produces nociceptive behavior and edema by release of histamine and 5-hydroxytryptamine. Eur J Pharmacol 333:1–7
Sawynok J, Reid AR, Fredholm BB (2008) Caffeine reverses antinociception by amitriptyline in wild type mice but not those lacking adenosine A_1 receptors. Neurosci Lett 440:181–184
Sawynok J, Reid AR, Fredholm BB (2010) Caffeine reverses antinociception by oxcarbazepine by inhibition of adenosine A_1 receptors: insights using knockout mice. Neurosci Lett 473:178–181
Schmidt AP, Böhmer AE, Antunes C, Schallenberger C, Porciúncula LO, Elisabetsky E, Lara DR, Souza DO (2009) Anti-nociceptive properties of the xanthine oxidase inhibitor allopurinal in mice: role of A_1 adenosine receptors. Br J Pharmacol 156:163–172
Scholz J, Woolf CJ (2007) The neuropathic pain triad: neurons, immune cells and glia. Nat Neurosci 10:1361–1368
Schulte G, Robertson B, Fredholm BB, DeLander GE, Shortland P, Molander C (2003) Distribution of antinociceptive adenosine A_1 receptors in the spinal cord dorsal horn, and relationship to primary afferents and neuronal subpopulations. Neuroscience 121:907–916
Sharma M, Mohta M, Chawla R (2006) Efficacy of intrathecal adenosine for postoperative pain relief. Eur J Anaesthesiol 23:449–453
Sneyd JR, Langton JA, Allan LG, Peacock JE, Rowbotham DJ (2007) Multicentre evaluation of the adenosine agonist GR79236X in patients with dental pain after third molar extraction. Br J Anaesth 98:672–676
Sowa NA, Vadakkan KI, Zylka MJ (2009) Recombinant mouse PAP has pH-dependent ectonucleotidase activity and acts through A_1-adenosine receptors to mediate antinociception. PLoS One 4:e4248

Sowa NA, Street SE, Vihko P, Zylka MJ (2010a) Prostatic acid phosphatase reduces thermal sensitivity and chronic pain sensitization by depleting phosphatidylinositol 4,5-bisphosphate. J Neurosci 30:10282–10293

Sowa NA, Taylor-Blake B, Zylka MJ (2010b) Ecto-5'-nucleotidase (CD73) inhibits nociception by hydrolyzing AMP to adenosine in nociceptive circuits. J Neurosci 30:2235–2244

Sowa NA, Voss M, Zylka MJ (2010c) Recombinant ecto-5'-nucleotidase (CD73) has long lasting antinociceptive effects that are dependent on adenosine A_1 receptor activation. Mol Pain 6:20

Suzuki R, Gale A, Dickenson AH (2000) Altered effects of an A_1 adenosine receptor agonist on the evoked responses of spinal dorsal horn neurons in a rat model of mononeuropathy. J Pain 1:99–110

Taiwo YO, Levine JD (1990) Direct cutaneous hyperalgesia induced by adenosine. Neuroscience 38:757–762

Tanase D, Baghdoyan HA, Lydic R (2002) Microinjection of an adenosine A_1 agonist into the medial pontine reticular formation increases tail flick latency to thermal stimulation. Anesthesiology 97:1597–1601

Taylor-Blake B, Zylka MJ (2010) Prostatic acid phosphatase is expressed in peptidergic and non-peptidergic nociceptive neurons of mice and rats. PLoS One 5:e8674

Tian L, Ji G, Wang C, Bai X, Lu Y, Xiong L (2010) Excitatory synaptic transmission in the spinal substantia gelatinosa is under an inhibitory tone of endogenous adenosine. Neurosci Lett 477:28–32

Tomić MA, Vučović SM, Stepanović-Petrović RM, Ugrešić N, Prostran MS, Bošković B (2004) The anti-hyperalgesic effects of carbamazepine and oxcarbazepine are attenuated by treatment with adenosine receptor antagonists. Pain 111:253–260

Wu W-P, Hao J-X, Halldner-Henricksson L, Xu XJ, Jacobson MA, Wiesenfeld-Hallin Z, Fredholm BB (2002) Decreased inflammatory pain due to reduced carrageenan-induced inflammation in mice lacking adenosine A_3 receptors. Neuroscience 114:523–527

Wu W-P, Hao J-X, Halldner L, Lövdahl C, DeLander GE, Wisenfeldl-Hallin Z, Fredholm BB, Xu X-J (2005) Increased nociceptive response in mice lacking the adenosine A_1 receptor. Pain 113:395–404

Yang D, Zhang Y, Nguyen HG, Koupenova M, Chauhan AK, Makitalo M, Jones MR, St. Hilaire C, Seldin DC, Toselli P, Lamperti E, Schreiber BM, Gavras H, Wagner DD, Ravid K (2006) The A_{2B} adenosine receptor protects against inflammation and excessive vascular adhesion. J Clin Invest 116:1913–1923

Zahn PK, Straub H, Wenk M, Pogatzki-Zahn EM (2007) Adenosine A_1 but not A_{2a} receptor agonist reduces hyperalgesia caused by a surgical incision in rats. A pertussis toxin-sensitive G protein-dependent process. Anesthesiology 107:797–806

Zhao ZQ (2008) Neural mechanism underlying acupuncture analgesia. Prog Neurobiol 85:355–375

Zhu CZ, Mikusa J, Chu K, Cowart M, Kowaluk EA, Jarvis MF (2001) A-134974, a novel ADO kinase inhibitor, relieves tactile allodynia via spinal sites of action in peripheral nerve injured rats. Brain Res 905:104–110

Zylka MJ (2010) Needling adenosine receptors for pain relief. Nat Neurosci 13:783–784

Zylka MJ (2011) Pain-relieving prospects for adenosine receptors and ectonucleotidases. Trends Mol Med 17:188–196

Zylka MJ, Sowa NA, Taylor-Blake B, Twomey MA, Herrala A, Voikar V, Vihko P (2008) Prostatic acid phosphatase is an ectonucleotidase and suppresses pain by generating adenosine. Neuron 60:111–122

Chapter 18
Symptomatic and Neuroprotective Effects of A_{2A} Receptor Antagonists in Parkinson's Disease

Annalisa Pinna*, Nicola Simola*, Lucia Frau, and Micaela Morelli

Abstract The motor symptoms of Parkinson's disease (PD) are due primarily to the degeneration of the dopaminergic nigrostriatal pathway; however other neurotransmitters besides dopamine are affected in the disease. Treatment of PD is largely unsatisfactory due to several side effects such as "on/off," "wearing-off," and dyskinesia, associated with dopaminergic chronic therapy. Therefore, new pharmacological approaches based on non-dopaminergic therapy have been recently called for a broadening of therapeutic options beyond traditional dopaminergic drugs. Adenosine A_{2A} receptors have a selective localization in richly dopamine-innervated areas and A_{2A} receptor antagonists can regulate GABA and glutamate release in basal ganglia offering a unique opportunity to modulate basal ganglia functions mediated by dopamine. Indeed, A_{2A} receptor antagonists have been shown to restore motor function and contrast parkinsonian tremor acutely, either alone or in combination with dopaminergic drugs, in experimental models of PD. Moreover, in clinical trials, adenosine A_{2A} receptor antagonists reduce "off" time in patients with PD receiving

*Annalisa Pinna and Nicola Simola have contributed equally to this work.

A. Pinna
Department of Biomedical Sciences - Toxicology, University of Cagliari, Via Ospedale, 72, Cagliari 09124, Italy

CNR, Institute of Neuroscience, Cagliari, Italy

N. Simola • L. Frau
Department of Biomedical Sciences - Toxicology, University of Cagliari, Via Ospedale, 72, Cagliari 09124, Italy

M. Morelli (✉)
Department of Biomedical Sciences - Toxicology, University of Cagliari, Via Ospedale, 72, Cagliari 09124, Italy

CNR, Institute of Neuroscience, Cagliari, Italy

Centre of Excellence for Neurobiology of Dependence, University of Cagliari, Cagliari, Italy
e-mail: morelli@unica.it

S. Masino and D. Boison (eds.), *Adenosine: A Key Link between Metabolism and Brain Activity*, DOI 10.1007/978-1-4614-3903-5_18,

optimal dopaminergic therapy without the exacerbation of dyskinesia. In addition preclinical data have shown that adenosine A_{2A} receptor antagonists help to prevent neurodegeneration in PD, raising the possibility of their use as disease-modifying agents. With their proposed symptomatic and neuroprotective efficacy, A_{2A} receptor antagonists might be realistic prospects to advance PD therapeutics.

Keywords Dyskinesia • L-DOPA • 6-hydroxydopamine • MPTP • Urate • Dopamine D_2 receptor • Basal ganglia • Caudate–putamen • Clinical trials • Tremor

Abbreviations

$A_{2A}R$	A_{2A} receptors
AMP	Adenosine monophosphate
ATP	Adenosine triphosphate
CPu	Caudate–putamen
CSF	Cerebrospinal fluid
DDS	Dopamine-dysregulation syndrome
FDA	Food and drug administration
GAD67	Glutamic acid decarboxylase 67
GP	Globus pallidus
6-OHDA	6-hydroxydopamine
L-DOPA	Levo-DOPA
MDMA	3,4-methylenedioxymethamphetamine
mGlu5	Metabotropic glutamate receptor 5
MPTP	1-methyl-4-phenyl-1,2,3,6-tetrahydropyridine
PD	Parkinson's disease
PET	Positron emission tomography
SNr	Substantia nigra pars reticulata
TJMs	Tremulous jaw movements
VEGF	Vascular endothelial growth factor
UPDRS	Unified PD Rating Scale

18.1 Parkinson's Disease and Its Treatment

Parkinson's disease (PD) is the second most common neurodegenerative disease after Alzheimer's disease and has an incidence that increases with age and a male predominance. PD is characterized by bradykinesia, rigidity, tremor often progressing to postural instability, gait problems, and "freezing." Although attracting considerably less attention, several non-motor symptoms related to autonomic disturbances and neuropsychiatric components, such as depression, anxiety, and cognitive decline, are manifested (Obeso et al. 2000).

The motor symptoms of PD are due primarily to the degeneration of the dopaminergic nigrostriatal pathway, leading to disruption of processing in the basal ganglia, motor cortex, and thalamus, which are responsible for the integration of sensorimotor information that controls voluntary movement whereas the mesolimbic/mesocortical dopaminergic pathways remain relatively spared.

Cell loss occurring in brain areas other than the substantia nigra and changes in several neurotransmitters, including norepinephrine, serotonin, glutamate, adenosine, and acetylcholine, contribute to the symptomatology of PD (Braak et al. 2003; Jellinger 2002).

Current therapy is dependent on dopamine-replacement strategies, with L-DOPA and dopamine agonists forming the primary tool for controlling motor symptoms. Several problems, however, are associated with the treatment of PD: loss of drug efficacy and onset of motor complications ("wearing-off," "on/off," dyskinesia) together with side effects, such as psychosis and dopamine-dysregulation syndrome (DDS) (e.g., compulsive gambling, hypersexuality), which can become treatment limiting and difficult to manage with currently available therapies (Fenu et al. 2009). Disease progression also remains untreated, with no drugs currently approved for use as disease-modifying agents.

18.2 Adenosine A_{2A} Receptor Antagonists as New Drugs for the Treatment of PD

Adenosine is a nucleoside derived from the degradation of ATP/AMP and functions as a signaling molecule in the nervous system through different receptors (A_1, A_{2A}, A_{2B}, and A_3) (Masino and Dulla 2005).

Among these receptors, adenosine A_{2A} receptors have emerged as a potential candidate for the treatment of PD for two major reasons: their selective localization in basal ganglia and their direct interaction with D_2 dopamine receptors and indirect interaction with D_1 dopamine receptors (Ferré et al. 1997) (Fig. 18.1).

The occurrence of A_{2A} receptors in basal ganglia and their scarce presence in other brain areas offer a unique opportunity to modulate motor functions without affecting unwanted non-motor behavior, as for example the occurrence of DDS or of other psychiatric complications. Adenosine A_{2A} receptors are, in fact, selectively localized in cortical glutamatergic terminals and GABAergic neurons of the indirect striatopallidal output pathway, containing enkephalin—the same pathway where D_2 dopamine receptors are expressed (Schiffmann et al. 2007) (Fig. 18.1). Moreover, A_{2A} receptors are observed primarily at asymmetric synapses, suggesting that adenosine may be important in modulating excitatory input to striatal neurons (Hettinger et al. 2001). In this respect, it is particularly interesting that dopamine depletion leads to a partial loss of spine and glutamatergic synapses on the indirect striatopallidal pathway (Day et al. 2006).

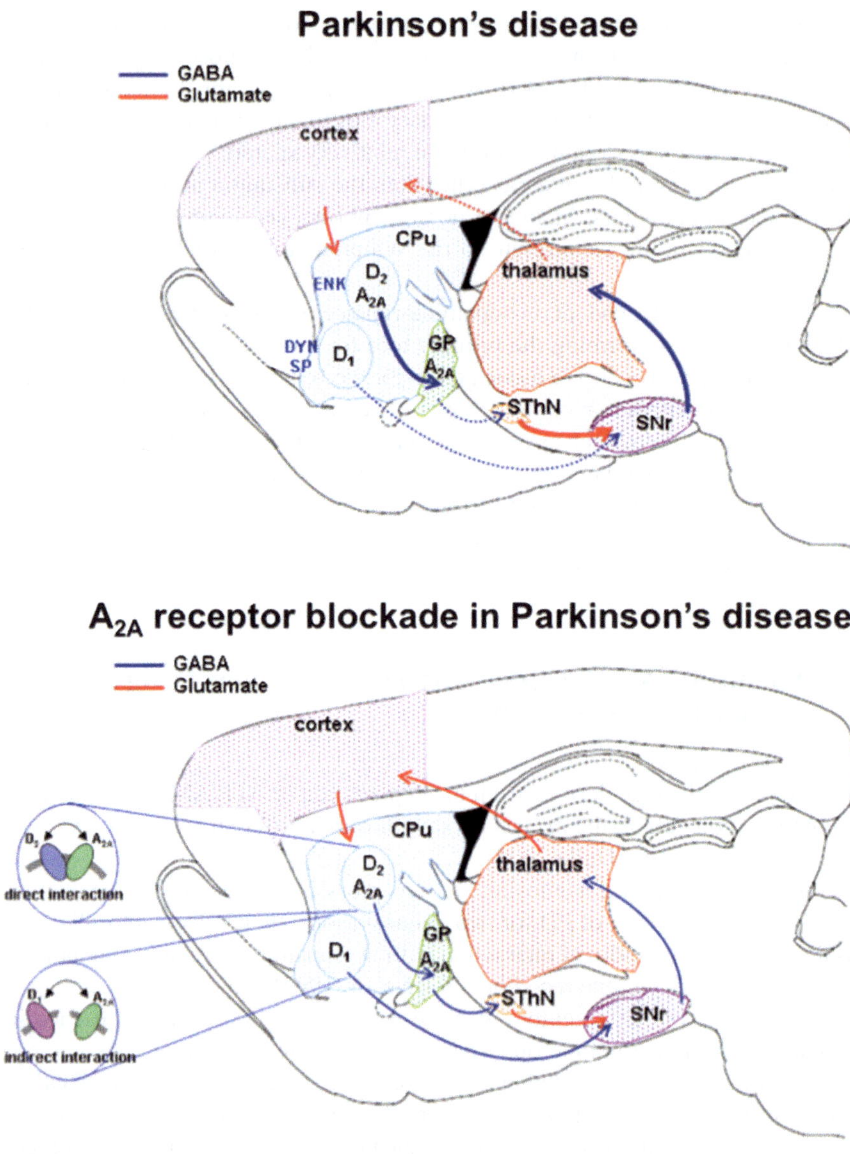

Fig. 18.1 Rationale for the use of A_{2A} receptor antagonists in PD. In PD, lack of dopamine generates an imbalance in the activity of the striatal output pathways. Striatonigral neurons become hypoactive, whereas striatopallidal neurons, losing the inhibitory effect of dopamine while undergoing the stimulatory influence of adenosine, become hyperactive, boosting their inhibitory influence on globus pallidus (GP) neurons. Such imbalanced activity leads to a markedly increased inhibitory output from substantia nigra pars reticulata (SNr) or to thalamo-cortical neurons, which produce hypokinetic symptoms in PD. A_{2A} receptor antagonist efficacy in PD relies on the blockade of A_{2A} receptors on striatopallidal neurons, which should dampen their excessive activity and restore, either directly (D_2) or indirectly (D_1), a balance between striatonigral and striatopallidal neurons, consequently relieving thalamo-cortical activity. Presynaptic A_{2A} receptors regulate neuronal activity at both the caudate–putamen (CPu) and GP levels by regulating glutamatergic and GABAergic synaptic transmissions, thus contributing to the restoration of GP activity and, in turn, subthalamic nucleus (SThN) activity

Adenosine A_{2A} receptor activation can produce direct inhibition of D_2 receptors and may also reduce D_2 receptor coupling to G_i proteins, suggesting that molecular interactions between A_{2A} and D_2 receptors may be important for the antiparkinsonian actions of A_{2A} receptor antagonists (Chen et al. 2001a). Moreover, studies at the molecular level have demonstrated that the A_{2A} receptor may form heteromers with the dopamine D_2 receptor (Table 18.1). A_{2A}/D_2 receptor interaction includes antagonistic interaction at the transcriptional level, as shown by synergistic *c-fos* activation in the indirect striatopallidal pathway by L-DOPA or D_2 receptor agonists and A_{2A} receptor antagonists (Fenu et al. 1997). Of interest is the dimerization of A_1/A_{2A} receptors for the tuning of glutamate release (Quarta et al. 2004) and co-localization and formation of heteromers observed between striatal A_{2A} and metabotropic glutamate receptor 5 (mGlu5) (Ferré et al. 2002) since a synergistic potentiation of motor activity has been reported upon combined administration of A_{2A} and mGlu5 receptor antagonists, together with a synergistic interaction at the level of the signal-transduction pathways (Coccurello et al. 2004; Kachroo et al. 2005) (Table 18.1).

Studies performed in human brain confirmed and supported results obtained in rodents and primates. Expression of the A_{2A} receptor and its mRNA is highly concentrated in the caudate–putamen (CPu) of human brain (Calon et al. 2004). Moreover, according to positron emission tomography (PET) studies, utilizing the A_{2A} receptor antagonists [^{11}C]istradefylline and [^{11}C]SCH442416, A_{2A} receptor binding is highly concentrated in the putamen>caudate nucleus>nucleus accumbens, and over 90 % receptor occupancy is achieved with low oral doses in humans (Brooks et al. 2009, 2010).

Alterations in the basal ganglia of patients who died with PD and who were receiving treatment with dopaminergic drugs were observed in the A_{2A} receptors. Increased levels of A_{2A} receptor and mRNA were found in the putamen of PD patients with dyskinesia versus non-dyskinetic L-DOPA-treated controls, suggesting that an increase in A_{2A} receptors is associated with either the development or the expression of dyskinesia (Calon et al. 2004).

Besides central actions, adenosine and A_{2A} receptors have been known to regulate several peripheral functions (Lukashev et al. 2004). The A_{2A} receptor is important in mediating vasodilatation, including the regulation of coronary blood flow, and supporting synthesis of new blood vessels via the generation of vascular endothelial growth factor (VEGF) and/or by other mechanisms. Most importantly, many cells of the immune system express a high density of adenosine receptors which influence their function (Deussen et al. 2006; Hershfield 2005; Adair 2005) (Fig. 18.2).

Therefore, in the clinical perspective of chronic A_{2A} receptor antagonist utilization, these aspects should be taken into great consideration, although, to date, there have been no reports of immune system impairment or inflammatory change in either toxicological studies or studies on the clinical development of the most advanced A_{2A} receptor antagonist candidates.

Table 18.1 Significant interactions between adenosine A_{2A} receptors and other receptors relevant to Parkinson's Disease

Receptors	Heteromer	Transduction	Mechanism or ion channels involved	Effects
A_{2A}–D_2 (opposite)	• In vitro[1] • Ex vivo[2]	• G proteins • cAMP[1,3] • PKA[1,4]	• NMDA receptor-associated Ca^{2+} channel[4]	• Glu release in CPu[4] • Motor activation[3]
A_{2A}–A_1 (opposite)	• In vitro[5] • Ex vivo[5]	• G proteins[5,10]	• Receptor–receptor interaction[5]	• Glu release in CPu[5]
A_{2A}–CB_1 (synergistic)	• In vitro[6] • Ex vivo[6]	• G proteins[6] • cAMP[6]	• Control of direct or indirect CPu-SN pathway	• Glu transmission in CPu[6] • Motor activation[6] • Addiction[6]
A_{2A}–mGlu5 (synergistic)	• In vitro[7] • Ex vivo[7]	• cAMP[7,8]	• DARPP-32[8]	• Glu transmission in CPu[8] • Motor activation[9]

A_{2A}–D_2: may form heteromeric complexes[1,2], colocalized in striatal enkephalinergic neurons, showing a functional and biochemical antagonistic interaction involving cAMP and protein kinase A (PKA)[1,3,4]. Postsynaptic A_{2A} receptor counteracts the inhibitory modulation of glutamatergic NMDA receptor activity mediated by D_2 receptors, which involves modulation of Ca^{2+} signaling[4] This interaction contributes to the locomotor depression and activation provoked by A_{2A} receptor agonists and antagonists, respectively

A_{2A}–A_1: may form heteromeric complexes[5], colocalized in the same striatal glutamatergic nerve terminals. The principal biochemical characteristic of the A_1–A_{2A} receptor heteromers is the ability of A_{2A} receptor activation to reduce the affinity of the A_1 receptors for agonists. This provides a control mechanism by which low and high concentrations of adenosine inhibit and stimulate, respectively, glutamate release[5,10]

A_{2A}–CB_1: may form heteromeric complexes and; are preferentially colocalized postsynaptically, in enkephalinergic dendrites, and presynaptically in glutamatergic terminals. Blockade of A_{2A} receptors counteracts the motor depressant effects produced by the intrastriatal administration of a cannabinoid CB_1 receptor agonist. Motor effects of cannabinoids depend on physical and functional positive interactions between striatal A_{2A} and CB_1 receptors. In the functional A_{2A}–CB_1 receptor interdependence, CB_1 receptor stimulation could produce a decrease in cAMP levels only if A_{2A} receptors were simultaneously co-activated. Activation of A_{2A} receptors in the CB_1–A_{2A} receptor heteromers allows the effective coupling of CB_1 receptor to G_i proteins[6]

A_{2A}–mGlu5: colocalize and may form heteromeric complexes in cells and striatal neurons[7]. mGlu5 receptors potentiate adenosine A_{2A}/DARPP-32 signaling by an adenosine A_{2A} receptor-mediated formation of cAMP in an extracellular signal-regulated kinase-dependent manner[8]. A synergistic potentiation of motor activity has been reported upon combined administration of A_{2A} and mGlu5 receptor antagonists, suggesting the rationale for developing dual antagonist strategies for symptomatic relief in PD[9]

[1]Hillion et al. (2002), [2]Soriano et al. (2009), [3]Ferré et al. (1997), [4]Higley and Sabatini (2010), [5]Ciruela et al. (2006), [6]Carriba et al. (2007), [7]Ferré et al. (2002), [8]Nishi et al. (2003), [9]Kachroo et al. (2005), [10]Lopes et al. (1999)

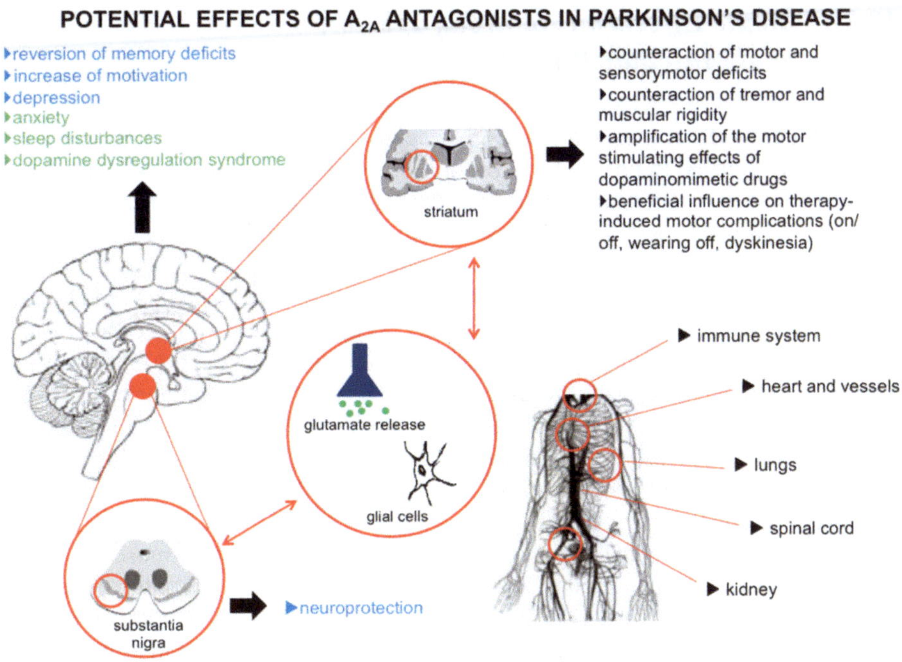

Fig. 18.2 Overview of the potential effects of adenosine A_{2A} receptor antagonists in PD and of their sites of action. The effects demonstrated in both experimental models of PD and clinical trials are shown in *black* and the effects which have been demonstrated in experimental animal models of PD alone so far are shown in *blue*. The non-motor symptoms of PD which could be affected by A_{2A} receptor antagonists, as hypothesized on the basis of preclinical evidence, are shown in *green*. By acting at the level of the dorsal CPu, A_{2A} receptor antagonists counteract many motor and sensorimotor deficits occurring in PD. In addition, they exert a beneficial effect on the motor complications caused by the pharmacological therapy used to manage PD-associated motor impairment. Moreover, A_{2A} receptor antagonists can attenuate the degeneration of dopaminergic neurons located in the substantia nigra pars compacta, with an action involving complex mechanisms, such as the reduction of neuroinflammatory events mediated by glial cells and the decrease of glutamate release. Finally, A_{2A} receptor antagonists have also proven to affect non-motor features of PD, by acting on extra basal ganglia areas, and the function of organs other than the brain, which may influence their effects on PD symptoms

18.3 Effects of Adenosine A_{2A} Receptor Antagonists in PD: Preclinical Studies

Several studies have demonstrated the beneficial effects exerted by antagonists of A_{2A} receptors in animal models of PD. These effects include counteraction of motor impairment not associated with worsening of motor complications, attenuation of tremor and muscular rigidity, and reversion of cognitive deficits. Moreover, some antagonists of A_{2A} receptors have been entered in clinical trials with encouraging results, demonstrating that they are good candidates for further development (see below).

18.3.1 Acute Effects on Akinesia, Tremor, and Muscular Rigidity

One of the first effects described for A_{2A} receptor antagonists is the counteraction of catalepsy induced in rats by blockade of dopamine D_2 receptors (e.g., with haloperidol), a model of parkinsonian-like akinesia. Acute administration of an A_{2A} receptor antagonist attenuates both duration and severity of catalepsy, and enhances the anticataleptic effects of L-DOPA when given concurrently (Pinna et al. 2005; Stasi et al. 2006). A_{2A} receptor antagonists have also proven effective in animal models of parkinsonian-like tremor and muscular rigidity. Regarding tremor, interesting results have been obtained in the rat paradigm of tremulous jaw movements (TJMs), a validated model for evaluating the antitremor properties of drugs (Salamone et al. 1998). Acute administration of A_{2A} receptor antagonists effectively suppresses TJMs in rats (Tronci et al. 2007), with an action mediated by striatal mechanisms, which makes this effect particularly relevant, since parkinsonian tremor stems from the CPu (Simola et al. 2004).

Next to their effects on TJMs, A_{2A} receptor antagonists attenuate muscular rigidity induced in rats by haloperidol, modifying both its electromyographic and mechanographic features (Wardas et al. 2001). The efficacy of A_{2A} receptor antagonists in both of these paradigms is particularly interesting, and envisions them as new drugs for the management of PD-associated tremor and muscular rigidity, for which there are hardly any effective treatments.

18.3.2 Acute Effects in Lesion Models of PD

Another important effect of A_{2A} receptor antagonists is the counteraction of motor impairment, which has been observed in both experimental rodents and nonhuman primates rendered parkinsonian through dopaminergic neurotoxins, such as 6-hydroxydopamine (6-OHDA) and 1-methyl-4-phenyl-1,2,3,6-tetrahydropyridine (MPTP). Degeneration of dopaminergic nigrostriatal neurons by unilateral infusion of 6-OHDA leads to motor impairment manifested as a tendency to rotate towards the side of the 6-OHDA infusion (ipsilateral turning). It is known that drugs bearing antiparkinsonian properties revert this impairment, by stimulating contralateral turning and/or amplifying that elicited by a dopaminomimetic drug (Dauer and Przedborski 2003; Lane et al. 2006). Notably, acute administration of A_{2A} receptor antagonists amplifies the contralateral turning stimulated in 6-OHDA-lesioned rats by either L-DOPA or apomorphine as well as by selective D_1 or D_2 dopamine receptor agonists (Pinna et al. 1996, 2010; Fenu et al. 1997). Moreover, A_{2A} receptor antagonists counteract both motor initiation slowness and sensorimotor deficits associated to 6-OHDA lesion and also amplify the effects of L-DOPA on these symptoms (Pinna et al. 2010).

In line with these results are those observed in MPTP-treated nonhuman primates, which display motor deficits closely mimicking those featured in human

PD, thus being the best preclinical model of the disease (Jenner and Marsden 1986). In both cynomolgus monkeys and common marmosets treated with MPTP, acute administration of A_{2A} receptor antagonists counteracts motor impairment, showing, in addition, a synergic interaction with either L-DOPA or dopaminergic agonists (Kanda et al. 1998a, b, 2000; Grondin et al. 1999; Rose et al. 2006). These results suggest that A_{2A} receptor antagonists could be of interest for the management of PD-associated motor impairment as both monotherapy and adjunct to other antiparkinsonian drugs. Since the dose of L-DOPA has been linked to the severity of side effects by several studies, the possibility of increasing L-DOPA antiparkinsonian activity by A_{2A} receptor antagonists opens the possibility of lowering L-DOPA dosage, decreasing the appearance of L-DOPA side effects.

18.3.3 Chronic Effects on Motor Impairment and Dyskinetic Potential

The beneficial impact of A_{2A} receptor antagonists on PD-associated motor impairment has been confirmed by preclinical studies examining the effects of long-term treatment with these drugs. Prolonged administration of A_{2A} receptor antagonists leads to a persistent counteraction of motor impairment in both rodent and primate models of PD, with no sign of tolerance. In 6-OHDA-lesioned rats, A_{2A} receptor antagonists retain their ability to potentiate L-DOPA-induced contralateral turning, even after subchronic administration (Pinna et al. 2001). Similarly, a persistent improvement of parkinsonian motor deficits was observed in MPTP-treated common marmosets chronically treated with an A_{2A} receptor antagonist (Kanda et al. 1998a, b). The finding that no tolerance develops to the motor-stimulating effects of A_{2A} receptor antagonists is remarkable. In fact, PD necessitates a pharmacological treatment which retains its motor-improving effects over a chronic administration regimen.

Interesting results have also been obtained in preclinical studies examining the effects of a chronic treatment with A_{2A} receptor antagonists in combination with L-DOPA on dyskinesia and motor fluctuations in parkinsonian rats and primates. In 6-OHDA-lesioned rats, chronic treatment with L-DOPA for 19 days leads to sensitization in the intensity of L-DOPA-stimulated contralateral turning, and this is deemed to model some features of L-DOPA-induced abnormal motor responses (Pinna et al. 2006). Using this paradigm, a comparison was made between the effects of a chronic treatment with a higher dose of L-DOPA and those of a chronic treatment with an A_{2A} receptor antagonist plus a lower dose of L-DOPA, given in combination to engender a motor stimulation equivalent to that of L-DOPA at the higher dose. Sensitization in the intensity of contralateral turning occurred only after L-DOPA alone, whereas a stable motor response was observed after the coadministration of an A_{2A} receptor antagonist and L-DOPA, suggesting that this drug combination bears a low dyskinetic potential (Pinna et al. 2001; Tronci et al. 2007). In contrast a study using a high dose of L-DOPA administered chronically for 21 days together with an A_{2A} receptor antagonist failed to observe such a difference (Lundblad et al. 2003). Studies in 6-OHDA-lesioned rats have also shown that a progressive reduction in the duration of contralateral turning takes place during chronic L-DOPA administration for 22 days, and this is thought to model the L-DOPA "wearing-off" observed in patients (Bibbiani et al. 2003; Oh and Chase 2002; Bové et al. 2006). This phenomenon is significantly attenuated when L-DOPA is coadministered with an A_{2A} receptor antagonist, suggesting that "wearing-off" might benefit from A_{2A} receptor blockade (Bibbiani et al. 2003).

Data from MPTP-treated primates confirm and extend these findings. In parkinsonian primates previously rendered dyskinetic through chronic L-DOPA administration for 21 days, A_{2A} receptor antagonists counteract motor impairment without exacerbating the already established dyskinesia (Kanda et al. 1998a, b, 2000; Grondin et al. 1999; Hodgson et al. 2010). Moreover, it was observed that parkinsonian cynomolgus monkeys chronically treated with apomorphine in combination with an A_{2A} receptor antagonist for 19 days displayed hardly any dyskinesia, and that this treatment delayed the manifestation of dyskinesia when the primates were maintained on apomorphine alone (Bibbiani et al. 2003).

Overall, data from animal models of PD indicate that A_{2A} receptor antagonists counteract motor impairment and beneficially impact motor complications associated with the pharmacological treatment used to manage movement deficits. A_{2A} receptor antagonists are not pro-dyskinetic drugs themselves, and combining A_{2A} receptor antagonists with low non-dyskinetic doses of L-DOPA yields a satisfactory motor stimulation accompanied by a less severe dyskinesia, as compared with L-DOPA alone at high doses.

18.3.4 Mechanisms of Action in PD Models

Regarding the mechanisms underlying the beneficial effects of A_{2A} receptor antagonists in PD experimental models, a major one appears to be the facilitation of dopamine transmission; however, several important interactions of adenosine with the cannabinoid and the glutamatergic systems have also been described (Table 18.1).

As discussed elsewhere in this book (see Chap. 9), adenosine A_{2A} receptors interact in an opposite fashion with both dopamine D_2 and D_1 receptors through direct and indirect basal ganglia network mechanisms, respectively (Fig. 18.1). Following the administration of an A_{2A} receptor antagonist, the net result of this interaction is the amplification of dopamine-mediated effects, which leads to motor stimulation. It must also be highlighted that coadministration of an A_{2A} receptor antagonist attenuates the neurochemical modifications induced in the basal ganglia by L-DOPA. This has been reported for neuropeptides (dynorphin, enkephalin), enzymes (GAD67), and phosphorylation state of receptors (AMPA glutamate receptors) (Bibbiani et al. 2003; Carta et al. 2002, 2003; Tronci et al. 2007; Chase et al. 2003). Furthermore, A_{2A} receptors can modulate the extracellular levels of glutamate, whose excessive elevation contributes to the abnormal functioning of basal ganglia taking place in PD. Blockade of A_{2A} receptors depresses glutamate release by an action involving both neuronal terminals and glial cells, in particular astrocytes, with important functional consequences on the cross talk between neurons and glia in the basal ganglia (Ciruela et al. 2006; Yu et al. 2008). Hence, A_{2A} receptors appear to be critically involved in the fine-tuning of basal ganglia functionality, and this may explain the beneficial effects of A_{2A} receptor antagonists on PD-associated movement deficits and motor complications.

18.3.5 Effect on Non-motor Symptoms of PD: Cognition, Depression, and Sleep

The idea has recently emerged that A_{2A} receptor antagonists might be a new option for the management of cognitive decline and other non-motor features of PD which, as mentioned elsewhere in this chapter, are important and often underdiagnosed symptoms of this pathology, and for which hardly any effective therapies are available (Fig. 18.2). Adenosine A_{2A} receptors are critically involved in cognitive processes, as shown by several experiments. Thus, memory deficits are displayed by rats overexpressing the human A_{2A} receptors, whilst either blockade or genetic deletion of A_{2A} receptors improves cognition in rodents (Wang et al. 2006; O'Neill and Brown 2006; Higgins et al. 2007; Giménez-Llort et al. 2007; Wei et al. 2011). Experimental results also suggest that A_{2A} receptor antagonists may be effective on parkinsonian-like cognitive decline. In rats infused with MPTP in the substantia nigra, a model of early PD, caffeine, an A_1–A_{2A} receptor antagonist (Fredholm et al. 1999), counteracts the impaired performance in a two-way active avoidance task (Gevaerd et al. 2001). Moreover, in the rat reserpine model of parkinsonism, administration of either caffeine or selective A_{2A} receptor antagonists reverted the deficits in olfaction-mediated short-term memory (Takahashi et al. 2008). These encouraging results push for a thorough evaluation of A_{2A} receptor antagonists for the management of PD-associated cognitive decline. In this regard, it is worth mentioning that preclinical studies show that A_{2A} receptor antagonists also beneficially impact cognitive functions other than memory which are affected in the course of PD, such as motivation and attention (O'Neill and Brown 2006; Higgins et al. 2007; see Chap. 23). Moreover, epidemiological studies demonstrate that regular caffeine consumption inversely correlates with cognitive decline in the elderly (Ritchie et al. 2007), supporting, though indirectly, the possibility that A_{2A} receptor antagonists may positively impact cognitive impairment (Fig. 18.2).

Considerations similar to those on cognitive decline also apply to the potential efficacy of A_{2A} receptor antagonists on other non-motor symptoms of PD. For example, A_{2A} receptor blockade is beneficial in animal models of depression (El Yacoubi et al. 2001). It is also possible that A_{2A} receptor antagonists may be of interest for the management of other non-motor features observed in PD patients, for example, sleep disturbances. Thus, it is well acknowledged that adenosine is critically involved in the wake/sleep cycle, and that blockade of A_{2A} receptors stimulates arousal and counteracts sleepiness. In this connection, A_{2A} receptor antagonists might be helpful in counteracting PD-associated sleepiness, although potential untoward effects of these agents on the wake/sleep cycle cannot be ruled out at this stage (Huang et al. 2005; Lazarus et al. 2011). Moreover, it has been suggested that A_{2A} receptor blockade might be helpful to add to dopamine agonists in order to lower the dosage of the latter drugs, the administration of which at high doses is believed to be at the origin of the impulse-control disorders represented by the DDS and other neuropsychiatric diseases (Fenu et al. 2009). Therefore, there is a need for a thorough investigation of the efficacy of A_{2A} receptor antagonists on the wide array of non-motor symptoms associated with PD, as these drugs may represent a promising opportunity to simultaneously target both motor and non-motor features of the disease. This is particularly important as drugs currently used to control PD-associated motor impairment are scarcely effective on non-motor symptoms and, at times, can aggravate them.

18.4 Adenosine and Neurodegeneration in PD

At present, there are no drugs available that are capable of counteracting the neuronal degeneration underlying PD. A large number of molecules from different classes, including antioxidants, dopamine agonists, metabolic enhancers, and others, are continuously being screened for their neuroprotective potential in experimental models of PD: in this connection, very interesting results have been obtained with A_{2A} receptor antagonists.

Adenosine A_{2A} receptors critically modulate the survival of neurons in response to a range of noxious insults, including ischemia, excitotoxicity, and neurodegeneration (Chen et al. 1999; Popoli et al. 2004). With regard to PD, marked neuroprotective effects of A_{2A} receptors antagonists have been reported in both rats infused with 6-OHDA in the brain and mice treated with MPTP (Ikeda et al. 2002; Bové et al. 2005; Pierri et al. 2005; Chen et al. 2001a, b; Frau et al. 2011). In these paradigms, A_{2A} receptor antagonists significantly attenuated the loss of tyrosine hydroxylase content in the substantia nigra. Moreover, in MPTP-treated mice, A_{2A} receptor antagonists counteracted the drop in the levels of dopamine in the CPu triggered by MPTP (Chen et al. 2001a, b; Pierri et al. 2005). These results indicate that A_{2A} receptor antagonists exert a protective effect on both cell bodies and terminals of dopaminergic neurons. Remarkably, the beneficial influence of A_{2A} receptor antagonism on parkinsonian-like neurodegeneration has been substantiated by studies in mice bearing a deletion of the gene encoding for the A_{2A} receptor, since a markedly reduced dopaminergic neurotoxicity by MPTP was observed in these mice compared with wild-type animals (Chen et al. 2001a, b; Carta et al. 2009) (Fig. 18.2).

The molecular mechanisms mediating neuroprotection by A_{2A} receptor antagonists in animal models of PD are not completely understood, although they appear distinct from those underlying motor stimulation (Yu et al. 2008). It has been suggested that A_{2A} receptor blockade may afford neuroprotection by dampening the glutamatergic transmission, based on the modulation of glutamate outflow by A_{2A} receptors located on presynaptic terminals and glial cells, in particular astrocytes (Fig. 18.2). Since the increase of glutamate release might promote the degeneration of dopaminergic neurons, it could be hypothesized that both neuronal and glial elements might be involved in dopamine neuron degeneration (Lange and Riederer 1994; Popoli et al. 2004) (Fig. 18.2). Alternative mechanisms have also been explored. In this connection, it is conceivable that an interplay between adenosine A_{2A} receptors and P2 purinoreceptors, which have been shown to regulate neurodegeneration, may exist. Moreover, in recent years a great deal of attention has been paid to the involvement of A_{2A} receptors in neuroinflammation, which is increasingly being envisioned as a key step in the onset of PD-associated neurodegeneration (Hunot and Hirsch 2003; Kerschensteiner et al. 2009; Ohta and Sitkovsky 2001). In this regard, a study in MPTP-treated mice has demonstrated that either blockade or genetic deletion of A_{2A} receptors not only attenuates the demise of dopaminergic neurons, but also dampens the activation of microglia and astroglia induced by MPTP in the substantia nigra and CPu (Carta et al. 2009). Finally, evidence exists which demonstrates that antagonism of adenosine receptors can counteract the loss of striatal dopamine and the degeneration of nigral neurons in an animal model of mitochondrial dysfunction, which is regarded as a potential cause of PD (Alfinito et al. 2003). On these bases, it can be hypothesized that A_{2A} receptor antagonists may protect dopaminergic neurons by acting at different stages of a complex cascade of events, including neuroinflammation and actual neurotoxic insults, which may trigger the demise of the dopaminergic neurons underlying PD (Fig. 18.2).

Up to now, no clinical studies have been performed to evaluate the protective effects of A_{2A} receptor antagonists in PD patients, but clinical trials aimed at conclusively addressing this issue are currently under way (see below). However, it must be mentioned that epidemiological studies have suggested blockade of adenosine receptors as a factor which might beneficially impact the degeneration of dopaminergic neurons underlying PD. Thus, independent investigators have observed the existence of a negative correlation between the incidence of idiopathic PD and a regular lifetime intake of the nonselective adenosine antagonist, caffeine (Ross et al. 2000; Ascherio et al. 2001). Obviously, this is indirect evidence and only suggests, but does not prove, that caffeine and/or selective A_{2A} receptor antagonists are able to prevent the loss of dopaminergic mesencephalic neurons. Nevertheless, this finding is remarkable, and has paved the path for the clinical evaluation of A_{2A} receptor antagonists as new protective agents for the neurodegeneration underlying PD.

Overall, preclinical results depicting A_{2A} receptor antagonists as neuroprotective drugs for PD are very encouraging (Chen et al. 1999; Pierri et al. 2005; Carta et al. 2009). These results acquire particular importance in the consideration of the beneficial effects of A_{2A} receptor antagonists on motor symptoms of PD (see above), and suggest that A_{2A} receptor antagonists might possess a dual beneficial effect on motor impairment and disease progression, thus being an ideal class of drugs for further development. Nevertheless, caution should be exercised when considering A_{2A} receptor antagonists as new tools for neuroprotection. In fact, the preclinical data on this issue suggest, on the whole, that the actions of A_{2A} receptors on neuronal survival are complex, and may be critically dependent on the duration of the insult and the specific areas which degenerate. Therefore, there is a need for a detailed characterization of the specific conditions under which A_{2A} receptor antagonists are effective on PD neurodegeneration and how they can be properly used as therapeutic tools.

18.4.1 Detrimental Effects of Adenosine Antagonism: Caffeine

As mentioned above, caffeine and A_{2A}receptor antagonists show neuroprotective effects in animal models of the disease. However, depending on the experimental protocol and the type of drug concomitantly administered, caffeine can have a detrimental effect on neuron survival.

Caffeine is frequently used in combination with amphetamine-like stimulants that are often abused orally in the form of tablets for recreational purposes. Because of its stimulant effect in the central nervous system, beverages containing caffeine, such as coffee, tea, and soft drinks, are intentionally taken mostly by young people in association with drugs of abuse in order to avoid drowsiness and fatigue (Fredholm et al. 1999).

Different preclinical studies in the rat have shown that caffeine, when administered at doses similar to those that may be taken for recreational use, may enhance the acute toxicity and lethality of 3,4-methylenedioxymethamphetamine (MDMA) (McNamara et al. 2006; Camarasa et al. 2006) and may increase mortality caused by acute amphetamine or cocaine administration (Derlet et al. 1992). At the preclinical level, caffeine intake potentiates toxicity of MDMA, increasing MDMA-induced hyperthermia in group-housed rats (McNamara et al. 2006; Camarasa et al. 2006), inducing profound tachycardia (McNamara et al. 2007), potentiating the MDMA-induced decrease of serotonin transporter sites in the cortex, CPu, and hippocampus (Camarasa et al. 2006), and enhancing the permeation and the absorption of MDMA in intestinal epithelial cells (Kuwayama et al. 2007). Caffeine enhances MDMA-induced dopamine release, provoking a greater response than that obtained following either caffeine or MDMA alone (Vanattou-Saïfoudine et al. 2011). Finally, combined use of caffeine and MDMA induces neuroinflammation characterized by an increased microglial and astroglial reactivity in different brain areas (Khairnar et al. 2010). Since neuroinflammation is one of the factors responsible for the pathogenesis of neurodegenerative disorders (Kerschensteiner et al. 2009), this effect may facilitate neurodegenerative processes such as in PD.

18.5 Clinical Studies on A_{2A}R Antagonists in Parkinsonian Patients

The promising antiparkinsonian responses of xanthine and non-xanthine highly selective A_{2A} receptor antagonists in rodent and nonhuman primate models of PD, have provided a strong motivation to initiate clinical programs assessing the safety and the therapeutic potential of these drugs in PD patients.

To date, several A_{2A} receptor antagonists have been entered into clinical trials: istradefylline (KW-6002) from Kyowa Hakko Kogo; preladenant (SCH-420814) from Merck & Co Inc. following its acquisition of Schering-Plough Corp; SYN115 from Synosia Therapeutics, acquired by Biotie Therapies and UCB Pharma;

vipadenant (BIIB014/V2006) from Vernalis plc-Biogen Idec; and ST-1535 from Sigma-Tau. Other companies working in this field have not yet disclosed significant information related to specific drug development programs (for reviews see Pinna 2009; Shah and Hodgson 2010).

The first A_{2A} receptor antagonist evaluated in clinical trials in PD patients was the xanthine istradefylline (Jenner 2005), which has progressed through phase II and III studies. The vast majority of clinical data on this drug have been collected in patients with moderate-to-severe PD, with overt motor complications, under treatment with L-DOPA either alone or in combination with other PD medications. Although clinical findings obtained with istradefylline are not always statistically significant, they clearly indicate that this drug, at dose of 20–80 mg/day, reduces the time spent in the "off" state by about 1 h (Bara-Jimenez et al. 2003; LeWitt et al. 2008; Stacy et al. 2008; Hauser et al. 2008; Mizuno et al. 2010). Moreover, this A_{2A} receptor antagonist significantly increased L-DOPA "on" time with non-troublesome dyskinesia, whereas L-DOPA "on time" with troublesome dyskinesia remained unchanged (LeWitt et al. 2008; Stacy et al. 2008; Hauser et al. 2008). Interestingly, evidence of consistent and sustained effects of istradefylline has been supported by a long phase III study in PD patients, in which istradefylline showed a stable efficacy in decreasing the "off" time (Factor et al. 2010). Recently, istradefylline (40 mg/day) was compared with placebo as a monotherapy in early PD patients (Fernandez et al. 2010). Although istradefylline provided numerical improvement of motor Unified PD Rating Scale (UPDRS) scores at each time point and significantly greater improvement at week 2 in treated patients compared with placebo, the difference across groups was not statistically significant (Fernandez et al. 2010). This discrepancy may be due to the low dose used in this clinical trial; indeed, when administered alone, higher doses of istradefylline may be needed to elicit beneficial effects on parkinsonian symptoms. All clinical trials with istradefylline have provided evidence for a suitable tolerability and safety, with discontinuation due to adverse effects being infrequent and comparable in placebo and istradefylline groups (Bara-Jimenez et al. 2003; Hauser et al. 2008; LeWitt et al. 2008; Stacy et al. 2008; Factor et al. 2010). Despite these findings, istradefylline did not receive approval by the Food and Drug Administration (FDA), as the agency expressed concern about the adequacy of overall efficacy of the drug and asked for more thorough clinical investigations (http://www.istradefylline.com/fda.html). Considering the positive clinical results obtained in PD patients, Kyowa company decided to assess an additional and detailed examination.

Another very promising A_{2A} receptor antagonist is the non-xanthine derivative preladenant, with good oral bioavailability, high A_{2A} receptor selectivity, good pharmacokinetic properties, and in vivo activity against parkinsonian symptoms (Hodgson et al. 2009). The phase I PET study in healthy humans confirmed the possibility of development of preladenant under a twice-per-day administration regimen (Brooks et al. 2009). Preladenant, at doses of 5 and 10 mg, in moderate-to-severe PD patients on a stable regimen with L-DOPA and other adjunctive medications, was significantly more effective in reducing the "off" time than placebo. Besides, preladenant, at both doses, significantly increased the "on" time without producing a proportional overall increase in any dyskinesias (troublesome or non-troublesome) compared with placebo

(Hauser et al. 2011). Moreover, in these clinical studies, preladenant was shown to be well tolerated and safe at all doses. The incidence of adverse effects and discontinuation rates was similar between preladenant- and placebo-treated groups (Hauser et al. 2011). These encouraging results have motivated the company to plan two more extensive phase II and three phase III trials with different doses of preladenant, in moderate-to-severe PD patients, as adjunctive therapies with L-DOPA or dopamine agonists (Pinna 2009; http://www.clinicaltrials.gov/ct2/results?term=preladenant). Notably, one more phase III trial has been planned to evaluate the efficacy and safety of preladenant in early PD patients as monotherapy for 1 year (http://www.clinicaltrials.gov/ct2/results?term=preladenant).

Regarding the clinical program of the other non-xantine A_{2A} receptor antagonists, vipadenant has been discontinued (http://www.vernalis.com/media-centre/latest-releases/2010-releases/584), whereas SYN115 and ST-1535 are at an early stage of development (phase IIb and phase I trials, respectively). Phase IIa study, with SYN115 alone or in combination with an infusion of low dose of L-DOPA, showed that this drug significantly improved tapping speed as compared with placebo, either in the presence or not of a subtherapeutic infusion of L-DOPA, in patients with mild-to-moderate PD (Black et al. 2010). Moreover, the total UPDRS motor score was 20 % lower with SYN115 compared with placebo when administered with L-DOPA (Black et al. 2010). In particular, improvement in two UPDRS measures of bradykinesia achieved statistical significance; SYN115 was also well tolerated (Black et al. 2010). On the basis of these promising results, the company planned to start a phase IIb study to investigate four doses of SYN115 as adjunctive therapy in L-DOPA-treated PD patients experiencing "wearing-off" (http://www.biotie.com/en/recearch_and_development/central_nervous_system_disorders/syn115).

The Sigma-Tau company has started a clinical phase I program, with the non-xanthine A_{2A} receptor antagonist, ST-1535, after the promising results of preclinical studies in animal models of PD, in which ST-1535 displayed a clear efficacy as antiparkinsonian drug (Tronci et al. 2007; Rose et al. 2006). The initial phase I study showed that all single doses of ST-1535 (between 50 and 450 mg) were generally well tolerated. Furthermore, the company has started to investigate the antiparkinsonian activity of two metabolites of ST-1535 which have shown good efficacy in rodent models of PD.

The encouraging clinical results obtained with A_{2A} receptor antagonists strongly support the utilization of these drugs in the management of parkinsonian patients. Indeed, the observed temporal reduction of L-DOPA "off" time by these drugs is of notable importance, "wearing-off" being one of the major untoward effects manifested during long-term use of this drug. Conversely, to ameliorate the management with A_{2A} receptor antagonists, additional investigation is also needed, as suggested by preclinical studies, in combination with a suboptimal dose of L-DOPA or dopamine agonists, rather than with the optimal doses usually utilized so far in clinical trials. Besides, further data from the studies assessed in patients with early-stage PD are expected to elucidate whether A_{2A} receptor antagonists are suitable for administration as monotherapy against motor symptoms. Finally, evaluation of these drugs in patients devoid of L-DOPA motor complications should be performed in order to evaluate in detail the efficacy of A_{2A} receptor antagonists on the onset and progression of human dyskinesia.

18.6 Adenosine, Urate, and PD

Urate (uric acid) is the "end product" generated during the metabolisms of purines, such as adenosine. This product, in physiologic concentrations, is an important natural antioxidant, as effective as ascorbate, in reducing oxidative stress through its actions as a scavenger of free radicals and a chelator of iron and copper (Cipriani et al. 2010). This neuroprotective feature of urate has been established in several cellular and animal models of numerous diseases, including PD (Cipriani et al. 2010). Recently, epidemiological findings showed that cerebrospinal fluid (CSF) and serum concentrations of urate may predict clinical progression of PD; indeed, humans with high plasma urate levels have a markedly reduced incidence of developing PD (Ascherio et al. 2009; Schwarzschild et al. 2008). Similarly, individuals with gout, a rheumatic disease resulting from hyperuricemia, have a significantly reduced risk of developing PD (Alonso et al. 2007; De Vera et al. 2008). These data have been supported by the study in which dietary interventions produced an increase of blood urate concentrations reducing the risk of PD (Gao et al. 2008). On the other hand, lower concentrations of urate were observed in serum, CSF, and postmortem nigrostriatal tissue samples of PD patients compared with controls (Church and Ward 1994; Morelli et al. 2010). Notably, the relationship between plasma urate and cognitive impairments in PD patients has been investigated in two studies (Annanmaki et al. 2008; Wang et al. 2009). Both these reports demonstrated a positive correlation between urate levels and preserved neuropsychological performance, supporting the view that serum urate may possess an additional feature as a diagnostic biomarker for cognitive dysfunction of PD.

Overall, these lines of evidence involve adenosine as a key factor for disease development and suggest urate as the first biomarker of the risk, development, and clinical progression of PD. Epidemiological and clinical data have been rapidly translated into clinical trials that investigate the safety and ability of inosine (a precursor of urate as well as the deamination product of adenosine) to increase serum and CSF urate levels, in early PD patients, in order to evaluate its potential for further development as a novel strategy to delay progression of the disease (http://www.clinicaltrials.gov/ct2/show/NCT00833690).

18.7 Conclusions

The management of PD has two important phases: early management, which includes initiation of therapy after diagnosis, and management of late, complicated PD, when the response to L-DOPA is altered and "on/off," "wearing-off," and dyskinesia are present. In the first, uncomplicated, phase, either L-DOPA or dopamine receptor agonists efficiently counteract motor dysfunctions without the emergence of motor side effects. This phase, otherwise called "honey moon," lasts about 5–10 years, until motor and non-motor complications appear (see Sect. 18.1 above). The management of this first phase has important consequences on the appearance of the side effects that characterize the second, complicated, phase. Introduction of A_{2A} receptor antagonists in this early phase may allow the dose of L-DOPA to be lowered, preventing the appearance of side effects and might also delay the dopamine neuron degeneration through the disease-modifying properties of these drugs.

Similarly to the first, uncomplicated, phase, in the second, complicated, phase of PD, the lowering of L-DOPA dosage by addition of A_{2A} receptor antagonists may increase the efficacy of this drug and reduce the "off" phase, without increasing the associated dyskinesia.

In conclusion, the role of adenosine and adenosine receptors so far described has raised the possibility that using A_{2A} receptor antagonists might be a uniquely effective strategy for counteracting the motor dysfunctions and neurodegeneration that characterize PD. Adenosine participates in several important physiological processes, from energy supply through ATP, to the regulation of anti-inflammatory and immune system responses. Therefore more detailed studies should be undertaken in both experimental animals and humans in order to clarify under which specific limitations A_{2A} receptor antagonists might be used in the treatment of PD.

References

Adair TH (2005) Growth regulation of the vascular system: an emerging role for adenosine. Am J Physiol Regul Integr Comp Physiol 289:R283–R296

Alfinito PD, Wang SP, Manzino L, Rijhsinghani S, Zeevalk GD, Sonsalla PK (2003) Adenosinergic protection of dopaminergic and GABAergic neurons against mitochondrial inhibition through receptors located in the substantia nigra and striatum, respectively. J Neurosci 23:10982–10987

Alonso A, Rodríguez LA, Logroscino G, Hernán MA (2007) Gout and risk of Parkinson disease: a prospective study. Neurology 69:1696–1700

Annanmaki T, Pessala-Driver A, Hokkanen L, Murros K (2008) Uric acid associates with cognition in Parkinson's disease. Parkinsonism Relat Disord 14:576–578

Ascherio A, Zhang SM, Hernán MA, Kawachi I, Colditz GA, Speizer FE, Willett WC (2001) Prospective study of caffeine consumption and risk of Parkinson's disease in men and women. Ann Neurol 50:56–63

Ascherio A, LeWitt PA, Xu K, Eberly S, Watts A, Matson WR, Marras C, Kieburtz K, Rudolph A, Bogdanov MB et al (2009) Urate as a predictor of the rate of clinical decline in Parkinson disease. Arch Neurol 66:1460–1468

Bara-Jimenez W, Sherzai A, Dimitrova T, Favit A, Bibbiani F, Gillespie M, Morris MJ, Mouradian MM, Chase TN (2003) Adenosine $A_{(2A)}$ receptor antagonist treatment of Parkinson's disease. Neurology 61:293–296

Bibbiani F, Oh JD, Petzer JP, Castagnoli N Jr, Chen JF, Schwarzschild MA, Chase TN (2003) A_{2A} antagonist prevents dopamine agonist-induced motor complications in animal models of Parkinson's disease. Exp Neurol 184:285–294

Black KJ, Campbell MC, Dickerson W, Creech ML, Koller JM, Chung S, Bandak SI (2010) A randomized, double-blind, placebo-controlled cross-over trial of the adenosine A_{2A} antagonist SYN115 in Parkinson disease. Neurology 74(Suppl 2):A317

Bové J, Serrats J, Mengod G, Cortés R, Tolosa E, Marin C (2005) Neuroprotection induced by the adenosine A_{2A} antagonist CSC in the 6-OHDA rat model of parkinsonism: effect on the activity of striatal output pathways. Exp Brain Res 165:362–374

Bové J, Serrats J, Mengod G, Cortés R, Aguilar E, Marin C (2006) Reversion of levodopa-induced motor fluctuations by the A_{2A} antagonist CSC is associated with an increase in striatal preprodynorphin mRNA expression in 6-OHDA-lesioned rats. Synapse 59:435–444

Braak H, Del Tredici K, Rüb U, de Vos RA, Jansen Steur EN, Braak E (2003) Staging of brain pathology related to sporadic Parkinson's disease. Neurobiol Aging 24:197–211

Brooks DJ, Warrington S, Tendolkar A, Cutler DL, Hunter J (2009) Positron emission tomography (PET) study of preladenant in healthy male subjects. Mov Disord 24(suppl 1):S257

Brooks DJ, Papapetropoulos S, Vandenhende F, Tomic D, He P, Coppell A, O'Neill G (2010) An open-label, positron emission tomography study to assess adenosine A2A brain receptor occupancy of vipadenant (BIIB014) at steady-state levels in healthy male volunteers. Clin Neuropharmacol 33:55–60

Calon F, Dridi M, Hornykiewicz O, Bedard PJ, Rajput AH, Di Paolo T (2004) Increased adenosine A_{2A} receptors in the brain of Parkinson's disease patients with dyskinesias. Brain 127:1075–1084

Camarasa J, Pubill D, Escubedo E (2006) Association of caffeine to MDMA does not increase antinociception but potentiates adverse effects of this recreational drug. Brain Res 1111:72–82

Carriba P, Ortiz O, Patkar K, Justinova Z, Stroik J, Themann A, Müller C, Woods AS, Hope BT, Ciruela F et al (2007) Striatal adenosine A_{2A} and cannabinoid CB_1 receptors form functional heteromericcomplexesthatmediatethemotoreffectsofcannabinoids.Neuropsychopharmacology 32:2249–2259

Carta AR, Pinna A, Cauli O, Morelli M (2002) Differential regulation of GAD67, enkephalin and dynorphin mRNAs by chronic-intermittent L-DOPA and A_{2A} receptor blockade plus L-DOPA in dopamine-denervated rats. Synapse 44:166–174

Carta AR, Fenu S, Pala P, Tronci E, Morelli M (2003) Selective modifications in GAD67 mRNA levels in striatonigral and striatopallidal pathways correlate to dopamine agonist priming in 6-hydroxydopamine-lesioned rats. Eur J Neurosci 18:2563–2572

Carta AR, Kachroo A, Schintu N, Xu K, Schwarzschild MA, Wardas J, Morelli M (2009) Inactivation of neuronal forebrain A_{2A} receptors protects dopaminergic neurons in a mouse model of Parkinson's disease. J Neurochem 111:1478–1489

Chase TN, Bibbiani F, Bara-Jimenez W, Dimitrova T, Oh-Lee JD (2003) Translating A_{2A} antagonist KW6002 from animal models to parkinsonian patients. Neurology 61:S107–S111

Chen JF, Huang Z, Ma J, Zhu J, Moratalla R, Standaert D, Moskowitz MA, Fink JS, Schwarzschild MA (1999) $A_{(2A)}$ adenosine receptor deficiency attenuates brain injury induced by transient focal ischemia in mice. J Neurosci 19:9192–9200

Chen JF, Moratalla R, Impagnatiello F, Grandy DK, Cuellar B, Rubinstein M, Beilstein MA, Hackett E, Fink JS, Low MJ et al (2001a) The role of the $D_{(2)}$ dopamine receptor ($D_{(2)}R$) in $A_{(2A)}$ adenosine receptor ($A_{(2A)}R$)-mediated behavioral and cellular responses as revealed by $A_{(2A)}$ and $D_{(2)}$ receptor knockout mice. Proc Natl Acad Sci USA 98:1970–1975

Chen JF, Xu K, Petzer JP, Staal R, Xu YH, Beilstein M, Sonsalla PK, Castagnoli K, Castagnoli N Jr, Schwarzschild MA (2001b) Neuroprotection by caffeine and $A_{(2A)}$ adenosine receptor inactivation in a model of Parkinson's disease. J Neurosci 21:RC143

Church WH, Ward VL (1994) Uric acid is reduced in the substantia nigra in Parkinson's disease: effect on dopamine oxidation. Brain Res Bull 33:419–425

Cipriani S, Chen X, Schwarzschild MA (2010) Urate: a novel biomarker of Parkinson's disease risk, diagnosis and prognosis. Biomark Med 4:701–712

Ciruela F, Casadó V, Rodrigues RJ, Luján R, Burgueño J, Canals M, Borycz J, Rebola N, Goldberg SR, Mallol J et al (2006) Presynaptic control of striatal glutamatergic neurotransmission by adenosine A_1-A_{2A} receptor heteromers. J Neurosci 26:2080–2087

Coccurello R, Breysse N, Amalric M (2004) Simultaneous blockade of adenosine A_{2A} and metabotropic glutamate mGlu5 receptors increase their efficacy in reversing Parkinsonian deficits in rats. Neuropsychopharmacology 29:1451–1461

Dauer W, Przedborski S (2003) Parkinson's disease: mechanisms and models. Neuron 39:889–909

Day M, Wang Z, Ding J, An X, Ingham CA, Shering AF, Wokosin D, Ilijic E, Sun Z, Sampson AR et al (2006) Selective elimination of glutamatergic synapses on striatopallidal neurons in Parkinson disease models. Nat Neurosci 9:251–259

De Vera M, Rahman MM, Rankin J, Kopec J, Gao X, Choi H (2008) Gout and the risk of Parkinson's disease: a cohort study. Arthritis Rheum 59:1549–1554

Derlet RW, Tseng JC, Albertson TE (1992) Potentiation of cocaine and d-amphetamine toxicity with caffeine. Am J Emerg Med 10:211–216

Deussen A, Brand M, Pexa A, Weichsel J (2006) Metabolic coronary flow regulation—current concepts. Basic Res Cardiol 101:453–464

El Yacoubi M, Ledent C, Parmentier M, Bertorelli R, Ongini E, Costentin J, Vaugeois JM (2001) Adenosine A_{2A} receptor antagonists are potential antidepressants: evidence based on pharmacology and A_{2A} receptor knockout mice. Br J Pharmacol 134:68–77

Factor S, Mark MH, Watts R, Struck L, Mori A, Ballerini R, Sussman NM (2010) A long-term study of istradefylline in subjects with fluctuating Parkinson's disease. Parkinsonism Relat Disord 16:423–426

Fenu S, Pinna A, Ongini E, Morelli M (1997) Adenosine A_{2A} receptor antagonism potentiates L-DOPA-induced turning behaviour and c-fos expression in 6-hydroxydopamine-lesioned rats. Eur J Pharmacol 321:143–147

Fenu S, Wardas J, Morelli M (2009) Impulse control disorders and dopamine dysregulation syndrome associated with dopamine agonist therapy in Parkinson's disease. Behav Pharmacol 20:363–379

Fernandez HH, Greeley DR, Zweig RM, Wojcieszek J, Mori A, Sussman NM (2010) Istradefylline as monotherapy for Parkinson disease: results of the 6002-US-051 trial. Parkinsonism Relat Disord 16:16–20

Ferré S, Fredholm BB, Morelli M, Popoli P, Fuxe K (1997) Adenosine-dopamine receptor-receptor interactions as an integrative mechanism in the basal ganglia. Trends Neurosci 20:482–487

Ferré S, Karcz-Kubicha M, Hope BT, Popoli P, Burgueno J, Gutierrez MA, Casadó V, Fuxe K, Goldberg SR, Lluis C et al (2002) Synergistic interaction between adenosine A_{2A} and glutamate mGlu5 receptors: implications for striatal neuronal function. Proc Natl Acad Sci USA 99:11940–11945

Frau L, Borsini F, Wardas J, Khairnar AS, Schintu N, Morelli M (2011) Neuroprotective and anti-inflammatory effects of the adenosine $A_{(2A)}$ receptor antagonist ST1535 in a MPTP mouse model of Parkinson's disease. Synapse 65:181–188

Fredholm BB, Bättig K, Holmén J, Nehlig A, Zvartau EE (1999) Actions of caffeine in the brain with special reference to factors that contribute to its widespread use. Pharmacol Rev 51(1):83–133

Gao X, Chen H, Choi HK, Curhan G, Schwarzschild MA, Ascherio A (2008) Diet, urate, and Parkinson's disease risk in men. Am J Epidemiol 167:831–838

Gevaerd MS, Takahashi RN, Silveira R, Da Cunha C (2001) Caffeine reverses the memory disruption induced by intra-nigral MPTP-injection in rats. Brain Res Bull 55:101–106

Giménez-Llort L, Schiffmann SN, Shmidt T, Canela L, Camón L, Wassholm M, Canals M, Terasmaa A, Fernández-Teruel A, Tobeña A et al (2007) Working memory deficits in transgenic rats overexpressing human adenosine A_{2A} receptors in the brain. Neurobiol Learn Mem 87:42–56

Grondin R, Bédard PJ, Hadj Tahar A, Grégoire L, Mori A, Kase H (1999) Antiparkinsonian effect of a new selective adenosine A_{2A} receptor antagonist in MPTP-treated monkeys. Neurology 52:1673–1677

Hauser RA, Shulman LM, Trugman JM, Roberts J, Mori A, Ballerini R, Sussman NM (2008) Study of istradefylline in patients with Parkinson's disease on levodopa with motor fluctuations. Mov Disord 23:2177–2185

Hauser RA, Cantillon M, Pourcher E, Micheli F, Mok V, Onofrj M, Huyck S, Wolski K (2011) Preladenant in patients with Parkinson's disease and motor fluctuations: a phase 2, double-blind, randomised trial. Lancet Neurol 10:221–229

Hershfield MS (2005) New insights into adenosine-receptor-mediated immunosuppression and the role of adenosine in causing the immunodeficiency associated with adenosine deaminase deficiency. Eur J Immunol 35:25–30

Hettinger BD, Lee A, Linden J, Rosin DL (2001) Ultrastructural localization of adenosine A_{2A} receptors suggests multiple cellular sites for modulation of GABAergic neurons in rat striatum. J Comp Neurol 431:331–346

Higgins GA, Grzelak ME, Pond AJ, Cohen-Williams ME, Hodgson RA, Varty GB (2007) The effect of caffeine to increase reaction time in the rat during a test of attention is mediated through antagonism of adenosine A_{2A} receptors. Behav Brain Res 185:32–42

Higley MJ, Sabatini BL (2010) Competitive regulation of synaptic Ca^{2+} influx by D_2 dopamine and A_{2A} adenosine receptors. Nat Neurosci 13:958–966

Hillion J, Canals M, Torvinen M, Casado V, Scott R, Terasmaa A, Hansson A, Watson S, Olah ME, Mallol J et al (2002) Coaggregation, cointernalization, and codesensitization of adenosine A_{2A} receptors and dopamine D_2 receptors. J Biol Chem 77:18091–18097

Hodgson RA, Bertorelli R, Varty GB, Lachowicz JE, Forlani A, Fredduzzi S, Cohen-Williams ME, Higgins GA, Impagnatiello F, Nicolussi E et al (2009) Characterization of the potent and highly selective A_{2A} receptor antagonists preladenant and SCH 412348 in rodent models of movement disorders and depression. J Pharmacol Exp Ther 330:294–303

Hodgson RA, Bedard PJ, Varty GB, Kazdoba TM, Di Paolo T, Grzelak ME, Pond AJ, Hadjtahar A, Belanger N, Gregoire L et al (2010) Preladenant, a selective $A_{(2A)}$ receptor antagonist, is active in primate models of movement disorders. Exp Neurol 225:384–390

Huang ZL, Qu WM, Eguchi N, Chen JF, Schwarzschild MA, Fredholm BB, Urade Y, Hayaishi O (2005) Adenosine A_{2A}, but not A_1, receptors mediate the arousal effect of caffeine. Nat Neurosci 8:858–859

Hunot S, Hirsch EC (2003) Neuroinflammatory processes in Parkinson's disease. Ann Neurol 53:S49–S58

Ikeda K, Kurokawa M, Aoyama S, Kuwana Y (2002) Neuroprotection by adenosine A_{2A} receptor blockade in experimental models of Parkinson's disease. J Neurochem 80:262–270

Jellinger KA (2002) Recent developments in the pathology of Parkinson's disease. J Neural Transm Suppl 62:347–376

Jenner P (2005) Istradefylline, a novel adenosine A_{2A} receptor antagonist, for the treatment of Parkinson's disease. Expert Opin Investig Drugs 14:729–738

Jenner P, Marsden CD (1986) The actions of 1-methyl-4-phenyl-1,2,3,6-tetrahydropyridine in animals as a model of Parkinson's disease. J Neural Transm Suppl 20:11–39

Kachroo A, Orlando LR, Grandy DK, Chen JF, Young AB, Schwarzschild MA (2005) Interactions between metabotropic glutamate 5 and adenosine A_{2A} receptors in normal and parkinsonian mice. J Neurosci 25:10414–10419

Kanda T, Jackson MJ, Smith LA, Pearce RK, Nakamura J, Kase H, Kuwana Y, Jenner P (1998a) Adenosine A_{2A} antagonist: a novel antiparkinsonian agent that does not provoke dyskinesia in parkinsonian monkeys. Ann Neurol 43:507–513

Kanda T, Tashiro T, Kuwana Y, Jenner P (1998b) Adenosine A_{2A} receptors modify motor function in MPTP-treated common marmosets. Neuroreport 9:2857–2860

Kanda T, Jackson MJ, Smith LA, Pearce RK, Nakamura J, Kase H, Kuwana Y, Jenner P (2000) Combined use of the adenosine $A_{(2A)}$ antagonist KW-6002 with L-DOPA or with selective D_1 or D_2 dopamine agonists increases antiparkinsonian activity but not dyskinesia in MPTP-treated monkeys. Exp Neurol 162:321–327

Kerschensteiner M, Meinl E, Hohlfeld R (2009) Neuro-immune crosstalk in CNS diseases. Neuroscience 158:1122–1132

Khairnar A, Plumitallo A, Frau L, Schintu N, Morelli M (2010) Caffeine enhances astroglia and microglia reactivity induced by 3,4-methylenedioxymethamphetamine ('ecstasy') in mouse brain. Neurotox Res 17:435–439

Kuwayama K, Inoue H, Kanamori T, Tsujikawa K, Miyaguchi H, Iwata Y, Miyauchi S, Kamo N, Kishi T (2007) Interactions between 3,4-methylenedioxymethamphetamine, methamphetamine, ketamine, and caffeine in human intestinal Caco-2 cells and in oral administration to rats. Forensic Sci Int 170:183–188

Lane EL, Cheetham SC, Jenner P (2006) Does contraversive circling in the 6-OHDA-lesioned rat indicate an ability to induce motor complications as well as therapeutic effects in Parkinson's disease? Exp Neurol 197:284–290

Lange KW, Riederer P (1994) Glutamatergic drugs in Parkinson's disease. Life Sci 55: 2067–2075

Lazarus M, Shen HY, Cherasse Y, Qu WM, Huang ZL, Bass CE, Winsky-Sommerer R, Semba K, Fredholm BB, Boison D, Hayaishi O, Urade Y, Chen JF (2011) Arousal effect of caffeine depends on adenosine A_{2A} receptors in the shell of the nucleus accumbens. J Neurosci 6:10067–10075

LeWitt PA, Guttman M, Tetrud JW, Tuite PJ, Mori A, Chaikin P, Sussman NM (2008) Adenosine A_{2A} receptor antagonist istradefylline (KW-6002) reduces OFF time in Parkinson's disease: a double-blind, randomized, multicenter clinical trial (6002-US-005). Ann Neurol 63:295–302

Lopes LV, Cunha RA, Ribeiro JA (1999) Cross talk between A(1) and A(2A) adenosine receptors in the hippocampus and cortex of young adult and old rats. J Neurophysiol 82:3196–3203

Lukashev D, Ohta A, Apasov S, Chen JF, Sitkovsky M (2004) Cutting edge: physiologic attenuation of proinflammatory transcription by the Gs protein-coupled A_{2A} adenosine receptor in vivo. J Immunol 173:21–24

Lundblad M, Vaudano E, Cenci MA (2003) Cellular and behavioural effects of the adenosine A_{2A} receptor antagonist KW-6002 in a rat model of l-DOPA-induced dyskinesia. J Neurochem 84:1398–1410

Masino SA, Dulla CG (2005) Adenosine, glutamate and pH: interactions and implications. Neurol Res 27:149–152

McNamara R, Kerans A, O'Neill B, Harkin A (2006) Caffeine promotes hyperthermia and serotonergic loss following co-administration of the substituted amphetamines, MDMA ("Ecstasy") and MDA ("Love"). Neuropharmacology 50:69–80

McNamara R, Maginn M, Harkin A (2007) Caffeine induces a profound and persistent tachycardia in response to MDMA ("Ecstasy") administration. Eur J Pharmacol 555:194–198

Mizuno Y, Hasegawa K, Kondo T, Kuno S, Yamamoto M (2010) Clinical efficacy of istradefylline (KW-6002) in Parkinson's disease: a randomized, controlled study. Mov Disord 25:1437–1443

Morelli M, Carta AR, Kachroo A, Schwarzschild MA (2010) Pathophysiological roles for purines: adenosine, caffeine and urate. Prog Brain Res 183:183–208

Nishi A, Liu F, Matsuyama S, Hamada M, Higashi H, Nairn AC, Greengard P (2003) Metabotropic mGlu5 receptors regulate adenosine A_{2A} receptor signaling. Proc Natl Acad Sci USA 100:1322–1327

O'Neill M, Brown VJ (2006) The effect of the adenosine A(2A) antagonist KW-6002 on motor and motivational processes in the rat. Psychopharmacology (Berl) 184:46–55

Obeso JA, Rodríguez-Oroz MC, Rodríguez M, Lanciego JL, Artieda J, Gonzalo N, Olanow CW (2000) Pathophysiology of the basal ganglia in Parkinson's disease. Trends Neurosci 23:S8–S19

Oh JD, Chase TN (2002) Glutamate-mediated striatal dysregulation and the pathogenesis of motor response complications in Parkinson's disease. Amino Acids 23:133–139

Ohta A, Sitkovsky M (2001) Role of G-protein-coupled adenosine receptors in downregulation of inflammation and protection from tissue damage. Nature 414:916–920

Pierri M, Vaudano E, Sager T, Englund U (2005) KW-6002 protects from MPTP induced dopaminergic toxicity in the mouse. Neuropharmacology 48:517–524

Pinna A (2009) Novel investigational adenosine A_{2A} receptor antagonists for Parkinson's disease. Expert Opin Investig Drugs 18:1619–1631

Pinna A, Di Chiara G, Wardas J, Morelli M (1996) Blockade of A_{2A} adenosine receptors positively modulates turning behaviour and c-Fos expression induced by D_1 agonists in dopamine-denervated rats. Eur J Neurosci 8:1176–1181

Pinna A, Fenu S, Morelli M (2001) Motor stimulant effects of the adenosine A_{2A} receptor antagonist SCH 58261 do not develop tolerance after repeated treatments in 6-hydroxydopamine-lesioned rats. Synapse 39:233–238

Pinna A, Volpini R, Cristalli G, Morelli M (2005) New adenosine A_{2A} receptor antagonists: actions on Parkinson's disease models. Eur J Pharmacol 512:157–164

Pinna A, Pontis S, Morelli M (2006) Expression of dyskinetic movements and turning behaviour in subchronic L-DOPA 6-hydroxydopamine-treated rats is influenced by the testing environment. Behav Brain Res 171:175–178

Pinna A, Tronci E, Schintu N, Simola N, Volpini R, Pontis S, Cristalli G, Morelli M (2010) A new ethyladenine antagonist of adenosine $A_{(2A)}$ receptors: behavioral and biochemical characterization as an antiparkinsonian drug. Neuropharmacology 58:613–623

Popoli P, Minghetti L, Tebano MT, Pintor A, Domenici MR, Massotti M (2004) Adenosine A_{2A} receptor antagonism and neuroprotection: mechanisms, lights, and shadows. Crit Rev Neurobiol 16:99–106

Quarta D, Ferré S, Solinas M, You ZB, Hockemeyer J, Popoli P, Goldberg SR (2004) Opposite modulatory roles for adenosine A_1 and A_{2A} receptors on glutamate and dopamine release in the shell of the nucleus accumbens: effects of chronic caffeine exposure. J Neurochem 88:1151–1158

Ritchie K, Carrière I, de Mendonca A, Portet F, Dartigues JF, Rouaud O, Barberger-Gateau P, Ancelin ML (2007) The neuroprotective effects of caffeine: a prospective population study (the Three City Study). Neurology 69:536–545

Rose S, Jackson MJ, Smith LA, Stockwell K, Johnson L, Carminati P, Jenner P (2006) The novel adenosine A_{2A} receptor antagonist ST1535 potentiates the effects of a threshold dose of L-DOPA in MPTP treated common marmosets. Eur J Pharmacol 546:82–87

Ross GW, Abbott RD, Petrovitch H, Morens DM, Grandinetti A, Tung KH, Tanner CM, Masaki KH, Blanchette PL, Curb JD et al (2000) Association of coffee and caffeine intake with the risk of Parkinson disease. JAMA 283:2674–2679

Salamone JD, Mayorga AJ, Trevitt JT, Cousins MS, Conlan A, Nawab A (1998) Tremulous jaw movements in rats: a model of parkinsonian tremor. Prog Neurobiol 56:591–611

Schiffmann SN, Fisone G, Moresco R, Cunha RA, Ferré S (2007) Adenosine A_{2A} receptors and basal ganglia physiology. Prog Neurobiol 83:277–292

Schwarzschild MA, Schwid SR, Marek K, Watts A, Lang AE, Oakes D, Shoulson I, Ascherio A, Hyson C, Parkinson Study Group PRECEPT Investigators et al (2008) Serum urate as a predictor of clinical and radiographic progression in Parkinson disease. Arch Neurol 65:716–723

Shah U, Hodgson R (2010) Recent progress in the discovery of adenosine $A_{(2A)}$ receptor antagonists for the treatment of Parkinson's disease. Curr Opin Drug Discov Devel 13:466–480

Simola N, Fenu S, Baraldi PG, Tabrizi MA, Morelli M (2004) Blockade of adenosine A_{2A} receptors antagonizes parkinsonian tremor in the rat tacrine model by an action on specific striatal regions. Exp Neurol 189:182–188

Soriano A, Ventura R, Molero A, Hoen R, Casadó V, Cortés A, Fanelli F, Albericio F, Lluís C, Franco R et al (2009) Adenosine A_{2A} receptor-antagonist/dopamine D_2 receptor-agonist bivalent ligands as pharmacological tools to detect A_{2A}-D_2 receptor heteromers. J Med Chem 52:5590–5602

Stacy M, Silver D, Mendis T, Sutton J, Mori A, Chaikin P, Sussman NM (2008) A 12-week, placebo-controlled study (6002-US-006) of istradefylline in Parkinson disease. Neurology 70:2233–2240

Stasi MA, Borsini F, Varani K, Vincenzi F, Di Cesare MA, Minetti P, Ghirardi O, Carminati P (2006) ST 1535: a preferential A_{2A} adenosine receptor antagonist. Int J Neuropsychopharmacol 9:575–584

Takahashi RN, Pamplona FA, Prediger RD (2008) Adenosine receptor antagonists for cognitive dysfunction: a review of animal studies. Front Biosci 13:2614–2632

Tronci E, Simola N, Borsini F, Schintu N, Frau L, Carminati P, Morelli M (2007) Characterization of the antiparkinsonian effects of the new adenosine A_{2A} receptor antagonist ST1535: acute and subchronic studies in rats. Eur J Pharmacol 566:94–102

Vanattou-Saïfoudine N, Gossen A, Harkin A, Neuropsychopharmacology Research Group (2011) A role for adenosine A(1) receptor blockade in the ability of caffeine to promote MDMA "Ecstasy"-induced striatal dopamine release. Eur J Pharmacol 650:220–228

Wang JH, Ma YY, van den Buuse M (2006) Improved spatial recognition memory in mice lacking adenosine A_{2A} receptors. Exp Neurol 199:438–445

Wang XJ, Luo WF, Wang LJ, Mao CJ, Wang L, Liu CF (2009) Study on uric acid and the related factors associated with cognition in the patients with Parkinson's disease. Zhonghua Yi Xue Za Zhi 89:1633–1635

Wardas J, Konieczny J, Lorenc-Koci E (2001) SCH 58261, an $A_{(2A)}$ adenosine receptor antagonist, counteracts parkinsonian-like muscle rigidity in rats. Synapse 41:160–171

Wei CJ, Singer P, Coelho J, Boison D, Feldon J, Yee BK, Chen JF (2011) Selective inactivation of adenosine $A(_{2A})$ receptors in striatal neurons enhances working memory and reversal learning. Learn Mem 18:459–474

Yu L, Shen HY, Coelho JE, Araújo IM, Huang QY, Day YJ, Rebola N, Canas PM, Rapp EK, Ferrara J et al (2008) Adenosine A_{2A} receptor antagonists exert motor and neuroprotective effects by distinct cellular mechanisms. Ann Neurol 63:338–346

Chapter 19
Adenosine Receptors and Alzheimer's Disease

David Blum, Ursula Sandau, Cyril Laurent, Vânia Batalha, Antoine Leboucher, Malika Hamdane, Florence Pasquier, Detlev Boison, Luc Buée, and Luísa V. Lopes

Abstract Alzheimer's disease (AD) is the most prevalent form of dementia in the aged population. Definitive diagnosis of AD is based on the presence of senile plaques and neurofibrillary tangles that are identified in post-mortem brain specimens. The formation of these AD-specific lesions is attributed to the pathological accumulation of either extracellular amyloid beta (Aβ) peptide or intraneuronal hyperphosphorylated Tau protein. The AD brain is also characterized by astrogliosis and inflammation. Sporadic AD results from multiple genetic and environmental risk factors. Prospective, retrospective epidemiological studies and experimental

D. Blum (✉) • M. Hamdane • L. Buée
Inserm, U837, 1 place de Verdun, Lille Cedex 59045, France

UDSL, Jean-ierre Aubert Research Centre, Institut de Médecine Prédictive et de Recherche Thérapeutique, Université Lille-Nord de France, Lille 59045, France

CHRU-Lille, Lille 59037, France
e-mail: david.blum@inserm.fr

U. Sandau • D. Boison
Legacy Research Institute, Portland, OR 97232, USA

C. Laurent • A. Leboucher
Inserm, U837, 1 place de Verdun, Lille Cedex 59045, France

UDSL, Jean-ierre Aubert Research Centre, Institut de Médecine Prédictive et de Recherche Thérapeutique, Université Lille-Nord de France, Lille 59045, France

V. Batalha • L.V. Lopes (✉)
Neurosciences Unit, Instituto de Medicina Molecular, Av. Prof. Egas Moniz, Lisbon 1640-028, Portugal
e-mail: lvlopes@fm.ul.pt

F. Pasquier
CHRU-Lille, Lille 59037, France

Department Neurology, CHRU-Lille, Lille, France

S. Masino and D. Boison (eds.), *Adenosine: A Key Link between Metabolism and Brain Activity*, DOI 10.1007/978-1-4614-3903-5_19,

findings have identified chronic caffeine consumption as a protective factor. Caffeine effects would essentially result from modulation of the adenosine system. In this frame, the present review aims to discuss the role of adenosinergic system and in particular involvement of the $A_{2A}R$ in AD pathology and associated cognitive impairments. We also discuss the apparent paradox in regards to $A_{2A}R$ blockade and the aforementioned protective effects versus the disruption of specific biochemical processes that influence hippocampal synaptic plasticity, BDNF/TrkB signaling and acetylcholine release, all being associated with AD physiopathology. As an alternative to targeting specific pathways a more effective option to treat AD may be achieved by utilizing novel treatment strategies that restore adenosine homeostasis in the diseased brain. Thus, prior to exploring the efficacy of $A_{2A}R$ blockade as a therapeutic option for AD, we conclude that a better understanding of adenosine signaling in AD is needed.

Keywords Amyloid plaque • Brain-derived neurotrophic factor • Caffeine • Inflammation • Neurofibrillary tangle • Tau protein

19.1 Alzheimer's Disease

19.1.1 Generalities, Symptoms, and Lesions

Alzheimer's disease (AD) was first described one hundred years ago. With the increase in life expectancy, its incidence has increased dramatically and current forecasts speak in terms of a doubling of the number of persons affected every 20 years (Wittchen and Jacobi 2005). AD is the most frequently encountered form of dementia (about 70 % of cases of dementia). The earliest and most frequent manifestations are benign memory disorders relating to recent facts (Amieva et al. 2005; Ritchie et al. 2001). There is then a slow evolution of the symptoms, which will gradually spread to organizational and programming disorders, aphasia, apraxia, and agnosia. In most cases, AD appears as a sporadic multifactorial disease resulting from the interaction of different environmental, epigenetic, and genetic factors that might facilitate its onset. Various epidemiologic studies have identified "risk factors" and "protective factors." Not only aging but also cardiovascular factors like high blood pressure, diabetes and obesity are examples of the former (Ballard et al. 2011); while a healthy lifestyle (physical and intellectual activities, fish consumption, among other factors) would seem to have protective effects (Belarbi et al. 2011; La Rue 2010; and references herein). More recently, genome-wide association studies have allowed for the identification of genes associated to the disease (Lambert and Amouyel 2011).

The definitive diagnosis of AD is based on the observation of characteristic brain lesions, usually found during a post-mortem examination: senile plaques and neurofibrillary tangles. Neurofibrillary degeneration is due to the pathological accumulation in the neuron of a naturally present protein, the Tau protein; while the amyloid pathology is characterized by the extracellular accumulation of the amyloid-beta peptide (Aβ), which is normally present in low concentrations. Aβ peptides derive from a precursor called βAPP (β-Amyloid Precursor Protein): Aβ peptides result from the combined action of two distinct proteolytic enzymatic activities, β- and γ-secretase, respectively releasing the N- and C-terminal extremities of Aβ peptides (Checler 1995; De Strooper et al. 2010). This constitutes the amyloidogenic pathway. The activity of β-secretase is now well characterized. It is an acid protease called BACE1 (β-site-APP Cleaving Enzyme 1); γ-secretase is the enzymatic activity which releases the C-terminal extremity of the amyloid peptides. Numerous studies suggest that presenilins 1 and 2 (PS1 and PS2) are themselves carriers of the γ-secretase activity. The γ-secretase activity depending on the presenilins appears to be carried by a multiprotein complex of high molecular weight implicating at least three other proteins: nicastrin (NCT), Aph-1 (Anterior pharynx defective 1 homolog), and Pen-2 (Presenilin enhancer 2 homolog) (De Strooper 2010).

Neurofibrillary degeneration consists of the intraneuronal accumulation of proteinaceous fibrils into paired helical filaments (PHF) (for reviews see Buee et al. 2000; Sergeant et al. 2008). The major antigen of PHF was shown to correspond to Tau protein, which is phosphorylated. Tau is a neuronal protein essentially located within the axonal compartment. Its structure makes it essential for the organization, stabilization and dynamics of microtubules (Buee et al. 2000; Sergeant et al. 2008), but recent data also emphasize that Tau has other important neuronal functions at the dendritic and nuclear levels (Ittner et al. 2010; Sultan et al. 2011). The physiologic and pathologic functions of Tau are also regulated by posttranslational modifications such as phosphorylation. In AD and related disorders, aggregated Tau proteins are always found hyperphosphorylated. These changes in phosphorylation may affect a number of Tau functions and facilitate Tau aggregation (Buee et al. 2000; Sergeant et al. 2008). Notably, the progression of the neurofibrillary lesions in the cortex (entorhinal cortex, then hippocampus, and lastly neocortex), corresponds to the progression of the symptoms (Duyckaerts et al. 1997; Grober et al. 1999). In line with these findings, and supporting the instrumental role of Tau pathology in cognitive alterations, data stress that both long-term potentiation (LTP) and depression (LTD), two well-known manifestations of hippocampal synaptic plasticity, are reduced in transgenic models mimicking AD-like Tau pathology (Polydoro et al. 2009; Van der Jeugd et al. 2011).

19.1.2 Astrogliosis and Inflammation and AD

Besides Aβ load and Tau pathology, both inflammatory processes and astrogliosis appear to be integral components of AD onset and progression (Heneka and O'Banion 2007).

Affirmation of the role that reactive astrocytes play in the pathophysiology of AD stems from several observations. Post-mortem characterization of brain specimens from human AD patients reveals that the marker for glial activation, S100β, is elevated (Sheng et al. 1994; Van Eldik and Griffin 1994) and that activated astrocytes are found juxtaposed to amyloid plaques (Wisniewski and Wegiel 1991). Additionally, hippocampal sclerosis, a common pathology of medial Temporal Lobe Epilepsy (mTLE) characterized by astrogliosis and severe loss of neurons, is a common finding in autopsies from AD patients (Attems and Jellinger 2006). Within the AD brain, a tight association between astrogliosis and either aberrant Aβ or Tau function has been established. Exogenous application of the $A\beta_{1-42}$ peptide to the rodent cortex causes a rapid activation of astrocytes. Importantly, the astroglial response occurs in the absence of senile plaques indicative of a direct Aβ-mediated effect on astrocyte function (Perez et al. 2010). Considering that Aβ, acting as an acute toxin, is a potent glial activator, it remains possible that astrocyte activation could be an early event in disease onset, occurring even in the absence of amyloid deposition (Moreira et al. 2008; Nunomura et al. 2000). Similarly, glial activation was detected at very early stages in the AD brain (Cagnin et al. 2001), and in a recent population-based study, increased gliosis has been found before the development of AD-lesions (Wharton et al. 2009). The presence of reactive astrocytes in early AD is of great importance considering that cognitive deficits and neurodegeneration precede the formation of senile plaques in multiple transgenic mouse models of AD (Hsia et al. 1999; Westerman et al. 2002). Astrocytes are also endowed with the capability of promoting amyloid plaque formation (Nagele et al. 2004). Within the AD brain there is intracellular accumulation of Aβ in the astrocyte processes that are in close proximity of amyloid deposits (Kurt et al. 1999). Of note, alterations in astrocyte function are not only limited to Aβ toxicity but also associated with Tau pathology (Belarbi et al. 2011; Schindowski et al. 2006). Astrocyte dysfunction may thus have a central role in regulating disease progression and the behavioral outcomes, and modulating astrogliosis function could be of importance.

Inflammatory processes are an important hallmark of AD (Heneka and O'Banion 2007; Wyss-Coray 2006; for reviews). A recent study points out that both acute and chronic systemic inflammation, characterized with increases in serum TNFα, is associated with an increase in cognitive decline in Alzheimer disease (Holmes et al. 2009). Microglial cells seem to play an important role in AD-related central inflammatory processes. Whether microglial activation is protective or deleterious in AD is still a matter of debate and has been discussed elsewhere (e.g., Wyss-Coray 2006 for review). Some data indicate that microglia may promote Aβ clearance. For instance, removing microgial CX3CR1 receptor-mediated signaling between neurons and microglia results in reduction of β-amyloid levels and deposition in APP mouse models while microglial reaction is increased (Lee et al. 2010b). In accordance, intra-hippocampal injection of lipopolysaccharide (LPS), a well-known promoter of inflammation, can reduce Aβ load in vivo (Herber et al. 2004; Wyss-Coray et al. 2001). Besides resident microglia, it has also been shown that peripheral monocytes could infiltrate the brain and contribute to amyloid deposit reduction (Simard et al. 2006). Depletion of the chemokine receptor CCR2, known to be involved in the accumulation of the mononuclear phagocyte in the brain, results in an increase of Aβ burden and memory defects in APP mouse models (El Khoury et al. 2007; Naert and Rivest 2011). However, in opposition to the aforementioned results some studies have concluded that microglia are unable to remove Aβ deposits but instead produce cytotoxic damage (Fang et al. 2010; Wegiel et al. 2003). Puzzlingly, a recent study points out that microglia ablation did not modulate plaque formation in an APP transgenic model (Grathwohl et al. 2009). Therefore, the precise role of innate immunity towards the amyloid side of AD remains unclear.

Little is known about the relationships between inflammation and the Tau side of AD. Previous works readily described neuroinflammatory processes in several models of AD-Like Tau pathology (Bellucci et al. 2004; Sasaki et al. 2008; Zilka et al. 2009). It has been particularly stressed that microglial activation would even precede tangle formation (Yoshiyama et al. 2007). Interestingly, microglial activation would be detrimental towards Tau pathology. Indeed, LPS administration in rTg4510 mice has been shown to exacerbate hippocampal and cortical Tau phosphorylation (Lee et al. 2010a). This is in line with other data showing that LPS significantly induced Tau hyperphosphorylation through activation of cyclin-dependent kinase 5 in the 3xTg transgenic model, without affecting APP processing (Kitazawa et al. 2005). Involvement of microglia in processes underlying Tau pathology has been further emphasized by demonstrating that removal of CX3CR1 leads to enhanced Tau pathology (Bhaskar et al. 2010). These observations fit well with the ability of pro-inflammatory mediators, known to be released by microglial cells, as IL1β or TNFα, to promote Tau phosphorylation and even its neuritic aggregation (Gorlovoy et al. 2009; Li et al. 2003).

Overall, all the above data stress that astrogliosis and neuroinflammatory events, especially those mediated by microglial cells, may have an instrumental role in AD while their respective contribution to the amyloid and Tau sides remain unclear so far. This, however, indicates that modulation of brain glial cells is prone to affect mechanisms underlying AD pathogenesis.

19.2 Targeting Adenosine Receptors in AD

19.2.1 A Strong Rationale: The Effect of Caffeine in Memory Loss During Aging and AD

Caffeine is one of the most consumed psychoactive drugs. At high millimolar concentrations, irrelevant for normal consumption, caffeine can act at the level of ryanodine receptors and cyclic nucleotide phosphodiesterases, but it is now well established that under normal physiological conditions, the effects exerted in the brain by caffeine depend on its ability to block adenosine A_1 and A_{2A} receptors (A_1R and $A_{2A}R$) (Fredholm et al. 1999, 2005). Notably, in humans, caffeine is largely demethylated to its dimethyl metabolic intermediates, with over 80 % of orally delivered caffeine metabolized to paraxanthine (1,7-dimethyl-xanthine), and about 16 % is converted to theophylline (3,7-dimethylxanthine) and theobromine (1,3-dimethylxanthine), both being potent antagonists of adenosine receptors.

Adenosine receptors have a crucial neuromodulatory role and regulate both synaptic transmission and plasticity (Linden 2001) either by directly modulating synaptic responses or by interfering with other receptors (Ribeiro and Sebastiao 2010). Adenosine receptors are G-protein coupled receptors that mostly regulate, in opposite directions, the second messenger cAMP. While A_1R are inhibitory $G_{i/o}$ coupled, thus decreasing cAMP levels, $A_{2A}R$ are excitatory G_s coupled, increasing cAMP levels. Activation of adenosine receptors interferes also with Ca^{2+} channels and the phospholipase C pathway. Through these actions and by modulating both the release and the uptake of different neurotransmitters, the balance between adenosine A_1R and $A_{2A}R$ activation allows the fine-tuning of synaptic transmission and plasticity in the hippocampus (Sebastiao and Ribeiro 2009). Upon aging, as well as under chronic noxious brain conditions, the levels of adenosine increase, associated to an increased density and effects of $A_{2A}R$ and a decrease in A_1R (Cunha et al. 2001; Lopes et al. 1999; Rebola et al. 2003). Thus, by blocking adenosine receptors, caffeine emerges as a potential drug to modulate brain function and cognition (de Mendonca and Cunha 2010).

The widespread consumption of caffeine underlies the importance of studying the consequences of its chronic intake to the aging-associated cognitive decline, particularly in pathological conditions. Caffeine consumption increases alertness and improves cognition in physiological conditions (Fisone et al. 2004; Haskell et al. 2005; Sawyer et al. 1982; Smit and Rogers 2000). Caffeine also improves attention and information processing (Lorist and Tops 2003). In rodents, evidence from the past few years supports the cognitive enhancer properties of caffeine in a variety of behavioral tasks that evaluate learning and memory (Takahashi et al. 2008; Marques et al. 2011). However, only recently, the relationship between its consumption and the pathophysiology of many neurodegenerative diseases has been firmly established (Rahman 2009). Even if some controversy may exist (Eskelinen and Kivipelto 2010), caffeine consumption has beneficial effects not only in aging, reducing the associated cognitive decline in healthy subjects (Johnson-Kozlow et al. 2002; Ritchie et al. 2007; van Gelder et al. 2007), but also in AD patients. Indeed, different epidemiological reports highlight the inverse correlation between caffeine consumption and the incidence or severity of AD. In the Canadian Study of Health and Aging (CSHA), daily coffee drinking decreased the risk of AD by 31 % during a 5-year follow-up (Lindsay et al. 2002). Similarly, another study from the CAIDE population reports that coffee drinkers at midlife had lower risk of dementia and AD later in life compared with those drinking no or only little coffee. The lowest risk (65 % decreased) was found in people who drank 3–5 cups per day (Eskelinen et al. 2009). Finally, a retrospective study has shown inverse correlation between coffee intake and the occurrence of AD later on in life. Indeed, while age-matched controls had an average daily caffeine intake of 198.7 ± 135.7 mg during the corresponding 20 years of their lifetimes, patients with AD had a lower average daily caffeine intake of 73.9 ± 97.9 mg during the 20 years that preceded diagnosis of AD (Maia and de Mendonca 2002). Thus, caffeine consumption is able to slow down the cognitive decline in the elderly and even appears to reduce the risk to develop AD.

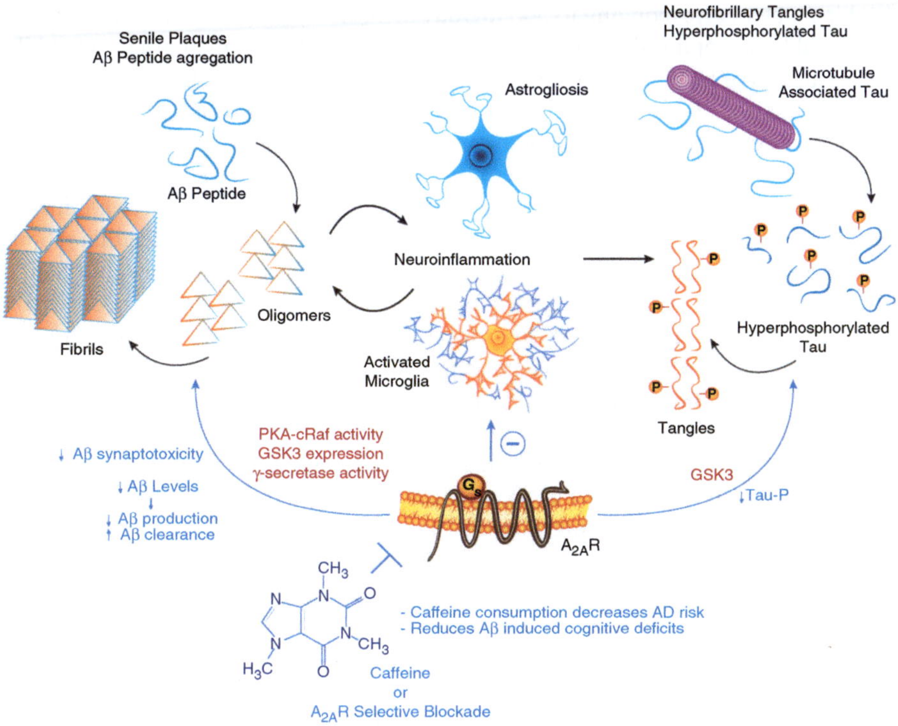

Fig. 19.1 Potential pathways involved in protective effects provided by caffeine and adenosine $A_{2A}R$ blockade against Alzheimer's pathophysiology. Alzheimer's disease is characterized by the presence of two brain lesions. Senile plaques, composed of aggregated Aβ peptide, and neurofibrillary tangles, composed of hyperphosphorylated Tau proteins. Recent data point out that neuroinflammation has also a crucial role in Alzheimer pathophysiology. Caffeine consumption decreases the incidence and severity of AD and mitigates Aβ-induced cognitive deficits. Caffeine was shown to decrease Aβ-induced synaptotoxicity as well as to modulate Aβ production and clearance. Modulation of BACE1 expression and reduction of c-Raf1 activity possibly through PKA pathway, may explain the enhancement of Aβ degradation. Caffeine also reduces GSK3 expression and/or levels influencing Aβ production. A possible role in γ-secretase activity cannot be excluded. Finally caffeine modulation of GSK3 is also able to decrease Tau phosphorylation. Other data support that caffeine could also reduce brain inflammation. Part of the effects of caffeine are thought to relate to the blockade of $A_{2A}R$. Indeed, $A_{2A}R$ blockade was so far shown to decrease neuroinflammation and mitigates Aβ neuronal toxicity (Figure by Carla Batalha)

19.2.2 Modulation of AD-Related Pathology by Caffeine and Adenosine Receptor Targeting

Caffeine has been shown to mitigate cognitive decline induced by Aβ as well as to reduce amyloid burden in transgenic mice overexpressing mutated APP (APPsw) not only in preventive but also therapeutic paradigms (Fig. 19.1). Indeed, APPsw mice chronically treated from 4 to 9.5 months of age with caffeine (300 mg/L by drinking water corresponding to 500 mg/day in humans) were cognitively improved in several behavioral tasks that evaluated working and spatial memories and exhibited reduction of hippocampal $A\beta_{1-40}$ and $A\beta_{1-42}$ (Arendash et al. 2006). Importantly, a similar treatment of APPsw mice at late pathological stages (18–19 months) for 4–5 weeks reversed memory deficits and reduced amyloid deposition as well as soluble Aβ levels in both entorhinal cortex and hippocampus (Arendash et al. 2009). Such beneficial effects of caffeine against Aβ production has recently been confirmed by another group in an experimental model of sporadic AD based on feeding rabbits with cholesterol-enriched diet that elevates both Aβ levels and Tau phosphorylation in the brain (Prasanthi et al. 2010). In this study, rabbits fed with cholesterol-enriched diet were treated with low doses of caffeine (0.5–30 mg/day) through drinking water, corresponding to a maximal 60 mg/day consumption in humans. In this paradigm, caffeine significantly decreased Aβ production in accordance with Arendash's results (Prasanthi et al. 2010). Interestingly, reduced production of $A\beta_{1-40}$ and $A\beta_{1-42}$ was also observed in a neuroblastoma model overexpressing mutated APP following treatment with concentrations of caffeine below 10 μM (Arendash et al. 2006), further supporting that caffeine impacts on mechanisms underlying amyloid peptide production. In accordance, chronic caffeine treatment of APPsw mice has been associated with decreased PS1 and BACE1 protein expression as well as increased insulin-degrading enzyme (IDE) levels, the latter presumably contributing to enhanced Aβ degradation (Arendash et al. 2006; Prasanthi et al. 2010). Effects of caffeine on BACE1 expression could relate to its ability to reduce c-Raf1 activity, possibly through PKA activation (Arendash et al. 2009). In addition, caffeine would reduce GSK3 expression and/or activity and thereby influence Aβ production (Arendash et al. 2009). However, a direct effect of caffeine on γ-secretase activity remains elusive, and mechanisms linking caffeine and Aβ production/clearance deserve further evaluations. It is finally noteworthy that although the beneficial effects of coffee on cognitive decline and decreased AD risk in human has been mostly ascribed to caffeine, other coffee constituents may play an important role. Indeed, two recent studies support that non-caffeine components of coffee display neuroprotective effects in Drosophila and *C. elegans* amyloid models through activation of the Nrf2 detoxification pathway (Dostal et al. 2010; Trinh et al. 2010).

Doses used in experimental studies describing beneficial effects of caffeine achieve plasmatic concentrations of caffeine and metabolites below 10 μM (Georgiev et al. 1993). Taking into account the brain-to-plasma ratio (Kaplan et al. 1989), the brain levels achieved by caffeine are compatible with adenosine receptor inhibition (Fredholm et al. 1999; Muller and Jacobson 2011) even if the involvement of other targets cannot be excluded (Fredholm et al. 1999; Guerreiro et al. 2008). Therefore beneficial effects of caffeine could then be mostly ascribed to its effects upon adenosine receptors. Previous studies indicate that A_1R activation increases soluble APP production, a non-amyloidogenic form of APP (Angulo et al. 2003). Thus, A_1R blockade by caffeine could reduce the non-amyloidogenic pathway, an effect considered as deleterious in the AD context. However, the involvement of A_1R blockade by caffeine in the modulation of amyloid production and toxicity remains to be thoroughly evaluated, as well as the potential role of adaptive changes of A_1R promoted by chronic caffeine (Jacobson et al. 1996; Johansson et al. 1997). Indeed, adaptive increases in A_1R expression are expected to promote the neuroprotective potential of *endogenous* adenosine (Cunha 2008) and have been suggested to be involved in protection afforded by caffeine in ischemia or traumatic brain injury (Jacobson et al. 1996; Johansson et al. 1997; Li et al. 2008). Alternatively, caffeine could mediate its long-term effects through $A_{2A}R$ blockade. Interestingly, besides Aβ production, convergent data indicate that caffeine protects against the synapto-neurotoxicity induced by Aβ through blockade of $A_{2A}R$. Caffeine was indeed shown protective against the death of rat primary hippocampal neurons induced by adenovirus carrying mutated APP gene (Stoppelkamp et al. 2011). These data are in accordance with former works showing that, in primary cultures of cerebellar granule cells, low doses of caffeine (1–25 μM) are able to counteract $A\beta_{25\text{-}35}$ toxicity, an effect mimicked by ZM241385, an $A_{2A}R$ antagonist but not CPT, a selective A_1R antagonist (Dall'Igna et al. 2003). These protective effects were confirmed in vivo. Subchronic treatment with caffeine 30 mg/kg was shown protective against aversive and working memory deficits induced by intra-cerebroventricular (icv) injection of $A\beta_{25\text{-}35}$ in mice (Dall'Igna et al. 2007) and mimicked by administration of SCH58261, a selective $A_{2A}R$ antagonist. $A_{2A}R$ blockade, through intraperitoneal injection of SCH58261 or KW6002 or genetic knockout, was also shown to prevent working memory impairment as well as synaptic loss induced by icv injection of $A\beta_{1\text{-}42}$ (Canas et al. 2009; Cunha et al. 2008). Working memory improvement observed following $A_{2A}R$ blockade was thought to be related to prevention of synaptotoxicity promoted by Aβ through modulation of p38 MAPK and mitochondrial function (Canas et al. 2009). Interestingly, it has been demonstrated that memory improvement promoted by $A_{2A}R$ blockade following icv injection of Aβ was not observed in amnestic conditions induced by MK801 or scopolamine (Cunha et al. 2008). This reinforces the idea that $A_{2A}R$ blockade targets specific synaptic mechanisms rather than general mechanisms controlling memory (Cunha et al. 2008 and Gomes et al. 2011 for discussion). However, this possibility should be tempered since physiologically, $A_{2A}R$ blockade is pro-cognitive and particularly favors working memory but apparently not reference memory needing memory consolidation (Wei et al. 2011; Zhou et al. 2009). This raises the possibility that the overall cognitive effect observed in the paradigms discussed above may not entirely be related to normalization of synaptic defects induced by Aβ, but involve more complex regulations. In line with this view, beneficial effects of $A_{2A}R$ blockade in the hippocampus are not exclusive to Aβ toxicity. $A_{2A}R$ antagonism also prevents acute stress-induced effects (Cunha et al. 2006), convulsive episodes (Cognato et al. 2010), and even age-associated memory impairments (Costa et al. 2008).

Contrasting with the available encouraging data supporting that caffeine and $A_{2A}R$ blockade impacts on the amyloid side of AD, much less is known about impact of adenosine receptor modulation on Tau pathology. So far, only few data are available, and most studies focused on the effects of caffeine. Caffeine was found protective against the death of rat primary hippocampal neurons expressing a mutated Tau protein (Stoppelkamp et al. 2011). Furthermore, caffeine was found able to mitigate Tau hyperphosphorylation induced by cholesterol-rich diet in rabbits (Prasanthi et al. 2010). Similarly, caffeine, at relatively high doses, was recently shown to reduce Tau phosphorylation in rat primary cortical neurons (Currais et al. 2011). Caffeine effects would rely at least on GSK3 modulation, but the underlying mechanisms remain largely unknown. Effects of chronic caffeine consumption, either as prevention or cure, have not been evaluated in a reliable model exhibiting Tau pathology. Furthermore, available data concerning adenosine receptors are largely insufficient to estimate the potential of adenosine receptor targeting towards Tau pathology. The only study available so far demonstrates that A_1R activation promotes Tau phosphorylation in a neuroblastoma model (Angulo et al. 2003) supporting that the effect of caffeine towards Tau may be mediated through this receptor subtype.

Data from Dall'Igna et al. (2003) and Canas et al. (2009) support that protection afforded by $A_{2A}R$ blockade against Aβ is ascribed to neuron-autonomous mechanisms (Canas et al. 2009). However, nonneuronal autonomous mechanisms are also expected. Recent observations indicate that inhibitory effect of Aβ upon astrocytic glutamate uptake is prevented by $A_{2A}R$ blockade (Agostinho et al. 2011) supporting involvement of nonautonomous neuronal mechanisms. As discussed earlier in this review, astroglial and microglial reactions are AD hallmarks and both type of cells express $A_{2A}R$. Also interesting is the observation that increased expression of $A_{2A}R$ observed in the AD brain would be particularly related to its microglial upregulation (Angulo et al. 2003). Together, this would suggest that modulating astroglial and microglial functions by $A_{2A}R$ may also impact AD-related neuropathologies. The role of $A_{2A}R$ expressed by astro- and microglial cells is far from understood (for review see Hasko et al. 2005). $A_{2A}R$ may promote activation and proliferation of astroglial cells (Hindley et al. 1994; Ke et al. 2009) as well as their ability to release glutamate (Nishizaki 2004), while $A_{2A}R$ blockade counteracts the astrocyte induction driven by b-FGF (Brambilla et al. 2003). However, other studies indicate that $A_{2A}R$ stimulation inhibits nitric oxide and TNFα release by astrocytes (Brodie et al. 1998). Concerning microglial cells, it has been shown that $A_{2A}R$ stimulation causes microglial activation (Orr et al. 2009) and potentiates the release of nitric oxide as well as prostaglandin E2 release from these cells (Fiebich et al. 1996; Saura et al. 2005). Also, different experimental evidence supports the anti-inflammatory effect of $A_{2A}R$ blockade in different neuropathological situations (Dai et al. 2010a; Rebola et al. 2011; Yu et al. 2008). In particular, it has recently been shown that $A_{2A}R$ blockade mitigates LTP defects as well as microglial activation and IL1β release following LPS administration (Rebola et al. 2011). Then, the role of $A_{2A}R$ towards glial and microglial function remains unclear. Of interest, the role of $A_{2A}R$ may vary depending on the pathophysiological context and particularly with regard to the local environment. Indeed, although $A_{2A}R$ may have anti-inflammatory functions in low glutamate conditions, their function shifts toward pro-inflammatory function under high glutamate concentrations following a transduction shift (Dai et al. 2010b). This raises the possibility that depending on the neuropathological context, $A_{2A}R$ function may vary in microglial cells, and possibly astroglial cells, providing a different effect. This remains to be studied in vivo using appropriate cell-specific models.

19.3 Is There an Adenosine Paradox in AD?

Data concerning caffeine and adenosine receptors, reviewed above, support that adenosine receptor blockade—especially of the $A_{2A}R$ subtype—could be of therapeutical interest in AD. However, different sets of data also indicate that adenosine receptor activation, especially of the $A_{2A}R$ subtype, is physiologically important towards synaptic plasticity, acetylcholine release, neurotrophin function, and seizure occurrence. In addition to the well-investigated adenosine receptor-mediated effects of adenosine, it must be stressed that adenosine, as homeostatic bioenergetic network regulator, exerts important adenosine receptor-independent functions as key regulator of basic biochemical reactions, as regulator of epigenetic functions, and by its ability to directly couple to mitochondrial bioenergetics (Boison 2011). Thus, adenosine is most likely able to regulate the pathophysiology of AD on a homeostatic network level, adding to the complexity of its well-known adenosine receptor-dependent functions.

19.3.1 Is There an $A_{2A}R$-Based Paradox in AD?

19.3.1.1 Hippocampal Synaptic Plasticity

Although brain adenosine $A_{2A}R$ are classically recognized to be most abundant in the striatum, where they have a predominant postsynaptic localization, in the hippocampus they are predominantly localized at presynaptic nerve terminals, therefore, in a privileged position to control neurotransmitter release and synaptic function (Rebola et al. 2005). Data from in vitro studies indicate that adenosine $A_{2A}R$ activated by endogenous adenosine are not usually involved in basal hippocampal synaptic transmission. However, upon exogenous activation with selective $A_{2A}R$ agonists there is enhanced glutamatergic synaptic transmission in the hippocampus (Cunha et al. 1994, 1995; Jin and Fredholm 1997; Sebastiao and Ribeiro 1992). Remarkably, the enhancing effects of adenosine $A_{2A}R$ on basal transmission are more pronounced in old animals (Rebola et al. 2003) most likely due to the increased expression of these receptors (Cunha et al. 1995; Lopes et al. 1999; Rodrigues et al. 2008). This information is relevant because cognitive deficits, and particularly AD, are essentially found in aged people, although researchers commonly use young animals as experimental models for AD. Endogenous activation of adenosine $A_{2A}R$ appears to become particularly relevant in conditions of high-frequency or theta-burst stimulation leading to LTP (de Mendonca and Ribeiro 1994; Dias et al. 2010). In particular, post-synaptic $A_{2A}R$ seem to be essential for a form of LTP at mossy fiber synapses (Rebola et al. 2008). These synaptic effects resulting from $A_{2A}R$ activation on LTP and glutamate transmission contrast with the reported cognitive benefits achieved by using either caffeine or $A_{2A}R$ antagonists. Whether this means that the LTP recorded in those particular synapses does not necessarily reflect enhanced performance in memory tasks remains to be clarified

(Costenla et al. 2010; Diogenes et al. 2011). In vivo recordings of LTP in freely moving animals (which would capture the whole synaptic activity), under chronic intake of caffeine or selective receptor antagonists, are needed in order to solve this apparent paradox and to further strengthen the rationale of blocking $A_{2A}R$ as a way to enhance cognition (see also Lopes et al. 2011).

19.3.1.2 Neurotrophic Signaling

Compelling evidence indicates that $A_{2A}R$ plays an important role for Brain-Derived Neurotrophic Factor (BDNF) function (Tebano et al. 2010). BDNF is a neurotrophin highly expressed in the hippocampus where it plays, through the activation of its cognate TrkB receptor, a critical role in synaptic plasticity processes as well as learning and memory (Cunha et al. 2010). The pathophysiological context of AD, Aβ, and Tau lesions could thus act synergistically to impair the BDNF/TrkB system, thereby contributing to cognitive dysfunction. Aβ promotes the down-regulation of BDNF and TrkB expression (Espana et al. 2010); while Tau pathology impacts on the hippocampal synaptic response to BDNF (Blum et al. 2011) supporting that both AD-related lesions impair BDNF functions. Current evidence indicates that $A_{2A}R$ are prone to modulate BDNF/TrkB functions by regulating TrkB activity, synaptic effects of BDNF, and BDNF release itself. First, $A_{2A}R$ have been shown to modulate the activity of the BDNF/TrkB signaling system. Indeed, activation of $A_{2A}R$ on primary hippocampal neurons leads to TrkB receptor activation, even in absence of the neurotrophic factor BDNF (Lee and Chao 2001). This also seems to occur in vivo, at least in motor neurons (Wiese et al. 2007). Interestingly, it has been recently demonstrated that $A_{2A}R$ activation promotes the recruitment of TrkB receptors to lipid rafts and potentiates BDNF-induced TrkB activation (Assaife-Lopes et al. 2010). Further, several electrophysiological studies now support that $A_{2A}R$ activation is important for BDNF facilitatory effects at hippocampal synapses. Activation of adenosine $A_{2A}R$ with CGS 21680, or the increase in extracellular adenosine levels induced by 5-iodotubercidin, triggered the excitatory action of BDNF, a process prevented by an $A_{2A}R$ antagonist and by a protein kinase A inhibitor (Diogenes et al. 2004). Furthermore, the facilitatory effect of BDNF on LTP requires adenosine $A_{2A}R$ activation by endogenous adenosine (Diogenes et al. 2011; Fontinha et al. 2008). Thus, activation of $A_{2A}R$ facilitates BDNF modulation of synaptic transmission at hippocampal synapses. These observations have been strengthened by data showing that BDNF-induced synaptic facilitation in hippocampal slices was lost by $A_{2A}R$ pharmacologic and genetic blockade (Tebano et al. 2010). Interestingly, the reduced functional ability of BDNF to promote hippocampal plasticity driven by $A_{2A}R$ blockade was accompanied by a reduction in hippocampal BDNF levels (Tebano et al. 2010). However, consequences of such $A_{2A}R$-based regulations on memory processes remain unclear. The importance of BDNF/TrkB system in the occurrence of different forms of memory together with the apparent ability of $A_{2A}R$ to promote BDNF/TrkB system function would indicate that memory functions would be altered by $A_{2A}R$ knockout. However, $A_{2A}R$ invalidation promotes enhanced spatial memory (Wang et al. 2006) and $A_{2A}R$ overexpression in rat leads to working memory alterations with limited effects on spatial memory (Gimenez-Llort et al. 2007). Together, these observations raise the question of the contribution of the $A_{2A}R$-BDNF/TrkB signaling in AD processes and of the mechanisms underlying protective processes promoted by $A_{2A}R$ blockade.

19.3.1.3 Acetylcholine Release

The high abundance of $A_{2A}R$ in striatal areas (Bruns et al. 1986), along with their ability to tightly regulate dopaminergic transmission (Brown et al. 1990) and the low levels of these receptors in other brain regions, render $A_{2A}R$ particularly interesting targets in Parkinson's disease (PD) (Schwarzschild et al. 2006). The administration of $A_{2A}R$ antagonists attenuates the motor impairments in different animal models of PD (Kalda et al. 2006; Morelli et al. 2007; Xu et al. 2005) and recent clinical trials are probing $A_{2A}R$ antagonists as an alternative to traditional therapeutic approaches (Pinna 2009). Besides striatum, there is now evidence that exogenous activation of $A_{2A}R$ in physiological conditions with selective agonists facilitates the release of acetylcholine (Cunha et al. 1994, 1995; Rodrigues et al. 2008; Sebastiao and Ribeiro 1992). This observation is of importance since cholinergic system defect is another hallmark of AD thought to participate in the cascade that leads to memory impairment in the disease. Basal forebrain cholinergic neurons provide the major cholinergic innervation to the cortex and the hippocampus and play a role in memory behavior, and loss of these neurons is consistently found in postmortem brains of subjects with endstage AD (Gold 2003). Thus, regulation afforded by $A_{2A}R$ towards acetylcholine release is precisely the opposite of what would be expected to underline the beneficial effects of $A_{2A}R$ antagonists on memory performance in AD.

19.3.2 Is There a Homoeostasis-Based Paradox in AD?

19.3.2.1 Contribution of Adenosine and Adenosine Kinase to AD-Related Seizures

Seizures and epilepsy are a common comorbidity of AD (Menendez 2005; Palop and Mucke 2009). Within the sporadic AD population there is an 87 % increase in the incidence of unprovoked clinical seizures in comparison to the age-matched reference population (Amatniek et al. 2006; Menendez 2005). In addition to these behavioral seizures, it is postulated that the numerous episodes of severe confusion reported by AD patients is the occurrence of "silent" (i.e., subclinical, electrographic) seizures (Thomas 1997). So far, non-convulsive seizure occurrence in AD has been ascribed to the amyloid side through the use of several APP transgenic mouse models (Minkeviciene et al. 2009; Palop and Mucke 2009). Similarly, seizure threshold in the hAPP transgenic mice was reduced in response to pentylentetrazol (Del Vecchio et al. 2004; Palop and Mucke 2009). The spontaneous and chemically induced seizure phenotypes in the APP mutant were attenuated by genetically removing endogeneous Tau protein, thereby restoring synaptic function and reducing hyperexcitability (Roberson et al. 2007). These data raise the possibility that amyloid may be instrumental in seizure occurrence observed in AD patients, and also that these seizures may relate to Aβ-Tau interaction. Interestingly, seizure occurrence has been observed in different Tauopathies including fronto-temporal dementia linked to chromosome 17 (FTDP-17) and progressive supranuclear palsy (Beach et al. 2003; Nygaard et al. 1989). However, whether AD-like Tau pathology is also able to promote seizure occurrence has yet to be established.

Dysregulation of the homeostatic role of adenosine could be impaired in AD and instrumentally involved in seizure occurrence observed in AD. Indeed, astrogliosis and the associated increase in adenosine kinase (ADK) leading to focal adenosine deficiency, is sufficient to trigger seizures (Boison 2008). In addition, adenosine deficiency as a result of transgenic overexpression of ADK is sufficient to promote severe learning deficits in the Morris water maze and in Pavlovian conditioning (Yee et al. 2007). The role of adenosine deficiency and ADK on the generation of AD-related seizures and memory deficits remains to be established in reliable models and human brain tissue. Importantly, this hypothesis suggests that reduction of the adenosinergic tone by increasing its metabolic clearance through overexpressed ADK is deleterious. Although this notion apparently contrasts with the possible protective effect expected from adenosine receptor blockade, by caffeine or $A_{2A}R$ antagonists, it needs to be stressed that the net-effects of disrupted adenosine homeostasis affect bioenergetic network regulation and encompass adenosine receptor-dependent, as well as adenosine receptor-independent, aspects of homeostatic network control by adenosine. Future work needs to address adenosine receptor-independent and epigenetic functions of adenosine within the context of AD, integrate those functions with the better characterized adenosine receptor-dependent functions of adenosine, and to ascribe specific roles of adenosine-dependent homeostatic network disruption to cognitive and seizure phenotypes of AD. Eventually, it might be possible to treat or possibly cure AD synergistically by reconstructing homostatic bioenergetic network functions that depend on correct adenosine homeostasis. Therefore, carefully tailored focal adenosine augmentation strategies, perhaps in combination with $A_{2A}R$ blockade might hold promise for the future therapy of AD.

19.4 Conclusion

Several lines of evidence indicate that caffeine administration and $A_{2A}R$ blockade constitute promising therapeutical approaches for AD. Despite epidemiological data on the effects of caffeine in aged and AD subjects, and data from animal studies, no clinical trials have been performed to date to evaluate the extent by which caffeine can slow down disease progression in patients who have already developed

AD. Although caffeine consumption is considered safe (Fredholm et al. 1999), pilot studies are mandatory given potential side effects, particularly at the cardiovascular level (Riksen et al. 2009). Although encouraging preliminary data have been obtained related to the role of $A_{2A}R$ in AD, the major interest towards $A_{2A}R$ antagonists stems from encouraging safety and tolerability trials in PD (for reviews see Morelli et al. 2009; Pinna 2009; Simola et al. 2008). Nevertheless, the major drawback for the use of $A_{2A}R$ antagonists in AD is the limited experimental studies and the lack of pilot clinical studies assessing their effects on cognitive impairment during aging or in AD patients. The only indirect evidence is epidemiological and solely regarding caffeine consumption and this is clearly insufficient. Our poor knowledge of the overall effect of $A_{2A}R$ blockade in AD models is also a limit for the development of $A_{2A}R$-based clinical trials. Several questions are still pending: so far no studies have evaluated preventive or curative effects of $A_{2A}R$ blockade in models of AD that mimic both amyloid pathology and Tau aggregation. In addition to the evaluation of $A_{2A}R$ modulation on neuropathology and cognition, one crucial point remains in regard to the effect of these therapeutic interventions towards inflammatory processes and their influence on both amyloid burden and Tau pathology. Further, the impact of $A_{2A}R$ blockade towards the cholinergic pathway remains below our current knowledge in the AD context. This translates our ignorance regarding the consequences of $A_{2A}R$ modulation at the different levels of regulation (neuronal/nonneuronal) and the lack of data of cell-specific dynamics of $A_{2A}R$ during pathophysiological development. This may be of crucial interest given that $A_{2A}R$ dynamics are thought to play an important role in other disorders such as Huntington's disease (see Blum et al., this issue). Finally, it remains to be determined to what extent network regulation mediated by adenosine through its homeostatic role in addition to adenosine receptor-mediated regulations are interconnected and whether reconstitution of adenosine-based homeostatic network functions would be of therapeutical interest. Overall, modulation of the adenosinergic system is certainly important to consider for the treatment of AD, but precise evaluations of its importance towards pathological and memory processes await further experimental work in the future.

References

Agostinho PM, Matos M, Augusto E, Rodrigues A, Cunha RA (2011) Amyloid-beta inhibits the astrocytic glutamate uptake through a mechanism involving adenosine A2A receptor activation. AD/PD meeting, 2011, Barcelona

Amatniek JC, Hauser WA, DelCastillo-Castaneda C, Jacobs DM, Marder K, Bell K, Albert M, Brandt J, Stern Y (2006) Incidence and predictors of seizures in patients with Alzheimer's disease. Epilepsia 47:867–872

Amieva H, Jacqmin-Gadda H, Orgogozo JM, Le Carret N, Helmer C, Letenneur L, Barberger-Gateau P, Fabrigoule C, Dartigues JF (2005) The 9 year cognitive decline before dementia of the Alzheimer type: a prospective population-based study. Brain 128:1093–1101

Angulo E, Casado V, Mallol J, Canela EI, Vinals F, Ferrer I, Lluis C, Franco R (2003) A1 adenosine receptors accumulate in neurodegenerative structures in Alzheimer disease and mediate both amyloid precursor protein processing and tau phosphorylation and translocation. Brain Pathol 13:440–451

Arendash GW, Schleif W, Rezai-Zadeh K, Jackson EK, Zacharia LC, Cracchiolo JR, Shippy D, Tan J (2006) Caffeine protects Alzheimer's mice against cognitive impairment and reduces brain beta-amyloid production. Neuroscience 142:941–952

Arendash GW, Mori T, Cao C, Mamcarz M, Runfeldt M, Dickson A, Rezai-Zadeh K, Tane J, Citron BA, Lin X et al (2009) Caffeine reverses cognitive impairment and decreases brain amyloid-beta levels in aged Alzheimer's disease mice. J Alzheimers Dis 17:661–680

Assaife-Lopes N, Sousa VC, Pereira DB, Ribeiro JA, Chao MV, Sebastiao AM (2010) Activation of adenosine A2A receptors induces TrkB translocation and increases BDNF-mediated phospho-TrkB localization in lipid rafts: implications for neuromodulation. J Neurosci 30:8468–8480

Attems J, Jellinger KA (2006) Hippocampal sclerosis in Alzheimer disease and other dementias. Neurology 66:775

Ballard C, Gauthier S, Corbett A, Brayne C, Aarsland D, Jones E (2011) Alzheimer's disease. Lancet 377:1019–1031

Beach TG, Sue L, Scott S, Layne K, Newell A, Walker D, Baker M, Sahara N, Yen SH, Hutton M et al (2003) Hippocampal sclerosis dementia with tauopathy. Brain Pathol 13:263–278

Belarbi K, Burnouf S, Fernandez-Gomez FJ, Laurent C, Lestavel S, Figeac M, Sultan A, Troquier L, Leboucher A, Caillierez R et al (2011) Beneficial effects of exercise in a transgenic mouse model of Alzheimer's disease-like Tau pathology. Neurobiol Dis 43:486–494

Bellucci A, Westwood AJ, Ingram E, Casamenti F, Goedert M, Spillantini MG (2004) Induction of inflammatory mediators and microglial activation in mice transgenic for mutant human P301S tau protein. Am J Pathol 165:1643–1652

Bhaskar K, Konerth M, Kokiko-Cochran ON, Cardona A, Ransohoff RM, Lamb BT (2010) Regulation of tau pathology by the microglial fractalkine receptor. Neuron 68:19–31

Blum D, Burnouf S, Martire A, Belarbi K, Laurent C, Muhr-Tailleux A, Van der Jeugd A, Leboucher A, Fernandez-gomez F, Troquier L et al (2011) Impairment of BDNF-mediated hippocampal synaptic facilitation in a transgenic model of Alzheimer's disease-like Tau pathology. Society for neuroscience meeting, San Diego, 2010

Boison D (2008) The adenosine kinase hypothesis of epileptogenesis. Prog Neurobiol 84:249–262

Boison D (2011) Modulators of nucleoside metabolism in the therapy of brain diseases. Curr Top Med Chem 11:1068–1086

Brambilla R, Cottini L, Fumagalli M, Ceruti S, Abbracchio MP (2003) Blockade of A2A adenosine receptors prevents basic fibroblast growth factor-induced reactive astrogliosis in rat striatal primary astrocytes. Glia 43:190–194

Brodie C, Blumberg PM, Jacobson KA (1998) Activation of the A2A adenosine receptor inhibits nitric oxide production in glial cells. FEBS Lett 429:139–142

Brown SJ, James S, Reddington M, Richardson PJ (1990) Both A1 and A2a purine receptors regulate striatal acetylcholine release. J Neurochem 55:31–38

Bruns RF, Lu GH, Pugsley TA (1986) Characterization of the A2 adenosine receptor labeled by [3H]NECA in rat striatal membranes. Mol Pharmacol 29:331–346

Buee L, Bussiere T, Buee-Scherrer V, Delacourte A, Hof PR (2000) Tau protein isoforms, phosphorylation and role in neurodegenerative disorders. Brain Res Brain Res Rev 33:95–130

Cagnin A, Brooks DJ, Kennedy AM, Gunn RN, Myers R, Turkheimer FE, Jones T, Banati RB (2001) In-vivo measurement of activated microglia in dementia. Lancet 358:461–467

Canas PM, Porciuncula LO, Cunha GM, Silva CG, Machado NJ, Oliveira JM, Oliveira CR, Cunha RA (2009) Adenosine A2A receptor blockade prevents synaptotoxicity and memory dysfunction caused by beta-amyloid peptides via p38 mitogen-activated protein kinase pathway. J Neurosci 29:14741–14751

Checler F (1995) Processing of the beta-amyloid precursor protein and its regulation in Alzheimer's disease. J Neurochem 65:1431–1444

Cognato GP, Agostinho PM, Hockemeyer J, Muller CE, Souza DO, Cunha RA (2010) Caffeine and an adenosine A(2A) receptor antagonist prevent memory impairment and synaptotoxicity in adult rats triggered by a convulsive episode in early life. J Neurochem 112:453–462

Costa MS, Botton PH, Mioranzza S, Souza DO, Porciuncula LO (2008) Caffeine prevents age-associated recognition memory decline and changes brain-derived neurotrophic factor and tirosine kinase receptor (TrkB) content in mice. Neuroscience 153:1071–1078

Costenla AR, Cunha RA, de Mendonca A (2010) Caffeine, adenosine receptors, and synaptic plasticity. J Alzheimers Dis 20(Suppl 1):S25–S34

Cunha RA (2008) Caffeine, adenosine receptors, memory and Alzheimer disease. Med Clin (Barc) 131:790–795

Cunha RA, Johansson B, van der Ploeg I, Sebastiao AM, Ribeiro JA, Fredholm BB (1994) Evidence for functionally important adenosine A2a receptors in the rat hippocampus. Brain Res 649:208–216

Cunha RA, Constantino MC, Sebastiao AM, Ribeiro JA (1995) Modification of A1 and A2a adenosine receptor binding in aged striatum, hippocampus and cortex of the rat. Neuroreport 6:1583–1588

Cunha RA, Almeida T, Ribeiro JA (2001) Parallel modification of adenosine extracellular metabolism and modulatory action in the hippocampus of aged rats. J Neurochem 76:372–382

Cunha GM, Canas PM, Oliveira CR, Cunha RA (2006) Increased density and synapto-protective effect of adenosine A2A receptors upon sub-chronic restraint stress. Neuroscience 141:1775–1781

Cunha GM, Canas PM, Melo CS, Hockemeyer J, Muller CE, Oliveira CR, Cunha RA (2008) Adenosine A2A receptor blockade prevents memory dysfunction caused by beta-amyloid peptides but not by scopolamine or MK-801. Exp Neurol 210:776–781

Cunha C, Brambilla R, Thomas KL (2010) A simple role for BDNF in learning and memory? Front Mol Neurosci 3:1

Currais A, Kato K, Canuet L, Ishii R, Tanaka T, Takeda M, Soriano S (2011) Caffeine modulates tau phosphorylation and affects Akt signaling in postmitotic neurons. J Mol Neurosci 43:326–332

Dai SS, Li W, An JH, Wang H, Yang N, Chen XY, Zhao Y, Li P, Liu P, Chen JF et al (2010a) Adenosine A2A receptors in both bone marrow cells and non-bone marrow cells contribute to traumatic brain injury. J Neurochem 113:1536–1544

Dai SS, Zhou YG, Li W, An JH, Li P, Yang N, Chen XY, Xiong RP, Liu P, Zhao Y et al (2010b) Local glutamate level dictates adenosine A2A receptor regulation of neuroinflammation and traumatic brain injury. J Neurosci 30:5802–5810

Dall'Igna OP, Porciuncula LO, Souza DO, Cunha RA, Lara DR (2003) Neuroprotection by caffeine and adenosine A2A receptor blockade of beta-amyloid neurotoxicity. Br J Pharmacol 138:1207–1209

Dall'Igna OP, Fett P, Gomes MW, Souza DO, Cunha RA, Lara DR (2007) Caffeine and adenosine A(2a) receptor antagonists prevent beta-amyloid (25-35)-induced cognitive deficits in mice. Exp Neurol 203:241–245

de Mendonca A, Cunha RA (2010) Therapeutic opportunities for caffeine in Alzheimer's disease and other neurodegenerative disorders. J Alzheimers Dis 20(Suppl 1):S1–S2

de Mendonca A, Ribeiro JA (1994) Endogenous adenosine modulates long-term potentiation in the hippocampus. Neuroscience 62:385–390

De Strooper B (2010) Proteases and proteolysis in Alzheimer disease: a multifactorial view on the disease process. Physiol Rev 90:465–494

De Strooper B, Vassar R, Golde T (2010) The secretases: enzymes with therapeutic potential in Alzheimer disease. Nat Rev Neurol 6:99–107

Del Vecchio RA, Gold LH, Novick SJ, Wong G, Hyde LA (2004) Increased seizure threshold and severity in young transgenic CRND8 mice. Neurosci Lett 367:164–167

Dias RB, Ribeiro JA, Sebastiao AM (2010) Enhancement of AMPA currents and GluR1 membrane expression through PKA-coupled adenosine A(2A) receptors. Hippocampus 22:276–291

Diogenes MJ, Fernandes CC, Sebastiao AM, Ribeiro JA (2004) Activation of adenosine A2A receptor facilitates brain-derived neurotrophic factor modulation of synaptic transmission in hippocampal slices. J Neurosci 24:2905–2913

Diogenes MJ, Costenla AR, Lopes LV, Jeronimo-Santos A, Sousa VC, Fontinha BM, Ribeiro JA, Sebastiao AM (2011) Enhancement of LTP in aged rats is dependent on endogenous BDNF. Neuropsychopharmacology 36:1823–1836

Dostal V, Roberts CM, Link CD (2010) Genetic mechanisms of coffee extract protection in a Caenorhabditis elegans model of beta-amyloid peptide toxicity. Genetics 186:857–866

Duyckaerts C, Bennecib M, Grignon Y, Uchihara T, He Y, Piette F, Hauw JJ (1997) Modeling the relation between neurofibrillary tangles and intellectual status. Neurobiol Aging 18:267–273

El Khoury J, Toft M, Hickman SE, Means TK, Terada K, Geula C, Luster AD (2007) Ccr2 deficiency impairs microglial accumulation and accelerates progression of Alzheimer-like disease. Nat Med 13:432–438

Eskelinen MH, Kivipelto M (2010) Caffeine as a protective factor in dementia and Alzheimer's disease. J Alzheimers Dis 20(Suppl 1):S167–S174

Eskelinen MH, Ngandu T, Tuomilehto J, Soininen H, Kivipelto M (2009) Midlife coffee and tea drinking and the risk of late-life dementia: a population-based CAIDE study. J Alzheimers Dis 16:85–91

Espana J, Valero J, Minano-Molina AJ, Masgrau R, Martin E, Guardia-Laguarta C, Lleo A, Gimenez-Llort L, Rodriguez-Alvarez J, Saura CA (2010) beta-Amyloid disrupts activity-dependent gene transcription required for memory through the CREB coactivator CRTC1. J Neurosci 30:9402–9410

Fang F, Lue LF, Yan S, Xu H, Luddy JS, Chen D, Walker DG, Stern DM, Schmidt AM, Chen JX et al (2010) RAGE-dependent signaling in microglia contributes to neuroinflammation, Abeta accumulation, and impaired learning/memory in a mouse model of Alzheimer's disease. FASEB J 24:1043–1055

Fiebich BL, Biber K, Lieb K, van Calker D, Berger M, Bauer J, Gebicke-Haerter PJ (1996) Cyclooxygenase-2 expression in rat microglia is induced by adenosine A2a-receptors. Glia 18:152–160

Fisone G, Borgkvist A, Usiello A (2004) Caffeine as a psychomotor stimulant: mechanism of action. Cell Mol Life Sci 61:857–872

Fontinha BM, Diogenes MJ, Ribeiro JA, Sebastiao AM (2008) Enhancement of long-term potentiation by brain-derived neurotrophic factor requires adenosine A2A receptor activation by endogenous adenosine. Neuropharmacology 54:924–933

Fredholm BB, Battig K, Holmen J, Nehlig A, Zvartau EE (1999) Actions of caffeine in the brain with special reference to factors that contribute to its widespread use. Pharmacol Rev 51:83–133

Fredholm BB, Chen JF, Cunha RA, Svenningsson P, Vaugeois JM (2005) Adenosine and brain function. Int Rev Neurobiol 63:191–270

Georgiev V, Johansson B, Fredholm BB (1993) Long-term caffeine treatment leads to a decreased susceptibility to NMDA-induced clonic seizures in mice without changes in adenosine A1 receptor number. Brain Res 612:271–277

Gimenez-Llort L, Schiffmann SN, Shmidt T, Canela L, Camon L, Wassholm M, Canals M, Terasmaa A, Fernandez-Teruel A, Tobena A et al (2007) Working memory deficits in transgenic rats overexpressing human adenosine A2A receptors in the brain. Neurobiol Learn Mem 87:42–56

Gold PE (2003) Acetylcholine modulation of neural systems involved in learning and memory. Neurobiol Learn Mem 80:194–210

Gomes CV, Kaster MP, Tome AR, Agostinho PM, Cunha RA (2011) Adenosine receptors and brain diseases: neuroprotection and neurodegeneration. Biochim Biophys Acta 1808:1380–1399

Gorlovoy P, Larionov S, Pham TT, Neumann H (2009) Accumulation of tau induced in neurites by microglial proinflammatory mediators. FASEB J 23:2502–2513

Grathwohl SA, Kalin RE, Bolmont T, Prokop S, Winkelmann G, Kaeser SA, Odenthal J, Radde R, Eldh T, Gandy S et al (2009) Formation and maintenance of Alzheimer's disease beta-amyloid plaques in the absence of microglia. Nat Neurosci 12:1361–1363

Grober E, Dickson D, Sliwinski MJ, Buschke H, Katz M, Crystal H, Lipton RB (1999) Memory and mental status correlates of modified Braak staging. Neurobiol Aging 20:573–579

Guerreiro S, Toulorge D, Hirsch E, Marien M, Sokoloff P, Michel PP (2008) Paraxanthine, the primary metabolite of caffeine, provides protection against dopaminergic cell death via stimulation of ryanodine receptor channels. Mol Pharmacol 74:980–989

Haskell CF, Kennedy DO, Wesnes KA, Scholey AB (2005) Cognitive and mood improvements of caffeine in habitual consumers and habitual non-consumers of caffeine. Psychopharmacology (Berl) 179:813–825

Hasko G, Pacher P, Vizi ES, Illes P (2005) Adenosine receptor signaling in the brain immune system. Trends Pharmacol Sci 26:511–516

Heneka MT, O'Banion MK (2007) Inflammatory processes in Alzheimer's disease. J Neuroimmunol 184:69–91

Herber DL, Roth LM, Wilson D, Wilson N, Mason JE, Morgan D, Gordon MN (2004) Time-dependent reduction in Abeta levels after intracranial LPS administration in APP transgenic mice. Exp Neurol 190:245–253

Hindley S, Herman MA, Rathbone MP (1994) Stimulation of reactive astrogliosis in vivo by extracellular adenosine diphosphate or an adenosine A2 receptor agonist. J Neurosci Res 38:399–406

Holmes C, Cunningham C, Zotova E, Woolford J, Dean C, Kerr S, Culliford D, Perry VH (2009) Systemic inflammation and disease progression in Alzheimer disease. Neurology 73:768–774

Hsia AY, Masliah E, McConlogue L, Yu GQ, Tatsuno G, Hu K, Kholodenko D, Malenka RC, Nicoll RA, Mucke L (1999) Plaque-independent disruption of neural circuits in Alzheimer's disease mouse models. Proc Natl Acad Sci USA 96:3228–3233

Ittner LM, Ke YD, Delerue F, Bi M, Gladbach A, van Eersel J, Wolfing H, Chieng BC, Christie MJ, Napier IA et al (2010) Dendritic function of tau mediates amyloid-beta toxicity in Alzheimer's disease mouse models. Cell 142:387–397

Jacobson KA, von Lubitz DK, Daly JW, Fredholm BB (1996) Adenosine receptor ligands: differences with acute versus chronic treatment. Trends Pharmacol Sci 17:108–113

Jin S, Fredholm BB (1997) Adenosine A2A receptor stimulation increases release of acetylcholine from rat hippocampus but not striatum, and does not affect catecholamine release. Naunyn Schmiedebergs Arch Pharmacol 355:48–56

Johansson B, Georgiev V, Lindstrom K, Fredholm BB (1997) A1 and A2A adenosine receptors and A1 mRNA in mouse brain: effect of long-term caffeine treatment. Brain Res 762:153–164

Johnson-Kozlow M, Kritz-Silverstein D, Barrett-Connor E, Morton D (2002) Coffee consumption and cognitive function among older adults. Am J Epidemiol 156:842–850

Kalda A, Yu L, Oztas E, Chen JF (2006) Novel neuroprotection by caffeine and adenosine A(2A) receptor antagonists in animal models of Parkinson's disease. J Neurol Sci 248:9–15

Kaplan GB, Greenblatt DJ, Leduc BW, Thompson ML, Shader RI (1989) Relationship of plasma and brain concentrations of caffeine and metabolites to benzodiazepine receptor binding and locomotor activity. J Pharmacol Exp Ther 248:1078–1083

Ke RH, Xiong J, Liu Y, Ye ZR (2009) Adenosine A2a receptor induced gliosis via Akt/NF-kappaB pathway in vitro. Neurosci Res 65:280–285

Kitazawa M, Oddo S, Yamasaki TR, Green KN, LaFerla FM (2005) Lipopolysaccharide-induced inflammation exacerbates tau pathology by a cyclin-dependent kinase 5-mediated pathway in a transgenic model of Alzheimer's disease. J Neurosci 25:8843–8853

Kurt MA, Davies DC, Kidd M (1999) beta-Amyloid immunoreactivity in astrocytes in Alzheimer's disease brain biopsies: an electron microscope study. Exp Neurol 158:221–228

La Rue A (2010) Healthy brain aging: role of cognitive reserve, cognitive stimulation, and cognitive exercises. Clin Geriatr Med 26:99–111

Lambert JC, Amouyel P (2011) Genetics of Alzheimer's disease: new evidences for an old hypothesis? Curr Opin Genet Dev 21:295–301

Lee FS, Chao MV (2001) Activation of Trk neurotrophin receptors in the absence of neurotrophins. Proc Natl Acad Sci USA 98:3555–3560

Lee DC, Rizer J, Selenica ML, Reid P, Kraft C, Johnson A, Blair L, Gordon MN, Dickey CA, Morgan D (2010a) LPS- induced inflammation exacerbates phospho-tau pathology in rTg4510 mice. J Neuroinflammation 7:56

Lee S, Varvel NH, Konerth ME, Xu G, Cardona AE, Ransohoff RM, Lamb BT (2010b) CX3CR1 deficiency alters microglial activation and reduces beta-amyloid deposition in two Alzheimer's disease mouse models. Am J Pathol 177:2549–2562

Li Y, Liu L, Barger SW, Griffin WS (2003) Interleukin-1 mediates pathological effects of microglia on tau phosphorylation and on synaptophysin synthesis in cortical neurons through a p38-MAPK pathway. J Neurosci 23:1605–1611

Li W, Dai S, An J, Li X, Chen X, Xiong R, Liu P, Wang H, Zhao Y, Zhu M, Liu X, Zhu P, Chen JF, Zhou Y (2008) Chronic but not acute treatment with cafeine attenuates traumatic brain injury in the mouse cortical impact model. Neuroscience 151:1198–1207

Linden J (2001) Molecular approach to adenosine receptors: receptor-mediated mechanisms of tissue protection. Annu Rev Pharmacol Toxicol 41:775–787

Lindsay J, Laurin D, Verreault R, Hebert R, Helliwell B, Hill GB, McDowell I (2002) Risk factors for Alzheimer's disease: a prospective analysis from the Canadian Study of Health and Aging. Am J Epidemiol 156:445–453

Lopes LV, Cunha RA, Ribeiro JA (1999) Cross talk between A(1) and A(2A) adenosine receptors in the hippocampus and cortex of young adult and old rats. J Neurophysiol 82:3196–3203

Lopes LV, Sebastiao AM, Ribeiro JA (2011) Adenosine and related drugs in brain diseases: present and future in clinical trials. Curr Top Med Chem 11:1087–1101

Lorist MM, Tops M (2003) Caffeine, fatigue, and cognition. Brain Cogn 53:82–94

Maia L, de Mendonca A (2002) Does caffeine intake protect from Alzheimer's disease? Eur J Neurol 9:377–382

Marques S, Batalha VL, Lopes LV, Outeiro TF (2011) Modulating Alzheimer's disease through caffeine: a putative link to epigenetics. J Alzheimers Dis 24:161–171

Menendez M (2005) Down syndrome, Alzheimer's disease and seizures. Brain Dev 27:246–252

Minkeviciene R, Rheims S, Dobszay MB, Zilberter M, Hartikainen J, Fulop L, Penke B, Zilberter Y, Harkany T, Pitkanen A et al (2009) Amyloid beta-induced neuronal hyperexcitability triggers progressive epilepsy. J Neurosci 29:3453–3462

Moreira PI, Santos MS, Oliveira CR, Shenk JC, Nunomura A, Smith MA, Zhu X, Perry G (2008) Alzheimer disease and the role of free radicals in the pathogenesis of the disease. CNS Neurol Disord Drug Targets 7:3–10

Morelli M, Di Paolo T, Wardas J, Calon F, Xiao D, Schwarzschild MA (2007) Role of adenosine A2A receptors in parkinsonian motor impairment and l-DOPA-induced motor complications. Prog Neurobiol 83:293–309

Morelli M, Carta AR, Jenner P (2009) Adenosine A2A receptors and Parkinson's disease. Handb Exp Pharmacol 589–615

Muller CE, Jacobson KA (2011) Recent developments in adenosine receptor ligands and their potential as novel drugs. Biochim Biophys Acta 1808:1290–1308

Naert G, Rivest S (2011) CC chemokine receptor 2 deficiency aggravates cognitive impairments and amyloid pathology in a transgenic mouse model of Alzheimer's disease. J Neurosci 31:6208–6220

Nagele RG, Wegiel J, Venkataraman V, Imaki H, Wang KC (2004) Contribution of glial cells to the development of amyloid plaques in Alzheimer's disease. Neurobiol Aging 25:663–674

Nishizaki T (2004) ATP- and adenosine-mediated signaling in the central nervous system: adenosine stimulates glutamate release from astrocytes via A2a adenosine receptors. J Pharmacol Sci 94:100–102

Nunomura A, Perry G, Pappolla MA, Friedland RP, Hirai K, Chiba S, Smith MA (2000) Neuronal oxidative stress precedes amyloid-beta deposition in Down syndrome. J Neuropathol Exp Neurol 59:1011–1017

Nygaard TG, Duvoisin RC, Manocha M, Chokroverty S (1989) Seizures in progressive supranuclear palsy. Neurology 39:138–140

Orr AG, Orr AL, Li XJ, Gross RE, Traynelis SF (2009) Adenosine A(2A) receptor mediates microglial process retraction. Nat Neurosci 12:872–878

Palop JJ, Mucke L (2009) Epilepsy and cognitive impairments in Alzheimer disease. Arch Neurol 66:435–440

Perez JL, Carrero I, Gonzalo P, Arevalo-Serrano J, Sanz-Anquela JM, Ortega J, Rodriguez M, Gonzalo-Ruiz A (2010) Soluble oligomeric forms of beta-amyloid (Abeta) peptide stimulate Abeta production via astrogliosis in the rat brain. Exp Neurol 223:410–421

Pinna A (2009) Novel investigational adenosine A2A receptor antagonists for Parkinson's disease. Expert Opin Investig Drugs 18:1619–1631

Polydoro M, Acker CM, Duff K, Castillo PE, Davies P (2009) Age-dependent impairment of cognitive and synaptic function in the htau mouse model of tau pathology. J Neurosci 29:10741–10749

Prasanthi JR, Dasari B, Marwarha G, Larson T, Chen X, Geiger JD, Ghribi O (2010) Caffeine protects against oxidative stress and Alzheimer's disease-like pathology in rabbit hippocampus induced by cholesterol-enriched diet. Free Radic Biol Med 49:1212–1220

Rahman A (2009) The role of adenosine in Alzheimer's disease. Curr Neuropharmacol 7:207–216

Rebola N, Sebastiao AM, de Mendonca A, Oliveira CR, Ribeiro JA, Cunha RA (2003) Enhanced adenosine A2A receptor facilitation of synaptic transmission in the hippocampus of aged rats. J Neurophysiol 90:1295–1303

Rebola N, Rodrigues RJ, Lopes LV, Richardson PJ, Oliveira CR, Cunha RA (2005) Adenosine A1 and A2A receptors are co-expressed in pyramidal neurons and co-localized in glutamatergic nerve terminals of the rat hippocampus. Neuroscience 133:79–83

Rebola N, Lujan R, Cunha RA, Mulle C (2008) Adenosine A2A receptors are essential for long-term potentiation of NMDA-EPSCs at hippocampal mossy fiber synapses. Neuron 57:121–134

Rebola N, Simoes AP, Canas PM, Tome AR, Andrade GM, Barry CE, Agostinho PM, Lynch MA, Cunha RA (2011) Adenosine A2A receptors control neuroinflammation and consequent hippocampal neuronal dysfunction. J Neurochem 117:100–111

Ribeiro JA, Sebastiao AM (2010) Modulation and metamodulation of synapses by adenosine. Acta Physiol (Oxf) 199:161–169

Riksen NP, Rongen GA, Smits P (2009) Acute and long-term cardiovascular effects of coffee: implications for coronary heart disease. Pharmacol Ther 121:185–191

Ritchie K, Artero S, Touchon J (2001) Classification criteria for mild cognitive impairment: a population-based validation study. Neurology 56:37–42

Ritchie K, Carriere I, de Mendonca A, Portet F, Dartigues JF, Rouaud O, Barberger-Gateau P, Ancelin ML (2007) The neuroprotective effects of caffeine: a prospective population study (the Three City Study). Neurology 69:536–545

Roberson ED, Scearce-Levie K, Palop JJ, Yan F, Cheng IH, Wu T, Gerstein H, Yu GQ, Mucke L (2007) Reducing endogenous tau ameliorates amyloid beta-induced deficits in an Alzheimer's disease mouse model. Science 316:750–754

Rodrigues RJ, Canas PM, Lopes LV, Oliveira CR, Cunha RA (2008) Modification of adenosine modulation of acetylcholine release in the hippocampus of aged rats. Neurobiol Aging 29:1597–1601

Sasaki A, Kawarabayashi T, Murakami T, Matsubara E, Ikeda M, Hagiwara H, Westaway D, George-Hyslop PS, Shoji M, Nakazato Y (2008) Microglial activation in brain lesions with tau deposits: comparison of human tauopathies and tau transgenic mice TgTauP301L. Brain Res 1214:159–168

Saura J, Angulo E, Ejarque A, Casado V, Tusell JM, Moratalla R, Chen JF, Schwarzschild MA, Lluis C, Franco R et al (2005) Adenosine A2A receptor stimulation potentiates nitric oxide release by activated microglia. J Neurochem 95:919–929

Sawyer DA, Julia HL, Turin AC (1982) Caffeine and human behavior: arousal, anxiety, and performance effects. J Behav Med 5:415–439

Schindowski K, Bretteville A, Leroy K, Begard S, Brion JP, Hamdane M, Buee L (2006) Alzheimer's disease-like tau neuropathology leads to memory deficits and loss of functional synapses in a novel mutated tau transgenic mouse without any motor deficits. Am J Pathol 169:599–616

Schwarzschild MA, Agnati L, Fuxe K, Chen JF, Morelli M (2006) Targeting adenosine A2A receptors in Parkinson's disease. Trends Neurosci 29:647–654

Sebastiao AM, Ribeiro JA (1992) Evidence for the presence of excitatory A2 adenosine receptors in the rat hippocampus. Neurosci Lett 138:41–44

Sebastiao AM, Ribeiro JA (2009) Tuning and fine-tuning of synapses with adenosine. Curr Neuropharmacol 7:180–194

Sergeant N, Bretteville A, Hamdane M, Caillet-Boudin ML, Grognet P, Bombois S, Blum D, Delacourte A, Pasquier F, Vanmechelen E et al (2008) Biochemistry of Tau in Alzheimer's disease and related neurological disorders. Expert Rev Proteomics 5:207–224

Sheng JG, Mrak RE, Griffin WS (1994) S100 beta protein expression in Alzheimer disease: potential role in the pathogenesis of neuritic plaques. J Neurosci Res 39:398–404

Simard AR, Soulet D, Gowing G, Julien JP, Rivest S (2006) Bone marrow-derived microglia play a critical role in restricting senile plaque formation in Alzheimer's disease. Neuron 49:489–502

Simola N, Morelli M, Pinna A (2008) Adenosine A2A receptor antagonists and Parkinson's disease: state of the art and future directions. Curr Pharm Des 14:1475–1489

Smit HJ, Rogers PJ (2000) Effects of low doses of caffeine on cognitive performance, mood and thirst in low and higher caffeine consumers. Psychopharmacology (Berl) 152:167–173

Stoppelkamp S, Bell HS, Palacios-Filardo J, Shewan DA, Riedel G, Platt B (2011) In vitro modelling of Alzheimer's disease: degeneration and cell death induced by viral delivery of amyloid and tau. Exp Neurol 229:226–237

Sultan A, Nesslany F, Violet M, Begard S, Loyens A, Talahari S, Mansuroglu Z, Marzin D, Sergeant N, Humez S et al (2011) Nuclear tau, a key player in neuronal DNA protection. J Biol Chem 286:4566–4575

Takahashi RN, Pamplona FA, Prediger RD (2008) Adenosine receptor antagonists for cognitive dysfunction: a review of animal studies. Front Biosci 13:2614–2632

Tebano MT, Martire A, Chiodi V, Ferrante A, Popoli P (2010) Role of adenosine A(2A) receptors in modulating synaptic functions and brain levels of BDNF: a possible key mechanism in the pathophysiology of Huntington's disease. ScientificWorldJournal 10:1768–1782

Thomas RJ (1997) Seizures and epilepsy in the elderly. Arch Intern Med 157:605–617

Trinh K, Andrews L, Krause J, Hanak T, Lee D, Gelb M, Pallanck L (2010) Decaffeinated coffee and nicotine-free tobacco provide neuroprotection in Drosophila models of Parkinson's disease through an NRF2-dependent mechanism. J Neurosci 30:5525–5532

Van der Jeugd A, Ahmed T, Burnouf S, Belarbi K, Hamdame M, Grosjean ME, Humez S, Balschun D, Blum D, Buee L et al (2011) Hippocampal tauopathy in tau transgenic mice coincides with impaired hippocampus-dependent learning and memory, and attenuated late-phase long-term depression of synaptic transmission. Neurobiol Learn Mem 95:296–304

Van Eldik LJ, Griffin WS (1994) S100 beta expression in Alzheimer's disease: relation to neuropathology in brain regions. Biochim Biophys Acta 1223:398–403

van Gelder BM, Buijsse B, Tijhuis M, Kalmijn S, Giampaoli S, Nissinen A, Kromhout D (2007) Coffee consumption is inversely associated with cognitive decline in elderly European men: the FINE Study. Eur J Clin Nutr 61:226–232

Wang JH, Ma YY, van den Buuse M (2006) Improved spatial recognition memory in mice lacking adenosine A2A receptors. Exp Neurol 199:438–445

Wegiel J, Imaki H, Wang KC, Wronska A, Osuchowski M, Rubenstein R (2003) Origin and turnover of microglial cells in fibrillar plaques of APPsw transgenic mice. Acta Neuropathol 105:393–402

Wei CJ, Li W, Chen JF (2011) Normal and abnormal functions of adenosine receptors in the central nervous system revealed by genetic knockout studies. Biochim Biophys Acta 1808: 1358–1379

Westerman MA, Cooper-Blacketer D, Mariash A, Kotilinek L, Kawarabayashi T, Younkin LH, Carlson GA, Younkin SG, Ashe KH (2002) The relationship between Abeta and memory in the Tg2576 mouse model of Alzheimer's disease. J Neurosci 22:1858–1867

Wharton SB, O'Callaghan JP, Savva GM, Nicoll JA, Matthews F, Simpson JE, Forster G, Shaw PJ, Brayne C, Ince PG (2009) Population variation in glial fibrillary acidic protein levels in brain ageing: relationship to Alzheimer-type pathology and dementia. Dement Geriatr Cogn Disord 27:465–473

Wiese S, Jablonka S, Holtmann B, Orel N, Rajagopal R, Chao MV, Sendtner M (2007) Adenosine receptor A2A-R contributes to motoneuron survival by transactivating the tyrosine kinase receptor TrkB. Proc Natl Acad Sci USA 104:17210–17215

Wisniewski HM, Wegiel J (1991) Spatial relationships between astrocytes and classical plaque components. Neurobiol Aging 12:593–600

Wittchen HU, Jacobi F (2005) Size and burden of mental disorders in Europe—a critical review and appraisal of 27 studies. Eur Neuropsychopharmacol 15:357–376

Wyss-Coray T (2006) Inflammation in Alzheimer disease: driving force, bystander or beneficial response? Nat Med 12:1005–1015

Wyss-Coray T, Lin C, Yan F, Yu GQ, Rohde M, McConlogue L, Masliah E, Mucke L (2001) TGF-beta1 promotes microglial amyloid-beta clearance and reduces plaque burden in transgenic mice. Nat Med 7:612–618

Xu K, Bastia E, Schwarzschild M (2005) Therapeutic potential of adenosine A(2A) receptor antagonists in Parkinson's disease. Pharmacol Ther 105:267–310

Yee BK, Singer P, Chen JF, Feldon J, Boison D (2007) Transgenic overexpression of adenosine kinase in brain leads to multiple learning impairments and altered sensitivity to psychomimetic drugs. Eur J Neurosci 26:3237–3252

Yoshiyama Y, Higuchi M, Zhang B, Huang SM, Iwata N, Saido TC, Maeda J, Suhara T, Trojanowski JQ, Lee VM (2007) Synapse loss and microglial activation precede tangles in a P301S tauopathy mouse model. Neuron 53:337–351

Yu L, Shen HY, Coelho JE, Araujo IM, Huang QY, Day YJ, Rebola N, Canas PM, Rapp EK, Ferrara J et al (2008) Adenosine A2A receptor antagonists exert motor and neuroprotective effects by distinct cellular mechanisms. Ann Neurol 63:338–346

Zhou SJ, Zhu ME, Shu D, Du XP, Song XH, Wang XT, Zheng RY, Cai XH, Chen JF, He JC (2009) Preferential enhancement of working memory in mice lacking adenosine A(2A) receptors. Brain Res 1303:74–83

Zilka N, Stozicka Z, Kovac A, Pilipcinec E, Bugos O, Novak M (2009) Human misfolded truncated tau protein promotes activation of microglia and leukocyte infiltration in the transgenic rat model of tauopathy. J Neuroimmunol 209:16–25

Chapter 20
Adenosine Receptors in Huntington's Disease

David Blum, Alberto Martire, Sylvie Burnouf, Bernard Sablonnière, Pierre Krystkowiak, Catherine Ledent, Luisa V. Lopes, and Patrizia Popoli

Abstract Huntington's disease is a devastating hereditary neurodegenerative disorder caused by CAG mutation within the IT15 gene encoding Huntingtin protein. Even though mutant and normal Huntingtin are ubiquitously expressed, the degenerative processes primarily occur within the striatum and particularly hit the GABAergic enkephalin neuronal subpopulation of medium spiny neurons particularly enriched with adenosine A_{2A}Rs, suggesting that the latter might play a role in HD. In agreement, variants in the ADORA2A gene influence the age at onset in HD

D. Blum (✉) • S. Burnouf • B. Sablonnière
Inserm, U837, Place de Verdun, Lille, 59045, France

Jean-Pierre Aubert Reserach Centre, Institut de Médecine Prédictive et de Recherche Thérapeutique, Université Lille-Nord de France, UDSL, Lille, France

CHRU-Lille, Lille, France
e-mail: david.blum@inserm.fr

A. Martire • P. Popoli (✉)
Section of Central Nervous System Pharmacology, Department of Therapeutic Research and Medicine Evaluation, Istituto Superiore di Sanità, Viale Regina Elena, Rome 299, Italy
e-mail: patrizia.popoli@iss.it

P. Krystkowiak
Department of Neurology, Amiens University Hospital, Amiens, France

CNRS UMR 8160, Amiens, France

C. Ledent
IRIBHM, Université Libre de Bruxelles, Brussels, Belgium

L.V. Lopes
Department of Neurology and Institute of Pharmacology and Neurosciences, Institute of Molecular Medicine and Faculty of Medicine of Lisbon, University of Lisbon, Lisbon, Portugal

S. Masino and D. Boison (eds.), *Adenosine: A Key Link between Metabolism and Brain Activity*, DOI 10.1007/978-1-4614-3903-5_20,

and $A_{2A}R$ dynamics is largely altered by mutated Huntingtin. Adenosine receptors are involved in a number of processes critical for neuronal function and homeostasis, such as modulation of synaptic activity and excitotoxicity, the control of neurotrophin levels and functions as well as the regulation of protein degradation mechanisms. In the present review, we critically reviewed the current knowledge involving adenosine receptors in HD and discussed whether they represent a suitable therapeutic target.

Keywords Brain-derived neurotrophic factor • Excitotoxicity • Glia • Huntingtin • Mutation • Striatum

20.1 Huntington's Disease: Generalities and Physiopathological Mechanisms

20.1.1 Overview

Huntington's disease (HD) is a monogenic autosomal dominantly inherited neurodegenerative disorder generally affecting young adults and characterized by involuntary abnormal movements and postures (chorea, dyskinesia, dystonia), psychiatric disturbances, and cognitive alterations (for review see Ross and Tabrizi 2011). Prevalence is 4–10/100.000. This disorder is fatal within 15–20 years after onset of symptoms. Although several cerebral regions (cerebral cortex, layers III, V, and VI; pallidum; subthalamic nucleus; cerebellum) show signs of neurodegeneration, primary and prominent neuronal loss is found in the caudate and putamen (Vonsattel and DiFiglia 1998). Within the striatum, the subpopulation of striato-pallidal medium spiny neurons expressing enkephalin, dopamine D_2, and adenosine A_{2A} receptors appear more vulnerable in accordance with the predominance of chorea in the early course of the disease (see Reiner et al. 1998; Schiffmann and Vanderhaeghen 1993).

HD is caused by a mutation in the gene IT15 encoding the protein Huntingtin (Htt; The Huntington's Disease Collaborative Research Group 1993). The mutation consists in a CAG triplet repeat expansion translated into an abnormal polyglutamine (polyQ) tract within the N-terminal region of the protein. Penetrance is full with a repeat CAG length above 40 and longer CAG repeats are associated with earlier onset. It is important to stress that although the number of CAG repetitions is the primary determinant of disease onset, it accounts for only ~60 % (see Walker 2007 for review). This supports that other genetic and environmental factors are disease modifier components to take into account.

Htt is a ubiquitous and large protein of about 350 kDa involved in an important number of cellular functions reviewed elsewhere (for review see Bantubungi and Blum 2007a, b; Popoli et al. 2008; Ross and Tabrizi 2011; Zuccato et al. 2010). In the central nervous system, distribution of Huntingtin is large with neuronal (prominent in cortical pyramidal neurons, Purkinje cells, and striatal interneurons; Gourfinkel-An et al. 1997; Trottier et al. 1995) as well as nonneuronal expressions

(Shin et al. 2005). The cellular functions of Htt remain incompletely understood (see Ross and Tabrizi 2011 and references therein). Among others, physiological Htt function is involved in early embryonic development (Dragatsis et al. 1998), fate of cortical progenitors (Godin et al. 2010), axonal transport (Colin et al. 2008; Gauthier et al. 2004), and BDNF expression/transport (Zuccato and Cattaneo 2009). Mutated Htt (mHtt) is thus prone to impair several mechanisms important for neuronal activity and survival by promoting a toxic gain-of-function and also impairing the normal Htt function through loss-of-function mechanisms (Zuccato et al. 2010). In the following, we will specifically address pathways impaired by mutated Huntingtin and prone to be modulated by adenosine receptors: namely excitotoxicity as well as glial and BDNF functions.

20.1.2 Excitotoxicity

Excitotoxicity is primarily defined as cell death ensuing from the toxic action of glutamate through excessive activation of glutamate receptors. Whereas it remains unclear whether glutamate is over-released or not by cortical afferents in HD, there is clear evidence that dysfunctions of the glutamatergic system in the striatum account for the toxic effect of mHtt (Fan and Raymond 2007; Stack et al. 2007). Several neuronal impairments contribute to excitotoxicity such as deficient glial reuptake of glutamate and/or NMDA receptor hypersensitivity, the latter being itself dependent on mitochondrial function (Brouillet et al. 2005; Fan and Raymond 2007; Jacquard et al. 2006). Interestingly, several previous works involved abnormal NMDA receptor activity in HD. R6/2 transgenic HD mice exhibit an enhanced response to NMDA (Cepeda et al. 2001). Intraneuronally, mHtt increases the activity of the NR2B subunit of the NMDA receptor—preferentially expressed by the medium spiny neurons within the striatum—possibly through a dysfunction of PSD95, a docking protein whose interaction with NR2B is reduced in the presence of the mutated protein (Chen et al. 1999; Sun et al. 2001; Zeron et al. 2002). These abnormal interactions lead to increased NMDA currents accompanied by an altered NMDA receptor trafficking (Fan et al. 2007) and neuronal vulnerability to NMDA (Shehadeh et al. 2006; Zeron et al. 2002). Recent data particularly support that the latter is related to an increased expression of extrasynaptic NR2B-containing NMDA receptors (Milnerwood et al. 2010).

In HD, increased NMDA response is probably favored by environmental modulation of NMDA receptor. Indeed, an early endogenous increase of quinolinic acid (QA) in the striatum from HD patients and models has been demonstrated (Guidetti et al. 2004, 2006). QA is an NMDA receptor agonist, derived from the kynurenin pathway, a major route of tryptophan degradation, able to produce lesions reminiscent of HD in animals (Beal et al. 1986) and favoring glutamate release from cortico-striatal endings (Blum et al. 2003a; Popoli et al. 2002). In addition, mHtt has been shown to impair expression of glutamate transporters and glutamate handling by astrocytes (Bradford et al. 2009; Faideau et al. 2010; Lievens et al. 2001; Shin et al. 2005), favoring its increase in the synaptic cleft.

In HD, excitotoxicity is also thought to be indirectly favored by mitochondrial alterations. Several imaging studies have revealed an early metabolic dysfunction in the striatum of HD patients (Brouillet et al. 1999, 2005). Importantly, the severity of metabolic alterations correlates with the size of the CAG expansion (Jenkins et al. 1998). Several post-mortem studies point a significant reduction of complex II-III activity (to which succinate dehydrogenase belongs) in the caudate nucleus of HD patients (Browne et al. 1997; Gu et al. 1996; Tabrizi et al. 1999). Such alterations could be related to an altered expression of the complex II subunits induced by mHtt (Benchoua et al. 2006) and be favored by dopamine (Benchoua et al. 2008). An instrumental role of complex II inhibition in the striatal degeneration in HD is also suggested by the specific profile of degeneration in animals treated by the irreversible complex II inhibitor 3-nitropropionic acid (3NP; Brouillet et al. 2005). Accordingly, several other works have reported strong mitochondrial alterations promoted by mHtt. Indeed, the latter, found localized in the neuronal mitochondrial membrane (Panov et al. 2002), has been shown to impair mitochondrial biogenesis, fission (Kim et al. 2010; Weydt et al. 2006), axonal transport (Shirendeb et al. 2011), and membrane potential (Panov et al. 2002) as well as ATP production (Milakovic and Johnson 2005; Seong et al. 2005) and calcium handling (Choo et al. 2004; Panov et al. 2002). Such mitochondrial defects represent one of the events contributing to the emergence of neuronal excitotoxicity in HD (Brouillet et al. 2005; Jacquard et al. 2006).

20.1.3 Glial Dysfunctions

Besides promoting striatal excitotoxicity, several recent studies point out that expression of mHtt in astroglial cells has a profound impact on HD pathophysiology. Indeed, expression of mHtt in glial cells promotes motor dysfunction (Bradford et al. 2009) and exacerbates the neurological phenotype in a transgenic mouse model of HD (Bradford et al. 2010). Furthermore, genes regulating cholesterol synthesis and efflux are downregulated in astrocytes from HD mice (Valenza et al. 2010). This observation is crucial since astroglial cells are a primary site for the brain production of cholesterol (Benarroch 2008), the latter being mandatory for neuronal function and survival (Valenza and Cattaneo 2006). On the other hand, several other reports indicate that microglial reaction also plays a role in HD. The involvement of inflammation has been reported in HD (Schwab et al. 2010 for review). Particularly, activated microglial cells have been reported in affected area of HD patients (McGeer et al. 1988; Sapp et al. 2001). Accordingly, increased binding of the peripheral benzodiazepine receptor ligand PK11195 has been observed both by ex vivo binding (Messmer and Reynolds 1998) and tomographic imaging in living patients (Politis et al. 2011). Studies in patients particularly stress that microglial activation could be an early event in HD (Politis et al. 2011; Tai et al. 2007) and correlated with severity later on (Pavese et al. 2006). Microglial activation is also observed in various genetic HD models (Schwab et al. 2010 and references therein). Recent data emphasize the instrumental role of microglial activation in HD. Indeed, genetic ablation of cannabinoid receptor 2, the expression of which is markedly increased in microglial cells from HD patients and models, enhances symptomatology and reduces life span (Palazuelos et al. 2009). Altogether, these data support that modulation of the function of both astroglial and microglial cells may affect significantly physiopathological outcomes in HD.

20.1.4 BDNF

BDNF is an abundant neurotrophin in the mammalian brain involved in a variety of brain processes such as development, differentiation, neuronal plasticity, and synaptic activity (Chao 2003). In the striatum, BDNF essentially comes from the cerebral cortex. Thus, BDNF is anterogradely transported to cortical nerve endings to be released in the striatum (Zuccato and Cattaneo 2007). In HD, mutated Huntingtin alters BDNF transcription (Zuccato et al. 2001), trafficking and axonal transport (Gauthier et al. 2004). It has been particularly established that mHtt perturbs the negative influence exerted by wild-type Huntingtin on the silencing activity of the RE1/NRSE silencer, favoring the downregulation of a set of genes, including the one coding BDNF (Zuccato et al. 2003). Mutated Huntingtin also alters the axonal transport of BDNF vesicles (Dompierre et al. 2007) as well as the post-Golgi trafficking of this factor (del Toro et al. 2006). More recently, alteration of BDNF transport in HD has been suggested to involve abnormal interactions between pro-BDNF and Huntingtin-associated protein 1 (Wu et al. 2010). Importantly, loss of striatal BDNF may preferentially affect the function of the subpopulation of medium spiny neurons expressing $A_{2A}R$ and enkephalin (Canals et al. 2004), known to be impaired early in HD. These latter observations support that BDNF impairment is crucially involved in the early vulnerability of striato-pallidal neurons. In accordance with this important role of BDNF in HD, its increase, by gene overexpression and pharmacological or environmental modulation has been shown to be beneficial in several experimental models of HD (Borrell-Pages et al. 2006; Gharami et al. 2008; Giralt et al. 2010; Lynch et al. 2007; Peng et al. 2008; Simmons et al. 2009; Xie et al. 2010).

20.2 Huntington's Disease Pathophysiology and Adenosine Receptors

20.2.1 Single Nucleotide Polymorphisms

As stated above, mutation in the IT15 gene statistically predicts for the age at onset (AAO) in HD patients, but on an individual basis, repeat length accounts for about 60 % of its variance (Walker 2007). For the most common expansions between 41 and 45 CAG repeats, the size of the CAG expansions even explains only 31 % of the variance. These observations have led to a search for genetic factors that influence AAO. In this context, several associations between AAO and polymorphisms of various genes (*GRIK2*, *TCERG1*, *UCLH1*, *TP53*, *DFFB*, *GRIN2A*, *GRIN2B*, *CA150*, *BDNF*, *Atg7*, *pPGC1a*…) have been reported, with several being subject to controversy (Alberch et al. 2005; Andresen et al. 2007; Di Maria et al. 2006; Kishikawa et al. 2006; Metzger et al. 2010; Taherzadeh-Fard et al. 2009; see also Zuccato et al. 2010). Currently, there are no data on the association between polymorphism in the ADORA1 gene and HD. Nonetheless, some data support association of ADORA1 variants with posttraumatic seizures (Wagner et al. 2010) or schizophrenia (Gotoh et al. 2009) supporting possible disease modifying effects in other neurological conditions. This remains to be determined in HD.

In contrast, some recent data points out that variants in the ADORA2A gene may influence AAO in cohorts of HD patients. ADORA2A polymorphisms have been previously associated with caffeine sensitivity/-induced anxiety, sleep disturbances, response to amphetamine, panic disorders, etc.(Alsene et al. 2003; Childs et al. 2008; Hohoff et al. 2005; Yang et al. 2010). We have recently shown in a population of almost 800 HD patients from the French Speaking HD Network that one exonic variant (1976 C>T, formerly 1083 C>T; rs5751876) previously associated with the anxiety response to amphetamine or caffeine (Alsene et al. 2003; Childs et al. 2008; Hohoff et al. 2005) or panic disorder (Hamilton et al. 2004) was a genetic modifier influencing AAO with a significant deleterious influence of the T/T genotype, anticipating AAO by 3.8 years (Dhaenens et al. 2009). Association of the rs5751876 polymorphism with HD AAO has been replicated in an independent population of 419 German patients (Taherzadeh-Fard et al. 2010). Furthermore, the latter study interestingly demonstrated that other ADORA2A polymorphisms are also associated with AAO in this German cohort. Together, these genetic data support a pathophysiological involvement of A_{2A}Rs in HD. Interestingly, two recent studies support that the rs5751876 variant in the ADORA2A gene may be associated with habitual caffeine consumption (Cornelis et al. 2007; Rogers et al. 2010). Given caffeine is a widely consumed psychoactive drug that primarily targets adenosine receptors within the normal consumption range (Fredholm et al. 1999), it could be hypothesized that association between some ADORA2A polymorphisms and habitual caffeine consumption may impact on AAO in HD (see below).

20.2.2 Dynamics of Adenosine Receptor Expression in HD

Dynamics of A_1R in HD has not been strictly evaluated in HD models and patients (but see Varani et al. 2003). The present section will be thus entirely related to the dynamics of A_{2A}R expression in HD patients and experimental models.

Former studies in the field have described A_{2A}Rs as largely down-regulated. Indeed, Cha et al. (1999) have observed in the R6/2 transgenic mouse model an early decrease of A_{2A}R binding and mRNA expression from 4 weeks of age, i.e., before motor symptom occurrence, indicating possible early dysfunction of striato-pallidal neurons. These data were later consistently reproduced in the same or other transgenic (Luthi-Carter et al. 2000; Mievis et al. 2011; Tarditi et al. 2006) and phenotypic models (Blum et al. 2002a). In HD patients, only post-mortem observations are available, waiting for reliable longitudinal PET studies, and demonstrate a substantial loss of $A_{2A}R$ binding as early as grade 0 (Glass et al. 2000). Alteration of $A_{2A}R$ expression in the presence of mHtt seems mediated through prevention of CREB binding to an $A_{2A}R$ core promoter (Chiang et al. 2005) consistent with the observation that altered function of CREB is one of the earliest dysfunctional events at the transcriptional level in HD (Sugars et al. 2004). However, given the large alterations of transcriptional mechanisms seen in HD (see Cha 2007 for review; Zuccato et al. 2010) and the complex structure of the ADORA2A gene (Fredholm et al. 2007), other transcriptional systems may be involved in the alterations of $A_{2A}R$ expression in HD. Recent findings indicate that the transcription factor YY-1 regulates $A_{2A}R$ expression in vitro and in the human brain (Buira et al. 2010a). It is thus a potential candidate to investigate further. Epigenetic involvement remains also possible because other recent findings suggest modulation of $A_{2A}R$ expression by DNA methylation at least in vitro (Buira et al. 2010b). Molecular basis of $A_{2A}R$ expression dysregulation in HD remains thus to be further investigated.

Early loss of $A_{2A}R$ expression in HD would suggest they represent a weak target. However, $A_{2A}R$ physiology appears far more complex in HD and cannot be only recapitulated by expression changes. Actually, mHtt exerts a strong influence on $A_{2A}R$ activity. Data from Varani et al. (2001) have demonstrated aberrant $A_{2A}R$ function in striatal cells expressing mHtt. Particularly, they reported an aberrant amplification of $A_{2A}R$-stimulated adenylyl cyclase in striatal-derived cells engineered to express mHtt (Varani et al. 2001), a phenomenon later described in the striatum of R6/2 mice (Chou et al. 2005; Tarditi et al. 2006). Thus, despite loss of expression, $A_{2A}Rs$ remain functionally active in transducing cAMP signaling in the presence of mHtt. Whether these changes represent a compensatory mechanism to the loss of mRNA expression remain to be determined, but this raises the possibility that $A_{2A}R$ can be pharmacologically targeted in HD.

Besides the latter canonical transduction pathway mediated through $A_{2A}R$, it remains also to be established how mHtt impacts on noncanonical pathways. Indeed, $A_{2A}Rs$ are constitutively active and stimulate their cognate G protein G_s to activate adenylyl cyclase leading to cAMP accumulation, yet there is a second signaling pathway that does not rely on coupling to heterotrimeric G proteins but rather promotes activation of MAP kinase through the recruitment of ARNO and ARF6 (Fredholm et al. 2007; Gsandtner et al. 2005). Relative modulation of both pathways may have functional consequences but the influence of mHtt on noncanonical MAPK-dependent transduction pathway remains unclear. Signaling by the $A_{2A}Rs$ is dramatically affected by membrane cholesterol content: coupling to G_s is blunted in the absence of cholesterol while signaling via ARNO to MAP kinase is not affected (Charalambous et al. 2008). Interestingly, the cholesterol biosynthetic pathway is altered in HD (Valenza et al. 2005, 2007a, b): mHtt induces an accumulation of cholesterol and alters its cellular distribution (del Toro et al. 2010), and depletion of the brain-generated cholesterol metabolite 24-hydroxycholesterol has been recently demonstrated in HD patients (Leoni et al. 2008). These cholesterol alterations may promote changes in membrane fluidity, themselves affecting $A_{2A}R$ function (Varani et al. 2007). How these changes impact on the relative canonical versus noncanonical $A_{2A}R$ transduction pathways has been undercharacterized so far.

Another unsolved point is the relative expression of pre- vs. post-synaptic receptors in the striatum of HD patients. Actually, and as indicated later in this review, pharmacological modulation of pre- vs. post-synaptic receptors may lead to differential clinical outcomes in HD (Blum et al. 2003a, b; Popoli et al. 2002). Thus, determining the precise dynamics of both receptor subpopulations is of interest, despite being challenging. Comparing at least the binding of ligands, preferential for each population (Orru et al. 2011) could provide some valuable information. Finally, another pending question is the validity of $A_{2A}R$ binding parameters as a peripheral biomarker. Indeed a change in $A_{2A}R$ expression and function has been consistently found in the peripheral blood cells of human subjects carrying the mHtt gene. $A_{2A}R$ B(max) is significantly increased in blood cells (including platelets, lymphocytes, and neutrophils) of HD patients (Maglione et al. 2005, 2006; Varani et al. 2003, 2007). However, a controversy exists as to whether peripheral $A_{2A}R$ binding correlates with age at onset in HD patients. Indeed, Varani et al. (2007) demonstrated an increase of $A_{2A}R$ B(max) at all HD stages including pre-symptomatic patients supporting, as stated above, that these alterations represent a toxic effect of polyglutamine expansions but could not be taken as an index of HD progression. This is in contrast with data from Maglione et al. (2005, 2006) that reported a correlation between $A_{2A}R$ B(max) and CAG repeat number as with AAO of patients with predominant motor chorea. Thus, whereas $A_{2A}R$ B(max) is a reliable and robust peripheral marker of mHtt toxic effects, its predictive value towards clinical evolution remains to be established. However, it could represent a reliable marker to monitor effects of pharmacological compounds aimed at counteracting toxic effects of mHtt (Varani et al. 2007).

20.3 Modulating Adenosine Receptors in Huntington's Disease

The hypothesis that adenosine receptors could be targeted in HD originates from data indicating ability of A_1Rs and $A_{2A}Rs$ to modulate excitotoxicity. Indeed, compelling evidence exists that A_1Rs exert an inhibitory presynaptic activity on the release of excitatory amino acids, restraining, in particular, the activation of NMDA receptors (Corradetti et al. 1984; Fredholm and Dunwiddie 1988; Palmer and Stiles 1995). A_1Rs could also promote membrane hyperpolarization, inhibiting NMDA-receptor activation (see Blum et al. 2003b). Finally, recent data point that A_1Rs may control EAAT2 expression (Wu et al. 2011) supporting that A_1Rs would control glutamate uptake by astrocytes. In accordance, A_1R activation was

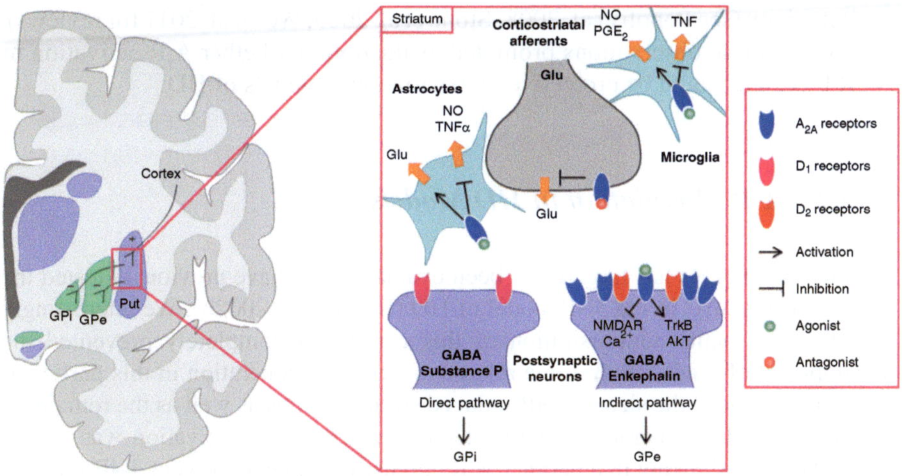

Fig. 20.1 Schematic representation of adenosine A_{2A}Rs localization and functions in the striatum. A_{2A}Rs are found both pre- and postsynaptically in the striatum. Presynaptic corticostriatal glutamatergic neurons express low levels of A_{2A}Rs, blockade of which may exert beneficial effects through inhibition of glutamate release. At the postsynaptic level, A_{2A}Rs are highly expressed on the GABAergic striatopallidal neurons projecting to the Gpe, which contain the peptide enkephalin and are enriched with D_2Rs. A_{2A}R activation at this level leads to inhibition of NMDA receptor currents and activation of TrkB receptors, both being potentially beneficial effects. A_{2A}Rs are also expressed on nonneuronal cells. In astrocytes, A_{2A}R stimulation can induce both deleterious effects by an increase in glutamate outflow and beneficial effects through an inhibition of nitric oxide and TNFα release. This latter beneficial effect has also been observed in microglial cells. However, stimulation of A_{2A}Rs in this cell population can also induce potentially deleterious effects through enhancement of nitric oxide and PGE_2 release. Altogether, the various and sometimes opposite functions of A_{2A}Rs in the striatum make the use of modulators more complex in the context of Huntington's disease. *Put* putamen, *GPe* external globus pallidus, *GPi* internal globus pallidus

shown to be protective in various neuropathological conditions involving excitotoxicity (de Mendonca and Ribeiro 2000; Dunwiddie and Masino 2001; Rudolphi et al. 1992; Stone et al. 2009; Wei et al. 2011). Similar functions are shared by A_{2A}Rs (see Fig. 20.1). Indeed, the A_{2A}R agonist CGS21680 was able to significantly increase extracellular glutamate in the rat striatum (Corsi et al. 1999; Popoli et al. 1995). The regulation of glutamate release through A_{2A}Rs may be not tonically operative, since basal glutamate levels were unaffected by pharmacological blockade of A_{2A}Rs (Pintor et al. 2001). However, when glutamate outflow was stimulated by high K^+, the A_{2A}R antagonist SCH58261 reduced it (Corsi et al. 2000; Pintor et al. 2001). In addition, increased glutamate release promoted by adenosine from cultured hippocampal astrocytes is mimicked by the A_{2A}R agonist CGS21680 and inhibited by A_{2A}R antagonists (Li et al. 2001; Nishizaki et al. 2002). These observations are in accordance with the prevention of extracellular glutamate outflow induced by a blocker of glial glutamate uptake by the A_{2A}R antagonist ZM241385 (Pintor et al. 2004). Accordingly, like for A_1R activation, A_{2A}R blockade has been consistently shown to be protective in various neuropathological models involving excitotoxicity (see Popoli et al. 2008; Stone et al. 2009; Wei et al. 2011 for reviews). Altogether, these observations prompted evaluation of whether A_1R activation or $A_{2A}R$ blockade would be protective in experimental models of HD.

20.3.1 A_1R Modulation in HD Models

Modulation of A_1Rs has largely not been evaluated. We have previously tested the A_1R agonist ADAC in the 3NP model of HD (Blum et al. 2002b). We interestingly observed that acute administration of this compound completely prevented the development of hindlimb dystonia related to striatal degeneration in this particular model and reduced the size of 3NP-induced striatal lesions as well as the remaining ongoing striatal degeneration. The protective effect of A_1R activation was in line with other data showing that another A_1R agonist was able to prevent 3NP-induced seizures in mice (Zuchora et al. 2001) and that A_1R blockade was deleterious in another metabolic model of HD induced by malonate (Alfinito et al. 2003). The protective effects of ADAC were ascribed to its presynaptic ability to reduce glutamate release within the striatum (Blum et al. 2002b). However, so far, it remains difficult to extrapolate the real therapeutic potential of A_1R activation in HD. Indeed, the number of experimental studies in the field is quite limited. In addition, the central dynamics of A_1R expression and function is not known in the case of HD. Besides the lack of experimental information, in the end, chronic A_1R activation may finally remain unable to provide protection due to receptor desensitization. In addition, potential side effects associated with A_1R activation, at the cardio-vascular level for instance, may limit a clinical interest to chronically activate A_1Rs in HD (see Blum et al. 2003b for review). Thus, it is doubtful that A_1R can be considered as a druggable target in HD.

20.3.2 $A_{2A}R$ Modulation in HD Models

In sharp contrast, more extensive work explored the potential of $A_{2A}R$ modulation in phenotypic and genetic HD models (see Table 20.1). One can assume that this greater interest was motivated by their enrichment in the striatum (especially in the most vulnerable medium spiny neurons), the encouraging experimental results obtained in the Parkinson's disease field, as well as availability of modulating drugs suitable for clinical trials. For instance, the $A_{2A}R$ antagonist istradefylline has proven safe, well tolerated, and effective at improving end-of-dose wearing off in PD patients taking L-DOPA (LeWitt et al. 2008), while other $A_{2A}R$ antagonists (e.g., BIIB014, preladenant and ST-1535) are currently in Phase I and II clinical trials and have been proven safe and well tolerated so far (for reviews see Morelli et al. 2009; Pinna 2009; Simola et al. 2008).

Table 20.1 Protective and deleterious effects of A_{2A} receptors in Huntington's disease

		QA MODEL DIRECT EXCITOTOXICITY	3NP MODEL INDIRECT EXCITOTOXICITY	TRANSGENIC MODELS	HD PATIENTS
A_{2A} BLOCKADE	SCH58261 ZM241385 MSX-3	• 0.01 mg/kg SCH58261 i.p. ↘ Striatal degeneration ↘ Motor deficits ↘ Non-motor behavioural alterations (Popoli et al., 2002; Scattoni et al., 2007) • 1 mg/kg SCH58261 i.p. ↗ Striatal degeneration ↗ Motor deficits	• 1 mg/kg MSX-3 i.p. ↘ Striatal degeneration (Blum et al., 2003a) • 5 mg/kg MSX-3 i.p. ↗ Striatal lesions (Blum et al., 2003a)	• 0.01 mg/kg SCH58261 i.p. – early treatment ↘ Non-motor behavioural alterations ↘ Vulnerability to NMDA ↗ Motor deficits (Domenici et al., 2007; Cipriani et al., 2008)	
	CAFFEINE		↘ Striatal damage (Blum et al., 2003a; Fink et al., 2004)		• Low consumption ↗ Neuroprotection (Seward et al., 2011) • High consumption ↘ AAO (Duru et al., 2011; Seward et al., 2011)
	GENETIC INVALIDATION		↘ Striatal damage (Blum et al., 2003a; Fink et al., 2004)	↗ Motor deficits ↘ Survival (Mievis et al., 2011)	
A_{2A} ACTIVATION	CGS21680 T1-11		• 1mg/kg CGS21680 i.p. ↗ Striatal degeneration (Blum et al., 2003a)	• 2.5 - 5 mg/kg CGS21680 i.p. – chronic treatment ↘ Motor deficits ↘ Brain atrophy ↘ Striatal lesions (Chou et al., 2005) • 0.05mg/mL T1-11 drinking water – chronic treatment ↘ Motor deficits ↘ Striatal lesions (Huang et al., in press)	

Blue: Beneficial effects; Red: Detrimental effects

Consistent with the ability of $A_{2A}R$ to modulate glutamate effects and the excitotoxicity hypothesis in HD, the first studies in the field were focused on the potential of $A_{2A}R$ agonists and antagonists in the QA and 3NP models, mimicking direct and indirect excitotoxicity seen in HD, respectively (Brouillet et al. 2005 for review). In the QA model, the pre-treatment with a single dose of the $A_{2A}R$ antagonist SCH58261 (0.01 mg/kg i.p.) protected striatal neurons from QA-induced cell death and fully antagonized QA-induced motor hyperactivity (Popoli et al. 2002). Extending these findings, SCH58261 prevented alterations of wall rearing behavior and emotional changes in the same model (Scattoni et al. 2007). SCH58261-induced neuroprotection was ascribed to its ability to reduce QA-induced glutamate outflow in the striatum (Popoli et al. 2002), a mechanism confirmed using another antagonist, ST1575 (Galluzzo et al. 2008). Both $A_{2A}R$ activation and blockade have also been evaluated in phenotypic models mimicking complex II inhibition seen in HD. In protocols of acute 3NP intoxication leading to the emergence of striatal lesions and, thus, involving glutamate excitotoxicity, both the pharmacological blockade by 8-(3-chlorostyryl)caffeine (CSC) and the global genetic inactivation of $A_{2A}Rs$ strongly mitigated striatal damage (Blum et al. 2003a; Fink et al. 2004). In accordance, in a chronic 3NP model, stimulation of $A_{2A}Rs$ using the agonist CGS21680 (1 mg/kg) led to a dramatic increase of striatal degeneration whereas a low dose of the MSX-3 antagonist was rather protective (Blum et al. 2003a). Finally, in rats, administration of the antagonist 3,7-dimethyl-1-propargylxanthine (DMPX; 5 mg/kg i.p. or 100 ng intrastriatal) reversed the drop of striatal GABA content induced by malonate, the striatal toxicity of which partly depends on glutamate release from cortico-striatal inputs (Alfinito et al. 2003). Together, these data supported that $A_{2A}R$ blockade may be neuroprotective against HD-related excitotoxicity. Later work then moved to the study of $A_{2A}R$ blockade in the transgenic HD models, essentially the R6/2 strain. An interesting starting point was the observation that in microdialysis experiments, SCH58261 was able to reduce basal glutamate outflow in R6/2 mice (Gianfriddo et al. 2004).

We later showed that 1-week treatment with SCH58261 (0.01 mg/kg i.p.) performed in a pre-symptomatic stage, i.e., between the 5th and the 6th week of age, prevented emotional/anxious alterations at later stages (Domenici et al. 2007). Furthermore, the increased vulnerability to NMDA in corticostriatal slices from R6/2 mice at 12 weeks of age was completely prevented by SCH58261 (Domenici et al. 2007), a result that can be considered an index of neuroprotection. In accordance with a protective effect of SCH58261, treatment of R6/2 mice at late stages led to an increased number of nNOS-positive striatal neurons, taken as a protective effect (Cipriani et al. 2008).

Together, all these data were in favor of a protective effect of $A_{2A}R$ blockade in HD. However, other observations made in parallel revealed that the story is far more complex than initially thought. Indeed, the neuroprotective effect of SCH58261 against QA as well as the effects on motor activity and glutamate outflow were achieved only by treating rats with 0.01 mg/kg i.p. and not with a higher dose (1 mg/kg i.p.) of the antagonist. Such an inverse dose relationship was also observed after intrastriatal perfusion with another $A_{2A}R$ antagonist, ZM 241385, the ability of which to reduce QA-induced glutamate outflow was evident at 5 nM but not at 50 nM (Tebano et al. 2004). A paradoxic dose-dependent effect of $A_{2A}R$ blockade was also observed in the chronic rat 3NP model with a low dose of MSX-3 (1 mg/kg i.p.) being protective whereas a higher dose (5 mg/kg i.p.) induced a more severe striatal lesion than in vehicle-treated animals (Blum et al. 2003a). Involvement of some local mechanisms, for instance, blockade of $A_{2A}Rs$ located on cerebral vessels, modulation of early inflammatory reactions, and different degree of pre-synaptic versus post-synaptic neuronal receptor blockade, were invoked. Our data (Blum et al. 2003a; Popoli et al. 2002) strongly supported the last possibility. In particular, blockade of $A_{2A}Rs$ was deleterious against QA toxicity in primary neuronal cultures (Popoli et al. 2002) whereas activation of PKA was protective against 3NP toxicity in a similar neuronal system (Blum et al. 2003a). Together, all the data obtained from QA and 3NP models supported the existence of a balance between presynaptic receptors, the blockade of which is beneficial by preventing excitotoxicity, and postsynaptic receptors the blockade of which alters survival signaling (for reviews see Blum et al. 2003b; Popoli et al. 2007, 2008). Then, *in vivo*, increasing antagonist doses might override the benefit afforded by presynaptic blockade by increasing the blockade of postsynaptic receptors. Notably, this balance was supported to be anatomically determined, expression of pre- and postsynaptic receptors being nonhomogeneously distributed in the striatum, at least in the rat (Blum et al. 2003a). Some studies performed in transgenic mouse models of HD led to contrasting results as well. Indeed, while improving striatal sensitivity to NMDA, early treatment with the $A_{2A}R$ antagonist SCH58261 did not allow recovery, and even worsened deficits in motor coordination showed by R6/2 mice using the rotarod test (Cipriani et al. 2008; Domenici et al. 2007). On the other hand, chronic CGS21680 (Ferrante et al. 2010) and SCH58261 (Martire et al. 2010) both rescued the reduced NR2A/NR2B ratio observed in R6/2 mice. It is worth noting that chronic treatment with CGS21680 modulated the NR2A/NR2B ratio in opposite ways in R6/2 versus WT mice. In particular, CGS21680 decreased the above ratio in the striatum of WT mice (a kind of "pro-excitotoxic" effect) whereas it increased it in HD mice (an "anti-excitotoxic effect").

Finally, to add further complexity, in addition to the relative contribution of pre- and postsynaptic and time window considerations, A_{2A}Rs localized in the non-neuronal compartment may also play a role. Regulation by A_{2A}R of glial and particularly microglial cell functions remain complex and unclear so far (see Dai et al. 2010; Milne and Palmer 2011; Orr et al. 2009; Rebola et al. 2011 and references therein). The local pathophysiological context may particularly influence the function of A_{2A}R in microglial cells (Dai et al. 2010). We previously found that SCH58261 differentially modulates astrogliosis and microglial activation in the striatum and cortex of QA-lesioned animals (Minghetti et al. 2007) suggesting that depending on the region studied and the associated degenerative context, A_{2A}Rs may play a differential role. Interestingly, it was observed in acute 3NP and malonate models, that selective inactivation of A_{2A}Rs on bone marrow-derived cells led to exacerbation of 3NP-induced striatal damage (Huang et al. 2006). However, in an excitotoxic context, A_{2A}R blockade was prone to reduce microglial activation (Dai et al. 2010) raising the possibility that A_{2A}R modulation of inflammatory processes may not only be ascribed to contextual and regional considerations but also to the inflammatory cell type involved.

Together the above data thus raised the following comments. Data obtained so far clearly underlined the complexity of A_{2A}R involvement in HD pathophysiology with nonlinearly altered dynamics (see Tarditi et al. 2006 in particular). The overall fate of blocking A_{2A}Rs thus depends on disease stage as well as the relative contribution of neuronal (pre/post) and nonneuronal compartments that can themselves vary over time. With regards to this, one can ask whether A_{2A}R blockade is really of therapeutic interest in HD and what is the ultimate effect of chronic modulation of A_{2A}Rs in HD transgenic models. Actually, daily treatment with CGS21680 (2.5–5 mg/kg i.p.) from early to frankly symptomatic stages delayed the progressive deterioration of motor performance, improved brain atrophy, reduced the size of striatal ubiquitin-positive neuronal intranuclear inclusions and normalized the over-activation of a major metabolic sensor (5'AMP-activated protein kinase, AMPK) in the striatum of R6/2 mice (Chou et al. 2005). Additionally, it has been recently described that chronic delivery of T1-11, a compound activating A_{2A}Rs and inhibiting ENT1, originally isolated from a Chinese medicinal herb, was able to improve progressive deterioration in motor coordination, to reduce the formation of striatal Htt aggregates and to increase the level of BDNF in the R6/2 model (Huang et al. 2011). All these data from Chern's group were clearly in accordance with our recent data showing that genetic deletion of A_{2A}Rs in N171-82Q transgenic HD mice worsens motor performances and survival as well as leads to a further decrease in striatal enkephalin expression (Mievis et al. 2011). Although these data await confirmation in other transgenic models, these data support that chronic blockade of A_{2A}Rs may be deleterious. In support of this notion, we have recently performed a retrospective caffeine study in a cohort of 80 HD patients from the French HD Speaking Network showing that—adjusted for CAG repeats, smoking, and other confounding factors—habitual consumption of caffeine above two cups significantly advanced AAO (Duru et al. 2011). These observations are in line with another cohort of data supporting that while low level caffeine intake may have neuroprotective effects in HD, moderate to high caffeine intake levels may contribute to earlier AAO and/or more aggressive rates of disease progression (Seward et al. 2010). Despite the limitation of such retrospective studies and even if caffeine also targets A_1 receptors, these data however further support that chronic blockade of $A_{2A}R$ may be deleterious in HD.

20.3.3 Adenosine Receptors and Modulation of Synaptic Transmission in HD

The corticostriatal pathway provides the excitatory glutamatergic input to the striatal medium spiny neurons (MSSNs). Direct pathway MSSNs (which project to the substantia nigra pars reticulata, SNr and to the internal segment of the globus pallidus, GPi) express mostly dopamine D_1-like receptors. Indirect pathway MSSNs (which project to GPi and SNr through to the external segment of the GP and then the subthalamic nucleus), preferentially express D_2-like receptors and $A_{2A}Rs$ (Bolam et al. 2000; Gerfen 1992; Schiffmann and Vanderhaeghen 1993). Complex alterations in corticostriatal transmission occur in experimental HD. Progressive reduction in the frequency of spontaneous EPSCs, progressive reduction in spontaneous synaptic activity in late-stage HD, and a progressive deafferentation that develops until the death of the HD animals have been described (Cepeda et al. 2003). Since the stimulation of striatal $A_{2A}Rs$ by CGS21680 increased the frequency of spontaneous EPSCs in cells from R6/2 mice (Cepeda et al. 2010), the restoration of excitatory synaptic activity by activation of (post-synaptic) $A_{2A}Rs$ could be beneficial in the late stages of the disease.

It should be noted, however, that glutamate transmission is increased at an early stage in HD. In early symptomatic YAC128 and BACHD mice, D_1R expressing cells—i.e., the direct pathway MSSNs—specifically display increased spontaneous EPSC frequencies and decreased paired-pulse ratios, both indices of increased presynaptic glutamate release (Andre et al. 2011). Interestingly, while postsynaptic $A_{2A}Rs$ are expressed in MSSNs of the indirect efferent pathway (where they heteromerize with D_2Rs), presynaptic $A_{2A}Rs$ are localized in corticostriatal glutamatergic terminals contacting MSSNs of the direct efferent pathway (where they heteromerize with adenosine A_1Rs) (Orru et al. 2011; Quiroz et al. 2009). This indicates that in the presymptomatic and early symptomatic stage, when a phasic increase in glutamate release does occur, either A_1R agonists or (presynaptic) $A_{2A}R$ antagonists may help restoring the normal corticostriatal transmission. Again, the influence of $A_{2A}Rs$ on the synaptic alterations occurring in HD is thus not unequivocal, being largely dependent on the stage of the disease.

To the best of our knowledge, no studies have explored so far the effects of A_1Rs on the corticostriatal synaptic transmission in HD models. As mentioned before, A_1Rs should promote membrane hyperpolarisation, inhibiting NMDA-receptor activation (see Blum et al. 2003b) and inhibit presynaptic glutamate release (Blum et al. 2002b). Such effects are expected to normalize both the early and late synaptic alterations observed in HD. However, whether A_1R functions and/or expression may be altered in genetic HD models—as it definitely happens to $A_{2A}Rs$—is currently unknown.

20.3.4 *Mechanisms Related to Protection by $A_{2A}R$ Activation in HD*

20.3.4.1 Effects of Adenosine $A_{2A}Rs$ on BDNF Levels and Functions

As stated above, striatal BDNF levels, and thereby activation of its cognate TrkB receptors, are important for striatal survival. The first link between BDNF and adenosine was provided by Lee and Chao in 2001, with the demonstration that activation of TrkA receptors in PC12 cells and TrkB in hippocampal neurons could be obtained in the absence of neurotrophins by treatment with adenosine (Lee and Chao 2001). These effects were reproduced by using the adenosine agonist CGS21680 and were counteracted with the antagonist ZM241385, indicating that this transactivation by adenosine involves the $A_{2A}R$ subtype. More recently, transactivation of TrkB receptors by $A_{2A}Rs$ has been also reported *in vivo* (Wiese et al. 2007). The close interaction between $A_{2A}Rs$ and the BDNF signaling pathway has been further exemplified by recent electrophysiological studies. In 2004, Diogenes and coworkers investigated how adenosine influences the action of BDNF on synaptic transmission in the CA1 area of rat hippocampal slices (Diogenes et al. 2004). These authors were able to show that activation of adenosine $A_{2A}Rs$ with CGS21680, or the increase in extracellular adenosine levels induced by 5-iodotubercidin (ITU), triggered the excitatory action of BDNF, a process prevented by an $A_{2A}R$ antagonist. Moreover, only the specific activation of the presynaptic cAMP–PKA transducing system seems to be critical for the excitatory action of BDNF; indeed, while a membrane-permeable cAMP analog, dibutyryl cAMP (dbcAMP) mimics the effect of ITU or CGS 21680 on the excitatory action of BDNF, in postsynaptic CA1 pyramidal neurons loaded with cAMP, BDNF did not produce any effect on EPSC amplitude. These data indirectly suggest that presynaptic, more than postsynaptic, $A_{2A}Rs$ trigger the action of BDNF. It was concluded that presynaptic activity-dependent release of adenosine, through activation of $A_{2A}Rs$, facilitates BDNF modulation of synaptic transmission at hippocampal synapses. A similar positive interaction has been more recently confirmed at the neuromuscular junction (Pousinha et al. 2006). The importance of $A_{2A}Rs$ in regulating BDNF has recently been strengthened by our demonstration that both BDNF levels and functions are significantly reduced in the brain of $A_{2A}R$ knock-out mice (Tebano et al. 2008), and that the systemic injection of SCH58261 dramatically reduced striatal BDNF levels in rats (Minghetti et al. 2007). These results suggest that the presence and the tonic activation of $A_{2A}Rs$ are necessary to allow BDNF-induced potentiation of synaptic transmission and to maintain normal BDNF levels. Despite mechanisms underlying the interplay between A_{2A}Rs and the BDNF physiology being far from understood, these observations support the view that reinforcing A_{2A}R signaling may be beneficial in HD.

20.3.4.2 Role of Adenosine A_{2A}Rs in the Regulation of Ubiquitin-Proteasome System in HD

One of the explanations of the impairment of abnormal protein degradation is the downregulation of the ubiquitin–proteasome system (UPS; de Pril et al. 2004; Petrucelli and Dawson 2004). UPS is a cellular process for the non-lysosomal protein degradation of abnormal, misfolded or oxidized proteins. Proteasome activities are generally inhibited in several brain regions and skin fibroblasts of HD patients (Seo et al. 2004), and enhancing UPS function by using proteasome activators can improve cell survival against glutamate toxicity in a cell culture model of HD (Seo et al. 2007). Proteasome activity is also reduced in HD striatal neurons containing mHtt (105 CAG repeats) and it could be improved by lentiviral gene transfer of a proteasome activator, suggesting a potential therapeutic strategy of HD through proteasome activation (Seo et al. 2007). Interestingly, UPS activity is also greatly reduced in the liver of HD mice, and chronic treatment with CGS21680 effectively rescued the urea cycle deficiency in the liver by enhancing the activity of the UPS through a cAMP/PKA-dependent pathway (Chiang et al. 2009). Thus, given its ability to stimulate the activity of UPS, A_{2A}R might be considered a novel drug target to improve the cellular capacity to remove misfolded proteins in neuronal cells and, as a consequence, to treat HD and other neurodegenerative diseases, where aggregate formation and/or UPS dysfunction are strongly involved.

20.4 Conclusion

Adenosine receptors, and especially the A_{2A}R subtype, are clearly linked to HD pathophysiology as attested by genetic, epidemiological and experimental studies. The role of A_1Rs in HD pathogenesis has been largely understudied and this field deserves consideration in the future. Conversely, it is undoubted that A_{2A}Rs play an important role in HD. Several aspects concerning its pathophysiological involvement remain to be deciphered. The pre/postsynaptic aspects deserve further investigation using specific ligands as well as genetic murine tools. Also, how A_{2A}Rs interact with glial dysfunction promoted by mutated Huntingtin has been largely understudied. Further, given that A_{2A}Rs interact with several other GPCRs that play a presumed role in striatal dysfunction and degeneration in HD (D_2R, cannabinoid receptor 1), one may consider that the A_{2A}R heterodimerization itself could be of importance and deserves further investigation. Then, from a fundamental perspective, much remains to be done to better understand the role of adenosine receptors and purines in HD. Nonetheless, in a therapeutic perspective, targeting adenosine receptors in

HD remains to be carefully considered. It is evident that this uncertainty relies on our partial knowledge of their pathophysiological function but it also relies on the apparent dynamic complexity of their role. Especially, targeting A_{2A}Rs may lead to completely different outcomes depending on the disease stage and the relative contribution of the different receptor subpopulation at the time of modulation, itself dependent on the disease progression. With our current background, the most promising approach would be activation of postsynaptic A_{2A}R coupled with presynaptic blockade. Thus, although HD may represent an interesting model to better understand the differential regulation afforded by A_{2A}Rs in neuropathological situations, adenosine-based therapeutical development in HD remains complex and demanding.

References

Alberch J, Lopez M, Badenas C, Carrasco JL, Mila M, Munoz E, Canals JM (2005) Association between BDNF Val66Met polymorphism and age at onset in Huntington disease. Neurology 65:964–965

Alfinito PD, Wang SP, Manzino L, Rijhsinghani S, Zeevalk GD, Sonsalla PK (2003) Adenosinergic protection of dopaminergic and GABAergic neurons against mitochondrial inhibition through receptors located in the substantia nigra and striatum, respectively. J Neurosci 23:10982–10987

Alsene K, Deckert J, Sand P, de Wit H (2003) Association between A2a receptor gene polymorphisms and caffeine-induced anxiety. Neuropsychopharmacology 28:1694–1702

Andre VM, Cepeda C, Fisher YE, Huynh M, Bardakjian N, Singh S, Yang XW, Levine MS (2011) Differential electrophysiological changes in striatal output neurons in Huntington's disease. J Neurosci 31:1170–1182

Andresen JM, Gayan J, Djousse L, Roberts S, Brocklebank D, Cherny SS, Cardon LR, Gusella JF, MacDonald ME, Myers RH et al (2007) The relationship between CAG repeat length and age of onset differs for Huntington's disease patients with juvenile onset or adult onset. Ann Hum Genet 71:295–301

Bantubungi K, Blum D (2007a) Mechanisms of neuronal death in Huntington's disease. First part: general considerations and histopathological features. Rev Med Brux 28:413–421

Bantubungi K, Blum D (2007b) Mechanisms of neuronal death in Huntington's disease. Second part: therapeutic challenges. Rev Med Brux 28:487–494

Beal MF, Kowall NW, Ellison DW, Mazurek MF, Swartz KJ, Martin JB (1986) Replication of the neurochemical characteristics of Huntington's disease by quinolinic acid. Nature 321:168–171

Benarroch EE (2008) Brain cholesterol metabolism and neurologic disease. Neurology 71: 1368–1373

Benchoua A, Trioulier Y, Diguet E, Malgorn C, Gaillard MC, Dufour N, Elalouf JM, Krajewski S, Hantraye P, Deglon N et al (2008) Dopamine determines the vulnerability of striatal neurons to the N-terminal fragment of mutant huntingtin through the regulation of mitochondrial complex II. Hum Mol Genet 17:1446–1456

Benchoua A, Trioulier Y, Zala D, Gaillard MC, Lefort N, Dufour N, Saudou F, Elalouf JM, Hirsch E, Hantraye P et al (2006) Involvement of mitochondrial complex II defects in neuronal death produced by N-terminus fragment of mutated huntingtin. Mol Biol Cell 17:1652–1663

Blum D, Galas MC, Gall D, Cuvelier L, Schiffmann SN (2002a) Striatal and cortical neurochemical changes induced by chronic metabolic compromise in the 3-nitropropionic model of Huntington's disease. Neurobiol Dis 10:410–426

Blum D, Galas MC, Pintor A, Brouillet E, Ledent C, Muller CE, Bantubungi K, Galluzzo M, Gall D, Cuvelier L et al (2003a) A dual role of adenosine A2A receptors in 3-nitropropionic acid-induced striatal lesions: implications for the neuroprotective potential of A2A antagonists. J Neurosci 23:5361–5369

Blum D, Gall D, Galas MC, d'Alcantara P, Bantubungi K, Schiffmann SN (2002b) The adenosine A1 receptor agonist adenosine amine congener exerts a neuroprotective effect against the development of striatal lesions and motor impairments in the 3-nitropropionic acid model of neurotoxicity. J Neurosci 22:9122–9133

Blum D, Hourez R, Galas MC, Popoli P, Schiffmann SN (2003b) Adenosine receptors and Huntington's disease: implications for pathogenesis and therapeutics. Lancet Neurol 2: 366–374

Bolam JP, Hanley JJ, Booth PA, Bevan MD (2000) Synaptic organisation of the basal ganglia. J Anat 196(Pt 4):527–542

Borrell-Pages M, Canals JM, Cordelieres FP, Parker JA, Pineda JR, Grange G, Bryson EA, Guillermier M, Hirsch E, Hantraye P et al (2006) Cystamine and cysteamine increase brain levels of BDNF in Huntington disease via HSJ1b and transglutaminase. J Clin Invest 116:1410–1424

Bradford J, Shin JY, Roberts M, Wang CE, Li XJ, Li S (2009) Expression of mutant huntingtin in mouse brain astrocytes causes age-dependent neurological symptoms. Proc Natl Acad Sci USA 106:22480–22485

Bradford J, Shin JY, Roberts M, Wang CE, Sheng G, Li S, Li XJ (2010) Mutant huntingtin in glial cells exacerbates neurological symptoms of Huntington disease mice. J Biol Chem 285:10653–10661

Brouillet E, Conde F, Beal MF, Hantraye P (1999) Replicating Huntington's disease phenotype in experimental animals. Prog Neurobiol 59:427–468

Brouillet E, Jacquard C, Bizat N, Blum D (2005) 3-Nitropropionic acid: a mitochondrial toxin to uncover physiopathological mechanisms underlying striatal degeneration in Huntington's disease. J Neurochem 95:1521–1540

Browne SE, Bowling AC, MacGarvey U, Baik MJ, Berger SC, Muqit MM, Bird ED, Beal MF (1997) Oxidative damage and metabolic dysfunction in Huntington's disease: selective vulnerability of the basal ganglia. Ann Neurol 41:646–653

Buira SP, Albasanz JL, Dentesano G, Moreno J, Martin M, Ferrer I, Barrachina M (2010a) DNA methylation regulates adenosine A(2A) receptor cell surface expression levels. J Neurochem 112:1273–1285

Buira SP, Dentesano G, Albasanz JL, Moreno J, Martin M, Ferrer I, Barrachina M (2010b) DNA methylation and Yin Yang-1 repress adenosine A2A receptor levels in human brain. J Neurochem 115:283–295

Canals JM, Pineda JR, Torres-Peraza JF, Bosch M, Martin-Ibanez R, Munoz MT, Mengod G, Ernfors P, Alberch J (2004) Brain-derived neurotrophic factor regulates the onset and severity of motor dysfunction associated with enkephalinergic neuronal degeneration in Huntington's disease. J Neurosci 24:7727–7739

Cepeda C, Ariano MA, Calvert CR, Flores-Hernandez J, Chandler SH, Leavitt BR, Hayden MR, Levine MS (2001) NMDA receptor function in mouse models of Huntington disease. J Neurosci Res 66:525–539

Cepeda C, Cummings DM, Hickey MA, Kleiman-Weiner M, Chen JY, Watson JB, Levine MS (2010) Rescuing the corticostriatal synaptic disconnection in the R6/2 mouse model of Huntington's disease: exercise, adenosine receptors and ampakines. PLoS Curr 2:1182

Cepeda C, Hurst RS, Calvert CR, Hernandez-Echeagaray E, Nguyen OK, Jocoy E, Christian LJ, Ariano MA, Levine MS (2003) Transient and progressive electrophysiological alterations in the corticostriatal pathway in a mouse model of Huntington's disease. J Neurosci 23:961–969

Cha JH (2007) Transcriptional signatures in Huntington's disease. Prog Neurobiol 83:228–248

Cha JH, Frey AS, Alsdorf SA, Kerner JA, Kosinski CM, Mangiarini L, Penney JB Jr, Davies SW, Bates GP, Young AB (1999) Altered neurotransmitter receptor expression in transgenic mouse models of Huntington's disease. Philos Trans R Soc Lond B Biol Sci 354:981–989

Chao MV (2003) Neurotrophins and their receptors: a convergence point for many signalling pathways. Nat Rev Neurosci 4:299–309

Charalambous C, Gsandtner I, Keuerleber S, Milan-Lobo L, Kudlacek O, Freissmuth M, Zezula J (2008) Restricted collision coupling of the A2A receptor revisited: evidence for physical separation of two signaling cascades. J Biol Chem 283:9276–9288

Chen N, Luo T, Wellington C, Metzler M, McCutcheon K, Hayden MR, Raymond LA (1999) Subtype-specific enhancement of NMDA receptor currents by mutant huntingtin. J Neurochem 72:1890–1898

Chiang MC, Chen HM, Lai HL, Chen HW, Chou SY, Chen CM, Tsai FJ, Chern Y (2009) The A2A adenosine receptor rescues the urea cycle deficiency of Huntington's disease by enhancing the activity of the ubiquitin-proteasome system. Hum Mol Genet 18:2929–2942

Chiang MC, Lee YC, Huang CL, Chern Y (2005) cAMP-response element-binding protein contributes to suppression of the A2A adenosine receptor promoter by mutant Huntingtin with expanded polyglutamine residues. J Biol Chem 280:14331–14340

Childs E, Hohoff C, Deckert J, Xu K, Badner J, de Wit H (2008) Association between ADORA2A and DRD2 polymorphisms and caffeine-induced anxiety. Neuropsychopharmacology 33:2791–2800

Choo YS, Johnson GV, MacDonald M, Detloff PJ, Lesort M (2004) Mutant huntingtin directly increases susceptibility of mitochondria to the calcium-induced permeability transition and cytochrome c release. Hum Mol Genet 13:1407–1420

Chou SY, Lee YC, Chen HM, Chiang MC, Lai HL, Chang HH, Wu YC, Sun CN, Chien CL, Lin YS et al (2005) CGS21680 attenuates symptoms of Huntington's disease in a transgenic mouse model. J Neurochem 93:310–320

Cipriani S, Bizzoco E, Gianfriddo M, Melani A, Vannucchi MG, Pedata F (2008) Adenosine A2A receptor antagonism increases nNOS-immunoreactive neurons in the striatum of Huntington transgenic mice. Exp Neurol 213:163–170

Colin E, Zala D, Liot G, Rangone H, Borrell-Pages M, Li XJ, Saudou F, Humbert S (2008) Huntingtin phosphorylation acts as a molecular switch for anterograde/retrograde transport in neurons. EMBO J 27:2124–2134

Cornelis MC, El-Sohemy A, Campos H (2007) Genetic polymorphism of the adenosine A2A receptor is associated with habitual caffeine consumption. Am J Clin Nutr 86:240–244

Corradetti R, Lo Conte G, Moroni F, Passani MB, Pepeu G (1984) Adenosine decreases aspartate and glutamate release from rat hippocampal slices. Eur J Pharmacol 104:19–26

Corsi C, Melani A, Bianchi L, Pedata F (2000) Striatal A2A adenosine receptor antagonism differentially modifies striatal glutamate outflow in vivo in young and aged rats. Neuroreport 11:2591–2595

Corsi C, Melani A, Bianchi L, Pepeu G, Pedata F (1999) Striatal A2A adenosine receptors differentially regulate spontaneous and K+-evoked glutamate release in vivo in young and aged rats. Neuroreport 10:687–691

Dai SS, Zhou YG, Li W, An JH, Li P, Yang N, Chen XY, Xiong RP, Liu P, Zhao Y et al (2010) Local glutamate level dictates adenosine A2A receptor regulation of neuroinflammation and traumatic brain injury. J Neurosci 30:5802–5810

de Mendonca A, Ribeiro JA (2000) Long-term potentiation observed upon blockade of adenosine A1 receptors in rat hippocampus is N-methyl-D-aspartate receptor-dependent. Neurosci Lett 291:81–84

de Pril R, Fischer DF, Maat-Schieman ML, Hobo B, de Vos RA, Brunt ER, Hol EM, Roos RA, van Leeuwen FW (2004) Accumulation of aberrant ubiquitin induces aggregate formation and cell death in polyglutamine diseases. Hum Mol Genet 13:1803–1813

del Toro D, Canals JM, Gines S, Kojima M, Egea G, Alberch J (2006) Mutant huntingtin impairs the post-Golgi trafficking of brain-derived neurotrophic factor but not its Val66Met polymorphism. J Neurosci 26:12748–12757

del Toro D, Xifro X, Pol A, Humbert S, Saudou F, Canals JM, Alberch J (2010) Altered cholesterol homeostasis contributes to enhanced excitotoxicity in Huntington's disease. J Neurochem 115:153–167

Dhaenens CM, Burnouf S, Simonin C, Van Brussel E, Duhamel A, Defebvre L, Duru C, Vuillaume I, Cazeneuve C, Charles P et al (2009) A genetic variation in the ADORA2A gene modifies age at onset in Huntington's disease. Neurobiol Dis 35:474–476

Di Maria E, Marasco A, Tartari M, Ciotti P, Abbruzzese G, Novelli G, Bellone E, Cattaneo E, Mandich P (2006) No evidence of association between BDNF gene variants and age-at-onset of Huntington's disease. Neurobiol Dis 24:274–279

Diogenes MJ, Fernandes CC, Sebastiao AM, Ribeiro JA (2004) Activation of adenosine A2A receptor facilitates brain-derived neurotrophic factor modulation of synaptic transmission in hippocampal slices. J Neurosci 24:2905–2913

Domenici MR, Scattoni ML, Martire A, Lastoria G, Potenza RL, Borioni A, Venerosi A, Calamandrei G, Popoli P (2007) Behavioral and electrophysiological effects of the adenosine A2A receptor antagonist SCH 58261 in R6/2 Huntington's disease mice. Neurobiol Dis 28:197–205

Dompierre JP, Godin JD, Charrin BC, Cordelieres FP, King SJ, Humbert S, Saudou F (2007) Histone deacetylase 6 inhibition compensates for the transport deficit in Huntington's disease by increasing tubulin acetylation. J Neurosci 27:3571–3583

Dragatsis I, Efstratiadis A, Zeitlin S (1998) Mouse mutant embryos lacking huntingtin are rescued from lethality by wild-type extraembryonic tissues. Development 125:1529–1539

Dunwiddie TV, Masino SA (2001) The role and regulation of adenosine in the central nervous system. Annu Rev Neurosci 24:31–55

Duru C, Simonin C, Salleron J, Hincker P, Charles P, Delval A, Youssov K, Burnouf S, Azulay JP, Verny C et al (2011) Caffeine is a modifier of age at onset in Huntington's disease. In: Movement disorders society meeting, Toronto

Faideau M, Kim J, Cormier K, Gilmore R, Welch M, Auregan G, Dufour N, Guillermier M, Brouillet E, Hantraye P et al (2010) In vivo expression of polyglutamine-expanded huntingtin by mouse striatal astrocytes impairs glutamate transport: a correlation with Huntington's disease subjects. Hum Mol Genet 19:3053–3067

Fan MM, Fernandes HB, Zhang LY, Hayden MR, Raymond LA (2007) Altered NMDA receptor trafficking in a yeast artificial chromosome transgenic mouse model of Huntington's disease. J Neurosci 27:3768–3779

Fan MM, Raymond LA (2007) N-methyl-D-aspartate (NMDA) receptor function and excitotoxicity in Huntington's disease. Prog Neurobiol 81:272–293

Ferrante A, Martire A, Armida M, Chiodi V, Pezzola A, Potenza RL, Domenici MR, Popoli P (2010) Influence of CGS 21680, a selective adenosine A(2A) receptor agonist, on NMDA receptor function and expression in the brain of Huntington's disease mice. Brain Res 1323:184–191

Fink JS, Kalda A, Ryu H, Stack EC, Schwarzschild MA, Chen JF, Ferrante RJ (2004) Genetic and pharmacological inactivation of the adenosine A2A receptor attenuates 3-nitropropionic acid-induced striatal damage. J Neurochem 88:538–544

Fredholm BB, Battig K, Holmen J, Nehlig A, Zvartau EE (1999) Actions of caffeine in the brain with special reference to factors that contribute to its widespread use. Pharmacol Rev 51:83–133

Fredholm BB, Chern Y, Franco R, Sitkovsky M (2007) Aspects of the general biology of adenosine A2A signaling. Prog Neurobiol 83:263–276

Fredholm BB, Dunwiddie TV (1988) How does adenosine inhibit transmitter release? Trends Pharmacol Sci 9:130–134

Galluzzo M, Pintor A, Pezzola A, Grieco R, Borsini F, Popoli P (2008) Behavioural and neurochemical characterization of the adenosine A2A receptor antagonist ST1535. Eur J Pharmacol 579:149–152

Gauthier LR, Charrin BC, Borrell-Pages M, Dompierre JP, Rangone H, Cordelieres FP, De Mey J, MacDonald ME, Lessmann V, Humbert S et al (2004) Huntingtin controls neurotrophic support and survival of neurons by enhancing BDNF vesicular transport along microtubules. Cell 118:127–138

Gerfen CR (1992) The neostriatal mosaic: multiple levels of compartmental organization. Trends Neurosci 15:133–139

Gharami K, Xie Y, An JJ, Tonegawa S, Xu B (2008) Brain-derived neurotrophic factor over-expression in the forebrain ameliorates Huntington's disease phenotypes in mice. J Neurochem 105:369–379

Gianfriddo M, Melani A, Turchi D, Giovannini MG, Pedata F (2004) Adenosine and glutamate extracellular concentrations and mitogen-activated protein kinases in the striatum of Huntington transgenic mice. Selective antagonism of adenosine A2A receptors reduces transmitter outflow. Neurobiol Dis 17:77–88

Giralt A, Friedman HC, Caneda-Ferron B, Urban N, Moreno E, Rubio N, Blanco J, Peterson A, Canals JM, Alberch J (2010) BDNF regulation under GFAP promoter provides engineered astrocytes as a new approach for long-term protection in Huntington's disease. Gene Ther 17:1294–1308

Glass M, Dragunow M, Faull RL (2000) The pattern of neurodegeneration in Huntington's disease: a comparative study of cannabinoid, dopamine, adenosine and GABA(A) receptor alterations in the human basal ganglia in Huntington's disease. Neuroscience 97:505–519

Godin JD, Colombo K, Molina-Calavita M, Keryer G, Zala D, Charrin BC, Dietrich P, Volvert ML, Guillemot F, Dragatsis I et al (2010) Huntingtin is required for mitotic spindle orientation and mammalian neurogenesis. Neuron 67:392–406

Gotoh L, Mitsuyasu H, Kobayashi Y, Oribe N, Takata A, Ninomiya H, Stanton VP Jr, Springett GM, Kawasaki H, Kanba S (2009) Association analysis of adenosine A1 receptor gene (ADORA1) polymorphisms with schizophrenia in a Japanese population. Psychiatr Genet 19:328–335

Gourfinkel-An I, Cancel G, Trottier Y, Devys D, Tora L, Lutz Y, Imbert G, Saudou F, Stevanin G, Agid Y et al (1997) Differential distribution of the normal and mutated forms of huntingtin in the human brain. Ann Neurol 42:712–719

Gsandtner I, Charalambous C, Stefan E, Ogris E, Freissmuth M, Zezula J (2005) Heterotrimeric G protein-independent signaling of a G protein-coupled receptor. Direct binding of ARNO/cytohesin-2 to the carboxyl terminus of the A2A adenosine receptor is necessary for sustained activation of the ERK/MAP kinase pathway. J Biol Chem 280:31898–31905

Gu M, Gash MT, Mann VM, Javoy-Agid F, Cooper JM, Schapira AH (1996) Mitochondrial defect in Huntington's disease caudate nucleus. Ann Neurol 39:385–389

Guidetti P, Bates GP, Graham RK, Hayden MR, Leavitt BR, MacDonald ME, Slow EJ, Wheeler VC, Woodman B, Schwarcz R (2006) Elevated brain 3-hydroxykynurenine and quinolinate levels in Huntington disease mice. Neurobiol Dis 23:190–197

Guidetti P, Luthi-Carter RE, Augood SJ, Schwarcz R (2004) Neostriatal and cortical quinolinate levels are increased in early grade Huntington's disease. Neurobiol Dis 17:455–461

Hamilton SP, Slager SL, De Leon AB, Heiman GA, Klein DF, Hodge SE, Weissman MM, Fyer AJ, Knowles JA (2004) Evidence for genetic linkage between a polymorphism in the adenosine 2A receptor and panic disorder. Neuropsychopharmacology 29:558–565

Hohoff C, McDonald JM, Baune BT, Cook EH, Deckert J, de Wit H (2005) Interindividual variation in anxiety response to amphetamine: possible role for adenosine A2A receptor gene variants. Am J Med Genet B Neuropsychiatr Genet 139B:42–44

Huang NK, Lin JH, Lin JT, Lin CI, Liu EM, Lin CJ, Chen WP, Shen YC, Chen HM, Chen JB et al (2011) A new drug design targeting the adenosinergic system for Huntington's disease. PLoS One 6:e20934

Huang QY, Wei C, Yu L, Coelho JE, Shen HY, Kalda A, Linden J, Chen JF (2006) Adenosine A2A receptors in bone marrow-derived cells but not in forebrain neurons are important contributors to 3-nitropropionic acid-induced striatal damage as revealed by cell-type-selective inactivation. J Neurosci 26:11371–11378

Jacquard C, Trioulier Y, Cosker F, Escartin C, Bizat N, Hantraye P, Cancela JM, Bonvento G, Brouillet E (2006) Brain mitochondrial defects amplify intracellular [Ca2+] rise and neurodegeneration but not Ca2+ entry during NMDA receptor activation. FASEB J 20:1021–1023

Jenkins BG, Rosas HD, Chen YC, Makabe T, Myers R, MacDonald M, Rosen BR, Beal MF, Koroshetz WJ (1998) 1H NMR spectroscopy studies of Huntington's disease: correlations with CAG repeat numbers. Neurology 50:1357–1365

Kim J, Moody JP, Edgerly CK, Bordiuk OL, Cormier K, Smith K, Beal MF, Ferrante RJ (2010) Mitochondrial loss, dysfunction and altered dynamics in Huntington's disease. Hum Mol Genet 19:3919–3935

Kishikawa S, Li JL, Gillis T, Hakky MM, Warby S, Hayden M, MacDonald ME, Myers RH, Gusella JF (2006) Brain-derived neurotrophic factor does not influence age at neurologic onset of Huntington's disease. Neurobiol Dis 24:280–285

Lee FS, Chao MV (2001) Activation of Trk neurotrophin receptors in the absence of neurotrophins. Proc Natl Acad Sci USA 98:3555–3560

Leoni V, Mariotti C, Tabrizi SJ, Valenza M, Wild EJ, Henley SM, Hobbs NZ, Mandelli ML, Grisoli M, Bjorkhem I et al (2008) Plasma 24S-hydroxycholesterol and caudate MRI in pre-manifest and early Huntington's disease. Brain 131:2851–2859

LeWitt PA, Guttman M, Tetrud JW, Tuite PJ, Mori A, Chaikin P, Sussman NM (2008) Adenosine A2A receptor antagonist istradefylline (KW-6002) reduces "off" time in Parkinson's disease: a double-blind, randomized, multicenter clinical trial (6002-US-005). Ann Neurol 63:295–302

Li XX, Nomura T, Aihara H, Nishizaki T (2001) Adenosine enhances glial glutamate efflux via A2a adenosine receptors. Life Sci 68:1343–1350

Lievens JC, Woodman B, Mahal A, Spasic-Boscovic O, Samuel D, Kerkerian-Le Goff L, Bates GP (2001) Impaired glutamate uptake in the R6 Huntington's disease transgenic mice. Neurobiol Dis 8:807–821

Luthi-Carter R, Strand A, Peters NL, Solano SM, Hollingsworth ZR, Menon AS, Frey AS, Spektor BS, Penney EB, Schilling G et al (2000) Decreased expression of striatal signaling genes in a mouse model of Huntington's disease. Hum Mol Genet 9:1259–1271

Lynch G, Kramar EA, Rex CS, Jia Y, Chappas D, Gall CM, Simmons DA (2007) Brain-derived neurotrophic factor restores synaptic plasticity in a knock-in mouse model of Huntington's disease. J Neurosci 27:4424–4434

Maglione V, Cannella M, Martino T, De Blasi A, Frati L, Squitieri F (2006) The platelet maximum number of A2A-receptor binding sites (Bmax) linearly correlates with age at onset and CAG repeat expansion in Huntington's disease patients with predominant chorea. Neurosci Lett 393:27–30

Maglione V, Giallonardo P, Cannella M, Martino T, Frati L, Squitieri F (2005) Adenosine A2A receptor dysfunction correlates with age at onset anticipation in blood platelets of subjects with Huntington's disease. Am J Med Genet B Neuropsychiatr Genet 139B:101–105

Martire A, Ferrante A, Potenza RL, Armida M, Ferretti R, Pezzola A, Domenici MR, Popoli P (2010) Remodeling of striatal NMDA receptors by chronic A(2A) receptor blockade in Huntington's disease mice. Neurobiol Dis 37:99–105

McGeer PL, Itagaki S, McGeer EG (1988) Expression of the histocompatibility glycoprotein HLA-DR in neurological disease. Acta Neuropathol 76:550–557

Messmer K, Reynolds GP (1998) Increased peripheral benzodiazepine binding sites in the brain of patients with Huntington's disease. Neurosci Lett 241:53–56

Metzger S, Saukko M, Van Che H, Tong L, Puder Y, Riess O, Nguyen HP (2010) Age at onset in Huntington's disease is modified by the autophagy pathway: implication of the V471A polymorphism in Atg7. Hum Genet 128:453–459

Mievis S, Blum D, Ledent C (2011) A2A receptor knockout worsens survival and motor behaviour in a transgenic mouse model of Huntington's disease. Neurobiol Dis 41:570–576

Milakovic T, Johnson GV (2005) Mitochondrial respiration and ATP production are significantly impaired in striatal cells expressing mutant huntingtin. J Biol Chem 280:30773–30782

Milne GR, Palmer TM (2011) Anti-inflammatory and immunosuppressive effects of the A2A adenosine receptor. ScientificWorldJournal 11:320–339

Milnerwood AJ, Gladding CM, Pouladi MA, Kaufman AM, Hines RM, Boyd JD, Ko RW, Vasuta OC, Graham RK, Hayden MR et al (2010) Early increase in extrasynaptic NMDA receptor signaling and expression contributes to phenotype onset in Huntington's disease mice. Neuron 65:178–190

Minghetti L, Greco A, Potenza RL, Pezzola A, Blum D, Bantubungi K, Popoli P (2007) Effects of the adenosine A2A receptor antagonist SCH 58621 on cyclooxygenase-2 expression, glial activation, and brain-derived neurotrophic factor availability in a rat model of striatal neurodegeneration. J Neuropathol Exp Neurol 66:363–371

Morelli M, Carta AR, Jenner P (2009) Adenosine A2A receptors and Parkinson's disease. Handb Exp Pharmacol 589–615

Nishizaki T, Nagai K, Nomura T, Tada H, Kanno T, Tozaki H, Li XX, Kondoh T, Kodama N, Takahashi E et al (2002) A new neuromodulatory pathway with a glial contribution mediated via A(2a) adenosine receptors. Glia 39:133–147

Orr AG, Orr AL, Li XJ, Gross RE, Traynelis SF (2009) Adenosine A(2A) receptor mediates microglial process retraction. Nat Neurosci 12:872–878

Orru M, Bakesova J, Brugarolas M, Quiroz C, Beaumont V, Goldberg SR, Lluis C, Cortes A, Franco R, Casado V et al (2011) Striatal pre- and postsynaptic profile of adenosine A(2A) receptor antagonists. PLoS One 6:e16088

Palazuelos J, Aguado T, Pazos MR, Julien B, Carrasco C, Resel E, Sagredo O, Benito C, Romero J, Azcoitia I et al (2009) Microglial CB2 cannabinoid receptors are neuroprotective in Huntington's disease excitotoxicity. Brain 132:3152–3164

Palmer TM, Stiles GL (1995) Adenosine receptors. Neuropharmacology 34:683–694

Panov AV, Gutekunst CA, Leavitt BR, Hayden MR, Burke JR, Strittmatter WJ, Greenamyre JT (2002) Early mitochondrial calcium defects in Huntington's disease are a direct effect of polyglutamines. Nat Neurosci 5:731–736

Pavese N, Gerhard A, Tai YF, Ho AK, Turkheimer F, Barker RA, Brooks DJ, Piccini P (2006) Microglial activation correlates with severity in Huntington disease: a clinical and PET study. Neurology 66:1638–1643

Peng Q, Masuda N, Jiang M, Li Q, Zhao M, Ross CA, Duan W (2008) The antidepressant sertraline improves the phenotype, promotes neurogenesis and increases BDNF levels in the R6/2 Huntington's disease mouse model. Exp Neurol 210:154–163

Petrucelli L, Dawson TM (2004) Mechanism of neurodegenerative disease: role of the ubiquitin proteasome system. Ann Med 36:315–320

Pinna A (2009) Novel investigational adenosine A2A receptor antagonists for Parkinson's disease. Expert Opin Investig Drugs 18:1619–1631

Pintor A, Galluzzo M, Grieco R, Pezzola A, Reggio R, Popoli P (2004) Adenosine A 2A receptor antagonists prevent the increase in striatal glutamate levels induced by glutamate uptake inhibitors. J Neurochem 89:152–156

Pintor A, Quarta D, Pezzola A, Reggio R, Popoli P (2001) SCH 58261 (an adenosine A(2A) receptor antagonist) reduces, only at low doses, K(+)-evoked glutamate release in the striatum. Eur J Pharmacol 421:177–180

Politis M, Pavese N, Tai YF, Kiferle L, Mason SL, Brooks DJ, Tabrizi SJ, Barker RA, Piccini P (2011) Microglial activation in regions related to cognitive function predicts disease onset in Huntington's disease: a multimodal imaging study. Hum Brain Mapp 32:258–270

Popoli P, Betto P, Reggio R, Ricciarello G (1995) Adenosine A2A receptor stimulation enhances striatal extracellular glutamate levels in rats. Eur J Pharmacol 287:215–217

Popoli P, Blum D, Domenici MR, Burnouf S, Chern Y (2008) A critical evaluation of adenosine A2A receptors as potentially "druggable" targets in Huntington's disease. Curr Pharm Des 14:1500–1511

Popoli P, Blum D, Martire A, Ledent C, Ceruti S, Abbracchio MP (2007) Functions, dysfunctions and possible therapeutic relevance of adenosine A2A receptors in Huntington's disease. Prog Neurobiol 81:331–348

Popoli P, Pintor A, Domenici MR, Frank C, Tebano MT, Pezzola A, Scarchilli L, Quarta D, Reggio R, Malchiodi-Albedi F et al (2002) Blockade of striatal adenosine A2A receptor reduces, through a presynaptic mechanism, quinolinic acid-induced excitotoxicity: possible relevance to neuroprotective interventions in neurodegenerative diseases of the striatum. J Neurosci 22:1967–1975

Pousinha PA, Diogenes MJ, Ribeiro JA, Sebastiao AM (2006) Triggering of BDNF facilitatory action on neuromuscular transmission by adenosine A2A receptors. Neurosci Lett 404:143–147

Quiroz C, Lujan R, Uchigashima M, Simoes AP, Lerner TN, Borycz J, Kachroo A, Canas PM, Orru M, Schwarzschild MA et al (2009) Key modulatory role of presynaptic adenosine A2A receptors in cortical neurotransmission to the striatal direct pathway. ScientificWorldJournal 9:1321–1344

Rebola N, Simoes AP, Canas PM, Tome AR, Andrade GM, Barry CE, Agostinho PM, Lynch MA, Cunha RA (2011) Adenosine A2A receptors control neuroinflammation and consequent hippocampal neuronal dysfunction. J Neurochem 117:100–111

Reiner A, Medina L, Veenman CL (1998) Structural and functional evolution of the basal ganglia in vertebrates. Brain Res Brain Res Rev 28:235–285

Rogers PJ, Hohoff C, Heatherley SV, Mullings EL, Maxfield PJ, Evershed RP, Deckert J, Nutt DJ (2010) Association of the anxiogenic and alerting effects of caffeine with ADORA2A and ADORA1 polymorphisms and habitual level of caffeine consumption. Neuropsychopharmacology 35:1973–1983

Ross CA, Tabrizi SJ (2011) Huntington's disease: from molecular pathogenesis to clinical treatment. Lancet Neurol 10:83–98

Rudolphi KA, Schubert P, Parkinson FE, Fredholm BB (1992) Adenosine and brain ischemia. Cerebrovasc Brain Metab Rev 4:346–369

Sapp E, Kegel KB, Aronin N, Hashikawa T, Uchiyama Y, Tohyama K, Bhide PG, Vonsattel JP, DiFiglia M (2001) Early and progressive accumulation of reactive microglia in the Huntington disease brain. J Neuropathol Exp Neurol 60:161–172

Scattoni ML, Valanzano A, Pezzola A, March ZD, Fusco FR, Popoli P, Calamandrei G (2007) Adenosine A2A receptor blockade before striatal excitotoxic lesions prevents long term behavioural disturbances in the quinolinic rat model of Huntington's disease. Behav Brain Res 176:216–221

Schiffmann SN, Vanderhaeghen JJ (1993) Adenosine A2 receptors regulate the gene expression of striatopallidal and striatonigral neurons. J Neurosci 13:1080–1087

Schwab C, Klegeris A, McGeer PL (2010) Inflammation in transgenic mouse models of neurodegenerative disorders. Biochim Biophys Acta 1802:889–902

Seo H, Sonntag KC, Isacson O (2004) Generalized brain and skin proteasome inhibition in Huntington's disease. Ann Neurol 56:319–328

Seo H, Sonntag KC, Kim W, Cattaneo E, Isacson O (2007) Proteasome activator enhances survival of Huntington's disease neuronal model cells. PLoS One 2:e238

Seong IS, Ivanova E, Lee JM, Choo YS, Fossale E, Anderson M, Gusella JF, Laramie JM, Myers RH, Lesort M et al (2005) HD CAG repeat implicates a dominant property of huntingtin in mitochondrial energy metabolism. Hum Mol Genet 14:2871–2880

Seward A, Longwell C, Young G, Kostyk S (2010) Caffeine Intake and Disease Characteristics in a Huntington's Disease Clinic Population. In Fourth Annual Huntington Disease, Clinical Research Symposium (San Diego)

Shehadeh J, Fernandes HB, Zeron Mullins MM, Graham RK, Leavitt BR, Hayden MR, Raymond LA (2006) Striatal neuronal apoptosis is preferentially enhanced by NMDA receptor activation in YAC transgenic mouse model of Huntington disease. Neurobiol Dis 21:392–403

Shin JY, Fang ZH, Yu ZX, Wang CE, Li SH, Li XJ (2005) Expression of mutant huntingtin in glial cells contributes to neuronal excitotoxicity. J Cell Biol 171:1001–1012

Shirendeb U, Reddy AP, Manczak M, Calkins MJ, Mao P, Tagle DA, Hemachandra Reddy P (2011) Abnormal mitochondrial dynamics, mitochondrial loss and mutant huntingtin oligomers in Huntington's disease: implications for selective neuronal damage. Hum Mol Genet 20:1438–1455

Simmons DA, Rex CS, Palmer L, Pandyarajan V, Fedulov V, Gall CM, Lynch G (2009) Up-regulating BDNF with an ampakine rescues synaptic plasticity and memory in Huntington's disease knockin mice. Proc Natl Acad Sci USA 106:4906–4911

Simola N, Morelli M, Pinna A (2008) Adenosine A2A receptor antagonists and Parkinson's disease: state of the art and future directions. Curr Pharm Des 14:1475–1489

Stack EC, Dedeoglu A, Smith KM, Cormier K, Kubilus JK, Bogdanov M, Matson WR, Yang L, Jenkins BG, Luthi-Carter R et al (2007) Neuroprotective effects of synaptic modulation in Huntington's disease R6/2 mice. J Neurosci 27:12908–12915

Stone TW, Ceruti S, Abbracchio MP (2009) Adenosine receptors and neurological disease: neuroprotection and neurodegeneration. Handb Exp Pharmacol 193:535–587

Sugars KL, Brown R, Cook LJ, Swartz J, Rubinsztein DC (2004) Decreased cAMP response element-mediated transcription: an early event in exon 1 and full-length cell models of Huntington's disease that contributes to polyglutamine pathogenesis. J Biol Chem 279: 4988–4999

Sun Y, Savanenin A, Reddy PH, Liu YF (2001) Polyglutamine-expanded huntingtin promotes sensitization of N-methyl-D-aspartate receptors via post-synaptic density 95. J Biol Chem 276:24713–24718

Tabrizi SJ, Cleeter MW, Xuereb J, Taanman JW, Cooper JM, Schapira AH (1999) Biochemical abnormalities and excitotoxicity in Huntington's disease brain. Ann Neurol 45:25–32

Taherzadeh-Fard E, Saft C, Andrich J, Wieczorek S, Arning L (2009) PGC-1alpha as modifier of onset age in Huntington disease. Mol Neurodegener 4:10

Taherzadeh-Fard E, Saft C, Wieczorek S, Epplen JT, Arning L (2010) Age at onset in Huntington's disease: replication study on the associations of ADORA2A, HAP1 and OGG1. Neurogenetics 11:435–439

Tai YF, Pavese N, Gerhard A, Tabrizi SJ, Barker RA, Brooks DJ, Piccini P (2007) Microglial activation in presymptomatic Huntington's disease gene carriers. Brain 130:1759–1766

Tarditi A, Camurri A, Varani K, Borea PA, Woodman B, Bates G, Cattaneo E, Abbracchio MP (2006) Early and transient alteration of adenosine A2A receptor signaling in a mouse model of Huntington disease. Neurobiol Dis 23:44–53

Tebano MT, Martire A, Potenza RL, Gro C, Pepponi R, Armida M, Domenici MR, Schwarzschild MA, Chen JF, Popoli P (2008) Adenosine A(2A) receptors are required for normal BDNF levels and BDNF-induced potentiation of synaptic transmission in the mouse hippocampus. J Neurochem 104:279–286

Tebano MT, Pintor A, Frank C, Domenici MR, Martire A, Pepponi R, Potenza RL, Grieco R, Popoli P (2004) Adenosine A2A receptor blockade differentially influences excitotoxic mechanisms at pre- and postsynaptic sites in the rat striatum. J Neurosci Res 77:100–107

The Huntington's Disease Collaborative Research Group (1993) A novel gene containing a trinucleotide repeat that is expanded and unstable on Huntington's disease chromosomes. Cell 72:971–983

Trottier Y, Devys D, Imbert G, Saudou F, An I, Lutz Y, Weber C, Agid Y, Hirsch EC, Mandel JL (1995) Cellular localization of the Huntington's disease protein and discrimination of the normal and mutated form. Nat Genet 10:104–110

Valenza M, Carroll JB, Leoni V, Bertram LN, Bjorkhem I, Singaraja RR, Di Donato S, Lutjohann D, Hayden MR, Cattaneo E (2007a) Cholesterol biosynthesis pathway is disturbed in YAC128 mice and is modulated by huntingtin mutation. Hum Mol Genet 16:2187–2198

Valenza M, Cattaneo E (2006) Cholesterol dysfunction in neurodegenerative diseases: is Huntington's disease in the list? Prog Neurobiol 80:165–176

Valenza M, Leoni V, Karasinska JM, Petricca L, Fan J, Carroll J, Pouladi MA, Fossale E, Nguyen HP, Riess O et al (2010) Cholesterol defect is marked across multiple rodent models of Huntington's disease and is manifest in astrocytes. J Neurosci 30:10844–10850

Valenza M, Leoni V, Tarditi A, Mariotti C, Bjorkhem I, Di Donato S, Cattaneo E (2007b) Progressive dysfunction of the cholesterol biosynthesis pathway in the R6/2 mouse model of Huntington's disease. Neurobiol Dis 28:133–142

Valenza M, Rigamonti D, Goffredo D, Zuccato C, Fenu S, Jamot L, Strand A, Tarditi A, Woodman B, Racchi M et al (2005) Dysfunction of the cholesterol biosynthetic pathway in Huntington's disease. J Neurosci 25:9932–9939

Varani K, Abbracchio MP, Cannella M, Cislaghi G, Giallonardo P, Mariotti C, Cattabriga E, Cattabeni F, Borea PA, Squitieri F et al (2003) Aberrant A2A receptor function in peripheral blood cells in Huntington's disease. FASEB J 17:2148–2150

Varani K, Bachoud-Levi AC, Mariotti C, Tarditi A, Abbracchio MP, Gasperi V, Borea PA, Dolbeau G, Gellera C, Solari A et al (2007) Biological abnormalities of peripheral A(2A) receptors in a large representation of polyglutamine disorders and Huntington's disease stages. Neurobiol Dis 27:36–43

Varani K, Rigamonti D, Sipione S, Camurri A, Borea PA, Cattabeni F, Abbracchio MP, Cattaneo E (2001) Aberrant amplification of A(2A) receptor signaling in striatal cells expressing mutant huntingtin. FASEB J 15:1245–1247

Vonsattel JP, DiFiglia M (1998) Huntington disease. J Neuropathol Exp Neurol 57:369–384

Wagner AK, Miller MA, Scanlon J, Ren D, Kochanek PM, Conley YP (2010) Adenosine A1 receptor gene variants associated with post-traumatic seizures after severe TBI. Epilepsy Res 90:259–272

Walker FO (2007) Huntington's disease. Lancet 369:218–228

Wei CJ, Li W, Chen JF (2011) Normal and abnormal functions of adenosine receptors in the central nervous system revealed by genetic knockout studies. Biochim Biophys Acta 1808: 1358–1379

Weydt P, Pineda VV, Torrence AE, Libby RT, Satterfield TF, Lazarowski ER, Gilbert ML, Morton GJ, Bammler TK, Strand AD et al (2006) Thermoregulatory and metabolic defects in Huntington's disease transgenic mice implicate PGC-1alpha in Huntington's disease neurodegeneration. Cell Metab 4:349–362

Wiese S, Jablonka S, Holtmann B, Orel N, Rajagopal R, Chao MV, Sendtner M (2007) Adenosine receptor A2A-R contributes to motoneuron survival by transactivating the tyrosine kinase receptor TrkB. Proc Natl Acad Sci USA 104:17210–17215

Wu J, Lee MR, Kim T, Johng S, Rohrback S, Kang N, Choi DS (2011) Regulation of ethanol-sensitive EAAT2 expression through adenosine A1 receptor in astrocytes. Biochem Biophys Res Commun 406:47–52

Wu LL, Fan Y, Li S, Li XJ, Zhou XF (2010) Huntingtin-associated protein-1 interacts with pro-brain-derived neurotrophic factor and mediates its transport and release. J Biol Chem 285: 5614–5623

Xie Y, Hayden MR, Xu B (2010) BDNF overexpression in the forebrain rescues Huntington's disease phenotypes in YAC128 mice. J Neurosci 30:14708–14718

Yang A, Palmer AA, de Wit H (2010) Genetics of caffeine consumption and responses to caffeine. Psychopharmacology (Berl) 211:245–257

Zeron MM, Hansson O, Chen N, Wellington CL, Leavitt BR, Brundin P, Hayden MR, Raymond LA (2002) Increased sensitivity to N-methyl-D-aspartate receptor-mediated excitotoxicity in a mouse model of Huntington's disease. Neuron 33:849–860

Zuccato C, Cattaneo E (2007) Role of brain-derived neurotrophic factor in Huntington's disease. Prog Neurobiol 81:294–330

Zuccato C, Cattaneo E (2009) Brain-derived neurotrophic factor in neurodegenerative diseases. Nat Rev Neurol 5:311–322

Zuccato C, Ciammola A, Rigamonti D, Leavitt BR, Goffredo D, Conti L, MacDonald ME, Friedlander RM, Silani V, Hayden MR et al (2001) Loss of huntingtin-mediated BDNF gene transcription in Huntington's disease. Science 293:493–498

Zuccato C, Tartari M, Crotti A, Goffredo D, Valenza M, Conti L, Cataudella T, Leavitt BR, Hayden MR, Timmusk T et al (2003) Huntingtin interacts with REST/NRSF to modulate the transcription of NRSE-controlled neuronal genes. Nat Genet 35:76–83

Zuccato C, Valenza M, Cattaneo E (2010) Molecular mechanisms and potential therapeutical targets in Huntington's disease. Physiol Rev 90:905–981

Zuchora B, Turski WA, Wielosz M, Urbanska EM (2001) Protective effect of adenosine receptor agonists in a new model of epilepsy–seizures evoked by mitochondrial toxin, 3-nitropropionic acid, in mice. Neurosci Lett 305:91–94

Chapter 21
Adenosine and Multiple Sclerosis

María Victoria Sánchez-Gómez, Estibaliz González-Fernández, Rogelio O. Arellano, and Carlos Matute

Abstract The major demyelinating disease of the CNS is multiple sclerosis (MS), which is the foremost disabling pathology among young adults. MS is a chronic, degenerative disease characterized by focal lesions with inflammation, demyelination, infiltration of immune cells, oligodendroglial death, and axonal degeneration. These cellular alterations are accompanied by neurological deficits, such as sensory disturbances, lack of motor coordination, and visual impairment. It is widely accepted that the etiology of this illness has autoimmune and inflammatory grounds, and that a derailment of the immune system leads to cell- and antibody-mediated attacks on myelin.

Both genetic and environmental factors contribute to MS susceptibility. Among them, primary and/or secondary alterations in neurotransmitter signaling including the glutamate and purinergic system lead to oligodendrocyte and myelin damage and contribute to MS pathology. In addition, recent data indicate that adenosine is involved in neuroinflammation, and that activation of adenosine receptors may contribute to tissue damage in experimental models of MS. Moreover, some alterations of adenosine metabolism occur in MS though it is not clear yet whether they are primary or secondary to the disease process. Finally, emerging evidence suggests that enhanced activity of the adenosinergic system may also contribute to the pathophysiology of MS and that this feature may open new therapeutic approaches beneficial for the treatment of MS.

M.V. Sánchez-Gómez (✉) • E. González-Fernández • C. Matute
CIBERNED and Departamento de Neurociencias,
Universidad del País Vasco (UPV/EHU), Leioa 48940, Spain

R.O. Arellano
Instituto de Neurobiología, Universidad Nacional Autónoma de México,
Boulevard Juriquilla 3001, Juriquilla Querétaro, Querétaro, Mexico

S. Masino and D. Boison (eds.), *Adenosine: A Key Link between Metabolism and Brain Activity*, DOI 10.1007/978-1-4614-3903-5_21,

Keywords Demyelination • EAE • Oligodendrocyte • Myelin • Neuroinflammation • P1 purinergic receptor • Tissue damage

21.1 Multiple Sclerosis

Multiple sclerosis (MS) is an inflammatory, demyelinating disease of the central nervous system (CNS), first clinically described by Charcot in 1868 (Charcot 1868), in which autoreactive, myelin-specific T cells cause extensive tissue damage resulting in neurological deficits. MS courses with immunological attacks, which may last from days to weeks with a posterior recovery phase; however, the antigenic stimuli that initiate or perpetuate this abnormal immune reactivity are still a matter of intense research and debate. Focal lesions showing perivascular inflammation, infiltration of immune cells, axonal degeneration, oligodendroglial death, and demyelination characterize this chronic and degenerative disease of the CNS (Prineas et al. 2002). These cellular alterations are accompanied by neurological deficits, such as sensory disturbances, lack of motor coordination, and visual impairment. However, the clinical presentation of MS varies greatly and correlates well with the multiplicity of lesions and their distribution at various anatomical sites within the brain and spinal cord.

MS is a major, chronic, disabling neurological disease among adults between 20 and 40 years of age, and affects women more often than it does men. Its incidence is approximately 3.6 cases among women and 2 cases among men per 100,000 individuals per year. Its prevalence varies geographically: higher MS incidence is classically associated with countries with cold climates and at high latitudes (averaging 90–100 cases per 100,000 individuals), although recent studies report that this gradient has become attenuated since 1980, apparently due to increased incidence in lower latitudes (Alonso and Hernan 2008).

21.1.1 Diagnostic Criteria for MS

The multiplicity of MS plaques and their location at various anatomic sites may account for the great variability of clinical symptoms and signs. Visual impairment to different degrees, orbital pain, and frontal headaches are often the presenting symptoms. However, any cerebral or spinal cord dysfunction may induce the disease. MS symptoms typically appear with an "attack" or "relapse," and disease diagnosis is based on clinical criteria supported by visualization of CNS white matter lesions by magnetic resonance imaging (MRI) following established standards (Polman et al. 2005). MRI can show the typical periventricular plaques, as well as plaques as small as 3–4 mm in diameter. Moreover, using contrast material, MRI identifies acute plaques and is helpful in monitoring therapeutic efficacy (Prineas et al. 2002).

21.1.2 Clinical Subtypes and Prognosis

MS patients are classified into different subtypes according to their clinical description and the course of their disease. Most patients (approximately 85–90 %) initially have a relapsing–remitting form of MS (RRMS), which is characterized by discrete clinical "attacks" or "relapses" followed by subsequent improvement. RRMS is the most typical presentation in younger patients. Within 10 years of disease onset, approximately 50 % of patients with RRMS will develop a slow, insidiously progressive, neurological deterioration (usually progressive gait impairment), with or without clinical attacks superimposed. This is termed secondary progressive MS (SPMS), and 90 % of RRMS patients advance into this stage within 30 years of disease onset. A minority of patients (approximately 10–15 %) have primary progressive MS (PPMS), which is characterized by a progressive course from onset, with occasional plateaus and absence of clinically evident relapses. In addition, PPMS presents less conspicuous inflammation on MRI. Due to its difference from RRMS, and particularly because of the absence of any relapse, it has been suggested that PPMS may represent a different disease (Thompson et al. 1997). Finally, approximately 5 % of patients develop progressive relapsing MS (PRMS), which is characterized by worsening from onset with clear, acute relapses (with or without recovery), and with periods between relapses with continuing progression (Lublin and Reingold 1996; Rovaris et al. 2006).

The prognosis of MS varies widely. The factors responsible for this variability are unclear and it therefore remains difficult to estimate prognosis in individual patients at the time of diagnosis. The outcome of an attack depends on the severity and extent of axonal injury, and this largely determines the degree of recovery or the persistence of neurological deficits. Usually, after inflammation has resolved and myelin debris has been removed, the conduction of the neural impulses is reestablished in the denuded nerve fibers. Although the resolution of inflammation allows a remission with full or partial recovery of neurologic deficits, the appearance of new plaques, which may develop at any time throughout the course of the disease, would cause a relapse. Disability becomes permanent when the structural continuity of the nerve fibers is disrupted and wallerian degeneration develops (reviewed in Prineas et al. 2002).

Clinical indicators of a relatively good prognosis are female gender, younger age of onset, optic neuritis, sensory attacks, complete recovery from attacks, few attacks, and long inter-attack interval. Relatively poor prognostic factors include male gender, predominant cerebellar and motor involvement, incomplete resolution of attacks, progressive course from onset, frequent early attacks, and short inter-attack interval (Keegan and Noseworthy 2002).

In general, the average duration of the disease is 25–30 years. One-third of patients have a benign course, remain fully functional, and show little disability for 15 years after disease onset. In contrast, malignant MS is defined by a rapid, progressive course, leading to significant neurological deficits and even death (Lublin and Reingold 1996; Prineas et al. 2002).

21.1.3 Etiology

Despite great interest and extensive pathological and experimental studies on MS, the etiology of this complex and multifactorial disease remains elusive. It is widely accepted that a combination of genetic susceptibility and environmental factors is implicated in the origin of MS. Familiar occurrences are recognized, and the disease has been reported in monozygotic and dizygotic twins; however, the genes that contribute to MS susceptibility are difficult to identify because they exert a relatively modest effect on disease risk.

In support of the genetic contribution, certain gene variants in the class II major histocompatibility complexes (MHCs) occur more frequently among MS patients than among the general population (Prineas et al. 2002). Two genetic studies have revealed an association of MS with particular human leukocyte antigen (HLA) haplotypes, particularly the DR15 haplotype (DRB1, DRB5, DQA1, DQB1) for the European and North American Caucasian population (Lincoln et al. 2005; Sawcer et al. 2005). Recently, this link was confirmed in the largest genome-wide association study conducted to date, which also revealed two other immune-related targets including the receptors for the interleukins IL-2 and IL-7 (Hafler et al. 2007; Lundmark et al. 2007).

Aside from the genetic component, the role of transmissible agents as a cause of MS is a relatively popular hypothesis. It is possible that exposure to an unidentified infectious agent may occur during the early years of life. Studies indicate that individuals who migrate from a cold climate to a warm climate after 15 years of age retain the higher risk associated with their native locality. Several pathogens have been postulated to trigger the immune reaction: measles virus, Epstein–Barr virus (EBV), human herpes virus 6, retroviruses, and *Chlamydia pneumoniae*. Among the infectious agent candidates, data associating EBV infection with MS remains strong. Thus, people with symptomatic EBV infection are at increased risk of developing MS, similar to people with high titers of anti-EBV antibodies. There are also observations supporting EBV-induced molecular mimicry as an underlying mechanism in MS autoimmunity. This association may be causative, or it may simply be a phenomenon required for disease onset (Giovannoni and Ebers 2007).

Moreover, and although there is evidence for both genetic (Niino et al. 2007) and environmental components (Ascherio and Munger 2007), many studies demonstrate that the immune system plays an integral role in the initiation and progression of MS (Hemmer et al. 2002; Keegan and Noseworthy 2002). In this view, MS can be seen as a disease in which genetically susceptible individuals, upon encountering an environmental stimulus such as an infection, generate an autoimmune attack against CNS myelin based on molecular mimicry between infectious and myelin antigens.

Today, autoimmnune-mediated reactions against myelin, which are based on the inappropriate immunological attack towards self-antigen, constitute a more widely accepted hypothesis for MS etiology, although its initial cause is still unknown. This autoimmune hypothesis is supported by similarities between the pathologies of MS and experimental allergic encephalomyelitis (EAE), the dominant MS animal model, which is induced by immunization with brain and spinal cord myelin extracts. In this model, myelin antigen-specific $CD4^+$ T cells can induce CNS inflammation, demyelination, and neurodegeneration, resulting in the loss of motor functions or paralysis (Brown and Sawchenko 2007).

21.1.4 Pathophysiology of Multiple Sclerosis

Pathophysiological models of MS should reproduce the generation of acute demyelinating lesions and their evolution into chronic sclerotic plaques, as well as an unpredictable clinical course that is characterized initially by recurrent relapses and later by steady progression. Although MS is regarded as a white matter disease, and the principal features of the disease are demyelination, perivascular inflammation, and the presence of plaques within the white matter, demyelination and oligodendrocyte or neuron/axon injury are also prominent and widespread in grey matter structures, such as the cerebral cortex, thalamus, and basal ganglia (Lassmann 2007a; Stadelmann et al. 2008). Thus, a particularly high prevalence of plaques in the cerebral cortex has been observed in progressive stages of the disease, and constitutes a significant proportion of the overall pathology of the brains of MS patients (Amadio et al. 2011).

The generally accepted pathophysiological model for MS is centered on an immune-mediated attack against CNS myelin antigens, in which several components of the immune system generate an inflammatory response that damages myelin and axons, leading to the formation of acute plaques (Frohman et al. 2006; Fontoura and Garren 2010). However, although MS is considered an autoimmune CNS inflammatory disease, it is widely accepted that neurodegeneration is the predominant pathophysiological substrate of disability. In this sense, distinct evidence suggests a strong correlation between inflammation and neurodegeneration (Zipp and Atkas 2006; Lassmann 2007b; Franciotta et al. 2008; Frischer et al. 2009; Lee et al. 2011); however, debate remains regarding whether inflammation is a primary or a secondary process during both the onset and the development of MS (Trapp and Nave 2008; Craner and Fugger 2011).

21.1.4.1 Neuroinflammation in MS

Inflammatory events in MS could be due to an autoimmune reaction against myelin antigens, which is more profound in actively demyelinating lesions and involves both cellular and humoral immunity with the participation of macrophages, T lymphocytes, and B cells. In the early stage of the disease process, T cells are primed in the periphery by antigen-presenting dendritic cells, and activated $CD4^+$ T lymphocytes cross the blood–brain barrier and react with myelin and/or oligodendroglia antigens. There is also evidence for humoral autoantibodies produced by local B cells binding myelin and other CNS components, as well as expression of immune-associated molecules, such as major histocompatibility antigens, adhesion molecules or pro-inflammatory cytokines, interleukin 12 (IL-12), and tumor necrosis factor-α (TNF-α). All of these factors contribute to the generation of a diffuse inflammatory process, with infiltrates mainly composed by lymphocytes and macrophages and with additional activation of the resident microglia.

In general, the inflammation process occurs during all stages of MS, although it changes as the disease progresses. In acute MS, infiltrating macrophages and activated T cells are predominant in focal demyelinated lesions. In addition, in chronic MS there is a diffuse inflammatory/degenerative process supported by microglia and macrophages that may be more or less independent of current focal inflammation (Kerschensteiner et al. 2009; Edan and Leray 2010). In this chronic phase, particularly oligodendrocytes, myelin, and axons degenerate in the CNS, causing a wide variety of symptoms that often progress to physical and cognitive disabilities (Amadio et al. 2011).

21.1.4.2 Neurodegeneration in MS

Although MS has been mainly considered an immune disease, there is an evolving concept where MS may be a primary neurodegenerative disease with secondary inflammatory demyelination. There is evidence for early axonal damage resulting from a direct interaction with immune cells, as well as a positive correlation between axonal transaction and degree of inflammation in brain white matter MS lesions undergoing demyelination (Zipp and Atkas 2006; Trapp and Nave 2008; Nikic et al. 2011). In addition, recent studies support the concept that neurodegeneration is an independent process in MS, which may explain why current disease-modifying therapies predominantly targeting immunomodulatory mechanisms have a reduced efficacy against the development of permanent physical disability in the later stages of MS (Craner and Fugger 2011).

In the vast majority of MS patients, MS develops into a progressive stage. Following the relapse–remitting stage of MS, patients develop "secondary progressive" MS, in which neurological deterioration continues in the absence of relapses (Dutta and Trapp 2007; Lassmann et al. 2007; Noseworthy et al. 2000). Alternatively, others are diagnosed with primary progressive MS, in which neurological dysfunction occurs without relapses from disease onset (Dutta and Trapp 2007; Lassmann et al. 2007; Noseworthy et al. 2000). In all instances, imaging, neuropathological studies, and animal studies of MS show that markers for neurodegeneration (primarily axonal damage and atrophy) appear during the progressive phase of the illness and correlate with neurological disability (Bjartmar et al. 2000; Lassmann et al. 2007). Although the mechanisms of axonal degeneration are uncertain, inflammation and demyelination appear to be major risk factors (Ferguson et al. 1997; Trapp et al. 1998; reviewed in Trapp and Nave 2008). With this in mind, numerous studies have focused on axonal protection as a major therapeutic goal in MS, both promoting remyelination and analyzing different aspects of spontaneous remyelination and axonal recovery (Black et al. 2006; Coman et al. 2006; Howell et al. 2006; Patrikios et al. 2006; Nikic et al. 2011).

There is also evidence suggesting a link between autoimmunity, clinical phenotype, and neurodegeneration in MS. MS patients develop antibodies to oligodendrocytes, myelin, and other neuronal antigens that cause neurodegeneration and contribute to the pathogenesis of the disease (Franciotta et al. 2008; Lee et al. 2011).

Together, the evidence summarized above strongly suggests that the molecular mechanisms responsible for axonal degeneration may differ between early and late stages of MS, and that there is an unclear link between the mechanisms of neurodegeneration and focal/diffuse inflammation and their relative contribution to the clinical deficits observed in different phases of the disease.

21.1.4.3 Oligodendroglial Injury in MS

As mentioned above, MS is considered a progressive disease of the CNS characterized by autoimmune attack on myelin and oligodendrocytes. Although the precise basis for this selective injury remains to be established, the depletion of oligodendrocytes is a recognized feature of MS lesions, becoming more apparent as the disease evolves (Raine 1994). In compliance with this idea, Fas expression is elevated in oligodendrocytes in chronic active and chronic silent MS lesions. Fas is a cell surface receptor that transduces cell death signals, and its upregulation in oligodendrocytes suggests that Fas-mediated signaling might contribute to immune-mediated oligodendrocyte injury and subsequent demyelination in MS (D'Souza et al. 1996). In turn, treatment of oligodendrocytes with antibodies against myelin-oligodendrocyte glycoprotein (MOG) leads to an increase in Ca^{2+} influx and activation of the MAPK/Akt pathways, a signaling cascade relevant to the initial steps of MOG-mediated demyelination (Marta et al. 2005).

A number of experimental studies have demonstrated a strong positive correlation between oligodendrocyte susceptibility to injury and the extent of CNS inflammation in EAE. In a knockout mouse system, absence of oligodendrocyte protective factors increases oligodendrocyte susceptibility to injury and augments the inflammatory reaction and the severity of symptoms (Butzkueven et al. 2002; Linker et al. 2002; Balabanov et al. 2007). In contrast, mice lacking proapoptotic genes or overexpressing antiapoptotic molecules, specifically in oligodendrocytes, display resistance to EAE and inflammatory demyelination (Hisahara et al. 2000; Hövelmeyer et al. 2005). A recent study has described that the oligodendrocytic overexpression of the dominant-negative form of interferon regulatory factor-1 (IRF-1), a severity factor for both MS and EAE, results in significant protection against EAE with a reduction of inflammatory demyelination and with oligodendrocyte and axonal preservation (Ren et al. 2011). These data suggest that oligodendrocytes are actively involved both in the regulation of EAE and in the network of neuroimmune responses. Therefore, exploring oligodendrocyte-related pathogenic mechanisms in addition to conventional immune-based mechanisms may have important therapeutic implications in MS.

However, early MS lesions have a prephagocytic nature and display primary oligodendrocyte injury in the absence of microglial activation and adaptive T cell or B cell responses (Barnett and Prineas 2004; Barnett et al. 2006; Matute et al. 2001; Matute and Pérez-Cerdá 2005). The trigger for that selective and apparently primary oligodendrocyte damage is unknown; however, it may include viruses, glutamate, and other agents known to be oligotoxic. These data correlate with previous pathological characterizations of MS lesions, such that some plaques are highly suggestive of a primary oligodendrocyte dystrophy rather than an autoimmunity process (Lucchinetti et al. 2000). Moreover, these findings suggest that all MS lesions may initiate with oligodendrocyte death of unknown origin in the absence of inflammatory activity, and that the heterogeneity observed in the neuropathology of the lesions within and among patients may be a reflection of the time point at which a given lesion is observed. Therefore, it is possible that primary oligodendrocyte injury leads to microglial activation, with the adaptive T cell and B cell response appearing as a secondary event. In this context, the innate immune system plays a much more fundamental role than previously thought in MS lesion pathogenesis, and therapies aimed at this side of the immune response will prove effective, including in progressive stages of the disease (Fontoura and Garren 2010).

Finally, in addition to changes to oligodendrocytes and neurons, astrocytes and microglia are also relevant to MS pathophysiology (He and Sun 2007). Reactive astrocytes contribute to the glial scar that fills the demyelinated plaque (Holley et al. 2003). In addition, astrocytes can promote inflammation and damage to oligodendrocytes and axons; however, at the same time they can also support the migration, proliferation, and differentiation of oligodendrocyte progenitors (Williams et al. 2007). Likewise, reactive microglia may be both deleterious and protective in MS pathogenesis (Muzzio et al. 2007; Sanders and De Keyser 2007). Consequently, both astrocytes and microglia play important roles in both the destructive and restorative phases of MS. Taken together, all of these findings outline the complexity of MS pathology, in which both the adaptive and innate immune systems and all the cells of the CNS participate in MS lesions and repair in a manner that may depend on the stage of MS (Fontoura and Garren 2010).

21.2 Adenosinergic Signaling in Neuroinflammation

Extracellular purine/pyrimidine nucleotides and nucleosides are among the most widespread exogenous signals, with important detrimental or protective roles in normal and injured brains (Inoue et al. 2007; Apolloni et al. 2009). Although little is known regarding purinergic signaling in MS, there is growing evidence that not only adenosine triphosphate(ATP), but also adenosine can directly modulate different aspects of the pathophysiology of MS, particularly neuroinflammation and demyelination (Bours et al. 2006; Sperlágh and Illes 2007; Amadio et al. 2011). Because neuroinflammation is commonly associated with the course of MS, we devote this section to describing the effects of adenosine in macrophages/monocytes, as well as in microglia.

21.2.1 Macrophages/Monocytes in MS and Their Regulation by Adenosine

MS is associated with the activation of immune cells, including macrophages, peripheral blood mononuclear cells (PBMCs), microglia, and astrocytes. Macrophages are key cellular components of the innate immune system. Although macrophage function is essential for an efficient immune response, failure to control macrophage activation or prolonged or inappropriate inflammatory processes will lead to unacceptable levels of collateral damage to surrounding cells. In addition, activated PBMCs release numerous inflammatory effector molecules, such as TNF-α or IL-6, which potentiate or abrogate demyelination through pro-inflammatory or immunosuppressive effects. Activated monocytes and lymphocytes secrete IL-6, which can activate B cells and the proliferation of T cells, and circulating levels of TNF-α, IL-6, and other cytokines are increased in the serum, cerebrospinal fluid, and brains of MS patients. In particular, TNF-α is related to the pathogenesis of MS models, as illustrated by the finding that neutralizing antibodies to TNF-α or soluble TNF-α receptors abrogate EAE development (Selmaj et al. 1991, 1995). In addition, overexpression of TNF-α in transgenic mice causes demyelination similar to the inflammatory demyelination observed in patients with MS (Probert et al. 1995). Nevertheless, TNF-α is not the only cytokine or mediator of MS, because TNF-α receptor knockout mice also develop EAE (Frei et al. 1997).

In addition to cytokines, ATP and adenosine are relevant to immunity and inflammation, although the signals they mediate in immune responses in vivo are extremely complex. Thus, the effects of ATP and adenosine may vary from immunostimulatory to immunoregulatory depending on their extracellular concentrations, as well as on the expression patterns of purinergic receptors and ecto-enzymes (Bours et al. 2006). Specifically, the neuromodulator adenosine regulates a variety of physiological processes, including immune activation and neuronal survival, through specific G-protein-coupled receptors expressed on macrophages and neurons (Cronstein 1994; Ribeiro et al. 2002). Under normal conditions, adenosine is tightly regulated to low nanomolar levels in tissues throughout the body (Geiger et al. 1997). Utilization of ATP during periods of high metabolic activity leads to an increased concentration of intracellular adenosine that can be secreted through nucleoside transporters. Another major pathway contributing to high extracellular adenosine concentration during metabolic stress is the release of precursor adenine nucleotides (ATP, ADP, and AMP) from cells, followed by extracellular degradation to adenosine. Neutrophils and endothelial cells release large amounts of adenosine at sites of inflammation, and activated macrophages can serve as a major source of extracellular adenosine via ATP production (Cronstein 1994).

All of these mechanisms lead to increased levels of adenosine, which acts at the cell surface through four G-protein-coupled adenosine receptors: A_1, A_{2a}, A_{2b}, and A_3. Monocytes and macrophages express all four adenosine receptors (Bours et al. 2006), and these receptors allow the cells to detect stressful conditions and modulate their cellular functions to adapt to their microenvironment.

Also relevant to the pathogenesis of MS is the observation that the activation of adenosine receptors in macrophages has been related to the anti-inflammatory effects of adenosine, including the downregulation of TNF-α production (Haskó et al. 1996, 2008; Haskó and Cronstein 2004; Sitkovsky and Lukashev 2005; Sitkovsky and Ohta 2005; Ruiz-García et al. 2011). In addition, the anti-inflammatory properties of adenosine and adenosine-related compounds have been confirmed in various animal disease models and in vivo genetic experiments (Ohta and Sitkovsky 2001; Lukashev et al. 2004; Sitkovsky et al. 2004).

In general, the activation of all four adenosine receptors on macrophages and monocytes modulates the synthesis and release of immunomodulatory molecules, including cytokines, matrix metalloproteinases, and reactive oxygen species (Boyle et al. 1996; Haskó et al. 1996). Elevated levels of cytokines in relapsing–remitting MS patients may be due to altered levels of adenosine and/or adenosine receptors (Mayne et al. 1999). Consistent with this hypothesis, a study including MS patients showed that plasma levels of TNF-α were higher and that adenosine levels were lower in MS cases than in control subjects. In addition, adenosine A_1 receptor stimulation in PBMCs from MS patients displayed differing effects on the regulation of TNF-α and IL-6 compared with PBMCs from controls (Mayne et al. 1999). Taken together, these results suggest that decreased levels of adenosine and its A_1 receptor can modulate TNF-α and IL-6 levels and contribute to the pathogenesis of MS.

21.2.2 Adenosine and the Modulation of Microglial Activation

Microglial cells originate from monocyte/macrophage precursors and are regarded as the major immunocompetent cell type of the nervous system, constituting approximately 10 % of all cells in the brain. The immune response of the brain is spatially segregated from the peripheral immune response by the blood–brain barrier and, together with astroglial cells and infiltrating peripheral immune cells, is predominantly executed by microglial cells. Thus, microglial cells, as brain-resident immune cells, are a sensor of pathological signals in the CNS and play a major role in host defense and tissue repair in the brain. They are rapidly activated and respond with morphological changes, transforming the resting ramified microglia into an amoeboid form with phagocytic activity, proliferation, and production of a wide array of inflammatory mediators (Kreutzberg 1996; Cuadros and Navascués 1998). Past studies have shown that exposure to different factors, such as lipopolysaccharide, interferon-γ, or β-amyloid, leads to microglial activation and induces the production of various pro-inflammatory mediators that are potentially neurotoxic (Meda et al. 1995; Zielasek and Hartung 1996). These mediators include nitric oxide, prostaglandin E2, pro-inflammatory cytokines (e.g., TNF-α, IL-1β, and IL-6), and reactive oxygen species (reviewed in Bi et al. 2011).

There is evidence that the constant activation and release of pro-inflammatory factors promote the development of neurodegenerative diseases and therefore, microglial activation plays an important role in the pathophysiology of these diseases (Moon et al. 2007a, b; Bi et al. 2011). Therefore, inhibition of pro-inflammatory mediators in microglia attenuates the severity of Alzheimer's disease, Parkinson's disease, trauma, multiple sclerosis, and cerebral ischemia (Eikelenboom and van Gool 2004; Koning et al. 2007; Krause and Müller 2010; Liu and Hong 2003; Qian et al. 2010).

As in macrophages and monocytes, purinergic signaling initiated by ATP released from both neurons and glial cells, as well as its breakdown product, adenosine, activates A_1, A_{2a}, and A_3 in microglia (Haskó et al. 2005) and contributes to the neuroimmune and neuroinflammatory response (reviewed in Bours et al. 2006; Sperlágh and Illes 2007). The pivotal role of A_1 adenosine receptors in microglia-mediated neuroinflammation is underscored by a study showing that A_1 receptor null mice develop severe demyelination and motor symptoms in chronic EAE, which indicates that endogenous adenosine might have a protective role in the defense against this disease (Tsutsui et al. 2004). These effects are mostly mediated by cells in the microglial lineage, and may be involved, at least in part, with promyelinating effects mediated by A_1 receptors, which may support the partial remyelination of tissue during chronic EAE.

Additionally, activation of A_{2a} adenosine receptors differentially regulates the transcription and *de novo* synthesis of various subtypes of K^+ channels via cAMP-mediated and PKC-mediated pathways in cultured rat microglial cells (Kust et al. 1999). This mechanism may participate in the transition of microglia from the resting state to the activated form, in which the function of K^+ channels seems to play a pivotal role (Illes et al. 1996; reviewed in Sperlágh and Illes 2007). Selective stimulation of A_3 adenosine receptors induces p38 and ERK1/2 phosphorylation in primary murine microglia and in the N13 microglial cell line (Hammarberg et al. 2003, 2004). This effect is mediated by G_i protein coupling and phosphatidylinositol-3′-kinase (PI3K) (Hammarberg et al. 2003), and is absent in microglial cells derived from A_3 receptor-deficient transgenic mice (Hammarberg et al. 2004). Finally, adenosinergic stimulation also influences the proliferation/survival, or even apoptosis, of microglial cells. However, the identity of the receptor subtype involved in these effects remains unknown (Sperlágh and Illes 2007).

21.2.3 Role of Adenosine Receptors in Inflammatory Processes

Macrophage activation is mediated by multiple mechanisms (Gilroy et al. 2004) including adenosine that modulates the inflammatory response by limiting macrophage activation (Haskó and Cronstein 2004; Sitkovsky and Ohta 2005). Thus, macrophages activated through Toll receptor increase the expression of the A_{2a} and A_{2b} adenosine receptors which in turn enhance the expression of genes related to their metabolic activity. In particular, endogenously generated adenosine in macrophages cooperates with bacterial components to increase PFKFB3 isozyme activity, resulting in greater fructose 2,6-bisphosphate accumulation. Meanwhile, A_{2a} receptor mediates, in part, the induction of *pfkfb3* by LPS, whereas the A_{2b} receptor, with lower adenosine affinity, cooperates when high adenosine levels are present. As a whole, these processes enhance the glycolytic flux and favors ATP generation helping to develop and maintain the long-term defensive and reparative functions of the macrophages (Ruiz-García et al. 2011).

In microglial cells, A_{2a} receptor activation increases the production and release of nerve growth factor (NGF), up-regulates COX-2 expression, and induces subsequent PGE2 release (reviewed in Sperlágh and Illes 2007). However, only simultaneous activation of A_1 and A_{2a} receptors promotes microglia proliferation (Gebicke-Haerter et al. 1996). Together, these data indicate that activation of adenosine receptors influences the proliferation/survival of microglial cells and inflammatory processes. However, further investigations are needed to unravel the identity of the receptor subtypes involved, the signaling pathways mediating the microglial response, and the mechanisms underlying the protective action of adenosine in the process of microglial activation.

21.3 Adenosine Receptor Regulation in MS

Because of the strong immunosuppressive and anti-inflammatory properties of adenosine, dysfunction of the A_1 receptors in the CNS has been implicated in the development of MS in humans and EAE in nonhuman animals. In particular, the A_1 receptor, which is expressed principally on cells of monocyte/macrophage lineage in both brain and blood cells, is selectively diminished in MS patients, potentially leading to increased macrophage activation and CNS inflammation. This phenomenon suggests that modulation of neuroinflammation by A_1 receptors may represent a mechanism that could provide new therapeutic opportunities for MS and other demyelinating diseases (Johnston et al. 2001).

Consistent with the aforementioned idea, A_1 receptor null mice develop a more severe progressive/relapsing form of EAE as compared with their wild-type littermates (Tsutsui et al. 2004). The absence of A_1 receptors negatively regulates the suppression of anti-inflammatory genes, and the enhanced activation of microglia/macrophages fostering oligodendrocyte cytotoxicity, demyelination, and axonal injury (Tsutsui et al. 2004). Conversely, chronic treatment with caffeine, a nonselective antagonist of adenosine receptors, increases A_1 receptor mRNA levels in microglia/macrophages, suppresses pro-inflammatory responses, diminishes oligodendrocyte cytotoxicity, and attenuates EAE severity (Tsutsui et al. 2004).

Levels of A_1 receptors are diminished at the onset of EAE, and chronic caffeine treatment decreases the incidence of EAE and alleviates the symptoms on the disease (Chen et al. 2010). Specifically, caffeine attenuates EAE pathology symptoms, inflammatory cell infiltration, levels of inflammatory cytokines, and demyelination. Caffeine's protective actions in EAE are possibly due to up-regulation of A_1 receptors and transforming growth factor-beta mRNAs and suppression of interferon-gamma mRNA, rather than due to inhibition of adenosine receptors (Chen et al. 2010).

On the other hand, the A_{2a} receptor-specific antagonist SCH58261 also protects mice from EAE induction by limiting the efficient entry of lymphocytes into the CNS during EAE development, (Mills et al. 2008). Moreover, the protective actions of A_{2a} receptor blockade may also directly prevent oligodendrocyte damage, since its antagonism reduces activation of JNK/MAPK signaling in oligodendrocytes (Melani et al. 2009).

Taken together, an imbalance between the A_1 and A_{2a} adenosine receptor activities may be involved in EAE development (Chen et al. 2010). However, no direct information is currently available regarding the contribution of the additional A_{2b} and A_3 adenosine receptor subtypes during MS in humans and EAE in nonhuman animals (Amadio et al. 2011).

The data described above indicates that A_1 and A_{2a} receptor activation have opposite effects (beneficial and deleterious, respectively) on the outcome of EAE by primarily influencing the onset of this experimental disease via interactions with the inflammatory and immune responses. However, manipulating these responses with therapeutic aims is complicated by the fact that adenosine can directly modulate migration, proliferation, and differentiation of oligodendrocyte progenitor cells (OPCs). Thus, adenosine inhibits OPC proliferation while promotes OPC differentiation and myelination (Stevens et al. 2002), and stimulates OPC's migration via A_1 receptors (Othman et al. 2003). Therefore, prolonged blockade of A_{2a} receptors may compromise remyelination naturally occurring in the MS brain and thus, limit its therapeutic potential.

21.4 Adenosine Metabolism and Its Involvement in MS

Adenosine levels are reduced in blood from MS patients (Mayne et al. 1999). Adenosine is generated from the breakdown of AMP by CD73 (ecto-5′-nucleotidase), a cell surface enzyme of the purine catabolic pathway. Curiously, CD73 is expressed in brain endothelial cells and the choroid plexus epithelium, which regulates lymphocyte immune surveillance between blood and cerebrospinal fluid. Moreover, CD73, with adenosine receptor signaling, is required for the efficient entry of lymphocytes into the CNS during EAE development (Mills et al. 2008). Therefore, $CD73^{-/-}$ mice, which display preserved T cell responsiveness, are resistant to EAE. In addition, because interferon-β is known to increase the expression of CD73 on endothelial cells, both in vitro and after systemic treatment of MS patients in vivo, it was postulated that CD73-derived adenosine might contribute in part to the therapeutic effects of interferon-β (Airas et al. 2007).

Furthermore, the activities of ectonucleotide pyrophosphatase/phosphodiesterase and adenosine deaminase (enzymes responsible for extracellular ATP/adenosine metabolism and for altering the circulatory levels of nucleotides and nucleosides) are decreased in both the lymphocytes and platelets of relapsing/remitting MS patients (Vivekanandhan et al. 2005; Spanevello et al. 2010), possibly as a compensatory mechanism to increase adenosine levels in MS patients (Spanevello et al. 2010). Meanwhile, adenosine kinase (ADK) is an enzyme involved in the adenosine metabolism reducing adenosine levels by forming AMP. Although a deficiency in ADK may result in adenosine accumulation (Boison et al. 2002), its role in the modulation of MS currently remains unknown.

Taken together, alterations of ATP, ADP, and AMP hydrolysis occur in EAE and MS. These alterations may be exploited to develop novel therapeutic strategies in immune-mediated diseases, including MS (Amadio et al. 2011).

21.5 Adenosine and Oligodendrocyte Injury in MS

Although oligodendrocyte loss is a major MS feature, there is little (and apparently contradictory) information regarding how purinergic signaling influences oligodendroglia. It has been suggested that adenosine is likely to inhibit oligodendrocyte precursor cell (OPC) proliferation, promote OPC differentiation, and increase myelinization through the activation of adenosine receptors in oligodendrocytes (Stevens et al. 2002). However, another study revealed that the stimulation of A_1 adenosine receptor induces OPC migration, but does not influence the proliferation or viability of OPCs (Othman et al. 2003). Instead, ATP triggers the proliferation, migration, and differentiation of OPCs, primarily via several different P2Y receptors (James and Butt 2001; Agresti et al. 2005). The activation of P2 receptors evokes Ca^{2+} signals in OPCs and oligodendrocytes in situ and in culture (James and Butt 2001; Alberdi et al. 2005; Agresti et al. 2005; Butt et al. 2005). Therefore, it appears that axons release adenosine and ATP during action potential propagation, thereby controlling oligodendrocyte development, with a role for adenosine in stimulating terminal differentiation, and for ATP in promoting myelination (Ishibashi et al. 2006; reviewed in Amadio et al. 2011).

Regarding neurodegenerative disorders, a number of experimental studies have demonstrated a strong positive correlation between the susceptibility of oligodendrocytes to injury and the extent of CNS inflammation in EAE. However, the molecular mechanisms involving oligodendrocytes in the regulation of EAE remain poorly understood.

Oligodendrocytes express adenosine A_1 (Othman et al. 2003) and adenosine A_{2a} receptors (Melani et al. 2003). In addition, work in progress carried out in our laboratory shows that cultured oligodendrocytes derived from rat optic nerves express the mRNAs of all four adenosine receptors, as well as equilibrative nucleoside transporters ENT1 and ENT2 and the adenosine-degrading enzymes adenosine deaminase and adenosine kinase (González-Fernández et al. 2010; Fig. 21.1). Notably, the activation of adenosine receptors causes oligodendroglial death in a dose-dependent

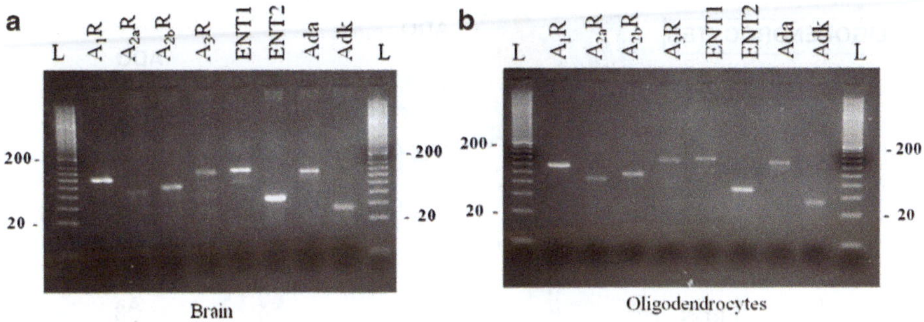

Fig. 21.1 Expression of the adenosinergic system in cultures of optic nerve oligodendrocytes. Reverse-transcription polymerase chain reaction (RT-PCR) reveals expression of the four adenosine receptors (A_1, A_{2a}, A_{2b}, and A_3), adenosine transporters (ENT1 and ENT2), and the enzymes adenosine deaminase and adenosine kinase in samples of rat brain, used as a positive control (**a**), and in cultured oligodendrocytes obtained from the optic nerves of 12-day-old rats (**b**). PCR fragments were of the expected size and sequence. Molecular standards in base pairs correspond to the ϕX174 HαεIII digest (L). A_1R A_1 adenosine receptor, $A_{2a}R$ A_{2a} adenosine receptor, $A_{2b}R$ A_{2b} adenosine receptor, A_3R A_3 adenosine receptor, *ENT1* equilibrative nucleoside transporter (member 1), *ENT2* equilibrative nucleoside transporter (member 2), *Ada* adenosine deaminase, *Adk* adenosine kinase

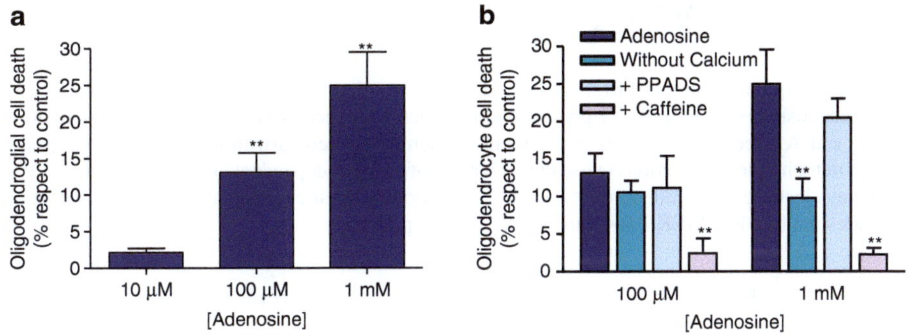

Fig. 21.2 Adenosine induces oligodendrocyte death. (**a**) Optic nerve oligodendrocytes were exposed to increasing concentrations of adenosine (10 μM to 1 mM) for 15 min, and cell viability was measured after 24 h using the calcein-AM assay as previously described in detail (Matute et al. 2007). Adenosine causes concentration-dependent oligodendrocyte death in comparison with control untreated cells ($n=4$–6; $^{**}p<0.01$). (**b**) Adenosine-mediated oligodendrocyte cell death is not reduced in the presence of the P2X receptor antagonist PPADS ($p>0.2$); however, blocking the adenosine receptors with caffeine significantly decreased oligodendroglial death, when compared with cells treated with adenosine alone ($n=4$–6; $^{**}p<0.005$)

manner (Fig. 21.2) that can be blocked by caffeine, as well as oxidative stress and the activation of both caspase-dependent and caspase-independent apoptosis (Fig. 21.3). Finally, levels of phosphorylated ERK1/2 and Akt are significantly reduced after treatment with adenosine receptor agonists. Although oligodendrocyte toxicity by adenosine is partly mediated by the A_3 receptor, the specific antagonist of this receptor, MRS1220, did not ameliorate EAE (Fig. 21.4).

In summary, these results indicate that adenosine receptor activation in oligodendrocytes triggers cell death. The findings also extend our previous observations

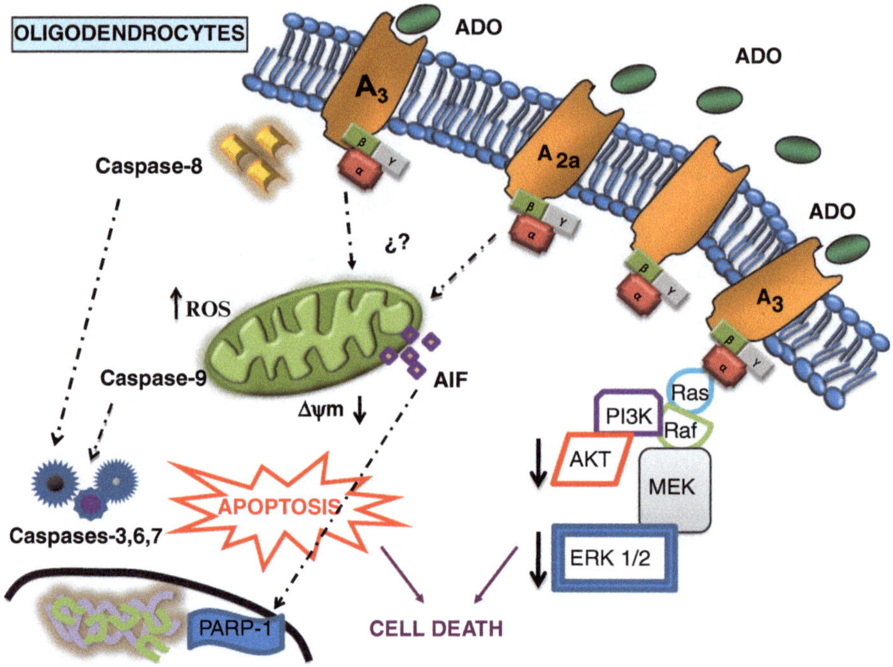

Fig. 21.3 Signaling cascades triggered by adenosine in oligodendrocyte toxicity. The oligotoxicity of adenosine is mediated by G-protein-coupled adenosine receptors. In particular, the activation of A_3 and A_{2a} receptors induces mitochondrial membrane depolarization and oxidative stress. These mitochondrial alterations cause the activation of caspase-dependent and caspase-independent pathways leading to oligodendrocyte death. In addition, downstream signaling after A_3 activation includes the Ras/Raf/MEK/ERK and PLC/PI3K/Akt pathways

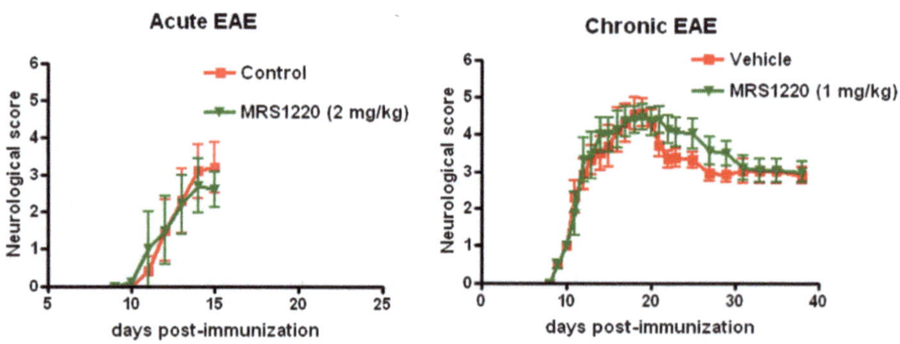

Fig. 21.4 Blockade of A_3 receptors does not change the course of acute or chronic EAE. Acute EAE was induced in Wistar male rats following immunization with myelin basic protein (Smith et al. 2000); rats were treated with vehicle or MRS1220 from post-immunization day 7 until the end of the experiment (2 mg/kg, i.p. once daily). Chronic EAE was induced in C57BL/6 mice with myelin-oligodendrocyte glycoprotein as described (Matute et al. 2007); vehicle or MRS1220 ($n=12$ each) was administered from post-immunization day 16 until the end of the experiment (1 mg/kg, i.p. once daily). Motor symptoms were scored from 0 to 8, as described previously (Matute et al. 2007). $p>0.5$ at all time points examined

showing that the activation of the purinergic receptor P2X7 in oligodendrocytes mediates cell death in vitro and in vivo, and contributes to tissue damage in EAE pathology (Matute et al. 2007). The vulnerability of oligodendrocytes to enhanced activation of the purinergic/adenosinergic system illustrates that a general shift in the dynamic equilibrium regulating the adenosinergic biomolecular network constitutes a pathogenic feature in MS, and also suggest that modulation of adenosine and adenosine-related molecules may be beneficial for the treatment of MS.

21.6 Conclusions

Multiple sclerosis is a chronic, immune-mediated disorder of the central nervous system, and the most common neurological disorder in young adults. Pathologically, prolonged chronic inflammation and widespread demyelination in the central nervous system lead to atrophy and progressive worsening of disease. The cause and etiology of MS are incompletely understood and current treatments are, by many measures, inadequate. In recent years, many insights have been acquired about the autoimmune mechanisms of MS and, in particular, the molecular targets involved, so several treatment approaches have emerged. In this regard, the role of adenosine in the control of immune and inflammatory systems has generated excitement regarding the potential use of adenosine receptor-based therapies in the treatment of different diseases, as degenerative pathologies.

Adenosine receptors, and especially A_1 and A_{2a} subtype, seem to be linked to MS pathophysiology as attested by experimental and epidemiological studies. In this regard, adenosine receptor antagonists represent an ideal target for the therapy of certain immune-related disorders because their action is selectively targeted to the site of injury, where endogenous adenosine is released. In this line, interfering with the function of A_{2a} adenosine receptor using specific antagonist allows local reduction of adenosine deleterious effects, providing an opportunity for interventions with limited side effects. Thus, modulation of neuroinflammation by the adenosine receptors represents a novel mechanism that provides new therapeutic opportunities for MS and other demyelinating diseases. Notwithstanding, the high abundance of adenosine receptors across tissues and opposite protective effects mediated by them warrant further preclinical studies before proving them of use in the treatment of MS.

References

Agresti C, Meomartini ME, Amadio S, Ambrosini E, Serafini B, Franchini L, Volonté C, Aloisi F, Visentin S (2005) Metabotropic P2 receptor activation regulates oligodendrocyte progenitor migration and development. Glia 50:132–144

Airas L, Niemela J, Yegutkin G, Jalkanen S (2007) Mechanism of action of IFN-beta in the treatment of multiple sclerosis: a special reference to CD73 and adenosine. Ann N Y Acad Sci 1110: 641–648

Alberdi E, Sánchez-Gómez MV, Matute C (2005) Calcium and glial cell death. Cell Calcium 38:417–425

Alonso A, Hernan MA (2008) Temporal trends in the incidence of multiple sclerosis: a systematic review. Neurology 71:129–135

Amadio S, Apollini S, D'Ambrosi N, Volonté C (2011) Purinergic signalling at the plasma membrane: a multipurpose and multidirectional mode to deal with amyotrophic lateral sclerosis and multiple sclerosis. J Neurochem 116:796–805

Apolloni S, Montilli C, Finocchi P, Amadio S (2009) Membrane compartments and purinergic signaling: P2X receptors in neurodegenerative and neuroinflammatory events. FEBS J 276:354–364

Ascherio A, Munger KL (2007) Environmental risk factors for multiple sclerosis. II. Noninfectious factors. Ann Neurol 61:504–513

Balabanov R, Strand K, Goswami R, McMahon E, Begolka W, Miller SD, Popko B (2007) Interferon-gamma-oligodendrocyte interactions in the regulation of experimental autoimmune encephalomyelitis. J Neurosci 27:2013–2024

Barnett MH, Prineas JW (2004) Relapsing and remitting multiple sclerosis: pathology of the newly forming lesion. Ann Neurol 55:458–468

Barnett MH, Henderson AP, Prineas JW (2006) The macrophage in MS: just a scavenger after all? Pathology and pathogenesis of the acute MS lesion. Mult Scler 12:121–132

Bi W, Zhu L, Wang C, Liang Y, Liu J, Shi Q, Tao E (2011) Rifampicin inhibits microglial inflammation and improves neuron survival against inflammation. Brain Res 1995:12–20

Bjartmar C, Kidd G, Mork S, Rudick R, Trapp BD (2000) Neurological disability correlates with spinal cord axonal loss and reduced N-acetyl aspartate in chronic multiple sclerosis patients. Ann Neurol 48:893–901

Black JA, Liu S, Hains BC, Saab CY, Waxman SG (2006) Long-term protection of central axons with phenytoin in monophasic and chronic-relapsing EAE. Brain 129:3196–3208

Boison D, Scheurer L, Zumsteg V, Rülicke T, Lityniski P, Fowler B, Brander S, Mohler H (2002) Neonatal hepatic steatosis by disruption of the adenosine kinase gene. Proc Natl Acad Sci U S A 99:6985–6999

Bours MJL, Swennen ELR, Di Virgilio F, Cronstein BN, Dagnelie PC (2006) Adenosine 5′-triphosphate and adenosine as endogenous signaling molecules in immunity and inflammation. Pharmacol Ther 112:358–404

Boyle DL, Sajjadi FG, Firestein GS (1996) Inhibition of synoviocyte collagenase gene-expression by adenosine receptor stimulation. Arthritis Rheum 39:923–930

Brown DA, Sawchenko PE (2007) Time course and distribution of inflammatory and neurodegenerative events suggest structural bases for the pathogenesis of experimental autoimmune encephalomyelitis. J Comp Neurol 502:236–260

Butt AM, Hamilton N, Hubbard P, Pugh M, Ibrahim M (2005) Synantocytes: the fifth element. J Anat 207:695–706

Butzkueven H, Zhang JG, Soilu-Hanninen M, Hochrein H, Chionh F, Shipham KA, Emery B, Turnley AM, Petratos S, Ernst M et al (2002) LIF receptor signaling limits immune mediated demyelination by enhancing oligodendrocyte survival. Nat Med 8:613–619

Charcot J (1868) Histologie de la sclérose en plaque. Gazette des Hôpitaux 41:554–566

Chen GQ, Chen YY, Wang XS, Wu SZ, Yang HM, Xu HQ, He JC, Wang XT, Chen JF, Zheng RY (2010) Chronic caffeine treatment attenuates experimental autoimmune encephalomyelitis induced by guinea pig spinal cord homogenates in Wistar rats. Brain Res 1309:116–125

Coman I, Aigrot MS, Seilhean D, Reynolds R, Girault JA, Zalc B, Lubetzki C (2006) Nodal, paranodal and juxtaparanodal axonal proteins during demyelination and remyelination in multiple sclerosis. Brain 129:3186–3195

Craner MJ, Fugger L (2011) Axonal injury in reverse. Nat Med 17:423–425

Cronstein BN (1994) Adenosine, an endogenous anti-inflammatory agent. J Appl Physiol 76:5–13

Cuadros MA, Navascués J (1998) The origin and differentiation of microglial cells during development. Prog Neurobiol 56:173–189

D'Souza SD, Bonetti B, Balasingam V, Cashman NR, Barker PA, Troutt AB, Raine CS, Antel JP (1996) Multiple sclerosis: Fas signaling in oligodendrocyte cell death. J Exp Med 184:2361–2370

Dutta R, Trapp D (2007) Pathogenesis of axonal and neuronal damage in multiple sclerosis. Neurology 68:S22–S31, discussion S43–S54

Edan G, Leray E (2010) A new treatment era in multiple sclerosis: clinical applications of new concepts. J Neurol Sci. doi:10.1016/j.jns.2010.09.018

Eikelenboom P, van Gool WA (2004) Neuroinflammatory perspectives on the two faces of Alzheimer's disease. J Neural Transm 111:281–294

Ferguson B, Matyszak MK, Esiri MM, Perry VH (1997) Axonal damage in acute multiple sclerosis lesions. Brain 120:393–399

Fontoura P, Garren H (2010) Multiple sclerosis therapies: molecular mechanisms and future. In: Martin R, Lutterotti A (eds) Results and problems in cell differentiation, vol 51. Springer, Berlin, pp 259–285

Franciotta D, Salvetti M, Lolli F, Serafini B, Aloisi F (2008) B cells and multiple sclerosis. Lancet Neurol 7:852–858

Frei K, Eugster HP, Bopst M, Constantinescu CS, Lavi E, Fontana A (1997) Tumor necrosis factor alpha and lymphotoxin alpha are not required for induction of acute experimental autoimmune encephalomyelitis. J Exp Med 185:2177–2182

Frischer JM, Bramow S, Dal-Bianco A, Lucchinetti CF, Rauschka H, Schmidbauer M, Laursen H, Sorensen PS, Lassmann H (2009) The relation between inflammation and neurodegeneration in multiple sclerosis brains. Brain 132:1175–1189

Frohman EM, Racke MK, Raine CS (2006) Multiple sclerosis—the plaque and its pathogenesis. N Engl J Med 354:942–955

Gebicke-Haerter PJ, Christoffel F, Timmer J, Northoff H, Berger M, Van Calker D (1996) Both adenosine A1- and A2-receptors are required to stimulate microglial proliferation. Neurochem Int 29:37–42

Geiger JD, Parkinson FE, Kowaluk EA (1997) Regulators of endogenous adenosine levels as therapeutic agents. In: Jacobson JA, Javis MA (eds) Purinergic approaches in experimental therapeutics. Wiley-Liss, New York, pp 55–84

Gilroy DW, Lawrence T, Perretti M, Rossi AG (2004) Inflammatory resolution: new opportunities for drug discovery. Nat Rev Drug Discov 3:401–416

Giovannoni G, Ebers G (2007) Multiple sclerosis: the environment and causation. Curr Opin Neurol 20:261–268

González-Fernández E, Sánchez-Gómez MV, Matute C (2010) Activation of adenosine receptors induces apoptosis in oligodendrocytes. Sixth Cajal Winter Conference, p 5

Hafler DA, Compston A, Sawcer S, Lander ES, Daly MJ, De Jager PL, de Bakker PI, Gabriel SB, Mirel DB, Ivinson AJ, International Multiple Sclerosis Genetics Consortium et al (2007) Risk alleles for multiple sclerosis identified by a genome wide study. N Engl J Med 357: 851–862

Hammarberg C, Schulte G, Fredholm BB (2003) Evidence for functional adenosine A3 receptors in microglia cells. J Neurochem 86:1051–1054

Hammarberg C, Fredholm BB, Schulte G (2004) Adenosine A3 receptor-mediated regulation of p38 and extracellular-regulated kinase ERK1/2 via phosphatidylinositol-3′-kinase. Biochem Pharmacol 67:129–134

Haskó G, Cronstein BN (2004) Adenosine: an endogenous regulator of innate immunity. Trends Immunol 25:33–39

Haskó G, Szabo C, Nemeth ZH, Kvetan V, Pastores SM, Vizi ES (1996) Adenosine receptor agonists differentially regulate IL-10, TNF-alpha, and nitric oxide production in RAW 264.7 macrophages and in endotoxemic mice. J Immunol 157:4634–4640

Haskó G, Pacher P, Vizi ES, Illes P (2005) Adenosine receptor signaling in the brain immune system. Trends Pharmacol Sci 26:511–516

Haskó G, Linden J, Cronstein B, Pacher P (2008) Adenosine receptors: therapeutic aspects for inflammatory and immune diseases. Nat Rev Drug Discov 7:759–770

He F, Sun YE (2007) Glial cells more than support cells? Int J Biochem Cell Biol 39:661–666
Hemmer B, Archelos JJ, Hartung HP (2002) New concepts in the immunopathogenesis of multiple sclerosis. Nat Rev Neurosci 3:291–301
Hisahara S, Araki T, Sugiyama F, Yagami K, Suzuki M, Abe K, Yamamura K, Miyazaki J, Momoi T, Saruta T et al (2000) Targeted expression of baculovirus p35 caspase inhibitor in oligodendrocytes protects mice against autoimmune-mediated demyelination. EMBO J 19: 341–348
Holley JE, Gveric D, Newcombe J, Cuzner ML, Gutowski NJ (2003) Astrocyte characterization in the multiple sclerosis glial scar. Neuropathol Appl Neurobiol 29:434–444
Hövelmeyer N, Hao Z, Kranidioti K, Kassiotis G, Buch T, Frommer F, von Hoch L, Kramer D, Minichiello L, Kollias G et al (2005) Apoptosis of oligodendrocytes via Fas and TNF-R1 is a key event in the induction of experimental autoimmune encephalomyelitis. J Immunol 175:5875–5884
Howell OW, Palser A, Polito A, Melrose S, Zonta B, Scheiermann C, Vora AJ, Brophy PJ, Reynolds R (2006) Disruption of neurofascin localisation reveals early changes preceding demyelination and remyelination in multiple sclerosis. Brain 129:3173–3185
Illes P, Norenberg W, Gebicke-Haerter PJ (1996) Molecular mechanisms of microglial activation. B. Voltage- and purinoceptor-operated channels in microglia. Neurochem Int 29:13–24
Inoue K, Koizumi S, Tsuda M (2007) The role of nucleotides in the neuron–glia communication responsible for the brain functions. J Neurochem 102:1447–1458
Ishibashi T, Dakin K, Stevens B, Lee P, Kozlov S, Stewart C, Fields R (2006) Astrocytes promote myelination in response to electrical impulses. Neuron 49:823–832
James G, Butt AM (2001) P2X and P2Y purinoreceptors mediate ATP-evoked calcium signalling in optic nerve glia in situ. Cell Calcium 30:251–259
Johnston JB, Silva C, Gonzalez G, Holden J, Warren KG, Metz LM, Power C (2001) Diminished adenosine A1 receptor expression on macrophages in brain and blood of patients with multiple sclerosis. Ann Neurol 49:650–658
Keegan BM, Noseworthy JH (2002) Multiple sclerosis. Annu Rev Med 53:285–302
Kerschensteiner M, Meinl E, Hohlfeld R (2009) Neuro-immune crosstalk in CNS diseases. Neuroscience 158:1122–1132
Koning N, Bö L, Hoek RM, Huitinga I (2007) Downregulation of macrophage inhibitory molecules in multiple sclerosis lesions. Ann Neurol 62:504–514
Krause DL, Müller N (2010) Neuroinflammation, microglia and implications for anti-inflammatory treatment in Alzheimer's disease. Int J Alzheimers Dis. doi:10.4061/2010/732806
Kreutzberg GW (1996) Microglia: a sensor for pathological events in the CNS. Trends Neurosci 19:312–318
Kust BM, Biber K, van Calker D, Gebicke-Haerter PJ (1999) Regulation of K^+ channel mRNA expression by stimulation of adenosine A2A receptors in cultured rat microglia. Glia 25:120–130
Lassmann H (2007a) Cortical, subcortical and spinal alterations in neuroimmunological diseases. J Neurol 254:15–17
Lassmann H (2007b) Multiple sclerosis: is there neurodegeneration independent from inflammation? J Neurol Sci 259:3–6
Lassmann H, Bruck W, Lucchinetti CF (2007) The immunopathology of multiple sclerosis: an overview. Brain Pathol 17:210–218
Lee LS, Xu L, Shin Y, Gardner L, Hartzes A, Dohan FC, Raine C, Homayouni R, Levin MC (2011) A potential link between autoimmunity and neurodegeneration in immune-mediated neurological disease. J Neuroimmunol 235:56–69
Lincoln MR, Monpetit A, Cader MZ, Saarela J, Dyment DA, Tiislar M, Fereti V, Tienari PJ, Sadovnick AD, Peltonen L et al (2005) A predominant role for the HLA class II region in the association of the MHC region with multiple sclerosis. Nat Genet 37:1108–1112
Linker RA, Mäuer M, Gaupp S, Martini R, Holtmann B, Giess R, Rieckmann P, Lassmann H, Toyka KV, Sendtner M, Gold R (2002) CNTF is a major protective factor in demyelinating

CNS disease: a neuroprotective cytokine as modulator in neuroinflammation. Nat Med 8: 620–624

Liu B, Hong JS (2003) Role of microglia in inflammation-mediated neurodegenerative diseases: mechanisms and strategies for therapeutic intervention. J Pharmacol Exp Ther 304:1–7

Lublin FD, Reingold SC (1996) Defining the clinical course of multiple sclerosis: results of an international survey. Neurology 46:907–911

Lucchinetti C, Bruck W, Parisi J, Scheithauer B, Rodríguez M, Lassmann H (2000) Heterogeneity of multiple sclerosis lesions: implications for the pathogenesis of demyelination. Ann Neurol 47:707–717

Lukashev D, Ohta A, Apasov S, Chen JF, Sitkovsky M (2004) Cutting edge: physiologic attenuation of proinflammatory transcription by the Gs protein-coupled A2A adenosine receptor in vivo. J Immunol 173:21–24

Lundmark F, Duvefelt K, Iacobaeus E, Kockum I, Wallstrom E, Khademi M, Oturai A, Ryder LP, Saarela J, Harbo HF et al (2007) Variation in interleukin 7 receptor alpha chain (IL7R) influences risk of multiple sclerosis. Nat Genet 39:1108–1113

Marta CB, Montano MB, Taylor CM, Taylor AL, Bansal R, Pfeiffer SE (2005) Signaling cascades activated upon antibody cross-linking of myelin oligodendrocyte glycoprotein: potential implications for multiple sclerosis. J Biol Chem 280:8985–8993

Matute C, Pérez-Cerdá F (2005) Multiple sclerosis: novel perspectives on newly forming lesions. Trends Neurosci 28:173–175

Matute C, Alberdi E, Domercq M, Pérez-Cerdá F, Pérez-Samartín A, Sánchez-Gómez MV (2001) The link between excitotoxic oligodendroglial death and demyelinating diseases. Trends Neurosci 24:224–230

Matute C, Torre I, Pérez-Cerdá F, Pérez-Samartín A, Alberdi E, Etxebarria E, Arranz AM, Rodríguez-Antigüedad A, Sánchez-Gómez MV, Domercq M (2007) P2X7 receptor blockade prevents ATP excitotoxicity in oligodendrocytes and ameliorates experimental autoimmune encephalomyelitis. J Neurosci 7:9525–9533

Mayne M, Shepel PN, Jiang Y, Geiger JD, Power C (1999) Dysregulation of adenosine A1 receptor-mediated cytokine expression in peripheral blood mononuclear cells from multiple sclerosis patients. Ann Neurol 45:633–639

Meda L, Cassatella MA, Szendrei GI, Otvos L Jr, Baron P, Villalba M (1995) Activation of microglial cells by β-amyloid protein and interferon-γ. Nature 373:647–650

Melani A, Pantoni L, Bordoni F, Gianfriddo M, Bianchi L, Vannucchi MG, Bertorelli R, Monopoli A, Pedata F (2003) The selective A2A receptor antagonist SCH 58261 reduces striatal transmitter outflow, turning behaviour and ischemic brain damage induced by permanent focal ischemia in the rat. Brain Res 959:243–250

Melani A, Cipriani S, Vannucchi MG, Nosi D, Donati C, Bruni P, Giovannini MG, Pedata F (2009) Selective adenosine A2a receptor antagonism reduces JNK activation in oligodendrocytes after cerebral ischaemia. Brain 132:1480–1495

Mills JH, Thompson LH, Mueller C, Waickman AT, Jalkanen S, Niemela J, Airas L, Bynoe MS (2008) CD73 is required for efficient entry of lymphocytes into the central nervous system during experimental autoimmune encephalomyelitis. Proc Natl Acad Sci U S A 105:9325–9330

Moon DO, Choi YH, Kim ND, Park YM, Kim GY (2007a) Anti-inflammatory effects of β-lapachone in lipopolysaccharide-stimulated BV2 microglia. Int Immunopharmacol 7:506–514

Moon DO, Kim KC, Jin CY, Han MH, Park C, Lee KJ, Park YM, Choi YH, Kim GY (2007b) Inhibitory effects of eicosapentaenoic acid on lipopolysaccharide-induced activation in BV2 microglia. Int Immunopharmacol 7:222–229

Muzzio L, Martino G, Furlan R (2007) Multifaceted aspects of inflammation in multiple sclerosis: the role of microglia. J Neuroimmunol 191:39–44

Niino M, Fukazawa T, Kikuchi S, Sasaki H (2007) Recent advances in genetic analysis of multiple sclerosis: genetic associations and therapeutic implications. Expert Rev Neurother 7:1175–1188

Nikic I, Merkler D, Sorbara C, Brinkoetter M, Kreutzfeldt M, Bareyre FM, Brück W, Bishop D, Missgeld T, Kerschensteiner M (2011) A reversible form of axon damage in experimental autoimmune encephalomyelitis and multiple sclerosis. Nat Med 17:495–500

Noseworthy J, Lucchinetti C, Rodriguez M, Weinshenker B (2000) Multiple sclerosis. N Engl J Med 343:938–952

Ohta A, Sitkovsky M (2001) Role of G-protein-coupled adenosine receptors in downregulation of inflammation and protection from tissue damage. Nature 414:916–920

Othman T, Yan H, Rivkees SA (2003) Oligodendrocytes express functional A1 adenosine receptors that stimulate cellular migration. Glia 44:166–172

Patrikios P, Stadelmann C, Kutzelnigg A, Rauschka H, Schmidbauer M, Laursen H, Sorensen PS, Brück W, Lucchinetti C, Lassmann H (2006) Remyelination is extensive in a subset of multiple sclerosis patients. Brain 129:3165–3172

Polman CH, Reingold SC, Edan G, Filippi M, Hartung HP, Kappos L, Lublin FD, Metz LM, McFarland HF, O'Connor PW et al (2005) Diagnosis criteria for multiple sclerosis: 2005 revisions to the "McDonald Criteria". Ann Neurol 58:840–846

Prineas JW, McDonald WI, Franklin RJM (2002) Demyelinating diseases. In: Graham DI, Lantos PL (eds) Greenfield's neuropathology, vol 2. Arnold, London, pp 471–550

Probert L, Akassoglou K, Pasparakis M, Kontogeorgos G, Kollias G (1995) Spontaneous inflammatory demyelinating disease in transgenic mice showing central nervous system-specific expression of tumor necrosis factor alpha. Proc Natl Acad Sci U S A 92:11294–11298

Qian L, Flood PM, Hong JS (2010) Neuroinflammation is a key player in Parkinson's disease and a prime target for therapy. J Neural Transm 117:971–979

Raine CS (1994) The Dale E McFarlin Memorial Lecture: the immunology of the multiple sclerosis lesion. Ann Neurol 36:561–572

Ren Z, Wang Y, Tao D, Liebenson D, Liggett T, Goswami R, Clarke R, Stefoski D, Balabanov R (2011) Overexpression of the dominant-negative form of interferon regulatory factor 1 in oligodendrocytes protects against experimental autoimmune encephalomyelitis. J Neurosci 31:8329–8341

Ribeiro JA, Sebastiao AM, de Mendoza A (2002) Adenosine receptors in the nervous system: pathophysiological implications. Prog Neurobiol 68:377–392

Rovaris M, Confavreux C, Furlan R, Kappos L, Comi G, Filippi M (2006) Secondary progressive multiple sclerosis: current knowledge and future challenges. Lancet Neurol 5:343–354

Ruiz-García A, Monsalve E, Novellasdemunt L, Navarro-Sabate A, Manzano A, Rivero S, Castrillo A, Casado M, Laborda J, Bartrons R, Díaz-Guerra MJ (2011) Cooperation of adenosine with macrophage Toll-4 receptor agonists leads to increased glycolytic flux through the enhanced expression of *PFKFB3* gene. J Biol Chem 286:19247–19258

Sanders P, De Keyser J (2007) Janus faces of microglia in multiple sclerosis. Brain Res Rev 54:274–285

Sawcer S, Ban M, Maranian M, Yeo TW, Compston A, Kirby A, Daly MJ, De Jager PL, Walsh E, Lander ES et al (2005) A high-density screen for linkage in multiple sclerosis. Am J Hum Genet 77:454–467

Selmaj K, Raine CS, Cross AH (1991) Anti-tumor necrosis factor therapy abrogates autoimmune demyelination. Ann Neurol 30:694–700

Selmaj K, Papierz W, Glabinski A, Kohno T (1995) Prevention of chronic relapsing experimental autoimmune encephalomyelitis by soluble tumor necrosis factor receptor I. J Neuroimmunol 56:135–141

Sitkovsky M, Lukashev D (2005) Regulation of immune cells by local-tissue oxygen tension: HIF1alpha and adenosine receptors. Nat Rev Immunol 5:712–721

Sitkovsky MV, Ohta A (2005) The 'danger' sensors that STOP the immune response: the A2 adenosine receptors? Trends Immunol 26:299–304

Sitkovsky MV, Lukashev D, Apasov S, Kojima H, Koshiba M, Caldwell C, Ohta A, Thiel M (2004) Physiological control of immune response and inflammatory tissue damage by hypoxia-inducible factors and adenosine A2A receptors. Annu Rev Immunol 22:657–682

Smith T, Groom A, Zhu B, Turski L (2000) Autoimmune encephalomyelitis ameliorated by AMPA antagonists. Nat Med 6:62–66

Spanevello RM, Mazzanti CM, Bagatini M, Correa M, Schmatz R, Stefanello N, Thomé G, Morsch VM, Becker L, Bellè L et al (2010) Activities of the enzymes that hydrolyze adenine nucleotides in platelets from multiple sclerosis patients. J Neurobiol 257:24–30

Sperlágh B, Illes P (2007) Purinergic modulation of microglial cell activation. Purinergic Signal 3:117–127

Stadelmann C, Albert M, Wegner C, Bruck W (2008) Cortical pathology in multiple sclerosis. Curr Opin Neurol 21:229–234

Stevens B, Porta S, Haak LL, Gallo V, Fields RD (2002) Adenosine: a neuron-glial transmitter promoting myelination in the CNS in response to action potentials. Neuron 36:855–868

Thompson AJ, Polman CH, Miller DH, McDonald WI, Brochet B, Filipi M, Montalban X, De Sá J (1997) Primary progressive multiple sclerosis. Brain 120:1085–1096

Trapp BD, Nave KA (2008) Multiple sclerosis: an immune or neurodegenerative disorder? Annu Rev Neurosci 31:247–269

Trapp B, Peterson J, Ransohoff R, Rudick R, Mork S, Bo L (1998) Axonal transection in the lesions of multiple sclerosis. N Engl J Med 338:278–285

Tsutsui S, Schnermann J, Noorbakhsh F, Henry S, Yong VW, Winston BW, Warren K, Power C (2004) A1 adenosine receptor upregulation and activation attenuates neuroinflammation and demyelination in a model of multiple sclerosis. J Neurosci 24:1521–1529

Vivekanandhan S, Soundararajan CC, Tripathi M, Maheshwari MC (2005) Adenosine deaminase and 5′nucleotidase activities in peripheral blood T cells of multiple sclerosis patients. Neurochem Res 30:453–456

Williams A, Piaton G, Lubetzki C (2007) Astrocytes-friends or foes in multiple sclerosis? Glia 55:1300–1312

Zielasek J, Hartung HP (1996) Molecular mechanisms of microglial activation. Adv Neuroimmunol 6:191–202

Zipp F, Atkas O (2006) The brain as a target of inflammation: common pathways link inflammatory and neurodegenerative diseases. Trends Neurosci 29:518–527

Chapter 22
Adenosinergic Perspectives on Schizophrenia: Opportunity for an Integrative Synthesis

Benjamin K. Yee, Philipp Singer, and Detlev Boison

Abstract Adenosine is a homeostatic modulator capable of influencing brain-wide network functioning through receptor-dependent interactions, and receptor-independent biochemical, bioenergetic and epigenetic mechanisms. In the brain, it directly modulates and interferes with dopaminergic and glutamatergic neuronal activities, the dysfunctions of which have been consistently implicated in the pathophysiology of schizophrenia. In addition, adenosinergic mechanisms may also be relevant to the aetiology of schizophrenia due to its importance in early brain development and its interaction with the brain immune response. This chapter aims to integrate these different perspectives on the neurobiology of schizophrenia in relation to the unique functional properties of adenosine within the central nervous system. The rationale of an adenosinergic hypothesis of schizophrenia is outlined with evidence supporting the suggestion that adenosinergic dysfunction might contribute to the emergence of multiple behavioural and cognitive dysfunctions characteristic of schizophrenia. Possible novel therapeutic strategies are explored, especially against negative and cognitive schizophrenia symptoms where current pharmacotherapy remains highly unsatisfactory. Multiple pharmacological targets within the adenosine system are identified, including the high-affinity adenosine receptors (A_1 receptor and A_{2A} receptor) and the major regulatory enzyme of adenosine metabolism in the brain, adenosine kinase. Preliminary clinical data are encouraging in highlighting the therapeutic potential of these new targets, but preclinical data have also yielded a more complex picture underlining the necessity to consider brain region and receptor subtype specificity in order to maximize benefits and minimize possible side-effects of any adenosine-based approach to treat schizophrenia.

B.K. Yee (✉) • P. Singer • D. Boison
Legacy Research Institute, 1225 NE 2nd Ave, Portland, OR 97232, USA
e-mail: byee@downneurobiology.orf

S. Masino and D. Boison (eds.), *Adenosine: A Key Link between Metabolism and Brain Activity*, DOI 10.1007/978-1-4614-3903-5_22,

22.1 Introduction

Schizophrenia is a debilitating mental disorder with a global prevalence of 1 % with considerable human and social costs. However, existing treatments are only effective in controlling its most florid psychotic symptoms characterized by functional excesses such as delusions, hallucinations, and disorganized thinking. In contrast to such positive symptoms, the negative and cognitive symptoms of the disease comprising chronic deterioration of mental functions such as memory, attention and executive functioning, as well as avolition and mood dysfunction, are often resistant to current pharmacotherapy (Coyle et al. 2010). Until this medical need is met, successful long-term rehabilitation and social re-integration of patients would remain difficult. The current stagnation in novel treatment development is a reflection of our current lack of a cohesive biological theory of the disease. Indeed, a comprehensive explanation of the disease from aetiology to symptom genesis is still lacking in spite of continual efforts in identifying a multitude of genetic and environmental risk factors.

Current pharmacotherapy for schizophrenia is mainly guided by the "dopamine" and "glutamate" hypotheses, which emphasize the contribution of excessive dopaminergic activities and deficient glutamatergic activity via *N*-methyl-D-aspartate receptor (NMDAR), respectively, to the generation of disease symptoms (Gordon 2010; Heinz and Schlagenhauf 2010; Inta et al. 2010; Javitt 2008). Thus, symptom-oriented drug development has been largely guided by these hypotheses, although the contribution of GABAergic mechanism to symptom production has also received increasing attention (Gonzalez-Burgos et al. 2010; Lewis et al. 1999, 2005).

While positive symptoms are generally responsive to drugs that effectively block neurotransmission via dopamine D_2 receptor (D_2R), these drugs have limited efficacy against negative and cognitive symptoms and they are associated with several undesirable side effects that often limit compliance. On the other hand, it has been hypothesized that effective treatment of negative and cognitive symptoms requires the augmentation of glutamatergic function, especially via the ionotropic glutamate receptor—NMDARs. Indeed, NMDAR blockers can produce psychotic-like symptoms, disrupt some forms of neural plasticity and impair learning and memory (Citri and Malenka 2008; Malenka and Nicoll 1993). However, direct NMDAR agonism is associated with excitotoxicity, and one favoured alternative approach is to target allosteric and other neuromodulators of glutamatergic transmission in order to achieve activity-dependent enhancement of glutamate receptor or NMDAR function. A new glycine transporter-1 inhibitor from Roche (RG1678) has recently been shown to add significant benefits when combined with conventional antipsychotic drugs in a phase II clinical trial (Pinard et al. 2010). The drug works by indirectly increasing the occupancy of the glycine-B site in NMDARs and thereby facilitates receptor channel opening upon activation by glutamate. This strategy has received considerable support from preclinical genetic and pharmacological animal models (Black et al. 2009; Yang and Svensson 2008; Yee et al. 2006). Hence, it seems that a functionally meaningful enhancement of the diffused and widespread cortical-subcortical glutamatergic network in conjunction with blockade of the more anatomically defined mesocorticolimbic dopaminergic system could yield highly desirable clinical outcome. Drug targets that can achieve normalization of both dopaminergic and glutamatergic transmissions in such manners therefore would be most promising and innovative.

Adenosine—a neuromodulator, which regulates both dopaminergic and glutamatergic signalling in the brain, represents one such candidate target. It is functionally positioned to incorporate elements of the dopamine and glutamate hypotheses. An adenosine hypothesis of schizophrenia posits that an imbalance in the ambient tone of adenosine may critically determine the susceptibility to schizophrenia and its restoration may be therapeutic (Lara 2002; Lara et al. 2006; Lara and Souza 2000). This chapter attempts to foster a collective perspective on schizophrenia based on the multiple functions of adenosine in the brain by offering a framework to integrate existing neurotransmitter hypotheses of schizophrenia and specific neurodevelopmental perspectives of the disease.

22.2 The Adenosine Cycle and Adenosine Receptors

The purine ribonucleoside adenosine directly affects a variety of signalling pathways including synaptic neurotransmission and assumes important regulatory functions in the CNS (Boison 2008; Boison et al. 2010; Fredholm et al. 2005; Sebastião and Ribeiro 2009; Stone et al. 2009). Unlike classical neurotransmitters, however, adenosine is neither stored in synaptic vesicles nor released by exocytosis, and its does not exclusively act on synapses (Boison et al. 2010; Fredholm et al. 2005). Its release and uptake is mediated by bidirectional nucleoside transporters with the direction of transport being solely dependent on the concentration gradient between the cytoplasm and the extracellular space (Boison et al. 2010; Gu et al. 1995). Adenosine is therefore considered a neuromodulator. It influences neural activity (1) presynaptically by controlling neurotransmitter release, (2) postsynaptically by hyper- or de-polarizing neurons, and (3) non-synaptically mainly via regulatory effects on glial cells (Boison et al. 2010).

As summarized in Fig. 22.1, extracellular levels of adenosine are largely controlled by an astrocyte-based adenosine cycle (Boison 2008; Boison et al. 2010; Halassa et al. 2007a, b; Haydon and Carmignoto 2006; Martin et al. 2007). The major source for synaptic adenosine is the release of its precursor 5′-adenosinetriphosphate (ATP) from astrocytes, which can occur via vesicular release (Pascual et al. 2005) or secretion through hemichannels (Kang et al. 2008; Kawamura et al. 2010). The ATP released into extracellular space is rapidly degraded to adenosine by ectonucleotidases (Pascual et al. 2005; Zimmermann 2000). In addition, adenosine can be directly released from astrocytes through nucleoside transporters when intracellular adenosine level is elevated (Geiger and Fyda 1991). In contrast to conventional neurotransmitters, the reuptake of adenosine does not depend on active transport. Instead, astrocytes contain two types of membrane-bound equilibrative nucleoside transporters for rapid exchange of adenosine between extra- and intracellular space (Baldwin

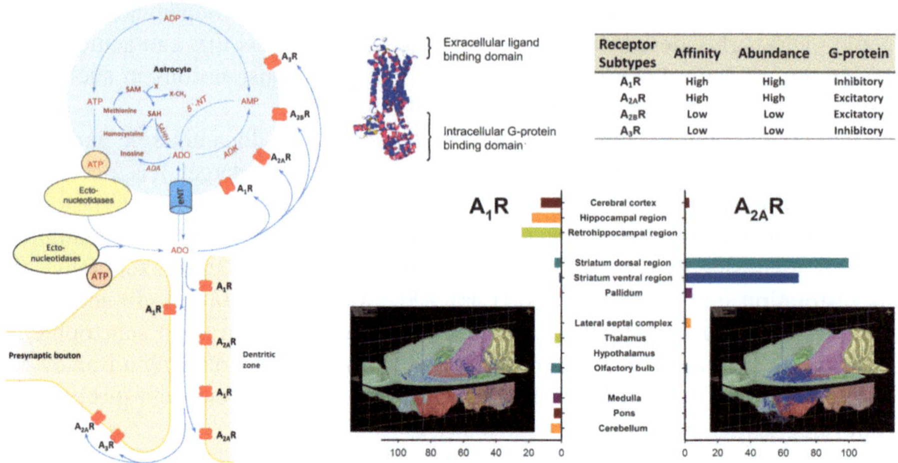

Receptor Subtypes	Affinity	Abundance	G-protein
A_1R	High	High	Inhibitory
$A_{2A}R$	High	High	Excitatory
$A_{2B}R$	Low	Low	Excitatory
A_3R	Low	Low	Inhibitory

Fig. 22.1 The adenosine metabolic cycle and adenosine receptor subtypes. Major metabolic pathways of adenosine take place in astrocytes: ecto-nucleotidase, adenosine kinase (ADK), *S*-adenosylhomocysteine hydrolase (SAHH), *S*-adenosylmethionine (SAM), equilibrative nucleoside transporter (eNT), adenosine deaminase (ADA), 5′-nucleotidase (5′-NT). Release into extracellular space is controlled by passive transporters. The four major receptors subtypes are A_1, A_{2A}, A_{2B} and A_3. They are coupled with G-protein mechanisms. Their relative affinity for adenosine and abundance in the brain is tabulated. In particular, the gene expression (as *blue dots*) of the high-affinity receptors, A_1Rs and A_{2A}Rs (genes symbols = *Adora1* and *Adora2a*), in the mouse (C57BL/6J strain) brain are plotted in 3D graphics illustrating their complementary regionally distribution. While A1Rs are preferentially expressed in the cerebral cortex and hippocampal/retrohippocampal regions, A_{2A}Rs are most highly enriched in the striatum. *Histograms* showing the gene expression density data are normalized to a range of [0,100]. Original data and images obtained from The Allen Institute for Brain Science (http://mouse.brain-map.org/welcome.do) for non-commercial use: *Adora1*: http://mouse.brain-map.org/brain/gene/73496905/ExpressionGraph.html; *Adora2*a: http://mouse.brain-map.org/brain/gene/69855739/ExpressionGraph.html

et al. 2004). Reuptake of adenosine into the astrocytes is driven by its efficient and rapid removal by intracellular adenosine kinase (ADK), a ribokinase, which phosphorylates adenosine into 5′-adenosine monophosphate (AMP) (Boison 2006; Etherington et al. 2009; Park and Gupta 2008; Pignataro et al. 2007; Studer et al. 2006). Lower levels of intracellular (astrocytic) adenosine would drive adenosine influx into astrocytes via bidirectional nucleoside transporters (Boison et al. 2010), thereby reducing extracellular adenosine. Hence, ADK expressed in astrocytes functions as a key regulator of extracellular adenosine, and manipulation targeting it bears predictable changes in ambient adenosinergic tone (Boison et al. 2010).

The central effects of adenosine are mediated via two high-affinity receptors—classified into A_1 and A_{2A}, and two low-affinity receptors—classified into A_{2B} and A_3 (Fig. 22.1). They are also dichotomized according to whether they are couple to inhibitory (A_1Rs and A_3Rs) or excitatory (A_{2A}Rs and A_{2B}Rs) metabotropic G-proteins (Fredholm et al. 2001, 2005; Linden 2001). The A_1R subtype is the most abundant adenosine receptor subtype; they are densely expressed throughout the CNS with the highest abundance in the hippocampal region, the adjoining retrohippocampal area and across the cerebral cortex, with moderate levels of expression in the midbrain and dorsal horn of the spinal cord. A_{2A}Rs are very strongly expressed in both dorsal and ventral striatal neurons, with much lower levels occurring in neurons outside of the striatum and in glial cells (Hettinger et al. 2001; Ribeiro et al. 2003; Svenningsson et al. 1999). The relative distribution between A_1Rs and A_{2A}Rs in the brain to a large extent complements each other.

A_1R activation inhibits the release of neurotransmitters such as dopamine and glutamate and decreases neural excitability by post-synaptic hyperpolarization (Dunwiddie and Masino 2001; Fredholm et al. 2005), whilst activation of A_{2A}R opposes that of A_1R and tends to promote neurotransmitter release (Ciruela et al. 2006a; Fredholm et al. 2005; Stone et al. 2009). More direct receptor–receptor interaction can be achieved via the formation of A_1R–A_{2A}R heterodimers (Ciruela et al. 2006a, b). Similarly, heterodimers are also formed by A_{2A}R and D_2R (Férre 1997; Fredholm and Svenningsson 2003; Fuxe et al. 2003). A_{2A}R activation is also linked to transactivation of BDNF Trk-B receptor (Diogenes et al. 2004, 2007) and therefore the modulation of BDNF-dependent neural plasticity, including long-term changes in synaptic strength (Sebastião et al. 2011).

Hence, adenosine's neural modulatory functions through the two high-affinity receptors (A_1R and A_{2A}R) alone are already extensive, complex and intricate. By activating receptors with opposing effects on neuronal excitability, adenosine is uniquely positioned to stabilize, fine-tune and integrate competing excitatory and inhibitory processes in the CNS. By contrast, expression of the lower-affinity A_{2B}Rs and A_3Rs is relatively weak and diffused (Boison 2007a; Fredholm et al. 2005; Jacobson and Gao 2006), and their influence on normal brain function may therefore be less impactful and critical. Their relevance to health and disease remains largely unknown since their possible interactions with other classical neurotransmitter systems are much less well characterized. However, their importance should not be ignored for the development of adenosine-based pharmacotherapy in the long run. In addition, adenosine is a bioenergetic regulator that can directly affect the equilibrium of enzymatic pathways including transmethylation reactions, e.g. DNA methylation (Boison et al. 2002) as well as basic biochemistry via adenosine receptor independent mechanisms (Cunha 2008; Férre et al. 2007; Fredholm et al. 2005; Sebastião and Ribeiro 2009). In summary, the diverse synergistic and antagonistic interactive functions of adenosine are in line with its role as a prime homeostatic regulator (Newby et al. 1985). The present chapter essentially focuses on the interplay of adenosine receptors with dopamine receptors and the NMDARs because of their relevance to schizophrenia.

22.3 The Clinical Perspective

A purinergic hypothesis of schizophrenia was first proposed by Lara and colleagues (Lara et al. 2006; Lara and Souza 2000). They speculated that a dysfunction in purinergic system would give rise to adenosinergic deficiency as a possible common explanation for the functional imbalance between dopaminergic and glutamatergic activities as well as the neuroimmunological dysregulation characteristic of schizophrenia The hypothesis gave rise directly to the initial attempt to treat schizophrenia patients with the purine derivative, allopurinol (Lara et al. 2000, 2001, 2003), which is efficacious in elevating brain adenosine in preclinical models (Schmidt et al. 2009). This is one of the few examples in which a hypothesis driven medication has been developed or discovered in psychiatry.

The xanthine oxidase inhibitor allopurinol has been studied as add-on therapy for schizophrenia (Akhondzadeh et al. 2005; Brunstein et al. 2004, 2005; Lara et al. 2001). By inhibiting a major degradation pathway of purines, allopurinol is believed to exert some of its beneficial effects by raising the endogenous pool of purines, including adenosine. Adjunctive allopurinol was moderately effective on positive and general symptoms, yielding 20 % improvement in a double-blind, placebo-controlled, crossover clinical trial of add-on allopurinol (300 mg b.i.d.) for poorly responsive schizophrenia or schizoaffective disorder (Brunstein et al. 2005). However, this study was only based on 22 patients who had completed the study and larger samples would be necessary to substantiate its findings, especially when a recent case report has described the relapse of symptoms in a schizophrenia patient with gout having been treated with allopurinol (Gomberg 2007). Nevertheless, in support of the double-blind allopurinol trial (Brunstein et al. 2005), the adenosine transport inhibitor dipyramidole (thereby raising adenosine) was also found to be beneficial in another study (Akhondzadeh et al. 2000).

Several lines of evidence have lent further credence to an adenosine hypothesis. First, a post mortem study has revealed an increase in the density of A_{2A}Rs within the striatum of schizophrenia patients (Kurumaji and Toru 1998), which may have developed as a compensatory response to reduced adenosinergic activity. This in turn could result in a functional hyperdopaminergia state as predicted by Lara and colleagues (Lara et al. 2006; Lara and Souza 2000). Second, the olfactory G-protein (G_{olf}) that is coupled to A_{2A}Rs (Kull et al. 2000) is a candidate gene for schizophrenia (Schwab et al. 1998a, b) although linkage studies suggest that the *A2AR* gene itself is not (Hong et al. 2005; Deckert et al. 1996, 1997). Third, a study of a cohort of Japanese schizophrenia patients has highlighted the possible relevance of A_1R polymorphisms in the pathophysiology of schizophrenia (Gotoh et al. 2009). Fourth, a variant of adenosine deaminase characterized by lower enzymatic activity has been shown to be rarer amongst schizophrenia patients, suggesting that adenosine clearance might be faster, leading to lower ambient adenosine levels than the general population (Dutra et al. 2010). Fifth, acute blockade of A_1R/A_{2A}R by oral theophylline can induce impairment of P50 suppression in healthy human subjects, suggesting that adenosine hypofunction may be related to sensory gating deficit in schizophrenia (Ghisolfi et al. 2002).

In summary, although some of the supportive findings still remain controversial, an adenosine hypothesis of schizophrenia is not without any merits, especially given that the disease is unlikely to be fully accounted by a singular dysfunction in a neurotransmitter or neuromodulator. Additional clinical studies with larger patient samples across multiple centres will help to clarify the limits and scope that any possible adenosinergic account of the disease would yield viable therapeutic options. Next, the rationale for linking adenosine to the dopamine and glutamate hypothesis of schizophrenia is outlined with reference to existing preclinical data.

22.4 The Dopaminergic Perspective

The dopamine hypothesis of schizophrenia is essentially based on the complementary actions of dopamine D_2R antagonists and agonists to suppress and promote psychotic symptoms, respectively. Whilst the efficacy of many antipsychotics correlates with their ability to block dopamine D_2Rs (Seeman 2006; Seeman et al. 2006), dopamine-releasing agents, such as amphetamine, can exacerbate psychotic symptoms in schizophrenia (Snyder 1976). The prolonged abuse of amphetamine may even induce psychotic-like symptoms in healthy subjects (Laruelle and Abi-Dargham 1999). Functional imaging studies confirm the presence of elevated basal occupation of D_2Rs by dopamine (Abi-Dargham et al. 2000), increased dopamine turnover (Lindstrom et al. 1999), and enhanced amphetamine-induced dopamine release (Laruelle and Abi-Dargham 1999) in schizophrenia patients. Notably, these clinical findings are compatible with a deficiency in adenosine.

Extensive behavioural and neurochemical studies have provided strong support for an antagonistic interaction between adenosine and dopamine receptors especially within the basal ganglia. $A_{2A}Rs$ are highly expressed in GABAergic striatopallidal neurons where they co-localize with dopamine D_2Rs. The reported increase in basal D_2R occupancy in schizophrenia patients (Abi-Dargham et al. 2000) could undermine the regulation of $A_{2A}R$ on D_2R and thereby increase D_2R's affinity for dopamine (Férre et al. 1991a, b, c). On the other hand, reduced activation of adenosine A_1Rs can weaken tonic inhibition (i.e. disinhibition) on dopamine release (Golembiowska and Zylewska 1998). Such disinhibition is expected to potentiate amphetamine-induced locomotion (e.g. Popoli et al. 1994b) and dopamine release in the nucleus accumbens as suggested by the effects of A_1R antagonists (Solinas et al. 2002). Thus, there exist possible A_1R–D_2R and $A_{2A}R$–D_2R mechanisms to predict a typical neuroleptic-like profile against hyperdopaminergia by adenosine receptor agonists.

Conversely, blockade of both A_1R and $A_{2A}R$ by non-selective adenosine receptor antagonist, caffeine, can induce hyperactivity resembling that seen following acute psychostimulant exposure (Chen et al. 2001; Kuzmin et al. 2000). Here, dopamine D_1R mechanism might allow further opportunity for interaction with A_1Rs, which are found in close proximity to D_1Rs in striatonigral GABAergic neurons (Férre et al. 1994, 2005; Fink et al. 1992; Schiffmann et al. 1991; Schiffmann and Vanderhaeghen 1993; Svenningsson et al. 1999). The formation of A_1R–D_1R as well as $A_{2A}R$–D_2R heteromers also provide additional mechanisms for adenosinergic regulation of overall dopaminergic basal ganglia functions, linked not only to motor functions but also motivation, reward and executive functions. At the circuitry level, the activation of $A_{2A}Rs$ located outside the striatum may augment glutamatergic signals to the striatum, and therefore counter the postsynaptic inhibition of striatopallidal neurons resulting from local $A_{2A}R$ activation (Shen and Chen 2009; Yu et al. 2009a, b). Hence, the global impact via adenosine receptors can indirectly influence basal ganglia function via dopamine receptor-independent mechanisms (see Schiffmann et al. 2007).

In vivo models are indispensable in gauging the effect of systemic adenosinergic drugs against hyperdopaminergia. The picture emerged clearly demarcated adenosine receptors agonists and antagonists (against either A_1R or $A_{2A}R$) with opposing effects (see review by Boison et al. 2012). As summarized in Table 22.1A, the evidence overwhelmingly suggests that the hyperlocomotor effect of indirect (amphetamine or cocaine) or direct dopamine receptor agonists (SKF38393 or bromocriptine) is attenuated by adenosine receptors agonists but exacerbated by adenosine receptors antagonists.

Another important translational model to assess schizophrenia-related endophenotypes is prepulse inhibition (PPI) of the acoustic startle reflex. PPI refers to the gating of sensory inputs whereby a weak auditory stimulus substantially weakens the response to a succeeding startle-eliciting acoustic pulse (Graham 1975; Hoffman and Searle 1968). Deficiency in PPI is believed to be linked to the perceptual dysfunction in schizophrenia leading to sensory flooding and susceptibility to intruding and distracting stimuli (Geyer et al. 2001; Swerdlow et al. 2008). Dopamine agonists readily induce PPI deficits in animals (e.g. Yee et al. 2004). The A_1R agonist, CPA (Koch and Hauber 1998), and the $A_{2A}R$ agonist, CGS21680 (Hauber and Koch 1997), are both effective in attenuating the impairment in PPI produced by systemic amphetamine or apomorphine. Hauber and Koch (1997) further localized the action of CGS21680 specifically to the nucleus accumbens (ventral striatum). Conversely, the mixed $A_1R/A_{2A}R$ antagonist, theophylline, exacerbated the impact of a non-disruptive dose of apomorphine on PPI (Koch and Hauber 1998). The opposing effects between agonists and antagonists therefore match those described above regarding hyperlocomotion induced by dopaminergic stimulation. Notably, adenosinergic drugs do not seem to have any effect on PPI function when administered alone (see review by Boison et al. 2012), which is ideal from a therapeutic perspective, supporting the interpretation that adenosine receptor agonists effectively correct the induced hyperdopaminergic state. Adenosine receptor antagonists are detrimental only when the dopamine system is already functionally overactive.

22.5 The Glutamatergic Perspective

The glutamate hypothesis of schizophrenia posits that deficient NMDAR function may contribute to cognitive and negative symptoms of schizophrenia (Coyle and Tsai 2004; Farber 2003), and thus, typical neuroleptic drugs providing solely targeted blockade of dopamine receptors are ineffective against such symptoms. Numerous empirical studies have shown that blockade of NMDARs can impair the induction of long-term potentiation (LTP) and produce severe learning deficits in

Table 22.1 Summary of adenosinergic manipulation against psychostimulant drugs reaction on locomotor activity

A. Summary of data in support of the suggestion that adenosinergic receptor agonists reduces the motor stimulating effect of dopamine receptor indirect and direct agonists

Hyperlocomotor induced by indirect and direct dopamine agonists		Attenuation by adenosine receptor agonists			Augmentation by adenosine receptor antagonists		
		A_1R	$A_{2A}R$	Mixed $A_1R/A_{2A}R$	A_1R	$A_{2A}R$	Mixed $A_1R/A_{2A}R$
Dopamine-releasers	Amphetamine	2-CLA (Heffner et al. 1989)	APEC (Turgeon et al. 1996)	NECA (Heffner et al. 1989; Shi et al. 1994)		DMPX (Poleszak and Malec 2002)	Caffeine (Poleszak and Malec 2002)
		CHA (Heffner et al. 1989; Turgeon et al. 1996)	CGS21680 (Poleszak and Malec 2002; Popoli et al. 1994a, b; Rimondini et al. 1997; Sill et al. 2001)				
		CPA (Heffner et al. 1989; Poleszak and Malec 2002)	CV-1808 (Heffner et al. 1989)				
		PIA (Heffner et al. 1989)					
	Cocaine	CPA (Poleszak and Malec 2002)	CGS2160 (Poleszak and Malec 2002)			DMPX (Poleszak and Malec 2002)	Caffeine (Poleszak and Malec 2002)
D_1R agonist	SKF38393	CPA (Férre et al. 1994)			8-CPT2 (Popoli et al. 1996)		
D_2R agonist	Bromocriptine	L-PIA (Férre et al. 1991b)		NECA (Férre et al. 1991b)			Caffeine (Férre et al. 1991a) Theophylline (Férre et al. 1991a)

(continued)

Table 22.1 (continued)

B. Summary of data in support of the suggestion that adenosinergic receptor agonists reduces the motor stimulating effect of NMDAR blockers

Hyperlocomotor induced by NMDA receptor antagonists	Attenuation by adenosine receptor agonists			Augmentation by adenosine receptor antagonists		
	A_1R	$A_{2A}R$	Mixed $A_1R/A_{2A}R$	A_1R	$A_{2A}R$	Mixed $A_1R/A_{2A}R$
MK-801(Dizocilpine)	CPA (Malec and Poleszak 2006)	CGS21680 (Malec and Poleszak 2006) DPMA (Fraser et al. 1997)	NECA (Malec and Poleszak 2006)		DPMX (Malec and Poleszak 2006)	Caffeine (Kuribara et al. 1992) Theophylline (Malec and Poleszak 2006)
PCP (Phencyclidine)	CPA (Gotoh et al. 2002)	CGS21680 (Rimondini et al. 1997)				
Ketamine					DPMX (Mandryk et al. 2005)	Caffeine (Mandryk et al. 2005)

C. Summary the psychostimulant-induced phenotypes by the major genetic engineered mouse models

Genetic manipulations	On the reaction to Dopamine receptor agonists		On the reaction to NMDA receptor blockade
	Amphetamine/cocaine	SKF81297/Quinpirole	MK-801 (Dizocilpine)/PCP (Phencyclidine)
Adktm1$^{-/-}$Tg(UbiAdk) mice	Attenuated (Yee et al. 2007)		Enhanced (Yee et al. 2007)
Constitutive $A_{2A}{}^{-/-}$	Attenuated (Chen et al. 2000)	Unaffected (Chen et al. 2000)	
CaMKIIα-Cre$^{(+)}$$A_{2A}R^{flox+/+}$ mice	Attenuated (Shen et al. 2008)		Attenuated (Shen et al. 2008)
Dlx5/6-Cre$^{(+)}$$A_{2A}R^{flox+/+}$ mice	Enhanced (Shen et al. 2008)		Enhanced (Shen et al. 2008)

2-CLA (2-chloroadenosine), *8-CPT2* (8-cyclopentyltheophylline), *APEC* (2-[(2-aminoethylamino)carbonylethylphenylethylamino]-50-*N*-ethylcarboxamidoadenosine), *caffeine* (1,3,7-trimethylxanthine), *CGS21680* (2-(4-(2-carboxyethyl)phenylethylamino)adenosine-50-*N*-ethylcarboxamideadenosine), *CHA* (*N*6-cyclohexyl-adenosine), *CPA* (*N*6-cyclopenty-ladenosine), *CV-1808* (2-(phenylamino)adenosine), *DP MA* (*N*6-[2-(3,5-dimethoxyphenyl)-2-(2-methylphenyl)-ethyl]adenosine), *DMPX* (3,7-dimethyl-1-propargylxanthine), *L-PIA* (L-*N*6-phenylisopropyl-adenosine), *NECA* (*N*6-phenylisopropyl-adenosine), *theophylline* (1,3-dimethyl-7*H*-purine-2,6-dione)

Reference: Sills et al. (2001)

animals (Davis et al. 1992; Morris et al. 1986; Morris 1989). Although not all forms of learning and memory are necessarily dependent on NMDAR-mediated plasticity, augmentation of NMDAR function can yield pro-cognitive effects, especially when this is achieved in an impulse-dependent manner, e.g. by increasing the availability of the obligatory co-agonists of NMDAR, such as D-serine and glycine—a strategy that has yielded some promising outcomes regarding cognition and negative symptoms in schizophrenia (Coyle and Tsai 2004; Javitt 2008). It is important to recognize that NMDAR blockade not only impairs learning and memory, but also gives rise to behavioural dysfunction such as impulsivity (Tonkiss et al. 1988) and psychotic-like behaviour (Farber 2003). For instance, phencyclidine (PCP) and ketamine (two non-competitive NMDAR blockers) are well-known psychomimetics when administered to humans (Farber 2003). Hence, augmentation of NMDAR function, such as that achieved by disruption of glycine transporter-1 (thereby elevating synaptic glycine levels) is associated with promnesic as well as antipsychotic effects in animal models (Möhler et al. 2008; Singer et al. 2007a, b; Yee et al. 2006). Similar opportunities may exist by targeting adenosine given its role as an endogenous modulator of glutamatergic activity, providing bidirectional control over neuronal excitation including that via NMDARs.

22.5.1 Striatum

In the striatum, adenosine can interact synergistically with NMDARs via the abundant $A_{2A}R$ located on striatal medium spiny neurons (Wardas et al. 2003), but activation of $A_{2A}Rs$ can also antagonistically inhibit NMDAR-evoked currents in the same neurons (Wirkner et al. 2000, 2004). Further synergism is also suggested by the observation that activation of either NMDAR or $A_{2A}R$ stimulates adenylyl cyclase, which in turn elevates cAMP and therefore activity of downstream second messenger system (Nash and Brotchie 2000). As summarized in Table 22.1B, A_1R and $A_{2A}R$ agonists can attenuate the motor stimulant effect of NMDAR blockers just as they do against that of dopamine receptor agonists (Table 22.1A). It is conceivable that the synergistic interactions between $A_{2A}R$ and NMDAR predominate in the striatum and are responsible at least partly for such antipsychotic-like outcomes, although antagonistic interaction between striatal $A_{2A}R$ and D_2R described before may also contribute (e.g. Kafka and Corbett 1996; Popoli et al. 1998) given that the motor effects of NMDAR blockade are mediated partly (but not exclusively) via an increase in striatal dopamine activity (Breier et al. 1998; Smith et al. 1998; Steinpreis 1996; Steinpreis and Salamone 1993).

The parallelism against the two classes of psychostimulant drugs, however, breaks down regarding the effects of adenosine receptor antagonists: Only antagonism against $A_{2A}R$ is capable in exacerbating the motor stimulant effect of NMDAR blockade as it does against dopamine stimulation (see Malec and Poleszak 2006). The two A_1R antagonists, CPT and DPCPX, were ineffective against the hyperlocomotor response induced by PCP and ketamine, respectively (Malec and Poleszak

2006; Mandryk et al. 2005). Yet, the A_1R agonist, CPA, was effective against PCP as well as MK-801 (see Table 22.1B; Gotoh et al. 2002; Malec and Poleszak 2006). The asymmetric effects via A_1R mechanisms suggest a separate mechanism independent of NMDARs interaction might be largely responsible—namely, via tonic inhibition of dopamine activity by A_1R activation (Golembiowska and Zylewska 1998; Popoli et al. 1994a, b; Solinas et al. 2002). This may further explain why A_1R antagonism is less effective in exacerbating the impact of NMDAR blockade in this regard.

The functional integrity of striatum is central to normal PPI function (see above), and $A_{2A}R$ agonism was also effective against PPI deficiency induced by NMDAR blockade as exemplified by the action of CGS21680 (Sills et al. 2001; Wardas et al. 2003), which was without any effect on PPI when administered alone (Sills et al. 2001; Hauber and Koch 1997). Unfortunately, no studies have examined whether A_1R agonist may share the same profile, and whether adenosine receptor antagonists might exacerbate the PPI-disruptive effect of NMDAR blockade.

22.5.2 Hippocampus

In the hippocampus, activation of A_1Rs inhibits glutamate release and the post-synaptic action of excitatory glutamatergic transmission (Dunwiddie and Masino 2001). Indeed, these form the basis for the neuroprotective effect of adenosine during hypoxia or against excitotoxicity due to prolonged NMDAR stimulation (Boeck et al. 2005; Dunwiddie and Masino 2001). However, the resulting tonic inhibition and suppression of NMDAR-mediated currents can inhibit NMDAR-dependent LTP and long-term depression (LTD) (de Mendonça and Ribeiro 1997, 2000; Rebola et al. 2008). This inhibition can be relieved via a positive feedback mechanism whereby NMDAR stimulation inhibits the suppressive action of A_1Rs located at the presynaptic terminals (Bartrup and Stone 1990; Bartrup et al. 1991; Nikbakht and Stone 2001). Thus, adenosine critically fine-tunes hippocampal synaptic plasticity—it effectively serves as a gate keeper that defends against some forms of (spurious) NMDAR stimulation but allows or even amplifies other (meaningful) signals. It may represent a molecular switch closely linked to the concept that effective learning must be coupled with appropriate task-dependent selectivity in order to achieve adaptive and flexible behavioural change that is the hallmark of learning (e.g. Rebola et al. 2008; Stone et al. 2009; Yu et al. 2009a). Hence, while memory trace that is too weak or labile does not support robust recall, memory trace that is too strong or resistant could also be interfering and detrimental, yielding inflexible and inappropriate behaviour as is often associated with schizophrenia negative and cognitive symptoms. Accordingly, A_1R antagonism in the hippocampus may facilitate signal processing by lowering the threshold of signal strength to activate NMDARs, and enhance related hippocampus-dependent cognitive functioning (Thummler and Dunwiddie 2000). Conversely, A_1R agonism is expected to yield the opposite effect.

Although $A_{2A}R$ expression is lower than A_1R expression in the hippocampus, it nonetheless is critical for various neural plastic mechanisms. At the mossy fibres postsynaptic junctions at CA3 pyramidal cells, $A_{2A}R$ activity is necessary for the induction of NMDAR-dependent LTP (Rebola et al. 2008), and $A_{2A}R$ activation can also induce a form of NMDAR-independent LTP in hippocampal CA1 (Kessey and Mogul 1997). In addition, through transactivation of BDNF Trk-B receptor (Diogenes et al. 2004, 2007), $A_{2A}R$ can also regulate BDNF-dependent neural plastic events, including long-term changes in synaptic strength and associated structural changes and growth (Sebastião et al. 2011). This would predict that A_2R agonism should facilitate synaptic plasticity in the hippocampus.

Hence, there exist parallel A_1R- and $A_{2A}R$-dependent homeostatic mechanisms that regulate and coordinate activities between disparate neuronal networks involving glutamatergic and dopaminergic activities, the disturbance of which are central to the genesis of schizophrenia symptoms. Accordingly, restoring such homeostatic control may yield potential benefits against multiple schizophrenia symptoms. Admittedly, however, A_1R- and $A_{2A}R$-dependent adenosine–glutamate interaction is both complex and extensive with considerable degrees of freedom. This could be a double-edge sword making adenosine a powerful and yet possibly unpredictable drug target. In vivo investigations become indispensable in delineating the functional impacts at the system level before attempts at any precise mechanistic accounts at the cellular, physiological and circuitry levels. This is particularly relevant in the domain of learning and memory where multiple dissociable memory systems exist.

A recent survey of systemic adenosinergic interventions in animals across a comprehensive spectrum of learning and memory tests has yielded some valuable empirical consensuses on the pro-cognitive potential of adenosinergic drugs (Boison et al. 2012) that could be instrumental to the formulation of any realistic mechanistic model.

Amongst the studies involving administration of drugs to normal animals, caffeine consistently leads to promnesic outcomes across tests. This is consistent with the well documented pro-cognitive effects of caffeine in humans (Lara 2010; Takahashi et al. 2008). The fact that caffeine improves active (Haraguchi and Kuribara 1991; Yonkov 1984) and inhibitory avoidance learning (Angelucci et al. 1999; Kopf et al. 1999) similarly confirms that the effects cannot be simply attributed to the drug's concomitant motor effects. However, the lack of an effect on working memory function may be disappointing with respect to schizophrenia cognitive symptoms. Yet, studies reporting improved spatial reference memory learning by the non-specific antagonists, caffeine or theophylline are supportive of enhanced hippocampus-dependent processing (Angelucci et al. 1999; Kopf et al. 1999; Haraguchi and Kuribara 1991; Hauber and Bareiss 2001; Prediger and Takahashi 2005). Although there is some evidence that selective A_1R or $A_{2A}R$ antagonists may also improve memory function (e.g. Kopf et al. 1999; Pereira et al. 2002), it is important to recognize that reports of null or even memory impairing effects by such antagonist drugs also exist, and sometimes emerged from the same studies (e.g. Fontinha et al. 2009; Kopf et al. 1999; Prediger and Takahashi 2005). By comparison, however, agonists more consistently yielded memory deficits across test paradigms regardless of $A_1R/A_{2A}R$ selectivity (see Boison et al. 2012), with the exception of one study reporting improved spatial learning performance in the water maze following CPA (an A_1R agonist) treatment (Von Lubitz et al. 1993). However, such inconsistent outcomes do not necessarily undermine our general thesis that adenosinergic mechanism is assuming a modulatory function over other neurotransmitter systems rather than being itself directly responsible for the neural processing and long-term plastic events underlying learning and memory.

Indeed, consistent with this view a much clearer consensus emerges when the survey of adenosinergic interventions focuses solely on animal models of memory impairment (see Boison et al. 2012). This survey gives a strong impression that A_1R or $A_{2A}R$ antagonism can ameliorate loss of memory functions across a variety of pathological models ranging from NMDAR blockade, anti-cholinergic treatment, aging, β-amyloid induced damage, to alcohol abuse. This is in agreement with the overview described above on the efficacy of caffeine to improve performance in normal animals. The consistency also extends across test paradigms similar to the survey of pharmacological studies conducted in normal non-pathological subjects. Only one study reported that caffeine exacerbated the social memory impairment associated with aging (Prediger et al. 2005) out of the 15 experiments on caffeine surveyed by Boison et al. (2012). However, there is a bias in the literature for studies of adenosine receptor antagonists with relatively few examining agonists. The mechanisms underlying the memory loss in some models are far from understood and are likely to involve multiple pathological causes (e.g. aging) beyond glutamatergic or NMDAR deficiency that is believed to underlie schizophrenia treatment-resistant negative/cognitive symptoms.

In this regard, it is important to note that the efficacy of $A_{2A}R$ antagonism to restore memory function is unexpected based on the synergistic interaction between $A_{2A}R$ and NMDAR in the hippocampus. According to this, selective $A_{2A}R$ antagonists should impair learning instead. There is, however, no pharmacological support for this, except one study examining compound SCH58261 on eyeblink conditioning in non-pathological animals (Fontinha et al. 2009). The pro-cognitive effects due to $A_{2A}R$ antagonism may not stem from its impact on hippocampal glutamatergic receptor/NMDAR activities, but rather its extensive interaction with the striatal dopamine system. Evidence in support of this would come from the lessons learned from mutant mouse models that examine the relevance of region-specific action of adenosinergic manipulations.

22.6 Perspectives from Mutant Mice Models

Genetic engineering allows the manipulation of a protein with a level of specificity that effectively avoids possible cross-target reactivity that would otherwise be hard to achieve by pharmacological manipulations alone. Together with the recent availability of many conditional knockout systems displaying different regional or even cell-type specificity, these tools represent an invaluable preclinical tool to explore mechanistically relevant disease and therapeutic models.

22.6.1 A Genetic Model of Adenosine Deficiency: Adktm1$^{-/-}$-Tg(UbiAdk) Mice

Central to the adenosinergic hypothesis is that adenosine deficiency should lead to both glutamatergic and dopaminergic dysfunction with consequential emergence of multiple schizophrenia endophenotypes (Lara 2002; Lara et al. 2006; Lara and Souza 2000). The engineered mouse line, Adktm1$^{-/-}$-Tg(UbiAdk), henceforth referred to as "*Adk-tg* mice", provides a unique model to evaluate the behavioural impact of a global adenosine deficiency. They were generated by breeding a loxP-flanked *Adk* transgene under the control of a human ubiquitin promoter into ADK knockout mice (Boison et al. 2002). ADK expression was elevated globally in Adk-tg mice, achieving a 47 % increase in the brain compared with wild type controls (Fedele et al. 2005; Pignataro et al. 2007; Yee et al. 2007). The resulting increase in adenosine clearance renders Adk-tg mice with a global brain adenosine deficiency (Li et al. 2007, 2008), expected to down-regulate both $A_{2A}R$ and A_1R-dependent adenosine signalling.

Adk-tg mice exhibited transient hyperlocomotion when exposed to a novel environment with activity rapidly resuming to normal levels demonstrating intact habituation (Fedele et al. 2005; Yee et al. 2007). Robust disturbed dopaminergic and glutamatergic homeostasis readily emerged when these animals were challenged with the psychostimulant drugs, amphetamine and MK-801 (Table 22.1C). The motor stimulating effect of amphetamine was nearly abolished, while that of MK-801 was clearly potentiated (Yee et al. 2007). These two seemingly opposing phenotypes (reducing versus enhancing psychostimuant reactions) readily exclude any explanation related to non-specific changes in general locomotor activity. The MK-801 induced phenotype is suggestive of NMDAR hypofunction, which is opposite to that seen following manipulations designed to augment activity-dependent NMDAR signalling (e.g. Yee et al. 2006; Singer et al. 2009). Adk-tg mice may therefore serve as a model to study NMDAR deficiency within the context of adenosine–NMDAR interaction and schizophrenia-related symptoms attributed to NMDAR hypofunction. The latter is further supported by the emergence of some severe learning deficits in these mutants across reference memory, working memory and associative learning (Yee et al. 2007).

The amphetamine-induced phenotype does not conform to an interpretation of hyperdopaminergia or psychotic-like phenotype. Instead it suggests that dopamine release was severely compromised. This phenotype may stem from an effect on glutamate–dopamine interaction, resembling the observation that systemic NMDAR blockade attenuated the hyperlocomotor response to cocaine (Uzbay et al. 2000). Regardless of whether this phenotype is causally related to NMDAR hypofunction, the amphetamine-induced phenotype is indicative of a hypo- rather than hyperdopamergic state. Instead of being directly linked to the production of psychotic-like symptoms often attributed to overactive striatal dopamine activity, this phenotype may instead contribute to the cognitive phenotypes in this mutant line. First of all, deficient frontal dopaminergic activity is critical to the network of associational cortices supporting effective and efficient working memory function (e.g. Simpson et al. 2010; Williams and Castner 2006). Second, dopamine-rich nucleus accumbens (ventral striatal) assumes a critical role in cognitive processes including sensorimotor gating and selective attention. In particular, disruption of dopaminergic activity in the nucleus accumbens shell can induce a specific form of attentional dysfunction representing the opposite pole of positive symptomatology (see Weiner 2003). This can be effectively reproduced also by NMDAR blockade and reversed by glycine transporter-1 inhibitor that indirectly enhances NMDAR-mediated neurotransmission (e.g. Black et al. 2009).

Here, it is worth noting that amphetamine has been shown to improve cognitive (Barch and Carter 2005) and negative symptoms (e.g. Angrist et al. 1982), particularly in patients with severe baseline negative symptomatology (Kirrane et al. 2000; Sanfilipo et al. 1996), and it has been suggested that the adjunctive use of glycine transporter-1 inhibitors might alleviate negative symptoms partly by increasing NMDAR-mediated tonic dopamine release (Grace 1991, 2000; Singer et al. 2009). Hence, the resulting severe cognitive deficits seen in Adk-tg mice are likely the result of a combination of NMDAR and related cortical dopaminergic dysfunction implicated in schizophrenia cognitive and negative psychopathology. The Adk-tg mouse model therefore does not seem to capture psychotic-like symptoms typical of positive symptomatology. This distinction can be tested by evaluating the relative efficacy of classical neuroleptic drugs like haloperidol and atypical neuroleptics such as clozapine to normalize behaviour in this mutant mouse model. A preferential response to the latter would lend support for the Adk-tg mouse might more closely approximate the pathophysiology of cognitive and negative schizophrenia symptoms.

22.6.2 Genetic Manipulations Targeting $A_{2A}R$

The severe phenotypes seen in Adk-tg mice readily suggest that reduction of brain adenosinergic tone is critical for normal behavioural functioning, and selective adenosine receptor subtype deletion and/or over-expression are instructive in dissecting the relative contributions between A_1R and $A_{2A}R$. The constitutive and conditional deletion of $A_{2A}R$ models (Chen et al. 1999; Shen et al. 2008; Yu et al. 2009a; Wei et al. 2011) and the rat model with brain $A_{2A}R$ over-expression (Giménez-Llort et al. 2002) have all been instrumental in examining the relevance of $A_{2A}R$-dependent pathway to schizophrenia-related behavioural dysfunction.

22.6.2.1 Psychostimulant-Induced Phenotypes

Constitutive $A_{2A}R^{-/-}$ mice were characterized by attenuated locomotor responses to amphetamine and cocaine (Chen et al. 2000), thus capturing a dopamine-dependent phenotype in Adk-tg mice (Yee et al. 2007). Interestingly, the response to either direct agonists against D_1R (locomotor stimulation and grooming) or D_2R

(motor-depression and stereotypy) was indistinguishable from wild-type controls (Chen et al. 2000). The distinction between direct and indirect agonists suggests that mechanisms other than tonic modulation of postsynaptic dopaminergic response were responsible for the attenuation of activity-dependent dopaminergic transmission triggered by amphetamine or cocaine. The critical mechanisms might instead involve local presynaptic A_{2A}R-dependent modulation or distal extra-striatal regulation by A_{2A}Rs upstream.

Comparisons between mutant mice with striatal-specific A_{2A}R deletion (Dlx5/6-Cre$^{(+)}$$A_{2A}R^{flox+/+}$ mice) or forebrain-wide A_{2A}R knockout (CaMKIIα-Cre$^{(+)}$$A_{2A}R^{flox+/+}$) have yielded further insights into functional differences between striatal and extra-striatal A_{2A}Rs within the forebrain in terms of psychostimulant response to cocaine and PCP (Table 22.1C). While deletion of striatal A_{2A}R enhanced the motor stimulant response to both drugs, an opposite effect (i.e. attenuated response to both drugs) was observed when the deletion was extended to the entire forebrain (Shen et al. 2008). The effect of complete forebrain knockout aligns well with that of constitutive knockout when challenged with a dopamine releasing psychostimulant (c.f., Chen et al. 2000). One interpretation is that the additional loss of extra-striatal A_{2A}Rs within the forebrain radically reverses the phenotype associated with loss of striatal A_{2A}Rs alone, and therefore the two subpopulations of A_{2A}Rs are functionally antagonistic, at least regarding their interaction with, or modulation of, psychostimulant drug action. Although we do not know whether constitutive knockout would also yield a similar phenotype as forebrain knockout when challenged with NMDAR blockers (because such an experiment has not been performed), it is tempting to predict so.

The radical component introduced by constitutive or conditional forebrain knockout could be the loss of presynaptic modulation by A_{2A}Rs discussed above. This should be spared in the striatal knockout mice because the deletion of A_{2A}Rs was restricted to striatal neurons, and their opposite phenotypes (exacerbating the response to dopaminergic stimulation or NMDAR blockade) are likely attributable solely to the disruption of the postsynaptic modulation by A_{2A}Rs on striatal principal neurons activity. One may therefore deduce that blockade of extra-striatal A_{2A}Rs may be particularly effective against behaviour associated with dopamine hyperfunction as well as NMDAR hypofunction. Genetic models with specific deletion of extra-striatal A_{2A}Rs can directly address this speculation that antipsychotic action might be maximized by targeting this subpopulation of extra-striatal A_{2A}Rs. It would provide an interesting contrast to the similar effects (antagonizing the response to dopaminergic stimulation or NMDAR blockade) by systemic pharmacological activation of A_{2A}Rs (Table 22.1A, B) that instead are likely derived primarily from their action on striatal A_{2A}Rs.

Surprisingly, amidst the complex interaction of between striatal and extra-striatal A_{2A}Rs, the observed effects of A_{2A}Rs antagonism (see Table 22.1A, B) resembles the psychostimulant-induced phenotypes revealed in the striatal specific knockout mice rather than that of constitutive or conditional forebrain knockout mice. Thus, systemic exposure to A_{2A}R antagonists displays a functional profile suggestive of a preferential blockade of striatal A_{2A}Rs, rather than that of constitutive homozygous knockout or complete forebrain knockout. This may either be because (1) $A_{2A}R$ expression in the striatum far exceeds that of any forebrain region, or (2) the modulation by striatal $A_{2A}Rs$ is normally very tightly regulated, rendering it very sensitive to pharmacological interference. Such insights could be valuable to future $A_{2A}R$-based drug development.

Lastly, the comparison would not be complete without recalling that global adenosine reduction in Adk-tg mice abolished response to amphetamine but potentiated that to MK-801 (Yee et al. 2007). The former resembles constitutive/forebrain knockout mice, but the latter drug-induced phenotype matches the striatal knockout mouse as well as systemic $A_{2A}R$ antagonist. Thus, the seemingly contradictory phenotypes against the two classes of psychostimulant drugs observed in Adk-tg mice may reflect the antagonistic contributions of striatal versus extra-striatal $A_{2A}R$ disruption. Although the precise pharmacological and biochemical mechanisms that give rise to this unique pattern of results have not been worked out, it demonstrates the potential of adenosine kinase as another possible target to manipulate adenosinergic signalling pathways relevant to schizophrenia.

22.6.2.2 Cognitive Phenotypes

As reviewed above, systemic caffeine robustly enhances performance across a wide range of cognitive assays in both animals and humans. Comparison between constitutive $A_{2A}R^{-/-}$ and $A_1R^{-/-}$ mice shows that $A_{2A}R$-dependent mechanism is solely responsible for the arousal effects of caffeine. The effects were completely abolished in $A_{2A}R^{-/-}$ mice but remained intact in $A_1R^{-/-}$ mice (Huang et al. 2005), and might once again point to the relevance of $A_{2A}R$-dependent adenosine–dopamine interaction given that arousal and alertness are both closely linked to dopaminergic control (Robbins and Arnsten 2009). Although it remains debatable whether the pro-cognitive effects of caffeine simply reflect the drug's effects on arousal and alertness, this emphasis is consistent with the recent findings that striatal-specific $A_{2A}R$ deletion (Dlx5/6-Cre$^{(+)}A_{2A}R^{flox+/+}$ mice) was sufficient to produce pro-cognitive effects seen in mutant mice (CaMKIIα-Cre$^{(+)}A_{2A}R^{flox+/+}$) with forebrain-wide $A_{2A}R$ knockout (Wei et al. 2011).

Wei et al. (2011) compared the two conditional knockout models from recognition memory, working memory, reference memory, to reversal learning. Both mutants were specifically associated with enhanced working memory function and facilitation in reversal learning, suggesting that loss of $A_{2A}R$-dependent signalling in the striatum was mainly responsible for these cognitive phenotypes. This finding contrasts with the null effects on either working or reference memory in constitutive $A_{2A}R$ knockout mice (Duan et al. 2009), but supports a similar study reporting tentative evidence for a promnesic effect (Zhou et al. 2009). Although it was not certain if the latter finding referred specifically to working or reference memory function because of an error in their classification of error types, the phenotypes revealed in the striatal and forebrain knockout animals (Wei et al. 2011) would favour a primary effect in working memory function. It is important to note that the working memory phenotype in these mutants appeared to be dependent on task difficulty—when the animals were required to hold the relevant information for longer and under conditions with increasing interference. The latter is in line with the reversal learning phenotype revealed in these mutant mice, and together the two phenotypes are suggestive of an underlying facilitation in cognitive flexibility in general. This interpretation is in keeping with the relevance of striatum in the execution of goal-directed behaviour, and is further substantiated by the finding that regulation of positively reinforced instrumental behaviour in the striatal knockout mice remained sensitive to the incentive value of the reinforcer in spite of overtraining (Yu et al. 2009a). Sufficient over-training of an operant task typically would render wild type mice insensitive to devaluation of the reinforcer as their response gradually becomes habitual and therefore less flexible. Hence, striatal A_{2A}Rs may regulate the formation and maintenance of habit in relation to changing reinforcement contingency, which could be relevant to some of the cognitive and negative symptoms of schizophrenia. An emphasis on the executive function rather than strength of memory trace as such also agrees with the general impression that systemic adenosine receptor antagonists are effective against the cognitive deficits across a spectrum of pathological models as comprehensively reviewed by Boison et al. (2012).

Finally, analysis of the rat model with brain A_{2A}R over-expression has provided some support that A_{2A}R stimulation might lead to cognitive deficiency (Giménez-Llort et al. 2007), although the reported deficits in object memory and working memory were subtle or transient in nature.

22.6.3 Genetic Manipulations Targeting A_1R

Existing studies of A_1R deletion are restricted to the use of homozygous and heterozygous constitutive knockout models (Giménez-Llort et al. 2002; Johansson et al. 2001). These models however might not be ideal for cognitive assays because life expectancy, anxiety as well as home cage activity are all clearly affected in a manner dependent on the dosage of gene knockout (Giménez-Llort et al. 2002). Pain threshold was also reduced in these mice (Johansson et al. 2001), and comparison between homozygous and heterozygous knockout mice also revealed some intriguing bidirectional effects on aggression and explorative behaviour (Giménez-Llort et al. 2002) making interpretation far from straightforward. In spite of these problematic confounds, there was no evidence for any cognitive effect based on a single study suggesting that A_1Rs are at least not essential for normal learning in the water maze (Giménez-Llort et al. 2002).

This data set is not easily assimilated with the preceding genetic models or data derived from systemic pharmacological intervention of A_1Rs. In particular, global reduction in adenosine signalling in Adk-tg mice is severely damaging to cognitive function in general. Seemingly, their phenotype profile cannot be simply understood as the sum of A_1R and A_{2A}R hypofunction. Comparison with double A_1R/A_{2A}R knockout mice (Xiao et al. 2011) might be highly instructive here. There is also an urgent need to examine the psychostimulant response in A_1R knockout mice, before any clear conclusion on the functional significance of brain A_1Rs can be made.

22.7 The Neurodevelopmental and Neuroimmunological Perspective

The relevance of neurodevelopment from the aetiological perspective of schizophrenia has long been recognized (e.g. Weinberger 1987; Harrison 1997). This emphasizes that the functional abnormalities in schizophrenia originate in early life, unfold over the course neurodevelopment (Benes et al. 1994) and accumulate into early adulthood leading to the emergence of overt symptoms. The "two-hit" perspective of schizophrenia aetiology, in particular, emphasizes the interplay between various genetic and environmental risk factors, which might mould and shift the path of disease development (Shenton et al. 2001). The developmental nature of the disease is substantiated by neuropathological examination of the relevant brain structures including recent imaging studies suggesting that a subset of brain abnormalities may change during pathogenesis, such as white matter alterations indicative of disturbance in connectivity between disparate brain regions (Kubicki et al. 2007; Shenton et al. 2001, 2010). One major early life factor singled out by epidemiological data is immune stress associated with maternal infection (Mednick et al. 1988; O'Callaghan et al. 1994; Hultman et al. 1999; Brown et al. 2000, 2004; Yolken et al. 2000; Pearce 2003) that might further be related to the immune dysfunctions identified amongst schizophrenia patients (Müller et al. 2000; Nawa et al. 2000; Rothermund et al. 2001; Hinze-Selch and Pollmächer 2000; Gaughran 2002).

Boison et al. (2012) have speculated that dysfunction in normal adenosine homeostasis during critical periods of pregnancy, as would be triggered by prenatal infection might set off a cascade of neurodevelopmental events implicated in schizophrenia aetiology. Adenosine provides critical homeostatic control over the brain immune system, and its dysfunction would offset the balance between pro- and anti-inflammatory cytokines that are critical for normal brain development (Hasko et al. 2005) as suggested by animal models of prenatal immune challenge (Meyer et al. 2008, 2009). Stressors, including viral and bacterial infections (Hasko et al. 2005) and injury (Clark et al. 1997; Pignataro et al. 2008) are associated with a surge of adenosine beyond normal physiological levels (Clark et al. 1997); and to protect the embryo against such fluctuations of systemic adenosine in the pregnant dams, the adenosine degrading enzyme, adenosine deaminase (ADA), is highly expressed in the placenta (Nagy et al. 1990). Indeed, inhibition ADA disrupts foetal development (Knudsen et al. 1992) and ADA knockout is linked to perinatal lethality (Blackburn et al. 1995). The other key adenosine degrading enzyme, ADK, undergoes a progressive switch from neuronal expression during perinatal life to exclusive astrocytic expression as the brain matures (Studer et al. 2006). Hence, a tight and well-coordinated regulation of brain adenosine is essential for brain development. This must be considered in any transgenic models, such as the Adk-tg mice with global brain disturbance of adenosine levels including the transgenic expression of Adk in neurons during adult life (Fedele et al. 2005; Pignataro et al. 2007; Yee et al. 2007).

Future studies and experimental validation of possible neurodevelopmental and neuroimmunological links are needed to explore this novel perspective of adenosinergic regulation of long-term brain development. Tentative evidence linking dysregulation in adenosine signalling with schizophrenia-relevant neurodevelopmental disturbances can be traced to a study of neonatal pregnenolone challenge (Muneoka et al. 2002). Exposure to this major precursor of the neurosteroidogenesis from postnatal day 3 to 7 increased dopamine turnover and stimulated open field locomotor activity as well as a deficiency in endogenous competitive binding to A_1R in later life (Muneoka et al. 2002). Separate evidence also suggested that neonatal pregnenolone-treatment reduced adenosine $A_{2A}R$ density in the brain (Shirayama et al. 2001). Further experiments are necessary to decide whether the deficient adenosine signalling caused by transient exposure to a neurosteroid during the critical time window of neuronal Adk expression (Studer et al. 2006) might be causally related to the concomitant signs of hyperdopaminergia.

22.8 Conclusions

This chapter has outlined the preclinical and clinical findings supporting an adenosine deficit in schizophrenia, and the basic neurobiological concepts behind the adenosine hypothesis as well as its compatibility with the existing neurochemical hypotheses of schizophrenia based primarily on dopamine and glutamate mechanism as summarized in Fig. 22.2. There are ample data showing that deficiency in adenosine can simultaneously affect dopaminergic and glutamatergic homeostatic regulation resulting in behavioural effects related to both psychotic/positive and cognitive/negative schizophrenia symptoms. To restore the global homeostatic balance would be the basis of any adenosine-based therapy. Although the chapter has explored only a limited subset of key adenosinergic biological functions within the context of schizophrenia, the complexity involved can already be quite formidable. Appreciating and delineating the mechanisms involved will be the next challenge, given some of the inconsistencies between physiological and pharmacological data, as well as data derived from genetic models (e.g. see Table 22.1). Multiple parallel bidirectional pathways have evolved to orchestrate the many roles of adenosine as a homeostatic bioenergetic network regulator in a highly supervised and structured manner. Thus, it may not be surprising that any disturbance in adenosine signalling could be related to some parts of the broad spectrum of schizophrenia symptoms. In vivo analysis will continue to play an important role in identifying and clarifying the behavioural/cognitive impacts by manipulations of specific adenosinergic targets (between different receptor subtypes and individual regulatory enzymes) in order to build testable hypotheses with sufficient specifications for successful and meaningful translation into clinically relevant drugs.

The distinction between receptor subtypes may be seemingly inadequate, such that the conventional strategy by pharmaceutical industries to achieve specific symptom control through biochemical selectivity alone may not be viable here. Instead, regional specific interaction with other neurotransmitter systems must be

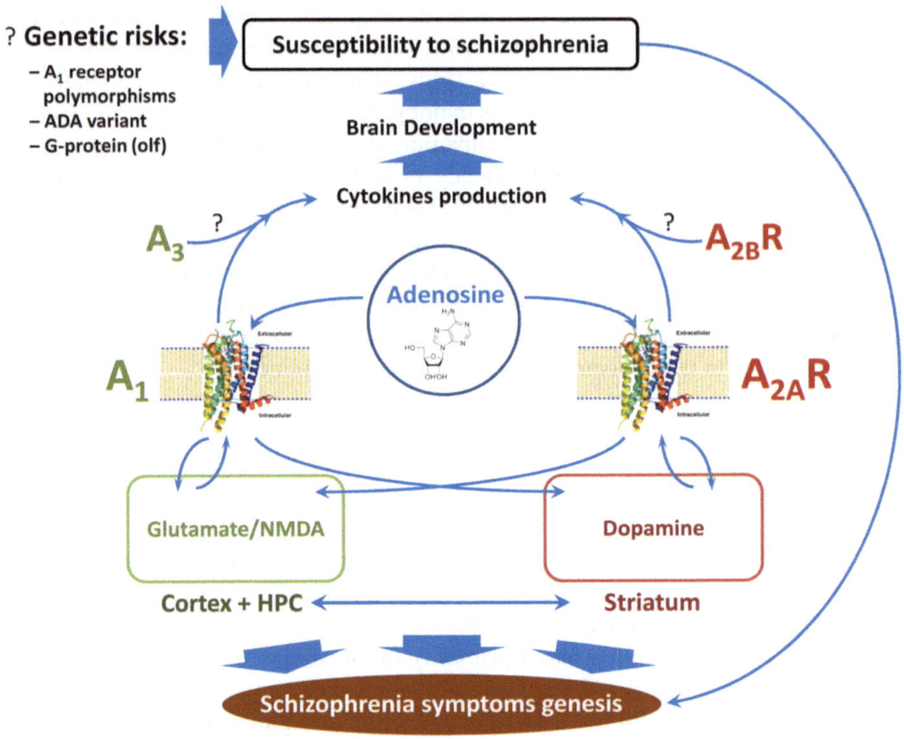

Fig. 22.2 Schematic summary of the links between adenosine and the pathophysiology/aetiology of schizophrenia. *ADA* adenosine deaminase, *G-protein* guanine nucleotide binding protein (G protein). "?" possible interaction with cytokine-mediated immune response by A_3Rs and ARs

taken into account. For example, the promising pro-cognitive effects demonstrated in the striatal A_{2A}R knockout mice were associated with enhanced response to both classes of psychostimulant drugs. And, the latter phenotypes were reversed when extra-striatal A_{2A}Rs within the forebrain were also disrupted, yielding the opposite phenotype, resembling the effects of systemic adenosine receptor antagonists. We also see that targeting enzymes, such as adenosine kinase or nucleoside transporters, which control ambient adenosinergic tone may well lead to outcomes not easily explicable in terms of receptor agonism/antagonism alone, as exemplified by the Adk-tg mice. Perhaps by affecting multiple adenosinergic pathways in parallel, the effects could be particularly impactful and sustainable due to existing synergism in resuming their balance back to their non-pathological range.

Adjunctive adenosinergic therapy could be an alternative option here—so as to enhance therapeutic efficacy as well as minimizing side effects associated with conventional antipsychotic drugs, which often introduce additional imbalance. This approach has proved feasible in the recent development of the novel glycine transporter-1 inhibitor, RG1678, by Hoffmann-La Roche (Pinard et al. 2010). Indeed, the limited clinical evidence reviewed here was based on adjunctive therapy

(Akhondzadeh et al. 2005; Brunstein et al. 2004, 2005; Lara et al. 2001). Another more challenging avenue deserves to be explored is perhaps focal adenosine therapy as in the treatment of epilepsy (Wilz et al. 2008) with appropriate technical advances in cell and gene therapies (Boison 2007b), thus enabling some of the region-specific effects discussed here to be translatable to the clinics.

Acknowledgement This work was supported by NIH grant R01MH083973 and the Legacy Foundations.

References

Abi-Dargham A, Rodenhiser J, Printz D, Zea-Ponce Y, Gil R, Kegeles LS, Weiss R, Cooper TB, Mann JJ, Van Heertum RL, Gorman JM, Laruelle M (2000) Increased baseline occupancy of D2 receptors by dopamine in schizophrenia. Proc Natl Acad Sci U S A 97:8104–8109

Akhondzadeh S, Shasavand E, Jamilian H, Shabestari O, Kamalipour A (2000) Dipyridamole in the treatment of schizophrenia: adenosine-dopamine receptor interactions. J Clin Pharm Ther 25:131–137

Akhondzadeh S, Safarcherati A, Amini H (2005) Beneficial antipsychotic effects of allopurinol as add-on therapy for schizophrenia: a double blind, randomized and placebo controlled trial. Prog Neuropsychopharmacol Biol Psychiatry 29:253–259

Angelucci ME, Vital MA, Cesario C, Zadusky CR, Rosalen PL, Da Cunha C (1999) The effect of caffeine in animal models of learning and memory. Eur J Pharmacol 373:135–140

Angrist BM, Peselow E, Rubinstein M, Corwin J, Rotrosen J (1982) Partial improvement in negative schizophrenic symptoms after amphetamine. Psychopharmacology (Berl) 78:128–130

Baldwin SA, Beal PR, Yao SY, King AE, Cass CE, Young JD (2004) The equilibrative nucleoside transporter family, SLC29. Pflugers Arch 447:735–743

Barch DM, Carter CS (2005) Amphetamine improves cognitive function in medicated individuals with schizophrenia and in healthy volunteers. Schizophr Res 77:43–58

Bartrup JT, Stone TW (1990) Activation of NMDA receptor-coupled channels suppresses the inhibitory action of adenosine on hippocampal slices. Brain Res 530:330–334

Bartrup JT, Addae JI, Stone TW (1991) Depression of purine induced inhibition during NMDA receptor mediated activation of hippocampal pyramidal cellsean Q3 iontophoretic study. Brain Res 564:323–327

Benes FM, Turtle M, Khan Y, Farol P (1994) Myelination of a key relay zone in the hippocampal formation occurs in the human brain during childhood, adolescence, and adulthood. Arch Gen Psychiatry 51:477–484

Black MD, Varty GB, Arad M, Barak S, De Levie A, Boulay D, Pichat P, Griebel G, Weiner I (2009) Procognitive and antipsychotic efficacy of glycine transport 1 inhibitors (GlyT1) in acute and neurodevelopmental models of schizophrenia: latent inhibition studies in the rat. Psychopharmacology (Berl) 202:385–396

Blackburn MR, Wakamiya M, Caskey CT, Kellems RE (1995) Tissue-specific rescue suggests that placental adenosine deaminase is important for fetal development in mice. J Biol Chem 270:23891–23894

Boeck CR, Kroth EH, Bronzatto MJ, Vendite D (2005) Adenosine receptors co-operate with NMDA preconditioning to protect cerebellar granule cells against glutamate neurotoxicity. Neuropharmacology 49:17–24

Boison D (2006) Adenosine kinase, epilepsy and stroke: mechanisms and therapies. Trends Pharmacol Sci 27:652–658

Boison D (2007a) Adenosine as a modulator of brain activity. Drug News Perspect 20:607–611

Boison D (2007b) Cell and gene therapies for refractory epilepsy. Curr Neuropharmacol 5:115–125

Boison D (2008) Adenosine as a neuromodulator in neurological diseases. Curr Opin Pharmacol 8:2–7

Boison D, Scheurer L, Zumsteg V, Rülicke T, Litynski P, Fowler B, Brandner S, Mohler H (2002) Neonatal hepatic steatosis by disruption of the adenosine kinase gene. Proc Natl Acad Sci U S A 99:6985–6990

Boison D, Chen JF, Fredholm BB (2010) Adenosine signalling and function in glial cells. Cell Death Differ 17:1071–1082

Boison D, Singer P, Shen HY, Feldon J, Yee BK (2012) Adenosine hypothesis of schizophrenia—Opportunities for pharmacotherapy. Neuropharmacology 62(3):1527–1543

Breier A, Adler C, Weisenfeld NI, Elman I, Picken L, Malhotra AK, Pickar D (1998) Effects of NMDA antagonism on striatal dopamine release in healthy subjects: application of a novel PET approach. Synapse 29:142–147

Brown AS, Schaefer CA, Wyatt RJ, Goetz R, Begg MD, Gorman JM, Susser ES (2000) Maternal exposure to respiratory infections and adult schizophrenia spectrum disorders: a prospective birth cohort study. Schizophr Bull 26:287–295

Brown AS, Begg MD, Gravenstein S, Schaefer CA, Wyatt RJ, Bresnahan M, Babulas VP, Susser ES (2004) Serologic evidence of prenatal influenza in the etiology of schizophrenia. Arch Gen Psychiatry 61:774–780

Brunstein MG, Ghisolfi ES, Ramos FL, Lara DR (2004) Clinical trial for allopurinol adjuvant therapy for poorly responsive schizophrenia. Schizophr Res S142 67

Brunstein MG, Ghisolfi ES, Ramos FL, Lara DR (2005) A clinical trial of adjuvant allopurinol therapy for moderately refractory schizophrenia. J Clin Psychiatry 66:213–219

Chen JF, Huang Z, Ma J, Zhu J, Moratalla R, Standaert D, Moskowitz MA, Fink JS, Schwarzschild MA (1999) A(2A) adenosine receptor deficiency attenuates brain injury induced by transient focal ischemia in mice. J Neurosci 19:9192–9200

Chen JF, Beilstein M, Xu YH, Turner TJ, Moratalla R, Standaert DG, Aloyo VJ, Fink JS, Schwarzschild MA (2000) Selective attenuation of psychostimulant induced behavioral responses in mice lacking A(2A) adenosine receptors. Neuroscience 97:195–204

Chen JF, Moratalla R, Impagnatiello F, Grandy DK, Cuellar B, Rubinstein M, Beilstein MA, Hackett E, Fink JS, Low MJ, Ongini E, Schwarzschild MA (2001) The role of the D(2) dopamine receptor (D(2)R) in A(2A) adenosine receptor (A(2A)R)-mediated behavioral and cellular responses as revealed by A(2A) and D(2) receptor knockout mice. Proc Natl Acad Sci U S A 98:1970–1975

Ciruela F, Casadó V, Rodrigues RJ, Luján R, Burgueño J, Canals M, Borycz J, Rebola N, Goldberg SR, Mallol J, Cortés A, Canela EI, López-Giménez JF, Milligan G, Lluis C, Cunha RA, Ferré S, Franco R (2006a) Presynaptic control of striatal glutamatergic neurotransmission by adenosine A1eA2A receptor heteromers. J Neurosci 26:2080–2087

Ciruela F, Férre S, Casado V, Cortes A, Cunha RA, Lluis C, Franco R (2006b) Heterodimeric adenosine receptors: a device to regulate neurotransmitter release. Cell Mol Life Sci 63:2427–2431

Citri A, Malenka RC (2008) Synaptic plasticity: multiple forms, functions, and mechanisms. Neuropsychopharmacology 33:18–41

Clark RS, Carcillo JA, Kochanek PM, Obrist WD, Jackson EK, Mi Z, Wisneiwski SR, Bell MJ, Marion DW (1997) Cerebrospinal fluid adenosine concentration and uncoupling of cerebral blood flow and oxidative metabolism after severe head injury in humans. Neurosurgery 41:1284–1293

Coyle JT, Tsai G (2004) NMDA receptor function, neuroplasticity, and the pathophysiology of schizophrenia. Int Rev Neurobiol 59:491–515

Coyle JT, Balu D, Benneyworth M, Basu A, Roseman A (2010) Beyond the dopamine receptor: novel therapeutic targets for treating schizophrenia. Dialogues Clin Neurosci 12:359–382

Cunha RA (2008) Different cellular sources and different roles of adenosine: A(1) receptor-mediated inhibition through astrocytic-driven volume transmission and synapse-restricted A(2A) receptor-mediated facilitation of plasticity. Neurochem. Int 52:65–72

Davis S, Butcher SP, Morris RGM (1992) The NMDA receptor antagonist D-2-amino-5-phosphonopentanoate (D-AP5) impairs spatial learning and LTP in vivo at intracerebral concentrations comparable to those that block LTP in vitro. J Neurosci 12:21–34

de Mendonça A, Ribeiro JA (1997) Adenosine and neuronal plasticity. Life Sci 60:245–251

de Mendonça A, Ribeiro JA (2000) Long-term potentiation observed upon blockade of adenosine A(1) receptors in rat hippocampus is N-methyl-Daspartate receptor-dependent. Neurosci Lett 291:81–84

Deckert J, Nöthen MM, Rietschel M, Wildenauer D, Bondy B, Ertl MA, Knapp M, Schofield PR, Albus M, Maier W, Propping P (1996) Human adenosine A2a receptor (A2aAR) gene: systematic mutation screening in patients with schizophrenia. J Neural Transm 103:1447–1455

Deckert J, Nothen MM, Bryant SP, Schuffenhauer S, Schofield PR, Spurr NK, Propping P (1997) Mapping of the human adenosine A2a receptor gene: relationship to potential schizophrenia loci on chromosome 22q and exclusion from the CATCH 22 region. Hum Genet 99:326–328

Diogenes MJ, Fernandes CC, Sebastião AM, Ribeiro JA (2004) Activation of adenosine A_{2A} receptor facilitates brain-derived neurotrophic factor modulation of synaptic transmission in hippocampal slices. J Neurosci 24:2905–2913

Diogenes MJ, Assaife-Lopes N, Pinto-Duarte A, Ribeiro JA, Sebastião AM (2007) Influence of age on BDNF modulation of hippocampal synaptic transmission: interplay with adenosine A(2A) receptors. Hippocampus 17:577–585

Duan W, Gui L, Zhou Z, Liu Y, Tian H, Chen JF, Zheng J (2009) Adenosine A_{2A} receptor deficiency exacerbates white matter lesions and cognitive deficits induced by chronic cerebral hypoperfusion in mice. J Neurol Sci 285:39–45

Dunwiddie TV, Masino SA (2001) The role and regulation of adenosine in the central nervous system. Annu Rev Neurosci 24:31–55

Dutra GP, Ottoni GL, Lara DR, Bogo MR (2010) Lower frequency of the low activity adenosine deaminase allelic variant (ADA1*2) in schizophrenic patients. Rev Bras Psiquiatr 32: 275–278

Etherington LA, Patterson GE, Meechan L, Boison D, Irving AJ, Dale N, Frenguelli B (2009) Astrocytic adenosine kinase regulates basal synaptic adenosine levels and seizure activity but not activity-dependent adenosine release in the hippocampus. Neuropharmacology 56:429–437

Farber NB (2003) The NMDA receptor hypofunction model of psychosis. Ann N Y Acad Sci 1003:119–130

Fedele DE, Gouder N, Gsüttinger M, Gabernet L, Scheurer L, Rulicke T, Crestani F, Boison D (2005) Astrogliosis in epilepsy leads to overexpression of adenosine kinase resulting in seizure aggravation. Brain 128:2383–2395

Férre S (1997) Adenosine-dopamine interactions in the ventral striatum. Implications for the treatment of schizophrenia. Psychopharmacology (Berl) 133:107–120

Férre S, Herrera-Marschitz M, Grabowska-Anden M, Casas M, Ungerstedt U, Anden NE (1991a) Postsynaptic dopamine/adenosine interaction: II Postsynaptic dopamine agonism and adenosine antagonism of methylxanthines in short-term reserpinized mice. Eur J Pharmacol 192:31–37

Férre S, Herrera-Marschitz M, Grabowska-Anden M, Ungerstedt U, Casas M, Anden NE (1991b) Postsynaptic dopamine/adenosine interaction. I. Adenosine analogues inhibit dopamine D2-mediated behaviour in short-term reserpinized mice. Eur J Pharmacol 192:25–30

Férre S, von Euler G, Johansson B, Fredholm BB, Fuxe K (1991c) Stimulation of high-affinity adenosine A2 receptors decreases the affinity of dopamine D2 receptors in rat striatal membranes. Proc Natl Acad Sci U S A 88:7238–7241

Férre S, Popoli P, Giménez-Liort L, Finnman UB, Martìnez E, Scotti de Carlos A, Fuxe K (1994) Postsynaptic antagonistic interaction between adenosine A1 and dopamine D1 receptors. Neuroreport 6:73–76

Férre S, Borycz J, Goldberg SR, Hope BT, Morales M, Lluis C, Franco R, Ciruela F, Cunha R (2005) Role of adenosine in the control of homosynaptic plasticity in striatal excitatory synapses. J Integr Neurosci 4:445–464

Férre S, Ciruela F, Woods AS, Lluis C, Franco R (2007) Functional relevance of neurotransmitter receptor heteromers in the central nervous system. Trends Neurosci 30:440–446

Fink JS, Weaver DR, Rivkees SA, Peterfreund RA, Psollack AE, Adler EM, Reppert SM (1992) Molecular cloning of the rat A2 adenosine receptor: selective co-expression with D2 dopamine receptors in rat striatum. Mol Brain Res 14:186–195

Fontinha BM, Delgado-García JM, Madroñal N, Ribeiro JA, Sebastião AM, Gruart A (2009) Adenosine A(2A) receptor modulation of hippocampal CA3-CA1 synapse plasticity during associative learning in behaving mice. Neuropsychopharmacology 34:1865–1874

Fraser CM, Fisher A, Cooke MJ, Thompson ID, Stone TW (1997) Purine modulation of dizocilpine effects on spontaneous alternation. Psychopharmacology (Berl) 130:334–342

Fredholm BB, Svenningsson P (2003) Adenosine-dopamine interactions: development of a concept and some comments on therapeutic possibilities. Neurology 61:S5–S9

Fredholm BB, Jacobson KA, Klotz KN, Linden J (2001) International Union of Pharmacology. XXV. Nomenclature and classification of adenosine receptors. Pharmacol Rev 53:527–552

Fredholm BB, Chen JF, Cunha RA, Svenningsson P, Vaugeois JM (2005) Adenosine and brain function. Int Rev Neurobiol 63:191–270

Fuxe K, Agnati LF, Jacobsen K, Hillion J, Canals M, Torvinen M, Tinner-Staines B, Staines W, Rosin D, Terasmaa A, Popoli P, Leo G, Vergoni V, Lluis C, Ciruela F, Franco R, Ferré S (2003) Receptor heteromerization in adenosine A2A receptor signaling: relevance for striatal function and Parkinson's disease. Neurology 61:S19–S23

Gaughran F (2002) Immunity and schizophrenia: autoimmunity, cytokines, and immune response. Int Rev Neurobiol 52:275–302

Geiger JD, Fyda DM (1991) Adenosine transport in nervous system tissues. In: Stone TW (ed) Adenosine in the nervous system. Academic, London

Geyer MA, Krebs-Thomson K, Braff DL, Swerdlow NR (2001) Pharmacological studies of prepulse inhibition models of sensorimotor gating deficits in schizophrenia: a decade in review. Psychopharmacology (Berl) 156:117–154

Ghisolfi ES, Prokopiuk AS, Becker J, Ehlers JA, Belmonte-de-Abreu P, Souza DO, Lara DR (2002) The adenosine antagonist theophylline impairs P50 auditory sensory gating in normal subjects. Neuropsychopharmacology 27:629–637

Giménez-Llort L, Fernandez-Teruel A, Escorihuela RM, Fredholm BB, Tobena A, Pekny M, Johansson B (2002) Mice lacking the adenosine A1 receptor are anxious and aggressive, but are normal learners with reduced muscle strength and survival rate. Eur J Neurosci 16:547–550

Giménez-Llort L, Schiffmann SN, Shmidt T, Canela L, Camón L, Wassholm M, Canals M, Terasmaa A, Fernández-Teruel A, Tobeña A, Popova E, Ferré S, Agnati L, Ciruela F, Martínez E, Scheel-Kruger J, Lluis C, Franco R, Fuxe K, Bader M (2007) Working memory deficits in transgenic rats overexpressing human adenosine A2A receptors in the brain. Neurobiol Learn Mem 87:42–56

Golembiowska K, Zylewska A (1998) Agonists of A1 and A2A adenosine receptors attenuate methamphetamine-induced overflow of dopamine in rat striatum. Brain Res 806:202–209

Gomberg R (2007) Possible case of allopurinol causing relapse of psychosis. Schizophr Res 93:409

Gonzalez-Burgos G, Hashimoto T, Lewis DA (2010) Alterations of cortical GABA neurons and network oscillations in schizophrenia. Curr Psychiatry Rep 12:335–344

Gordon JA (2010) Testing the glutamate hypothesis of schizophrenia. Nat Neurosci 13:2–4

Gotoh L, Kawanami N, Nakahara T, Hondo H, Motomura K, Ohta E, Kanchiku I, Kuroki T, Hirano M, Uchimura H (2002) Effects of the adenosine A(1) receptor agonist N(6)-cyclopentyladenosine on phencyclidine-induced behavior and expression of the immediate-early genes in the discrete brain regions of rats. Brain Res Mol Brain Res 100:1–12

Gotoh L, Mitsuyasu H, Kobayashi Y, Oribe N, Takata A, Ninomiya H, Stanton VP Jr, Springett GM, Kawasaki H, Kanba S (2009) Association analysis of adenosine A1 receptor gene (ADORA1) polymorphisms with schizophrenia in a Japanese population. Psychiatr Genet 19:328–335

Grace AA (1991) Phasic versus tonic dopamine release and the modulation of dopamine system responsivity: a hypothesis for the etiology of schizophrenia. Neuroscience 41:1–24

Grace AA (2000) Gating of information flow within the limbic system and the pathophysiology of schizophrenia. Brain Res Brain Res Rev 31:330–341

Graham FK (1975) The more or less startling effects of weak prestimulation. Psychophysiology 12:238–248

Gu JG, Foga IO, Parkinson FE, Geiger J (1995) Involvement of bidirectional adenosine transporters in the release of L-[3H]adenosine from rat brain synaptosomal preparations. J Neurochem 64: 2105–2110

Halassa MM, Fellin T, Haydon PG (2007a) The tripartite synapse: roles for gliotransmission in health and disease. Trends Mol Med 13:54–63

Halassa MM, Fellin T, Takano H, Dong JH, Haydon PG (2007b) Synaptic islands defined by the territory of a single astrocyte. J Neurosci 27:6473–6477

Haraguchi H, Kuribara H (1991) Behavioral effects of adenosine agonists: evaluation by punishment, discrete shuttle avoidance and activity tests in mice. Jpn J Pharmacol 55:303–310

Harrison PJ (1997) Schizophrenia: a disorder of neurodevelopment? Curr Opin Neurobiol 7:285–289

Hasko G, Pacher P, Vizi ES, Illes P (2005) Adenosine receptor signaling in the brain immune system. Trends Pharmacol Sci 26:511–516

Hauber W, Bareiss A (2001) Facilitative effects of an adenosine A1/A2 receptor blockade on spatial memory performance of rats: selective enhancement of reference memory retention during the light period. Behav Brain Res 118:43–52

Hauber W, Koch M (1997) Adenosine A2a receptors in the nucleus accumbens modulate prepulse inhibition of the startle response. Neuroreport 8:1515–1518

Haydon PG, Carmignoto G (2006) Astrocyte control of synaptic transmission and neurovascular coupling. Physiol Rev 86:1009–1031

Heffner TG, Wiley JN, Williams AE, Bruns RF, Coughenour LL, Downs DA (1989) Comparison of the behavioral effects of adenosine agonists and dopamine antagonists in mice. Psychopharmacology (Berl) 98:31–37

Heinz A, Schlagenhauf F (2010) Dopaminergic dysfunction in schizophrenia: salience attribution revisited. Schizophr Bull 36:472–485

Hettinger BD, Lee A, Linden J, Rosin DL (2001) Ultrastructural localization of adenosine A2A receptors suggests multiple cellular sites for modulation of GABAergic neurons in rat striatum. J Comp Neurol 431:331–346

Hinze-Selch D, Pollmächer T (2000) In vitro cytokine secretion in individuals with schizophrenia: results, confounding factors, and implications for further research. Brain Behav Immun 15: 282–318

Hoffman HS, Searle JL (1968) Acoustic and temporal factors in the evocation of startle. J Acoust Soc Am 43:269–282

Hong CJ, Liu HC, Liu TY, Liao DL, Tsai SJ (2005) Association studies of the adenosine A2a receptor (1976T>C) genetic polymorphism in Parkinson's disease and schizophrenia. J Neural Transm 112:1503–1510

Huang ZL, Qu WM, Eguchi N, Chen JF, Schwarzschild MA, Fredholm BB, Urade Y, Hayaishi O (2005) Adenosine A2A, but not A1, receptors mediate the arousal effect of caffeine. Nat Neurosci 8:858–859

Hultman CM, Sparén P, Takei N, Murray RM, Cnattingius S (1999) Prenatal and perinatal risk factors for schizophrenia, affective psychosis, and reactive psychosis of early onset: case–control study. BMJ 318:421–426

Inta D, Monyer H, Sprengel R, Meyer-Lindenberg A, Gass P (2010) Mice with genetically altered glutamate receptors as models of schizophrenia: a comprehensive review. Neurosci Biobehav Rev 34:285–294

Jacobson KA, Gao ZG (2006) Adenosine receptors as therapeutic targets. Nat Rev Drug Discov 5:247–264

Javitt DC (2008) Glycine transport inhibitors and the treatment of schizophrenia. Biol Psychiatry 63:6–8

Johansson B, Halldner L, Dunwiddie TV, Masino SA, Poelchen W, Giménez-Llort L, Escorihuela RM, Fernandez-Teruel A, Wiesenfeld-Hallin Z, Xu X-J, Hårdemark A, Betsholtz C, Herlenius E,

Fredholm BB (2001) Hyperalgesia, anxiety, and decreased hypoxic neuroprotection in mice lacking the adenosine A1 receptor. Proc Natl Acad Sci U S A 98:9407–9412
Kafka SH, Corbett R (1996) Selective adenosine A2A receptor/dopamine D2 receptor interactions in animal models of schizophrenia. Eur J Pharmacol 295:147–154
Kang J, Kang N, Lovatt D, Torres A, Zhao Z, Lin J, Nedergaard M (2008) Connexin 43 hemichannels are permeable to ATP. J Neurosci 28:4702–4711
Kawamura M Jr, Ruskin DN, Masino SA (2010) Metabolic autocrine regulation of neurons involves cooperation among pannexin hemichannels, adenosine receptors, and KATP channels. J Neurosci 30:3886–3895
Kessey K, Mogul DJ (1997) NMDA-Independent LTP by adenosine A2 receptormediated postsynaptic AMPA potentiation in hippocampus. J. Neurophysiol. 78:1965–1972
Kirrane RM, Mitropoulou V, Nunn M, New AS, Harvey PD, Schopick F, Silverman J, Siever LJ (2000) Effects of amphetamine on visuospatial working memory performance in schizophrenia spectrum personality disorder. Neuropsychopharmacology 22:14–18
Knudsen TB, Winters RS, Otey SK, Blackburn MR, Airhart MJ, Church JK, Skalko RG (1992) Effects of (R)-deoxycoformycin (pentostatin) on intrauterine nucleoside catabolism and embryo viability in the pregnant mouse. Teratology 45:91–103
Koch M, Hauber W (1998) Regulation of sensorimotor gating by interactions of dopamine and adenosine in the rat. Behav Pharmacol 9:23–29
Kopf SR, Melani A, Pedata F, Pepeu G (1999) Adenosine and memory storage: effect of A(1) and A(2) receptor antagonists. Psychopharmacology (Berl) 146:214–219
Kubicki M, McCarley R, Westin CF, Park HJ, Maier S, Kikinis R, Jolesz FA, Shenton ME (2007) A review of diffusion tensor imaging studies in schizophrenia. J Psychiatr Res 41:15–30
Kull B, Svenningsson P, Fredholm BB (2000) Adenosine A(2A) receptors are colocalized with and activate g(olf) in rat striatum. Mol Pharmacol 58:771–777
Kuribara H, Asami T, Ida I, Tadokoro S (1992) Characteristics of the ambulation-increasing effect of the noncompetitive NMDA antagonist MK-801 in mice: assessment by the coadministration with central-acting drugs. Jpn J Pharmacol 58:11–18
Kurumaji A, Toru M (1998) An increase in [3H] CGS21680 binding in the striatum of postmortem brains of chronic schizophrenics. Brain Res 808:320–323
Kuzmin A, Johansson B, Fredholm BB, Ogren SO (2000) Genetic evidence that cocaine and caffeine stimulate locomotion in mice via different mechanisms. Life Sci 66:L113–L118
Lara DR (2002) Inhibitory deficit in schizophrenia is not necessarily a GABAergic deficit. Cell Mol Neurobiol 3:239–247
Lara DR (2010) Caffeine, mental health, and psychiatric disorders. J Alzheimers Dis 20(Suppl 1):2348–2392
Lara DR, Souza DO (2000) Schizophrenia: a purinergic hypothesis. Med Hypotheses 54:157–166
Lara DR, Belmonte-de-Abreu P, Souza DO (2000) Allopurinol for refractory aggression and self-inflicted behaviour. J Psychopharmacol 14:81–83
Lara DR, Brunstein MG, Ghisolfi ES, Lobato MI, Belmonte-de-Abreu P, Souza DO (2001) Allopurinol augmentation for poorly responsive schizophrenia. Int Clin Psychopharmacol 16:235–237
Lara DR, Cruz MR, Xavier F, Souza DO, Moriguchi EH (2003) Allopurinol for the treatment of aggressive behaviour in patients with dementia. Int Clin Psychopharmacol 18:53–55
Lara DR, Dall'Igna OP, Ghisolfi ES, Brunstein MG (2006) Involvement of adenosine in the neurobiology of schizophrenia and its therapeutic implications. Prog Neuropsychopharmacol Biol Psychiatry 30:617–629
Laruelle M, Abi-Dargham A (1999) Dopamine as the wind of the psychotic fire: new evidence from brain imaging studies. J Psychopharmacol 13:358–371
Lewis DA, Pierri JN, Volk DW, Melchitzky DS, Woo TU (1999) Altered GABA neurotransmission and prefrontal cortical dysfunction in schizophrenia. Biol Psychiatry 46:616–626
Lewis DA, Hashimoto T, Volk DW (2005) Cortical inhibitory neurons and schizophrenia. Nat Rev Neurosci 6:312–324
Li T, Lan JQ, Fredholm BB, Simon RP, Boison D (2007) Adenosine dysfunction in astrogliosis: cause for seizure generation? Neuron Glia Biol 3:353–366

Li T, Ren G, Lusardi T, Wilz A, Lan JQ, Iwasato T, Itohara S, Simon RP, Boison D (2008) Adenosine kinase is a target for the prediction and prevention of epileptogenesis in mice. J Clin Invest 118:571–582
Linden J (2001) Molecular approach to adenosine receptors: receptor-mediated mechanisms of tissue protection. Annu Rev Pharmacol Toxicol 41:775–787
Lindstrom LH, Gefvert O, Hagberg G, Lundberg T, Bergstrom M, Hartvig P, Langstrom B (1999) Increased dopamine synthesis rate in medial prefrontal cortex and striatum in schizophrenia indicated by L-(beta-11 C) DOPA and PET. Biol Psychiatry 46:681–688
Malec D, Poleszak E (2006) Involvement of adenosine receptors in dizocilpine induced motor activity in mice. Pharmacol Rep 58:101–106
Malenka RC, Nicoll RA (1993) NMDA-receptor-dependent synaptic plasticity: multiple forms and mechanisms. Trends Neurosci 16:521–527
Mandryk M, Fidecka S, Poleszak E, Malec D (2005) Participation of adenosine system in the ketamine-induced motor activity in mice. Pharmacol Rep 57:55–60
Martin ED, Fernandez M, Perea G, Pascual O, Haydon PG, Araque A, Cena V (2007) Adenosine released by astrocytes contributes to hypoxia-induced modulation of synaptic transmission. Glia 55:36–45
Mednick SA, Machon RA, Huttunen MO, Bonett D (1988) Adult schizophrenia following prenatal exposure to an influenza epidemic. Arch Gen Psychiatry 45:189–192
Meyer U, Murray PJ, Urwyler A, Yee BK, Schedlowski M, Feldon J (2008) Adult behavioral and pharmacological dysfunctions following disruption of the fetal brain balance between pro-inflammatory and IL-10-mediated anti-inflammatory signaling. Mol Psychiatry 13:208–221
Meyer U, Feldon J, Yee BK (2009) A review of the fetal brain cytokine imbalance hypothesis of schizophrenia. Schizophr Bull 35:959–972
Möhler H, Rudolph U, Boison D, Singer P, Feldon J, Yee BK (2008) Regulation of cognition and symptoms of psychosis: focus on GABAA receptors and glycine transporter 1. Pharmacol Biochem Behav 90:58–64
Morris RG (1989) Synaptic plasticity and learning: selective impairment of learning rats and blockade of long-term potentiation in vivo by the N-methyl-D-aspartate receptor antagonist AP5. J Neurosci 9:3040–3057
Morris RG, Anderson E, Lynch GS, Baudry M (1986) Selective impairment of learning and blockade of long-term potentiation by an N-methyl-D-aspartate receptor antagonist, AP5. Nature 319:774–776
Müller N, Riedel M, Ackenheil M, Schwarz MJ (2000) Cellular and humoral immune system in schizophrenia: a conceptual re-evaluation. World J Biol Psychiatry 1:173–179
Muneoka KT, Shirayama Y, Minabe Y, Takigawa M (2002) Effects of a neurosteroid, pregnenolone, during the neonatal period on adenosine A1 receptor, dopamine metabolites in the fronto-parietal cortex and behavioral response in the open field. Brain Res 956:332–338
Nagy JI, Geiger JD, Staines WA (1990) Adenosine deaminase and purinergic neuroregulation. Neurochem Int 16:211–221
Nash JE, Brotchie JM (2000) A common signaling pathway for striatal NMDA and adenosine A2a receptors: implications for the treatment of Parkinson's disease. J Neurosci 20:7782–7789
Nawa H, Takahashi M, Patterson PH (2000) Cytokine and growth factor involvement in schizophrenia—support for the developmental model. Mol Psychiatry 5:594–603
Newby AC, Worku Y, Holmquist CA (1985) Adenosine formation. Evidence for a direct biochemical link with energy metabolism. Adv Myocardiol 6:273–284
Nikbakht MR, Stone TW (2001) Suppression of presynaptic responses to adenosine by activation of NMDA receptors. Eur J Pharmacol 427:13–25
O'Callaghan E, Sham PC, Takei N, Murray G, Glover G, Hare EH, Murray RM (1994) The relationship of schizophrenic births to 16 infectious diseases. Br J Psychiatry 165:353–356
Park J, Gupta RS (2008) Adenosine kinase and ribokinase—the RK family of proteins. Cell Mol Life Sci 65:2875–2896
Pascual O, Casper KB, Kubera C, Zhang J, Revilla-Sanchez R, Sul JY, Takano H, Moss SJ, McCarthy K, Haydon PG (2005) Astrocytic purinergic signalling coordinates synaptic networks. Science 310:113–116

Pearce BD (2003) Can a virus cause schizophrenia? Facts and hypotheses. Kluwer, New York
Pereira GS, Mello e Souza T, Vinadé ER, Choi H, Rodrigues C, Battastini AM, Izquierdo I, Sarkis JJ, Bonan CD (2002) Blockade of adenosine A1 receptors in the posterior cingulate cortex facilitates memory in rats. Eur J Pharmacol 437:151–154
Pignataro G, Simon RP, Boison D (2007) Transgenic overexpression of adenosine kinase aggravates cell death in ischemia. J Cereb Blood Flow Metab 27:1–5
Pignataro G, Maysami S, Studer FE, Wilz A, Simon RP, Boison D (2008) Downregulation of hippocampal adenosine kinase after focal ischemia as potential endogenous neuroprotective mechanism. J Cereb Blood Flow Metab 28:17–23
Pinard E, Alanine A, Alberati D, Bender M, Borroni E, Bourdeaux P, Brom V, Burner S, Fischer H, Hainzl D, Halm R, Hauser N, Jolidon S, Lengyel J, Marty HP, Meyer T, Moreau JL, Mory R, Narquizian R, Nettekoven M, Norcross RD, Puellmann B, Schmid P, Schmitt S, Stalder H, Wermuth R, Wettstein JG, Zimmerli D (2010) Selective GlyT1 inhibitors: discovery of [4-(3-fluoro-5-trifluoromethylpyridin-2-yl)piperazin-1-yl][5-methanesulfon yl-2-((S)-2,2,2-trifluoro-1-methylethoxy)phenyl]methanone (RG1678), a promising novel medicine to treat schizophrenia. J Med Chem 53:4603–4614
Poleszak E, Malec D (2002) Cocaine-induced hyperactivity is more influenced by adenosine receptor agonists than amphetamine-induced hyperactivity. Pol J Pharmacol 54:359–366
Popoli P, Pèzzola A, Reggio R, Caporali MG, Scotti de Carolis A (1994a) CGS21680 antagonizes motor hyperactivity in a rat model of Huntington's disease. Eur J Pharmacol 257:R5–R6
Popoli P, Pezzola A, de Carolis AS (1994b) Modulation of striatal adenosine A1 and A2 receptors induces rotational behaviour in response to dopaminergic stimulation in intact rats. Eur J Pharmacol 257:21–25
Popoli P, Giménez-Llort L, Pezzola A, Reggio R, Martínez E, Fuxe K, Ferré S (1996) Adenosine A1 receptor blockade selectively potentiates the motor effects induced by dopamine D1 receptor stimulation in rodents. Neurosci Lett 218:209–213
Popoli P, Reggio R, Pezzola A, Fuxe K, Ferre S (1998) Adenosine A1 and A2A receptor antagonists stimulate motor activity: evidence for an increased effectiveness in aged rats. Neurosci Lett 251:201–204
Prediger RD, Takahashi RN (2005) Modulation of short-term social memory in rats by adenosine A1 and A(2A) receptors. Neurosci Lett 376:160–165
Prediger RD, Da Cunha C, Takahashi RN (2005) Antagonistic interaction between adenosine A2A and dopamine D2 receptors modulates the social recognition memory in reserpine-treated rats. Behav Pharmacol 16:209–218
Rebola N, Lujan R, Cunha RA, Mulle C (2008) Adenosine A2A receptors are essential for long-term potentiation of NMDA-EPSCs at hippocampal mossy fiber synapses. Neuron 57:121–134
Ribeiro JA, Sebastião AM, de Mendonca A (2003) Adenosine receptors in the nervous system: pathophysiological implications. Prog Neurobiol 68:377–392
Rimondini R, Férre S, Ogren SO, Fuxe K (1997) Adenosine A2A agonists: a potential new type of atypical antipsychotic. Neuropsychopharmacology 17:82–91
Robbins TW, Arnsten AF (2009) The neuropsychopharmacology of fronto-executive function: monoaminergic modulation. Annu Rev Neurosci 32:267–287
Rothermund M, Arolt V, Bayer TA (2001) Review of immunological and immunopathological findings in schizophrenia. Brain Behav Immun 5:319–339
Sanfilipo M, Wolkin A, Angrist B, van Kammen DP, Duncan E, Wieland S, Cooper TB, Peselow ED, Rotrosen J (1996) Amphetamine and negative symptoms of schizophrenia. Psychopharmacology (Berl) 123:211–214
Schiffmann SN, Vanderhaeghen JJ (1993) Adenosine A2 receptors regulate the gene expression of striatopallidal and striatonigral neurons. J Neurosci 13:1080–1087
Schiffmann SN, Jacobs O, Vanderhaeghen JJ (1991) Striatal restricted adenosine A2 receptor (RDC8) is expressed by enkephalin but not by substance P neurons: an in situ hybridization histochemistry study. J Neurochem 57:1062–1067

Schiffmann SN, Fisone G, Moresco R, Cunha RA, Férre S (2007) Adenosine A(2A) receptors and basal ganglia physiology. Prog Neurobiol 83:277–292

Schmidt AP, Böhmer AE, Antunes C, Schallenberger C, Porciúncula LO, Elisabetsky E, Lara DR, Souza DO (2009) Anti-nociceptive properties of the xanthine oxidase inhibitor allopurinol in mice: role of A1 adenosine receptors. Br J Pharmacol 156:163–172

Schwab SG, Hallmayer J, Albus M, Lerer B, Hanses C, Kanyas K, Segman R, Borrman M, Dreikorn B, Lichtermann D, Rietschel M, Trixler M, Maier W, Wildenauer DB (1998a) Further evidence for a susceptibility locus on chromosome 10p14-p11 in 72 families with schizophrenia by nonparametric linkage analysis. Am J Med Genet 81:302–307

Schwab SG, Hallmayer J, Lerer B, Albus M, Borrmann M, Hönig S, Strauss M, Segman R, Lichtermann D, Knapp M, Trixler M, Maier W, Wildenauer DB (1998b) Support for a chromosome 18p locus conferring susceptibility to functional psychoses in families with schizophrenia, by association and linkage analysis. Am J Hum Genet 63:1139–1152

Sebastião AM, Ribeiro JA (2009) Tuning and fine-tuning of synapses with adenosine. Curr Neuropharmacol 7:180–194

Sebastião AM, Assaife-Lopes N, Diógenes MJ, Vaz SH, Ribeiro JA (2011) Modulation of brain-derived neurotrophic factor (BDNF) actions in the nervous system by adenosine A(2A) receptors and the role of lipid rafts. Biochim Biophys Acta 1808:1340–1349

Seeman P (2006) Targeting the dopamine D2 receptor in schizophrenia. Expert Opin Ther Targets 10:515–531

Seeman P, Schwarz J, Chen JF, Szechtman H, Perreault M, McKnight GS, Roder JC, Quirion R, Boksa P, Srivastava LK, Yanai K, Weinshenker D, Sumiyoshi T (2006) Psychosis pathways converge via D2high dopamine receptors. Synapse 60:319–346

Shen HY, Chen JF (2009) Adenosine A2A receptors in psychopharmacology: modulators of behavior, mood, and cognition. Curr Neuropharmacol 7:195–206

Shen HY, Coelho JE, Ohtsuka N, Canas PM, Day YJ, Huang QY, Rebola N, Yu L, Boison D, Cunha RA, Linden J, Tsien JZ, Chen JF (2008) A critical role of the adenosine A2A receptor in extra-striatal neurons in modulating psychomotor activity as revealed by opposite phenotypes of striatum- and forebrain-A2A receptor knockouts. J Neurosci 28:2970–2975

Shenton ME, Dickey CC, Frumin M, McCarley RW (2001) A review of MRI findings in schizophrenia. Schizophr Res 49:1–52

Shenton ME, Whitford TJ, Kubicki M (2010) Structural neuroimaging in schizophrenia: from methods to insights to treatments. Dialogues Clin Neurosci 12:317–332

Shi D, Nikodijevic O, Jacobson KA, Daly JW (1994) Effects of chronic caffeine on adenosine, dopamine and acetylcholine systems in mice. Arch Int Pharmacodyn Ther 328:261–287

Shirayama Y, Muneoka KT, Takigawa M, Minabe Y (2001) Adenosine A2A, 5-HT1A and 5-HT7 receptor in neonatally pregnenolone-treated rats. Neuroreport 12:3773–3776

Sills TL, Azampanah A, Fletcher PJ (2001) The adenosine A2A agonist CGS 21680 reverses the reduction in prepulse inhibition of the acoustic startle response induced by phencyclidine, but not by apomorphine and amphetamine. Psychopharmacology (Berl) 156:187–193

Simpson EH, Kellendonk C, Kandel E (2010) A possible role for the striatum in the pathogenesis of the cognitive symptoms of schizophrenia. Neuron 65:585–596

Singer P, Boison D, Möhler H, Feldon J, Yee BK (2007a) Enhanced recognition memory following glycine transporter 1 deletion in forebrain neurons. Behav Neurosci 121:815–825

Singer P, Feldon J, Yee BK (2007b) Interference of glycine transporter 1: modulation of cognitive functions via activation of glycine-B site of the NMDA receptor. Cent Nerv Syst Agents Med Chem 7:259–268

Singer P, Yee BK, Feldon J, Iwasato T, Itohara S, Grampp T, Prenosil G, Benke D, Möhler H, Boison D (2009) Altered mnemonic functions and resistance to N-METHYL-d-Aspartate receptor antagonism by forebrain conditional knockout of glycine transporter 1. Neuroscience 161:635–654

Smith GS, Schloesser R, Brodie JD, Dewey SL, Logan J, Vitkun SA, Simkowitz P, Hurley A, Cooper T, Volkow ND, Cancro R (1998) Glutamate modulation of dopamine measured in vivo

with positron emission tomography (PET) and 11Craclopride in normal human subjects. Neuropsychopharmacology 18:18–25
Snyder SH (1976) The dopamine hypothesis of schizophrenia: focus on the dopamine receptor. Am J Psychiatry 133:197–202
Solinas M, Ferre S, You ZB, Karcz-Kubicha M, Popoli P, Goldberg SR (2002) Caffeine induces dopamine and glutamate release in the shell of the nucleus accumbens. J Neurosci 22:6321–6324
Steinpreis R (1996) The behavioral and neurochemical effects of phencyclidine in humans and animals: some implications for modelling psychosis. Behav Brain Res 74:45–55
Steinpreis R, Salamone JD (1993) The role of nucleus accumbens dopamine in the neurochemical and behavioral effects of phencyclidine: a microdialysis and behavioral study. Brain Res 612:263–270
Stone TW, Ceruti S, Abbracchio MP (2009) Adenosine receptors and neurological disease: neuroprotection and neurodegeneration. Handb Exp Pharmacol 193:535–587
Studer FE, Fedele DE, Marowsky A, Schwerdel C, Wernli K, Vogt K, Fritschy J-M, Boison D (2006) Shift of adenosine kinase expression from neurons to astrocytes during postnatal development suggests dual functionality of the enzyme. Neuroscience 142:125–137
Svenningsson P, Le Moine C, Fisone G, Fredholm BB (1999) Distribution, biochemistry and function of striatal adenosine A2A receptors. Prog Neurobiol 59:355–396
Swerdlow NR, Weber M, Qu Y, Light GA, Braff DL (2008) Realistic expectations of prepulse inhibition in translational models for schizophrenia research. Psychopharmacology (Berl) 199:331–388
Takahashi RN, Pamplona FA, Schroder RU, Prediger RD (2008) Adenosine receptor antagonists for cognitive dysfunction: a review of animal studies. Front Biosci 13:2614–2632
Thummler S, Dunwiddie TV (2000) Adenosine receptor antagonists induce persistent bursting in the rat hippocampal CA3 region via an NMDA receptor dependent mechanism. J Neurophysiol 83:1787–1795
Tonkiss J, Morris RGM, Rawlins JNP (1988) Intra-ventricular infusion of the NMDA antagonist AP5 impairs performance on a non-spatial operant DRL task in the rat. Exp Brain Res 73:181–188
Turgeon SM, Pollack AE, Schusheim L, Fink JS (1996) Effects of selective adenosine A1 and A2a agonists on amphetamine-induced locomotion and c-Fos in striatum and nucleus accumbens. Brain Res 707:75–80
Uzbay IT, Wallis CJ, Lal H, Forster MJ (2000) Effects of NMDA receptor blockers on cocaine-stimulated locomotor activity in mice. Behav Brain Res 108:57–61
Von Lubitz DK, Paul IA, Bartus RT, Jacobson KA (1993) Effects of chronic administration of adenosine A1 receptor agonist and antagonist on spatial learning and memory. Eur J Pharmacol 249:271–280
Wardas J, Konieczny J, Pietraszek M (2003) Influence of CGS 21680, a selective adenosine A(2A) agonist, on the phencyclidine-induced sensorimotor gating deficit and motor behaviour in rats. Psychopharmacology (Berl) 168:299–306
Wei CJ, Singer P, Coelho J, Boison D, Feldon J, Yee BK, Chen JF (2011) Selective inactivation of adenosine A2A receptors in striatal neurons enhances working memory and reversal learning. Learn Mem 18:459–474
Weinberger DR (1987) Implications of normal brain development for the pathogenesis of schizophrenia. Arch Gen Psychiatry 44:660–669
Weiner I (2003) The "two-headed" latent inhibition model of schizophrenia: modeling positive and negative symptoms and their treatment. Psychopharmacology (Berl) 169:257–297
Williams GV, Castner SA (2006) Under the curve: critical issues for elucidating D1 receptor function in working memory. Neuroscience 139:263–276
Wilz A, Pritchard EM, Li T, Lan JQ, Kaplan DL, Boison D (2008) Silk polymer based adenosine release: therapeutic potential for epilepsy. Biomaterials 29:3609–3616
Wirkner K, Assmann H, Köles L, Gerevich Z, Franke H, Nörenberg W, Boehm R, Illes P (2000) Inhibition by adenosine A2A receptors of NMDA but not AMPA currents in striatal neurons. Br J Pharmacol 130:259–269

Wirkner K, Gerevich Z, Krause T, Göunther A, Köles L, Schneider D, Nörenberg W, Illes P (2004). Adenosine A2A receptor-induced inhibition of NMDA and GABAA receptor-mediated synaptic currents in a subpopulation of rat striatal neurons. Neuropharmacology 46:994–1007

Xiao D, Cassin JJ, Healy B, Burdett TC, Chen JF, Fredholm BB, Schwarzschild MA (2011) Deletion of adenosine A_1 or $A(_{2A})$ receptors reduces L-3,4-dihydroxyphenylalanine-induced dyskinesia in a model of Parkinson's disease. Brain Res 1367:310–318

Yang CR, Svensson KA (2008) Allosteric modulation of NMDA receptor via elevation of brain glycine and D-serine: the therapeutic potentials for schizophrenia. Pharmacol Ther 120:317–332

Yee BK, Russig H, Feldon J (2004) Apomorphine-induced prepulse inhibition disruption is associated with a paradoxical enhancement of prepulse stimulus reactivity. Neuropsychopharmacology 29:240–248

Yee BK, Balic E, Singer P, Schwerdel C, Grampp T, Gabernet L, Knuesel I, Benke D, Feldon J, Mohler H, Boison D (2006) Disruption of glycine transporter 1 restricted to forebrain neurons is associated with a pro-cognitive and antipsychotic phenotypic profile. J Neurosci 26:3169–3181

Yee BK, Singer P, Chen JF, Feldon J, Boison D (2007) Transgenic overexpression of adenosine kinase in brain leads to multiple learning impairments and altered sensitivity to psychomimetic drugs. Eur J Neurosci 26:3237–3252

Yolken RH, Karlsson H, Yee F, Johnston-Wilson NL, Torrey EF (2000) Endogenous retroviruses and schizophrenia. Brain Res Brain Res Rev 31:193–199

Yonkov DI (1984) Possible role of brain dopaminergic systems in the memory effects of central stimulants. Methods Find Exp Clin Pharmacol 6:235–239

Yu C, Gupta J, Chen JF, Yin HH (2009a) Genetic deletion of A2A adenosine receptors in the striatum selectively impairs habit formation. J Neurosci 29:15100–15103

Yu L, Coelho JE, Zhang X, Fu Y, Tillman A, Karaoz U, Fredholm BB, Weng Z, Chen JF (2009b) Uncovering multiple molecular targets for caffeine using a drug target validation strategy combining A_{2A} receptor knockout mice with microarray profiling. Physiol Genomics 37:199–210

Zhou SJ, Zhu ME, Shu D, Du XP, Song XH, Wang XT, Zheng RY, Cai XH, Chen JF, He JC (2009) Preferential enhancement of working memory in mice lacking adenosine A(2A) receptors. Brain Res 1303:74–83

Zimmermann H (2000) Extracellular metabolism of ATP and other nucleotides. Naunyn Schmiedebergs Arch Pharmacol 362:299–309

Chapter 23
The Role of Adenosine in the Ventral Striatal Circuits Regulating Behavioral Activation and Effort-Related Decision Making: Importance for Normal and Pathological Aspects of Motivation

John D. Salamone, Merce Correa, Patrick A. Randall, Eric J. Nunes, Marta Pardo, and Laura Lopez-Cruz

Abstract Motivated behavior can be characterized by a high degree of activity, vigor, and persistence. Brain dopamine (DA), particularly in the nucleus accumbens, has been implicated in activational aspects of motivation and effort-related processes. Accumbens DA depletions reduce the tendency of rats to work for food, and alter effort-related decision making, but leave aspects of food motivation such as appetite intact. Recent evidence indicates that the purine neuromodulator adenosine, largely through actions on adenosine A_{2A} receptors, also participates in regulating effort-related processes. Intra-accumbens injections of adenosine A_{2A} agonists produce effects that are similar to those produced by accumbens DA depletion or antagonism. Furthermore, systemic or intra-accumbens injections of adenosine A_{2A} receptor antagonists can reverse the effects of DA D_2 antagonists on effort-related choice behavior. In contrast, adenosine A_1 receptor antagonists fail to reverse these motivational effects of DA D_2 antagonism. These studies have implications for understanding the potential role of adenosine in the development and treatment of energy-related motivational symptoms such as anergia and fatigue in depression and other disorders.

Keywords Motivation • Accumbens • Adenosine A_{2A} • Dopamine • Behavioral activation • Anergia • Fatigue

---

J.D. Salamone, Ph.D. (✉) • P.A. Randall • E.J. Nunes
Division of Behavioral Neuroscience, Department of Psychology,
University of Connecticut, Storrs, CT 06269-1020, USA
e-mail: john.salamone@uconn.edu

M. Correa
Division of Behavioral Neuroscience, Department of Psychology,
University of Connecticut, Storrs, CT 06269-1020, USA

Department of Psicobiology, Universitat Jaume I, Castelló 12071, Spain

M. Pardo • L. Lopez-Cruz
Department of Psicobiology, Universitat Jaume I, Castelló 12071, Spain

S. Masino and D. Boison (eds.), *Adenosine: A Key Link between Metabolism and Brain Activity*, DOI 10.1007/978-1-4614-3903-5_23,

23.1 Behavioral Activation and Effort-Related Processes

23.1.1 Dopaminergic Involvement in Behavioral Activation and Exertion of Effort in Motivated Behavior

Motivation involves a complex and multifaceted set of processes. Motivated behavior occurs in phases that represent different degrees of physical or psychological distance from the primary motivational stimulus. For that reason it is sometimes said that there is a "consummatory" phase that represents a direct interaction with the motivational stimulus (e.g., food consumption), and another phase variously labeled as "approach," "appetitive," or "instrumental," which reflects the fact that the organism must engage in a pattern of behavior in order to obtain access to the motivational stimuli (e.g., forage, or run in a maze, to obtain food; Salamone 1992; Salamone and Correa 2002; van den Bos et al. 2006). Furthermore, theory and research on motivation has for years emphasized that there are "directional" and "activational" aspects of motivation (Salamone 1988, 1991, 1992). The behavior of animals is directed towards or away from particular motivational stimuli, but it also is evident that motivated behavior can be characterized by vigorous or persistent activity that involves high levels of work. Activational aspects of motivated behavior enable organisms to overcome challenges or work-related response costs that separate them from significant stimuli such as food (Salamone 1992; Salamone and Correa 2002; van den Bos et al. 2006; Salamone et al. 2009a). In humans, impairments in behavioral activation can display themselves as energy-related symptoms such as psychomotor slowing, anergia, apathy, and fatigue, which are core features of depression, and also are present in neurological disorders such as parkinsonism and multiple sclerosis (Salamone et al. 2006, 2009b).

In studying the neural basis of behavioral activation and exertion of effort, several brain areas and neurotransmitters (DA, GABA, adenosine) have been investigated. One of the neural systems most closely associated with behavioral activation is the dopamine (DA) innervation of the nucleus accumbens (Robbins and Everitt 2007; Salamone et al. 2007), which is a part of the striatal complex that also is known as ventral striatum. Several lines of evidence demonstrate that nucleus accumbens DA is involved generally in behavioral activation, and, more specifically, in effort-related aspects of food motivation. Accumbens DA depletions have been shown to reduce spontaneous and novelty-induced locomotor activity and rearing, as well as stimulant-induced activity (Koob et al. 1978; Correa et al. 2002). Behaviors such as wheel-running, polydipsia (excessive drinking), or locomotor activity, which are induced by scheduled presentation of food pellets to food-deprived animals, were shown to be reduced by accumbens DA depletions (Robbins and Koob 1980; Wallace et al. 1983; McCullough and Salamone 1992). Furthermore, the effects of nucleus accumbens DA depletions on food-reinforced instrumental behavior depend greatly upon the task requirements (e.g., the number of lever presses required to receive food reinforcement). Schedules of reinforcement with low work requirements are generally somewhat insensitive to the effects of accumbens DA depletions or antagonism (Aberman and Salamone 1999; Correa et al. 2002; Ishiwari et al. 2004; Wakabayashi et al. 2004; Mingote et al. 2005). However, a critical factor that makes a food-reinforced operant task sensitive to the effects of accumbens DA depletions is the size of the ratio requirement (i.e., the number of times they have to press the lever to receive food). Performance of an operant task with higher ratio requirements (i.e., as more lever presses are required to receive each unit of reinforcement) makes rats more sensitive to the response suppressing effects of accumbens DA depletions (Aberman et al. 1998; Aberman and Salamone 1999; Salamone et al. 2001; Ishiwari et al. 2004). For example, accumbens DA depletions have little effect on lever pressing in rats responding on a fixed ratio 1 schedule (FR1; 1 lever press required per reinforcer; Aberman and Salamone 1999; Ishiwari et al. 2004), whereas performance on FR5, FR16, FR64 schedules (or higher) is impaired substantially (Aberman et al. 1998; Salamone et al. 2001; Ishiwari et al. 2004). These effects of accumbens DA depletions are not simply dependent upon the degree of intermittence in the schedule (i.e., the time spent with no primary reinforcers). A series of studies showed that while accumbens DA depletions had little effect on performance of variable interval VI 30, 60, or 120 s schedules (i.e., an animal is reinforced for the first response after an interval elapses; the interval varies, but averages 30 or 60 s or 120 s), these DA depletions had a substantial effect when a ratio requirement (i.e., five or ten responses required after the interval) was added to the time interval (Correa et al. 2002; Mingote et al. 2005). Along with other experiments showing that accumbens DA depletions did not impair food intake, this overall pattern of results indicates that nucleus accumbens DA depletions leave fundamental aspects of appetite or primary food motivation intact (Salamone and Correa 2009), but reduce the tendency of the animals to work for food reinforcement.

23.1.2 Accumbens DA and Effort-Related Decision Making

At any one time, there may be several motivational stimuli available in a complex environment (e.g., other animals, water, different types of food), which can vary in terms of type or amount. Usually, there also would be different types of instrumental behaviors that are necessary for obtaining access to each stimulus, each having different response requirements that vary in terms of work, time, and other parameters. In order to deal with these conditions, organisms must make effort-related decisions, and allocate behavioral resources based upon multiple factors that involve response costs and reinforcement value. Along with other brain areas, nucleus accumbens DA is a critical participant in the process of effort-related choice behavior (Salamone et al. 1997, 2007, 2009a, b).

A number of behavioral procedures have been developed that allow for the assessment of how animals allocate resources based upon analyses of reinforcement value and response cost. One task that has been developed is an operant concurrent choice procedure that offers rats a choice between lever pressing to obtain a relatively preferred food (high carbohydrate operant pellets) vs. approaching and consuming a less preferred food (standard lab chow) that is concurrently available in the chamber. Under baseline or control conditions, rats responding on an FR5 schedule for the preferred pellets typically get most of their food by lever pressing, and they eat little chow (e.g., 1–2 g or less). Administration of moderate to low doses of DA antagonists, regardless of D_1 or D_2 receptor selectivity, produces a substantial alteration of response allocation; these drugs decrease lever pressing for the preferred pellets, but substantially increase intake of the concurrently available chow (Salamone et al. 1991, 2002; Sink et al. 2008). The nucleus accumbens, rather than the neostriatum, is the DA terminal region most closely associated with the effects of DA depletion or antagonism (Cousins et al. 1993; Koch et al. 2000; Nowend et al. 2001; Farrar et al. 2010). Thus, despite the fact that lever pressing is decreased by accumbens DA antagonism or depletions, the rats show a compensatory reallocation of behavior and select a new path to an alternative source of food. In contrast to control conditions, animals treated with low doses of DA antagonists or accumbens DA depletions obtain most of their food from the low cost option (i.e., chow intake) as opposed to the high cost option (lever pressing). However, these effects are not dependent upon changes in appetite or food preference (Salamone et al. 1991, 2002; Koch et al. 2000; Sink et al. 2008), indicating that changes in effort-related choice behavior are dissociable from changes in taste preference or "finickiness" (e.g., Dess 2000).

Another behavior test that is used to assess effort-related choice behavior is the T-maze barrier choice task (Salamone et al. 1994; Cousins et al. 1996; Pardo et al. submitted). With this task, the two choice arms of the maze can have different reinforcement densities (e.g., four vs. two food pellets, or four vs. zero). In order to provide an effort-related challenge, a large barrier can be placed in the arm with the higher density of food reinforcement. When there is no barrier in the arm with the high reinforcement density, or when both arms contain a barrier, rats and mice strongly prefer the arm with the high density of food, and DA antagonism or accumbens DA depletion does not alter their response choice (Salamone et al. 1994; Denk et al. 2005; Pardo et al. submitted). Under conditions in which the arm with the barrier contains four food pellets, but the no-barrier arm contains no pellets (thus, the only way to obtain food is by climbing the barrier), rats with accumbens DA depletions still choose the high-density arm, climb the barrier, and consume the pellets (Cousins et al. 1996). Moreover, pre-feeding the animals to reduce food motivation produces a dramatic increase in omissions, and a relative indifference to both arms (Pardo et al. 2012). Thus, the effects of DA antagonists or accumbens DA depletions on food-related tasks do not closely resemble those produced by pre-feeding or appetite-suppressant drugs (Aberman and Salamone 1999; Salamone et al. 1991; Sink et al. 2008). These control experiments indicate also that interference with DA transmission does not impair memory for which arm had the most pellets, and does not affect the discrimination of reinforcement density or the preference for the arm with the high density of food. Nevertheless, low doses of DA antagonists and accumbens DA depletions dramatically altered choice behavior when the high-density arm had the barrier in place, and the arm without the barrier contained an alternative food source. Under these conditions, rats and mice with impaired DA transmission showed decreased choice for the high-density arm that contained the barrier, and increased choice for the arm with less food that did not have a barrier (Salamone et al. 1994, 1997; Cousins et al. 1996; Mott et al. 2009; Pardo et al. 2012). The results of these studies are consistent with recent papers showing that DA antagonists also affect effort discounting in T-maze (Bardgett et al. 2009) and lever pressing (Floresco et al. 2008) choice tasks. Together with the findings from operant choice experiments, the results of these studies indicate that low doses of DA antagonists and interference with accumbens DA transmission cause animals to reallocate their instrumental responses based upon the response requirements of the task.

23.2 Adenosine A_{2A} Receptors Are Involved in Behavioral Activation and Effort-Related Processes

23.2.1 Adenosine A_{2A} Receptor Stimulation Affects Effort-Related Processes

As described above, considerable evidence indicates that DA antagonists and accumbens DA depletions profoundly affect functions such as behavioral activation, instrumental response output and allocation, and effort-related processes. Needless to say, accumbens DA does not regulate effort-related processes in isolation; several recent studies have shown that structures that project to nucleus accumbens, such as anterior cingulate cortex and basolateral amygdala, as well as output structures such as ventral pallidum, also participate in effort-related processes (Walton et al. 2002, 2003, 2006; Schweimer et al. 2005; Schweimer and Hauber 2005a, b; Floresco et al. 2008; Floresco and Ghods-Sharifi 2007; Farrar et al. 2008; Mingote et al. 2008; Hauber and Sommer 2009). In addition, recent work in this area has focused upon interactions between DA and the purine neuromodulator adenosine.

Four G-protein-coupled adenosine receptors have been identified, although the A_1 and A_{2A} subtypes are the most common ones in the brain (Svenningsson et al. 1999). It has been known for several years that nonselective adenosine antagonists, such as caffeine and theophylline, act as minor stimulants (Ferré 2008; Ferré et al. 2008a; Randall et al. 2011a; Lopez-Cruz et al. 2011). There has been a tremendous growth in research on adenosine receptor neurochemistry and pharmacology, and the A_{2A} receptor subtype has been the focus of considerable attention. Striatal subregions, including both caudate/putamen (neostriatum) and nucleus accumbens, are very rich in adenosine A_{2A} receptors (Schiffmann et al. 1991; Ferré et al. 2004; Vontell et al. 2010). There is a neurochemical interaction between striatal DA D_2 and adenosine A_{2A} receptors, which tend to be co-localized on the same enkephalin-positive striatopallidal medium spiny neurons (Fink et al. 1992; Ferré 1997; Ferré et al. 2001, 2004, 2008b; DeMet and Chicz-DeMet 2002; Hillion et al. 2002; Fuxe et al. 2003, 2007). DA D_2 and adenosine A_{2A} receptors form heteromeric complexes, and interact at the level of c-AMP-related signal transduction pathways (Ferré et al. 2001, 2004, 2008b; Fuxe et al. 2003, 2007; Azdad et al. 2009). This interaction has most typically been studied in relation to Parkinson's disease and neostriatal motor functions potentially related to parkinsonism (Wardas et al. 2001; Morelli and Pinna 2002; Correa et al. 2004; Simola et al. 2004; Salamone et al. 2008). Several adenosine A_{2A} receptor antagonists are being assessed for their potential antiparkinsonian actions (Jenner 2005; Pinna 2010; Salamone 2010).

In addition to these studies related to motor function, adenosine A_{2A} agonists and antagonists are being assessed for their actions on other behavioral functions as well. Drugs that act upon adenosine A_{2A} receptors also have been studied for their involvement in cognition (Takahashi et al. 2008) and anxiety-related processes (Correa and Font 2008). Over the last few years, the involvement of adenosine A_{2A} receptors in aspects of behavioral activation and effort-related processes also has become the focus of considerable research. Injections of the adenosine A_{2A} agonist CGS 21680 directly into nucleus accumbens have been shown to produce effects that resemble those of accumbens DA depletions or antagonism. Intra-accumbens injections of CGS 21680 were shown to reduce novelty-induced locomotor activity (Barraco et al. 1993). More recently, it was demonstrated that local injection of CGS 21680 into the core subregion of nucleus accumbens affected effort-related choice behavior (Font et al. 2008). Infusions 24.0 ng of CGS 21680 bilaterally into the accumbens core decreased lever pressing and increased chow intake in rats responding on the concurrent FR5/chow feeding choice task (Font et al. 2008), a pattern of effects similar to that produced by accumbens DA depletions and antagonism. In contrast, infusions of CGS 21680 into a control site dorsal to nucleus accumbens were ineffective (Font et al. 2008). A recent study (Mingote et al. 2008) also investigated the effects of intra-accumbens injections of CGS 21680 on performance of a VI 60-s operant schedule under conditions in which there was a low work requirement (FR1 attached to the interval) and a high work requirement (FR10 attached to the interval). Stimulation of accumbens adenosine A_{2A} receptors with local bilateral injections of CGS 21680 disrupted performance of an instrumental task with high work demands (i.e., the interval lever pressing schedule with a ratio requirement attached), but had little effect on a task with the lower work requirement. These actions were site specific, because infusions of CGS 21680 into a dorsal control site had no effect.

Immunohistochemical studies revealed that accumbens neurons that project to the ventral pallidum showed adenosine A_{2A} receptor immunoreactivity (Mingote et al. 2008). Moreover, activation of accumbens A_{2A} receptors by local injections of CGS 21680 increased extracellular GABA levels in the ventral pallidum as measured by microdialysis. Combined contralateral injections of CGS 21680 into the accumbens and the $GABA_A$ agonist muscimol into ventral pallidum (i.e., "disconnection" methods) also impaired response output, indicating that these structures are part of a common neural circuitry regulating the exertion of effort. Thus, accumbens adenosine A_{2A} receptors regulate exertion of effort in rats responding on operant tasks by modulating the activity of the ventral striatopallidal pathway (Mingote et al. 2008).

23.2.2 Adenosine A_{2A} Receptor Antagonists, but not A_1 Receptor Antagonists, Reverse the Effort-Related Effects of DA D_2 Receptor Blockade

The findings reviewed above indicate that an adenosine A_{2A} agonist can produce actions similar to those resulting from interference with DA transmission. Consistent with those observations, it also has been shown that adenosine A_{2A} receptor antagonists can reverse the effects of DA antagonists on effort-related choice behavior. Studies employing the T-maze barrier choice procedure demonstrated that the adenosine A_{2A} receptor antagonist MSX-3 could reverse the effects of the D_2 antagonist haloperidol in both rats (Mott et al. 2009) and mice (Pardo et al. submitted). MSX-3 also has been shown to attenuate the effects of the D_2 antagonists haloperidol and eticlopride in rats responding on the concurrent lever pressing FR5/chow feeding procedure (Farrar et al. 2007; Worden et al. 2009; Nunes et al. 2010). Similar effects have been produced by another adenosine A_{2A} receptor antagonist, istradefylline (KW 6002; Salamone et al. 2009b; Nunes et al. 2010). Furthermore, adenosine A_{2A} receptor knockout mice are resistant to the effects of haloperidol on effort-related choice behavior as assessed by the T-maze barrier task (Pardo et al. 2012). These studies indicate that there is a very specific interaction between DA D_2 and adenosine A_{2A} receptor subtypes. Although the adenosine A_{2A} receptor antagonist MSX-3 can reduce the effect of haloperidol in rats and mice performing on the T-maze task, the A_1 antagonists 8-cyclopentyl-1,3-dipropylxanthine (DPCPX) and 8-cyclopentyltheophylline (CPT) were ineffective (Mott et al. 2009; Pardo et al. 2012). Similarly, DPCPX and CPT were unable to reverse the effects of haloperidol and eticlopride in rats responding on the operant concurrent lever pressing FR5/chow intake choice task (Salamone et al. 2009b; Nunes et al. 2010). Although MSX-3 and istradefylline could substantially reverse the effects of D_2 antagonists such as haloperidol and eticlopride in rats responding on the operant concurrent choice procedure (Farrar et al. 2007; Worden et al. 2009; Salamone et al. 2009b), these drugs produced only a partial attenuation of the effects of the D_1 antagonist ecopipam (SCH 39166; Worden et al. 2009; Nunes et al. 2010). Furthermore, the adenosine A_1 antagonists DPCPX and CPT were ineffective at reversing the effort-related effects of the DA D_1 antagonists ecopipam and SCH 23390 on the concurrent lever pressing FR5/chow intake procedure (Nunes et al. 2010). Parallel studies focusing upon novelty-induced locomotion also yielded a similar pattern of effects in rats and mice (Collins et al. 2010; Pardo et al. 2010).

This overall pattern of results indicates that there is a selective interaction between antagonists of DA D_2 and adenosine A_{2A} receptors (see Fig. 23.1 for summary of results). Both systemic and local intra-accumbens injections of the adenosine A_{2A} receptor antagonist MSX-3 reversed the effects of intra-accumbens injections of the D_2 antagonist eticlopride on effort-related choice behavior (Farrar et al. 2010), demonstrating that nucleus accumbens is an important brain site for this D_2/A_{2A} interaction. Moreover, these results from studies of effort-related choice behavior are consistent with the large body of evidence demonstrating that A_{2A}

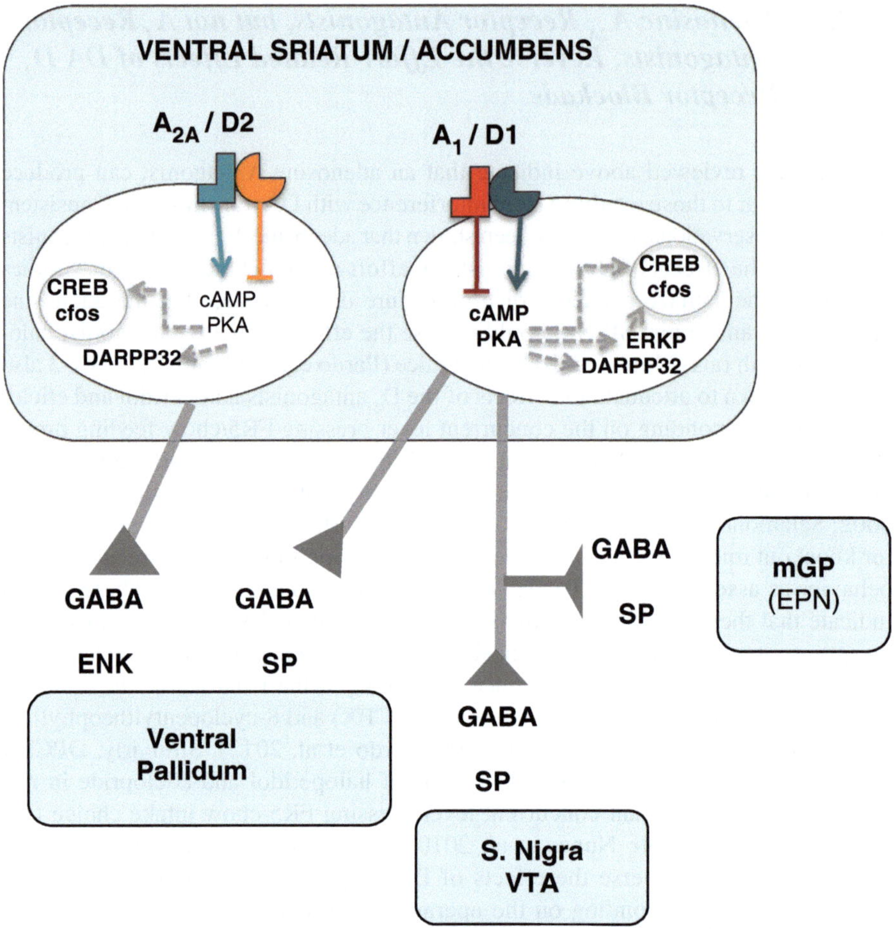

Fig. 23.1 Summary of results from studies employing the operant concurrent choice task, showing the ability of adenosine antagonists to reverse the effects of DA antagonists. This figure shows the effect sizes (R^2 values) that provide a marker of the magnitude of the reversal effect of each adenosine antagonist when co-administered with a DA antagonist. These calculations were performed on the lever pressing data from published papers (Salamone et al. 2009a, b; Nunes et al. 2010), as well as unpublished data (Santerre 2011, unpublished masters thesis, University of Connecticut). These analyses were conducted by removing the vehicle plus vehicle control data, and calculating the R^2 value for the four treatments that included a DA antagonist injection alone as well as the DA antagonist combined with an adenosine antagonist. With effect size calculations, the magnitude of the treatment effect is independent of the number of subjects, and is expressed as the proportion of total variance accounted for by treatment variance (for example $R^2=0.3$ reflects 30 % of the variance explained across experiments and measures; larger effect sizes mean greater reversal effects)

antagonism can generally reverse the effects of D_2 antagonism across a wide range of behavioral procedures, including tasks that involve functions related to dorsal as well as ventral striatum (Correa et al. 2004; Ishiwari et al. 2007; Salamone et al. 2008; Betz et al. 2009). The specific nature of this interaction is possibly related to

INTERACTION BETWEEN DA-ADENOSINE ANTAGONISTS.
LEVER PRESSING EFFECT SIZE (R2) FOR REVERSAL STUDIES

	D1	*D2*
A1	***ECOPIPAM-*** SCH23390 0.04 ***ECOPIPAM-*** DPCPX 0.14 CPT 0.11	***HALOPERIDOL-*** DPCPX 0.06
Non-selective	***ECOPIPAM-*** THEOPHYLLINE 0.10	***HALOPERIDOL-*** CAFFEINE 0.21
A2A	***ECOPIPAM-*** MSX3 0.12 ***ECOPIPAM-*** KW6002 0.26	***ETICLOPRIDE-*** MSX3 / MSX4 0.40 KW6002 0.43 ***HALOPERIDOL-*** KW6002 0.49

Data from: Salamone et al., 2009; Worden et al., 2009; Nunes et al., 2010; Santerre et al., in preparation

Fig. 23.2 Anatomical diagram depicting the pattern of adenosine (A_1 and A_{2A}) and DA (D_1 and D_2 family) receptor localization on medium spiny neurons in nucleus accumbens. See text for details. *mGP* medial globus pallidus, *epn* entopeduncular nucleus, *s. nigra* substantia nigra, *VTA* ventral tegmental area. Data from Salamone et al. 2009a, b; Worden et al. 2009; Nunes et al. 2010; Santerre et al. 2012 in preparation

the pattern of cellular localization of adenosine A_1 and A_{2A} receptors in striatal areas, including the nucleus accumbens (Fig. 23.2). As described above, adenosine A_{2A} receptors are located on enkephalin-positive striatal neurons that also express DA D_2 receptors. DA D_2 and adenosine A_{2A} receptors are able to form heteromeric complexes, in which the two receptors become linked and display alterations in their binding characteristics, and these receptors also converge onto the same cAMP-related signal transduction pathways. Thus, adenosine A_{2A} antagonists may be highly effective at reducing the effort-related effects of DA D_2 antagonists because they reverse the basic cellular manifestations of D_2 blockade. Consistent with this suggestion, recent studies have employed c-Fos immunoreactivity to provide a cellular marker of the interaction between DA D_2 and adenosine A_{2A} receptors. These experiments have demonstrated that doses of adenosine A_{2A} antagonists that are effective at reversing the actions of DA D_2 antagonists on effort-related choice behavior also can reverse the D_2-antagonist-induced enhancement of c-Fos expression in nucleus accumbens core (Farrar et al. 2010; Pardo et al. 2012).

23.2.3 Behavioral Stimulant Actions of Adenosine A_{2A} Receptor Antagonists: Studies of Operant Behavior

As reviewed above, adenosine A_{2A} antagonists can reverse many of the behavioral effects of DA antagonists, including actions on instrumental behavior and effort-related choice behavior. However, less is known about the effects of selective adenosine antagonists on operant behavior when these drugs are administered alone. A number of studies of the behavioral effects of adenosine antagonists in rodents have focused upon their ability to stimulate locomotor activity. Consistent with its classification as a minor stimulant, caffeine has been shown consistently to enhance locomotor activity in rats and mice (Garrett and Holtzman 1994; Daly and Fredholm 1998; Karcz-Kubicha et al. 2003; Antoniou et al. 2005). In addition, adenosine A_{2A} antagonists, including istradefylline (KW6002), MSX-3, and SCH 58261, also have been shown to increase locomotion (Popoli et al. 1998; Antoniou et al. 2005; Collins et al. 2010). The locomotor stimulant effects of A_1 antagonists reported in the literature appear to be more variable, and may vary depending upon the specific drug; CPT has been shown in several studies to induce locomotion in rodents (Marston et al. 1998; Popoli et al. 1998; Karcz-Kubicha et al. 2003), while DPCPX, which is more selective for A_1 receptors than CPT, generally does not stimulate locomotion (Marston et al. 1998; Collins et al. 2010).

Although there are large number of studies of the effects of adenosine A_{2A} antagonists on locomotion, and there is a growing body of evidence indicating that these drugs can reverse the suppression of lever pressing induced by DA antagonism, little is known about the effects of adenosine A_{2A} antagonists on operant behavior when they are administered in the absence of other drugs. One study reported that caffeine altered delay discounting, increasing choice of larger delayed rewards (Diller et al. 2008), but the effects of more selective adenosine antagonists are unknown. In terms of response rates, low to moderate doses of MSX-3 and istradefylline were reported as not having much effect on lever pressing rates in rats responding on either an FR5 schedule (Farrar et al. 2007; Salamone et al. 2009b) or the FR10 component of a drug discrimination procedure (Justinova et al. 2003). Justinova et al. (2009) observed that high doses of MSX-3 and CPT (i.e., 30.0 mg/kg) could decrease lever pressing rates on the FR10 component of a drug discrimination task in rats trained to discriminate nicotine (Justinova et al. 2009). However, in the behavioral pharmacology literature it is well known that the rate-enhancing effects of stimulant drugs depend greatly upon the baseline rate generated by the particular schedule of reinforcement (Dews 1955, 1958). For example, with major stimulants such as amphetamine, enhanced rates of lever pressing are most clearly shown when animals are responding on schedules that generate low rates of responding (e.g., fixed interval (FI) schedules), and response rates tend to be decreased when animals are responding on schedules that generate high baseline rates of responding (e.g., FR schedules; Dews 1958; Sanger and Blackman 1974). Studies of the rate-dependent effects of caffeine have yielded mixed results, with some papers indicating that caffeine can produce rate-dependent effects on operant responding (Davis et al. 1973; McKim 1980), and others failing to find this result (Harris et al. 1978).

Recent studies were undertaken to investigate the potential for rate-dependent stimulant effects of both selective and nonselective adenosine antagonists (Randall et al. 2011a). Six adenosine antagonists with varying patterns of selectivity were assessed: two nonselective adenosine antagonists (caffeine and theophylline), two adenosine A_1 antagonists (DPCPX and CPT), and two adenosine A_{2A} antagonists (istradefylline (KW6002) and MSX-3). In addition, two schedules of reinforcement were employed; a fixed interval 240-s (FI-240 s) schedule was used to generate low baseline rates of responding, and a fixed ratio 20 (FR20) schedule generated high rates. The methylxanthines caffeine and theophylline both produced rate-dependent effects on lever pressing, increasing responding on the FI-240 s schedule but decreasing responding on the FR20 schedule. However, the A_{2A} antagonists MSX-3 and istradefylline increased FI-240 s lever pressing but did not suppress FR20 lever pressing in the dose range tested. In fact, there was a tendency for istradefylline to increase FR20 responding at a moderate dose. In contrast to the effects of the adenosine A_{2A} antagonists, the A_1 receptor antagonists DPCPX and CPT failed to increase lever pressing rate, although DPCPX did act to decrease FR20 responding at higher doses. The results of the Randall et al. (2011a) paper indicate that adenosine A_{2A} antagonists enhance operant response rates, but A_1 antagonists do not. This pattern of results is consistent with the idea that nonselective and A_{2A} selective adenosine antagonists have behaviorally activating or arousing effects, which manifest themselves as increases in responding on schedules that generate low response rates, and which are consistent with the induction of locomotion by these drugs (Randall et al. 2011a).

The stimulant effects of MSX-3 also have been studied using a progressive ratio (PROG)/chow feeding concurrent choice procedure, in which rats can press on a PROG schedule reinforced by presentation of preferred high-carbohydrate pellets, or, alternatively, can approach and consume a less preferred chow available in the chamber (Randall et al. 2011b). The rat passes through each ratio level by completing 15 ratios, after which the ratio requirement is incremented by one additional ratio (e.g., FR1 15 times, FR2 15 times, etc.). When chow is not present, rats typically achieve an average ratio level of FR10-20. With chow present, a lower ratio is achieved (FR3-6), and the animals consume considerable chow (6–7 g). Thus, the rats eventually reach a point in the session at which the ratio value becomes too high, so they switch to less effort-requiring less preferred chow. Administration of MSX-3 increases PROG lever pressing and decreases chow intake. In contrast, the D_2 antagonist haloperidol decreases lever pressing (Randall et al. 2011b; Pardo et al. 2011).

23.3 Clinical Significance

As described above, adenosine A_{2A} antagonists are capable of reversing the effects of DA antagonists in various tasks related to effort-related choice behavior. In addition to contributing to our understanding of the basic neural mechanisms involved in motivation, this research also has considerable clinical significance. Psychologists and psychiatrists have come to emphasize the importance of energy-related dysfunctions, such as psychomotor slowing, apathy, anergia, and fatigue, in major depression, parkinsonism, and other disorders (Tylee et al. 1999; Stahl 2002; Demyttenaere et al. 2005; Salamone et al. 2006; Treadway and Zald 2011). Although anhedonia is often emphasized as a symptom of depression, many people with major depression have fundamental deficits in reward seeking, exertion of effort, and effort-related decision making that are over and above any problems that they have with experiencing pleasure (Treadway and Zald 2011). Moreover, it has been argued that many apparent self-reports of depressed mood or anhedonia in depressed patients may stem from core deficits in activational aspects of motivation and effort-related processes (Treadway and Zald 2011). In patients with major depression, the severity of these psychomotor or energy-related symptoms is correlated with problems with social function, employment, and treatment outcomes (Tylee et al. 1999; Stahl 2002). Energy-related symptoms are present in diverse disorders, including major depression, parkinsonism, organic brain disease, drug abstinence, and multiple sclerosis, and are thought to reflect a common set of dysfunctions (Caligiuri and Ellwanger 2000; Volkow et al. 2001; Salamone et al. 2006, 2009b; Friedman et al. 2007; Tellez et al. 2008), but the precise mechanisms underlying these symptoms remain unknown. Although the cardinal symptoms of Parkinson's disease involve movement dysfunction, this disease also can be associated with non-motor symptoms including cognitive dysfunction, depression, and fatigue. Fatigue is a widely reported problem in many patients with idiopathic or drug-induced Parkinsonism (Friedman et al. 2007; Salamone et al. 2009b). Fatigue symptoms are marked by subjective reports of a lack of energy, and also by changes in behavioral markers indicating reduced selection of high-effort activities (Friedman et al. 2007; Elbers et al. 2009). The neural basis of the impaired effort-related function seen in depression is still being characterized; nevertheless, considerable evidence implicates central DA, basal ganglia, and related cortical mechanisms (Rogers et al. 1987; Rampello et al. 1991; Brown and Gershon 1993; Hickie et al. 1999; Caligiuri and

Ellwanger 2000; Volkow et al. 2001; Schmidt et al. 2001; Brody et al. 2001; Salamone et al. 2006, 2007; Tellez et al. 2008; Treadway and Zald 2011).

Adenosine A_{2A} antagonists are currently being studied for their potential as antiparkinsonian agents (Salamone et al. 2008; Salamone 2010), but these compounds also could be useful for treating symptoms of depression, particularly the effort-related symptoms. Adenosine A_{2A} antagonists have been shown to have antidepressant-like effects in animal models (El Yacoubi et al. 2001, 2003; Minor et al. 2003, 2006; Cunha et al. 2008; Hodgson et al. 2009; Hanff et al. 2010). Adenosine A_{2A} antagonists can reverse the effects of IL-1β on forced swim test performance (Minor et al. 2003), and can interact with interleukins in the regulation of reserpine-induced alterations in performance on the forced swim test (Hanff et al. 2010). Because adenosine A_{2A} antagonists are so effective at reversing the effects of DA antagonism on tasks related to behavioral activation and effort-related processes, it has been suggested that these drugs could be useful for treating the motivational symptoms of depression (Farrar et al. 2007; Salamone et al. 2007, 2009a, b; Mingote et al. 2008; Nunes et al. 2010; Randall et al. 2011a). Future lines of translational research should focus on the potential utility of adenosine A_{2A} antagonism as a treatment for the psychomotor slowing, anergia, and fatigue that are debilitating for patients with depression and other disorders.

Acknowledgements This work was supported by a grant to J.D. Salamone from the National Institute of Mental Health (MH078023), and to M. Correa from Fundació Bancaixa-UJI (P1.1B2010-43), and Plan Nacional de Drogas, Ministerio de Política Social y Consumo (2010I024), Spain.

Financial and competing interests disclosure: For some studies reviewed in this manuscript, KW6002 was generously supplied by Lundbeck (Copenhagen, Denmark), and J.D. Salamone also has received a research grant from Pfizer and Merck.

References

Aberman JE, Salamone JD (1999) Nucleus accumbens dopamine depletions make rats more sensitive to high ratio requirements but do not impair primary food reinforcement. Neuroscience 92:545–552

Aberman JE, Ward SJ, Salamone JD (1998) Effects of dopamine antagonists and accumbens dopamine depletions on time-constrained progressive-ratio performance. Pharmacol Biochem Behav 61:341–348

Antoniou K, Papadopoulou-Daifoti Z, Hyphantis T, Papathanasiou G, Bekris E, Marselos M, Panlilio L, Müller CE, Goldberg SR, Ferré S (2005) A detailed behavioral analysis of the acute motor effects of caffeine in the rat: involvement of adenosine A1 and A2A receptors. Psychopharmacology (Berl) 183:154–162

Azdad K, Gall D, Woods AS, Ledent C, Ferré S, Schiffmann SN (2009) Dopamine D2 and adenosine A2A receptors regulate NMDA-mediated excitation in accumbens neurons through A2A-D2 receptor heteromerization. Neuropsychopharmacology 34:972–986

Bardgett ME, Depenbrock M, Downs N, Points M, Green L (2009) Dopamine modulates effort-based decision making in rats. Behav Neurosci 123:463–467

Barraco RA, Martens KA, Parizon M, Normile HJ (1993) Adenosine A2a receptors in the nucleus accumbens mediate locomotor depression. Brain Res Bull 31:397–404

Betz AJ, Vontell R, Valenta J, Worden L, Sink KS, Font L, Correa M, Sager TN, Salamone JD (2009) Effects of the adenosine A_{2A} antagonist KW-6002 (istradefylline) on pimozide-induced oral tremor and striatal c-Fos expression: comparisons with the muscarinic antagonist tropicamide. Neuroscience 163:97–108

Brody AL, Saxena S, Mandelkern MA, Fairbanks LA, Ho ML, Baxter LR (2001) Brain metabolic changes associated with symptom factor improvement in major depressive disorder. Biol Psychiatry 50:171–178

Brown AS, Gershon S (1993) Dopamine and depression. J Neural Transm Gen Sect 91:75–109

Caligiuri MP, Ellwanger J (2000) Motor and cognitive aspects of motor retardation in depression. J Affect Disord 57:83–93

Collins LE, Galtieri DJ, Collins P, Jones SK, Port RG, Paul NE, Hockemeyer J, Müller CE, Salamone JD (2010) Interactions between adenosine and dopamine receptor antagonists with different selectivity profiles: effects on locomotor activity. Behav Brain Res 211:148–155

Correa M, Font L (2008) Is there a major role for adenosine A2A receptors in anxiety? Front Biosci 13:4058–4070

Correa M, Carlson BB, Wisniecki A, Salamone JD (2002) Nucleus accumbens dopamine and work requirements on interval schedules. Behav Brain Res 137:179–187

Correa M, Wisniecki A, Betz A, Dobson DR, O'Neill MF, O'Neill MJ, Salamone JD (2004) The adenosine A2A antagonist KF17837 reverses the locomotor suppression and tremulous jaw movements induced by haloperidol in rats: possible relevance to parkinsonism. Behav Brain Res 148:47–54

Cousins MS, Sokolowski JD, Salamone JD (1993) Different effects of nucleus accumbens and ventrolateral striatal dopamine depletions on instrumental response selection in the rat. Pharmacol Biochem Behav 46:943–951

Cousins MS, Atherton A, Turner L, Salamone JD (1996) Nucleus accumbens dopamine depletions alter relative response allocation in a T-maze cost/benefit task. Behav Brain Res 74:189–197

Cunha RA, Ferré S, Vaugeois JM, Chen JF (2008) Potential therapeutic interest of adenosine A2A receptors in psychiatric disorders. Curr Pharm Des 14:1512–1524

Daly JW, Fredholm BB (1998) Caffeine-an atypical drug of dependence. Drug Alcohol Depend 51:199–206

Davis TR, Kensler CJ, Dews PB (1973) Comparison of behavioral effects of nicotine, d-amphetamine, caffeine and dimethylheptyl tetrahydrocannabinol in squirrel monkeys. Psychopharmacology (Berl) 32:51–65

DeMet EM, Chicz-DeMet A (2002) Localization of adenosine A2A-receptors in rat brain with [3 H]ZM-241385. Naunyn Schmiedebergs Arch Pharmacol 366:478–481

Demyttenaere K, De Fruyt J, Stahl SM (2005) The many faces of fatigue in major depressive disorder. Int J Neuropsychopharmacol 8:93–105

Denk F, Walton ME, Jennings KA, Sharp T, Rushworth MF, Bannerman DM (2005) Differential involvement of serotonin and dopamine systems in cost–benefit decisions about delay or effort. Psychopharmacology (Berl) 179:587–596

Dess NK (2000) Responses to basic taste qualities in rats selectively bred for high versus low saccharin intake. Physiol Behav 69:247–257

Dews PB (1955) Studies on behavior. I. Differential sensitivity to pentobarbital of pecking performance in pigeons depending on the schedule of reward. J Pharmacol Exp Ther 113:393–401

Dews PB (1958) Studies on behavior. IV. Stimulant actions of methamphetamine. J Pharmacol Exp Ther 122:137–147

Diller JW, Saunders BT, Anderson KG (2008) Effects of acute and repeated administration of caffeine on temporal discounting in rats. Pharmacol Biochem Behav 89:546–555

El Yacoubi M, Ledent C, Parmentier M, Bertorelli R, Ongini E, Costentin J, Vaugeois JM (2001) Adenosine A2A receptor antagonists are potential antidepressants: evidence based on pharmacology and A2A receptor knockout mice. Br J Pharmacol 134:68–77

El Yacoubi M, Costentin J, Vaugeois JM (2003) Adenosine A2A receptors and depression. Neurology 61(S6):S82–S87

Elbers R, van Wegen EEH, Rochester R, Hetherington V, Nieuwboer A, Willems AM, Jones D, Kwakkel G (2009) Is impact of fatigue an independent factor associated with physical activity in idiopathic Parkinson's disease. Mov Disord 24:1512–1518

Farrar AM, Pereira M, Velasco F, Hockemeyer J, Müller CE, Salamone JD (2007) Adenosine A(2A) receptor antagonism reverses the effects of dopamine receptor antagonism on instrumental output and effort-related choice in the rat: implications for studies of psychomotor slowing. Psychopharmacology (Berl) 191:579–586

Farrar AM, Font L, Pereira M, Mingote SM, Bunce JG, Chrobak JJ, Salamone JD (2008) Forebrain circuitry involved in effort-related choice: injections of the $GABA_A$ agonist muscimol into ventral pallidum alters response allocation in food-seeking behavior. Neuroscience 152:321–330

Farrar AM, Segovia KN, Randall PA, Nunes EJ, Collins LE, Stopper CM, Port RG, Hockemeyer J, Müller CE, Correa M, Salamone JD (2010) Nucleus accumbens and effort-related functions: behavioral and neural markers of the interactions between adenosine A2A and dopamine D2 receptors. Neuroscience 166:1056–1067

Ferré S (1997) Adenosine-dopamine interactions in the ventral striatum. Implications for the treatment of schizophrenia. Psychopharmacology (Berl) 133:107–120

Ferré S (2008) An update on the mechanisms of the psychostimulant effects of caffeine. J Neurochem 105:1067–1079

Ferré S, Popoli P, Gimenez-Llort L, Rimondini R, Muller CE, Stromberg I, Ogren SO, Fuxe K (2001) Adenosine/dopamine interaction: implications for the treatment of Parkinson's disease. Parkinsonism Relat Disord 7:235–241

Ferré S, Ciruela F, Canals M, Marcellino D, Burgueno J, Casado V, Hillion J, Torvinen M, Fanelli F, Benedetti PP, Goldberg SR, Bouvier M, Fuxe K, Agnati LF, Lluis C, Franco R, Woods A (2004) Adenosine A2A-dopamine D2 receptor-receptor heteromers. Targets for neuro-psychiatric disorders. Parkinsonism Relat Disord 10:265–271

Ferré S, Ciruela F, Borycz J, Solinas M, Quarta D, Antoniou K, Quiroz C, Justinova Z, Lluis C, Franco R, Goldberg SR (2008a) Adenosine A1-A2A receptor heteromers: new targets for caffeine in the brain. Front Biosci 13:2391–2399

Ferré S, Quiroz C, Woods AS, Cunha R, Popoli P, Ciruela F, Lluis C, Franco R, Azdad K, Schiffmann SN (2008b) An update on adenosine A2A-dopamine D2 receptor interactions: implications for the function of G protein-coupled receptors. Curr Pharm Des 14:1468–1474

Fink JS, Weaver DR, Rivkees SA, Peterfreund RA, Pollack AE, Adler EM, Reppert SM (1992) Molecular cloning of the rat A2 adenosine receptor: selective co-expression with D2 dopamine receptors in rat striatum. Brain Res Mol Brain Res 14:186–195

Floresco SB, Ghods-Sharifi S (2007) Amygdala-prefrontal cortical circuitry regulates effort-based decision making. Cereb Cortex 17:251–260

Floresco SB, Tse MT, Ghods-Sharifi S (2008) Dopaminergic and glutamatergic regulation of effort- and delay-based decision making. Neuropsychopharmacology 33:1966–1979

Font L, Mingote S, Farrar AM, Pereira M, Worden L, Stopper C, Port RG, Salamone JD (2008) Intra-accumbens injections of the adenosine A2A agonist CGS 21680 affect effort-related choice behavior in rats. Psychopharmacology (Berl) 199:515–526

Friedman JH, Brown RG, Comella C, Garber CE, Krupp LB, Lou JS, Marsh L, Nail L, Shulman L, Taylor CB (2007) Fatigue in Parkinson's disease: a review. Mov Disord 22:297–308

Fuxe K, Agnati LF, Jacobsen K, Hillion J, Canals M, Torvinen M, Tinner-Staines B, Staines W, Rosin D, Terasmaa A, Popoli P, Leo G, Vergoni V, Lluis C, Ciruela F, Franco R, Ferré S (2003) Receptor heteromerization in adenosine A2A receptor signaling: relevance for striatal function and Parkinson's disease. Neurology 61:S19–S23

Fuxe K, Ferré S, Genedani S, Franco R, Agnati LF (2007) Adenosine receptor-dopamine receptor interactions in the basal ganglia and their relevance for brain function. Physiol Behav 92:210–217

Garrett BE, Holtzman SG (1994) D1 and D2 dopamine receptor antagonists block caffeine-induced stimulation of locomotor activity in rats. Pharmacol Biochem Behav 47:89–94

Hanff TC, Furst SJ, Minor TR (2010) Biochemical and anatomical substrates of depression and sickness behavior. Isr J Psychiatry Relat Sci 47:64–71

Harris RA, Snell D, Loh HH (1978) Effects of d-amphetamine, monomethoxyamphetamines and hallucinogens on schedule-controlled behavior. J Pharmacol Exp Ther 204:103–117

Hauber W, Sommer S (2009) Prefrontostriatal circuitry regulates effort-related decision making. Cereb Cortex 19:2240–2247

Hickie I, Ward P, Scott E, Haindl W, Walker B, Dixon J, Turner K (1999) Neo-striatal rCBF correlates of psychomotor slowing in patients with major depression. Psychiatry Res 92:75–81

Hillion J, Canals M, Torvinen M, Casado V, Scott R, Terasmaa A, Hansson A, Watson S, Olah ME, Mallol J, Canela EI, Zoli M, Agnati LF, Ibanez CF, Lluis C, Franco R, Ferre S, Fuxe K (2002) Coaggregation, cointernalization, and codesensitization of adenosine A2A receptors and dopamine D2 receptors. J Biol Chem 277:18091–18097

Hodgson RA, Bertorelli R, Varty GB, Lachowicz JE, Forlani A, Fredduzzi S, Cohen-Williams ME, Higgins GA, Impagnatiello F, Nicolussi E, Parra LE, Foster C, Zhai Y, Neustadt BR, Stamford AW, Parker EM, Reggiani A, Hunter JC (2009) Characterization of the potent and highly selective A2A receptor antagonists preladenant and SCH 412348 [7-[2-[4-2,4-difluorophenyl]-1-piperazinyl]ethyl]-2-(2-furanyl)-7 H-pyrazolo[4,3-e][1,2,4]triazolo[1,5-c]pyrimidin-5-amine] in rodent models of movement disorders and depression. J Pharmacol Exp Ther 330:294–303

Ishiwari K, Weber SM, Mingote S, Correa M, Salamone JD (2004) Accumbens dopamine and the regulation of effort in food-seeking behavior: modulation of work output by different ratio or force requirements. Behav Brain Res 151:83–91

Ishiwari K, Madson LJ, Farrar AM, Mingote SM, Valenta JP, DiGianvittorio MD, Frank LE, Correa M, Hockemeyer J, Muller C, Salamone JD (2007) Injections of the selective adenosine A2A antagonist MSX-3 into the nucleus accumbens core attenuate the locomotor suppression induced by haloperidol in rats. Behav Brain Res 178:190–199

Jenner P (2005) Istradefylline, a novel adenosine A_{2A} receptor antagonist, for the treatment of Parkinson's disease. Expert Opin Investig Drugs 14:729–738

Justinova Z, Ferre S, Segal PN, Antoniou K, Solinas M, Pappas LA, Highkin JL, Hockemeyer J, Munzar P, Goldberg SR (2003) Involvement of adenosine A1 and A2A receptors in the adenosinergic modulation of the discriminative-stimulus effects of cocaine and methamphetamine in rats. J Pharmacol Exp Ther 307:977–986

Justinova Z, Ferré S, Barnes C, Wertheim CE, Pappas LA, Goldberg SR, Le Foll B (2009) Effects of chronic caffeine exposure on adenosinergic modulation of the discriminative-stimulus effects of nicotine, methamphetamine, and cocaine in rats. Psychopharmacology (Berl) 203:355–367

Karcz-Kubicha M, Antoniou K, Terasmaa A, Quarta D, Solinas M, Justinova Z, Pezzola A, Reggio R, Müller CE, Fuxe K, Goldberg SR, Popoli P, Ferré S (2003) Involvement of adenosine A1 and A2A receptors in the motor effects of caffeine after its acute and chronic administration. Neuropsychopharmacology 28:1281–1291

Koch M, Schmid A, Schnitzler HU (2000) Role of nucleus accumbens dopamine D1 and D2 receptors in instrumental and Pavlovian paradigms of conditioned reward. Psychopharmacology (Berl) 152:67–73

Koob GF, Riley SJ, Smith SC, Robbins TW (1978) Effects of 6-hydroxydopamine lesions of the nucleus accumbens septi and olfactory tubercle on feeding, locomotor activity, and amphetamine anorexia in the rat. J Comp Physiol Psychol 92:917–927

Lopez-Cruz L, Pardo M, Dosda A, Salamone JD, Correa M (2011) Comparison between high doses of caffeine and theophylline on motor and anxiogenic effects in CD1 mice: studies of acute and chronic administration. Behav Pharmacol 22:e71–e72

Marston HM, Finlayson K, Maemoto T, Olverman HJ, Akahane A, Sharkey J, Butcher SP (1998) Pharmacological characterization of a simple behavioral response mediated selectively by central adenosine A1 receptors, using in vivo and in vitro techniques. J Pharmacol Exp Ther 285:1023–1030

McCullough LD, Salamone JD (1992) Involvement of nucleus accumbens dopamine in the motor activity induced by periodic food presentation: a microdialysis and behavioral study. Brain Res 592:29–36

McKim WA (1980) The effect of caffeine, theophylline and amphetamine on operant responding of the mouse. Psychopharmacology (Berl) 68:135–138

Mingote S, Weber SM, Ishiwari K, Correa M, Salamone JD (2005) Ratio and time requirements on operant schedules: effort-related effects of nucleus accumbens dopamine depletions. Eur J Neurosci 21:1749–1757

Mingote S, Font L, Farrar AM, Vontell R, Worden LT, Stopper CM, Port RG, Sink KS, Bunce JG, Chrobak JJ, Salamone JD (2008) Nucleus accumbens adenosine A2A receptors regulate exertion of effort by acting on the ventral striatopallidal pathway. J Neurosci 28:9037–9046

Minor TR, Huang Q, Foley EA (2003) Cytokine-purine interactions in behavioral depression in rats. Integr Physiol Behav Sci 38:189–202

Minor TR, Huang Q, Witt AE (2006) Cytokine-purine interactions in traumatic stress, behavioral depression, and sickness. CNS Neurol Disord Drug Targets 5:547–560

Morelli M, Pinna A (2002) Interaction between dopamine and adenosine A_{2A} receptors as a basis for the treatment of Parkinson's disease. Neurol Sci 22:71–72

Mott AM, Nunes EJ, Collins LE, Port RG, Sink KS, Hockemeyer J, Müller CE, Salamone JD (2009) The adenosine A_{2A} antagonist MSX-3 reverses the effects of the dopamine antagonist haloperidol on effort-related decision making in a T-maze cost/benefit procedure. Psychopharmacology (Berl) 204:103–112

Nowend KL, Arizzi M, Carlson BB, Salamone JD (2001) D1 or D2 antagonism in nucleus accumbens core or dorsomedial shell suppresses lever pressing for food but leads to compensatory increases in chow consumption. Pharmacol Biochem Behav 69:373–382

Nunes EJ, Randall PA, Santerre JL, Given AB, Sager TN, Correa M, Salamone JD (2010) Differential effects of selective adenosine antagonists on the effort-related impairments induced by dopamine D1 and D2 antagonism. Neuroscience 170:268–280

Pardo M, Lopez-Cruz L, Hockemeyer J, Müller CE, Salamone JD, Correa M (2010) Interaction between dopamine D2 receptor and adenosine A1 and A2A receptors: studies of c-Fos immunoreactivity, behavioral activation and effort related decision making. Program No. 916.19. 2010. Neuroscience Meeting Planner, Society for Neuroscience, 2010, San Diego, CA. Online

Pardo M, Randall PA, Nunes E, Lopez-Cruz L, Correa M, Salamone JD (2011) A progressive ratio/chow feeding concurrent choice task as a measure of effort-related decision making: effects of dopamine and adenosine antagonism. Behav Pharmacol 22:e58

Pardo M, Lopez-Cruz L, Valverde O, Ledent C, Baqi Y, Müller CE, Salamone JD, Correa M (2012) Adenosine A_{2A} receptor antagonism and genetic deletion attenuate the effects of dopamine D_2 antagonism on effort-based decision making in mice. Neuropharmacology 62(5–6): 2068–2077

Pinna A (2010) Novel investigational adenosine A2A receptor antagonists for Parkinson's disease. Expert Opin Investig Drugs 18:1619–1631

Popoli P, Reggio R, Pèzzola A, Fuxe K, Ferré S (1998) Adenosine A1 and A2A receptor antagonists stimulate motor activity: evidence for an increased effectiveness in aged rats. Neurosci Lett 251:201–204

Rampello L, Nicoletti G, Raffaele R (1991) Dopaminergic hypothesis for retarded depression: a symptom profile for predicting therapeutical responses. Acta Psychiatr Scand 84:552–554

Randall PA, Nunes EJ, Janniere SL, Stopper CM, Farrar AM, Sager TN, Baqi Y, Hockemeyer J, Müller CE, Salamone JD (2011a) Stimulant effects of adenosine antagonists on operant behavior: differential actions of selective A2A and A1 antagonists. Psychopharmacology (Berl) 216:173–186

Randall PA, Pardo M, Nunes EJ, Lopez-Cruz L, Blodgett A, Lingiah K, Leser C, Vemuri VK, Mackriyannis A, Baqi Y, Müller CE, Correa M, Salamone JD (2011b) Effort-related choice behavior as assessed by a progressive ratio/chow feeding task: differential effects of DA D2 antagonism, adenosine A_{2A} antagonism, cannabinoid CB1 antagonism and pre-feeding. Program No. 732.03. 2011. Neuroscience Meeting Planner. Society for Neuroscience, 2011, Washington, DC. Online

Robbins TW, Everitt BJ (2007) A role for mesencephalic dopamine in activation: a commentary on Berridge (2007). Psychopharmacology (Berl) 191:433–437
Robbins TW, Koob GF (1980) Selective disruption of displacement behaviour by lesions of the mesolimbic dopamine system. Nature 285:409–412
Rogers D, Lees AJ, Smith E, Trimble M, Stern GM (1987) Bradyphrenia in Parkinson's disease and psychomotor retardation in depressive illness. An experimental study. Brain 110: 761–776
Salamone JD (1988) Dopaminergic involvement in activational aspects of motivation: effects of haloperidol on schedule induced activity, feeding and foraging in rats. Psychobiology 16:196–206
Salamone JD (1991) Behavioral pharmacology of dopamine systems: a new synthesis. In: Willner P, Scheel-Kruger J (eds) The mesolimbic dopamine system: from motivation to action. Cambridge University Press, Cambridge, pp 599–613
Salamone JD (1992) Complex motor and sensorimotor functions of striatal and accumbens dopamine: involvement in instrumental behavior processes. Psychopharmacology (Berl) 107:160–174
Salamone JD (2010) Preladenant, a novel adenosine A(2A) receptor antagonist for the potential treatment of parkinsonism and other disorders. IDrugs 13:723–731
Salamone JD, Correa M (2002) Motivational views of reinforcement: implications for understanding the behavioral functions of nucleus accumbens dopamine. Behav Brain Res 137:3–25
Salamone JD, Correa M (2009) Dopamine/adenosine interactions involved in effort-related aspects of food motivation. Appetite 53:422–425
Salamone JD, Steinpreis RE, McCullough LD, Smith P, Grebel D, Mahan K (1991) Haloperidol and nucleus accumbens dopamine depletion suppress lever pressing for food but increase free food consumption in a novel food choice procedure. Psychopharmacology (Berl) 104:515–521
Salamone JD, Cousins MS, Bucher S (1994) Anhedonia or anergia? Effects of haloperidol and nucleus accumbens dopamine depletion on instrumental response selection in a T-maze cost/benefit procedure. Behav Brain Res 65:221–229
Salamone JD, Cousins MS, Snyder BJ (1997) Behavioral functions of nucleus accumbens dopamine: empirical and conceptual problems with the anhedonia hypothesis. Neurosci Biobehav Rev 21:341–359
Salamone JD, Wisniecki A, Carlson BB, Correa M (2001) Nucleus accumbens dopamine depletions make animals highly sensitive to high fixed ratio requirements but do not impair primary food reinforcement. Neuroscience 105:863–870
Salamone JD, Arizzi M, Sandoval MD, Cervone KM, Aberman JE (2002) Dopamine antagonists alter response allocation but do not suppress appetite for food in rats: contrast between the effects of SKF 83566, raclopride and fenfluramine on a concurrent choice task. Psychopharmacology (Berl) 160:371–380
Salamone JD, Correa M, Mingote SM, Weber SM, Farrar AM (2006) Nucleus accumbens dopamine and the forebrain circuitry involved in behavioral activation and effort-related decision making: implications of understanding anergia and psychomotor slowing and depression. Curr Psychiatry Rev 2:267–280
Salamone JD, Correa M, Farrar A, Mingote SM (2007) Effort-related functions of nucleus accumbens dopamine and associated forebrain circuits. Psychopharmacology (Berl) 191:461–482
Salamone JD, Betz AJ, Ishiwari K, Felsted J, Madson L, Mirante B, Clark K, Font L, Korbey S, Sager TN, Hockemeyer J, Muller CE (2008) Tremorolytic effects of adenosine A2A antagonists: implications for parkinsonism. Front Biosci 13:3594–3605
Salamone JD, Correa M, Farrar AM, Nunes EJ, Pardo M (2009a) Dopamine, behavioral economics, and effort. Front Behav Neurosci 3:13
Salamone JD, Farrar AM, Font L, Patel V, Schlar DE, Nunes EJ, Collins LE, Sager TN (2009b) Differential actions of adenosine A1 and A2A antagonists on the effort-related effects of dopamine D2 antagonism. Behav Brain Res 201:216–222
Sanger DJ, Blackman DE (1974) Rate-dependent effects of drugs: a review of the literature. Pharmacol Biochem Behav 4:73–83

Santerre JL, Nunes EJ, Kovner R, Leser C, Randall PA, Collins-Praino LE, Correa M, Baqi Y, Müller CE, Salamone JD (2012) The novel adenosine A2A antagonist prodrug MSX-4 is effective in animal models related to motivational and motor function. Pharmacol Biochem Behav, in press

Schiffmann SN, Jacobs O, Vanderhaeghen JJ (1991) Striatal restricted adenosine A2 receptor (RDC8) is expressed by enkephalin but not by substance P neurons: an in situ hybridization histochemistry study. J Neurochem 57:1062–1067

Schmidt K, Nolte-Zenker B, Patzer J, Bauer M, Schmidt LG, Heinz A (2001) Psychopathological correlates of reduced dopamine receptor sensitivity in depression, schizophrenia, and opiate and alcohol dependence. Pharmacopsychiatry 34:66–72

Schweimer J, Hauber W (2005a) Involvement of the rat anterior cingulate cortex in control of instrumental responses guided by reward expectancy. Learn Mem 12:334–342

Schweimer J, Hauber W (2005b) Dopamine D1 receptors in the anterior cingulate cortex regulate effort-based decision making. Learn Mem 13:777–782

Schweimer J, Saft S, Hauber W (2005) Involvement of catecholamine neurotransmission in the rat anterior cingulate in effort-related decision making. Behav Neurosci 119:1687–1692

Simola N, Fenu S, Baraldi PG, Tabrizi MA, Morelli M (2004) Blockade of adenosine A2A receptors antagonizes parkinsonian tremor in the rat tacrine model by an action on specific striatal regions. Exp Neurol 189:182–188

Sink KS, Vemuri VK, Olszewska T, Makriyannis A, Salamone JD (2008) Cannabinoid CB1 antagonists and dopamine antagonists produce different effects on a task involving response allocation and effort-related choice in food-seeking behavior. Psychopharmacology (Berl) 196:565–574

Stahl SM (2002) The psychopharmacology of energy and fatigue. J Clin Psychiatry 63:7–8

Svenningsson P, Le Moine C, Fisone G, Fredholm BB (1999) Distribution, biochemistry and function of striatal adenosine A2A receptors. Prog Neurobiol 59:355–396

Takahashi RN, Pamplona FA, Prediger RD (2008) Adenosine receptor antagonists for cognitive dysfunction: a review of animal studies. Front Biosci 13:2614–2632

Tellez N, Alonso J, Rio M, Tintore M, Nos C, Montalban X, Rovira A (2008) The basal ganglia: a substrate for fatigue in multiple sclerosis. Neuroradiology 50:17–23

Treadway MT, Zald DH (2011) Reconsidering anhedonia in depression: lessons from translational neuroscience. Neurosci Biobehav Rev 35:537–555

Tylee A, Gastpar M, Lepine JP, Mendlewicz J (1999) DEPRES II (Depression Research in European Society II): a patient survey of the symptoms, disability and current management of depression in the community. DEPRES Steering Committee. Int Clin Psychopharmacol 14:139–151

van den Bos R, van der Harst J, Jonkman S, Schilders M, Spruijt B (2006) Rats assess costs and benefits according to an internal standard. Behav Brain Res 171:350–354

Volkow ND, Chang L, Wang GJ, Fowler JS, Leonido-Yee M, Franceschi D, Sedler MJ, Gatley SJ, Hitzemann R, Ding YS, Logan J, Wong C, Miller EN (2001) Association of dopamine transporter reduction with psychomotor impairment in methamphetamine abusers. Am J Psychiatry 158:377–382

Vontell R, Segovia KN, Betz AJ, Mingote S, Goldring K, Cartun RW, Salamone JD (2010) Immunocytochemistry studies of basal ganglia adenosine A2A receptors in rat and human tissue. J Histotechnol 33:41–47

Wakabayashi KT, Fields HL, Nicola SM (2004) Dissociation of the role of nucleus accumbens dopamine in responding to reward-predictive cues and waiting for reward. Behav Brain Res 154:19–30

Wallace M, Singer G, Finlay J, Gibson S (1983) The effect of 6-OHDA lesions of the nucleus accumbens septum on schedule-induced drinking, wheelrunning and corticosterone levels in the rat. Pharmacol Biochem Behav 18:129–136

Walton ME, Bannerman DM, Rushworth MF (2002) The role of rat medial frontal cortex in effort-based decision making. J Neurosci 22:10996–11003

Walton ME, Bannerman DM, Alterescu K, Rushworth MF (2003) Functional specialization within medial frontal cortex of the anterior cingulate for evaluating effort-related decisions. J Neurosci 23:6475–6479

Walton ME, Kennerley SW, Bannerman DM, Phillips PE, Rushworth MF (2006) Weighing up the benefits of work: behavioral and neural analyses of effort-related decision making. Neural Netw 19:1302–1314

Wardas J, Konieczny J, Lorenc-Koci E (2001) SCH 58261, an A(2A) adenosine receptor antagonist, counteracts parkinsonian-like muscle rigidity in rats. Synapse 41:160–171

Worden LT, Shahriari M, Farrar AM, Sink KS, Hockemeyer J, Muller CE, Salamone JD (2009) The adenosine A2A antagonist MSX-3 reverses the effort-related effects of dopamine blockade: differential interaction with D1 and D2 family antagonists. Psychopharmacology (Berl) 203:489–499

Chapter 24
Adenosine and Autism: Physiological Symptoms and Metabolic Opportunities

Julia Svedova, Inge-Marie Eigsti, and Susan A. Masino

Abstract Autism is characterized by impairments in social interactions, deficits in communication, and repetitive behaviors or stereotyped interests. People with autism also have an increased incidence of seizures, gastrointestinal disorders, sleep disruption, and psychological disorders such as anxiety and depression. Purine abnormalities have been associated with autism, but evidence has not typically linked purines to either the neurochemistry or behavioral hallmarks of the disorder. Here we review emerging and converging lines of evidence suggesting that a metabolic increase in the influence of the purine nucleoside adenosine has the potential to play a key role in reducing multiple behavioral and physiological symptoms of autism. *First*, results from a questionnaire study of parents of children with autism indicated that stimuli and activities predicted to increase adenosine are observed to reduce symptoms of autism. *Second*, a ketogenic diet—a high-fat, low-carbohydrate regimen used to treat epilepsy—can reduce electrographic seizures in mice via actions at adenosine A_1 receptors, and other research has shown that a ketogenic diet can improve symptoms of autism in children with autism and improve sleep in children with epilepsy. In animal studies, this established metabolic therapy has been shown to increase ATP levels and reduce levels of the intracellular adenosine-metabolizing enzyme adenosine kinase, suggesting possible mechanisms for increased extracellular adenosine. While the number of studies investigating each of these areas is limited to date, accumulating evidence suggests that metabolic strategies to increase adenosine could address some of the core and comorbid symptoms of autism. Further research is needed.

J. Svedova • S.A. Masino, Ph.D. (✉)
Neuroscience Program/Psychology Department, Life Sciences Center 210, Trinity College, 300 Summit St., Hartford, CT 06106, USA
e-mail: Susan.masino@trincoll.edu

I.-M. Eigsti
Psychology Department, University of Connecticut, Storrs, CT, USA

S. Masino and D. Boison (eds.), *Adenosine: A Key Link between Metabolism and Brain Activity*, DOI 10.1007/978-1-4614-3903-5_24,

Keywords Autism • Adenosine • Epilepsy • Ketogenic diet • Metabolism • Adenosine receptors • Autism spectrum disorders • Symptomatology • Behavior • Rett syndrome

24.1 Autism Spectrum Disorders: Core Symptoms, Prevalence, and Classification

Autism encompasses pervasive developmental disorders characterized by impairments in reciprocal social interactions and communication skills, along with the presence of restricted, stereotyped, and repetitive behaviors and interests (Buitelaar and Willemsen-Swinkels 2000). There is significant variability in the severity of autism spectrum disorders (ASDs), with IQs ranging from the severely impaired to the gifted range. ASD is more common in boys than in girls (average reported ratio of 4:1) and associated with mental retardation in up to 75 % of cases (Buitelaar and Willemsen-Swinkels 2000; Fombonne 2005).

The ASDs comprise three primary diagnoses: Autistic Disorder ("classic autism"), Pervasive Developmental Disorder—Not Otherwise Specified (PDD-NOS, sometimes known as "Autism Lite"), and Asperger's Disorder. Other developmental disorders on the spectrum, Child Disintegrative Disorder and Rett Syndrome (found exclusively in girls), are far less common but share some features of ASDs. Although these disorders vary in severity as well as physiological and developmental aspects, they share similar behavioral manifestations. In particular, the diagnostic criteria for ASD are classified into three major categories: qualitative impairments in reciprocal social interactions, qualitative impairments in verbal and nonverbal forms of communication, and stereotyped repetitive patterns of behavior (American Psychiatric Association 2000). Table 24.1 shows common symptoms in these primary categories.

According to data from the Center for Disease Control and Prevention's Autism and Developmental Disabilities Monitoring Network (2006),[1] ASD affects

Table 24.1 Behavioral symptoms of autistic disorder as listed in the DSM-IV

Primary category	Examples
Impairments in social interactions	Lack of eye contact Failure to develop relationships with others Lack of social or emotional reciprocity
Impairments in communication	Delayed development or a complete lack of language Inability to initiate or sustain conversation Stereotyped repetitive use of language
Stereotyped repetitive behavior	Inflexible adherence to routines or rituals Preoccupation with objects of interest Stereotyped and repetitive motor movements

[1]National Institute of Mental Health: Autism. http://www.nimh.nih.gov/statistics/1AUT_CHILD.shtml. Accessed 1/31/2012.

approximately 1 in 110 children (91 in 10,000; see also Lord (2011)). Although the reported rates of ASD have increased steadily in the past five decades (from 4.5 in 10,000 in 1966, according to Lotter (1966)), it is still unclear whether these represent secular increases and/or changes in surveillance. A recent study evaluating an entire community of 7–12 year-olds in Seoul, South Korea, found a startling incidence of 1 in 38; the significantly higher prevalence appeared to be largely because of high-functioning children in normal classrooms who were undiagnosed previously (Kim et al. 2011), suggesting actual increases through diagnostic procedures. However, a recent epidemiological study in England examining data from over 7,000 individuals in eight different age groups, stratified by decade, indicated that there was no change in rates over time (Brugha et al. 2011). These data support the possibility that changes in diagnostic criteria and greater public awareness may be primary contributors to reported changes in prevalence rates (Anonymous 2010).

24.1.1 Physiological and Behavioral Symptoms Associated with Autism

In addition to core diagnostic symptoms, children with autism tend to manifest multiple other behavioral, physiological, and structural abnormalities (Levy et al. 2010). Behavioral difficulties are likely related to core features of autism (e.g., obsessiveness, perseveration), sensory abnormalities, or other comorbid symptoms (e.g., self-injury, aggression). Neurological problems, such as epilepsy and sleep disturbances, and psychological disorders including depression and anxiety, are also prevalent in autism (Canitano 2007; Gurney et al. 2006; Newschaffer et al. 2007; Polimeni et al. 2005). In addition, compared to an unaffected population, individuals with autism tend to suffer from macrocephaly (Fidler et al. 2000), immunological abnormalities (Goines and Van de Water 2010; Weizman et al. 1982), gastrointestinal problems (Horvath et al. 1999), and food and skin allergies (Gurney et al. 2006). Recent structural and functional magnetic resonance imaging studies have consistently documented neurological changes including reduced volume of white matter tracts (Bode et al. 2011) such as the corpus callosum (Egaas et al. 1995) and decreased functional connectivity during cognitive tasks (e.g., Gepner and Deruelle 2008; Just et al. 2004; Kana et al. 2006; Mostofsky and Ewen 2011; see Eigsti and Shapiro (2003) for review).

24.1.2 Causes of Autism

In most cases, the cause of autism remains unknown, although there are several genetic and environmental factors considered likely to contribute to the pathology of ASD (Buitelaar and Willemsen-Swinkels 2000; Libbey et al. 2005). Twin studies have revealed concordance rates of 36–95 % in monozygotic twins and relative discordance in dizygotic twins (Folstein and Rutter 1977; Ritvo et al. 1985; Ronald and Hoekstra 2011). Although no single gene has been identified (Veenstra-VanderWeele and Cook 2004), a recent detailed study found copy number variations (CNVs) of several candidate genes in a linkage locus region on chromosome 7 called AUTS1 (Maestrini et al. 2010); recent work has suggested that CNVs are the primary genetic risk factor. Overall, however, a known genetic or structural disorder associated with autism explains only a minor part of the autistic population, possibly 10 % of all cases (Bailey et al. 1996).

Because of the known genetic association, many studies have searched for potential autism susceptibility genes, including genes involved in purine metabolism. In particular, a genetic polymorphism in adenosine deaminase (ADA) was investigated as a potential candidate for an autism susceptibility gene. ADA is a polymorphic enzyme with two codominant allelic forms which have different levels of activity (Battistuzi et al. 1981). In terms of influencing central nervous system (CNS) activity, ADA deaminates adenosine to inosine—thus reducing available adenosine—but there is evidence that it may also enhance adenosine binding at the widely distributed high-affinity inhibitory adenosine A_1 receptor (A_1R; Ciruela et al. 1996). The relative influence of these two mechanisms on synaptic transmission is unknown. Two studies of autistic subjects of Italian descent found an increased incidence of the less active ADA form (Bottini et al. 2001; Persico et al. 2000), whereas a study of autistic subjects in North America did not find any differences (Hettinger et al. 2008). Therefore, the role of ADA polymorphism in autism pathology remains unclear. Also related to genetic changes in adenosine signaling, a recent study by Freitag et al. suggested a link among adenosine A_{2A} receptor ($A_{2A}R$) variant ADORA2A, increased anxiety, and autistic symptoms in ASD (Freitag et al. 2010).

Purine metabolism disturbances have been implicated in the pathogenesis of autism. In addition to an early report connecting autistic symptoms to increased de novo purine synthesis (Nyhan et al. 1969), Page and Coleman found a fourfold increase in de novo purine synthesis and a decreased ratio of adenine to guanine in a small sample of autistic subjects with increased excretion of uric acid. The exact enzyme defect in purine metabolism, however, was not identified (Page and Coleman 2000). As noted above, further studies will be necessary to elucidate whether changes in purine metabolism and corresponding genetic variants contribute to pathogenesis of ASD.

Currently there is evidence that environmental factors are likely to be involved in the epidemiology of autism. Although indirect evidence (e.g., from twins; Gardener et al. 2009; Hallmayer et al. 2011; Newschaffer 2006) has suggested a role for environmental exposures to toxins such as metals, teratogens, perinatal insult, and prenatal infections such as rubella and cytomegalovirus, these account for few cases (Muhle et al. 2004). Specifically considering metabolism as an environmental factor, a recent preliminary report—part of the Child Autism Risks from Genetic and the Environment (CHARGE) initiative—noted that diabetes, maternal obesity, or gestational/type II diabetes during pregnancy were found to be associated with a higher incidence of autism (Krakowiak et al. 2011). These initial findings suggest that research should continue to examine the role of metabolic factors and mitochondrial dysfunction in ASD.

The aggregate contributions of epigenetic changes to ASD are not well understood but the possibilities are staggering. Epigenetic mechanisms integrate environmental and genetic factors, and help explain the relatively limited success in finding clear genetic causes for autism despite known genetic susceptibility. Regarding purines, and specifically adenosine, one epigenetic mechanism potentially linking adenosine to autism is altered DNA methylation and thus altered gene expression. Changes in DNA methylation, particularly in X-linked genes, have been postulated to account for autism's sporadic occurrence and its prevalence in males (Jones et al. 2008). Children with autism were determined to have a reduced ratio of *S*-adenosylmethionine (SAM) to *S*-adenosylhomocysteine (SAH), an indicator of methylation capacity (James et al. 2006). Previously, changes in the adenosine system have been shown to produce epigenetic changes: a loss of adenosine kinase, the main adenosine-regulating enzyme in the CNS, has been shown to alter DNA transmethylation in vivo (Boison et al. 2002).

24.1.3 Neurochemistry and Autism

It is widely believed that disparate factors which cause autism result in ongoing neurochemical changes in the brain. A wide array of neurotransmitter systems has been studied, including serotonin, dopamine, norepinephrine, acetylcholine, oxytocin, endogenous opioids, cortisol, glutamate, and gamma-aminobutyric acid (Lam et al. 2006). While many results have been compromised by small sample sizes and a lack of appropriate comparison samples, findings to date suggest little support for a role of neurotransmitter differences in the symptoms or etiology of ASD.

One notable exception is the serotonin system, in which differences have consistently been found in individuals with ASD. Serotonin is involved in regulating mood, eating, body temperature, pain sensitivity, hormone release, and general arousal levels. Over 30 % of subjects with ASD exhibit elevated blood concentrations of serotonin (for review see Buitelaar and Willemsen-Swinkels (2000)). However, the importance of hyperserotonemia is unknown. To date, it has not been linked to specific social or communicative deficits in ASD, or to repetitive behaviors (Cook et al. 1990; Cuccaro et al. 1993); furthermore, hyperserotonemia is linked to multiple medical and neuropsychiatric disorders (see Lam et al. (2006) for discussion).

To date there is some limited evidence for the role of the noradrenergic system in ASD (Beversdorf 2010); animal studies indicate that the dopaminergic system could be involved in hyperactivity and stereotyped behavior common in autism. Related to this, abnormalities in monoamine oxidase, particularly monamine oxidase A, have been considered (Yoo et al. 2009), with consequent effects on multiple monoamine transmitters. Other lines of evidence suggest that there could be decreased GABAergic or enhanced glutamatergic transmission (Fatemi et al. 2009; Purcell et al. 2001).

Beyond classical transmitters, oxytocin has been implicated in the symptomatology of autism. Whereas children with autism tend to have significantly lower levels of oxytocin in plasma compared to controls, adults with autism have higher levels of oxytocin than controls (Bartz and Hollander 2008). Excessive opioid system activity in the brain has been hypothesized to contribute to the social impairments characteristic of autism. Despite these and other hypotheses regarding underlying neurochemistry, there is currently no biologically validated neurochemical explanation for symptoms of ASD. Medications can be prescribed to treat some of the comorbid physiological and behavioral features, including depression, anxiety, aggression, hyperactivity/impulsivity, seizures, or sleep problems (Findling 2005). Studies estimated that 46.7 % of autistic children and adolescents (Witwer and Lecavalier 2005) and around 60 % of autistic adolescents and adults (Esbensen et al. 2009) were taking at least one psychotropic medication. Even though psychotropic drugs remove some of the hindrances to other therapeutic efforts, they are also associated with adverse side effects such as fatigue, irritability, abnormal involuntary movements, or weight gain (Esbensen et al. 2009; Witwer and Lecavalier 2005). There are *no* current pharmacological treatments which address the core symptoms of ASD.

24.1.4 Metabolism and Autism

Gastrointestinal problems and metabolic disturbances appear to be closely related to ASD. Recent reviews have highlighted mitochondrial dysfunction and metabolism as underlying multiple aspects of ASD (Frye and Rossignol 2011; Rossignol and Frye 2012). In addition, there are several known metabolic impairments associated with autistic symptoms. These include phenylketonuria, adenylosuccinate lyase deficiency, histidinemia, or hyperuricosuric autism (Page and Coleman 2000). The incidence of other metabolic abnormalities, including food allergies (Gurney et al. 2006), and the fact that about 40–70 % of children with autism tend to suffer from gastrointestinal problems, such as chronic diarrhea or constipation (Valicenti-McDermott et al. 2006; Wang et al. 2011), suggest that some types of ASD may be treated with dietary and/or metabolic approaches.

In the past few decades, several diets have been described in terms of their effects on symptoms of autism, including the gluten-free, casein-free diet, specific carbohydrate diet, body ecology diet, or ketogenic diet (Srinivasan 2009). A survey by Witwer and Lecavalier (Witwer and Lecavalier 2005) estimated that 15.5 % of autistic children or adolescents were on a modified diet for treatment of autism. Overall, these strategies have yielded disappointing or, at best, mixed results.

The most prominent type of diet attempted as a treatment for autism is the gluten-free, casein-free diet. The diet addresses the possibility that undigested opioid peptides may pass into the CNS and alter behavior (Srinivasan 2009). Initially promising results from a single blind study by Knivsberg et al. (2002) showed some improvements after 1 year of treatment in 20 children with autism. However, a better-controlled double-blind placebo-controlled clinical trial showed no statistically significant findings in 15 children over 12 weeks (Elder et al. 2006). Thus, specific aspects of this dietary regimen have not been supported.

Taken as a whole, the gluten-free, casein-free diet and other diets intended for treatment of autism are not supported by empirical evidence. Exceptions in the case of the ketogenic diet are discussed in detail below. Nevertheless, data from families with a variety of developmental disorders indicate significant enthusiasm for alternative and complementary therapies and interventions, and factors such as (1) characteristics of alternative practitioners and (2) increased invasiveness and costs of intervention contribute to greater adoption of these therapies. Because of the naturally waxing and waning course of ASD, and because many families make use of multiple interventions simultaneously, it is difficult to distinguish between change(s) due to one specific intervention versus another, and change(s) that is part of a broader developmental trajectory (see Levy and Hyman 2008 for review). Nevertheless, more targeted and hypothesis-driven metabolic therapies may offer some promise in ASD.

Recently the adenosine system has been considered as a possible therapeutic target for reducing symptoms of autism, and dietary and metabolic approaches may be particularly promising (Masino et al. 2009, 2011a, b; Ghanizadeh 2010; Lara 2010; Ribeiro and Sebastião 2010). Adenosine is a purine nucleoside, involved in metabolism, neuronal signaling, and gene regulation, and can be considered a homeostatic bioenergetic network regulator in the CNS (Boison et al. 2011). As described below, and illustrated in Fig. 24.1, we believe that targeting the adenosine system with metabolic approaches is a hypothesis-driven strategy which can ameliorate multiple core and comorbid symptoms of ASD.

24.2 Purines and Autism

Multiple reports over several decades have noted purine abnormalities associated with autism diagnoses (Bottini et al. 2001; Marie et al. 2002; Nyhan et al. 1969; Page and Coleman 2000). Uric acid is the end product of purine metabolism, and a high proportion of children diagnosed with autism excrete excessive uric acid in their urine; estimates range from 11 to 28 % (Coleman et al. 1976; Rosenberger-Debiesse and Coleman 1986). These purine abnormalities have not been linked directly to neural mechanisms underlying core symptoms of the disorder, and this aspect of autism has not been exploited for practical clinical benefits. Here we do not catalogue purine abnormalities in ASD, nor do we attempt to prove underlying purine or adenosine dysregulation as a metabolic underpinning of the core symptoms of ASD. Rather, we outline potential clinical benefits for purines, and particularly adenosine, for multiple core symptoms and serious comorbidities in autism.

24.2.1 Physiological and Behavioral Evidence of Decreased Adenosine

Adenosine is found throughout the extracellular space of the CNS and acts via G protein-coupled receptors. Acting at A_1Rs, adenosine is an endogenous antiseizure molecule (Dunwiddie 1999). In general, adenosine is a sleep-promoter, and antagonizing adenosine receptors with caffeine has known psychomotor effects (Burnstock

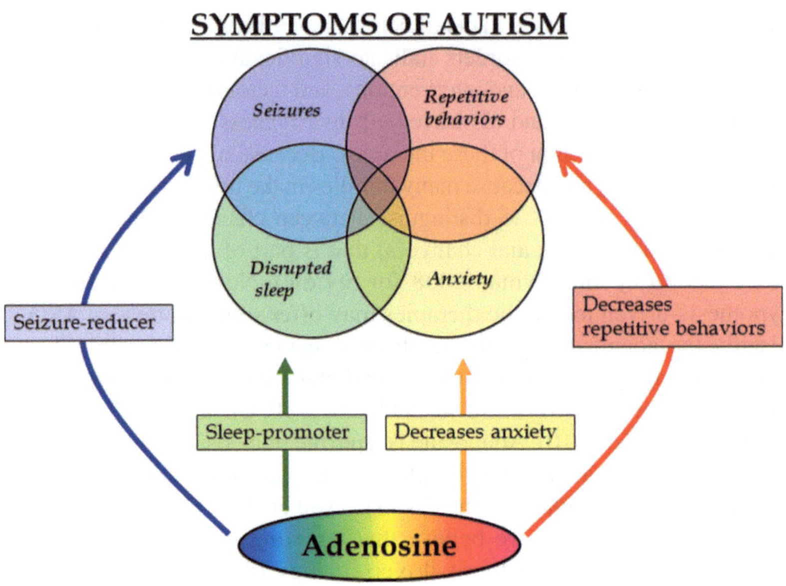

Fig. 24.1 Common comorbidities of autism, such as disrupted sleep, anxiety, seizures, and repetitive behaviors, would be expected to be reduced by an increased influence of adenosine. Adapted from Masino et al. (2011c). Available from http://www.intechopen.com/articles/show/title/adenosine-and-autism-recent-research-and-a-new-perspective

et al. 2011). Notably, people with autism have an increased incidence of epilepsy and/or abnormal EEG, and tend to suffer from poor sleep (Malow 2004). Thus it is likely that at least a subset of serious comorbidities of autism—sleep disruption, increased incidence of seizures, and perhaps anxiety as well—could benefit from increased adenosine or altered actions at A_1R or $A_{2A}R$ subtypes (Freitag et al. 2010; Masino et al. 2011a, c). Based on animal research, core symptoms such as perseverative behavior may also be reduced by increasing adenosine (Tanimura et al. 2010).

24.2.2 *A Parental Questionnaire Investigating a Link Between Adenosine and Autism*

Our group conducted an exploration into the relationship between adenosine and autism via a customized questionnaire offered to parents included in a national autism database. Over 2,000 parents with a child with a confirmed autism diagnosis were invited to participate. This questionnaire focused specifically on core symptoms of autism in their child, and queried whether these symptoms improved, did not change, or worsened relative to stimuli predicted a priori to increase, or decrease/have no effect on adenosine. After validation and analysis of responses, the set of stimuli predicted to increase adenosine was associated with the most positive impact on symptoms of autism, and this effect was highly significant. The relationship between increased adenosine and decreased symptoms of autism held across all developmental ages (2–18 years old) (Masino et al. 2011a).

Limitations of this approach are that the questionnaire relied on parent report, and could not correlate changes in behavior directly with changes in adenosine. Because the data were indirect and somewhat noisy we did not have sufficient power in our final data set to separately analyze individual behaviors. Some of these limitations were unavoidable: adenosine measurements in blood, urine, etc. are not informative, and invasive manipulations or measurements are neither practical nor ethically justified at this time. A future study could however be designed with more power and thus more specificity in the analyses.

The strengths of this approach are that all parents were naïve to study hypotheses, and all stimuli were predetermined regarding their predicted effects on adenosine. Furthermore, it is the first study to make a link between the neuromodulator adenosine and core symptoms of autism. Overall, these results comport with previous findings which each found that a subset of these stimuli reduced core and comorbid symptoms of autism (Edelson et al. 1999; Franke et al. 2006; Goldman et al. 2010). Regardless of whether or not a dysregulation in adenosine is measurable in persons with ASD, increasing adenosine would still seem to offer benefits. Decades of animal research have provided consistent proof-of-concept for adenosine-based therapeutic approaches for multiple neurological disorders (i.e., pain, brain injury, stroke, epilepsy; (Lopes et al. 2011)).

Typically, adenosine-based drug development has targeted adenosine receptors or enzymatic removal of adenosine; thus far this approach has not yielded success in the form of an approved pharmacological option. One notable exception is the development of adenosine $A_{2A}R$ antagonist-based approaches for Parkinson's disease (Schwarzschild et al. 2006), yet these have also not been approved to date. Nevertheless, based on accumulated evidence for physiological and metabolic stimuli that increase adenosine in the CNS (Edelson et al. 1999; Franke et al. 2006; Goldman et al. 2010; Kawamura et al. 2010; Masino et al. 2011b), there may be opportunities to mobilize increased adenosine, either acutely or perhaps chronically. Respectively, these stimuli could offer short- or long-term reductions in or relief from multiple symptoms, and perhaps enhance possibilities for targeted learning and improved management of ASD symptoms.

24.2.3 Ketogenic Diet, Adenosine, and Autism

Increasingly, metabolic dysfunction is recognized as a key component of neurological dysfunction; autism is no exception. However, metabolic therapies for the most part remain in their infancy as a treatment option. Notable exceptions include highly successful metabolic therapies for specific inborn errors of fatty acid or amino acid metabolism, such as phenylketonuria, and for hypercholesterolemia (Collins and Leonard 1985). Another exception, in use for over 90 years, and applied more broadly (either with or without a specific metabolic indication) is ketogenic diet therapy for epilepsy. Recent work described below has started to link a ketogenic diet to increased actions of adenosine in the nervous system. This emerging work suggests that a ketogenic diet has significant promise in treating ASD.

The ketogenic diet is a high-fat, low-carbohydrate formula which has been in use primarily for pediatric epilepsy. It is also considered the therapy of choice for glucose transporter-1 deficiency (Harris et al. 2008; Ito et al. 2011). In the absence of such a specific metabolic indication in epilepsy, it is often considered only as a last resort—after a child has failed two or more drugs. Nevertheless, it is consistently a highly effective treatment: more than 50 % of patients experience more than a 50 % reduction in seizures; a significant subset become seizure free and can discontinue the diet (and medication) after 6 months to 2 years of treatment (Keene 2006; Martinez et al. 2007). Perhaps most importantly, the ketogenic diet can sometimes resolve seizures that are refractory to all drugs available. This is particularly relevant for autism due to (1) the high comorbidity between epilepsy and autism: a conservative estimate is 25 % (Canitano 2007; Spence and Schneider 2009), and (2) the enormous treatment gap: children co-diagnosed with ASD and refractory epilepsy tend to have a seizure disorder of increased severity and suffer with particularly poor treatment outcomes (Sansa et al. 2011).

As described previously, it has been suggested that gene defects or environmental influences may result in metabolic dysfunctions that can eventually cause behavioral abnormalities associated with ASD, and different types of restrictive diets, including a ketogenic diet, have been utilized in attempting to treat autism and related disorders. As outlined below, currently there are few studies regarding the effects of ketogenic diets on autistic behavior; nevertheless, results appear consistently promising thus far.

Rett syndrome is a specific type of ASD. Unlike typical gender ratios of 4:1 boys:girls in autism, Rett syndrome specifically affects girls. It presents with dementia, repetitive hand movements, lack of speech, and severe seizures (Haas et al. 1986). Thus far there have been very positive results with vastly improved seizure control in two reports of children with Rett syndrome placed on the ketogenic diet (Haas et al. 1986; Liebhaber et al. 2003). In one of these studies the ketogenic diet appeared to improve behavior, sociability, and mood in five out of seven children (Haas et al. 1986) and six out of seven gained weight (a positive effect; previously their weights were below the 5th percentile). The other report, a case study, also noted behavioral improvements (Liebhaber et al. 2003). Complementing these clinical studies of positive effects of the ketogenic diet in children with Rett syndrome, a study by Mantis et al. (2009) described improvements in emotional withdrawal and hesitance and also a reduction in anxiety in the $MeCP2^{308/y}$ mouse model of Rett syndrome when fed a calorically restricted ketogenic diet. The positive effects in this mouse study were determined to depend on caloric restriction.

There has been only one prospective study investigating whether the ketogenic diet affects behavior of children with autism. This pilot study by Evangeliou et al.

(2003) was conducted on 30 children 4–10 years old. Rather than the standard ongoing diet administration typical for pediatric epilepsy (Neal et al. 2008), the diet was applied intermittently for 6 months; the parents were asked to evaluate their child's behavior using the Childhood Autism Rating Scale. Of the group that maintained the diet (18 of 30 children = 60 %), two children reported a significant improvement (>12 units on the scale), eight patients an average improvement (8–12 units), and eight children a mild improvement (2–8 units). Recently an online survey of epilepsy treatments in patients with autism and epilepsy noted that the ketogenic diet appeared to improve seizures as well as other clinical factors such as sleep, communication, attention, mood, and behavior (Frye et al. 2011). Taken together, the results of these studies suggest that the ketogenic diet could potentially be used as a new therapeutic option for children with autism.

24.2.4 *Ketogenic Diet and Adenosine*

Recent research in vitro and in vivo suggests that a ketogenic diet may reduce seizures by chronically increasing the influence of adenosine acting at the A_1 receptor subtype (Kawamura et al. 2010; Masino et al. 2011b). An in vitro electrophysiological study mimicking the key aspects of ketogenic diet metabolism in hippocampal area CA3 showed that sufficient or high intracellular ATP and reduced extracellular glucose leads to adenosine-based inhibition (Kawamura et al. 2010). The reduced glucose activated ATP release via pannexin channels, and ATP was then dephosphorylated into adenosine. Adenosine activated A_1Rs leading to opening of K^+ channels and membrane hyperpolarization/reduced excitability.

In an in vivo study using transgenic adenosine kinase-overexpressing mice (which exhibit spontaneous electrographic seizures in hippocampus), a ketogenic diet reduced seizure frequency by roughly 90 % and shortened seizure duration by almost 50 %, effects that were reversible by systemic injection of either glucose or an A_1R antagonist, suggesting that the ketogenic diet enhances adenosine levels and that its effects depend on reduced glucose levels (Masino et al. 2011b). In addition, in another transgenic mouse—with spontaneous electrographic seizures due to deletion of A_1Rs—the ketogenic diet was totally ineffective against seizure activity (Masino et al. 2011b). Thus genetic manipulations demonstrate that reduced adenosine, and thus reduced activation of A_1Rs, leads to CNS hyperexcitability—as expressed by spontaneous electrographic seizures. Furthermore, this study provides genetic, pharmacologic, and metabolic evidence that the ketogenic diet promotes adenosinergic inhibition via A_1Rs, possibly through the autocrine mechanism described above (Kawamura et al. 2010). This CNS regulation has a potential to increase seizure threshold and promote neuronal survival and homeostasis not only in epilepsy but also in other conditions, such as various types of brain injuries, inflammatory and neuropathic pain, behavioral disorders—such as autism, and neurodegenerative disorders (Masino et al. 2009).

To expand on this further, these insights into metabolic regulation of A_1R activity could offer significant potential in treating autism. It is a hypothesis-based metabolic approach with specific expected effects on an array of core and comorbid symptoms of autism. As noted above, Masino et al. (2011a) focused exclusively on the relationship between adenosine and autism based on core symptoms quantified via a customized questionnaire. Regarding symptoms comorbid with autism, adenosine is well established as an anticonvulsant, a sleep-promoter, and an anxiolytic. While ketogenic diets can have side effects (see Table 24.2), and are certainly less convenient than typical pharmacotherapy, peripheral side effects typically associated with adenosine—reduced heart rate, reduced blood pressure, reduced temperature—are not typically prohibitive side effects with ketogenic diets.

24.2.5 *Summary of Published Research*

Table 24.2 summarizes the published connections among adenosine, autism, and the ketogenic diet. We compiled research on ten major symptoms associated with autism, including repetitive behavior, sleep, anxiety, immune system deficiencies, and gastrointestinal problems—to determine if there was any published evidence relating these symptoms to the ketogenic diet and adenosine. Overall, diverse clinical research suggests that the ketogenic diet can improve behavior, cognition, mood, or social life; this compilation of published literature in Table 24.2 suggests strongly that many diverse symptoms of autism could be treated with the ketogenic diet, perhaps due in part to its effects on adenosine.

24.3 Conclusions and Implications

At this time, physiological and metabolic treatments that increase adenosine should be taken seriously as part of an emerging, hypothesis-driven approach for reducing core and comorbid symptoms of autism. Short-term physiological strategies may offer benefits, which, while temporary in their effects on adenosine, may open an improved window of opportunity for learning, sleep, and similar positive outcomes with few drawbacks. By inducing an increase in A_1R activity, the ketogenic diet may offer broad therapeutic possibilities. Overall, the ketogenic diet has known benefits for epilepsy, there is some evidence for benefits in autism, and the common side effects of this diet are minor, short term, or rare with good medical management. Thus, the ketogenic diet is a well-established pediatric therapy which—under medical supervision—could be considered for children with autism, and may be particularly useful in children with comorbid autism and epilepsy.

Table 24.2 Possible associations between ketogenic diet (KD), autism, and adenosine receptors

	Connections to:		
	Autism	Ketogenic diet	Adenosine receptors
Stereotypical behavior	Repetitive and stereotypical behaviors comprise one of the three core features of ASD (Lewis and Bodfish 1998; Newschaffer et al. 2007)	The KD significantly improved autistic symptoms, including stereotypy, in children with autism (Evangeliou et al. 2003) and motor control in a few girls with Rett syndrome (Haas et al. 1986). However, there is currently no direct evidence that the KD improves stereotypy	Stimulation of A_1 and A_2 adenosine receptors decreases amphetamine-induced stereotypical behavior in rats (Poleszak and Malec 2000) Conversely, constitutive loss of A_{2A}Rs reduces amphetamine- and cocaine-induced stereotypical behavior in mice (Chen et al. 2000)
Sociability deficits	Sociability deficits, such as lack of eye contact or failure to develop relationships, comprise one of the three core features of ASD (Newschaffer et al. 2007)	The KD improved behavior, including social problems, in children with epilepsy (Pulsifer et al. 2001) and sociability in children with autism (Evangeliou et al. 2003)	Mice lacking A_1Rs have heightened aggression (Gimenez-Llort et al. 2002)
Communication deficits	Impairments in verbal and nonverbal communication constitute one of the three core features of ASD (Newschaffer et al. 2007)	Three children suffering from epileptic aphasia experienced improved language abilities after being on the KD (Bergqvist et al. 1999)	–
Anxiety and depression	Children with autism experience a significantly higher incidence of anxiety and depression problems compared to normal children (Gurney et al. 2006)	The KD reduced symptoms of depression in rats undergoing the Porsolt test (Murphy et al. 2004)	A_1R knockout mice showed increased anxiety-related behavior (Johansson et al. 2001). In addition, an A_1R agonist had anxiolytic effects on CD1 mice (Florio et al. 1998) A_{2A}R antagonists protect against depression in mice (El Yacoubi et al. 2004)
Sleep	Children with autism have a higher incidence of sleep problems, including prolonged sleep onset times, sleep latency, frequent arousals during sleep, or sluggish and disoriented behavior during waking hours (Malow 2005; Polimeni et al. 2005)	The KD reduced the symptoms of narcolepsy, i.e., sleepiness and fatigue during the day, in nine subjects (Husain et al. 2004). In addition, the KD improved sleep quality in children with intractable epilepsy by decreasing the total night and daytime sleep and by increasing REM sleep (Hallböök et al. 2007)	Adenosine is a known sleep-promoting neuromodulator; adenosine levels increase during prolonged wakefulness and decrease during subsequent recovery sleep (Dunwiddie and Masino 2001; Huang et al. 2005; Porkka-Heiskanen 1999; Satoh et al. 1998)

(continued)

Table 24.2 (continued)

	Connections to:		
	Autism	Ketogenic diet	Adenosine receptors
Epilepsy and seizures	A conservative estimate of comorbidity of epilepsy in autism is 20–25 % (Canitano 2007) Medically refractory epilepsy is common in autism (Sansa et al. 2011)	The KD is at least as effective as anticonvulsant medications (Hemingway et al. 2001) The KD significantly reduces the frequency of seizures in children with intractable epilepsy (Keene 2006; Neal et al. 2008)	Adenosine exhibits anticonvulsant effects in experimental models of epilepsy probably by stimulating A_1Rs (Dunwiddie 1999; Dunwiddie and Masino 2001; Masino et al. 2011b)
Pain	Children with autism have altered pain responses. Contrary to previous anecdotal reports, children with autism may not have decreased sensitivity to pain (Tordjman et al. 2009). Children with autism displayed greater facial activity to pain instigated by venipuncture than normal children (Nader et al. 1998)	Rats on the KD showed reduced pain in the hotplate test (thermal nociception) compared to rats on the control diet (Ruskin et al. 2009) Beneficial effects of a KD on self-reported general bodily pain (not "pain" patients per se) were at the threshold of statistical significance (Yancy et al. 2009)	In general, stimulation of A_1Rs has inhibitory effects on pain (Sawynok 2007). Release of adenosine and subsequent activation of A_1Rs may mediate the anti-nociceptive effects of acupuncture (Goldman et al. 2010)
Gastrointestinal system	Children with autism have a high prevalence of gastrointestinal problems, including inflammation of upper and lower intestinal tract, increased intestinal permeability, decreased digestive enzyme activities, increased secretion after secretin injection, or decreased sulfation capacity of the liver (Horvath et al. 1999)	Gastrointestinal problems, including vomiting, constipation, or diarrhea, are sometimes reported shortly after administration of the KD (Freeman et al. 2007; Hartman and Vining 2007; Keene 2006), probably related to the dramatic change in diet composition. A risk of kidney stones is also a concern. Long-term effects of the KD on the GI tract, however, are not known	Adenosine receptors modulate enteric neural reflexes that control secretomotor and motor functions of the gastrointestinal tract (Christofi 2001). The adenosine system also acts as "a sensor apparatus" during inflammation and represents a novel target for inflammatory bowel disease (Antonioli et al. 2008). Inosine, a metabolite of adenosine, has protective effects during induced gut hyperpermeability (Garcia Soriano et al. 2001)

Immune system and inflammation	Children with autism have an altered immune system. Recent findings point to skewed cytokine profiles, decreased levels of IgG and IgM, and altered responses of immune cells, such as macrophages. Individuals with ASD also have active inflammation in the CNS (Goines and Van de Water 2010)	Rats on the KD have reduced inflammatory response to experimental peripheral inflammation compared to rats fed with a control diet (Ruskin et al. 2009) In overweight adults, a KD had overall greater anti-inflammatory effects (such as greater decreases in TNF-α, IL-6, and IL-8) than a low-fat diet (Forsythe et al. 2008)	Adenosine receptors can exert powerful immunomodulatory actions. For example, adenosine receptor activation can limit TNF-α production following macrophage activation, promote the recruitment of immature dendritic cells to sites of inflammation, and regulate neutrophil and lymphocyte function (Hasko et al. 2008)
Hormonal alternations	Previous studies showed that individuals with autism tend to have increased cortisol levels and altered oxytocin levels (Bartz and Hollander 2008; Spratt et al. 2012)	The current knowledge of hormonal changes due to the KD is limited. Some studies found increased levels of cortisol (Fraser et al. 2003), whereas others did not find any differences (Volek and Sharman 2004)	A_1Rs are located in the hypothalamic-pituitary-adrenal axis where they could potentially inhibit secretion of hormones (Stojilkovic 2009)

Note: If there are no published data the cell is left blank

Acknowledgements We acknowledge the support of the National Institutes of Health and Trinity College.

References

American Psychiatric Association (2000) Diagnostic and statistical manual of mental disorders, 4th edn. American Psychiatric Association, Washington DC

Anonymous (2010) Are rates of autism spectrum disorders increasing? Expanded diagnostic criteria and greater public awareness may explain why. Harv Ment Health Lett 26:6

Antonioli L, Fornai M, Colucci R, Ghisu N, Tuccori M, Del Tacca M, Blandizzi C (2008) Regulation of enteric functions by adenosine: pathophysiological and pharmacological implications. Pharmacol Ther 120:233–253

Bailey A, Phillips W, Rutter M (1996) Autism: towards an integration of clinical, genetic, neuropsychological, and neurobiological perspectives. J Child Psychol Psychiatry 37:89–126

Bartz JA, Hollander E (2008) Oxytocin and experimental therapeutics in autism spectrum disorders. Prog Brain Res 170:451–462

Battistuzi G, Iudicone P, Santalomazza P, Petrucci R (1981) Activity of adenosine deaminase allelic forms in intact erythrocytes and in lymphocytes. Ann Hum Genet 45:15–19

Bergqvist AGC, Chee CM, Lutchka LM, Brooks-Kayal AR (1999) Treatment of acquired epileptic aphasia with the ketogenic diet. J Child Neurol 14:696–701

Beversdorf DQ (2010) The role of the noradrenergic system in autism spectrum disorders. In: Blatt GJ (ed) Neurochemical basis of autism: from molecules to minicolumns. Springer, New York, pp 175–184

Bode MK, Mattila ML, Kiviniemi V, Rahko J, Moilanen I, Ebeling H, Tervonen O, Nikkinen J (2011) White matter in autism spectrum disorders—evidence of impaired fiber formation. Acta Radiol 52:1169–1174

Boison D, Scheurer L, Zumsteg V, Rülicke T, Litynski P, Fowler B, Brandner S, Mohler H (2002) Neonatal hepatic steatosis by disruption of the adenosine kinase gene. Proc Natl Acad Sci U S A 99:6985–6990

Boison D, Masino SA, Geiger JD (2011) Homeostatic bioenergetic network regulation: a novel concept to avoid pharmacoresistance in epilepsy. Expert Opin Drug Discov 6:713–724

Bottini N, De Luca D, Saccucci P, Fiumara A, Elia M, Porfirio MC, Lucarelli P, Curatolo P (2001) Autism: evidence of association with adenosine deaminase genetic polymorphism. Neurogenetics 3:111–113

Brugha TS, McManus S, Bankart J, Scott F, Purdon S, Smith J, Bebbington P, Jenkins R, Meltzer H (2011) Epidemiology of autism spectrum disorders in adults in the community in England. Arch Gen Psychiatry 68:459–465

Buitelaar JK, Willemsen-Swinkels SH (2000) Medication treatment in subjects with autistic spectrum disorders. Eur Child Adolesc Psychiatry 9(suppl 1):85–97

Burnstock G, Krügel U, Abbracchio MP, Illes P (2011) Purinergic signalling: from normal behaviour to pathological brain function. Prog Neurobiol 95:229–274

Canitano R (2007) Epilepsy in autism spectrum disorders. Eur Child Adolesc Psychiatry 16:61–66

Chen JF, Beilstein M, Xu YH, Turner TJ, Moratalla R, Standaert DG, Aloyo VJ, Fink JS, Schwarzschild MA (2000) Selective attenuation of psychostimulant-induced behavioral responses in mice lacking A(2A) adenosine receptors. Neuroscience 97:195–204

Christofi FL (2001) Unlocking mysteries of gut sensory transmission: is adenosine the key? News Physiol Sci 16:201–207

Ciruela F, Saura C, Canela EI, Mallol J, Lluis C, Franco R (1996) Adenosine deaminase affects ligand-induced signalling by interacting with cell surface adenosine receptors. FEBS Lett 380:219–223

Coleman M, Landgrebe MA, Landgrebe AR (1976) Purine autism. Hyperuricosuria in autistic children: does this identify a subgroup of autism? In: Coleman M (ed) The autistic syndromes. Elsevier, New York, pp 183–214

Collins JE, Leonard JV (1985) The dietary management of inborn errors of metabolism. Hum Nutr Appl Nutr 39:255–272

Cook EH, Leventhal BL, Heller W, Metz J, Wainwright M, Freedman DX (1990) Autistic children and their first-degree relatives: relationships between serotonin and norepinephrine levels and intelligence. J Neuropsychiatry Clin Neurosci 2:268–274

Cuccaro ML, Wright HH, Abramson RK, Marsteller FA, Valentine A (1993) Whole-blood serotonin and cognitive functioning in autistic individuals and their first-degree relatives. J Neuropsychiatry Clin Neurosci 5:94–101

Dunwiddie TV (1999) Adenosine and suppression of seizures. Adv Neurol 79:1001–1010

Dunwiddie TV, Masino SA (2001) The role and regulation of adenosine in the central nervous system. Annu Rev Neurosci 24:31–55

Edelson SM, Edelson MG, Kerr DC, Grandin T (1999) Behavioral and physiological effects of deep pressure on children with autism: a pilot study evaluating the efficacy of Grandin's Hug Machine. Am J Occup Ther 53:145–152

Egaas B, Courchesne E, Saitoh O (1995) Reduced size of corpus callosum in autism. Arch Neurol 52:794–801

Eigsti I-M, Shapiro T (2003) A systems neuroscience approach to autism: biological, cognitive and clinical perspectives. Ment Retard Dev Disabil Res Rev 9:205–215

El Yacoubi M, Costentin J, Vaugeois JM (2004) Adenosine A2A receptors and depression. Neurology 61(suppl 6):82–87

Elder JH, Shankar M, Shuster J, Theriaque D, Burns S, Sherrill L (2006) The gluten-free, casein-free diet in autism: results of a preliminary double blind clinical trial. J Autism Dev Disord 36:413–420

Esbensen AJ, Greenberg JS, Seltzer MM, Aman MG (2009) A longitudinal investigation of psychotropic and non-psychotropic medication use among adolescents and adults with autism spectrum disorders. J Autism Dev Disord 39:1339–1349

Evangeliou A, Vlachonikolis I, Mihailidou H, Spilioti M, Skarpalezou A, Makaronas N, Prokopiou A, Christodoulou P, Liapi-Adamidou G, Helidonis E et al (2003) Application of a ketogenic diet in children with autistic behavior: pilot study. J Child Neurol 18:113–118

Fatemi SH, Reutiman TJ, Folsom TD, Thuras PD (2009) $GABA_A$ receptor downregulation in brains of subjects with autism. J Autism Dev Disord 39:223–230

Fidler DJ, Bailey JN, Smalley SL (2000) Macrocephaly in autism and other pervasive developmental disorders. Dev Med Child Neurol 42:737–740

Findling RL (2005) Pharmacologic treatment of behavioral symptoms in autism and pervasive developmental disorders. J Clin Psychiatry 66(suppl 10):26–31

Florio C, Prezioso A, Papaioannou A, Vertua R (1998) Adenosine A1 receptors modulate anxiety in CD1 mice. Psychopharmacology (Berl) 136:311–319

Folstein S, Rutter M (1977) Infantile autism: a genetic study of 21 twin pairs. J Child Psychol Psychiatry 18:297–321

Fombonne E (2005) Epidemiology of autistic disorder and other pervasive developmental disorders. J Clin Psychiatry 66(suppl 10):3–8

Forsythe CE, Phinney SD, Fernandez ML, Quann EE, Wood RJ, Bibus DM, Kraemer WJ, Feinman RD, Volek JS (2008) Comparison of low fat and low carbohydrate diets on circulating fatty acid composition and markers of inflammation. Lipids 43:65–77

Franke H, Krügel U, Illes P (2006) P2 receptors and neuronal injury. Pflugers Arch 452:622–644

Fraser DD, Whiting S, Andrew RD, Macdonald EA, Musa-Veloso K, Cunnane SC (2003) Elevated polyunsaturated fatty acids in blood serum obtained from children on the ketogenic diet. Neurology 60:1026–1029

Freeman JM, Kossoff EH, Hartman AL (2007) The ketogenic diet: one decade later. Pediatrics 119:535–543

Freitag CM, Agelopoulos K, Huy E, Rothermundt M, Krakowitzky P, Meyer J, Deckert J, von Gontard A, Hohoff C (2010) Adenosine A(2A) receptor gene (ADORA2A) variants may increase autistic symptoms and anxiety in autism spectrum disorder. Eur Child Adolesc Psychiatry 19:67–74

Frye RE, Rossignol DA (2011) Mitochondrial dysfunction can connect the diverse medical symptoms associated with autism spectrum disorders. Pediatr Res 69:41R–47R

Frye RE, Sreenivasula S, Adams JB (2011) Traditional and non-traditional treatments for autism spectrum disorder with seizures: an on-line survey. BMC Pediatr 11:37

Garcia Soriano F, Liaudet L, Marton A, Hasko G, Batista Lorigados C, Deitch EA, Szabo C (2001) Inosine improves gut permeability and vascular reactivity in endotoxic shock. Crit Care Med 29:703–708

Gardener H, Spiegelman D, Buka SL (2009) Prenatal risk factors for autism: comprehensive meta-analysis. Br J Psychiatry 195:7–14

Gepner B, Deruelle C (2008) Abnormal cerebral effective connectivity during explicit emotional processing in adults with autism spectrum disorder. Soc Cogn Affect Neurosci 3:135–143

Ghanizadeh A (2010) Possible role of caffeine in autism spectrum disorders, a new testable hypothesis. J Food Sci 75:ix

Gimenez-Llort L, Fernandez-Teruel A, Escorihuela RM, Fredholm BB, Tobena A, Pekny M, Johansson B (2002) Mice lacking the adenosine A1 receptor are anxious and aggressive, but are normal learners with reduced muscle strength and survival rate. Eur J Neurosci 16:547–550

Goines P, Van de Water J (2010) The immune system's role in the biology of autism. Curr Opin Neurol 23:111–117

Goldman N, Chen M, Fujita T, Xu Q, Peng W, Liu W, Jensen TK, Pei Y, Wang F, Han X et al (2010) Adenosine A1 receptors mediate local anti-nociceptive effects of acupuncture. Nat Neurosci 13:883–888

Gurney JG, McPheeters ML, Davis MM (2006) Parental report of health conditions and health care use among children with and without autism. Arch Pediatr Adolesc Med 160:825–830

Haas RH, Rice MA, Tauner DA, Merritt TA (1986) Therapeutic effects of a ketogenic diet in Rett syndrome. Am J Med Genet Suppl 1:225–246

Hallböök T, Lundgren J, Rosén I (2007) Ketogenic diet improves sleep quality in children with therapy-resistant epilepsy. Epilepsia 48:59–65

Hallmayer J, Cleveland S, Torres A, Phillips J, Cohen B, Torigoe T, Miller J, Fedele A, Collins J, Smith K et al (2011) Genetic heritability and shared environmental factors among twin pairs with autism. Arch Gen Psychiatry 68:1095–1102

Harris ML, Patel H, Garg BP (2008) Intractable seizures, developmental delay, and the ketogenic diet. Semin Pediatr Neurol 15:209–211

Hartman AL, Vining EP (2007) Clinical aspects of the ketogenic diet. Epilepsia 48:31–42

Hasko G, Linden J, Cronstein B, Pacher P (2008) Adenosine receptors: therapeutic aspects for inflammatory and immune diseases. Nat Rev Drug Discov 7:759–770

Hemingway C, Freeman JM, Pillas DJ, Pyzik PL (2001) The ketogenic diet: a 3- to 6-year follow-up of 150 children enrolled prospectively. Pediatrics 108:898–905

Hettinger JA, Liu X, Holden JJ (2008) The G22A polymorphism of the ADA gene and susceptibility to autism spectrum disorders. J Autism Dev Disord 38:14–19

Horvath K, Papadimitriou JC, Rabsztyn A, Drachenberg C, Tildon JT (1999) Gastrointestinal abnormalities in children with autistic disorder. J Pediatr 135:559–563

Huang Z-L, Qu W-M, Eguchi N, Chen J-R, Schwartzschild MA, Fredholm BB, Urade Y, Hayaishi O (2005) Adenosine A_{2A}, but not A_1, receptors mediate the arousal effect of caffeine. Nat Neurosci 8:858–859

Husain AM, Yancy WS Jr, Carwile ST, Miller PP, Westman EC (2004) Diet therapy for narcolepsy. Neurology 62:2300–2302

Ito Y, Oguni H, Ito S, Oguni M, Osawa M (2011) A modified Atkins diet is promising as a treatment for glucose transporter type 1 deficiency syndrome. Dev Med Child Neurol 53:658–663

James SJ, Melnyk S, Jernigan S, Cleves MA, Halsted CH, Wong DH, Cutler P, Bock K, Boris M, Bradstreet JJ et al (2006) Metabolic endophenotype and related genotypes are associated with oxidative stress in children with autism. Am J Med Genet B Neuropsychiatr Genet 141: 947–956

Johansson B, Halldner L, Dunwiddie TV, Masino SA, Poelchen W, Giménez-Llort L, Escorihuela LM, Fernández-Teruel A, Wiesenfeld-Hallin Z, Xu X-J et al (2001) Hyperalgesia, anxiety, and decreased hypoxic neuroprotection in mice lacking the adenosine A_1 receptor. Proc Natl Acad Sci U S A 98:9407–9412

Jones JR, Skinner C, Friez MJ, Schwartz CE, Stevenson RE (2008) Hypothesis: dysregulation of methylation of brain-expressed genes on the X chromosome and autism spectrum disorders. Am J Med Genet A 146:2213–2220

Just MA, Cherkassky VL, Keller TA, Minshew NJ (2004) Cortical activation and synchronization during sentence comprehension in high-functioning autism: evidence of underconnectivity. Brain 127:1811–1821

Kana RK, Keller TA, Cherkassky VL, Minshew NJ, Just MA (2006) Sentence comprehension in autism: thinking in pictures with decreased functional connectivity. Brain 129:2484–2493

Kawamura M Jr, Ruskin DN, Masino SA (2010) Metabolic autocrine regulation of neurons involves cooperation among pannexin hemichannels, adenosine receptors and K_{ATP} channels. J Neurosci 30:3886–3895

Keene DL (2006) A systematic review of the use of the ketogenic diet in childhood epilepsy. Pediatr Neurol 35:1–5

Kim YS, Leventhal BL, Koh YJ, Fombonne E, Laska E, Lim EC, Cheon KA, Kim SJ, Kim YK, Lee H et al (2011) Prevalence of autism spectrum disorders in a total population sample. Am J Psychiatry 168:904–912

Knivsberg AM, Reichelt KL, Hoien T, Nodland M (2002) A randomised, controlled study of dietary intervention in autistic syndromes. Nutr Neurosci 5:251–261

Krakowiak P, Bremer AA, Baker AS, Walker CK, Hansen RL, Hertz-Picciotto I (2011) The role of maternal diabetes and related conditions in autism and other developmental delays. International Meeting for Autism Research (IMFAR) 10th Anniversary Meeting, abstract 131.007

Lam KS, Aman MG, Arnold LE (2006) Neurochemical correlates of autistic disorder: a review of the literature. Res Dev Disabil 27:254–289

Lara DR (2010) Caffeine, mental health, and psychiatric disorders. J Alzheimers Dis 20(suppl 1):S239–S248

Levy SE, Hyman SL (2008) Complementary and alternative medicine treatments for children with autism spectrum disorders. Child Adolesc Psychiatr Clin N Am 17:803–820

Levy SE, Giarelli E, Lee LC, Schieve LA, Kirby RS, Cunniff C, Nicholas J, Reaven J, Rice CE (2010) Autism spectrum disorder and co-occurring developmental, psychiatric, and medical conditions among children in multiple populations of the United States. J Dev Behav Pediatr 31:267–275

Lewis MH, Bodfish JW (1998) Repetitive disorders in autism. Ment Retard Dev Disabil Res Rev 4:80–89

Libbey JE, Sweeten TL, McMahon WM, Fujinami RS (2005) Autistic disorder and viral infection. J Neurovirol 11:1–10

Liebhaber GM, Riemann E, Baumeister FA (2003) Ketogenic diet in Rett syndrome. J Child Neurol 18:74–75

Lopes LV, Sebastião AM, Ribeiro JA (2011) Adenosine and related drugs in brain diseases: present and future in clinical trials. Curr Top Med Chem 11:1087–1101

Lord C (2011) Epidemiology: how common is autism? Nature 474:166–168

Lotter V (1966) Epidemiology of autistic conditions in young children. Soc Psychiatry 1: 124–137

Maestrini E, Pagnamenta AT, Lamb JA, Bacchelli E, Sykes NH, Sousa I, Toma C, Barnby G, Butler H, Winchester L et al (2010) High-density SNP association study and copy number variation analysis of the AUTS1 and AUTS5 loci implicate the IMMP2L-DOCK4 gene region in autism susceptibility. Mol Psychiatry 15:954–968

Malow BA (2004) Sleep disorders, epilepsy, and autism. Ment Retard Dev Disabil Res Rev 10:122–125

Malow BA (2005) Sleep and epilepsy. Neurol Clin 23:1127–1147

Mantis JG, Fritz CL, Marsh J, Heinrichs SC, Seyfried TN (2009) Improvement in motor and exploratory behavior in Rett syndrome mice with restricted ketogenic and standard diets. Epilepsy Behav 15:133–141

Marie S, Race V, Nassogne MC, Vincent MF, Van den Berghe G (2002) Mutation of a nuclear respiratory factor 2 binding site in the 5′ untranslated region of the ADSL gene in three patients with adenylosuccinate lyase deficiency. Am J Hum Genet 71:14–21

Martinez CC, Pyzik PL, Kossof EH (2007) Discontinuing the ketogenic diet in seizure-free children: recurrence and risk factors. Epilepsia 48:187–190

Masino SA, Kawamura M Jr, Wasser CD, Pomeroy LT, Ruskin DN (2009) Adenosine, ketogenic diet and epilepsy: the emerging therapeutic relationship between metabolism and brain activity. Curr Neuropharmacol 7:257–268

Masino SA, Kawamura M Jr, Plotkin LM, Svedova J, DiMario FJ Jr, Eigsti I-E (2011a) The relationship between the neuromodulator adenosine and behavioral symptoms of autism. Neurosci Lett 500:1–5

Masino SA, Li T, Theofilas P, Ruskin DN, Fredholm BB, Geiger JD, Aronica E, Boison D (2011b) A ketogenic diet suppresses seizures in mice through adenosine A_1 receptors. J Clin Invest 121:2679–2683

Masino SA, Svedova J, Kawamura M Jr, DiMario FD Jr, Eigsti I-M (2011c) Adenosine and autism—recent research and a new perspective. In: Eapen V (ed) Autism—a neurodevelopmental journey from genes to behaviour. InTech, Rijeka, Croatia, pp 103–122

Mostofsky SH, Ewen JB (2011) Altered connectivity and action model formation in autism is autism. Neuroscientist 5:5

Muhle R, Trentacoste SV, Rapin I (2004) The genetics of autism. Pediatrics 113:e472–e486

Murphy P, Likhodii S, Nylen K, Burnham WM (2004) The antidepressant properties of the ketogenic diet. Biol Psychiatry 56:981–983

Nader R, Forghani R, Krnjevic K (1998) In rat hippocampal slices, NMDA receptor-mediated EPSPs are more sensitive to hypoxia than AMPA receptor-mediated EPSPs. Neurosci Lett 245:135–138

Neal EG, Chaffe H, Schwartz RH, Lawson MS, Edwards N, Fitzsimmons G, Whitney A, Cross JH (2008) The ketogenic diet for the treatment of childhood epilepsy: a randomised controlled trial. Lancet Neurol 7:500–506

Newschaffer CJ (2006) Investigating diagnostic substitution and autism prevalence trends. Pediatrics 117:1436–1437

Newschaffer CJ, Croen LA, Daniels J, Giarelli E, Grether JK, Levy SE, Mandell DS, Miller LA, Pinto-Martin J, Reaven J et al (2007) The epidemiology of autism spectrum disorders. Annu Rev Public Health 28:235–258

Nyhan WL, James JA, Teberg AJ, Sweetman L, Nelson LG (1969) A new disorder of purine metabolism with behavioral manifestations. J Pediatr 74:20–27

Page T, Coleman M (2000) Purine metabolism abnormalities in a hyperuricosuric subclass of autism. Biochim Biophys Acta 1500:291–296

Persico AM, Militerni R, Bravaccio C, Schneider C, Melmed R, Trillo S, Montecchi F, Palermo MT, Pascucci T, Puglisi-Allegra S et al (2000) Adenosine deaminase alleles and autistic disorder: case-control and family-based association studies. Am J Med Genet 96:784–790

Poleszak E, Malec D (2000) Influence of adenosine receptor agonists and antagonists on ampheramine-induced stereotypy in rats. Pol J Pharmacol 52:423–429

Polimeni MA, Richdale AL, Francis AJ (2005) A survey of sleep problems in autism, Asperger's disorder and typically developing children. J Intellect Disabil Res 49:260–268

Porkka-Heiskanen T (1999) Adenosine in sleep and wakefulness. Ann Med 31:125–129

Pulsifer MB, Gordon JM, Brandt J, Vining EPG, Freeman JM (2001) Effects of ketogenic diet on development and behavior: preliminary report of a prospective study. Dev Med Child Neurol 43:301–306

Purcell AE, Jeon OH, Zimmerman AW, Blue ME, Pevsner J (2001) Postmortem brain abnormalities of the glutamate neurotransmitter system in autism. Neurology 57:1618–1628

Ribeiro JA, Sebastião AM (2010) Caffeine and adenosine. J Alzheimers Dis 20(suppl 1):S3–S15

Ritvo ER, Freeman BJ, Mason-Brothers A, Mo A, Ritvo AM (1985) Concordance for the syndrome of autism in 40 pairs of afflicted twins. Am J Psychiatry 142:74–77

Ronald A, Hoekstra RA (2011) Autism spectrum disorders and autistic traits: a decade of new twin studies. Am J Med Genet B Neuropsychiatr Genet 156:255–274

Rosenberger-Debiesse J, Coleman M (1986) Preliminary evidence for multiple etiologies in autism. J Autism Dev Disord 16:385–392

Rossignol DA, Frye RE (2012) Mitochondrial dysfunction in autism spectrum disorders: a systematic review and meta-analysis. Mol Psychiatry 17(3):290–314

Ruskin DN, Kawamura M Jr, Masino SA (2009) Reduced pain and inflammation in juvenile and adult rats fed a ketogenic diet. PLoS One 4:e8349

Sansa G, Carlson C, Doyle W, Weiner HL, Bluvstein J, Barr W, Devinsky O (2011) Medically refractory epilepsy in autism. Epilepsia 52:1071–1075

Satoh S, Matsumura H, Hayaishi O (1998) Involvement of adenosine A2A receptor in sleep promotion. Eur J Pharmacol 351:155–162

Sawynok J (2007) Adenosine and ATP receptors. Handb Exp Pharmacol 177:309–328

Schwarzschild MA, Agnati L, Fuxe K, Chen JF, Morelli M (2006) Targeting adenosine A2A receptors in Parkinson's disease. Trends Neurosci 29:647–654

Spence SJ, Schneider MT (2009) The role of epilepsy and epileptiform EEGs in autism spectrum disorders. Pediatr Res 65:599–606

Spratt EG, Nicholas JS, Brady KT, Carpenter LA, Hatcher CR, Meekins KA, Furlanetto RW, Charles JM (2012) Enhanced cortisol response to stress in children in autism. J Autism Dev Disord 42(1):75–81

Srinivasan P (2009) A review of dietary interventions in autism. Ann Clin Psychiatry 21:237–247

Stojilkovic SS (2009) Purinergic regulation of hypothalamopituitary functions. Trends Endocrinol Metab 20:460–468

Tanimura Y, Vaziri S, Lewis MH (2010) Indirect basal ganglia pathway mediation of repetitive behavior: attenuation by adenosine receptor agonists. Behav Brain Res 210:116–122

Tordjman S, Anderson GM, Botbol M, Brailly-Tabard S, Perez-Diaz F, Graignic R, Carlier M, Schmit G, Rolland AC, Bonnot O, Trabado S et al (2009) Pain reactivity and plasma beta-endorphin in children and adolescents with autistic disorder. PLoS One 4:e5289

Valicenti-McDermott M, McVicar K, Rapin I, Wershil BK, Cohen H, Shinnar S (2006) Frequency of gastrointestinal symptoms in children with autistic spectrum disorders and association with family history of autoimmune disease. J Dev Behav Pediatr 27(suppl 2):128–136

Veenstra-VanderWeele J, Cook EH Jr (2004) Molecular genetics of autism spectrum disorder. Mol Psychiatry 9:819–832

Volek JS, Sharman MJ (2004) Cardiovascular and hormonal aspects of very-low-carbohydrate ketogenic diets. Obes Res 12(suppl 2):115–123

Wang LW, Tancredi DJ, Thomas DW (2011) The prevalence of gastrointestinal problems in children across the United States with autism spectrum disorders from families with multiple affected members. J Dev Behav Pediatr 32:351–360

Weizman A, Weizman R, Szekely GA, Wijsenbeek H, Livni E (1982) Abnormal immune response to brain tissue antigen in the syndrome of autism. Am J Psychiatry 139:1462–1465

Witwer A, Lecavalier L (2005) Treatment incidence and patterns in children and adolescents with autism spectrum disorders. J Child Adolesc Psychopharmacol 15:671–681

Yancy WS Jr, Almirall D, Maciejewski ML, Kolotkin RL, McDuffie JR, Westman EC (2009) Effects of two weight-loss diets on health-related quality of life. Qual Life Res 18:281–289

Yoo HJ, Lee SK, Park M, Cho IH, Hyun SH, Lee JC, Yang SY, Kim SA (2009) Family- and population-based association studies of monoamine oxidase A and autism spectrum disorders in Korean. Neurosci Res 63:172–176

Chapter 25
Stress, Brain Adenosine Signaling, and Fatigue-Related Behavioral Processes

Traci N. Plumb, Sarah R. Sterlace, Kelly A. Cavanaugh, and Thomas R. Minor

Abstract This chapter reviews the role of adenosine in stress-induced behaviors such as fatigue and behavioral depression, as characterized by the state of conservation-withdrawal. Evidence is provided that adenosine A_{2A} receptors mediate these behaviors as seen in a variety of animal models of depression: learned helplessness, behavioral despair, reserpine-induced depression, sickness behavior, and cytokine-induced behavioral depression. A_{2A} receptors are colocalized with dopamine D_2 receptors on the indirect pathway of the striatum where they modulate the binding affinity of dopamine to the D_2 receptor. The striatum is segregated into functionally distinct regions, ranging from control of sensorimotor behaviors in the dorsal striatum to the motivational aspects of ongoing behavior in the ventral striatum. Preliminary evidence suggests that adenosine A_{2A} receptors in the core of the ventral striatum mediate behavioral depression in the learned helplessness paradigm. Activation of A_{2A} receptors in this region blocks the dopaminergic signal from the ventral tegmental area, resulting in the uncoupling of motivation from ongoing behavior.

Keywords Adenosine A_{2A} receptor • Behavioral depression • Conservation-withdrawal • Dopamine D_2 receptor • Indirect pathway • Learned helplessness • Striatum • Traumatic stress

The discovery of the adenosine receptor and its eventual link to action of methylxanthine stimulants (Sattin and Rall 1970; Snyder et al. 1981) foreshadowed a significant role of this signaling mechanism in behavior, particularly in fatigue-like processes. Caffeine and theophylline, the active ingredients in coffee and tea, respectively, are widely used to elevate mood, combat fatigue, and reverse the

T.N. Plumb • S.R. Sterlace • K.A. Cavanaugh • T.R. Minor (✉)
Department of Psychology and the Brain Research Institute,
University of California, Los Angeles, CA, USA
e-mail: minor@psych.ucla.edu

S. Masino and D. Boison (eds.), *Adenosine: A Key Link between Metabolism and Brain Activity*, DOI 10.1007/978-1-4614-3903-5_25,

effects of sleep. These compounds derive their stimulant properties by acting as nonselective antagonists at brain adenosine receptors.

Adenosine signaling is closely linked to cellular energy homeostasis (Fredholm and Dunwiddie 1988; Newby 1984; Meghji 1991; Phillis et al. 1987; Stone 1981; Van Wylen et al. 1986). The process is engaged whenever the rate of ATP (adenosine triphosphate) utilization exceeds the rate of synthesis. In brain neurons, this type of imbalance in the energy supply/demand ratio can result from excessive neural activation or from a shortage in brain glucose or oxygen. The nucleoside is produced in nanomolar concentrations as cellular work increases via *S*-adenosyl-L-homocysteine (SAH) metabolism and is extruded into extracellular space via bidirectional transporters (Geiger and Fyda 1991; Parkinson et al. 2011). Adenosine also can be rapidly hydrolyzed from extracellular nucleotides by a family of ectonucleotidases (Vorhoff et al. 2005). Extracellular adenosine acts as an energy-dependent neuromodulator of several functions in the brain, including neuronal viability, neuronal membrane potential, propagation of action potentials, astrocytic functions, microglia reactivity, primary metabolism in both neurons and astrocytes, and blood flow (Fredholm et al. 2005; Sebastião and Ribeiro 2000).

Adenosine exerts its homeostatic and regulatory actions by interacting with four G-protein coupled stereospecific receptors: A_1, A_{2A}, A_{2B}, and A_3 (see Haas and Selbach 2000; Phillis 2004 for reviews). A_1 receptors are widely distributed in the brain and mediate adenosine's inhibitory actions by coupling with a G_i protein that inhibits adenylyl cyclase (Linden 1991). A_2 receptors mediate adenosine's excitatory actions by coupling a G_s protein that excites adenylyl cyclase (Bruns et al. 1986; Sebastião and Ribeiro 1996). The A_{2B} subtype is a low-affinity receptor that is widely distributed in most brain regions. The high-affinity A_{2A} subtype has a much more limited distribution, being localized primarily on enkephaline-containing GABAergic neurons in the striatopallidal tract of the striatum (Calon et al. 2004; Svenningsson et al. 1999). Limited concentrations of A_{2A} receptors also are found in the thalamus (Ishiwata et al. 2005; Mishina et al. 2007; Weaver 1993), nucleus tractus solitarius (Castillo-Meléndez et al. 1994; Scislo and O'Leary 2006), and cholinergic neurons of the pontine reticular formation (Coleman et al. 2006; Ferré et al. 2007). A_3 receptors are found primarily in the periphery, with high concentrations in testes and mast cells, and are not heavily expressed in the brain. These receptors play an important role in regulating inflammatory reactions (Gessi et al. 2008; Linden et al. 1993).

25.1 Metabolic Fatigue and Conservation-Withdrawal

This molecular mechanism has a natural affinity with the behavioral concept of conservation-withdrawal. Engel and Schmale (1972) originally coined this term in describing an exaggerated withdrawal response of a psychiatric patient to emotional challenges. The reaction unconditionally follows periods of intense catabolic output. The sensory unresponsiveness, cognitive dullness, and behavioral depression that characterize this state serve as adaptive mechanisms for husbanding limited resources and facilitating the recovery of metabolic homeostasis. The term is used more broadly in modern parlance to refer to enervated states associated with physical or psychological stress.

Conservation-withdrawal is an integral component of major depression and related mood disorders (Engel and Schmale 1972; Field and Reite 1984; Weiner and Lovitt 1979). The state most closely corresponds to the affect-less, fatigue components of depression, rather than subsuming the entirety of the behavioral, cognitive, emotional, and motivational symptoms that comprise the disorder. It also represents the aspects of affective disorders that are most accurately modeled in animals. Conservation-withdrawal is also a key component of the after-reaction to physical and psychological stress. Symptoms of conservation-withdrawal are seen after a patient leaves the intensive care unit following a serious injury and are often confused with major depression (Mohta et al. 2003; Weiner 1983; Weiner and Lovitt 1979). These same symptoms also are the hallmark of the after-reaction to traumatic uncontrollable stress that has been variously termed learned helplessness (Overmier and Seligman 1967), behavioral despair (Porsolt et al. 1977), behavioral depression (Weiss et al. 1981), and the distress syndrome (Minor et al. 1991). Finally, conservation-withdrawal is a critical component of *sickness behavior*—the lethargy, hypoactivity, decreased libido, anorexia, anhedonia, and increased sleep that accompanies infectious disease (Dantzer 2001; Hestad et al. 2003; Maes 1995). This dramatic shift in ongoing activity, along with the induction of fever, is a highly adaptive strategy for fighting infection (Dantzer 2001). The overlap among mood disorders, the after-reaction to traumatic stress, recuperation from injury, and sickness behavior (Anisman et al. 2005; Dantzer et al. 1999; Yirmiya 1996) suggests a common biological mechanism underlying these enervated states. We have argued that the overlap is well accommodated by the concept of conservation-withdrawal (Minor et al. 1994a, b, 2006, 2010).

This chapter reviews the data relating brain adenosine signaling to conservation-withdrawal and other stress-related processes. There is now considerable data from several laboratories indicating that adenosine activation is critically associated with states of conservation-withdrawal. Recent data suggests that the behavior depression component of this reaction is organized in the striatum, where adenosine regulates the motivational effects of dopamine via an interaction at a D_2-A_{2A}-mGLU$_5$ heteromeric receptor complex in the striatopallidal pathway.

25.2 Adenosine Signaling in Animal Models of Conservation-Withdrawal

25.2.1 Learned Helplessness

The learned helplessness procedure is a traditional method for assessing the ability of psychological variables, such as prediction and control, to modulate the impact of traumatic stress. The paradigm also is a leading animal model of Posttraumatic Stress Disorder (PTSD) and comorbid depression.

The experiment is a simple two-phase procedure. Sets of three rats are restrained in wheel-turn chambers in the pretreatment phase. The first rat in each triad is exposed to a series of controllable shock escape trials. The rat must complete a 360° turn of the wheel with its paws following shock onset in order to terminate the aversive event on each trial. A second rat receives yoked inescapable (uncontrollable) shock on each trial. Shock comes on at the same time for both rats on a trial and terminates when the first rat completes the escape response. Wheel-turn responses by the yoked rats are ineffective in altering shock. Thus, both rats receive the same intensity pattern and duration of shock on each trial, but differ in the extent to which they can exert behavioral control over the stressor. These rats receive 100 such trials over a period of 2 h. A third rat is simply restrained in the chamber for the same period of time and receives no shock. The restrained rat provides a behavioral and physiological baseline from which any differential effects of stressor controllability can be assessed during later testing.

Although the nature of the test varies with the interests of the experimenter, the traditional measure of helplessness has been a performance in a shuttle-escape task conducted 24 h after stress pretreatment (Maier et al. 1973). The shuttle-escape task consists of five trials (or FR-1 trials) during which a single shuttle crossing is required to terminate shock, followed by 25 trials (or FR-2 trials) during which a rat must run from one side of the shuttlebox to the other, then return to terminate shock. Both trial types are presented with an average intertrial interval of 60 s; however, 3 min intervene between trial types, which maximize escape deficits in inescapably shocked rats without adversely affecting control subjects (Minor and LoLordo 1984). Trials terminate automatically if the appropriate response requirement is not met within 40 s of shock onset.

Groups typically do not differ in mean escape latencies during FR-1 trials. However, large performance differences occur when the response requirement is made more difficult (FR-2 trials). Rats preexposed to escapable shock in the pretreatment phase perform as efficiently as restrained controls during escape testing. By contrast, rats preexposed to yoked inescapable shock show severe impairment, with near-maximum escape latencies during the FR-2 trials. This general pattern among groups holds for a wide variety of behavioral and biological stress indexes. Moreover, because escapably and inescapably shocked rats receive the same pattern, intensity, and durations of shock during pretreatment, the differential performance of these two groups in the test phase provides unequivocal evidence that some psychological variable related to behavioral control, or lack thereof, modulates the impact of the shock stressor (Overmier and Seligman 1967).

A simpler version of this procedure is often used when testing the efficacy of experimental compounds at alleviating stress-induced pathology. Because rats exposed to escapable shock do not show behavioral or physiological impairment, this condition adds an unnecessary expense to basic pharmaceutical research. Thus, much of the research discussed here compares rats exposed to variable-duration inescapable shocks during the pretreatment phase with the restrained control.

Research over the past decade has yielded a consensus on a number of critical issues concerning the psychological and neurobiological determinants of helplessness.

Behavioral disturbances following inescapable shock are a consequence of the induction and prolonged maintenance of fear during exposure to the uncontrollable stressor (Drugan et al. 1984; Jackson and Minor 1988; Mineka et al. 1984). Controllable stress is less debilitating because it is less fear provoking (see Minor and Hunter 2002 for a review). The consequence of maintaining an intense catabolic reaction during inescapable shock is that a number of neural systems associated with fear and stress are rendered hyperresponsive and highly vulnerable to subsequent stress for 24–72 h.

Inescapably shocked rats enter the test phase of the helplessness experiment in an anxious, agitated state. Exposure to comparatively mild stress during the first few shuttle escape trials provokes exaggerated behavioral activation to the shock and excessive fear during the interval between trials (Maier 1990; Minor 1990). This behavioral reaction is correlated with sustaining glucocorticoid secretion (Orr and Mann 1992), rapid turnover of brain biogenic amines (Anisman and Zacharko 1986), depletion of forebrain norepinephrine (Anisman and Sklar 1979; Minor et al. 1988; Weiss et al. 1981), dopamine (Hranilovic et al. 2008; Li et al. 2011) and γ-aminobutyric acid (GABA) (Petty and Sherman 1980), and excessive release of excitatory amino acid transmitters (Hunter et al. 2003) and *N*-methyl-D-aspartate (NMDA) receptor activation (Shors and Servatius 1995). The initial response is short-lived and the animal rapidly transitions from a highly catabolic state to one of behavioral quiescence, dissociation, or conservation-withdrawal.

This general pattern of excessive neural activation is reminiscent of the conditions that compromise brain metabolic homeostasis and result in potent compensatory inhibition by adenosine. Although little was known about the behavioral effects of adenosine at the time we started this research, there was evidence that adenosine analogs suppress spontaneous motor activity and can produce a semi-hypnotic state (Barraco et al. 1983, 1984; Dunwiddie and Worth 1982; Heffner et al. 1989). These properties of receptor agonists suggest adenosine signaling as a plausible mechanism for the behavioral depression and cognitive dysfunction associated with conservation-withdrawal and helplessness.

Thus, we hypothesized that the excessive and unregulated neural activation that characterizes inescapably shocked rats during the first few minutes of escape testing rapidly compromises neural energy homeostasis, resulting in potent compensatory adenosine regulation. Because overactivation occurs in neural pathways associated with fear, stress and active coping, adenosine's inhibitory effect on these systems necessarily interferes with active commerce with the environment. The animal transitions into a state of conservation-withdrawal and performance deficits ensue.

The least intuitive prediction of this hypothesis, at least from a psychological perspective, is that treatment with methylxanthine stimulants just prior to testing in inescapably shocked rats should dramatically improve escape performance. The conceptual difficulty here is that methylxanthines have rather potent anxiogenic effects, particularly at high doses. Thus the prediction is that a condition that stems from too much fear (Jackson and Minor 1988; Minor et al. 1991) will be improved by increasing fear.

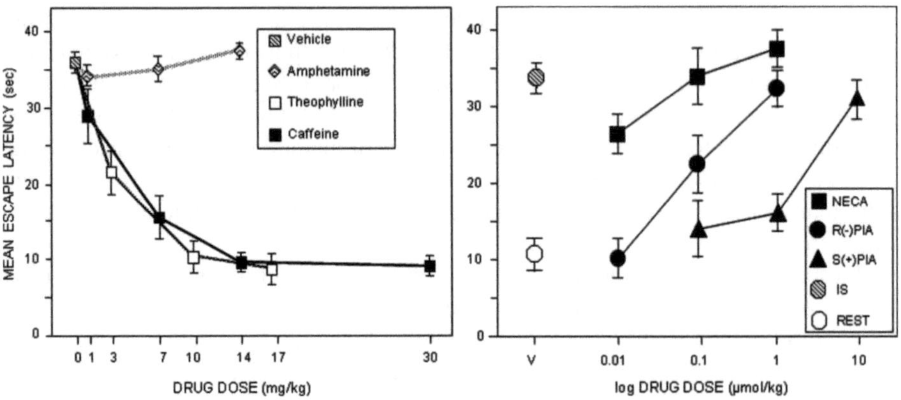

Fig. 25.1 (*Left panel*) Grand mean FR-2 shuttle-escape latencies as a function of drug type and dose. All rats were exposed to unsignaled inescapable shock on day 1. Shuttle-escape testing occurred 24 h later. Groups of rats received an i.p. injection of vehicle or various doses of amphetamine, caffeine or theophylline 15 min before testing. Each point in the figure is the mean for a group of eight rats. (*Right panel*) Grand mean FR-2 shuttle-escape latencies as a function of drug type and dose. One group of rats was exposed to inescapable shock (IS) and one group was restrained (REST) in tubes 24 h prior to shuttle escape testing. These groups were treated with vehicle 15 min before testing and their performance sets the boundaries for the learned helplessness effect. All other groups were restrained in tubes 24 h before receiving an i.p. injection of the nonselective adenosine agonists NECA, the highly selective A_1 receptor agonist R(−)PIA, or its enantiomer S(+)PIA. Escape testing occurred 15 min later

Despite this conundrum, the straightforward prediction is well substantiated. The left panel in Fig. 25.1 shows the effects of pretest treatment with various drugs on shuttle-escape performance in rats previously exposed to inescapable shock. Each point represents the mean FR-2 escape latencies for groups of eight rats. The nonselective adenosine receptor antagonists caffeine and theophylline reversed the effects of inescapable shock on later escape performance in a dose-dependent manner relative to the vehicle control. Amphetamine, a nonxanthine psychomotor stimulant, had no beneficial effect at any dose under study. The ability of the methylxanthines to reverse the helplessness effect was not due to drug state dependency and these drugs had no effect on the performance of nonshocked restrained controls.

Adenosine analogs also have the predicted effect on escape performance. The right panel of Fig. 25.1 shows the FR-2 shuttle escape latencies of groups of rats receiving a pretest injection of various adenosine analogs. One group of rats was exposed to inescapable shock (IS) and one group was restrained (REST) during the pretreatment phase of the experiment. Twenty-four hours later, these groups received vehicle (V) 15 min before escape testing and defined the boundaries of the helplessness effect. Other restrained groups were injected with various doses of either: (a) 5′-*N*-Ethylcarboxamidoadenosine (NECA), a high affinity, but nonselective adenosine receptor agonist; (b) R(−)-Phenylethyladenosine (R(−)PIA), a high-affinity, highly selective agonist of the A_1 receptor; or (c) S(+)-Phenylethyladenosine (S(+)PIA), the relatively inactive enantiomer of R-PIA. Later escape testing of these groups clearly indicated that the effects of inescapable shock are mimicked by activation of adenosine receptors. The comparison of NECA and R(−)PIA provided the first evidence that this effect may be mediated at the A_2 receptor. Moreover, the deleterious effects of pretreatment with inescapable shock or pretest treatment with NECA on escape performance is not reversed by peripheral treatment with the polarized methylxanthine stimulant 8-(*p*-Sulfopenyl)-theophylline (8-SPT), which does not cross the blood–brain barrier, but is completely reversed by theophylline, which acts both centrally and peripherally (not shown). These data suggest that the site of NECA's action is the central nervous system (CNS). NECA and inescapable shock also interact synergistically. Exposure to an ineffective number of inescapable shocks during pretreatment combines with the administration of subthreshold doses of NECA just prior to testing to maximize performance deficits in the shuttle-escape task.

The contribution of adenosine to the helplessness effect appears to be related to neural overactivation and subsequent metabolic failure. Overactivation of glutamate neurons in prefrontal cortex substantially impairs later escape performance (Petty et al. 1985). The deficit is completely reversed by adenosine receptor antagonists, suggesting that overactivation leads to compensatory adenosine signaling, which then impairs performance (Hunter et al. 2003). Escape performance also is impaired by systemic treatment with the glycolytic inhibitor 2-deoxy-D-glucose (2-DG) in a dose-dependent manner (Minor et al. 2001). This effect of 2-DG is not a direct consequence of glucoprivation or metabolic inhibition per se. Escape deficits are completely eliminated by peripheral and central administration of caffeine and theophylline, but are not reversed by the peripherally acting receptor antagonist 8-SPT. These data suggest that 2-DG compromises brain energy metabolism, resulting in compensatory adenosine regulation. Potent inhibition of brain substrates responsible for escape performance ultimately impairs performance.

Endogenous extracellular adenosine concentrations are regulated by two mechanisms. The nucleoside is converted to inactive inosine and eventually to uric acid via a degradation pathway involving adenosine deaminase (Geiger and Fyda 1991; Newby 1981; Noji et al. 2004). The nucleoside also is removed from the synaptic cleft via equilibrase nucleoside transporters (ENT1 and ENT2), as well as an active transport system involving Na^+ exchange (Parkinson et al. 2011). Adenosine is rapidly converted to 5′AMP via adenosine kinase once inside the cell, which stops active gradient transport of the nucleoside.

Disabling either of these regulatory mechanisms functionally increases synaptic adenosine concentrations and its action at extracellular receptors. Thus, if brain adenosine signaling mediates the escape deficits produced by initial exposure to inescapable shock, then enhancing the small and otherwise nondebilitating endogenous concentrations produced by prior restraint stress by either blocking degradation or uptake transport should disrupt test performance.

Woodson et al. (1998) demonstrated that inhibiting adenosine deaminase with erythro-9-(2-hydroxy-3-nonyl)adenine (EHNA) mimics the effects of inescapable shock on shuttle-escape performance in rats. Microinfusion of EHNA into the brain ventricles of previously restrained rats impaired test escape performance in a dose-dependent manner. This effect of EHNA, as well as the effect of earlier exposure to inescapable shock, was reversed by pretest treatment with adenosine receptor antagonist caffeine. A subthreshold dose of EHNA also interacted in synergy with preexposure to an ineffective number of inescapable shocks to maximally impair shuttle-escape performance at the time of testing.

Increasing endogenous adenosine concentrations by blocking the reuptake transporter yields a similar pattern of results (Minor et al. 2008). Microinfusion of the nucleoside transport blocker *S*-(4-nitrobenzyl)-6-theoinosine (NBTI) into the right lateral ventricle of rats undergoing restraint stress 24 h earlier produces a large deficit in shuttle-escape performance, which is comparable to that produced by prior exposure to inescapable shock. NBTI impairs escape performance in a dose-dependent manner. Moreover, a subthreshold dose of the adenosine uptake inhibitor acts synergistically with an ineffective number of inescapable preshocks to maximize deficits in test escape performance, suggesting that NBTI and inescapable shock are acting on the same neural mechanism.

Analysis of the receptor mediating shock-induced and NBTI-induced escape deficits using highly selective receptor antagonists supported earlier evidence for an A_2 receptor. Both types of deficits were reversed by the nonselective adenosine receptor antagonist caffeine, and the highly selective A_{2A} receptor antagonist CSC (8-(3-chloro-styrl)caffeine) in a dose-dependent manner. The highly selective A_1 (DPCPX: 8-Cyclopentyl-1,3-Dipropylxanthine) and A_{2B} (CGS 21680) receptor antagonists failed to improve performance in rats preexposed to inescapable shock or receiving intracerebral ventricular (i.c.v.) infusion of NBTI shortly before escape testing. These data strongly suggest that activation of A_{2A} receptors mediate deficits in escape performance, regardless of whether those deficits occur because of prior traumatic stress or by enhancing endogenous adenosine signaling through pharmacological means. As discussed in detail later, A_{2A} receptors have a limited and unique distribution in the CNS. Thus, these data have important implications for the anatomical locus and mechanism by which adenosine signaling mediates behavioral depression.

25.2.2 Behavioral Despair

Vaugeois and his colleagues have accumulated considerable evidence for a role of A_{2A} receptor signaling in animal models of behavioral despair (see El Yacoubi et al. 2003 for a review). They argue on the basis of these data that A_{2A} receptor antagonists may have potent antidepressant properties.

Much of the evidence for the role of adenosine signaling in behavioral despair comes from studying A_{2A} receptor knockout mice (El Yacoubi et al. 2001). These mice showed increased mobility relative to wild types in both forced swim and tail-suspension test, two widely used assessments of the antidepressant action of drugs (Porsolt et al. 1977; Stéru et al. 1987). Immobility scores in the forced swim task in CD1 mice (wild-type controls) are reduced in a dose-dependent manner by caffeine and the highly selective A_{2A} receptor antagonist SCH 58216. Similar results are obtained in the tail-suspension test. Moreover, the benefits of treatment with the A_{2A} receptor antagonist in mice genetically selected for spontaneous helplessness were only slightly less that the benefits of the antidepressant imipramine.

The benefits of an A_{2A} receptor antagonist in these procedures are likely to result from an interaction with striatal dopamine. The dopamine D_2 antagonist haloperidol increases immobility in both forced swim and tail-suspension tests (El Yacoubi et al. 2001, 2003). Moreover, cotreatment with haloperidol mitigates the behavioral activating effects of caffeine in these procedures. Interestingly, cotreatment with haloperidol and SCH 58216 decreases the activating effect of the A_{2A} receptor antagonist on spontaneous activity, but did not alter its benefits in behavioral despair paradigms.

25.2.3 Reserpine-Induced Depression

Reserpine was introduced in this country in the early 1950s as a treatment for hypertension (Cooper et al. 1978a; Gerber et al. 1990; Rech and Moore 1971). The extract reduces both cardiac output and peripheral vascular resistance by irreversibly binding to storage vesicles in monoaminergic neurons and thereby depletes stores of biogenic amines in the central and autonomic nervous systems (Norn and Shore 1999; Schwartz 1981). Cytoplasmic amines are either destroyed by intraneuronal monoamine oxidase or diffuse into the synaptic cleft. The end result is that little or no active transmitter is released at the synapse following depolarization. Recovery from the effects of reserpine requires synthesis of new storage vesicles, which can take several days to accomplish after discontinuing treatment (Cooper et al. 1978b; Ponzio et al. 1984).

Unfortunately, a significant portion of the population undergoing reserpine treatment for hypertension developed symptoms of major depression. These symptoms were severe enough to require antidepressant drug treatment and, at times, hospitalization. This observation, along with the findings a few years later that monoamine oxidase inhibitors and tricyclic antidepressants enhance brain biogenic amines, persevered as the empirical cornerstone of catecholamine (Schildkraut 1965) and later monoamine theories of depression (Akiskal and McKinney 1975; Bunney and Davis 1965).

One of the main problems in making a convincing case for a direct role of the biogenic amines in reserpine-induced depression is that the time course for their depletion does not fit the time course for behavioral impairment. For instance, Bean et al. (1989) reported that dopamine (DA) depletion in striatum, nucleus accumbens, and frontal cortex occurs rapidly and reaches a maximum about 6 h after an intraperitoneal (i.p.) injection of 6 mg/kg of reserpine. DA remained at floor levels for about 18 h and then increased thereafter such that DA baselines return to normal within 48 h. Behavioral deficits persist for a considerably longer period.

Huang and Minor (Huang et al. 2004; Minor et al. 2003, 2006) treated rats with an i.p. injection of 6 mg/kg of reserpine and then tested independent groups in a forced swim task for symptoms of conservation-withdrawal at various times thereafter. Large deficits in swim performance, as characterized by a large increase in floating time, were evident as early as 1 h post drug treatment, persisted for at least 72 h, and recovered within a week.

The determinants of reserpine-induced depression are complex and change over time. Moreover, the impairment is likely to be a "downstream" consequence of reserpine's effect on the monoamines rather than a direct result of their depletion per se. Adenosine plays a critical role in swim deficits as early as 1 h post drug treatment and continues to be a critical mediator at all time points thereafter. Caffeine and the highly selective A_{2A} receptor antagonist CSC reverse swim deficits at all time points under study (Minor and Hanff manuscript submitted). A_1 and A_{2B} receptor antagonists have no such benefit.

Additionally, the proinflammatory cytokine interleukin-1β (IL-1β) increases dramatically in the hypothalamus and to a lesser extent in the hippocampus 48–72 h after reserpine treatment and then returns to normal levels in the brain within 168 h. In this context, the IL-1β receptor antagonist has no effect on swim deficits 1 h after reserpine treatment, but substantially improves performance 48 h later.

Although there is only a small literature on the potential interactions between adenosine and IL-1β, it is clear that these pathways interact. For instance, application of endotoxin or IL-1β on PC12 or THP-1 cell upregulates the density of A_{2A} receptors and increases extracellular concentrations of adenosine (Bshesh et al. 2002; Trincavelli et al. 2002). Several investigators have argued on the basis of such data that adenosine (an anti-inflammatory agent) is induced as negative feedback on proinflammatory cytokines.

25.2.4 *Sickness Behavior and Cytokine-Induced Behavioral Depression*

The foregoing analysis of reserpine-induced depression indicates that brain IL-1β concentrations increase as a delayed consequence of drug treatment. Swim deficits at later time points are improved by blocking brain IL-1β receptors and completely eliminated by A_{2A} receptor antagonists (Minor and Hanff manuscript submitted; Hanff et al. 2010). The relationship between purine and cytokine signaling suggests that adenosine may play a role in sickness behavior. Indeed, the regulatory mechanism that is evident in adenosine's control of overt behavior may have evolved to suppress ongoing behavior at times of illness.

Sickness behavior most certainly has a conservation-withdrawal component. This behavioral syndrome is induced experimentally by systemic treatment of rats with endotoxin (e.g., Lipoplysacchrine; LPS), which indices the synthesis of proinflammatory cytokines in peripheral macrophages. Kuppfer cells in the liver recognize circulating IL-1β and serve as a primary immune-to-brain communication pathway. This signal is transferred via the afferent vagal nerve complex to the brain nucleus tractus solitarius (NTS) where IL-1β is expressed within 1–2 h of LPS treatment. The cytokine is also expressed relatively quickly thereafter in a variety of other brain nuclei, particularly in the hypothalamus (Anisman et al. 2005; Dantzer et al. 1999; Larson and Dunn 2001; Maes 1995). IL-1β binds to specific receptors throughout the brain to induce sickness behavior, consisting of lethargy, hypoactivity, decreased libido, anorexia, anhedonia, and increased sleep (Dantzer 2001; Larson and Dunn 2001). This dramatic shift in ongoing activity, along with the induction of fever, is assumed to be a highly adaptive strategy to fight infection.

Preliminary evidence clearly supports an interaction between central adenosine A_{2A} and cytokine signaling in the induction and regulation of sickness behavior (see Hanff et al. 2010; Minor et al. 2006, 2010 for reviews). A deficit in forced swim performance occurs 24 h after systemic treatment with LPS. The symptoms of behavioral depression are reversed by administration of the IL-1β receptor antagonist into the brain lateral ventricles. The swim deficit also is reversed by central or peripheral administration of caffeine, the selective A_{2A} antagonist CSC and the nonselective A_2 receptor antagonist DMPX (3,7-dimethyl-1-propargylxanthine). A_1 (DPCPX: 8-Cyclopentyl-1,3-Dipropylxanthine) and A_{2B} (AX: Alloxozine) receptor antagonists have no beneficial effects on performance. A deficit in swim performance also is evident 1 h after infusion of IL-1β into the brain lateral ventricles. Performance is again improved by peripheral treatment with caffeine, CSC, or AX, but not DPCPX or AX.

Overall, considerable evidence from leaned helplessness, behavioral despair, reserpine-induced depression, and cytokine-induced sickness behavior and depression paradigms, implicate adenosine A_{2A} receptor signaling (but not other forms of adenosine signaling) in conservation-withdrawal. Adenosine A_{2A} receptors have a limited distribution in the nervous system, being located primarily in the striatopallidal (indirect) tract of the striatum. These receptors have an important influence on both glutamate and dopamine signaling in this pathway. The molar effects of A_{2A} receptor activation in the striatum appears to increase uncertainty about appropriate courses of action and uncouples dopamine's motivational influence from ongoing behavior.

This uncoupling process receives empirical support from recent research with striatal A_{2A} knock-out mice demonstrating that habit formation requires A_{2A} receptor activation (Yu et al. 2009). Habits are formed when specific behaviors consistently lead to specific rewards. Under these circumstances the process of executing a response and the underlying learned association are simplified such that the individual is no longer aware of the consequences of his actions. In this sense, the response is uncoupled from its motivation consequences.

25.3 Adenosine, Dopamine, and Metabotropic Heteromers in the Striatum

Dopamine plays a complex role in the striatum's contribution to the learning and expression of behavior. This neurotransmitter is involved in both appetitive and aversive tasks, although two types of dopamine neurons appear to convey positive

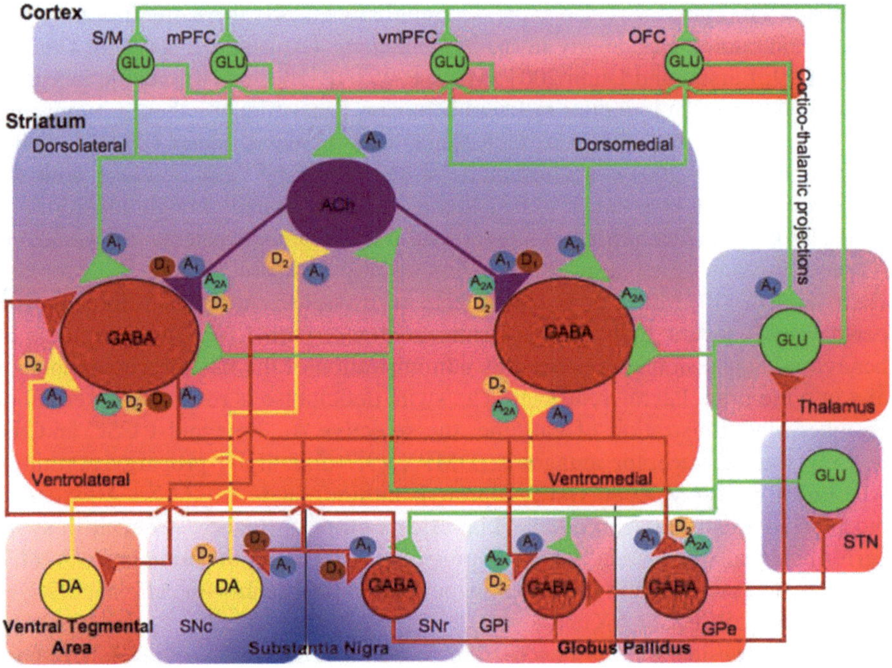

Fig. 25.2 This schematic represents the structural organization within the basal ganglia. The four functionally distinct sections of the striatum receive diverse inputs from various cortical regions and midbrain structures, such as the substantia nigra and ventral tegmental area, as well as the thalamus. The functional inputs and outputs of the striatum are *color coded*. *Blue* represents areas associated with sensorimotor function and transitions through associative regions into *red* limbic regions. Adenosine A_1 receptors are found on striatal GABAergic neurons containing dynorphin and substance P (nigrostriatal, striatopallidal neurons) and to a smaller extent GABAergic neurons containing enkephalin (striatopallidal neurons), as well as on cholinergic interneurons. Dopamine D_1 receptors are found mainly on GABAergic neurons that contain dynorphin and substance P, dopaminergic neurons, and cholinergic interneurons. The A_1 receptors are thought to modulate D_1 receptors even though they are not often colocalized. The adenosine A_{2A} receptors are found primarily on GABAergic neurons containing enkephalin and are often colocalized with dopamine D_2 receptors, though they are also found on dopaminergic neurons. *GLU* glutamate, *ACh* acetylcholine, *DA* dopamine, *GABA* γ-aminobutyric acid, *S/M* primary sensorimotor cortices, *mPFC* medial prefrontal cortex, *vmPFC* ventromedial prefrontal cortex, *OFC* orbitofrontal cortex, *SNc* substantia nigra pars compacta, *SNr* substantia nigra pars reticulata, *VTA* ventral tegmental area, *GPe* globus pallidus external, *GPi* globus pallidus internal, *STN* subthalamic nucleus

and negative motivational signals (Matsumoto and Hikosaka 2009). Adenosine modulates the role of dopamine by playing a significant role in regulating dopamine's action. The striatum exerts its function almost exclusively by medium-sized spiny GABAergic neurons. These neurons make up more than 90 % of the striatal neuronal population and are distributed among two efferent pathways, the direct and indirect pathways. (Herrero et al. 2002). See Fig. 25.2 for the inputs and outputs of the striatum.

Given the overwhelming evidence that the A_{2A} receptor is contributing to conservation withdrawal and other stress related pathology, the known trimeric receptor involving adenosine, dopamine, and metabotropic glutamate is of particular interest. Adenosine A_{2A} receptors are located mainly in the striatum on the dendritic spines of enkephalin-containing GABAergic neurons. These spines contain receptor complexes known as heteromers. The trimeric heteromer of interest contains receptors for adenosine (A_{2A}), dopamine (D_2), and glutamate ($mGLU_5$) (Ferré et al. 2007). These metabotropic heteromers are found postsynaptically at glutamatergic synapses. A_{2A} and D_2 receptors are G protein coupled and are involved in an antagonistic relationship where adenosine modulates the binding of dopamine (Schultz 2002). Adenosine bound to the A_{2A} receptor decreases the affinity of dopamine to bind to the D_2 receptor. There is also evidence that glutamate and adenosine have a synergistic relationship where the functional interaction of adenosine and glutamate is required to modulate the effects of dopamine in the striatum (Ferré et al. 2002, 2005; Anisman and Sklar 1979). Adenosine A_1 and A_{2A} receptors are also colocalized on presynaptic glutamate channels, but their function on the control of behavior is still unclear.

There are two main pathways within the striatum that contribute opposite influence on their targeted nuclei. The direct pathway, also known as the nigrostriatal pathway, facilitates movement through tonic input of this positive feedback loop to increase the activation of the thalamus, thereby increasing activation of the cortex and facilitating movement. Adenosine A_1 receptors and dopamine D_1 receptors are distributed throughout this pathway. The direct pathway has its origins in the substantia nigra pars compacta (SNc), where an excitatory dopaminergic signal is sent to the dorsal striatum innervating both GABAergic neurons and cholinergic interneurons. The dynorphin and substance P-containing GABAergic neurons of this area then project to GABAergic neurons in the globus pallidus interna (GPi), which in turn target glutamatergic thalamic nuclei via inhibitory GABAergic transmission. The result is an inhibition of the GPi GABAergic neurons by the striatal GABAergic neurons, producing an overall excitatory effect in the thalamus. Thalamic nuclei then send excitatory glutamate (GLU) to innervate areas of the cortex, the subthalamic nucleus (STN), and the pedunculopontine nucleus (PPN), as well as back to the substantia nigra.

The indirect, or striatopallidal, pathway normally wields an inhibitory influence on behavior. The indirect pathway contains a small number of both D_1 and A_1 receptors, but mostly consists of dopamine D_2 and adenosine A_{2A} receptors. Unlike the D_1 and A_1 receptors, the D_2 and A_{2A} are colocalized on the membrane in the striatopallidal enkephalin-containing GABAergic neurons. They are also colocalized on cholinergic interneurons and regulate acetylcholine (ACh) release in the striatum. The indirect pathway is considered to be the negative feedback loop within the striatum. Inputs from the SNc synapse on the dorsal striatum, which then sends inhibitory enkephalin-containing GABAergic projections to the globus pallidus externa (GPe). The GPe sends GABAergic projections to the STN, and the STN in turn sends glutamatergic projections to the GPi. The GPi projects to various glutamatergic thalamic nuclei via inhibitory GABAergic transmission, resulting in an inhibition of the thalamus. It is believed that the A_{2A} receptor is involved in regulating this inhibition by blocking D_2 receptor activation from the substantia nigra or the smaller dopaminergic signal from the ventral tegmental area (VTA) to the ventral striatum. Activation of the D_2 receptors results in a disinhibition of the indirect pathway, which facilitates movement by decreasing inhibition to the thalamus (for reviews see Gerfen and Surmeier 2011; Ferré et al. 2008; Haber 2008).

Fine-tune control of motor behavior requires interplay of these two pathways. To give a functional analogy, the direct pathway can be considered the accelerator of behavior, or what initiates movement. As animals become more certain of their expected outcomes, more force is applied to the accelerator. The indirect pathway, however, can be thought to serve as the brake on behavior when the expected outcomes are not forthcoming. In this sense, transitioning between behaviors requires disengaging the brake and initiating the accelerator. In conservation withdrawal, there appears to be potent activation of the brake.

This analogy is akin to Gray's (1982) analyses of behavioral inhibition which results from the discrepancy between predicted and obtained outcomes. Such a mismatch increases uncertainty about correct paths of action, and leads to the inhibition of behavior (also see Calhoun 1963). Gray et al. (1991) subsequently emphasized the dopaminergic function of the ventral striatum in behavioral inhibition.

The dopamine signal to the striatum from the substantia nigra (nigrostriatal pathway) and ventral tegmental area (mesoaccumbens pathway) has long been thought of as a prediction error signal (Ljungberg et al. 1992; Schultz 1998). Prediction error is described as a discrepancy between outcomes that are expected versus those that are actually observed. This can be commonly observed with a standard Pavlovian procedure, which is a method for conditioning hedonic or emotional value to stimuli. An influx of dopamine can be seen immediately following the presentation of a biological significant event. The phasic dopamine signal can then be seen to switch to the predictor of that event with repeated presentations. If the biologically significant event does not occur after the predictor as expected, a drop in phasic dopamine signaling is observed at the point where the expected event was to occur.

However, recent evidence suggests that the nigrostriatal and the mesoaccumbens pathways may serve different functions (Yin et al. 2008). The nigrostriatal pathway serves the dorsal striatum where its main function is the learning and performance of instrumental action. This dopamine signal may reflect the value of performing the action itself. The mesoaccumbens pathway serves mainly the ventral striatum which controls Pavlovian responses. This dopamine signal may reflect a different function where it encodes the value of the stimuli and of motivational states. It is suggested by Yin et al. (2008) that perhaps the phasic dopamine signal is encoding multiple prediction errors as opposed to a single prediction error. More recently, Ostlund et al. (2011) has suggested that tonic dopamine in the striatum may be encoding the costs and benefits of responding on an instrumental task, thereby controlling the performance of the instrumental response (also see Salamone et al. 2009). The complex function of dopamine in the striatum is still under much investigation.

25.4 The Function of the Striatum

The striatum, consisting of the caudate, putamen, nucleus accumbens, and olfactory tubercle, is the most prominent structure of the basal ganglia and is involved mainly in the performance of voluntary and involuntary motor movements. Movement disorders, such as Parkinson's disease and Huntington's disease, are associated with a malfunction of this structure or its output (Albin et al. 1989). Many discussions of striatal control of behavior greatly underestimate the complexity of behavioral reactions, the value of rewards, and the control of behavior by incentive stimuli. Overt behavior is not simply caused by one event or produced by one structure. Simple movement through space requires complex interplay among expected outcomes, the value of incentive stimuli that redirect behavior, and the level of motivation that constrains the range of responses (Bindra 1968). Likewise, the striatum is not a singular structure that simply produces behavior when told to. It is a structure of integration of cortical, thalamic, and limbic signals, all working together within the striatum to produce a wide array of complex behaviors.

The striatum can be broken down into four distinct areas, each with its own unique contribution to the overall performance of an action. Heimer and Wilson (1975) were the first to functionally distinguish the dorsal and ventral aspects of the striatum based on their connections to other structures. The traditional view assigned the dorsal striatum the sensorimotor behaviors and the ventral striatum the more associative and emotionally mediated behaviors. The current view of the functional distinctions of the striatum is similar to but more complex than the traditional view originally postulated.

The dorsal striatum, made up of the caudate–putamen complex, is involved in the learning and performance of an instrumental response, where one's action is necessary for receiving an outcome (Yin et al. 2008). It can be further divided into a dorsal medial segment and dorsal lateral segment with functional dissociation (Yin and Knowlton 2004; Yin et al. 2004). The dorsal medial striatum (DMS) is now known to be involved in the associated aspects of behavior and necessary for the learning of an instrumental response. The association made between the response and its consequence (or the outcome) is thought to occur in this area and is responsible for goal-directed actions. The dorsal lateral striatum (DLS) is involved in the development and performance of a habitual response. The learning of response–outcome associations takes place in the DMS and with overtraining, come under the control of the DLS where the behavior has now morphed into a habit (Yin et al. 2008). This sensorimotor region follows the traditional view of the dorsal striatum.

The ventral striatum, made up of the nucleus accumbens, plays a role in Pavlovian learning, where a stimulus acquires hedonic or incentive value because of its predictive relationship with a biologically significant event. Pavlovian conditioning is also a mechanism for controlling instinctive behavior, such as preparatory and consummatory behaviors. The ventral medial striatum (VMS) is the core of the accumbens and is involved in preparatory behaviors, commonly approach or exploration, which allow the organism to come into contact with a goal (Parkinson et al. 2000). The general emotional qualities of this outcome are also mediated by this area (Yin et al. 2008). The ventral lateral striatum (VLS) is the shell of the accumbens and is responsible for consummatory behaviors, those that allow an organism to carry out its goal—or consummate the act. This area is also implicated in the hedonics of a response and is thought to encode the specific sensory qualities of the outcome (Berridge and Robinson 1998; Yin et al. 2008). The ventral striatum as a whole is known as the limbic network because of inputs received to the core and shell from the limbic structures, allowing for emotional and motivational input to interact with ongoing behavior. The Fig. 25.2 color schematic outlines the projections of the striatum from blue sensorimotor regions transitioning through the associative network to the red limbic regions.

While each section of the striatum is functionally distinct, there is much interconnectivity between segments allowing for the integration of information in a hierarchical manner (Yin et al. 2008). For example, the dorsal medial striatum is necessary for the learning of a goal-directed action, but the ventral striatum is thought to mediate the performance of that action. The large number of projections from the ventral striatum to the dorsal segments allows for ventral striatal information to pass to the dorsal striatum, suggesting that the ventral striatum may be the starting place of this hierarchy.

25.5 Adenosine A_{2A} Receptor Signaling in Learned Helplessness

Adenosine A_{2A} receptors have been strongly implicated by a variety of paradigms related to conservation withdrawal. As A_{2A} receptors are mainly found colocalized with D_2 receptors in the indirect pathway of the striatum, this seems a likely region to study adenosine's influence on stress-induced fatigue.

The ventral striatum is the site of integration of dopaminergic inputs from the mesencephalon and excitatory glutamatergic inputs from the limbic cortices, amygdala, and hippocampus (Minor et al. 2006). A common feature of the ventral striatum is the dopaminergic pathway to and from the ventral tegmental area. This pathway is of particular importance to the aspect of motivation and reward-related behaviors (Meredith et al. 2008). The dopaminergic input to the accumbens shell and core provides for functionally strong associations to the motivational aspects of behavior by combining emotional information from the limbic system with motor information (Mogenson et al. 1980). Ferré et al. (2007) built upon this idea and described the ventral striatum as the conversion of motivation and action. As such, this region is a plausible center for motivational disorders caused by a dysfunction of this conversion, most commonly the development of behavioral depression.

Traci Plumb is currently assessing the role of adenosine signaling in the striatum, specifically A_{2A} receptors in the indirect pathway, and its relationship to the development of behavioral depression as measured in the learned helplessness procedure. The antagonistic interaction of A_{2A} and D_2 receptors is likely to be responsible for

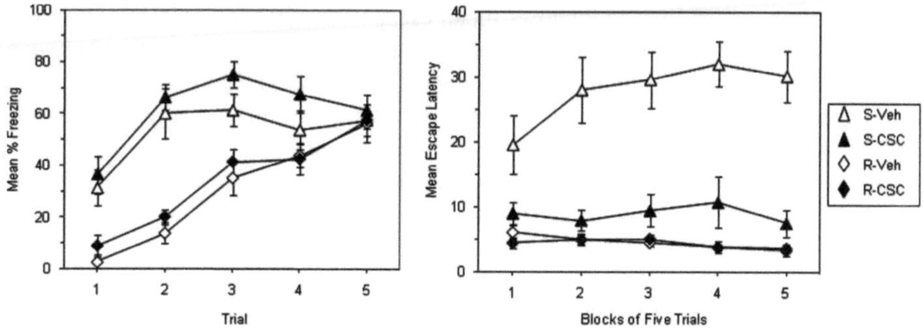

Fig. 25.3 (*Left panel*) Mean percent freezing in all groups during initial FR-1 trials. Rats were exposed to inescapable shock (S) or restraint (R) as a measure of traumatic stress and mild stress respectively. They were then injected with CSC, a highly selective adenosine A_{2A} receptor antagonist, or a vehicle (V). Rats exposed to traumatic stress exhibited freezing at a higher rate during the initial trials than those exposed to mild stress. CSC had little effect on time spent freezing. (*Right panel*) Mean escape latencies in all groups during FR-2 trials. The standard learned helplessness effect is observed between Groups S[V] and R[V] where performance was drastically impaired in those exposed to inescapable shock. Injection of CSC in the ventral striatum eliminated the learned helplessness effect

the uncoupling of motivation from ongoing behavior often seen in behavioral depression (Minor et al. 2006). Therefore, blocking adenosine's action in this area should allow the motivational signal to be transmitted to performance, thereby eliminating helplessness.

In a preliminary study, a guide cannula was implanted into the ventral medial striatum (accumbens core) during stereotaxic surgery. Rats were then exposed to a traumatic stress 7 days later in the form of 100, 1 mA tail shocks or simple restraint in tubes. Escape performance was measured 24 h later in a shuttle box. Immediately before testing, animals were infused with a highly selective adenosine A_{2A} receptor antagonist or vehicle. Subsequent escape latencies were measured as a form of behavioral depression.

The left panel of Fig. 25.3 shows mean percent freezing for each group during the initial FR-1 trials. Freezing is an excellent measure for the amount of fear a rat is experiencing at any given time as it is a species-typical defense response (Bolles 1971). Groups did not differ in amount of time spent freezing.

A comparison of shocked (S[V]) and restrained (R[V]) vehicle controls shows the exaggerated fearfulness that is typically observed following traumatic stress. Rats exposed to inescapable shock spend more time freezing during the initial trials than restrained rats. Blocking adenosine A_{2A} receptors in the core of the striatum has no effect on freezing behavior. This suggests that while the core is involved in the coupling of motivation and action, it is not specifically associated with fear behaviors. This is not surprising as there is considerable evidence suggesting fear related behaviors such as freezing are organized in the ventral periaqueductal gray (De Oca et al. 1998).

The right panel of Fig. 25.3 shows mean escape latencies for each group during 25 FR-2 trials. The standard learned helplessness effect can be observed between the shock and restraint groups given vehicle (S[V] and R[V]). Performance was dramatically impaired in the shock control group with the shift to the more complex FR-2 trials compared to restraint controls. More importantly, infusion of the adenosine A_{2A} receptor antagonist CSC in the core of rats exposed to inescapable shock was able to completely reverse the helplessness effect. CSC-treated rats performed similarly to restraint controls.

Further experiments are underway to determine the role of A_{2A} receptors in the remaining three areas of the striatum, but this data clearly suggests that adenosine from the core of the striatum plays a role in mediating the escape deficits seen in the learned helplessness procedure. When A_{2A} receptors were blocked, the rats were able to perform the escape task with no more difficulty than those that received a milder form of stress the day preceding test, providing evidence that those that receive traumatic shock stress show enhanced adenosine signaling on the day of test. This enhanced adenosine signaling may result from enhanced glutamate signaling with the overactivation of descending glutamatergic projections from the frontal cortex. Simultaneous activation of A_{2A} and $mGLU_5$ receptors within the heteromeric complexes of the striatum dramatically reduces the binding affinity of D_2 receptors. This loss of dopaminergic input from the VTA into the core essentially uncouples the motivational signal from ongoing behavior. The symptoms of helplessness, behavioral depression or conservation-withdrawal ensue.

25.6 Overview

Adenosine plays a significant role in stress-induced behaviors. The sensory unresponsiveness, cognitive dullness, and behavioral depression that characterize conservation-withdrawal are produced by adenosine receptor activation (Minor et al. 2006, 2010). Specifically, A_{2A} receptors have been implicated in a wide array of behavioral paradigms relating to behavioral depression, including exposure to uncontrollable shock, conditions that produce despair, and pharmacological manipulations.

A large number of adenosine receptors are organized in the direct and indirect pathways of the striatum. Adenosine A_{2A} receptors are colocalized with dopamine D_2 receptors on the indirect pathway, where adenosine serves to modulate activation of D_2 receptors to produce inhibition of behavior. Activating A_{2A} receptors blocks dopamine from the substantia nigra and ventral tegmental area to bind to D_2 receptors, resulting in behavioral inhibition. The functional consequence of this action is to increase uncertainty about the correct course of action and to block dopamine's motivational signal. Certain investigators have suggested on the basis of this relationship that adenosine A_{2A} receptor antagonists may be effective in treating certain aspects of depression. The affect-less, fatigue components of major depressive disorder are the most likely candidates (Minor et al. 2006; Woodson et al. 1998;

Bougarel et al. 2011; El Yacoubi et al. 2001, 2003). In this regard, the motivational components of major depression may not stem from a lack of drive per se, but rather from a failure to translate motivation into appropriate action.

Adenosine's contribution to ongoing behavior in the striatum is complex. The nucleoside is thought to regulate general motivation of behavior, the error signal critical for learning consequences of actions, and the transition from learned responses to habits. Each of these functions is organized in an anatomically distinct way in the striatum. Adenosine's contribution to this tripartite organization should be determined experimentally. Only then can we get an accurate picture of adenosine's influence on behavior.

References

Akiskal HS, McKinney WT Jr (1975) Overview of recent research in depression. Integration of ten conceptual models into a comprehensive clinical frame. Arch Gen Psychiatry 32:285–305

Albin RL, Young AB, Penney JB (1989) The functional anatomy of basal ganglia disorders. Trends Neurosci 12:366–375

Anisman H, Sklar LS (1979) Catecholamine depletion in mice upon reexposure to stress: mediation of the escape deficits produced by inescapable shock. J Comp Physiol Psychol 93:610–625

Anisman H, Zacharko RM (1986) Behavioral and neurochemical consequences associated with stressors. Ann N Y Acad Sci 467:205–225

Anisman H, Merali Z, Poulter MO, Hayley S (2005) Cytokines as a precipitant of depressive illness: animal and human studies. Curr Pharm Des 11:963–972

Barraco RA, Coffin VL, Altman HJ, Phillis JW (1983) Central effects of adenosine analogs on locomotor activity in mice and antagonism of caffeine. Brain Res 272:392–395

Barraco RA, Aggarwal AK, Phillis JW, Moron MA, Wu PH (1984) Dissociation of the locomotor and hypotensive effects of adenosine analogues in the rat. Neurosci Lett 48:139–144

Bean AJ, Adrian TE, Modlin IM, Roth RH (1989) Dopamine and neurotensin storage in colocalized and noncolocalized neuronal populations. J Pharmacol Exp Ther 249:681–687

Berridge KC, Robinson TE (1998) What is the role of dopamine in reward: hedonic impact, reward learning, or incentive salience? Brain Res Rev 28:309–369

Bindra D (1968) Neuropsychological interpretation of the effects of drive and incentive-motivation on general activity and instrumental behavior. Psychol Rev 75:1–22

Bolles RC (1971) Species-specific defense reactions. In F. R. Brush (Ed.), Aversive Conditioning and Learning. New York: Academic Press

Bougarel L, Guitton J, Zimmer L, Vaugeois JM, El Yacoubi M (2011) Behaviour of a genetic mouse model of depression in the learned helplessness paradigm. Psychopharmacology (Berl) 215:595–605

Bruns RF, Lu GH, Pugsley TA (1986) Characterization of the A2 adenosine receptor labeled by [3H]NECA in rat striatal membranes. Mol Pharmacol 29:331–346

Bshesh K, Zhao B, Spight D, Biaggioni I, Feokistov I, Denenberg A, Wong HR, Shanley TP (2002) The A2A receptor mediates an endogenous regulatory pathway of cytokine expression in THP-1 cells. J Leukoc Biol 72:1027–1036

Bunney WE Jr, Davis JM (1965) Norepinephrine in depressive reactions. A review. Arch Gen Psychiatry 13:483–494

Calhoun JB (1963) The ecology and sociology of the Norway rat. United States Department of Health, Education, and Welfare, Maryland

Calon F, Dridi M, Hornykiewicz O, Bédard PJ, Rajput AH, Di Paolo T (2004) Increased adenosine A2A receptors in the brain of Parkinson's disease patients with dyskinesias. Brain 127: 1075–1084

Castillo-Meléndez M, Krstew E, Lawrence AJ, Jarrott B (1994) Presynaptic adenosine A2a receptors on soma and central terminals of rat vagal afferent neurons. Brain Res 652:137–144

Coleman CG, Baghdoyan HA, Lydic R (2006) Dialysis delivery of an adenosine A2A agonist into the pontine reticular formation of C57BL/6J mouse increases pontine acetylcholine release and sleep. J Neurochem 96:1750–1759

Cooper JR, Bloom FE, Roth RH (1978a) The biochemical basis of neuropharmacology. Oxford University Press, New York

Cooper BR, Konkol RJ, Breese JR (1978b) Effects of catecholamine depleting drugs and d-amphetamine on self-stimulation of the substantia nigra and locus coeruleus. J Pharmacol Exp Ther 204:592–605

Dantzer R (2001) Cytokine-induced sickness behavior: where do we stand? Brain Behav Immun 15:7–24

Dantzer R, Wollman E, Vitkovic L, Yirmiya R (1999) Cytokines and depression: fortuitous or causative association? Mol Psychiatry 4:328–332

De Oca BM, DeCola JP, Maren S, Fanselow MS (1998) Distinct regions of the periaqueductal gray are involved in the acquisition and expression of defense responses. J Neurosci 18:3426–3432

Drugan RC, Ryan SM, Minor TR, Maier SF (1984) Librium prevents the analgesia and shuttlebox escape deficit typically observed following inescapable shock. Pharmacol Biochem Behav 21:749–754

Dunwiddie TV, Worth T (1982) Sedative and anticonvulsant effects of adenosine analogs in mouse and rat. J Pharmacol Exp Ther 220:70–76

El Yacoubi M, Ledent C, Parmentier M, Ongini E, Costentin J, Vaugeois JM (2001) Adenosine A2A receptor antagonists are potential antidepressants: evidence based on pharmacology and A2A receptor knockout mice. Br J Pharmacol 134:68–77

El Yacoubi M, Costenin J, Vaugeois JM (2003) Adenosine A2A receptors and depression. Neurology 61:S82–S87

Engel GL, Schmale AH (1972) Conservation-withdrawal: a primary regulatory process for organic homeostasis. In: Ciba Foundation Symposium on Physiology, Emotion and Psychosomatic Illness 8:57–75

Ferré S, Karcz-Kubicha M, Hope BT, Popoli P, Burgueno J, Gutierrez MA, Casado V, Fuxe K, Goldberg SR, Lluis C et al (2002) Synergistic interaction between adenosine A2A and glutamate mGlu5 receptors: implications for striatal neuronal function. Proc Natl Acad Sci U S A 99:11940–11945

Ferré S, Borycz J, Goldberg SR, Hope BT, Morales M, Lluis C, Franco R, Ciruela F, Cunha R (2005) Role of adenosine in the control of homosynaptic plasticity in striatal excitatory synapses. J Integr Neurosci 4:445–464

Ferré S, Diamond I, Goldberg SR, Yao L, Hourani SMO, Huang ZL, Urade Y, Kitchen I (2007) Adenosine A_{2A} receptors in ventral striatum, hypothalamus and nociceptive circuitry: implications for drug addiction, sleep and pain. Prog Neurobiol 8:332–347

Ferré S, Ciruela F, Borycz J, Solinas M, Quarta D, Antoniou K, Quiroz C, Justinova Z, Lluis C, Franco R, Goldberg SR (2008) Adenosine A1-A2A receptor herteromers: new targets for caffeine in the brain. Front Biosci 13:2391–2399

Field T, Reite M (1984) Children's responses to separation from mother during the birth of another child. Child Dev 55:1308–1316

Fredholm BB, Dunwiddie TV (1988) How does adenosine inhibit transmitter release? Trends Pharmacol Sci 9:130–134

Fredholm BB, Chen JF, Masino SA, Vaugeois JM (2005) Actions of adenosine at its receptors in the CNS: insights from knockouts and drugs. Annu Rev Pharmacol Toxicol 45:385–412

Geiger JD, Fyda DM (1991) Adenosine transport in nervous tissues. In: Stone TW (ed) Adenosine in the central nervous system. Academic, London, pp 1–23

Gerber JG, LoVerde M, Byyny RL, Nies AS (1990) The antihypertensive efficacy of hydrochlorothiazide is not prostacyclin dependent. Clin Pharmacol Ther 48:424–430

Gerfen C, Surmeier D (2011) Modulation of striatal projection systems by dopamine. Annu Rev Neurosci 34:441–466

Gessi S, Merighi S, Varani K, Leung E, Mac Lennan S, Borea PA (2008) The A3 adenosine receptor: an enigmatic player in cell biology. Pharmacol Ther 117:123–140

Gray JA (1982) The neuropsychology of anxiety. Oxford University Press, New York

Gray JA, Feldon J, Rawlins JNP, Hemsley DR, Smith AD (1991) The neuropsychology of schizophrenia. Behav Brain Sci 14:1–20

Haas HL, Selbach O (2000) Functions of neuronal adenosine receptors. Naunyn Schmiedebergs Arch Pharmacol 362:375–381

Haber S (2008) Parallel and integrative processing through the basal ganglia reward circuit: lessons from addiction. Biol Psychiatry 64(3):173–174

Hanff TC, Furst SJ, Minor TR (2010) Biochemical and anatomical substrates of depression and sickness behavior. Isr J Psychiatry Relat Sci 47:64–71

Heffner TG, Wiley JN, Williams AE, Bruns RF, Coughenour LL, Downs DA (1989) Comparison of the behavioral effects of adenosine agonists and dopamine antagonists in mice. Psychopharmacology (Berl) 98:31–37

Heimer L, Wilson RD (1975) The subcortical projections of the allocortex: similarities in the neural associations of the hippocampus, the piriform cortex, and the neocortex. In: Santinni M (ed) Golgi centennial symposium. Raven, New York, pp 177–193

Herrero MT, Barcia C, Navarro JM (2002) Functional anatomy of thalamus and basal ganglia. Childs Nerv Syst 18:386–404

Hestad KA, Tønseth S, Støen CD, Ueland T, Aukrust P (2003) Raised plasma levels of tumor necrosis factor alpha in patients with depression: normalization during electroconvulsive therapy. J ECT 19:183–188

Hranilovic D, Bucan M, Wang Y (2008) Emotional response in dopamine D2L receptor-deficient mice. Behav Brain Res 195:246–250

Huang QJ, Jiang H, Hao XL, Minor TR (2004) Brain IL-1 beta was involved in reserpine-induced behavioral depression in rats. Acta Pharmacol Sin 25:293–296

Hunter AM, Balleine BW, Minor TR (2003) Helplessness and escape performance: glutamate-adenosine interactions in the frontal cortex. Behav Neurosci 17:123–135

Ishiwata K, Mishina M, Kimura Y, Oda K, Sasaki T, Ishii K (2005) First visualization of adenosine A(2A) receptors in the human brain by positron emission tomography with [11C]TMSX. Synapse 55:133–136

Jackson RL, Minor TR (1988) Effects of signaling inescapable shock on subsequent escape learning: implications for theories of coping and "learned helplessness". J Exp Psychol Anim Behav Process 14:390–400

Larson SJ, Dunn AJ (2001) Behavioral effects of cytokines. Brain Behav Immun 15:371–387

Li B, Piriz J, Mirrione M, Chung C, Proulx CD, Schulz D, Henn F, Malinow R (2011) Synaptic potentiation onto habenula neurons in the learned helplessness model of depression. Nature 470:535–539

Linden J (1991) Structure and function of A1 adenosine receptors. FASEB J 5:2668–2676

Linden J, Taylor HE, Robeva AS, Tucker AL, Stehle JH, Rivkees SA, Fink JS, Reppert SM (1993) Molecular cloning and functional expression of a sheep A3 adenosine receptor with widespread tissue distribution. Mol Pharmacol 44:524–532

Ljungberg T, Apicella P, Schultz W (1992) Responses of monkey dopamine neurons during learning of behavioral reactions. J Neurophysiol 67:145–163

Maes M (1995) Evidence for an immune response in major depression: a review and hypothesis. Prog Neuropsychopharmacol Biol Psychiatry 19:11–38

Maier SF (1990) Role of fear in mediating shuttle escape learning deficit produced by inescapable shock. J Exp Psychol Anim Behav Process 16:137–149

Maier SF, Albin RW, Testa TJ (1973) Failure to learn to escape in rats previously exposed to inescapable shock depends on the nature of the escape response. J Comp Physiol Psychol 85:581–592

Matsumoto M, Hikosaka O (2009) Two types of dopamine neuron distinctly convey positive and negative motivational signals. Nature 459:837–842
Meghji P (1991) Adenosine production and metabolism. In: Stone TW (ed) Adenosine in the nervous system. Academic, California, pp 25–39
Meredith GE, Baldo BA, Andrezjewski ME, Kelley AE (2008) The structural basis for mapping behavior onto the ventral striatum and its subdivisions. Brain Struct Funct 213:17–27
Mineka S, Cook M, Miller S (1984) Fear conditioned with escapable and inescapable shock: effects of a feedback stimulus. J Exp Psychol Anim Behav Process 10:307–324
Minor TR (1990) Conditioned fear and neophobia following inescapable shock. Anim Learn Behav 18:212–226
Minor TR, Hunter AM (2002) Stressor controllability and learned helplessness research in the United States: sensitization and fatigue processes. Integr Physiol Behav Sci 37:44–58
Minor TR, LoLordo VM (1984) Escape deficits following inescapable shock: the role of contextual odor. J Exp Psychol Anim Behav Process 10:168–181
Minor TR, Pelleymounter MA, Maier SF (1988) Uncontrollable shock, forebrain norepinephrine, and stimulus selection during choice-escape leaning. Psychobiology 16:135–145
Minor TR, Dess NK, Overmier JB (1991) Inverting the traditional view of "learned helplessness". In: Denny MR (ed) Fear, avoidance, and phobias: a fundamental analysis. Lawrence-Erlbaum Associates, New Jersey, pp 87–133
Minor TR, Chang WC, Winslow JL (1994a) Stress and adenosine I: effects of methylxanthine and amphetamine stimulants on learned helplessness in rats. Behav Neurosci 108:254–264
Minor TR, Winslow JL, Chang WC (1994b) Stress and adenosine II: adenosine analogs mimic the effect of inescapable shock in shuttle-escape performance in rats. Behav Neurosci 108:265–276
Minor TR, Rowe MK, Soames Job RF, Ferguson EC (2001) Escape deficits induced by inescapable shock and metabolic stress are reversed by adenosine receptor antagonists. Behav Brain Res 120:203–212
Minor TR, Huang Q, Foley EA (2003) Cytokine-purine interactions in behavioral depression in rats. Integr Physiol Behav Sci 38:189–202
Minor TR, Huang Q, Witt AE (2006) Cytokine-purine interactions in traumatic stress, behavioral depression, and sickness. CNS Neurol Disord Drug Targets 5:547–560
Minor TR, Rowe M, Cullen PK, Furst S (2008) Enhancing brain adenosine signaling with the nucleoside transport blocker NBTI (S-(4-nitrobenzyl)-6-theoinosine) mimics the effects of inescapable shock on later shuttle-escape performance in rats. Behav Neurosci 122:1236–1247
Minor TR, Plumb TN, Schell CJ, Pham AK (2010) Brain adenosine signaling in psychological trauma and comorbid depression. In: Sher L, Vilens A (eds) Neurobiology of post-traumatic stress disorder. Nova Science Publishers, Inc., New York, pp 229–257
Mishina M, Ishiwata K, Kimura Y, Naganawa M, Oda K, Kobayashi S, Katayama Y, Ishii K (2007) Evaluation of distribution of adenosine A2A receptors in normal human brain measured with [11C]TMSX PET. Synapse 61:778–784
Mogenson GJ, Jones DL, Yim CY (1980) From motivation to action: functional interface between the limbic system and the motor system. Prog Neurobiol 14:69–97
Mohta M, Sethi AK, Tyagi A (2003) Psychological care in trauma patients. Injury 34:17–25
Newby AC (1981) The interaction of inhibitors with adenosine metabolizing enzymes in intact isolated cells. Biochem Pharmacol 30:2611–2615
Newby AC (1984) Adenosine and the concept of 'retaliatory metabolites'. Trends Biomed Sci 9:42–44
Noji T, Karasawa A, Ksuka H (2004) Adenosine uptake inhibitors. Eur J Pharmacol 495:1–16
Norn S, Shore PA (1999) Further studies on the nature of persistent reserpine binding: evidence for reversible and irreversible binding. Biochem Pharmacol 20:1291–1295
Orr TE, Mann DR (1992) Role of glucocorticoids in the stress-induced suppression of testicular steroidogenesis in adult male rats. Horm Behav 26:350–363

Ostlund SB, Wassum KM, Murphy NP, Balleine BW, Maidment NT (2011) Extracellular dopamine levels in striatal subregions track shifts in motivation and response cost during instrumental conditioning. J Neurosci 5:200–207
Overmier JB, Seligman ME (1967) Effects of inescapable shock upon subsequent escape and avoidance responding. J Comp Physiol Psychol 63:28–33
Parkinson JA, Willoughby PJ, Robbins TW, Everitt BJ (2000) Disconnection of the anterior cingulated cortex and nucleus accumbens core impairs Pavlovian approach behavior: further evidence for limbic cortical-ventral striatopallidal systems. Behav Neurosci 114:42–63
Parkinson FE, Damaraju VL, Graham K, Yao SY, Baldwin SA, Cass CE, Young JD (2011) Molecular biology of nucleoside transporters and their distributions and functions in the brain. Curr Top Med Chem 11:948–972
Petty F, Sherman AD (1980) Regional aspects of the prevention of learned helplessness by desiprimine. Life Sci 26:1447–1452
Petty F, McChesney C, Kramer G (1985) Intracortical glutamate injection produces helpless-like behavior in the rat. Pharmacol Biochem Behav 22:531–533
Phillis JW (2004) Adenosine and adenine nucleotides as regulators of cerebral blood flow: roles of acidosis, cell swelling, and KATP channels. Crit Rev Neurobiol 16:237–270
Phillis JW, Walter GA, O'Regan MH, Stair RE (1987) Increases in cerebral cortical perfusate adenosine and inosine concentrations during hypoxia and ischemia. J Cereb Blood Flow Metab 7:679–686
Ponzio F, Achilli G, Calderini G, Ferretti P (1984) Depletion and recovery of neuronal monoamine storage in rats of different ages treated with reserpine. Neurobiol Aging 5:101–104
Porsolt RD, Bertin A, Jalfre M (1977) Behavioral despair in mice: a primary screening test for antidepressants. Arch Int Pharmacodyn Ther 229:327–336
Rech RH, Moore KE (1971) An introduction to psychopharmacology. Raven, New York
Salamone JD, Correa M, Farrar AM, Nunes EJ, Pardo M (2009) Dopamine, behavioral economics, and effort. Front Behav Neurosci 3. doi:10.3389/neuro.08.013.2009
Sattin A, Rall TW (1970) The effect of adenosine and adenine nucleotides on the cyclic adenosine 3′-5′-phosphate content of guinea pig cerebral cortex slices. Mol Pharmacol 6:13–23
Schildkraut JJ (1965) The catecholamine hypothesis of affective disorders: a review of supporting evidence. Am J Psychiatry 122:509–522
Schultz W (1998) Predictive reward signal of dopamine neurons. J Neurophysiol 80:1–27
Schultz W (2002) Getting formal with dopamine and reward. Neuron 36:241–263
Schwartz JH (1981) Chemical basis of synaptic transmission: vesicles store and release. In: Kandel ER, Schwartz JH (eds) Principles of neural sciences. E. Arnold, London, pp 113–115
Scislo TJ, O'Leary DS (2006) Vasopressin V1 receptors contribute to hemodynamic and sympathoinhibitory responses evoked by stimulation of adenosine A2a receptors in NTS. Am J Physiol Heart Circ Physiol 290:H1889–H1898
Sebastião AM, Ribeiro JA (1996) Adenosine A2 receptor-mediated excitatory actions on the nervous system. Prog Neurobiol 48:167–189
Sebastião AM, Ribeiro JA (2000) Fine-tuning neuromodulation by adenosine. Trends Pharmacol Sci 21:341–346
Shors TJ, Servatius RJ (1995) Stress-induced sensitization and facilitated learning require NMDA receptor activation. Neuroreport 6:677–680
Snyder SH, Katims JJ, Annau Z, Bruns RF, Daly JW (1981) Adenosine receptors and behavioral actions of methylxanthines. Proc Natl Acad Sci U S A 78:3260–3264
Stéru L, Chermat R, Thierry B, Mico JA, Lenegre A, Stéru M, Simon P, Porsolt RD (1987) The automated tail suspension test: a computerized device which differentiates psychotropic drugs. Prog Neuropsychopharmacol Biol Psychiatry 11:659–671
Stone TW (1981) Physiological roles for adenosine and adenosine 5′-triphosphate in the nervous system. Neuroscience 6:523–555
Svenningsson P, Le Moine C, Fisone G, Fredholm BB (1999) Distribution, biochemistry and function of striatal adenosine A2A receptors. Prog Neurobiol 59:355–396

Trincavelli ML, Costa B, Tuscano D, Lucacchini A, Martini C (2002) Up-regulation of A(2A) adenosine receptors by proinflammatory cytokines in rat PC12 cells. Biochem Pharmacol 64:625–631

Van Wylen DGL, Park TS, Rubio R, Berne RM (1986) Increases in cerebral interstitial fluid adenosine concentrations during hypoxia, local potassium infusion, and ischemia. J Cereb Blood Flow Metab 6:522–528

Vorhoff T, Zimmermann H, Pelletier J, Sevigny J, Braun N (2005) Cloning and characterization of the ecto-nucleotidase NTPDase3 from rat brain: predicted and secondary structure and relation to other members of the E-NTPDase family and actin. Purinergic Signal 1:259–270

Weaver DR (1993) A2a adenosine receptor gene expression in developing rat brain. Brain Res Mol Brain Res 20:313–327

Weiner MF (1983) Conservation-withdrawal and mental retardation in medical and surgical patients. Psychosomatics 24:41–43

Weiner MF, Lovitt R (1979) Conservation-withdrawal versus depression. Gen Hosp Psychiatry 1:347–349

Weiss JM, Goodman PA, Losito BG, Corrigan S, Charry JM, Bailey WH (1981) Behavioral depression produced by an uncontrollable stressor: relationship to norepinephrine, dopamine, and serotonin levels in various regions of rat brain. Brain Res Rev 3:167–205

Woodson JC, Minor TR, Job RFS (1998) Inhibition of adenosine deaminase by erythro-9-(2-hydroxy-3-nonyl) adenine (EHNA) mimics the effect of inescapable shock on escape learning in rats. Behav Neurosci 112:399–409

Yin HH, Knowlton BJ (2004) Contributions of striatal subregions to place and response learning. Learn Mem 11:459–463

Yin HH, Knowlton BJ, Balleine BW (2004) Lesions of dorsolateral striatum preserve outcome expectancy but disrupt habit formation in instrumental learning. Eur J Neurosci 19:181–189

Yin HH, Ostlund SB, Balleine BW (2008) Reward-guided learning beyond dopamine in the nucleus accumbens: the integrative functions of cortico-basal ganglia networks. Eur J Neurosci 28:1437–1448

Yirmiya R (1996) Endotoxin produces a depressive-like episode in rats. Brain Res 711:163–174

Yu C, Gupta J, Chen JF, Yin HH (2009) Genetic deletion of A2A adenosine receptors in the striatum selectively impairs habit formation. J Neurosci 2:15100–15103

Part IV
Adenosine-Based Therapies

Chapter 26
Disruption of Adenosine Homeostasis in Epilepsy and Therapeutic Adenosine Augmentation

Detlev Boison

Abstract Despite the development of new classes of antiepileptic drugs during the past 40 years, about a third of all patients with epilepsy continue to be refractory to conventional treatment. It becomes evident that a disorder of complex network dysfunction, such as epilepsy, cannot be treated in a comprehensive manner with traditional treatment approaches. In contrast, the homeostatic bioenergetic network regulator adenosine is uniquely suited to affect several different pathways and mechanisms synergistically on multiple different levels. Adenosine is a known endogenous anticonvulsant of the brain. Its levels rise during seizures, and this increase in adenosine is part of an endogenous control mechanism to terminate seizures. However, disrupted adenosine homeostasis and resulting adenosine deficiency is a pathological hallmark of the epileptic brain. Adenosine deficiency as such was recently shown to be sufficient to trigger seizures. Thus, adenosine augmentation therapy (AAT) is a rational intervention to treat epilepsy. Unfortunately, systemic AAT is not a therapeutic option due to unacceptable side effects. To circumvent this problem, focal AATs have been developed based on the rationale to reconstruct normal adenosine homeostasis within an epileptogenic brain.

Keywords Adenosine kinase • Epileptogenesis • Silk cell therapy • Gene therapy • Ketogenic diet

---

D. Boison, Ph.D. (✉)
RS Dow Neurobiology Labs, Legacy Research Institute, 1225 NE 2nd Ave, Portland, OR 97232, USA
e-mail: dboison@downeurobiology.org

S. Masino and D. Boison (eds.), *Adenosine: A Key Link between Metabolism and Brain Activity*, DOI 10.1007/978-1-4614-3903-5_26,

26.1 Introduction

Epilepsy, a heterogeneous syndrome characterized by spontaneous recurrent seizures, is one of the most frequent neurological conditions and affects around 1 % of the population. Seizures can be brief periods of altered consciousness or absences, but can also generalize and involve motor functions leading to convulsions. Seizures are generated by excessive electrical discharges in clusters of neurons; however the cellular and molecular basis of epilepsy is still largely unknown. Today it is widely accepted that epilepsy is more complex than merely being a condition of increased neuronal excitation and decreased neuronal inhibition, and in particular the contribution of inflammatory, glial, bioenergetic, and epigenetic changes to the pathophysiology of epilepsy has gained much attention recently (Boison et al. 2011a; Pan et al. 2005; Qureshi and Mehler 2010; Ravizza et al. 2008; Seifert et al. 2010; Vezzani et al. 2008; Williamson et al. 2005). Consequently, epilepsy presents a disorder of complex network dysfunction, which also becomes evident in a wide spectrum of histopathological changes commonly found in the epileptic brain. Thus, temporal lobe epilepsy (TLE) is characterized by neural cell loss in select areas of the hippocampal formation (e.g., in CA1 and hilus), by granule cell dispersion, by mossy fiber sprouting, by ectopic neurons and aberrant or recurrent circuitry, but also by glial changes that control homeostatic functions of brain activity. In many patients, seizures can be suppressed by antiepileptic drugs (AEDs), but one-third of all patients, or over 20 million patients with epilepsy worldwide, cannot be treated with those AEDs. Findings from the past 15 years suggest that dysfunction of adenosine-dependent homeostatic brain functions contributes to the epileptic phenotype, and—consequently—adenosine augmentation therapies (AATs) constitute a novel and promising rational approach for seizure control in epilepsy.

26.2 Limitations of Conventional Epilepsy Therapy

Although most currently available AEDs effectively suppress seizures in many patients, 30 % of all epilepsies, in particular those of focal origin within the temporal lobe, have been described as being pharmacoresistant (Loscher and Schmidt 2011; Nadkarni et al. 2005). Furthermore, severe side effects such as drowsiness, sleepiness, or effects on cognition limit the most effective use of those AEDs (Nadkarni et al. 2005). Therefore alternative treatment options have been developed for clinical applications; those include neurosurgical resection of an epileptogenic focus (Wieser 1996) or invasive electrical stimulation (Boon et al. 2009). In addition, alternative treatment options such as cell and gene therapy (Loscher et al. 2008) or hormonal and dietary interventions (Herzog 2009; Kossoff and Rho 2009) are being explored.

26.2.1 Pharmacotherapy

A major caveat of the current use of systemic drug formulations are brain-wide side effects that are largely due to the unspecific blockade of neurotransmission in general (e.g., by sodium channel blockers), while conventional AEDs typically do not specifically target the hypersynchronous bursting that is a characteristic of epilepsy. Regarding effective drug treatment, the network complexity of epilepsy needs to be considered. A major limitation is the fact that the majority of AEDs have been designed using the same model systems (i.e., suppression of maximal electroshock-induced acute seizures) (Loscher and Schmidt 2011); in addition, currently available AEDs act on neuronal downstream targets, such as neuronal sodium or potassium channels, or enhancement of GABAergic neurotransmission and therefore are unlikely to affect the complex pathophysiology of epilepsy on a network level (Loscher and Schmidt 2011). Not surprisingly, AEDs at best provide symptom control (i.e., suppression of seizures; anti-ictogenesis), but do not affect the underlying pathogenetic processes (i.e., epileptogenesis). Therefore, requirements for more effective therapies include (1) efficacy in pharmacoresistant epilepsy; (2) prevention of epileptogenesis and targeting of underlying disease processes; and (3) limitation of side effects by focal mode of action. New concepts and strategies are needed to improve treatment options.

26.2.2 Surgery

Surgical resection of an epileptogenic brain region—if feasible—might be used as a last resort for some patients, but is accompanied by risks and the potential for loss of function. Although epilepsy surgery can significantly reduce the seizure burden in some patients and improve social outcome (Hamiwka et al. 2011), about 35 % of patients in a cohort of 416 patients with epilepsy involving hippocampal sclerosis experienced seizure recurrence during the first postoperative year (Ramesha et al. 2011) and cognitive changes after epilepsy surgery are common (Sherman et al. 2011). In addition, the disruption of mesial temporal structures that play a role in mood control may contribute to the development of new-onset major depression following mesial temporal lobectomy (Wrench et al. 2011). Alternative surgical treatment options include electrical stimulation of either the vagus nerve or deep brain structures using implanted electrodes (Boon et al. 2009). Although these approaches have demonstrated benefits in a subset of patients, invasive strategies are associated with inherent risks. Of note, intracranial electrode implantation is associated with significant risks for brain hemorrhage and infection (5 % risk, each) (Theodore and Fisher 2007).

26.3 An Astrocytic Basis of Epilepsy

Astrogliosis is a characteristic hallmark of epilepsy (Fig. 26.1); however the important role of astrocytes in the control of neuronal excitability has only recently been appreciated. Astrocytes couple synaptic transmission with the neuro-vasculature via the formation of tripartite synapses (Haydon and Carmignoto 2006). Importantly, both purinergic (Pascual et al. 2005) as well as glutamatergic (Tian et al. 2005) signaling from astrocytes was shown to influence neuronal networks and an astrocytic basis of epilepsy has been proposed (Tian et al. 2005). Several lines of evidence now demonstrate that disruption of astrocyte-based homeostatic functions critically contributes to the pathophysiology of epilepsy. Thus, calcium homeostasis in astrocytes from the epileptic brain is perturbed and calcium waves have been shown to propagate within the astrocyte network affecting gliotransmission (Halassa et al. 2007; Halassa and Haydon 2010; Haydon and Carmignoto 2006). Impaired K^+ buffering in epilepsy is further aggravated by reduced expression of astrocytic Kir channels, which play a fundamental role in astroglial K^+-homeostasis (Hinterkeuser et al. 2000). Further, failure of the astrocytic glutamine–glutamate cycle, caused by downregulation of glutamine synthetase, has been implicated in the pathophysiology of astrogliosis-based seizures (Ortinski et al. 2010). As is discussed in more

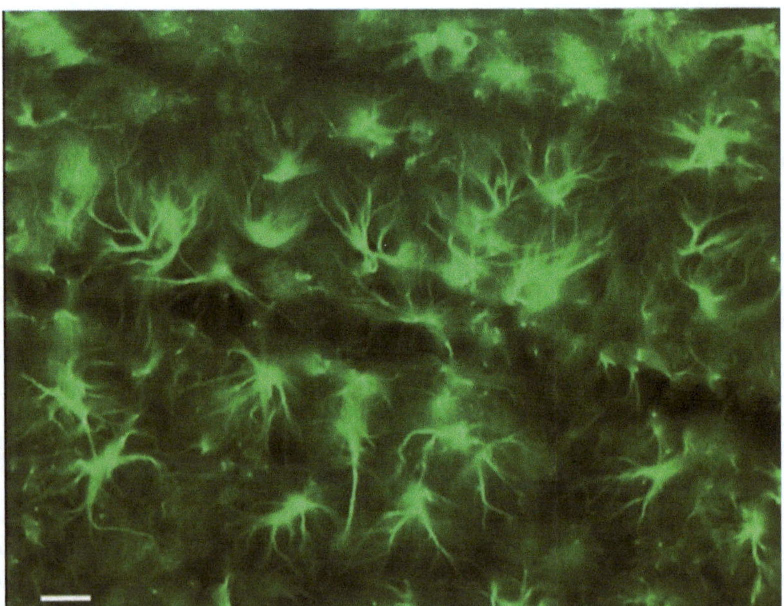

Fig. 26.1 Astrogliosis is a pathological hallmark of the epileptic brain. The image shows hypertrophied astrocytes from area CA1 of the hippocampal formation of an epileptic rat 21 weeks after systemic kainic acid-induced status epilepticus. Astrocytes are stained in green for glial fibrillary acid protein (GFAP) using immunofluorescence techniques. Scale bar: 100 μm

detail below, astrocytes play a key role in regulating the homeostatic functions of the brain's endogenous anticonvulsant adenosine (Aronica et al. 2011; Boison 2008; Boison et al. 2010). In addition to the disruption of homeostatic functions of astrocytes in epilepsy, changes in the blood–brain barrier (Heinemann et al. 2006) and glial-based immune functions (Vezzani 2008; Vezzani et al. 2008) contribute to the development of the epileptic phenotype.

26.4 Adenosine Deficiency: A Pathological Hallmark of the Epileptic Brain

26.4.1 Astrocyte Control of Adenosine Homeostasis

The physiological tone of adenosine is largely controlled by an astrocyte-based adenosine cycle (Boison 2008; Boison et al. 2010). ATP, released by astrocytes via regulated vesicular transport (Pascual et al. 2005) or via hemichannels (Stout et al. 2002), and rapidly cleaved by a cascade of extracellular ectonucleotidases (Zimmermann 2000), is a major source of synaptic adenosine. The astrocytic release of ATP plays a crucial role in regulating neuronal excitability via adenosine as was demonstrated using transgenic mice with an engineered inducible deficiency of astrocytic ATP release (Pascual et al. 2005). Astrocyte membranes contain two types of equilibrative nucleoside transporters that rapidly equilibrate extra- and intracellular levels of adenosine (Baldwin et al. 2004). Intracellular adenosine is rapidly phosphorylated by the astrocyte-based enzyme adenosine kinase (ADK), a mechanism creating a sink for adenosine and driving the influx of adenosine into the astrocyte resulting in the clearance of extracellular adenosine (Boison 2006, 2008; Boison et al. 2010).

26.4.2 Regulation of Seizures by Adenosine Kinase

The "ADK hypothesis of epileptogenesis" has recently been proposed to highlight a biphasic response of changes in adenosine homeostasis that are thought to contribute the development of seizures in epilepsy (Boison 2008). It appears that any type of injury or stress to the brain results in an acute surge of up to micromolar levels of adenosine (Clark et al. 1997; Gouder et al. 2004; Pignataro et al. 2008). This acute injury-associated surge of adenosine can further be potentiated by acute downregulation of ADK as shown during the first hours after experimental stroke or status epilepticus (Boison 2008; Gouder et al. 2004; Pignataro et al. 2008). Whereas the acute surge in adenosine provides an important neuroprotective function, it might also trigger several downstream pathways linked to epileptogenesis. Downstream mechanisms that can be influenced by an acute surge of adenosine include inflammatory processes (Hasko et al. 2005; Yu et al. 2004) and stimulation of microglia (Gebicke-Haerter et al. 1996). Increased levels of adenosine can induce downregulation of A_1Rs, but also upregulation of A_{2A}Rs on astrocytes (Cunha 2005). This is of interest because increased occupancy of astrocytic A_{2A}Rs was demonstrated to promote astrocyte proliferation and activation (Brambilla et al. 2003; Hindley et al. 1994). Based on those mechanisms, the acute injury-induced surge in adenosine might contribute to trigger subsequent astrogliosis. Therefore, it is no surprise that astrogliosis is common after a variety of insults to the brain, such as traumatic brain injury, status epilepticus, or stroke. As is discussed below, one major consequence of astrogliosis is overexpression of ADK, resulting in excessive adenosine clearance and spontaneous electrographic seizures. Evidence that astrocytic ADK is a key molecular link between astrogliosis and neuronal dysfunction in epilepsy has been obtained in a variety of experimental paradigms.

26.4.2.1 Rodent Models of Acquired Epilepsy

To dissect consequences of astrogliosis from other histopathological changes common in MTLE (e.g., mossy fiber sprouting, granule cell dispersion, changes in neuronal circuitry) a mouse model was investigated that develops focal astrogliosis in the absence of other histopathological changes (Li et al. 2008). The model is based on a focal onset status epilepticus, which is triggered by unilateral injection of the excitotoxin kainic acid into the basolateral amygdala. As a consequence of this injury two focal areas of neuronal cell loss, located in the ipsilateral CA3 and within the injured amygdala, are generated (Li et al. 2011, 2008). Within 3 weeks both focal areas that were initially injured developed prominent astrogliosis and overexpression of ADK (Li et al. 2011, 2008). Frequent spontaneous electrographic seizures originating from the two focal areas of ADK overexpression were identified (Li et al. 2011, 2008). Remarkably, seizure activity coincided temporally with overexpression of ADK (Li et al. 2007a), occurred independently in both foci, and remained focal (i.e., did not generalize) (Li et al. 2011). These findings demonstrate spatial and temporal colocalization of astrogliosis, overexpressed ADK, and spontaneous seizures, and suggest that astrogliosis and/or overexpression of ADK per se and in the absence of any other contributing factor for epileptogenesis is a minimum requirement to trigger electrographic seizures.

26.4.2.2 Genetically Engineered Mice

To address the question whether the isolated overexpression of ADK (i.e., in the absence of astrogliosis) might be sufficient to trigger seizures, transgenic mice were engineered with a global brain-wide overexpression of an *Adk*-transgene (Adk-tg mice). These animals are characterized by spontaneous hippocampal seizures in the absence of astrogliosis or any other histopathological alteration (Li et al. 2007a). Thus, overexpression of ADK per se is sufficient to trigger electrographic seizures. These findings were further substantiated using a viral gene expression approach based on an adeno-associated virus (AAV) overexpressing a cDNA of *Adk* selectively in astrocytes. Intrahippocampal injection of this viral vector into mice was shown to trigger the same type of spontaneous seizures, whereas the injection of an AAV expressing an *Adk* antisense construct was able to almost completely abolish spontaneous seizures in Adk-tg mice (Theofilas et al. 2011). Together these data define ADK as rational target for therapeutic intervention.

26.4.2.3 Overexpression of ADK in Human Epilepsy

The contribution of overexpressed ADK to seizure generation in epilepsy might be of wider relevance, since ADK was found to be overexpressed not only in a variety of rodent models of TLE (Aronica et al. 2011; Gouder et al. 2004), but also in epileptogenic tissue resected from human TLE patients (Aronica et al. 2011; Masino et al. 2011). In addition, microdialysis studies in human TLE patients have identified lower levels of adenosine in the epileptogenic compared to the contralateral healthy hippocampus (During and Spencer 1992). Thus overexpression of ADK and resulting disruption of adenosine homeostasis seems to be a pathological hallmark of the epileptic brain.

26.4.2.4 Pharmacological Approaches for Seizure Suppression

If overexpressed ADK and resulting disruption of adenosine homeostasis play a major role in seizure generation, then AATs should be highly effective in seizure control. Indeed, adenosine A_1R agonists effectively suppress seizures in a variety of models including one of pharmacoresistant epilepsy (Gouder et al. 2003; Jacobson and Gao 2006). However, systemic application of A_1R agonists leads to profound cardiovascular and sedative side effects (Monopoli et al. 1994) that might be related to the sleep-promoting effects of A_1R activation (Bjorness et al. 2009). Alternatively, seizure suppression can be achieved by ADK inhibitors such as ABT-702 (Jarvis et al. 2000; Kowaluk et al. 2000), which are able to amplify the site- and event-specific increase in adenosine resulting from injury or stress to the brain. Based on this rationale those agents could provide antiseizure activity comparable to adenosine receptor agonists (Kowaluk and Jarvis 2000; McGaraughty et al. 2005), however with a more favorable side effect profile. Indeed, pharmacological inhibition of ADK was shown to be highly effective in inhibiting epileptic seizures (Gouder et al. 2004; Kowaluk and Jarvis 2000) but systemic ADK inhibition might not be a long-term therapeutic option for epilepsy due to interference with methionine metabolism in liver (Boison et al. 2002; Mato et al. 2008) and the risk of brain hemorrhage (Erion et al. 2000; McGaraughty and Jarvis 2006).

26.5 Focal Adenosine Augmentation Therapy

To circumvent side effects of systemic adenosine augmentation and to translate adenosine-based therapies into clinical practice focal treatment approaches (Fig. 26.2) might become a necessity. The feasibility for those translational approaches is based on initial clinical data with adenosine, pharmacological findings, availability of focal treatment modalities, and availability of adenosine augmenting strategies that are amenable to focal application.

26.5.1 Clinical Use and Trials with Adenosine

Adenosine as such is an FDA-approved drug, which is clinically used to treat supraventricular tachycardia. In addition, adenosine administration is currently tested in a variety of clinical trials to study therapeutic effects on inflammation, cardioprotection, and pain (Fredholm et al. 2011). Intrathecal adenosine infusions have been performed in human subjects to study pain responses (Eisenach et al. 2002a, b). Based on clinical evidence to date, the direct application of adenosine to a human subject appears to be generally safe. However, since adenosine only poorly penetrates the blood–brain barrier and is rapidly metabolized within the circulation, systemic adenosine applications might not be a therapeutic option for patients with epilepsy.

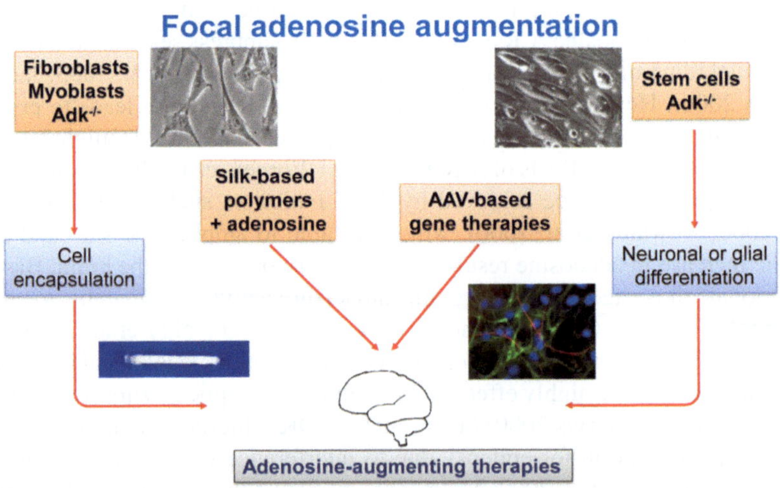

Fig. 26.2 Focal adenosine augmentation strategies. Local levels of adenosine in brain can be raised by a variety of strategies illustrated here

26.5.2 Clinical Use of Pharmacological Adenosine Augmentation

Adenosine-augmenting pharmacological agents have been and are currently being tested in a limited number of clinical studies (Fredholm et al. 2011). Dipyridamole augments adenosine signaling by blocking adenosine transport and thus reuptake of extracellular adenosine; the short half-life of the drug in vivo has spawned interest in its use as coronary vasodilator. The drug is also currently tested as an add-on treatment for schizophrenia. Adenosine augmentation is thought to be a promising approach for the treatment of schizophrenia (Boison et al. 2011b) and clinical studies with the adenosine-augmenting agent allopurinol have demonstrated some benefit in patients with refractory schizophrenia (Brunstein et al. 2004, 2005). Although a variety of adenosine receptor agonists are in clinical development for a variety of conditions (Fredholm et al. 2011) those approaches seem to be limited to highly specific peripheral applications and none of those pharmacological approaches has been pursued for the clinical treatment of epilepsy, most likely based on the need of focal applications.

26.5.3 Rationale for Focal AAT

Spatially restricted therapies are considered to be safe and feasible alternatives for systemic drug application and due to the focal nature of many epilepsies, focal interventions might be preferable (Nilsen and Cock 2004). Neurochemical deficits that are specific for an epileptogenic focus provide a direct rationale for focal intervention. Accordingly, endogenous antiepileptic compounds such as GABA (Gernert et al. 2002; Nolte et al. 2008; Thompson 2005), adenosine (Anschel et al. 2004; Boison 2009a; Boison and Stewart 2009), galanin (Kanter-Schlifke et al. 2007; Mazarati 2004; McCown 2009), or NPY (Noe et al. 2009; Sorensen et al. 2008) are therefore logical therapeutic candidates. Tools for focal drug delivery include polymeric brain implants (Kokaia et al. 1994; Wilz et al. 2008), cell therapy (Boison 2007; Loscher et al. 2008; Raedt et al. 2007; Shetty and Hattiangady 2007), or gene therapy (McCown 2004; Riban et al. 2009; Vezzani 2007). Opposed to the systemic use of AEDs, the focal use of an endogenous anticonvulsant and neuroprotectant, such as adenosine, is uniquely posed to reconstitute normal signaling within an epileptogenic brain area with the prospect to not only suppress seizures but also to affect the underlying neuropathology of epilepsy directly (Fig. 26.3). In addition, the focal delivery of adenosine to the limbic system might be a strategy to limit off-site effects of adenosine, such as interfering with A_{2A}R-dependent control of psychomotor functions in the striatum (Shen et al. 2008).

26.5.4 Engineering of Adenosine-Releasing Cells

The most effective strategy to increase adenosine signaling is disruption of metabolic adenosine clearance. Two molecular strategies have been used to augment adenosine release from cells: (1) gene targeting to disrupt the endogenous *Adk* gene

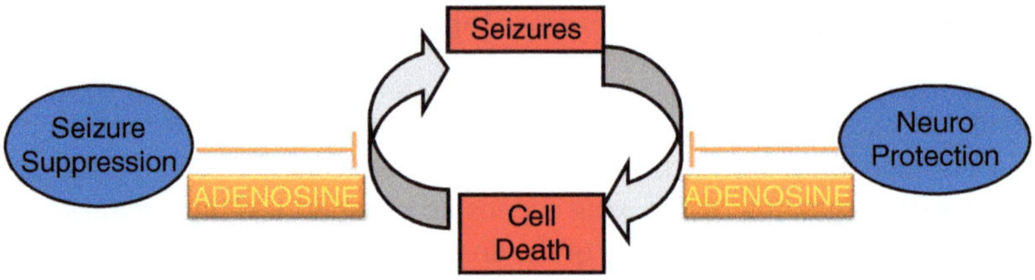

Fig. 26.3 Disease-modifying effects of adenosine. Combining antiepileptic with neuroprotective properties, adenosine might be ideally suited to disrupt a vicious cycle thought to be implicated in disease progression of epilepsy

(Fedele et al. 2004) and (2) RNA interference (RNAi) to knock down ADK expression (Boison 2010). Disruption of the endogenous *Adk* gene by homologous recombination in murine embryonic stem cells (ESCs) and subsequent selection for ADK deficiency resulted in ESCs with biallelic disruption of the endogenous *Adk*-gene (Fedele et al. 2004). After directed differentiation into neural cell populations, 10^5 ADK-deficient cells released a therapeutically relevant dose of up to 40 ng adenosine per hour. Adenosine release under those conditions is most likely due to passive diffusion through equilibrative nucleoside transporters. Similarly, therapeutic adenosine release in adult stem cells was induced using lentiviral RNAi to downregulate ADK in human mesenchymal stem cells (hMSCs) (Ren et al. 2007).

26.6 Seizure Suppression and Prevention by Focal Adenosine Augmentation

26.6.1 Proofs of Principle

The first demonstration that focal adenosine augmentation might effectively control seizures was obtained in rats that were electrically kindled in the hippocampus. Kindling is based on repetitive administration of sub-convulsive electrical stimulation to the limbic system of the brain, which results in a permanent increase in susceptibility and severity of electrically induced seizures (Racine and Burnham 1984). The rat kindling model is of high predictive value for the clinical efficacy of AEDs (Loscher 2002). Synthetic ethylene vinyl acetate copolymers were engineered to release an amount of around 20–50 ng adenosine per day (Boison et al. 1999) and were implanted into the lateral brain ventricles of rats that had previously been kindled in the hippocampus. Recipients of adenosine-releasing polymers were characterized by a strong reduction of convulsive seizures for at least 7 days whereas control implants had no effects. This was the first published demonstration that the focal release of adenosine can suppress epileptic seizures (Boison et al. 1999). Likewise, intraventricular implants of encapsulated fibroblasts engineered to release adenosine provided robust but transient seizure suppression in kindled rats (Huber et al. 2001). The proof of principle that focal adenosine delivery can be of therapeutic benefit was subsequently confirmed by an independent research group using an approach that was based on intracranial adenosine injection in a rat seizure model (Anschel et al. 2004).

26.6.2 Stem Cell-Based Adenosine Delivery

Stem cell therapies are a logical choice for regeneration and repair of the epileptic hippocampus (Hattiangady et al. 2008). In addition, adenosine-releasing stem cell-derived brain implants may exert direct anticonvulsant activity via paracrine effects (Güttinger et al. 2005), which can effectively be achieved by injecting stem cells or their derivatives into the infrahippocampal fissure of rodents, thereby avoiding interference with hippocampal circuitry (Li et al. 2007b). Infrahippocampal implants of neural precursor cells derived from murine or human ESCs, as well as hMSCs, survive within the infrahippocampal fissure for at least several weeks (Boison 2009b; Li et al. 2009, 2008, 2007b; Ren et al. 2007). In therapeutic cell transplantation experiments infrahippocampal grafts of adenosine-releasing stem cells provided potent seizure control in several experimental paradigms. ESC-derived neural precursor cells injected into rats prior to the onset of kindling provided robust suppression of kindling epileptogenesis (Li et al. 2007b). When the same cells were injected after intraamygdaloid kainic acid-induced status epilepticus in mice, they prevented the development of spontaneous seizures 3 weeks after the injury. In contrast, after the same time recipients of control cells had developed recurrent electrographic CA3-seizures at a rate of about four seizures per hour (Li et al. 2008).

26.6.3 Transient Focal Adenosine Delivery via Silk

To provide a platform for clinical implementation of focal AAT, the natural biopolymer silk was evaluated for the therapeutic delivery of adenosine (Pritchard et al. 2010; Wilz et al. 2008). Purified silk fibroin presents a unique option for therapeutic adenosine delivery as outlined in chapter 28 of this volume. When implanted after completion of kindling into fully kindled epileptic rats, implants engineered to release 1,000 ng adenosine for only 10 days effectively suppressed seizures for exactly those 10 days in fully kindled rats. After expiration of adenosine release from the polymers, convulsive stage 4 or 5 seizure activity recurred (Szybala et al. 2009). To assess a potential antiepileptogenic effect of focal adenosine delivery, unilateral (ipsilateral to side of kindling) implantations with the same adenosine-releasing or control polymers were performed prior to the onset of kindling. In a dose escalation study devices releasing defined doses of up to 1,000 ng adenosine per day were implanted into the infrahippocampal fissure of rats prior to the onset of electrical kindling. Focal adenosine release from silk-based polymers dose dependently retarded kindling epileptogenesis (Wilz et al. 2008). In a more refined study designed to gauge possible antiepileptogenic effects of focal adenosine delivery, rats were kindled partly in the presence of a silk-based polymer releasing a stable but temporally restricted daily dose of 1,000 ng adenosine for only 10 days. During the initial kindling recipients of adenosine-releasing implants failed to express any convulsive seizures, whereas all control rats reproducibly developed generalized convulsive seizure activity. Lack of seizures in the adenosine group could either reflect suppression of epileptogenesis or merely the suppression of seizures by active augmentation of adenosine signaling. To distinguish between those two possibilities, kindling was continued after expiration of adenosine from the polymer; i.e., kindling was resumed in the absence of polymer-derived adenosine. At the initial kindling stimulations following the washout period none of the rats from the adenosine group displayed convulsive seizures and kindling developed gradually with increasing number of stimulations. In contrast, all animals from the control group resumed kindling at the generalized convulsive seizure stage. This experiment demonstrates a disease-modifying and at least partial antiepileptogenic effect of transient focal adenosine release (Szybala et al. 2009). Pharmacological control experiments with the adenosine A_1 receptor antagonist DPCPX further corroborated disease-modifying effects of focal adenosine delivery: Whereas a dose of 1 mg/kg DPCPX has no effects in non-kindled control animals, DPCPX restores stage 5 seizures in fully kindled rats that are protected from transient adenosine delivery (Huber et al. 2001). Importantly, DPCPX did not trigger seizures in kindled animals that received adenosine-releasing implants prior to the onset of kindling, a result that further substantiated a possible antiepileptogenic effect of focal adenosine delivery (Szybala et al. 2009).

26.6.4 Gene Therapy

As outlined above, high expression levels of ADK have been linked with epileptic seizures in animal models of epilepsy (Aronica et al. 2011; Gouder et al. 2004; Li et al. 2011), and overexpression of ADK has been found in hippocampus resected from patients with TLE (Aronica et al. 2011; Masino et al. 2011). Likewise, transgenic overexpression of ADK in the brain of mice is sufficient to trigger electrographic seizures (Li et al. 2007a). ADK therefore emerges as a rational target for therapeutic intervention and downregulation of ADK in a gene therapy approach should therefore constitute a rational strategy for long-term seizure control (Boison 2010). In order to assess whether seizure activity can be controlled by modulating ADK expression in astrocytes, an adeno-associated virus serotype 8 (AAV8)-based expression system was used to permanently alter ADK expression in astrocytes based on the expression of an *Adk*-cDNA in either sense (to overexpress ADK) or antisense (to knock down ADK) orientation (Theofilas et al. 2011). Cell-type specificity of ADK expression for astrocytes was achieved by using an astrocyte-selective pGfa promoter. As predicted, AAV8-based overexpression of ADK in astrocytes of the hippocampus of wild-type mice resulted in spontaneous electrographic seizures, whereas the knockdown of ADK in astrocytes of the hippocampus of Adk-tg mice almost completely abrogated any seizure activity (Theofilas et al. 2011). These data provide a proof of feasibility to use gene therapy approaches aimed at downregulating ADK expression as possible therapeutic avenue for long-term seizure control.

26.7 Possible Benefit of Focal AAT for Epilepsy

Focal AAT is based on a solid rationale and as such is highly suited and efficient in controlling epileptic seizures. In addition, newer data suggest that adenosine might also have novel disease-modifying and possibly antiepileptogenic effects.

26.7.1 Seizure Suppression

As outlined above solid evidence exists for the capability of adenosine augmentation to control epileptic seizures. Importantly, augmentation of adenosine signaling was shown to be effective in a model of pharmacoresistant epilepsy (Gouder et al. 2003, 2004). Focal adenosine augmentation was shown to effectively suppress seizures in two different species (mice and rats) and in models of electrically induced kindled seizures, as well as in models of spontaneous recurrent seizures that resulted as a consequence of kainic acid-induced status epilepticus (Huber et al. 2001; Li et al. 2008). Focal AAT was effective irrespective of the technology used to augment adenosine signaling. Robust seizure suppression was achieved by infrahippocampal implants of adenosine-releasing silk, adenosine-releasing cells, or by gene therapy targeting adenosine metabolism (Huber et al. 2001; Szybala et al. 2009; Theofilas et al. 2011). As will be discussed in a separate chapter of this volume a ketogenic diet was recently shown to suppress seizures in mice by augmentation of adenosine signaling (Masino et al. 2011).

26.7.2 Antiepileptogenesis and Disease Modification

Apart from seizure control, therapeutic adenosine augmentation might also modify the phenotype and expression of epilepsy, and might have novel antiepileptogenic effects. It is important to realize that epilepsy is not just a seizure disorder but the expression of a complex pathophysiology that also includes cognitive impairment, depression, and sleep disorders, all of which can be linked to disrupted adenosine homeostasis. Therefore reconstitution of adenosine-based homeostatic functions might not only be of benefit for seizure suppression, but also for the treatment of epilepsy-associated comorbidities. Thus, therapeutic adenosine augmentation emerges as a holistic approach to treat the complex spectrum of endophenotypes associated with epilepsy.

Recent findings support an antiepileptogenic role of focal adenosine augmentation: (1) Adenosine-releasing implants located close to the hippocampal formation suppressed kindling epileptogenesis in rats (Li et al. 2007b; Wilz et al. 2008). (2) In rats that were kindled in the presence of a transient source of adenosine, lack of convulsive seizures was maintained even after washout of the adenosine (Szybala et al. 2009). (3) Adenosine-releasing stem cell-derived infrahippocampal implants in mice prevented the development of hippocampal chronic seizures 3 weeks after status epilepticus triggered by an intraamygdaloid injection of kainic acid. Remarkably, in recipients of adenosine-releasing cells astrogliosis was markedly attenuated and ADK expression levels were almost normal (Li et al. 2008). (4) Finally, transgenic mice with increased adenosine levels in forebrain (Shen et al. 2011) appeared to be completely resistant to epileptogenesis.

26.8 Impact and Outlook

As outlined above, focal AATs may combine anticonvulsive, neuroprotective, disease-modifying, and possibly antiepileptogenic properties. It is important to point out that the conceptual rationale for focal AAT development differs drastically from classical AED development, which is based on a *neuro*centric concept. It is unlikely that the development of new AEDs that all act on a limited number of neuronal targets, which are all isolated downstream components of complex homeostatic networks, will lead to any significant improvement in antiepileptic therapy. In contrast, augmentation of adenosine as an upstream modulator of several downstream pathways is uniquely suited to affect neuronal excitability on the network level. As an *endogenous* anticonvulsant, adenosine is subject to physiological clearance. Rather than leading to toxic accumulations of adenosine, adenosine augmentation is likely to restore the adenosinergic equilibrium, thereby avoiding undue side effects.

Before clinical implementation of focal AATs several issues need to be resolved. These include(1) determination of ED_{50}s and TD_{50}s and the respective therapeutic index; (2) detailed toxicology or toxicogenomic studies; (3) differentiation of antiepileptic efficacy in mechanistically different animal models; (4) determination of appropriate time points for therapeutic intervention; (5) determination of appropriate therapeutic targets; and (6) demonstration of long-term efficacy.

AATs make rational use of an endogenous seizure-control system of the brain. It is important to note that adenosine is already FDA approved for the treatment of supraventricular tachycardia and that it has already been used as intrathecal infusion in phase I human pain studies (Eisenach et al. 2002a). One possibility for first safety and feasibility studies in a clinical setting could be the infusion of adenosine into an epileptic temporal lobe during its surgical removal. When coupled to synchronous EEG recordings, a proof of principle could be established that adenosine is effective in pharmacoresistant human epilepsy.

Acknowledgement The work of the author is supported by grants R01NS058780, R01NS061844, and R01MH083973 from the National Institutes of Health (NIH).

References

Anschel DJ, Ortega EL, Kraus AC, Fisher RS (2004) Focally injected adenosine prevents seizures in the rat. Exp Neurol 190:544–547

Aronica E, Zurolo E, Iyer A, de Groot M, Anink J, Carbonell C, van Vliet EA, Baayen JC, Boison D, Gorter JA (2011) Upregulation of adenosine kinase in astrocytes in experimental and human temporal lobe epilepsy. Epilepsia. doi:10.1111/j.1528-1167.2011.03115.x, Epub ahead of print

Baldwin SA, Beal PR, Yao SY, King AE, Cass CE, Young JD (2004) The equilibrative nucleoside transporter family, SLC29. Pflugers Arch 447:735–743

Bjorness TE, Kelly CL, Gao T, Poffenberger V, Greene RW (2009) Control and function of the homeostatic sleep response by adenosine A1 receptors. J Neurosci 29:1267–1276

Boison D (2006) Adenosine kinase, epilepsy and stroke: mechanisms and therapies. Trends Pharmacol Sci 27:652–658

Boison D (2007) Cell and gene therapies for refractory epilepsy. Curr Neuropharmacol 5:115–125

Boison D (2008) The adenosine kinase hypothesis of epileptogenesis. Prog Neurobiol 84:249–262

Boison D (2009a) Adenosine augmentation therapies (AATs) for epilepsy: prospect of cell and gene therapies. Epilepsy Res 85:131–141

Boison D (2009b) Engineered adenosine-releasing cells for epilepsy therapy: human mesenchymal stem cells and human embryonic stem cells. Neurotherapeutics 6:278–283

Boison D (2010) Inhibitory RNA in epilepsy: research tool and therapeutic perspectives. Epilepsia 51:1659–1668

Boison D, Stewart K-A (2009) Therapeutic epilepsy research: from pharmacological rationale to focal adenosine augmentation. Biochem Pharmacol 78:1428–1437

Boison D, Scheurer L, Tseng JL, Aebischer P, Mohler H (1999) Seizure suppression in kindled rats by intraventricular grafting of an adenosine releasing synthetic polymer. Exp Neurol 160:164–174

Boison D, Scheurer L, Zumsteg V, Rülicke T, Litynski P, Fowler B, Brandner S, Mohler H (2002) Neonatal hepatic steatosis by disruption of the adenosine kinase gene. Proc Natl Acad Sci U S A 99:6985–6990

Boison D, Chen JF, Fredholm BB (2010) Adenosine signalling and function in glial cells. Cell Death Differ 17:1071–1082

Boison D, Masino MA, Geiger JD (2011a) Homeostatic bioenergetic network regulation—a novel concept to avoid pharmacoresistance in epilepsy. Expert Opin Drug Discov 6:713–724

Boison D, Singer P, Shen HY, Feldon J, Yee BK (2011b) Adenosine hypothesis of schizophrenia—opportunities for pharmacotherapy. Neuropharmacology 62(3):1527–43

Boon P, Raedt R, de Herdt V, Wyckhuys T, Vonck K (2009) Electrical stimulation for the treatment of epilepsy. Neurotherapeutics 6:218–227

Brambilla R, Cottini L, Fumagalli M, Ceruti S, Abbracchio MP (2003) Blockade of A2A adenosine receptors prevents basic fibroblast growth factor-induced reactive astrogliosis in rat striatal primary astrocytes. Glia 43:190–194

Brunstein MG, Ghisolfi ES, Ramos FL, Lara DR (2004) Clinical trial for allopurinol adjuvant therapy for poorly responsive schizophrenia. Schizophr Res 67:S142

Brunstein MG, Ghisolfi ES, Ramos FL, Lara DR (2005) A clinical trial of adjuvant allopurinol therapy for moderately refractory schizophrenia. J Clin Psychiatry 66:213–219

Clark RS, Carcillo JA, Kochanek PM, Obrist WD, Jackson EK, Mi Z, Wisneiwski SR, Bell MJ, Marion DW (1997) Cerebrospinal fluid adenosine concentration and uncoupling of cerebral blood flow and oxidative metabolism after severe head injury in humans. Neurosurgery 41:1284–1292, discussion 1292–1293

Cunha RA (2005) Neuroprotection by adenosine in the brain: from A1 receptor activation to A2A receptor blockade. Purinergic Signal 1:111–134
During MJ, Spencer DD (1992) Adenosine: a potential mediator of seizure arrest and postictal refractoriness. Ann Neurol 32:618–624
Eisenach JC, Hood DD, Curry R (2002a) Phase I safety assessment of intrathecal injection of an American formulation of adenosine in humans. Anesthesiology 96:24–28
Eisenach JC, Hood DD, Curry R (2002b) Preliminary efficacy assessment of intrathecal injection of an American formulation of adenosine in humans. Anesthesiology 96:29–34
Erion MD, Wiesner JB, Rosengren S, Ugarkar BG, Boyer SH, Tsuchiya M (2000) Therapeutic potential of adenosine kinase inhibitors as analgesic agents. Drug Dev Res 50:S14–06
Fedele DE, Koch P, Brüstle O, Scheurer L, Simpson EM, Mohler H, Boison D (2004) Engineering embryonic stem cell derived glia for adenosine delivery. Neurosci Lett 370:160–165
Fredholm BB, Ijzerman AP, Jacobson KA, Linden J, Muller CE (2011) International Union of Basic and Clinical Pharmacology. LXXXI. Nomenclature and classification of adenosine receptors—an update. Pharmacol Rev 63:1–34
Gebicke-Haerter PJ, Christoffel F, Timmer J, Northoff H, Berger M, Van Calker D (1996) Both adenosine A1- and A2-receptors are required to stimulate microglial proliferation. Neurochem Int 29:37–42
Gernert M, Thompson KW, Loscher W, Tobin AJ (2002) Genetically engineered GABA-producing cells demonstrate anticonvulsant effects and long-term transgene expression when transplanted into the central piriform cortex of rats. Exp Neurol 176:183–192
Gouder N, Fritschy JM, Boison D (2003) Seizure suppression by adenosine A_1 receptor activation in a mouse model of pharmacoresistant epilepsy. Epilepsia 44:877–885
Gouder N, Scheurer L, Fritschy J-M, Boison D (2004) Overexpression of adenosine kinase in epileptic hippocampus contributes to epileptogenesis. J Neurosci 24:692–701
Güttinger M, Fedele DE, Koch P, Padrun V, Pralong W, Brüstle O, Boison D (2005) Suppression of kindled seizures by paracrine adenosine release from stem cell derived brain implants. Epilepsia 46:1–8
Halassa MM, Haydon PG (2010) Integrated brain circuits: astrocytic networks modulate neuronal activity and behavior. Annu Rev Physiol 72:335–355
Halassa MM, Fellin T, Haydon PG (2007) The tripartite synapse: roles for gliotransmission in health and disease. Trends Mol Med 13:54–63
Hamiwka L, Macrodimitris S, Tellez-Zenteno JF, Metcalfe A, Wiebe S, Kwon CS, Jette N (2011) Social outcomes after temporal or extratemporal epilepsy surgery: a systematic review. Epilepsia 52:870–879
Hasko G, Pacher P, Vizi ES, Illes P (2005) Adenosine receptor signaling in the brain immune system. Trends Pharmacol Sci 26:511–516
Hattiangady B, Rao MS, Shetty AK (2008) Grafting of striatal precursor cells into hippocampus shortly after status epilepticus restrains chronic temporal lobe epilepsy. Exp Neurol 212:468–481
Haydon PG, Carmignoto G (2006) Astrocyte control of synaptic transmission and neurovascular coupling. Physiol Rev 86:1009–1031
Heinemann U, Kann O, Remy S, Beck H (2006) Novel mechanisms underlying drug resistance in temporal lobe epilepsy. Adv Neurol 97:85–95
Herzog AG (2009) Hormonal therapies: progesterone. Neurotherapeutics 6:383–391
Hindley S, Herman MA, Rathbone MP (1994) Stimulation of reactive astrogliosis in vivo by extracellular adenosine diphosphate or an adenosine A2 receptor agonist. J Neurosci Res 38:399–406
Hinterkeuser S, Schroder W, Hager G, Seifert G, Blumcke I, Elger CE, Schramm J, Steinhauser C (2000) Astrocytes in the hippocampus of patients with temporal lobe epilepsy display changes in potassium conductances. Eur J Neurosci 12:2087–2096
Huber A, Padrun V, Deglon N, Aebischer P, Mohler H, Boison D (2001) Grafts of adenosine-releasing cells suppress seizures in kindling epilepsy. Proc Natl Acad Sci U S A 98:7611–7616
Jacobson KA, Gao ZG (2006) Adenosine receptors as therapeutic targets. Nat Rev Drug Discov 5:247–264

Jarvis MF, Yu H, Kohlhaas K, Alexander K, Lee CH, Jiang M, Bhagwat SS, Williams M, Kowaluk EA (2000) ABT-702 (4-amino-5-(3-bromophenyl)-7-(6-morpholinopyridin-3-yl)pyrido[2, 3-d]pyrimidine), a novel orally effective adenosine kinase inhibitor with analgesic and anti-inflammatory properties: I. In vitro characterization and acute antinociceptive effects in the mouse. J Pharmacol Exp Ther 295:1156–1164

Kanter-Schlifke I, Toft Sorensen A, Ledri M, Kuteeva E, Hokfelt T, Kokaia M (2007) Galanin gene transfer curtails generalized seizures in kindled rats without altering hippocampal synaptic plasticity. Neuroscience 150:984–992

Kokaia M, Aebischer P, Elmér E, Bengzon J, Kalén P, Kokaia Z, Lindvall O (1994) Seizure suppression in kindling epilepsy by intracerebral implants of GABA- but not by noradrenaline-releasing polymer matrices. Exp Brain Res 100:385–394

Kossoff EH, Rho JM (2009) Ketogenic diets: evidence for short- and long-term efficacy. Neurotherapeutics 6:406–414

Kowaluk EA, Jarvis MF (2000) Therapeutic potential of adenosine kinase inhibitors. Expert Opin Investig Drugs 9:551–564

Kowaluk EA, Mikusa J, Wismer CT, Zhu CZ, Schweitzer E, Lynch JJ, Lee CH, Jiang M, Bhagwat SS, Gomtsyan A et al (2000) ABT-702 (4-amino-5-(3-bromophenyl)-7-(6-morpholino-pyridin- 3-yl) pyrido[2,3-d]pyrimidine), a novel orally effective adenosine kinase inhibitor with analgesic and anti-inflammatory properties. II. In vivo characterization in the rat. J Pharmacol Exp Ther 295:1165–1174

Li T, Lan JQ, Fredholm BB, Simon RP, Boison D (2007a) Adenosine dysfunction in astrogliosis: cause for seizure generation? Neuron Glia Biol 3:353–366

Li T, Steinbeck JA, Lusardi T, Koch P, Lan JQ, Wilz A, Segschneider M, Simon RP, Brustle O, Boison D (2007b) Suppression of kindling epileptogenesis by adenosine releasing stem cell-derived brain implants. Brain 130:1276–1288

Li T, Ren G, Lusardi T, Wilz A, Lan JQ, Iwasato T, Itohara S, Simon RP, Boison D (2008) Adenosine kinase is a target for the prediction and prevention of epileptogenesis in mice. J Clin Invest 118:571–582

Li T, Ren G, Kaplan DL, Boison D (2009) Human mesenchymal stem cell grafts engineered to release adenosine reduce chronic seizures in a mouse model of CA3-selective epileptogenesis. Epilepsy Res 84:238–241

Li T, Lytle N, Lan J-Q, Sandau US, Boison D (2011) Local disruption of glial adenosine homeostasis in mice triggers focal electrographic seizures: a first step in epileptogenesis? Glia 60(1):83–95

Loscher W (2002) Animal models of epilepsy for the development of antiepileptogenic and disease-modifying drugs. A comparison of the pharmacology of kindling and post-status epilepticus models of temporal lobe epilepsy. Epilepsy Res 50:105–123

Loscher W, Schmidt D (2011) Modern antiepileptic drug development has failed to deliver: ways out of the current dilemma. Epilepsia 52:657–678

Loscher W, Gernert M, Heinemann U (2008) Cell and gene therapies in epilepsy—promising avenues or blind alleys? Trends Neurosci 31:62–73

Masino SA, Li T, Theofilas P, Sandau US, Ruskin DN, Fredholm BB, Geiger JD, Aronica E, Boison D (2011) A ketogenic diet suppresses seizures in mice through adenosine A1 receptors. J Clin Invest 121:2679–2683

Mato JM, Martinez-Chantar ML, Lu SC (2008) Methionine metabolism and liver disease. Annu Rev Nutr 28:273–293

Mazarati AM (2004) Galanin and galanin receptors in epilepsy. Neuropeptides 38:331–343

McCown TJ (2004) The clinical potential of antiepileptic gene therapy. Expert Opin Biol Ther 4:1771–1776

McCown TJ (2009) Adeno-associated virus vector-mediated expression and constitutive secretion of galanin suppresses limbic seizure activity. Neurotherapeutics 6:307–311

McGaraughty S, Jarvis MF (2006) Purinergic control of neuropathic pain. Drug Dev Res 67:376–388

McGaraughty S, Cowart M, Jarvis MF, Berman RF (2005) Anticonvulsant and antinociceptive actions of novel adenosine kinase inhibitors. Curr Top Med Chem 5:43–58

Monopoli A, Conti A, Dionisotti S, Casati C, Camaioni E, Cristalli G, Ongini E (1994) Pharmacology of the highly selective A1 adenosine receptor agonist 2-chloro-N6-cyclopentyladenosine. Arzneimittelforschung 44:1305–1312

Nadkarni S, LaJoie J, Devinsky O (2005) Current treatments of epilepsy. Neurology 64:S2–S11

Nilsen KE, Cock HR (2004) Focal treatment for refractory epilepsy: hope for the future? Brain Res Brain Res Rev 44:141–153

Noe F, Frasca A, Balducci C, Carli M, Sperk G, Ferraguti F, Pitkanen A, Bland R, Fitzsimons H, During M et al (2009) Neuropeptide Y overexpression using recombinant adeno-associated viral vectors. Neurotherapeutics 6:300–306

Nolte MW, Loscher W, Herden C, Freed WJ, Gernert M (2008) Benefits and risks of intranigral transplantation of GABA-producing cells subsequent to the establishment of kindling-induced seizures. Neurobiol Dis 314:342–354

Ortinski PI, Dong J, Mungenast A, Yue C, Takano H, Watson DJ, Haydon PG, Coulter DA (2010) Selective induction of astrocytic gliosis generates deficits in neuronal inhibition. Nat Neurosci 13:584–591

Pan JW, Kim JH, Cohen-Gadol A, Pan C, Spencer DD, Hetherington HP (2005) Regional energetic dysfunction in hippocampal epilepsy. Acta Neurol Scand 111:218–224

Pascual O, Casper KB, Kubera C, Zhang J, Revilla-Sanchez R, Sul JY, Takano H, Moss SJ, McCarthy K, Haydon PG (2005) Astrocytic purinergic signaling coordinates synaptic networks. Science 310:113–116

Pignataro G, Maysami S, Studer FE, Wilz A, Simon RP, Boison D (2008) Downregulation of hippocampal adenosine kinase after focal ischemia as potential endogenous neuroprotective mechanism. J Cereb Blood Flow Metab 28:17–23

Pritchard EM, Szybala C, Boison D, Kaplan DL (2010) Silk fibroin encapsulated powder reservoirs for sustained release of adenosine. J Control Release 144:159–167

Qureshi IA, Mehler MF (2010) Epigenetic mechanisms underlying human epileptic disorders and the process of epileptogenesis. Neurobiol Dis 39:53–60

Racine RJ, Burnham WM (1984) The kindling model. In: Schwartz-kroin P, Wheal H (eds) Electrophysiology of epilepsy. Academic, London, pp 153–171

Raedt R, Van Dycke A, Vonck K, Boon P (2007) Cell therapy in models for temporal lobe epilepsy. Seizure 16:565–578

Ramesha KN, Mooney T, Sarma PS, Radhakrishnan K (2011) Long-term seizure outcome and its predictors in patients with recurrent seizures during the first year aftertemporal lobe resective epilepsy surgery. Epilepsia 52:917–924

Ravizza T, Gagliardi B, Noe F, Boer K, Aronica E, Vezzani A (2008) Innate and adaptive immunity during epileptogenesis and spontaneous seizures: evidence from experimental models and human temporal lobe epilepsy. Neurobiol Dis 29:142–160

Ren G, Li T, Lan JQ, Wilz A, Simon RP, Boison D (2007) Lentiviral RNAi-induced downregulation of adenosine kinase in human mesenchymal stem cell grafts: a novel perspective for seizure control. Exp Neurol 208:26–37

Riban V, Fitzsimons HL, During MJ (2009) Gene therapy in epilepsy. Epilepsia 50:24–32

Seifert G, Carmignoto G, Steinhauser C (2010) Astrocyte dysfunction in epilepsy. Brain Res Rev 63:212–221

Shen H-Y, Coelho JE, Ohtsuka N, Canas PM, Day Y-J, Huang Q-Y, Rebola N, Yu L, Boison D, Cunha RA et al (2008) A critical role of the adenosine A2A receptor in extra-striatal neurons in modulating psychomotor activity as revealed by opposite phenotypes of striatum- and forebrain-A2A receptor knockouts. J Neurosci 28:2970–2975

Shen HY, Lusardi TA, Williams-Karnesky RL, Lan JQ, Poulsen DJ, Boison D (2011) Adenosine kinase determines the degree of brain injury after ischemic stroke in mice. J Cereb Blood Flow Metab 31:1648–1659

Sherman EM, Wiebe S, Fay-McClymont TB, Tellez-Zenteno J, Metcalfe A, Hernandez-Ronquillo L, Hader WJ, Jette N (2011) Neuropsychological outcomes after epilepsy surgery: systematic review and pooled estimates. Epilepsia 52:857–869

Shetty AK, Hattiangady B (2007) Concise review: prospects of stem cell therapy for temporal lobe epilepsy. Stem Cells 25:2396–2407

Sorensen AT, Kanter-Schlifke I, Carli M, Balducci C, Noe F, During MJ, Vezzani A, Kokaia M (2008) NPY gene transfer in the hippocampus attenuates synaptic plasticity and learning. Hippocampus 18:564–574

Stout CE, Costantin JL, Naus CC, Charles AC (2002) Intercellular calcium signaling in astrocytes via ATP release through connexin hemichannels. J Biol Chem 277(12):10482–10488

Szybala C, Pritchard EM, Wilz A, Kaplan DL, Boison D (2009) Antiepileptic effects of silk-polymer based adenosine release in kindled rats. Exp Neurol 219:126–135

Theodore WH, Fisher R (2007) Brain stimulation for epilepsy. Acta Neurochir Suppl 97:261–272

Theofilas P, Brar S, Stewart K-A, Shen H-Y, Sandau US, Poulsen DJ, Boison D (2011) Adenosine kinase as a target for therapeutic antisense strategies in epilepsy. Epilepsia 52:589–601

Thompson KW (2005) Genetically engineered cells with regulatable GABA production can affect afterdischarges and behavioral seizures after transplantation into the dentate gyrus. Neuroscience 133:1029–1037

Tian GF, Azmi H, Takano T, Xu QW, Peng WG, Lin J, Oberheim N, Lou NH, Wang XH, Zielke HR et al (2005) An astrocytic basis of epilepsy. Nat Med 11:973–981

Vezzani A (2007) The promise of gene therapy for the treatment of epilepsy. Expert Rev Neurother 7:1685–1692

Vezzani A (2008) Epileptogenic role of astrocyte dysfunction. Epilepsy Curr 8:46–47

Vezzani A, Ravizza T, Balosso S, Aronica E (2008) Glia as a source of cytokines: implications for neuronal excitability and survival. Epilepsia 49(Suppl 2):24–32

Wieser HG (1996) Epilepsy surgery. Baillieres Clin Neurol 5:849–875

Williamson A, Patrylo PR, Pan J, Spencer DD, Hetherington H (2005) Correlations between granule cell physiology and bioenergetics in human temporal lobe epilepsy. Brain 128:1199–1208

Wilz A, Pritchard EM, Li T, Lan JQ, Kaplan DL, Boison D (2008) Silk polymer-based adenosine release: therapeutic potential for epilepsy. Biomaterials 29:3609–3616

Wrench JM, Rayner G, Wilson SJ (2011) Profiling the evolution of depression after epilepsy surgery. Epilepsia 52:900–908

Yu L, Huang Z, Mariani J, Wang Y, Moskowitz M, Chen JF (2004) Selective inactivation or reconstitution of adenosine A2A receptors in bone marrow cells reveals their significant contribution to the development of ischemic brain injury. Nat Med 10:1081–1087

Zimmermann H (2000) Extracellular metabolism of ATP and other nucleotides. Naunyn Schmiedebergs Arch Pharmacol 362:299–309

Chapter 27
Ketogenic Diet and Epilepsy: The Role of Adenosine

Jong M. Rho, Beth Zupec-Kania, and Susan A. Masino

Abstract The ketogenic diet (KD) debuted in the 1920s as a metabolic treatment for epilepsy, based on the historical observation that fasting could prevent seizures. Similar to fasting, the high-fat, low-carbohydrate KD restricts glucose and favors oxidation of fatty acids, which in turn generates ketones for energy. Despite nearly a century of clinical use, proof of efficacy was only recently established, and little is known about how the KD works. The lure of a continually growing armamentarium of pharmacological options and the inherent challenges in implementing a dietary treatment versus a drug have combined to relegate the KD to a therapy of last resort. Indeed, better knowledge of how the KD exerts broad-spectrum clinical activity would be required to develop enhanced metabolism-based treatments, and perhaps even a "diet in a pill." Recent evidence strongly implicates adenosine as a mediator of KD action, as it is well known that adenosine is a fundamental link between metabolism and neuronal membrane excitability. Through a greater mechanistic understanding of how the KD—and adenosine in particular—works to dampen aberrant excitation in the brain, novel insights and molecular targets are bound to emerge. However, given the complexity of metabolic pathways in both normal and disease states, it will be important to determine specific cause-and-effect relationships. At present, given growing interest in metabolic dysfunction as a major pathophysiological substrate for a multiplicity of disease states, as well as urgent

J.M. Rho
Departments of Pediatrics and Clinical Neurosciences, Alberta Children's Hospital, University of Calgary, Calgary, AB, Canada

B. Zupec-Kania
Ketogenic Seminars, The Charlie Foundation, Santa Monica, CA, USA

S.A. Masino (✉)
Neuroscience Program and Psychology Department, Trinity College, Life Sciences Center 210 300 Summit Street, Hartford, CT 06016, USA
e-mail: Susan.Masino@trincoll.edu

S. Masino and D. Boison (eds.), *Adenosine: A Key Link between Metabolism and Brain Activity*, DOI 10.1007/978-1-4614-3903-5_27,

concerns regarding unbridled healthcare costs worldwide, we predict that research in the area of "translational metabolism" will gain further momentum. Here we highlight current research on mechanism(s) of KD action, with a focus on adenosine, and promote the concept that the KD is both an accessible and affordable therapy by describing an international case study of recent KD training and initial outcomes in the Republic of Georgia.

27.1 Introduction: Ketogenic Diet as a Metabolic Therapy

The ketogenic diet (KD) enjoys a long-standing track record as an effective treatment for epilepsy (Freeman et al. 1998, 2006, 2009; Vining et al. 1998; Neal et al. 2008, 2009). Historically, the observation that fasting, or a "water diet," could reduce seizure activity spurred the quest for a metabolism-based therapy that could mimic the biochemical and therapeutic effects of fasting, yet retain adequate caloric and nutritional intake (Wheless 2008). Accordingly, a KD was developed more than 90 years ago to treat seizures, and it has remained on the treatment landscape ever since. Its high-fat low-carbohydrate formulation restricts the amount of glucose available to generate adenosine triphosphate (ATP) through glycolysis; this fundamental metabolic shift favors fatty acid oxidation that then generates ketones as the predominant bioenergetic substrate (Bough and Rho 2007). Despite an early and broad understanding of the major biochemical changes induced by the KD, the critical mechanisms responsible for its anticonvulsant effects have been difficult to identify (Vamecq et al. 2005; Bough and Rho 2007; Maalouf et al. 2009; Masino and Rho 2012).

Notwithstanding the lack of mechanistic information, the KD has experienced periods of waxing and waning enthusiasm by clinicians and awareness by the public at large (Wheless 2008). Shortly after its initial implementation the KD fell out of favor, due primarily to the advent of newer antiepileptic drugs (AEDs) such as phenytoin which were easier to administer and were regarded as being equally efficacious (Wheless 2008). This pharmacological bias remained for many decades, despite the fact that a significant fraction of patients with epilepsy typically fail to respond to AEDs—with either traditional or newer agents (Mattson et al. 1985; Kwan and Brodie 2000). While not yet proven, there is a clinical impression that the KD may be more effective than drugs, especially since there can be impressive response rates in patients with medically-refractory epilepsy (Freeman et al. 2006; Kossoff and Rho 2009).

Numerous clinical reports note 50 % responder rates in at least half the children administered the KD, and approximately one-third achieve more than a 90 % reduction in their seizures, with 7–10 % becoming seizure free (Kossoff and Rho 2009; Payne et al. 2011). In further support of the KD's efficacy in the most challenging of cases, there is growing evidence that the KD treats refractory status epilepticus (Nabbout et al. 2010; Nam et al. 2011)—a devastating condition of unrelenting seizure activity that can ensue for days and weeks. While used most

Table 27.1 Comparison of energy distribution and fat:nonfat ratios of therapeutic ketogenic diets (KD), a modified (Mod) Atkins diet, and a regular diet

Energy distribution	Classic KD		Liberal KD		Mod Atkins	Regular[a]
Ratio (fat:nonfat)	4:1	3:1	2:1	1:1	1:1	0.2:1
Fat: kilocalories	90 %	87 %	80 %	70 %	64 %	30 %
Protein: kilocalories	10 %	13 %	20 %	30 %	30 %	15–25 %
Carbohydrate: kilocalories					6 %	45–55 %

[a] Dietary guidelines for Americans; US Department of Health & Human Services, 2011

often in children, the KD and related dietary formulations are also effective in adolescents and adults (Payne et al. 2011). A recent case report describes how initiation with a KD, and maintenance with a modified Atkins diet, was able to control medically and surgically refractory epilepsy in an adult in a neurocritical care unit (Cervenka et al. 2011).

Against this historically rich backdrop, there is a current resurgence in consideration of metabolism-based treatments. In addition to the traditional KD and its well-known medium-chain triglyceride (MCT) variation, there are also low-carbohydrate diets for weight loss (such as the Atkins and South Beach/low-glycemic index treatments) which have become increasingly popular as these related approaches are associated with more liberal and palatable foodstuffs (Kossoff et al. 2006; Muzykewicz et al. 2009). Importantly, the KD is now considered a truly international treatment for epilepsy, even in underdeveloped countries (Kossoff and McGrogan 2005; Kossoff et al. 2011).

Table 27.1 outlines the energy composition of several diets of varying stringencies—including a classic KD and a modified Atkins diet as compared to a "standard" diet. A KD is customized to the individual, and takes into account his or her energy and protein requirements prior to determining the carbohydrate and fat content of his or her diet (hence the ratio, defined by weight as fats to carbohydrate plus protein). For example, to provide sufficient protein for an inactive adult, a 1:1 ratio may be prescribed. To achieve the very high fat content of a KD, typical meals include a small serving of meat, fish, or protein and a small serving of vegetables. Butter, mayonnaise, vegetable or coconut oil, and heavy cream are incorporated based on preferences and in amounts required to satisfy the ratio of fat:(protein + carbohydrate) (Table 27.1).

In many ways, the tremendous growth of KD centers worldwide, and scientific attention to this non-pharmacological treatment, would not have been possible without the sustained efforts of the Charlie Foundation to Cure Pediatric Epilepsy (http://www.charliefoundation.org; Santa Monica, California, USA), created by Jim and Nancy Abrahams, which brought substantial media attention to the benefits of the KD for epilepsy. For example, a 1997 made-for-television movie—"*First Do No Harm,*" starring Meryl Streep, documented the KD's "miraculous" effects on a young boy with catastrophic epilepsy which was unresponsive to numerous AEDs. Over the past decade and a half, the Charlie Foundation has been an invaluable resource for parents, patients, and professionals worldwide—and has been the driving force for both clinical and scientific activity surrounding the KD. More recently, the organization Matthew's Friends (http://www.matthewsfriends.org), spearheaded by Emma Williams and based in the United Kingdom, has quickly evolved to represent another major international resource and a voice for advocacy and education. Perhaps the greatest testament to the remarkable scientific growth of the KD is the fact that compared to the 15 years prior to its renaissance in the mid 1990s, there has been a tenfold increase in citations indexed on PubMed (http://www.ncbi.nlm.nih.gov/pubmed/) over the subsequent 15 years. Part of this phenomenal growth is likely due to expanding interest in metabolism and its dysfunction relative to chronic diseases (WHO Special Issue 2004; Uauy et al. 2008).

27.2 Mechanisms Mobilized by Ketogenic Diet Therapy

Ultimately, to produce anticonvulsant effects, the KD must reduce neuronal activity and/or excitability. Accordingly, there has been intense interest in a detailed understanding of how the metabolic effects of a KD translate into reduced excitation and/or a hyperpolarized membrane potential, altered neurotransmitter release or receptor affinities, or myriad other mechanisms that would result in fewer seizures (Vamecq et al. 2005; Bough and Rho 2007; Masino and Rho 2012).

One cardinal feature is the profound metabolic shift induced by the KD. As noted earlier, the KD is high in fat and low in carbohydrates. The enhanced fatty acid oxidation and restricted carbohydrate content switch metabolism from the preferred ATP-generating pathway—i.e., glycolysis—to intermediary metabolism that results in increased production of ketone bodies, decreased glucose, and increased levels of circulating fatty acids (Kim do and Rho 2008). Proximal metabolic changes are shown in Fig. 27.1 and have been described in more detail elsewhere (Kim do and Rho 2008).

Regarding the collective mechanisms proposed as relevant or critical to KD therapy, there are a number of excellent reviews and the reader is referred to these sources for more comprehensive coverage (Bough and Rho 2007; Gasior et al. 2006; Kim do and Rho 2008; Maalouf et al. 2009; Rho and Stafstrom 2011; Masino and Rho 2012). As an overview, these postulated mechanisms include (1) acute and chronic biochemical changes observed with KD administration in vivo (e.g., increased ketone production, decreased serum glucose levels, a reduction in the generation of reactive oxygen species (ROS), increased fatty acid levels (perhaps importantly, polyunsaturated fatty acids [PUFAs] which possess membrane-stabilizing properties), increased bioenergetic reserves (consisting of increased levels of ATP and ADP), and a reduction in levels of adenosine kinase, the major adenosine-metabolizing enzyme) and (2) cellular mechanisms with potentially direct effects on neuronal excitability such as opening of ATP-sensitive potassium (K_{ATP}) channels that cause membrane hyperpolarization (Ma et al. 2007; Kawamura et al. 2010) and acetoacetate-mediated presynaptic release of excitatory neurotransmitters via an

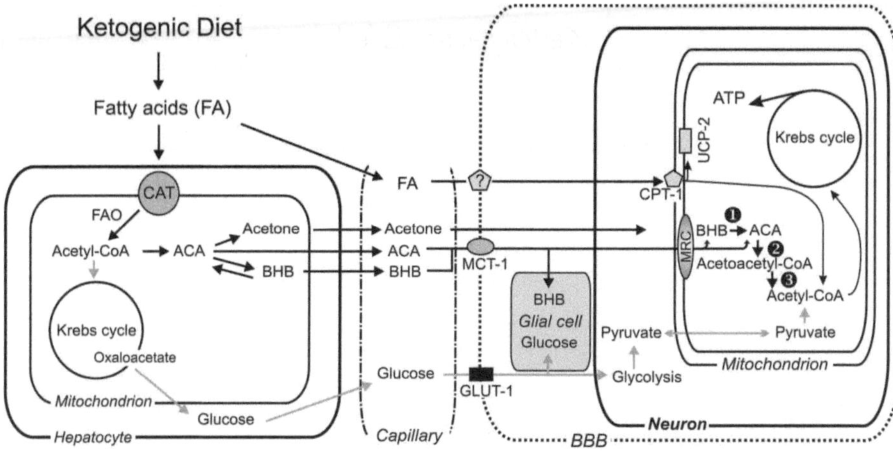

Fig. 27.1 Metabolic pathways involved in ketogenic diet (KD) treatment. In the liver (hepatocyte, *left*), fatty acids are ordinarily converted into acetyl-CoA which enters the tricarboxylic acid (TCA) cycle. When fatty acid levels exceed the metabolic capacity of the TCA cycle, acetyl-CoA is shunted to ketogenesis. Two acetyl-CoAs can combine through a thiolase enzyme to produce acetoacetyl-CoA, which is a precursor for the synthesis of acetoacetate (ACA) and β-hydroxybutyrate (BHB). Acetone, the other major ketone body, is produced primarily from spontaneous decarboxylation of ACA. These ketones are transported from the vascular lumen (capillary, *middle*) to the brain interstitial space, and to both glia and neurons. Within neurons (*right*), both ACA and BHB are transported directly into mitochondria, and ultimately converted to acetyl-CoA (through several enzymatic steps) which then enters the TCA cycle. Abbreviations: *CAT* carnitine-acylcarnitine translocase, *FAO* fatty acid oxidation, *ACA* acetoacetate, *BHB* β-hydroxybutyrate, *MCT-1* monocarboxylate transporter-1, *GLUT-1* glucose transporter-1, *BBB* blood–brain barrier, *CPT-1* carnitine palmitoyl transferase-1, *UCP* uncoupling protein, *ATP* adenosine triphosphate, (1) 3-hydroxybutyrate dehydrogenase, (2) succinyl-CoA:(CoA:3)3-oxoacid CoA transferase, (3) mitochondrial acetoacetyl-CoA thiolase. *MRC* mitochondrial respiratory complex. Reprinted with permission from Kim do Y, Rho JM, The ketogenic diet and epilepsy. Curr Opin Clin Nutr Metab Care 2008;11(2):113–120

interaction with vesicular glutamate transporters (VGLUTs) (Juge et al. 2010). A schematic summarizing these mechanisms is shown in Fig. 27.2, and mechanisms related to adenosine are discussed in more detail below.

27.3 Ketogenic Diet and Adenosine

Against this pleiotropic backdrop, and the many parallel and potentially synergistic pathways toward a reduction in membrane hyperexcitability, the question arises as to where adenosine fits in. Adenosine is in a unique position to translate KD-induced changes in metabolism into altered neuronal activity, and may be a critical mechanism for mediating the protective effects of the KD in epilepsy (and perhaps other neurological disorders). Similar to the KD, adenosine has been shown to be

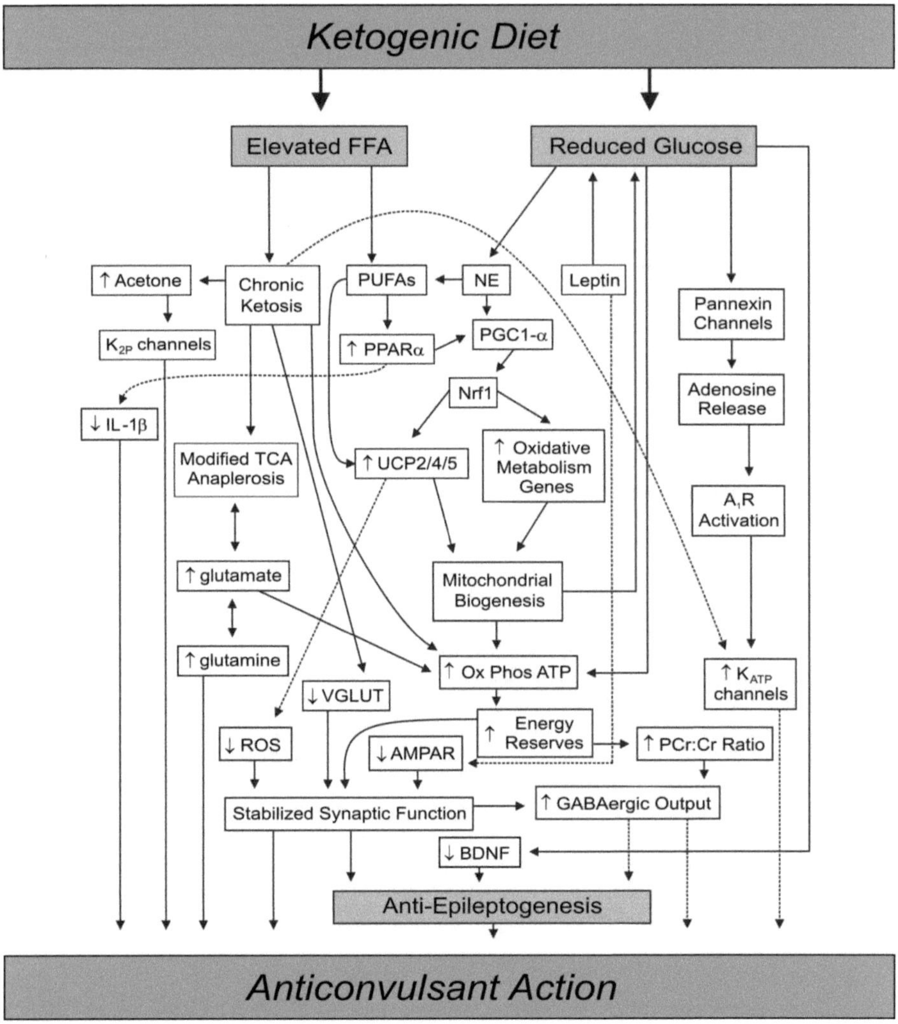

Fig. 27.2 Hypothetical pathways leading to the anticonvulsant effects of the ketogenic diet (KD). A KD elevates free fatty acids (FFAs) and reduces glucose (*top*), and each of these consequences mobilizes a host of mechanisms which could lead to anticonvulsant actions (*bottom*). Mechanisms related to adenosine are shown on the *right*. Elevated FFAs lead to chronic ketosis and increased concentrations of polyunsaturated fatty acids (PUFAs) in the brain. Chronic ketosis is anticipated to lead to increased levels of acetone; this might activate K_{2P} channels to hyperpolarize neurons and limit neuronal excitability. Chronic ketosis is also anticipated to modify the tricarboxylic acid (TCA) cycle, as would the presence of anaplerotic substrates such as triheptanoin. This would increase glutamate and, subsequently, GABA (γ-aminobutyric acid) synthesis in brain. Among several direct inhibitory actions, PUFAs boost the activity of brain-specific uncoupling proteins (UCPs). This is expected to limit reactive oxygen species (ROS) generation, neuronal dysfunction, and resultant neurodegeneration. Acting via the nuclear transcription factor peroxisome proliferator-activated receptor-α (PPARα) and its co-activator peroxisome proliferator-activated receptor γ co-activator-1 (PGC-1α), PUFAs would induce the expression of UCPs and coordinately up-regulate several dozen genes related to oxidative energy metabolism. PPARα expression is inversely correlated

effective in controlling seizures (likely via actions at the adenosine A_1 receptor (A_1R) subtype) (Dunwiddie 1999; Boison et al. 2011). However, systemic side effects, rapid metabolism and promiscuity with other adenosine receptor subtypes, and signaling mechanisms have hampered efforts over the past two decades to develop adenosine-based therapies for epilepsy. This scenario—in part fueled by linking adenosine to the KD—is now changing, and there is growing interest in reevaluating purinergic neurotransmission in epilepsy (Boison 2008; Masino and Geiger 2008; Masino et al. 2011a, b).

With respect to the relationship between metabolic changes and adenosine, diverse lines of evidence suggest that both the KD and ketones enhance brain energy metabolism. De Vivo and colleagues (1978) were the first to demonstrate that a KD increased ATP levels (and indeed all major measures of cellular bioenergetic reserves), and subsequent studies have generally confirmed these observations (Nakazawa et al. 1983; Bough et al. 2006; Nylen et al. 2009; Kim do et al. 2010).

In addition to increased levels of bioenergetic substrates, investigators have also shown that the KD increases mitochondrial numbers and profiles (Bough et al. 2006; Nylen et al. 2009), and preferentially up-regulates genes involved in

Fig. 27.2 (continued) with interleukin-1β (IL-1β) cytokine expression; given the role of IL-1β in hyperexcitability and seizure generation, diminished expression of IL-1β cytokines during KD treatment could lead to improved seizure control. Ultimately, PUFAs would stimulate mitochondrial biogenesis. Mitochondrial biogenesis is predicted to increase adenosine triphosphate (ATP) production capacity and enhance energy reserves, leading to stabilized synaptic function and improved seizure control. In particular, an elevated phosphocreatine:creatine (PCr:Cr) energy-reserve ratio is predicted to enhance GABAergic output, perhaps in conjunction with the ketosis-induced elevated GABA production, leading to diminished hyperexcitability. Reduced glucose coupled with elevated free fatty acids is proposed to reduce glycolytic flux during KD, which would further be feedback inhibited by high concentrations of citrate and ATP produced during KD treatment. This would activate metabolic ATP-sensitive potassium (K_{ATP}) channels. Ketones may also directly activate K_{ATP} channels. Reduced glucose alone, under conditions of adequate or enhanced energy levels, activates pannexins on CA3 pyramidal neurons, releasing ATP into the extracellular space; ATP is converted via ectonucleotidases to adenosine which subsequently activates adenosine receptors (A_1R). A_1R activation is also coupled to K_{ATP} channels. Ultimately, opening of K_{ATP} channels would hyperpolarize neurons and diminish neuronal excitability to contribute to the anticonvulsant (and perhaps neuroprotective actions of the KD). Increased leptin, seen with KD treatment, can reduce glucose levels and inhibit α-amino-3-hydroxy-5-methyl-4-isoxazolepropionic acid (AMPA) receptor-mediated synaptic excitation. Reduced glucose is also expected to down-regulate brain-derived neurotrophic factor (BDNF) and tropomyosin receptor kinase B (TrkB, a tyrosine kinase) signaling in brain. As activation of TrkB pathways by BDNF has been shown to promote hyperexcitability and kindling, these potential KD-induced effects would be expected to limit the symptom (seizures) as well as epileptogenesis. Boxed variables depict findings described from KD studies; *up* or *down arrows* indicate the direction of the relationship between variables as a result of KD treatment. *Dashed lines* are used to clarify linkages and are not meant to suggest either magnitude or relative importance compared to *solid lines*. Published previously as Fig: 78-4, p. 1015 from 'Mechanisms of Ketogenic Diet Action' by Susan A. Masino and Jong M. Rho in "Jaspers Basic: Mechanisms of the Epilepsies, 4E" edited by Noebels, J, Avoli M, Rogawski MA, Olsen RW, and Delgado-Escueta AV (2012). By permission of Oxford University Press, Inc

mitochondrial metabolism (Noh et al. 2004; Bough et al. 2006). Typically, such effects have been observed after maintenance on a KD for 2–4 weeks, but in published reports to date, a detailed time course for these effects has not been established. The biochemical, ultrastructural, and gene profiling from microarray expression studies all comport with animal research and clinical observations to indicate that in most cases, the anticonvulsant effects of the KD may take two or more weeks to develop (Freeman et al. 2006; Bough and Rho 2007; Kossoff and Rho 2009).

The compelling evidence that the KD increases ATP levels places the mechanistic spotlight squarely on adenosine. ATP is dephosphorylated rapidly to adenosine (via action of ectonucleotidases), and any increase in extracellular ATP will ultimately result in greater activation of adenosine receptors (Dunwiddie et al. 1997; Cunha et al. 1998; Dunwiddie 1999; Masino et al. 2002). There is direct evidence that the adenosine produced via dephosphorylation of ATP acts on inhibitory A_1Rs, which would lead to an anticonvulsant effect (Dunwiddie et al. 1997; Cunha et al. 1998; Masino et al. 2002). This fundamental link between metabolism and neuronal activity is discussed in an expanded fashion in an initial review postulating the potential mechanistic role of adenosine in KD therapy (Masino and Geiger 2008).

27.4 Evidence Linking the Ketogenic Diet to Adenosine

Two primary lines of evidence have bolstered support for adenosine's role in mediating the anticonvulsant effects of the KD. These include (1) an in vitro model of the KD—in particular, mirroring two essential features, namely, reduced glucose and normal or elevated ATP levels (Kawamura et al. 2010); and (2) assessment of electrographic seizure activity in cohorts of transgenic mice exhibiting spontaneous seizures, with and without KD treatment (Masino et al. 2011a, b). While these complementary studies have begun to address basic mechanisms of KD action, there are many questions that remain unanswered.

Kawamura and colleagues employed cellular electrophysiological techniques to address the question of whether a reduction in glucose (along with adequate or increased ATP levels) would result in membrane hyperpolarization, a *sine qua non* of anticonvulsant activity (Kawamura et al. 2010). This study was designed with the assumption that clinically, seizure control relates weakly to blood ketone levels, but perhaps more consistently with blood glucose (Freeman et al. 2006). Certainly, earlier studies have supported the concept that calorie restriction (with or without KD treatment) exerts anticonvulsant (and potentially anti-epileptogenic) effects (Todorova et al. 2000; Greene et al. 2001), and that reduced glucose may be more important than increased ketosis (Greene et al. 2003), but perhaps not in brain extracellular fluid (Samala et al. 2011). In animal models, the relationship between blood glucose levels and seizure frequency is not consistent (Mantis et al. 2004; Hartman et al. 2010), but the underlying idea behind reduced glucose is that cellular energy is compromised and as such, neurons are simply unable to reach high levels of repetitive firing required to sustain seizure activity. This concept, however, is in striking contrast

Table 27.2 Predicted and observed effects of the ketogenic diet in mice with adenosine-based electrographic seizures

Mouse model	A_1R expression	Predicted change in seizure frequency	Observed change in seizure frequency	Glucose-induced change in seizure frequency
Wild type: (C57BL/6)	Unaltered	N/A (no seizures)	N/A	No change
Transgenic: *Adk-tg*	Unaltered	Robust suppression	88 % decrease ($p<0.001$)	Reversed to 85 % of baseline ($p<0.001$)
Transgenic: $A_1R^{+/-}$	50 % normal	Partial suppression	53 % decrease ($p<0.001$)	Reversed to 89 % of baseline ($p<0.001$)
Transgenic: $A_1R^{-/-}$	No receptors	No suppression	4 % decrease (N.S.)	No change (N.S.)

Abbreviations: *Adk-tg* adenosine kinase overexpressing transgenic mice, $A_1R^{+/-}$ adenosine receptor subtype 1 heterozygous mutant, $A_1R^{-/-}$ adenosine receptor subtype 1 homozygous (null) mutant, *N.S.* not significant. From Masino et al. (2011b)

to several lines of evidence demonstrating increased, not decreased, bioenergetics (DeVivo et al. 1978; Bough et al. 2006; Nylen et al. 2009; Kim do et al. 2010).

The combination of increased intracellular ATP and decreased extracellular glucose revealed a novel adenosine-mediated autocrine mechanism in hippocampal CA3 neurons (Kawamura et al. 2010). The CA3 subfield of the hippocampus was chosen because it is a highly seizure-prone area. In short, Kawamura and colleagues found that ATP—released directly into the extracellular space through pannexin-1 channels and dephosphorylated to adenosine—led to activation of A_1Rs, which under these conditions hyperpolarized the membrane potential via coupling to the opening of K_{ATP} channels (Kawamura et al. 2010). This autocrine inhibition in CA3 did not require any direct exposure to ketones; rather, a combination of reduced glucose and sufficient or increased ATP was all that appeared necessary for this effect.

The second major line of evidence invoking adenosine in KD action emerged from experiments involving transgenic mice exhibiting spontaneous electrographic seizures due to deficient adenosine signaling (Masino et al. 2011a, b). After feeding a KD for 3–4 weeks, Masino and colleagues found that the diet reduced seizures by 90 % in mice overexpressing adenosine kinase (and hence deficient in extracellular adenosine) but with intact A_1Rs. In contrast, a KD was only partially effective in reducing electrographic seizures in mice with a partial complement of A_1Rs and was completely ineffective in mice lacking A_1Rs altogether (see Table 27.2). Similar to clinical reports, where anticonvulsant effects reverse rapidly with glucose infusion (Huttenlocher 1976), a systemic injection of glucose restored seizures in mice that experienced a KD-induced reduction in seizures. Further supporting the role of adenosine in KD action, adenosine kinase was down-regulated in normal mouse brain after 3–4 weeks of KD feeding, and levels of this enzyme were increased in tissue obtained from human patients with epilepsy, consistent with a deficiency in adenosine signaling (Masino et al. 2011a, b). Taken together, these data indicate that the anticonvulsant effects of the KD may be in part due to increased adenosine acting at inhibitory A_1Rs, and perhaps further augmented by deficiencies in adenosine kinase activity. At present, whether the KD induces changes in A_1R expression or affinity, or alters other aspects of the regulation of adenosine, is unknown. It is clear, however, that the relationship between the KD and adenosine signaling must be explored in other clinically relevant seizure models. Finally, it would be of great interest to determine whether augmentation of adenosine signaling might influence epileptogenesis, as there are reports suggesting a disease-modifying effect of the KD (Gasior et al. 2006; Maalouf et al. 2009; Stafstrom and Rho 2012).

27.5 Current Recommendations and Emerging Applications of Ketogenic Diet Therapy

The KD demonstrates an overall success rate equivalent to or perhaps superior to available AEDs (Kossoff and Rho 2009). However, with the exception of a few tertiary epilepsy centers (and a select few epileptic conditions), the KD has been relegated to the status of therapeutic last resort. It is becoming increasingly clear that this practice should be reevaluated (Nordli 2009). The most recent and comprehensive set of recommendations for implementation of the KD was published in 2009 as an international consensus statement commissioned by the Charlie Foundation (Kossoff et al. 2009b). This document addresses the following clinical management considerations: patient selection, pretreatment counseling and evaluation, specific dietary therapy selection (indications and contraindications), short-term and long-term implementation, supplementation with vitamins and minerals, follow-up visits and management, adverse event monitoring, concomitant use of AEDs, and eventual KD discontinuation. The consensus recommendations were made on the basis of best available evidence, considered areas of agreement and controversy, and touched upon unanswered questions and future research opportunities.

Based on anecdotal clinical observations wherein patients maintain a seizure-free state even after discontinuation of the KD (Hemingway et al. 2001; Marsh et al. 2006; Freeman et al. 2006), and a rapidly expanding experimental literature attesting to the neuroprotective effects of such high-fat, low-carbohydrate diets (both in vivo and in vitro) (Gasior et al. 2006), the notion that the KD may be effective for other neurological disorders arose—particularly those conditions associated with neurodegeneration (Maalouf et al. 2009; Stafstrom and Rho 2012). To date, dietary and metabolic therapies have been attempted in either experimental models or patients for the following conditions other than epilepsy: headache, neurotrauma, Alzheimer disease, Parkinson disease, sleep disorders, brain cancer, autism, and pain (Stafstrom and Rho 2012). The general impetus for using various diets to treat—or at least ameliorate symptoms of—these disorders stems from both a lack of effectiveness of pharmacological therapies and the intrinsic appeal of implementing a more "natural" treatment.

Whether adenosine is critically involved in the neuroprotective effects of the KD against any or all of these conditions remains to be determined. Certainly, there is existing evidence that adenosine acting at A_1Rs can produce neuroprotective actions

(Dunwiddie and Masino 2001; Masino and Geiger 2008; Tozaki-Saitoh et al. 2011; Gomes et al. 2011). Additionally, chronic adenosine exposure may lead to diverse epigenetic effects on DNA/RNA methylation (Skinner et al. 1986; Boison et al. 2002; Boison 2011). Adenosine and homocysteine are formed from S-adenosylhomocysteine, which is produced from S-adenosylmethionine via the action of methyltransferases. Thus, altered adenosine would influence the S-adenosylmethionine cycle. Adenosine's role in methylation reactions has long been recognized (Henderson 1979), and dysregulated adenosine metabolism (putatively increased due to a genetic loss of adenosine kinase) has been shown to inhibit DNA transmethylation. Specifically, a lack of adenosine kinase resulted in decreased adenine nucleotides and increased S-adenosylhomocysteine (a potent inhibitor of transmethylation reactions) in the liver (Boison et al. 2002). Yet these aspects of adenosine metabolism and regulation have not been the focus of adenosine-based therapeutics in recent decades—the goals of drug discovery have been primarily to influence adenosine receptor signaling. It may be time to revisit the potential biochemical and epigenetic roles of adenosine in determining cellular homeostasis and dysfunction in health and disease, respectively (Boison et al. 2011). Clearly epigenetic changes have enormous potential for disease-modifying effects, and further studies are necessary to clarify the long-term consequences of both the KD and adenosine on epigenetic nucleic acid modifications.

27.6 Use of the Ketogenic Diet in a Developing Country

Although the KD originated in the 1920s in the United States, variations on this metabolism-based therapy have been established in numerous academic and non-academic centers in over three dozen countries worldwide (Kossoff and McGrogan 2005; Kossoff et al. 2011), and there are anecdotal reports of several centers exploring uses of the KD outside of epilepsy, including brain cancer (A. Evangeliou, personal communication; T. Seyfried, personal communication). One of the authors (B.Z.-K.) has provided hands-on training to clinical staff at dozens of regional and international medical centers. In developing countries, these training workshops have been sponsored by The Charlie Foundation and/or provided free of charge in exchange for travel and accommodations. Because of the necessary coordination among neurologists, dietitians, and hospital staff, preparation for such visits often takes an entire year or more. KD meals are planned using local food supplies and sources, cultural preferences, and practices.

Box 27.1 highlights KD training and outcomes in the Republic of Georgia, a former member of the Soviet Union. The Republic of Georgia suffered from economic crisis and civil unrest since the collapse of the Soviet Union, and the healthcare system which had provided complete services for all citizens dissolved under the new democracy and is still struggling to recover. Healthcare is no longer free, with serious implications for patient care. Whereas in the United States prior laboratory studies rule out disorders in fatty acid metabolism and evaluate blood chemistries,

Box 27.1 Recent Training and Initial Outcomes: Republic of Georgia

In 2006, based on his work with pediatric epilepsy, Dr. Gia Milekshevelli, a neurologist working in the capital city Tbilisi, Republic of Georgia, received the Bernard D'Sousa International Fellowship Award. Dr. Milekshevelli's award included an all-expense, 2-week training at the Children's Hospital at Scottish Rite in Atlanta, Georgia (USA). Dr. Milekshevelli learned about the treatments available in the United States for pediatric epilepsy which included medication therapy, surgery, and the ketogenic diet. With limited access to antiseizure medications in Republic of Georgia, he knew dietary treatment was a feasible alternative.

With support from the Charlie Foundation, KD training was arranged in 2007. Dr. Milekshevelli's daughter, a medical student, served as an interpreter for two neurologists and pharmacist during the training and diet initiation in several patients by the author (B.Z.-K.). Families brought everything needed to the hospital to start the diet over 3 days, including food, dishes, utensils, soap, bedding, towels, etc., as well as ice and coolers to store their food.

One adult and two pediatric patients diagnosed with medication-resistant epilepsy of unknown etiology were identified as candidates for the KD (no EEG equipment was available to assist with further classification). All had severe developmental and cognitive delays. None had received special education and physical or speech therapy, nor did they have assistive devices such as special feeding utensils or wheelchairs. Their constant care was provided by their families. A KD was calculated for each patient and translated into Georgian for the family. The pharmacist reviewed and minimized the carbohydrate content of each patient's medications; she also provided the vitamin and mineral supplements.

A 4:1 KD ratio (90 % fat) was used for the children and a 1:1 ratio (70 % fat) was used for the adult. A rapid initiation method involved one ketogenic meal the first day, two the second day, and three the third day. Between meals each received a ketogenic beverage to provide the goal of calories for each patient. Within 3 days, all of the patients were producing strong urine ketones. The families ensured that every bit of food and drop of beverage was consumed, and worked together to prepare meals and share food and ideas to make the diet palatable for their children.

Case 1: Patient MR was 20 months of age. Her seizures started at 2 months of age and were described as "polymorphic" (characterized by two or more seizure types). She was trialed on an antiseizure medication that cost her family an entire month's salary. Her seizures did not improve, and the family discontinued it after 2 months. At the first meeting she lay flaccid in her mother's

(continued)

Box 27.1 (continued)

arms for the entire hour. Because she had such poor head control, she was spoon-fed thickened ketogenic drinks between meals. She was very weak and feeding her took hours each day.

Three weeks after starting the diet she was holding her head erect on her own and her mother reported that she was cooing for the first time. Her seizures declined dramatically over the next several months and she eventually became seizure free. She has been off of the diet for over a year and now, 4 years later, she remains seizure and medication free.

Case 2: Patient SK was 3 years old. He had generalized tonic-clonic seizures, had been trialed on all seven medications available at the time in Republic of Georgia, and was diagnosed with medication-resistant epilepsy. He appeared to be at an infant level physically and cognitively. He could sit by himself if positioned appropriately, but could not crawl or pull himself up.
SK had many seizures during the hospitalization for diet initiation. He had a history of constipation and, unfortunately, the KD did not help—constipation is a common side effect. He remained on the diet for 3 months.

Case 3: Patient AH was a 34-year-old lean, physically fit man. He had generalized seizures, and was diagnosed with medication-resistant epilepsy. He had the cognitive function of a child and could only speak in brief phrases.
AH had a prolonged seizure during the second day of diet initiation. However, he recovered without rescue medication. Over the 4 days of diet initiation his language skills improved markedly and he began speaking in full sentences. He remained on this very liberal KD for a year. Although he continued to have generalized tonic-clonic seizure every few months, his family was satisfied with the improvement in cognition, especially his verbal skills. When the medication Keppra became available in Georgia, he was started on this and discontinued the diet.

To date, Dr. Milekshevelli remains in contact with the author (B.Z-K.) and has placed a total of 12 patients on the diet: seven (58 %) have experienced significant improvement in seizure control, including three (25 %) who are seizure free. Although a small cohort, these statistics are impressive. Dr. Milekshevelli continues to advance his treatment of people with epilepsy, and has invited experts from Europe to visit him and review and offer advice on his most challenging patients. In 2009 he received certification in analyzing EEG data and he now has an EEG machine. Most recently, Dr. Milekshevelli trained his colleague, a neurologist in Bulgaria, on KD therapy and continues to mentor him.

(continued)

Box 27.1 (continued)

Ketogenic diet team members in the Republic of Georgia: Left to right; three mothers, author (B-.Z.K.), and Dr. Gia Milekshevelli

One of the mothers in the Republic of Georgia preparing her child's diet

these tests may be unavailable or unreliable in some countries; close monitoring with in-patient diet initiation is recommended under these circumstances.

Results in multiple countries have replicated published efficacy with pharmacoresistant epilepsy as well as success when drugs are not available or not economically viable. Western diets contain approximately 50 % of energy as carbohydrate. KDs limit carbohydrate-rich foods to less than 10 % of energy, and protein is consumed in moderate amounts—much lower than the typical Western diet. Fat is the KD's main source of energy and the least expensive food group when compared in terms of energy density to carbohydrate (i.e., bread, cereal, grains, and fruit) and protein-rich foods (i.e., meat and cheese). In the Republic of Georgia, success with a small cohort of patients yielded promising results—25 % became seizure free—underscoring that a dietary approach is accessible and affordable in countries with limited resources. In India, picture books and simple measuring devices allow successful application of the KD available to patients who cannot read (J. Nathan, personal communication). The KetoCalculator program (https://www.ketocalculator.com) is written in English and uses the metric system for diet calculations; this tool has been used successfully to help administer the KD in many countries.

27.7 Summary and Conclusions

Even though the KD demonstrates an overall success rate equivalent or potentially superior to available AEDs, aside from a few treatment centers it has been consistently relegated as a therapy of last resort for epilepsy. However, there is growing recognition that the KD should be considered in the treatment algorithm at a much earlier stage, and perhaps for a multiplicity of other neurological disorders (Nordli 2009; Stafstrom and Rho 2012) with specific predictions related to its effects on adenosine (Masino et al. 2009). Certainly, it is now well appreciated that the KD and its variants constitute both an affordable and accessible metabolic therapy. At a scientific level, the positive bioenergetic approach represented by the KD may be a general paradigm for restoring brain homeostasis, and is less likely to produce treatment resistance and significant untoward side effects (Boison et al. 2011). As the KD approaches its 100th anniversary (i.e., 2021), a critical reflection of its fascinating saga and its late maturation as a valid clinical and scientific entity places the KD on an exciting platform for further advances during the next century, especially as the leading metabolism-based treatment for a variety of neurological disorders.

Acknowledgements The authors would like to acknowledge the National Institutes of Health, National Science Foundation, Trinity College, the Alberta Children's Hospital Research Institute, and the Charlie Foundation to Cure Pediatric Epilepsy for support.

References

Boison D (2008) The adenosine kinase hypothesis of epileptogenesis. Prog Neurobiol 84(3):249–262
Boison D (2011) Adenosine: a novel regulator of DNA methylation in epilepsy. Presented as part of "The "Methylation Hypothesis:" Does Epigenetic Chromalin Modification Play a Role in Epileptogenesis?", organized by I. Blumcke, American Epilepsy Society 65th annual meeting, 2011, IW.8. Baltimore, MD
Boison D, Scheurer L, Zumsteg V, Rülicke T, Litynski P, Fowler B, Brandner S, Mohler H (2002) Neonatal hepatic steatosis by disruption of the adenosine kinase gene. Proc Natl Acad Sci USA 99(10):6985–6990
Boison D, Masino SA, Geiger JD (2011) Homeostatic bioenergetic network regulation—a novel concept to avoid pharmacoresistance in epilepsy. Expert Opin Drug Discov 6(7):713–724
Bough KJ, Wetherington J, Hassel B, Pare JF, Gawryluk JW, Greene JG, Shaw R, Geiger JD, Dingledine RJ (2006) Mitochondrial biogenesis in the anticonvulsant mechanism of the ketogenic diet. Ann Neurol 60:223–235
Bough KJ, Rho JM (2007) Anticonvulsant mechanisms of the ketogenic diet. Epilepsia 48(1):43–58
Cervenka MC, Hartman AL, Venkatesan A, Geocadin RG, Kossoff EH (2011) The ketogenic diet for medically and surgically refractory status epilepticus in the neurocritical care unit. Neurocrit Care 15(3):519–524
Cunha RA, Sebastião AM, Ribeiro JA (1998) Inhibition by ATP of hippocampal synaptic transmission requires localized extracellular catabolism by ecto-nucleotidases into adenosine and channeling to adenosine A1 receptors. J Neurosci 18(6):1987–1995
DeVivo DC, Leckie MP, Ferrendelli JS, McDougal DB Jr (1978) Chronic ketosis and cerebral metabolism. Ann Neurol 3(4):331–337
Dunwiddie TV (1999) Adenosine and suppression of seizures. Adv Neurol 79:1001–1010
Dunwiddie TV, Masino SA (2001) The role and regulation of adenosine in the central nervous system. Annu Rev Neurosci 24:31–55
Dunwiddie TV, Diao L, Proctor WR (1997) Adenine nucleotides undergo rapid, quantitative conversion to adenosine in the extracellular space in rat hippocampus. J Neurosci 17(20):7673–7682
Freeman JM, Vining EP, Pillas DJ, Pyzik PL, Casey JC, Kelly LM (1998) The efficacy of the ketogenic diet-1998: a prospective evaluation of intervention in 150 children. Pediatrics 102(6):1358–1363
Freeman J, Veggiotti P, Lanzi G, Tagliabue A, Perucca E, Institute of Neurology IRCCS C. Mondino Foundation (2006) The ketogenic diet: from molecular mechanisms to clinical effects. Epilepsy Res 68(2):145–180
Freeman JM, Vining EP, Kossoff EH, Pyzik PL, Ye X, Goodman SN (2009) A blinded, crossover study of the efficacy of the ketogenic diet. Epilepsia 50(2):322–325
Gasior M, Rogawski MA, Hartman AL (2006) Neuroprotective and disease-modifying effects of the ketogenic diet. Behav Pharmacol 17(5–6):431–439
Gomes CV, Kaster MP, Tomé AR, Agostinho PM, Cunha RA (2011) Adenosine receptors and brain diseases: neuroprotection and neurodegeneration. Biochim Biophys Acta 1808(5):1380–1399
Greene AE, Todorova MT, McGowan R, Seyfried TN (2001) Caloric restriction inhibits seizure susceptibility in epileptic EL mice by reducing blood glucose. Epilepsia 42(11):1371–1378
Greene AE, Todorova MT, Seyfried TN (2003) Perspectives on the metabolic management of epilepsy through dietary reduction of glucose and elevation of ketone bodies. J Neurochem 86(3):529–537
Hartman AL, Zheng X, Bergbower E, Kennedy M, Hardwick JM (2010) Seizure tests distinguish intermittent fasting from the ketogenic diet. Epilepsia 51(8):1395–1402
Hemingway C, Freeman JM, Pillas DJ, Pyzik PL (2001) The ketogenic diet: a 3- to 6-year follow-up of 150 children enrolled prospectively. Pediatrics 108(4):898–905
Henderson JF (1979) Regulation of adenosine metabolism. In: Baer HP, Drummond GI (eds) Physiological and regulatory function of adenosine and adenine nucletides. Raven, New York, pp 315–322
Huttenlocher PR (1976) Ketonemia and seizures: metabolic and anticonvulsant effects of two ketogenic diets in childhood epilepsy. Pediatr Res 10(5):536–540

Juge N, Gray JA, Omote H, Miyaji T, Inoue T, Hara C, Uneyama H, Edwards RH, Nicoll RA, Moriyama Y (2010) Metabolic control of vesicular glutamate transport and release. Neuron 68(1):99–112
Kawamura M Jr, Ruskin DN, Masino SA (2010) Metabolic autocrine regulation of neurons involves cooperation among pannexin hemichannels, adenosine receptors, and KATP channels. J Neurosci 30(11):3886–3895
Kim do Y, Rho JM (2008) The ketogenic diet and epilepsy. Curr Opin Clin Nutr Metab Care 11(2):113–120
Kim do Y, Vallejo J, Rho JM (2010) Ketones prevent synaptic dysfunction induced by mitochondrial respiratory complex inhibitors. J Neurochem 114(1):130–141
Kossoff EH, McGrogan JR (2005) Worldwide use of the ketogenic diet. Epilepsia 46(2):280–289
Kossoff EH, Rho JM (2009) Ketogenic diets: evidence for short- and long-term efficacy. Neurotherapeutics 6(2):406–414
Kossoff EH, McGrogan JR, Bluml RM, Pillas DJ, Rubenstein JE, Vining EP (2006) A modified Atkins diet is effective for the treatment of intractable pediatric epilepsy. Epilepsia 47(2):421–424
Kossoff EH, Zupec-Kania BA, Rho JM (2009a) Ketogenic diets: an update for child neurologists. J Child Neurol 24(8):979–988
Kossoff EH, Zupec-Kania BA, Amark PE, Ballaban-Gil KR, Christina-Bergqvist AG, Blackford R, Buchhalter JR, Caraballo RH, Helen Cross J, Dahlin MG, Donner EJ, Klepper J, Jehle RS, Kim HD, Christiana Liu YM, Nation J, Nordli DR Jr, Pfeifer HH, Rho JM, Stafstrom CE, Thiele EA, Turner Z, Wirrell EC, Wheless JW, Veggiotti P, Vining EP, Charlie Foundation, Practice Committee of the Child Neurology Society, Practice Committee of the Child Neurology Society, International Ketogenic Diet Study Group (2009b) Optimal clinical management of children receiving the ketogenic diet: recommendations of the International Ketogenic Diet Study Group. Epilepsia 50(2):304–317
Kossoff EH, Caraballo RH, du Toit T, Kim HD, Mackay MT, Nathan JK, Philip SG (2011) Dietary therapies: a worldwide phenomenon. Epilepsy Res [Epub Aug 17]
Kwan P, Brodie MJ (2000) Early identification of refractory epilepsy. N Engl J Med 342(5):314–319
Ma W, Berg J, Yellen G (2007) Ketogenic diet metabolites reduce firing in central neurons by opening K_{ATP} channels. J Neurosci 27(14):3618–3625
Maalouf M, Rho JM, Mattson MP (2009) The neuroprotective properties of calorie restriction, the ketogenic diet, and ketone bodies. Brain Res Rev 59(2):293–315
Mantis JG, Centeno NA, Todorova MT, McGowan R, Seyfried TN (2004) Management of multifactorial idiopathic epilepsy in EL mice with caloric restriction and the ketogenic diet: role of glucose and ketone bodies. Nutr Metab (Lond) 1(1):11
Marsh EB, Freeman JM, Kossoff EH, Vining EP, Rubenstein JE, Pyzik PL, Hemingway C (2006) The outcome of children with intractable seizures: a 3- to 6-year follow-up of 67 children who remained on the ketogenic diet less than one year. Epilepsia 47(2):425–430
Masino SA, Geiger JD (2008) Are purines mediators of the anticonvulsant/neuroprotective effects of ketogenic diets? Trends Neurosci 31(6):273–278
Masino SA, Rho JM (2012) Mechanisms of ketogenic diet action. In: Noebels JL, Avoli M, Rogawski MA, Olsen RW, Delgado-Escueta AV (eds) Jasper's basic mechanisms of the epilepsies, 4th edn. Oxford University Press, New York, pp 1001–1022
Masino SA, Diao L, Illes P, Zahniser NR, Larson GA, Johansson B, Fredholm BB, Dunwiddie TV (2002) Modulation of hippocampal glutamatergic transmission by ATP is dependent on adenosine A1 receptors. J Pharmacol Exp Ther 303(1):356–363
Masino SA, Kawamura M, Wasser CD, Pomeroy LT, Ruskin DN (2009) Adenosine, ketogenic diet and epilepsy: the emerging therapeutic relationship between metabolism and brain activity. Curr Neuropharmacol 7(3):257–268
Masino SA, Li T, Theofilas P, Sandau US, Ruskin DN, Fredholm BB, Geiger JD, Aronica E, Boison D (2011a) A ketogenic diet suppresses seizures in mice through adenosine A1R receptors. J Clin Invest 121(7):2679–2683
Masino SA, Kawamura M Jr, Ruskin DN, Geiger JD, Boison D (2011b) Purines and neuronal excitability: links to the ketogenic diet. Epilepsy Res [Epub Aug 29]
Mattson RH, Cramer JA, Collins JF, Smith DB, Delgado-Escueta AV, Browne TR, Williamson PD, Treiman DM, McNamara JO, McCutchen CB et al (1985) Comparison of carbamazepine,

phenobarbital, phenytoin, and primidone in partial and secondarily generalized tonic-clonic seizures. N Engl J Med 313(3):145–151

Muzykewicz DA, Lyczkowski DA, Memon N, Conant KD, Pfeifer HH, Thiele EA (2009) Efficacy, safety, and tolerability of the low glycemic index treatment in pediatric epilepsy. Epilepsia 50(5):1118–1126

Nabbout R, Mazzuca M, Hubert P, Peudennier S, Allaire C, Flurin V, Aberastury M, Silva W, Dulac O (2010) Efficacy of ketogenic diet in severe refractory status epilepticus initiating fever induced refractory epileptic encephalopathy in school age children (FIRES). Epilepsia 51(10):2033–2037

Nakazawa M, Kodama S, Matsuo T (1983) Effects of ketogenic diet on electroconvulsive threshold and brain contents of adenosine nucleotides. Brain Dev 5(4):375–380

Nam SH, Lee BL, Lee CG, Yu HJ, Joo EY, Lee J, Lee M (2011) The role of ketogenic diet in the treatment of refractory status epilepticus. Epilepsia 52(11):e181–e184

Nathan JK, Purandare AS, Parekh ZB, Manohar HV (2009) Ketogenic diet in Indian children with uncontrolled epilepsy. Indian Pediatr 46(8):669–673

Neal EG, Chaffe H, Schwartz RH, Lawson MS, Edwards N, Fitzsimmons G, Whitney A, Cross JH (2008) The ketogenic diet for the treatment of childhood epilepsy: a randomised controlled trial. Lancet Neurol 7(6):500–506

Neal EG, Chaffe H, Schwartz RH, Lawson MS, Edwards N, Fitzsimmons G, Whitney A, Cross JH (2009) A randomized trial of classical and medium-chain triglyceride ketogenic diets in the treatment of childhood epilepsy. Epilepsia 50(5):1109–1117

Noh HS, Lee HP, Kim DW, Kang SS, Cho GJ, Rho JM, Choi WS (2004) A cDNA microarray analysis of gene expression profiles in rat hippocampus following a ketogenic diet. Brain Res Mol Brain Res 129(1–2):80–87

Nordli DR Jr (2009) The ketogenic diet, four score and seven years later. Nat Clin Pract Neurol 5(1):12–13

Nylen K, Velazquez JL, Sayed V, Gibson KM, Burnham WM, Snead OC 3rd (2009) The effects of a ketogenic diet on ATP concentrations and the number of hippocampal mitochondria in Aldh5a1(-/-) mice. Biochim Biophys Acta 1790(3):208–212

Payne NE, Cross JH, Sander JW, Sisodiya SM (2011) The ketogenic and related diets in adolescents and adults—a review. Epilepsia 52(11):1941–1948

Rho JM, Stafstrom CE (2011) The ketogenic diet: what has science taught us? Epilepsy Res [Epub Aug 29]

Samala R, Klein J, Borges K (2011) The ketogenic diet changes metabolite levels in hippocampal extracellular fluid. Neurochem Int 58(1):5–8

Skinner MA, Ho HJ, Chan VL (1986) Inhibition of methylation of DNA and tRNA by adenosine in an adenosine-sensitive mutant of the baby hamster kidney cell line. Arch Biochem Biophys 246(2):725–732

Stafstrom CE, Rho JM (2012) The ketogenic diet as a treatment paradigm for diverse neurological disorders. Front Pharmacol 3:59

Todorova MT, Tandon P, Madore RA, Stafstrom CE, Seyfried TN (2000) The ketogenic diet inhibits epileptogenesis in EL mice: a genetic model for idiopathic epilepsy. Epilepsia 41(8):933–940

Tozaki-Saitoh H, Tsuda M, Inoue K (2011) Role of purinergic receptors in CNS function and neuroprotection. Adv Pharmacol 61:495–528

Uauy R, Kain J, Mericq V, Rojas J, Corvalán C (2008) Nutrition, child growth, and chronic disease prevention. Ann Med 40(1):11–20

Vamecq J, Vallée L, Lesage F, Gressens P, Stables JP (2005) Antiepileptic popular ketogenic diet: emerging twists in an ancient story. Prog Neurobiol 75(1):1–28

Vining EP, Freeman JM, Ballaban-Gil K, Camfield CS, Camfield PR, Holmes GL, Shinnar S, Shuman R, Trevathan E, Wheless JW (1998) A multicenter study of the efficacy of the ketogenic diet. Arch Neurol 55(11):1433–1437

Wheless JW (2008) History of the ketogenic diet. Epilepsia 49(Suppl 8):3–5

World Health Organization (WHO) (2004) Special issue on diet, nutrition and prevention of chronic diseases. Public Health Nutr 7(1A):Suppl 1001

Chapter 28
Silk: A Biocompatible and Biodegradable Biopolymer for Therapeutic Adenosine Delivery

Eleanor M. Pritchard, Detlev Boison, and David L. Kaplan

Abstract No system currently exists for controllable, sustained, long-term drug delivery via fully degradable implants. While there are a number of biomaterials that have been proposed to meet this need, purified silk fibroin protein is a unique material well suited to controlled release applications, and particularly suited to the delivery of neurological therapeutics. Implants derived from silk exhibit the requisite biocompatibility and degradation profile for implantable applications, but also possess the necessary material properties to provide a diffusion barrier even for small molecule drugs and highly controllable material features that can in turn be used to tailor drug release behavior. Adenosine has therapeutic potential in a wide range of neurological diseases, including pharmacoresistant epilepsy. However, the use of adenosine or its analogs is limited by unacceptable systemic side effects. Silk implants for controlled, sustained, local adenosine delivery were developed in an effort to overcome these limitations. The therapeutic potential of controlled release silk implants was demonstrated in the rat model of kindling epileptogenesis. Controlled, local adenosine delivery suppressed seizures in a dose-dependent manner and partially delayed the progression of epilepsy, while also avoiding the adverse effects of systemic administration and safety risks of xenotransplantation. Silk microspheres and silk encapsulated adenosine reservoirs were also investigated. Increasing coating thickness of silk encapsulated reservoirs, by increasing either the silk solution concentration, crystallinity or the number of silk coatings, delayed burst release of adenosine and increased the duration of release. The ability to achieve zero-order release profiles suggests these tunable implants will be able to

E.M. Pritchard • D.L. Kaplan (✉)
Department for Biomedical Engineering, Tufts University, Medford, MA 01255, USA
e-mail: David.Kaplan@tufts.edu

D. Boison
Robert S. Dow Neurobiology Laboratories, Legacy Research, 1225 NE 2nd Avenue, Portland, OR 97232, USA

S. Masino and D. Boison (eds.), *Adenosine: A Key Link between Metabolism and Brain Activity*, DOI 10.1007/978-1-4614-3903-5_28,

maintain therapeutic drug concentrations within narrow windows. Silk-based adenosine delivery systems represent a safe and efficient strategy to suppress seizures. Further, these degradable, implantable biomaterials could potentially be applied to a range of adenosine related pathologies.

Keywords Silk fibroin • Adenosine • Biomaterials • Drug delivery • Controlled release • Degradable polymers • Epilepsy • Epileptogenesis

28.1 Focal Drug Delivery for Adenosine Delivery in the CNS

Systemic periodic administration of therapeutics, such as via oral delivery, bolus injection, or intravenous infusion, has been the primary mode of drug delivery for decades, despite several significant disadvantages. Systemic administration relies on the circulatory system to bring drugs to their target site of action by distributing them throughout the body. Not only does this systemic exposure waste a large fraction of the initial drug dose, it can also result in adverse side effects (Langer 1980). Further, achieving the sustained drug levels necessary for long-term therapeutic efficacy can require repeated administration. Repeated injections can be inconvenient and painful (Danckwerts and Fassih 1991), and scheduled self-administered oral formulations can suffer from poor patient compliance (Lebovits et al. 1990).

The therapeutic administration of neuroactive molecules to treat neurological disorders is particularly challenging. For therapeutics intended to reach the brain, there is the added difficulty of penetrating the blood brain barrier (BBB). The BBB is a physiological and pharmacological barrier comprised of tight junctions between the endothelial cells of CNS capillaries which prevent the influx of molecules from the bloodstream into the brain (Wang et al. 2002). Limited drug passage through the BBB can require intolerably high systemic doses to achieve sufficient therapeutic doses within the CNS.

Despite robust evidence of the involvement of adenosine receptors in several pathophysiologies, therapeutically useful manipulation of their activation has been limited. Systemic administration of adenosine induced severe systemic side effects, including suppression of cardiac function (Dunwiddie 1999). Adenosine is also cleared rapidly and has a short half-life and low BBB permeability (Pagonopoulou et al. 2006). A system for site-specific, implantable, sustained adenosine delivery is therefore desired to enhance its efficacy and to reduce side effects.

Focal drug delivery offers many advantages over traditional systemic administration, including enhanced efficacy and cost-efficiency and reduction or elimination of unwanted side-effects. The advantages of local drug delivery are especially attractive for the administration of neurological therapeutics. Local delivery in the brain would eliminate the need to penetrate the BBB while maintaining high concentrations of therapeutic agents at the target site without causing systemic toxicity (Maysinger and Morinville 1997; Menei et al. 1993; Gutman et al. 2000).

Previous attempts to restrict adenosine delivery to target areas in the brain have predominantly fallen into three categories: (1) infusion from nondegradable pumps, (2) grafts of neuroactive substance-releasing cells, either encapsulated in semipermeable membranes or directly transplanted, and (3) drug eluting nondegradable polymer implants. A number of reviews have been published on strategies for local adenosine delivery (Van Dycke et al. 2011; Boison and Stewart 2009). While local infusion of adenosine reduced seizures (Van Dycke et al. 2010) and wakefulness (Portas et al. 1997), this approach may be limited by complications related to the implantable pumps, including mechanical failure, obstruction, and infection (Wang et al. 2002; Barcia and Gallego 2009). Intrathecal or intraventricular catheters suffer from drug instability in aqueous solution and fail to deliver drugs directly into the parenchyma (Harbaugh et al. 1988).

Numerous studies have demonstrated the potential of cell-based delivery strategies using various cell sources, including myoblasts (Güttinger et al. 2005b), fibroblasts (Huber et al. 2001), baby hamster kidney (BHK) cells (Boison et al. 2002), embryonic stem (ES) cells (Li et al. 2007; Güttinger et al. 2005a; Uebersax et al. 2006), human mesenchymal stem cells (hMSCs) (Ren et al. 2007), astrocytes derived from fetal neural proginetor cells (Van Dycke et al. 2010), and ES-cell derived glia (Fedele et al. 2004). A review of adenosine-releasing stem cells has been published (Boison 2009). While these cell-based approaches are theoretically capable of achieving the long-term release critical for chronic disorders like epilepsy, they must be carefully balanced against the possible risks of implanting biologically active cells into the brain (Barcia and Gallego 2009). In addition, further research to ensure long-term cell survival and stable adenosine secretion is needed to effectively implement graft-based local delivery. Although long delivery durations are comparably more difficult to achieve with polymer implants, polymeric delivery bypasses some of these safety concerns.

Controlled, sustained release implants offer the advantages of local delivery, with the added benefits of reduced frequency of administration, improved patient convenience and compliance, and drug levels that are continuously maintained in a therapeutically desirable range without peaks and valleys (Langer 1980). Ethylene vinyl acetate copolymer (EVAc) polymer rods (1 mm height × 0.4 mm diameter) prepared with 20 % adenosine (w/w) released approximately 20–50 ng of adenosine per day, and when implanted into kindled rat hippocampii, induced a reduction of epileptic seizures (Boison et al. 1999). Stage 5 seizures were reduced for at least 7 days and as the adenosine release rate diminished, the antiseizure effects gradually decreased and expired after 14 days (Boison et al. 1999). EVAc cylinders directly implanted in rat brains (4 mm height × 5 mm diameter) also effectively delivered the anticonvulsant phenytoin locally, resulting in significant seizure suppression compared with empty control polymer implants (Tamargo et al. 2002). However, EVAc is nondegradable and must therefore be retrieved at the completion of therapy. Though the effects of retrieval on the brain have not (to our knowledge) been investigated, previous attempts with nondegradable reservoirs have demonstrated that the retrieval operation can be much more traumatic than the original insertion. The hormone-releasing implantable contraceptive device Norplant, for example, was approved for use by the Food and

Drug Administration (FDA) in 1990, but was withdrawn in 2002 due to complications associated with implant removal (Practice Committee of the American Society for Reproductive Medicine 2004). Biodegradable implants would be preferable as they alleviate the need for surgical removal of the materials after the conclusion of therapy (Danckwerts and Fassih 1991).

Though these methods have some limitations, these studies demonstrated the therapeutic potential of local adenosine delivery. The success of these approaches has led to interest in the development of implantable, biocompatible biomaterials for controlled site-specific adenosine delivery.

28.2 Silk Biomaterials for Adenosine Delivery in the CNS

28.2.1 Biodegradable Polymers for CNS Delivery

Numerous biodegradable materials have been investigated for controlled release drug delivery, but these degradable biomaterials frequently lack the full scope of properties required for implantable drug delivery applications, including biocompatibility, nonimmunogenicity, nontoxicity, high mechanical, thermal and chemical stability, mild processing to preserve drug bioactivity during encapsulation, and stabilizing effects on the incorporated drug (Puppi et al. 2010; Wenk et al. 2011). Due to the broad spectrum of potential applications, the ability to process the material into a wide array of material formats and carrier morphologies is also considered advantageous. In addition, to ensure that constructs can perform dual functions (for example, as a combined tissue scaffold and drug carrier), materials should possess multiple control points through which both implant morphology/material properties (such as elasticity, porosity, crystallinity) and release behavior can be dialed in (Wenk et al. 2011).

Several systems for local intracerebral drug delivery based on biodegradable polymers have been reported, particularly for delivering chemotherapy to treat brain cancers. Brem et al. prepared wafers from carmustine (BCNU) and a copolymer matrix consisting of 1,3-bis(*p*-carboxyphenoxy)propane (CPP) (GLIADEL® wafers) to be implanted into the resected tumors of patients with recurring malignant glioma (Brem et al. 1995). Carboplatin-releasing PLGA microspheres injected into the tumors of glioma-bearing rats produced elevated carboplatin levels in the tumor tissue compared with the surrounding brain tissue. Median survival increased and weight loss decreased for animals receiving drug-eluting PLGA microspheres compared with animals that received systemic doses of carboplatin or empty PLGA microspheres (Chen et al. 1998). Fung et al. implanted polyanhydride pellets in monkey brains, and reported high drug concentrations within 3 mm of the implant site, and significant concentrations within 5 cm up to 30 days post-implantation (Fung et al. 1998). Reviews of biomaterials-based drug delivery for neurologic disease (Whittlesey and Shea 2004); and delivery to the CNS (Patel et al. 2009) have been published.

In the field of drug delivery, implants prepared from the synthetic polymer polylactide-co-glycolide (PLGA) have received considerable attention due to their ability to degrade in vivo and their tunability in terms of degradation lifetime. The degradation rate, rate of drug release, and physical and mechanical properties of PLGA implants are adjustable though manipulation of the polymer molecular weight or copolymer ratio (PLA–PGA) (Mundargi et al. 2008). However, PLGA suffers from numerous critical disadvantages. Degradation of PLGA implants proceeds via a random, bulk hydrolysis of ester bonds in the polymer chain, which can lead to dose dumping. The acidic by-products of PLGA degradation have been shown to incite local inflammation (Puppi et al. 2010; Daugherty et al. 1997) and instability of the incorporated therapeutic (Varde and Pack 2007), such as acid sensitive peptides and proteins. Further, many PLGA material formats require harsh processing conditions, such as high heat or the use of organic solvents (Jain 2000), which may inactivate sensitive drugs during device loading. Though the effects of low pH and harsh processing conditions on adenosine stability have not been extensively investigated, mild processing options are desirable for many therapeutics.

Naturally derived biodegradable biopolymers generally offer superior biocompatibility and cell support (Puppi et al. 2010). However, compared with PLGA, naturally derived materials often lack the tight control of properties needed for tunable drug delivery. Degradation lifetimes in vivo for unmodified naturally derived materials tend to be too short for long-term sustained release applications. Chemical cross-linking can be used to slow the degradation rate but can compromise the biological response to the material, impacting safety or biocompatibility. In addition, drug carriers prepared from naturally derived materials tend to become highly permeable (especially to small molecules) or disintegrate upon exposure to aqueous body fluids, releasing drug loads over short time frames (Puppi et al. 2010; Siepmann et al. 2008; Abletshauser et al. 1993).

To effectively achieve sustained, local adenosine delivery, biodegradable implants must be prepared which exhibit both the biocompatibility and mild processing associated with naturally derived biopolymers and the tunability and slow degradation associated with synthetic polymers like PLGA.

28.2.2 Properties of Silk Biomaterials for Drug Delivery

Silk fibroin is a biologically derived protein polymer isolated from the cocoons of the domestic silkworm (*Bombyx mori*) that possesses unique properties for implantable and sustained drug delivery applications. Silk possesses excellent biocompatibility (Leal-Egaña and Scheibel 2010; Tang et al. 2009; Meinel et al. 2005; Seo et al. 2009; Panilaitis et al. 2003), robust mechanical strength (Altman et al. 2003) and has been shown to support cell growth, proliferation, and differentiation (Acharya et al. 2008a; Wang et al. 2006). Silk degrades to nontoxic products in vivo and the degradation time course of silk implants can be controlled from weeks to years via regulation of beta sheet content (crystallinity) during processing

(Horan et al. 2005; Wang et al. 2008a; Numata and Kaplan 2010). Silk can be processed entirely in aqueous systems using mild, ambient conditions of temperature and pressure (Vepari and Kaplan 2007; Lawrence et al. 2008), avoiding the use of harsh manufacturing conditions that can degrade incorporated therapeutics (such as shear, heat, organic solvents, or extreme pH). Stable, physical cross-linking of silk can be achieved during the crystallization process to form beta sheets, negating any need for chemical cross-linking and thereby avoiding potentially toxic chemicals. Further, silk has been found to exert a significant stabilizing effect on encapsulated enzymes and antibodies, even at elevated storage temperatures (Lu et al. 2009, 2010a; Guziewic et al. 2011).

Silk also exhibits excellent versatility and tunability in terms of material features. Silk fibroin has been processed into a variety of useful material formats including porous tissue scaffolds that structurally mimic the extracellular matrix (ECM) (Wang et al. 2006), optical devices (Lawrence et al. 2008) and drug carriers including sponges, nano- and microspheres, films, and hydrogels (Wenk et al. 2011; Pritchard et al. 2011). As a biodegradable carrier, release rates from silk fibroin are dependent on diffusion of drug through the silk, degradation of the silk polymer carrier via proteases, or a combination of both mechanisms. Diffusivity can be controlled through carrier morphology (geometry, number of silk coatings, porosity) and manipulation of the silk properties, including average molecular weight and crystallinity/β-sheet content.

The combined diversity of material formats available for drug delivery, tight control of the various drug carriers and unique properties of the silk material (biocompatibility, biodegradation, stabilization, aqueous processing) result in a broad range of silk-based systems available for clinical and scientific applications. Reviews of drugs delivered using silk carriers have been published (Wenk et al. 2011; Pritchard et al. 2011; Wang et al. 2010; Numata and Kaplan 2010).

28.2.3 Fundamentals of Silk Structure and Self-assembly

Bombyx mori cocoon silk consists of two main components: hydrophilic "glue-like" sericin proteins and the hydrophobic structural protein fibroin. For use in biomedical applications, silk fibroin must be separated from the sericin proteins to prevent an inflammatory response in vivo (Panilaitis et al. 2003). The degumming process typically relies on exposure to denaturing conditions (such as boiling in an alkaline sodium carbonate solution), which removes the sericin protein (Altman et al. 2003; Vepari and Kaplan 2007).

Fibroin is comprised of a heavy chain (approximately 390 kDa) and a light chain (approximately 25 kDa) bonded together via a disulfide bond (Kaplan et al. 1997; Matsumoto et al. 2006). The heavy chain is a natural block copolymer composed of large hydrophobic blocks (domains), much smaller hydrophilic blocks and two large hydrophilic blocks at the chain ends (Kim et al. 2004; Matsumoto et al. 2006). The primary structure of silk fibroin is dominated by the amino acids glycine, alanine, serine, valine, and tyrosine with characteristic repetitive sequences of GAGAGS, GAGAGY, and GAGAGVGY (Matsumoto et al. 2006). This structure (large hydrophobic domains consisting of short side chain amino acids) permits the tight packing of stacked sheets of hydrogen bonded antiparallel chains, which forms characteristic antiparallel beta-sheet secondary structure that gives silk fibroin its strength and resilience (Vepari and Kaplan 2007).

Degummed silk fibroin fibers can be used to prepare fiber biomaterials, including yarns, sutures, rope, and woven fabrics (Horan et al. 2006). However, most silk biomaterials are prepared from regenerated silk fibroin, which is prepared by solubilizing degummed fibroin in a hot salt solution (typically ranging from 50 to 70 °C), then dialyzing out the salt to obtain an aqueous solution. For solvent processing, regenerated silk fibroin solution can be lyophilized, then dissolved in solvent (usually 1,1,1,3,3,3-hexafluoro-2-propanol [HFIP]).

Silk in its prespun water soluble state (silk I) is relatively unstable, and the transition to the more energetically stable, β-sheet rich silk II is considered essentially irreversible from a thermodynamic perspective (Kaplan et al. 1997). Beta-sheet stacking occurs such that methyl groups and hydrogen groups of opposing sheets interact to form intersheet stacking in the crystals. The crystal structure is thermodynamically stabilized by van der Waals forces and strong hydrogen bonds that form between amino acids perpendicular to the axis of the chains and the fiber (Vepari and Kaplan 2007; Kaplan et al. 1997).

In vivo, glandular silk is assembled into highly ordered, β-sheet rich fibers via water extraction, changes in pH and salt concentration and mechanical stresses during fiber spinning that induce chain alignment (Hofmann et al. 2006). These processes, along with others that manipulate the factors involved in the assembly process, have been biomimetically exploited in vitro. For example, silk films that are dried overnight at ambient conditions retain their silk I conformation (and water solubility) (Kaplan et al. 1997), but once the films are treated with methanol, localized dehydration removes the ordered water molecules that surround the fibroin hydrophobic moieties in the solution state, resulting in β-sheet formation (Matsumoto et al. 2006).

28.2.4 Biocompatibility and Biodegradation of Silk in the Brain

The biocompatibility of silk fibroin for CNS therapies has been extensively investigated. Silk fibroin supported the survival and growth of hippocampal neurons and neurons cultured in silk fibroin extract exhibited no significant difference in morphology or cell viability compared to plain neuronal culture medium (Tang et al. 2009). Moreover, no significant difference in mRNA or protein levels of growth-associated molecules (determined via immunohistochemistry, real-time PCR, and Western Blot) was detected between hippocampal neurons cultured in plain neuronal culture medium and medium containing silk fibroin extract, suggesting that silk fibroin has no significant cytotoxic effects on the hippocampal neuron phenotype or function (Tang et al. 2009). Similar studies have demonstrated the biocompatibility of silk fibroin with dorsal root ganglia and Schwann cells (Yang et al. 2007a). These biocompatibility studies, combined with the frequent use of silk as a suture material in brain and nerve tissue (Dehdashti et al. 2001 and Schmutz et al. 1997) and studies showing positive outcomes for silk fibroin-based nerve grafts (Yang et al. 2007b; Uebersax et al. 2007; Madduri et al. 2010), demonstrate that silk is a safe, supportive biomaterial for CNS applications.

As a protein, silk fibroin is susceptible to biological degradation by proteolytic enzymes. The final degradation products of silk digestion are amino acids which are easily absorbed in vivo (Cao and Wang 2009; Numata and Kaplan 2010). The degradation behavior of silk is dependent on many factors, including silk processing (organic solvent vs. aqueous processing related to crystalline content, silk concentration and porosity (Wang et al. 2008a)), silk material format (Arai et al. 2004; Yang et al. 2009), type and concentration of enzyme (Arai et al. 2004), and immune system response (Wang et al. 2008a). For reviews of silk degradation, see Altman et al. 2003; Cao and Wang 2009, and Pritchard et al. 2011.

To assess degradation of silk sponges in the brain, silk sponge porosity was compared prior to and 4 weeks post-implantation in a rat brain. Prior to implantation, the average silk implant porosity based on surface area analysis was 41.1 % and this increased to 50.9 % after implantation (Szybala et al. 2009). This increase in porosity suggested that silk-degrading proteases were present in the brain, which is also supported by the literature on brain proteases. Chymotrypsin has been shown to degrade silk (Li et al. 2003) and numerous chymotrypsin-like proteases have been identified in the rat brain, such as caldecrin (Tomomura et al. 2002). The biodegradability of silk implants in the brain constitutes an advantage for their use in drug delivery for neurological applications.

28.2.5 Silk Drug Delivery Systems and Adenosine Release Studies

28.2.5.1 Silk Films

To prepare drug loaded silk films, regenerated silk fibroin is prepared as previously described (Sofia et al. 2001) and mixed with an aqueous solution of the drug of interest. The solution is cast into films of the desired thickness and surface area, dried at ambient conditions and then treated to produce the desired material and release properties.

Processing options to physically cross-link silk films include controlled slow drying (Lu et al. 2010b), water annealing (Jin et al. 2005), stretching, compressing (Demura et al. 1989), and solvent immersion (including methanol (Hofmann et al. 2006) and ethanol (Miyairi et al. 1978)). Silk films that receive no further treatment post-drying retain water solubility and dissolve when brought into contact with aqueous body tissue (Kim et al. 2010). If desired, chemical cross-linking can be used, such as glutaraldehyde (Acharya et al. 2008b), or 1-ethyl-3-(3-dimethyl aminopropyl) carbodiimide (EDC) (Bayraktar et al. 2005). However, it is not clear how this impacts degradability in long term.

Recently, ultrathin electronics have been mounted onto dissolvable silk film substrates and these patterned films exhibited spontaneous conformal wrapping when applied to the soft, curvilinear surface of the brain tissue. The benefit of the conformal coverage was demonstrated in neural mapping in a feline animal model (Kim et al. 2010). Rapidly dissolving, highly conformational silk films could also be used for drug delivery to improve tissue contact without damaging the soft brain tissue. Dissolvable silk films loaded with 0.5, 0.25, or 0.125 mg of adenosine per 0.2 mm^2 film released the majority of the drug load (approximately 80 %) within 15 min of exposure to 37 °C phosphate buffered saline (PBS) (unpublished data).

Silk films are also attractive as coating materials (Pritchard and Kaplan 2011). Due to the anti-inflammatory and neuroprotective properties of adenosine, adenosine-eluting silk coatings could potentially improve therapeutic outcomes for neurological implants.

28.2.5.2 Silk Microspheres

Compared with implantable silk material formats, such as films, sponges, and coated dry powder reservoirs, microspheres offer the advantage of being injectable for minimally invasive delivery. If spatial stability of the injected particles is a concern, microspheres can be imbedded in silk hydrogels to restrict movement to the delivery site. Various processes have been used to produce drug loaded silk microspheres, including lipid templates (Wang et al. 2007c), spray-drying (Hino et al. 2000, 2003; Yeo et al. 2003), ethanol and freezing-induced self-assembly (Cao et al. 2007; Bessa et al. 2010a, b), salting out (Lammel et al. 2010), water-in-oil emulsification (Imsombut et al. 2010), laminar jet breakup of an aqueous silk solution (Wenk et al. 2008), and casting and dissolution of PVA-silk blend films (Wang et al. 2010).

Adenosine containing microspheres have been prepared according to the MeOH-based lipid protocol (Wang et al. 2007c). Briefly, 200 mL of 10 mg/mL adenosine stock solution was mixed with 1 mL of 8 % (w/v) silk aqueous solution, which was then added to 200 mg of 1,2-dioleoyl-*sn*-glycero-3-phosphocholine (DOPC) phospholipid that had been dissolved in 1 mL of chloroform, then dried under N_2 to a film on the interior of a glass test tube. The solution is diluted, then repeatedly freeze-thawed, and then lyophilized. Later, the microspheres are treated with methanol (MeOH) to remove the lipids and induce β-sheet silk physical cross-links to stabilize the structures. The resulting microspheres were less than 2 μm in diameter (Wang et al. 2007c). Using this protocol, up to 85 μg of adenosine has been loaded per mg of silk microspheres (unpublished data).

When suspended in PBS, adenosine-loaded silk microspheres exhibit a burst release with approximately 75 % of the total drug load released within the first 24 h, with no significant release after 3 days. Suspending microspheres in a sonication-induced 1 % (w/v) or 3 % (w/v) silk hydrogel (Wang et al. 2008b) did not impact release behavior in PBS, suggesting that for hydrophilic small molecule drugs the hydrogels provide structural support but no additional diffusion barrier. In contrast, suspending microspheres in aqueous-derived porous silk sponges (as described by Kim et al. 2005) extended the release in PBS from 3 to 7 days and resulted in constant, zero-order adenosine release for the first 3 days (unpublished data). The addition of silk coatings to adenosine-loaded silk microspheres (Wang et al. 2007b) also delayed burst and increased release duration in PBS, as measured by an increase in the fitted value of the Siepmann–Peppas release exponent (n) in the semiempirical power law equation (Ritger and Peppas 1987) with increased number of silk coatings (unpublished data).

Adenosine-releasing silk microspheres show promise both for minimally invasive, injectable sustained release drug delivery and as a component in composite implants, such as microsphere-loaded "plum-pudding" hydrogels and microsphere-loaded porous sponges.

28.2.5.3 Silk Encapsulated Adenosine Reservoirs

A critical challenge in designing sustained release implants is the ability to achieve zero-order release rates and maximize drug loading relative to implant volume in order to maximize release duration. Reservoir systems are comprised of a drug core surrounded by polymer film (Langer and Peppas 1981). Use of dry powder drug reservoirs allows reservoir loading far above the drug's solubility. Drug concentration at the internal wall is continuously maintained above the saturation concentration of the drug, resulting in near zero-order release (Langer and Peppas 1981) followed by a phase of first-order release once the concentration of drug in the core drops below the saturation concentration (Zhou et al. 2010). Studies have shown that a particular, desired release profile adapted to the target drug delivery application can be obtained by varying the polymer coating formulation and processing parameters (Siepmann et al. 2008).

Several natural and synthetic degradable polymers have been studied as coating materials, but they generally fail to produce zero-order release kinetics or sustained release durations due to their material properties. Most studies that investigate natural polymers for pill encapsulation focus on delivery of total drug load within short time frames corresponding to residence times in the target gastrointestinal organs (on the order of hours) for improved oral delivery. Silk fibroin encapsulated reservoirs were investigated to assess their ability to achieve controllable, zero-order, sustained release behavior while maintaining biocompatibility and degrading after the completion of therapy. Solid adenosine powder reservoirs approximately 10 mm in diameter and 2 mm in height were prepared using a manual pill press and coated by dipping in aqueous silk fibroin solution, drying, and then methanol treating. Material properties of the silk coating including thickness, crystallinity and morphology were investigated to assess the relationships between silk coating biomaterial features and adenosine release from silk encapsulated reservoirs. Coating thickness was varied through either manipulation of the silk coating solution

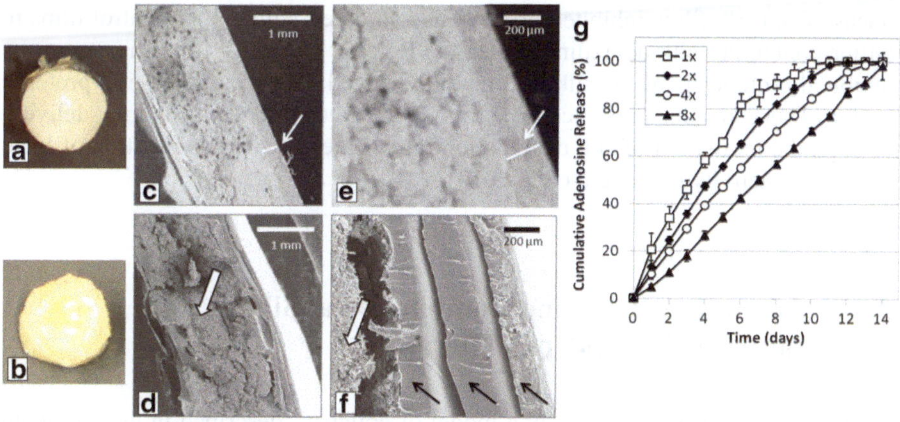

Fig. 28.1 Comparison between an encapsulated reservoir coated with single 8 % (w/v) silk coating with an encapsulated reservoir coated with multiple 8 % (w/v) coatings. (**a**) Gross morphology of single 8 % (w/v) silk coated reservoir. (**b**) Gross morphology of reservoir coated with eight 8 % (w/v) silk coatings. (**c**) and (**e**) SEM images of cross section of single 8 % (w/v) silk coated reservoir. (**d** and **f**) SEM cross section of multiple 8 % (w/v) silk layer coated reservoir. (**g**) Cumulative adenosine release from encapsulated adenosine reservoir coated in varied number of 8 % (w/v) silk coatings in PBS at 37 °C. 1×=one coating, 2×=two coatings, 4×=four coatings, 8×=8 coatings. $N=3$, error bars represent standard deviations (where error bars aren't shown they fall into background). Scale bars in (**c**) and (**d**) are 1 mm. Scale bars in (**e**) and (**f**) are 200 μm. (Modified from Pritchard et al. 2010)

concentration (2, 4, 8, 16, and 20 % (w/v)) or number of coatings applied (1, 2, 4, or 8 coatings of 8 % (w/v) (Fig. 28.1)). Single reservoir coatings exhibited thicknesses ranged from ~0.1 to ~0.8 mm depending on silk concentration. Release studies were performed in PBS at 37 °C and were also performed in proteinase type XIV to model the effects of degradation. Increasing the barrier to diffusion, either by increasing coating thickness or coating crystallinity was found to delay adenosine burst, decrease average daily release rate, and increase duration of release. Release behavior from encapsulated reservoirs was also related to proteolytic degradation rate. In the case of encapsulated reservoirs coated with eight layers of 8 % (w/v) silk, a linear release profile was observed and adenosine release was sustained for 14 days (Fig. 28.1g). The ability to achieve nearly constant release of a small molecule suggests these encapsulated reservoirs represent a novel system for local adenosine delivery. The relationships elucidated between release rate and coating thickness and crystallinity demonstrated the tunability of these systems (Pritchard et al. 2010).

To investigate a potential strategy for controlling the rate of degradation of the reservoir's coating, silk fibroin-encapsulated solid powder reservoirs were also prepared to corelease both adenosine and ethylenediamine tetraacetic acid (EDTA) and the effect of EDTA release on the rates of degradation and drug release was studied. EDTA use in humans is considered to be safe and EDTA has been used for decades as a common additive of the food industry to preserve freshness (Furia 1964). EDTA release from silk films inhibited proteolysis compared with empty control films in buffer containing silk-degrading proteases. Release of EDTA from the reservoirs inhibited proteolysis of the silk coating and delayed adenosine release in protease buffer. These results introduced a promising strategy to control drug delivery through the regulation of silk degradation rate, achieved via manipulation of local proteolytic activity (Pritchard et al. 2010).

28.3 Adenosine-Releasing Combinatorial Silk Implants for Epilepsy Therapy

Using the rat hippocampal kindling model of epilepsy (described in greater detail elsewhere (Wilz et al. 2008; Szybala et al. 2009)), the therapeutic potential of silk-based polymers engineered to release the anticonvulsant adenosine was investigated. Silk implants were designed to administer varying doses of adenosine to determine effective local dose and to test the hypothesis that different target release rates would suppress seizures in a dose-dependent manner. Following this dose-dependence study, the effect of adenosine delivery from silk implants on the progression of epilepsy was evaluated.

28.3.1 Implant Design and Fabrication

Silk implants with different target adenosine doses were prepared by imbedding adenosine loaded silk microspheres into nanofilm coated silk fibroin scaffolds. For dose dependence studies, implants were prepared to deliver target doses of 0 (control), 40, 200, and 1,000 ng of adenosine with the total drug load split roughly evenly between three systems: microspheres, macroscale films, and nanofilms. Microspheres were prepared according to the methanol-based lipid template protocol (Wang et al. 2007a). Porous scaffolds were prepared as previously described (Kim et al. 2005) using the water-based, NaCl-porogen protocol and a suspension of adenosine-loaded silk microspheres in 6 % (w/v) silk solution so that the microspheres were imbedded in the final porous scaffold. The macroscale adenosine-loaded silk film coating applied was based on a modified version of the bulk loaded films described previously (Hofmann et al. 2006). Nanofilm coatings were applied based on a previously described protocol of repeated dipping, washing and drying steps (Wang et al. 2007b) modified to accommodate the coating of a 3D porous scaffold. These three systems were integrated into a single implantable silk rod approximately 0.6–0.7 mm in diameter and 3–4 mm in length (as shown in Fig. 28.2). Target doses were controlled via manipulation of (1) the concentration of microspheres suspended in the porous scaffold, (2) the concentration of adenosine in the macroscale coating film, and (3) the number of drug loaded silk nanofilms deposited on the

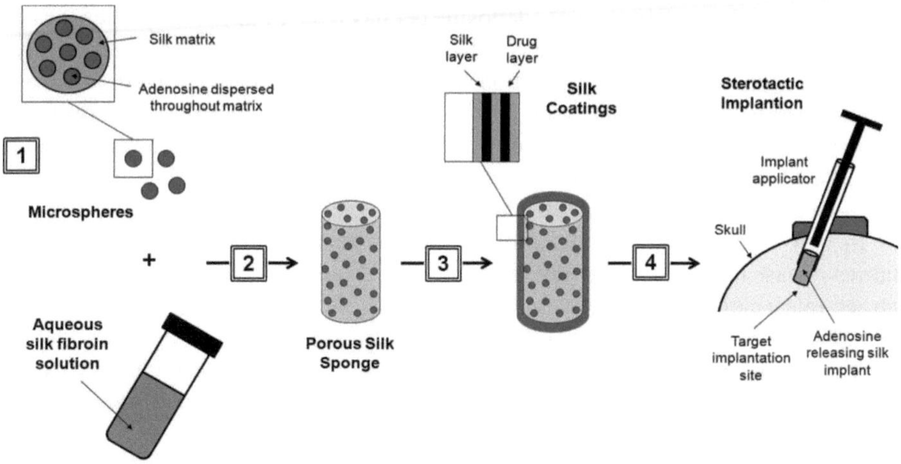

Fig. 28.2 Schematic of implant fabrication showing the individual silk-based drug delivery components. (1) Adenosine loaded silk microspheres are prepared according to the phospholipid template protocol (2) Microspheres are mixed with silk solution, and then porous scaffolds embedded with microspheres are fabricated using NaCl particulate leeching. (3) Porous scaffold is soaked in a silk + adenosine solution, reinforcing the implants and coating the implant with a macroscale drug-loaded film. Additional adenosine is loaded with alternating layer-by-layer (LbL) nanofilm coatings via alternate soakings in silk and adenosine solutions. Silk capping layers are added to increase the barrier to diffusion (4) Adenosine releasing silk rod is implanted by placing the implant in an applicator rod, inserting the applicator rod into the brain tissue and retracting first the outer shaft then the inner piston, leaving the silk implant at the target site of drug delivery (Modified from Wilz et al. 2008 and Boison et al. 2002)

system. After drug loading, all implants were coated with three capping layers (three subsequent silk dips) to delay burst.

Following the dose-dependence study, second generation implants designed to deliver target doses of 0 (empty/control) or 1,000 ng adenosine per day over a period of 10 days. These implants were designed and fabricated as described above, with some slight modifications. Briefly, implants were designed to split the target drug load evenly between microspheres and multiple macroscale films that were integrated into a single implant and capped with additional burst-delaying silk films.

28.3.2 Release Studies

Release studies were carried out by incubating implants in 1 mL of phosphate buffered saline (PBS) at 37 °C and removing and replacing the buffer at desired time points. Adenosine content in the PBS samples removed from the system was measured using a modified fluorescence assay based on conversion of adenosine to the fluorescent derivative 1,N^6-ethenoadenosine as previously described (Wojcik and Neff 1982). In PBS release studies, the respective polymers released an average of 0, 33.4, 170.5, and 819.0 ng of adenosine per day over 14 days (daily release rates are variable; average daily release is based on total adenosine released over 14 days divided by the 14 day release duration (Wilz et al. 2008)). Though average release rate is lower than the original target, this in vitro release data confirmed that implants designed to deliver varied doses were achieving this goal. Previous dose estimates from the literature suggested that an effective adenosine dose would be in the range of about 200 ng/day (Huber et al. 2001; Boison et al. 1999).

In vitro release study results showed that the second generation implants sustained release longer than previous implants and achieved release rates closer to the target: release of approximately 1,000 ng/day was maintained from day 4 to day 10 (Fig. 28.3a) (Szybala et al. 2009). The extended release duration compared with first generation silk implants is ascribed to the increase in the diffusional path length achieving by increasing the thickness of both the adenosine-loaded and implant-capping silk films.

28.3.3 *In Vivo Results*

28.3.3.1 Dose Dependent Seizure Suppression in a Rat Kindling Model

The therapeutic potential of adenosine-releasing silk implants was validated in a dose response study in the rat model of kindling epileptogenesis. Polymers were implanted using a stereotactic implantation device (internal diameter 0.7 mm, external diameter 1 mm) as previously described (Boison et al. 2002). The polymer-loaded device was stereotactically inserted into the brain using a drill hole. Upon reaching the target site the 3 mm long polymer was released and deposited within the infrahippocampal cleft by slowly retracting the outer tube of the device. Finally, the device was fully retracted as described previously (Boison et al. 2002). Thus, the implanted polymers were deposited within a formed cavity of 3 mm length within the right infrahippocampal cleft and adjacent to the electrode implantation site (Fig. 28.1). For each dose tested, either five or six animals received implants.

Four days prior to the onset of kindling, adenosine releasing polymers were implanted into the infrahippocampal cleft and progressive acquisition of kindled seizures was monitored over a total of 48 stimulations. Each rat received a total of eight kindling sessions (each session consisting of six stimulations delivered every 30 min), with sessions 1–8 corresponding to days 4, 6, 8, 11, 13, 15, 18, and 20 following implantation. Behavioral seizures were scored on a scale from 0 to 5 according to the scale of Racine (Racine 1978). Animals receiving empty/control silk implants exhibited first partial seizures during session 2 (around stimulation 10), while first emergence of partial seizures was delayed with increasing doses of released adenosine. In recipients of polymers releasing the highest adenosine dose (average release rate = total cumulative release divided by total release duration = 819 ng adenosine per day), kindling epileptogenesis was delayed by 1 week corresponding to 18 kindling stimulations, and these animals required 2.5 times as many stimulations to reach the same kindling state as the control

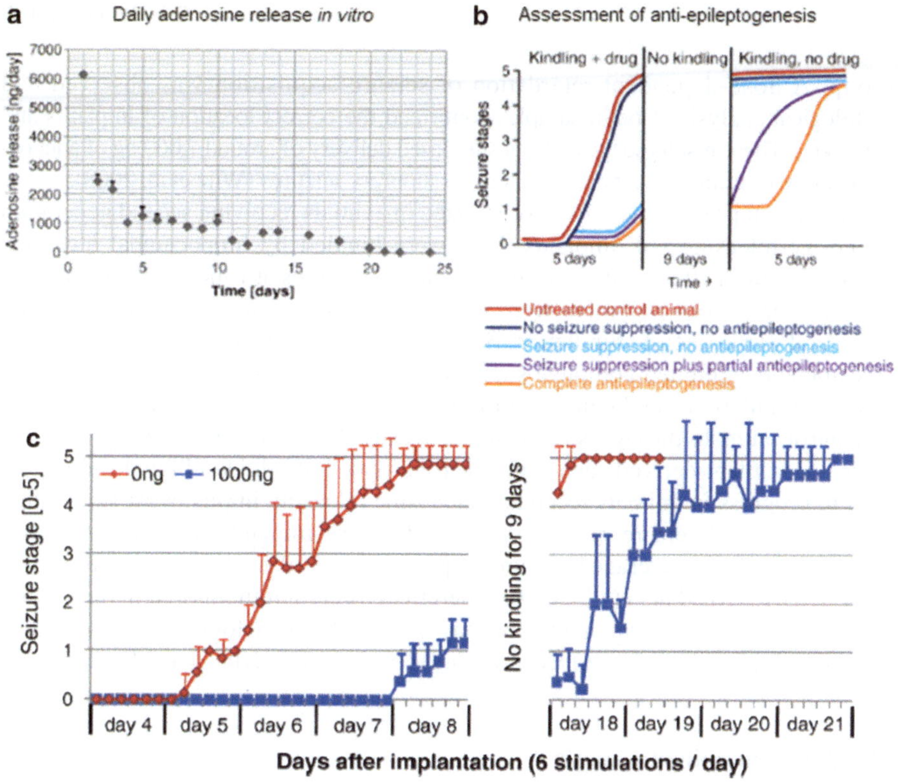

Fig. 28.3 Antiepileptogenesis study design and results (**a**) Daily release of adenosine from silk-based polymers. Adenosine release in vitro was determined for each day shown, based on averaged values from $N=3$ polymers. Note the stable release rate of around 1,000 ng adenosine per day from day 4 to 10, corresponding to the predesigned target release rate. Errors are given as ±SD. (**b**) Assessment of antiepileptogenesis in the rat kindling model according to Silver et al. (1991): kindling is initiated during drug delivery followed by a washout period of the drug; subsequently, kindling is resumed in the absence of the drug. Five potential kindling outcomes are shown: *red*, normal kindling development of an untreated or sham-treated animal; *dark blue*, kindling in the presence of a drug with no effects on seizure expression and epileptogenesis; *light blue* kindling development under the influence of a drug that suppresses seizures, but not epileptogenesis; *violet* kindling development under the influence of a drug that exerts partial antiepileptogenic effects; *orange* kindling under the influence of a drug that completely suppresses epileptogenesis. Note that the criterion for complete suppression of epileptogenesis compared with partial epileptogenesis is a shift of the kindling curve to the *right*; the number of drug-free kindling stimulations needed to trigger a specific seizure stage should be the same as in control animals. (**c**) Four days after implantation of silk-based polymers with daily target release rates for adenosine of 0 ng ($N=7$, *red*), or 1,000 ng ($N=5$, *blue*) kindling stimulations were delivered at a rate of six stimulations per day on days 4, 5, 6, 7, and 8 following implantation. A total of 30 kindling stimulations were delivered. After the 30th kindling stimulation, kindling was discontinued for 9 days to allow silk polymers to exhaust their adenosine release. Kindling stimulations were resumed at day 18. Seizure stages were averaged across animals from each group for each individual stimulus. Note that recipients of a target dose of 1,000 ng adenosine per day resumed kindling at day 18 at a level at which kindling was discontinued at day 8. After seven consecutive stage 5 seizures kindling was discontinued in control animals due to animal welfare considerations. Errors are given as ±SD (Modified from Szybala et al. 2009)

group. A dose-dependent retardation of seizure acquisition was observed and histological analysis of brain samples confirmed the correct location of implants and electrodes. These results suggested that silk-based delivery of around 1,000 ng of adenosine per day is a safe and efficient strategy to suppress seizures (Wilz et al. 2008).

Seizure suppression was also investigated in animals receiving second generation adenosine-releasing silk implants. Rats were kindled by hippocampal electrical stimulation until all animals reacted with stage 5 seizures at which point adenosine-releasing or control polymers were implanted into the infrahippocampal fissure ipsilateral to the site of stimulation. Subsequently, animals receiving 1,000 ng/day silk implants exhibited complete suppression of kindling development during the first 13 stimulations, and continued to display a suppression of seizures compared to control animals through stimulation 24. Only at day 11, when in vitro results indicated implant-derived adenosine release dropped off, did kindling development start to progress in animals receiving adenosine releasing implants. At stimulation 24, the averaged seizure response of the adenosine implant recipients was stage 1.3 ± 0.7, while the averaged seizure response observed in the control group was stage 3.3 ± 1.2 (Szybala et al. 2009). Side-effects were not monitored in these studies as focal adenosine augmentation has previously been found to be devoid of systemic and sedative side effects (Güttinger et al. 2005b; Nilsen and Cock 2004).

28.3.3.2 Antiepileptogenesis

To monitor seizure development in the presence of adenosine, adenosine releasing or control polymers were implanted prior to kindling. Animals receiving adenosine-releasing silk implants showed robust suppression of seizures during the first 5 days of stimulation and exhibited protection from convulsive seizures through 30 stimulations (average seizure score $= 1.3 \pm 0.5$). Recipients of control silk implants that received the same kindling stimulation exhibited an average seizure score of 4.9 ± 0.4 at stimulation 30 (day 8 post implantation).

After stimulation 30/day 8, kindling was discontinued for 9 days to allow the silk implants to exhaust their adenosine release and resumed at day 18. When kindling was resumed, all control animals continued to display stage 4 and 5 seizures, while recipients of adenosine-releasing silk implants resumed at stages similar to the levels recorded prior to kindling discontinuation (average score of 0.5 ± 0.6), then gradually increased in seizure severity until stage 4 and 5 seizures were reached at days 20 and 21 post implantation. The kindling curve of animals who received approximately 1,000 ng/day for 10 days following implant exhaustion resembled the kindling curve of recipients of control implants, but the number of stimulations required to produce comparable seizure responses to animals receiving control implants was lower than in those control rats (Fig. 28.3c). This suggests that focal adenosine delivery exerts partial antiepileptogenic effects (Fig. 28.3b) (Silver et al. 1991). This is significant because currently available antiepileptic drugs are largely limited to seizure suppression (antiictogenesis), failing to prevent disease progression (antiepileptogenesis) (Boison and Stewart 2009). The traditional symptomatic treatment approach has little prospect to affect the underlying causes of the disease. The focal adenosine delivery from silk implants has potential not only for seizure suppression, but also for preventative use in patients with high risk of developing epilepsy. For example, following traumatic brain injury, implantation of adenosine releasing silk scaffolds into the traumatized brain area could improve therapeutic outcomes for these patients.

28.4 Conclusions and Future Directions

These studies demonstrate that silk systems can deliver therapeutically relevant drug loads in vivo. Polymers prepared in different doses induced dose-dependent retardation of seizure acquisition, validating the therapeutic potential of implants and the ability to control dosing. Adenosine-releasing silk implants suppressed seizures in a rat kindling model of epilepsy, and also induced partial antiepileptogenesis. These studies demonstrated excellent agreement between in vitro release study predictions and in vivo response. Taken together, these results confirm the safety and efficacy of silk drug delivery implants for local delivery of neurological drugs, and suggest that these degradable, implantable biomaterial systems could potentially be applied to a range of therapeutics.

Adenosine is involved in nearly every aspect of cell function, including neuromodulation and neurotransmission (Gan and Habib 2007). Despite the abundance of evidence that manipulation of adenosine receptor activation might have significant therapeutic impact, the side effects of systemic adenosine administration have thus far limited adenosine's clinical usefulness. The epilepsy therapy based on adenosine-releasing silk biomaterials that has recently been reported (Wilz et al. 2008; Szybala et al. 2009) could potentially be extended to other adenosine-related pathophysiological applications, many of which have been reviewed (Ribeiro et al. 2003; Boison 2008).

References

Abletshauser CB, Schneider R, Rupprecht H (1993) Film coating of pellets with insoluble polymers obtained in situ crosslinking in the fluidized bed. J Control Release 21:149–156

Acharya C, Ghosh S, Kundu S (2008a) Silk fibroin protein from mulberry and non-mulberry silkworms: cytotoxicity, biocompatibility and kinetics of L929 murine fibroblast adhesion. J Mater Sci Mater Med 19:2827–2836

Acharya C, Kumar V, Sen R, Kundu SC (2008b) Performance evaluation of a silk protein-based matrix for the enzymatic conversion of tyrosine to L-DOPA. Biotechnol J 3:226–233

Altman GH, Diaz F, Jakuba C, Calabro T, Horan RL, Chen J, Lu H, Richmond J, Kaplan DL (2003) Silk-based biomaterials. Biomaterials 24:401–416

Arai T, Freddi G, Innocenti R, Tsukada M (2004) Biodegradation of *Bombyx mori* silk fibroin fibers and films. J Appl Polym Sci 91:2383–2390

Barcia JA, Gallego JM (2009) Intraventricular and intracerebral delivery of anti-epileptic drugs in the kindling model. Neurotherapeutics 6:337–343

Bayraktar O, Malay O, Özgarip Y, Batigün A (2005) Silk fibroin as a novel coating material for controlled release of theophylline. Eur J Pharm Biopharm 60:373–381

Bessa PC, Balmayor ER, Azevedo HS, Nürnberger S, Casal SM, van Griensven M, Reis RL, Redl H (2010a) Silk fibroin microparticles as carriers for delivery of human recombinant BMPs physical characterization and drug release. J Tissue Eng Regen Med 4:349–355

Bessa PC, Balmayor ER, Hartinger J, Zanoni G, Dopler D, Meinl A, Banerjee A, Casal M, Redl H, Reis RL, van Griensven M (2010b) Silk fibroin microparticles as carriers for delivery of human recombinant bone morphogenetic protein-2: *in vitro* and *in vivo* bioactivity. Tissue Eng Part C Methods 16:937–945

Boison D (2008) Adenosine as a neuromodulator in neurological diseases. Curr Opin Pharmacol 8:2–7

Boison D (2009) Engineered adenosine-releasing cells for epilepsy therapy: human mesenchymal stem cells and human embryonic stem cells. Neurotherapeutics 6:278–283

Boison D, Stewart K-A (2009) Therapeutic epilepsy research: from pharmacological rationale to focal adenosine augmentation. Biochem Pharmacol 78:1428–1437

Boison D, Scheurer L, Tseng JL, Aebischer P, Mohler H (1999) Seizure suppression in kindled rats by intraventricular grafting of an adenosine releasing synthetic polymer. Exp Neurol 160:164–174

Boison D, Huber A, Padrun V, Déglon N, Aebischer P, Möhlet H (2002) Seizure suppression by adenosine-releasing cells is independent of seizure frequency. Epilepsia 43:788–796

Brem H, Piantadosi S, Burger PC, Walker M, Selker R, Vick NA, Black K, Sisti M, Brem S, Mohr G et al (1995) Placebo-controlled trial of safety and efficacy of intraoperative controlled delivery by bio-degradable polymers of chemotherapy for recurrent gliomas. Lancet 345:1008–1012

Cao Y, Wang B (2009) Biodegradation of silk biomaterials. Int J Mol Sci 10:1514–1524

Cao Z, Chen X, Yao J, Huang L, Shao Z (2007) The preparation of regenerated silk fibroin microspheres. Soft Matter 3:910–915

Chen W, He J, Olson JJ, Lu DR (1998) Direct intracerebral delivery of carboplatin from PLGA microspheres against experimental malignant glioma in rats. Drug Deliv 5:101–110

Danckwerts M, Fassih A (1991) Implantable controlled release drug delivery systems: a review. Drug Dev Ind Pharm 17:1465–1502

Daugherty AL, Cleland JL, Duenas EM, Mrsny RJ (1997) Pharmacological modulation of the tissue response to implanted polylactic-co-glycolic acid microspheres. Eur J Pharm Biopharm 44:89–102

Dehdashti AR, Muster M, Reverdi A, de Tribolet V, Ruefenacht DA (2001) Preoperative silk suture embolization of cerebral and dural arteriovenous malformations. Neurosurg Focus 11:e6

Demura M, Asakura T, Kuroo T (1989) Immobilization of biocatalysts with *Bombyx mori* silk fibroin by several kinds of physical treatment and its application to glucose sensors. Biosensors 4:361–372

Dunwiddie TV (1999) Adenosine and suppression of seizures. Adv Neurol 79:1001–1010

Fedele DE, Koch P, Scheurer L, Simpson EM, Möhler H, Brüstle O, Boison D (2004) Engineering embryonic stem cell derived glia for adenosine delivery. Neurosci Lett 370:160–165

Fung LK, Ewend MG, Sills A, Sipos EP, Thompson R, Watts M, Colvin OM, Brem H, Saltzman WM (1998) Pharmacokinetics of interstitial delivery of carmustine, 4-hydroperoxycyclophosphamide, and paclitaxel from a biodegradable polymer implant in the monkey brain. Cancer Res 58:672–684

Furia T (1964) EDTA in foods—a technical review. Food Technol 18:1874–1882

Gan TJ, Habib AS (2007) Adenosine as a non-opioid analgesic in the perioperative setting. Anesth Analg 105:487–494

Gutman RL, Peacock G, Lu DR (2000) Targeted drug delivery for brain cancer treatment. J Control Release 65:31–41

Güttinger M, Fedele D, Koch P, Padrun V, Pralong WF, Brüstle O, Boison D (2005a) Suppression of kindled seizures by paracrine adenosine release from stem cell-derived brain implants. Epilepsia 46:1162–1169

Güttinger M, Padrun V, Pralong WF, Boison D (2005b) Seizure suppression and lack of adenosine A1 receptor desensitization after focal long-term delivery of adenosine by encapsulated myoblasts. Exp Neurol 193:53–64

Guziewic N, Best A, Perez-Ramirez B, Kaplan DL (2011) Lyophilized silk fibroin hydrogels for the sustained local delivery of therapeutic monoclonal antibodies. Biomaterials 32:2642–2645

Harbaugh RE, Saunders RL, Reeder RF (1988) Use of implantable pumps for central nervous system drug infusions to treat neurological disease. Neurosurgery 23:693–698

Hino T, Shimabayashi S, Nakai A (2000) Silk microspheres prepared by spray-drying of an aqueous system. Pharm Pharmacol Commun 6:335–339

Hino T, Tanimoto M, Shimabayashi S (2003) Change in secondary structure of silk fibroin during preparation of its microspheres by spray-drying and exposure to humid atmosphere. J Colloid Interface Sci 266:68–73

Hofmann S, Wong Po Foo CT, Rossetti F, Textor M, Vunjak-Novakovic G, Kaplan DL, Merkle HP, Meinel L (2006) Silk fibroin as an organic polymer for controlled drug delivery. J Control Release 111:219–227

Horan RL, Antle K, Collette AL, Wang Y, Huang J, Moreau JE, Volloch V, Kaplan DL, Altman GH (2005) *In vitro* degradation of silk fibroin. Biomaterials 26:3385–3393

Horan RL, Collette AL, Lee C, Antle K, Chen J, Altman GH (2006) Yarn design for functional tissue engineering. J Biomech 39:2232–2240

Huber A, Padrun V, Déglon N, Aebischer P, Möhler H, Boison D (2001) Grafts of adenosine-releasing cells suppress seizures in kindling epilepsy. Proc Natl Acad Sci U S A 98:7611–7616

Imsombut T, Srisuwan Y, Srihanam P, Baimark Y (2010) Genipin-cross-linked silk fibroin microspheres prepared by the simple water-in-oil emulsion solvent diffusion method. Powder Technol 203:603–608

Jain RA (2000) The manufacturing techniques of various drug loaded biodegradable poly(lactide-*co*-glycolide) (PLGA) devices. Biomaterials 21:2475–2490

Jin H-J, Park J, Karageorgiou V, Kim UJ, Valluzzi R, Cebe P, Kaplan DL (2005) Water-insoluble silk films with reduced β-sheet content. Adv Funct Mater 15:1241–1247

Kaplan DL, Mello CM, Arcidiacono S, Fossey S, Senecal K, Muller W (1997) Silk. In: McGrath K, Kaplan DL (eds) Protein-based materials. Birkhäuser, Boston, pp 103–131

Kim U-J, Park J, Li C, Jin H-J, Valluzzi R, Kaplan DL (2004) Structure and properties of silk hydrogels. Biomacromolecules 5:786–792

Kim U-J, Park J, Kim HJ, Wada M, Kaplan DL (2005) Three dimensional aqueous-derived biomaterial scaffolds from silk fibroin. Biomaterials 26:2775–2785

Kim D-H, Viventi J, Amsden JJ, Xiao J, Vigeland L, Kim Y-S, Blanco JA, Panilaitis B, Frechette ES, Contreras D et al (2010) Dissolvable films of silk fibroin for ultrathin conformal bio-integrated electronics. Nat Mater 9:511–517

Lammel AS, Hu X, Park S-H, Kaplan DL, Scheibel TR (2010) Controlling silk fibroin particle features for drug delivery. Biomaterials 31:4583–4591

Langer R (1980) Polymeric delivery systems for controlled drug release. Chem Eng Commun 6:1–48

Langer RS, Peppas NA (1981) Present and future applications of biomaterials in controlled drug delivery systems. Biomaterials 2:201–214

Lawrence BD, Cronin-Golomb M, Georgakoudi I, Kaplan DL, Omenetto FG (2008) Bioactive silk protein biomaterial systems for optical devices. Biomacromolecules 9:1214–1220

Leal-Egaña A, Scheibel T (2010) Silk-based materials for biomedical applications. Biotechnol Appl Biochem 55:155–167

Lebovits AH, Strain JJ, Messe MR, Schleifer SJ, Tanaka JS, Bhardwaj S (1990) Patient noncompliance with self-administered chemotherapy. Cancer 65:17–22

Li M, Ogiso M, Minoura N (2003) Enzymatic degradation behavior of porous silk fibroin sheets. Biomaterials 24:357–365
Li T, Steinbeck JA, Lusardi T, Koch P, Lan JQ, Wilz A, Segschneider M, Simon RP, Brüstle O, Boison D (2007) Suppression of kindling epileptogenesis by adenosine releasing stem cell-derived brain implants. Brain 130:1276–1288
Lu S, Wang X, Lu Q, Hu X, Uppal N, Omenetto FG, Kaplan DL (2009) Stabilization of enzymes in silk films. Biomacromolecules 10:1032–1042
Lu Q, Wang X, Hu X, Cebe P, Omenetto F, Kaplan DL (2010a) Stabilization and release of enzymes from silk films. Macromol Biosci 10:359–368
Lu Q, Hu X, Wang X, Kluge JA, Lu S, Cebe P, Kaplan DL (2010b) Water-insoluble silk films with silk I structure. Acta Biomater 6:1380–1387
Madduri S, Papaloïzos M, Gander B (2010) Trophically and topographically functionalized silk fibroin nerve conduits for guided peripheral nerve regeneration. Biomaterials 31:2323–2334
Matsumoto A, Chen J, Collette AL, Kim U-J, Altman GH, Cebe P, Kaplan DL (2006) Mechanisms of silk fibroin sol-gel transitions. J Phys Chem B 110:21630–21638
Maysinger D, Morinville A (1997) Drug delivery to the nervous system. Trends Biotechnol 15:410–418
Meinel L, Hofmann S, Karageorgiou V, Kirker-Head C, McCool J, Gronowicz G, Zichner L, Langer R, Vunjak-Novakovic G, Kaplan DL (2005) The inflammatory responses of silk films *in vitro* and *in vivo*. Biomaterials 26:147–155
Menei P, Daniel V, Montero-Menei C, Brouillard M, Pouplard-Barthelaix A, Benoit JP (1993) Biodegradation and brain tissue reaction to poly(D, L-lactide-co-glycolide). Biomaterials 14:470–478
Miyairi S, Sugiura M, Fukui S (1978) Immobilized β-glucosidase in fibroin membrane. Agric Biol Chem 42:1661–1667
Mundargi RC, Babu VR, Rangaswamy V, Patel P, Aminabhavi TM (2008) Nano/micro technologies for delivering macromolecular therapeutics using poly(D, L-lactide-co-glycolide) and its derivatives. J Control Release 125:193–209
Nilsen KE, Cock HR (2004) Focal treatment for refractory epilepsy: hope for the future? Brain Res Rev 44:141–153
Numata K, Kaplan DL (2010) Silk-based delivery systems of bioactive molecules. Adv Drug Deliv Rev 62:1497–1508
Pagonopoulou O, Efthimiadou A, Asimakopoulos B, Nikolettos NK (2006) Modulatory role of adenosine and its receptors in epilepsy: possible therapeutic approaches. Neurosci Res 56:14–20
Panilaitis B, Altman G, Chen J, Jin HJ, Karageorgiou V, Kaplan DL (2003) Macrophage responses to silk. Biomaterials 24:3079–3085
Patel MM, Goyal BR, Bhadada SV, Bhatt JS, Amin AF (2009) Getting into the brain: approaches to enhance brain drug delivery. CNS Drugs 23:35–58
Portas CM, Thakkar M, Rainnie DG, Greene RW, McCarley RW (1997) Role of adenosine in behavioral state modulation: a microdialysis study in the freely moving cat. Neuroscience 79:225–235
Practice Committee of the American Society for Reproductive Medicine (2004) Hormonal contraception: recent advances and controversies. Fertil Steril 82:520–526
Pritchard EM, Kaplan DL (2011) Silk fibroin biomaterials for controlled release drug delivery. Expert Opin Drug Deliv 8:797–811
Pritchard EM, Szybala C, Boison D, Kaplan DL (2010) Silk fibroin encapsulated powder reservoirs for sustained release of adenosine. J Control Release 144:159–167
Pritchard EM, Valentin T, Boison D, Kaplan DL (2011) Incorporation of proteinase inhibitors into silk-based delivery devices for enhanced control of degradation and drug release. Biomaterials 32:909–918
Puppi D, Chiellini F, Piras AM, Chiellini E (2010) Polymeric materials for bone and cartilage repair. Prog Polym Sci 35:403–440
Racine R (1978) Kindling: the first decade. Neurosurgery 3:234–252

Ren G, Li T, Lan JQ, Wilz A, Simon RP, Boison D (2007) Lentiviral RNAi-induced downregulation of adenosine kinase in human mesenchymal stem cell grafts: a novel perspective for seizure control. Exp Neurol 208:26–37

Ribeiro JA, Sebastião AM, de Mendonça A (2003) Adenosine receptors in the nervous system: pathophysiological implications. Prog Neurobiol 68:377–392

Ritger PL, Peppas NA (1987) A simple equation for description of solute release I. Fickian and non-fickian release from non-swellable devices in the form of slabs, spheres, cylinders or discs. J Control Release 5:23–36

Schmutz F, McAuliffe W, Anderson DM, Elliott JP, Eskridge JM, Winn HR (1997) Embolization of cerebral arteriovenous malformations with silk: histopathologic changes and hemorrhagic complications. AJNR Am J Neuroradiol 18:1233–1237

Seo Y, Yoon H, Song K, Kwon S, Lee H, Park Y, Park J (2009) Increase in cell migration and angiogenesis in a composite silk scaffold for tissue-engineered ligaments. J Orthop Res 27:495–503

Siepmann F, Siepmann J, Walther M, MacRae RJ, Bodmeier R (2008) Polymer blends for controlled release coatings. J Control Release 125:1–15

Silver JM, Shin C, McNamara JO (1991) Antiepileptogenic effects of conventional anticonvulsants in the kindling model of epilespy. Ann Neurol 29:356–363

Sofia S, McCarthy MB, Gronowicz G, Kaplan DL (2001) Functionalized silk-based biomaterials for bone formation. J Biomed Mater Res 54:139–148

Szybala C, Pritchard EM, Lusardi TA, Li T, Wilz A, Kaplan DL, Boison D (2009) Antiepileptic effects of silk-polymer based adenosine release in kindled rats. Exp Neurol 219:126–135

Tamargo RJ, Rossell LA, Kossoff EH, Tyler BM, Ewend MG, Aryanpur JJ (2002) The intracerebral administration of phenytoin using controlled-release polymers reduces experimental seizures in rats. Epilepsy Res 48:145–155

Tang X, Ding F, Yang Y, Hu N, Gi X (2009) Evaluation of *in vitro* biocompatibility of silk fibroin-based biomaterials with primarily cultured hippocampal neurons. J Biomed Mater Res A 91A:166–174

Tomomura A, Yamada H, Itagaki K, Fujimoto K, Katoh S (2002) Rat brain expresses serum calcium-decreasing factor (caldecrin). Neurosci Lett 317:17–20

Uebersax L, Fedele DE, Schumacher C, Kaplan DL, Merkle HP, Boison D, Meinel L (2006) The support of adenosine release from adenosine kinase deficient ES cells by silk substrates. Biomaterials 27:4599–4607

Uebersax L, Mattotti M, Papaloïzos M, Merkle HP, Gander B, Meinel L (2007) Silk fibroin matrices for the controlled release of nerve growth factor (NGF). Biomaterials 28:4449–4460

Van Dycke A, Raedt R, Verstraete A, Theofilas P, Wadman W, Vonck K, Boison D, Boon P (2010) Astrocytes derived from fetal neural progenitor cells as a novel source for therapeutic adenosine delivery. Seizure 19:390–396

VanDycke A, Raedt R, Vonck K, Boon P (2011) Local delivery strategies in epilepsy; a focus on adenosine. Seizure 20:376–382

Varde NK, Pack DW (2007) Influence of particle size and antacid on release and stability of plasmid DNA from uniform PLGA microspheres. J Control Release 124:172–180

Vepari C, Kaplan DL (2007) Silk as a biomaterial. Prog Polym Sci 32:991–1007

Wang PP, Frazier J, Brem H (2002) Local drug delivery to the brain. Adv Drug Deliv Rev 54:987–1013

Wang Y, Kim H-J, Vunjak-Novakovic G, Kaplan DL (2006) Stem cell-based tissue engineering with silk biomaterials. Biomaterials 2:6064–6082

Wang X, Hu X, Daley A, Rabotyagova O, Cebe P, Kaplan DL (2007a) Nanolayer biomaterial coatings of silk fibroin for controlled release. J Control Release 121:190–199

Wang X, Wenk E, Hu X, Castro GR, Meinel L, Wang X, Li C, Merkle H, Kaplan DL (2007b) Silk coatings on PLGA and alginate microspheres for protein delivery. Biomaterials 28:4161–4169

Wang X, Wenk E, Matsumoto A, Meinel L, Li C, Kaplan DL (2007c) Silk microspheres for encapsulation and controlled release. J Control Release 117:360–370

Wang X, Kluge JA, Leisk GG, Kaplan DL (2008a) Sonication-induced gelation of silk fibroin for cell encapsulation. Biomaterials 29:1054–1064
Wang Y, Rudym DD, Walsh A, Abrahamsen L, Kim HJ, Kim HS (2008b) *In vivo* degradation of three-dimensional silk fibroin scaffolds. Biomaterials 29:3415–3428
Wang X, Yucel T, Lu Q, Hu X, Kaplan DL (2010) Silk nanospheres and microspheres from silk/pva blend films for drug delivery. Biomaterials 31:1025–1035
Wenk E, Wandrey AJ, Merkle HP, Meinel L (2008) Silk fibroin spheres as a platform for controlled drug delivery. J Control Release 132:26–34
Wenk E, Merkle HP, Meinel L (2011) Silk fibroin as a vehicle for drug delivery applications. J Control Release 150:128–141
Whittlesey KJ, Shea LD (2004) Delivery strategies for drugs, proteins, and DNA: the neuroscience/biomaterial interface. Exp Neurol 190:1–16
Wilz A, Pritchard EM, Li T, Lan J-Q, Kaplan DL, Boison D (2008) Silk polymer-based adenosine release: therapeutic potential for epilepsy. Biomaterials 29:3609–3616
Wojcik WJ, Neff NH (1982) Adenosine measurement by a rapid HPLC-fluorometric method: induced changes of adenosine content in regions of rat brain. J Neurochem 39:280–282
Yang Y, Chen X, Ding F, Zhang P, Liu J, Gu X (2007a) Biocompatibility evaluation of silk fibroin with peripheral nerve tissues and cells *in vitro*. Biomaterials 28:1643–1652
Yang Y, Ding F, Wu J, Hu W, Liu W, Liu J, Gu X (2007b) Development and evaluation of silk fibroin-based nerve grafts used for peripheral nerve regeneration. Biomaterials 28:5526–5535
Yang Y, Zhao Y, Gu Y, Yan X, Liu J, Ding F, Gu X (2009) Degradation behaviors of nerve guidance conduits made up of silk fibroin *in vitro* and *in vivo*. Polym Degrad Stabil 94:2213–2220
Yeo J-H, Lee K-G, Lee Y-W, Kim SY (2003) Simple preparation and characteristics of silk fibroin microsphere. Eur Polym J 39:1195–1199
Zhou Y, Chu JS, Li JX, Wu XY (2010) Theoretical analysis of release kinetics of coated tablets containing constant and non-constant drug reservoirs. Int J Pharm 385:98–103

Chapter 29
Anatomical Distribution of Nucleoside System in the Human Brain and Implications for Therapy

Zsolt Kovács and Arpád Dobolyi

Abstract Nucleosides have a wide range of physiological and pathophysiological roles in the human brain as modulators of a variety of neural functions. For example, adenosine, inosine, guanosine, and uridine participate in the mechanisms underlying memory, cognition, sleep, pain, depression, schizophrenia, epilepsy, Alzheimer's disease, Huntington's disease, and Parkinson's disease. Consequently, increasing attention is now being given to the specific role of nucleosides in physiological and pathological processes in the human brain. Different elements of nucleoside system, including nucleoside concentrations, metabolic enzyme activity, and expression of nucleoside transporters and receptors, may be changed under normal and pathological conditions. The alterations suggest that interlinked elements of the nucleoside system are functioning in a tightly concerted manner.

Nucleoside levels, activity of nucleoside metabolic enzymes, and expression of nucleoside transporters and receptors are unevenly distributed in the brain, suggesting that nucleosides have different roles in functionally distinct human brain areas. The aim of this chapter is to summarize our present knowledge of the anatomical distribution of nucleoside system in the human brain, placing emphasis on potential therapeutic pharmacological strategies.

Keywords Nucleosides • Anatomical distribution of nucleoside system • Human brain diseases and therapy

Z. Kovács (✉)
Department of Zoology, University of West Hungary, Savaria Campus, Károlyi Gáspár tér 4, Szombathely 9700, Hungary
e-mail: zskovacs@ttk.nyme.hu

A. Dobolyi
Neuromorphological and Neuroendocrine Research Laboratory, Department of Anatomy, Histology and Embryology, Semmelweis University and the Hungarian Academy of Sciences, Budapest, Hungary

S. Masino and D. Boison (eds.), *Adenosine: A Key Link between Metabolism and Brain Activity*, DOI 10.1007/978-1-4614-3903-5_29,

Abbreviations

5′NT	5′-Nucleotidases
A_1 receptor/A_{2A} receptor/ A_{2B} receptor/A_3 receptor	$A_1R/A_{2A}R/A_{2B}R/A_3R$ subtype of adenosine receptors
AC	Adenylate cyclase
ADA	Adenosine deaminase
Ade	Adenine
ADK	Adenosine kinase
Ado	Adenosine
AMP	Adenosine monophosphate
CDP-choline	Cytidine diphosphocholine
cN	Cytoplasmic 5′-nucleotidases
CNS	Central nervous system
CNT transporters	Concentrative nucleoside transporters
CNT1/CNT2/CNT3 transporters	CNT1/CNT2/CNT3 subtype of concentrative nucleoside transporters
Cyd	Cytidine
EC	Extracellular
ENT transporters	Equilibrative nucleoside transporters
ENT1/ENT2/ENT3/ENT4 transporters	ENT1/ENT2/ENT3/ENT4 subtype of equilibrative nucleoside transporters
"es" nucleoside transporters	Equilibrative, NBTI sensitive type of ENT transporters
GABA	γ-Aminobutyric acid
GDA	Guanine deaminase
GMP	Guanosine monophosphate
Gn	Guanine
Guo	Guanosine
Hyp	Hypoxanthine
IMP	Inosine monophosphate
Ino	Inosine
NBTI	*S*-(4-nitrobenzyl)-6-thioinosine
PLC	Phospholipase C
PNP	Purine nucleoside phosphorylase
Urd	Uridine
Xn	Xanthine

29.1 Introduction

Nucleosides such as adenosine (Ado), guanosine (Guo), inosine (Ino), and uridine (Urd) have a role in the regulation of neuronal and glial functions in the brain (Burnstock et al. 2011; Dobolyi et al. 2011; Fields and Burnstock 2006; Haskó et al. 2004; Schmidt et al. 2007). In addition, nucleosides participate in physiological and pathophysiological processes in the brain, such as the regulation of sleep and memory, epilepsy, Parkinson's disease, and Alzheimer's disease (Dobolyi et al. 2011; Huang et al. 2011; Lopes et al. 2011; Sperlágh and Vizi 2011). Increasingly, nucleoside derivatives and uptake or metabolic inhibitors are being used in clinical or preclinical drug development for the treatment of different diseases, ranging from viral infections to neurodegenerative disorders (Boison 2011; Lopes et al. 2011; Parkinson et al. 2011).

Regional differences occur in the nucleoside system of the human central nervous system (CNS). Nucleoside levels, metabolic enzymes, transporters, and receptors are unevenly distributed in the human brain (Baldwin et al. 2005; Barnes et al. 2006; Dawson 1971; Fredholm et al. 2001; Jennings et al. 2001; Kovács et al. 1998, 2010a; Nagata et al. 1984; Norstrand et al. 1984; Norstrand and Glantz 1980; Pennycooke et al. 2001; Phillips and Newsholme 1979; Ritzel et al. 2001). In addition, nucleoside concentrations are dependent on age and gender (Kovács et al. 2010b). These results suggest region-, age-, and gender-dependent functions of nucleosides in the human brain. Correlations have been observed between the (1) *S*-(4-nitrobenzyl)-6-thioinosine (NBTI) binding site and the density of adenosine deaminase (ADA) immunoreactive neurons (Geiger and Nagy 1986), (2) regional differences in nucleoside levels and the nucleoside metabolic enzyme activities and distribution of adenosine receptors (Kovács et al. 2010a), (3) ENT1 subtype of equilibrative nucleoside transporters (ENT1) and A_1 adenosine receptor subtype (A_1R) density (Jennings et al. 2001), and (4) A_1R density and 5′-nucleotidase (5′NTs) levels (Fastbom et al. 1987). Interactions have also been observed between ADA and A_1Rs, resulting in the facilitation of agonist binding to A_1Rs and the enhancement of receptor functionality in the human caudate nucleus (Gracia et al. 2008). These results strengthen the hypothesis that the so-called "purinome" groups nucleoside and nucleotide receptors, transporters, metabolic enzymes and ligands together to organize purinergic signaling (Kovács and Dobolyi 2011; Volonté and D'Ambrosi 2009). Complex anatomical, biochemical, and pharmacological analyses of the purinome are necessary to understand the functions of nucleoside system and to develop novel and safe drugs to treat various CNS diseases.

The aim of this chapter is to summarize the anatomical distribution of the nucleoside system in the human brain and to examine their potential for the development of pharmacological therapies. We focus on four nucleosides, Ado, Ino, Guo, and Urd. The available knowledge regarding the physiological and/or pathophysiological role of other nucleosides in the human brain is too limited for comprehensive evaluation. We briefly summarize some relevant features of the brain nucleoside system. Then we describe the anatomical distribution of nucleoside levels, metabolic enzymes, transporters, and receptors. Finally, we discuss their potential as targets of pharmacological therapeutics.

29.2 Nucleosides in the Human Brain: Metabolism, Transporters, and Receptors

29.2.1 Metabolism

Ribonucleic acids (RNA) and deoxyribonucleic acids (DNA) are synthesized from nucleotides that are composed of nucleosides and phosphate moieties. Nucleosides contain purine or pyrimidine bases connected to a pentose moiety. The major purine ribonucleosides are Ado, Guo, Ino, while the major pyrimidine ribonucleosides are cytidine (Cyd), Urd, and thymidine (Thd) (Linden and Rosin 2006). Nucleosides are synthesized de novo in the liver and can be partly obtained from food. They are transported into the brain and metabolized to their corresponding nucleotides. De novo synthesis of nucleosides in the adult brain is limited. Therefore, a salvage mechanism in the brain preserves the purine and pyrimidine nucleosides and bases. The main precursors of nucleotides in the brain are Ado, adenine (Ade), hypoxanthine (Hyp), guanine (Gn), Urd, and Cyd. To maintain the synthesis of ribo- and deoxyribonucleotides, hypoxanthine phosphoribosyltransferase (HGPRT; hypoxanthine-guanine phosphoribosyltransferase) catalyzes the conversion of Hyp-inosine monophosphate (IMP) and Gn-guanosine monophosphate (GMP; Fig. 29.1). Adenosine kinase (ADK) converts Ado to adenosine monophosphate (AMP), but Ado can also be metabolized to IMP in salvage reactions. Ade is metabolized to AMP by the adenine phosphoribosyltransferase (APRT) salvage enzyme. Cytidine deaminase (CDA) and uridine-cytidine kinase (UCK) salvage Cyd and Urd (Ipata et al. 2011).

The degradation pathway of adenine nucleotides in the brain can convert AMP to IMP-Ino-Hyp or Ado-Ino-Hyp (Fig. 29.1). These metabolic steps are catalyzed by cytoplasmic 5′-nucleotidases (cN, 5′NT), AMP deaminase (AMPDA), ADA, and purine nucleoside phosphorylase (PNP). *S*-adenosylhomocysteine (SAH) can be converted to Ado by adenosylhomocysteinase (SAHH, *S*-adenosylhomocysteine hydrolase). The main route of guanine-ribonucleotide catabolism is the GMP-Guo-Gn-xanthine (Xn) pathway catalyzed by cN, PNP, and guanine deaminase

→

Fig. 29.1 (continued) deaminase; I: Nucleoside transporters; II: ATP channels and transporters; III: K^+ channels; IV: Ca^{2+}-channels; A_1, A_{2A}, A_{2B} and A_3, A_4 Adenosine receptors types; *AC* Adenylate cyclase; *ADAi* Adenosine deaminase inhibitors; *Ade* Adenine; *AdeR* Adenine receptor; *ADKi* Adenosine kinase inhibitors; *Ado* Adenosine; *ADP* Adenosine diphosphate; *AMP* Adenosine monophosphate; *ATi* Adenosine transporter inhibitors; *ATP* Adenosine triphosphate; *cAMP* Cyclic adenosine monophosphate; *DAG* Diacylglycerol; G_i, G_0, G_s, G_q, G_{olf}: G-proteins (f.e. G_i: Inhibitory, G_s: Stimulatory); *GMP* Guanosine monophosphate; *Gn* Guanine; *GTP* Guanosine triphosphate; *Guo* Guanosine; *GuoR* Guo receptor; *Hyp* Hypoxanthine; *IMP* Inosine monophosphate; *Ino* Inosine; IP_3 Inositol 1,4,5-triphosphate; *MAPK* Mitogen-activated protein kinase; *MTA* 5′-deoxy-5′-methylthioadenosine; *PIP2* Phosphatidylinositol bisphosphate; *PKA* Protein kinase A; *PKC* Protein kinase C; *PLC* Phospholipase C; *SAH* *S*-adenosylhomocysteine; *sNUC* Synthetic nucleosides/nucleoside analogues; *UA* Uric acid; *UMP* Uridine monophosphate; *Ura* Uracil; *Urd* Uridine; *UrdR* Urd receptor; *UTP* Uridine triphosphate; *Xn* Xanthine; *XOi* Xanthine oxidase inhibitors

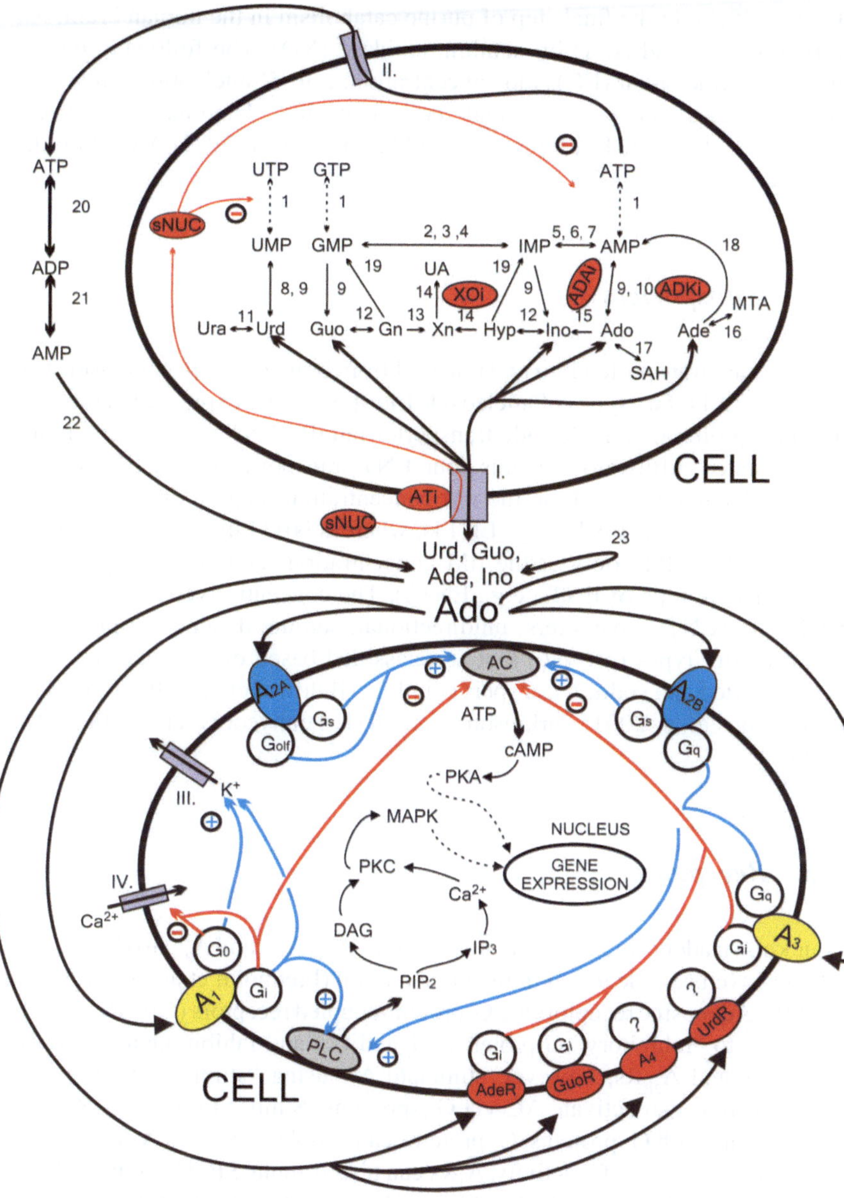

Fig. 29.1 Nucleoside production, transport and receptor signaling. *Abbreviations*: 1: Nucleoside mono- and diphosphate kinases and nucleoside di- and triphosphate phosphatases; 2: *GMPR* GMP reductase; 3: *GMPS* GMP synthetase; 4: *IMPDH* IMP dehydrogenase; 5: *AMPDA* AMP deaminase; 6: *ASL* Adenylosuccinate lyase; 7: *ASS* Adenylosuccinate synthetase; 8: *UCK* Uridine-cytidine kinase; 9: *5′NT* 5′-Nucleotidase; 10: *ADK* Adenosine kinase; 11: *UP* Uridine phosphorylase; 12: *PNP* Purine nucleoside phosphorylase; 13: *GDA* Guanine deaminase; 14: *XO* Xanthine oxidase; 15: *ADA* Adenosine deaminase; 16: *MTAP* 5′-deoxy-5′-methylthioadenosine phosphorylase; 17: *SAHH* S-adenosylhomocysteine hydrolase; 18: *APRT* Adenine phosphoribosyltransferase; 19: *HGPRT* Hypoxanthine phosphoribosyltransferase (hypoxanthine-guanine phosphoribosyltransferase); 20: ecto-ATPase; 21: ecto-ADPase; 22: *ecto-5′NT* ecto-5′-nucleotidase (eN); 23: *ecto-ADA* ecto-adenosine

(GDA; Fig. 29.1). In the final step of purine catabolism in the human brain, Xn is converted to uric acid (UA) by xanthine oxidase (XO). The following enzymes regulate the extracellular (EC) Ado concentration: ecto-5′-nucleotidase (eN), ecto-adenosine kinase (ecto-ADK), and ecto-adenosine deaminase (ecto-ADA) (Fernández et al. 2010; Firestein et al. 1999; Ipata et al. 2011; Yegutkin 2008; Zimmermann 1996) (Fig. 29.1).

29.2.2 *Transporters*

Nucleosides are transported into and released from brain cells via nucleoside transporters (Fig. 29.1). Two types of nucleoside transporters are expressed in the human brain. The equilibrative nucleoside transporter family (ENT transporters; bidirectional facilitated diffusion) contains four ENT transporter types: ENT1–ENT4. NBTI partially inhibits ENTs at the nM concentration range ("es": equilibrative, NBTI sensitive type of ENTs, e.g., ENT1), whereas NBTI insensitive transporters are inhibited by NBTI only at the μM concentration range ("ei": equilibrative, NBTI insensitive type of ENTs, e.g., ENT2). The concentrative nucleoside transporter family (CNT transporters; unidirectional, sodium-dependent) includes six CNT transporter types (N1–N6) that are classified based on the types of nucleosides transported and sodium transport coupling (Baldwin et al. 2005; Barnes et al. 2006; Jennings et al. 2001; Parkinson et al. 2011; Pennycooke et al. 2001; Ritzel et al. 2001).

29.2.3 *Receptors*

All four known adenosine receptor subtypes (A_1, A_{2A}, A_{2B}, and A_3: also known as P1 receptors) have been identified in the human brain (Fredholm et al. 2001; Jennings et al. 2001). Adenosine receptors are G-protein-coupled receptors (GPCR; Fig. 29.1). A_1Rs couple to "inhibitory" G-proteins (G_i and G_0) and inhibit adenylate cyclase (AC). A_{2A}Rs and A_{2B}Rs, however, stimulate AC using "stimulatory" G-proteins (G_S). A_{2A}Rs may also activate AC via G_{olf}-proteins. Similar to A_1Rs, A_3Rs inhibit AC by coupling with G_i-proteins. G_q proteins can couple to A_{2B} and A_3Rs and stimulate phospholipase C (PLC) activity. A_1Rs can also stimulate PLC and modulate the activity of K^+ and Ca^{2+} channels. In addition, the existence of yet unidentified nucleoside receptors cannot be excluded. For example, a novel subtype of adenosine receptors (A_4) has been proposed based on electrophysiological and pharmacological criteria in the brain (Cornfield et al. 1992; Luthin and Linden 1995; Tucker and Linden 1993). It is also conceivable that Urd, Guo, and Ade have their own receptors (UrdR, GuoR, AdeR, respectively; Fig. 29.1) that are used to execute certain functions in the nervous system (Bender et al. 2002; Borrmann et al. 2009; Kimura et al. 2001; Schulte and Fredholm 2003; Traversa et al. 2002).

29.3 Anatomical Distribution of the Nucleoside System in the Human Brain

29.3.1 Distribution of Nucleoside Levels

The concentration of nucleotide triphosphates, such as adenosine triphosphate (ATP), guanosine triphosphate (GTP), uridine triphosphate (UTP), and cytidine triphosphate (CTP), are 2–3 orders of magnitude higher (0.2–5 mM) in the human brain than that of nucleosides are. Consequently, the degradation of nucleotide triphosphates (Fig. 29.1) may increase the levels of corresponding nucleosides over baseline concentrations. For example, a 5–60 min period of ischemia was found to cause rapid degradation of nucleotide triphosphates and increase the concentrations of nucleosides and their metabolites (Ado, Guo, Ino, Hyp, and Xn) by 2–150 times that of baseline (Berne et al. 1974; Bjerring et al. 2010; Eells and Spector 1983; Hagberg et al. 1987; Kovács et al. 2010a; Melani et al. 2003; Traut 1994).

Both animal and human experiments have determined that nucleoside concentrations are unevenly distributed in different brain areas (Kékesi et al. 2006; Kovács et al. 2010a, 2011). Kovács and colleagues (Kovács et al. 2005) developed an extrapolation method that allows realistic estimates of the in vivo nucleoside levels from postmortem frozen and microwave-treated brain bank samples. Using this method, a nucleoside map of the human brain, consisting of 61 brain and 4 spinal cord areas, was constructed. High Ado (15.9–23.9 pmol/mg), Urd (44.1–66.2 pmol/mg), Ino (107.7–161.5 pmol/mg), and Guo (17.7–26.4 pmol/mg) concentrations were observed in several regions, including the cochlear nuclei, vestibular nuclei, cerebellar cortex, supraoptic nucleus, flocculonodular lobe, spinal trigeminal nucleus, temporal and occipital cortices, caudate nucleus, nucleus basalis, medial geniculate body, amygdala, spinal central gray, and ventral horn of the spinal cord (Table 29.1). The lowest concentrations of Ado (1.4–7.9 pmol/mg), Urd (15.7–22.0 pmol/mg), Ino (29.8–53.8 pmol/mg), and Guo (4.1–8.8 pmol/mg) were measured in the entorhinal cortex, septum, habenula, zona incerta, substantia nigra, locus coeruleus, preoptic area, pulvinar, and inferior colliculus (Table 29.1). Nucleoside metabolites such as Hyp, Xn, and uracil/Ura, (Fig. 29.1) were also unevenly distributed in the human brain (Kovács et al. 2010a).

Age and gender may modulate nucleoside expression. For example, the levels of Ino and Ado in the frontal cortex increase with age. Urd, Ino, and Guo concentrations are higher in the frontal cortex and white matter of middle-aged women when compared to middle-aged men, whereas Ado levels are lower in the frontal cortex of both middle-aged and elderly women when compared to men (Kovács et al. 2010b). These results suggest that the nucleoside microenvironment in the human brain may be an important factor in the aging processes and nucleosides might play a part in the reduced vulnerability of female brains to excitotoxic insults (Kovács et al. 2010b).

Table 29.1 Levels of nucleosides, activity of some nucleoside metabolizing enzymes, and relative density of nucleoside transporters and adenosine receptors in the human CNS

Anatomical distribution of nucleoside system	
Nucleosides in the CNS	
Nucleosides	*Nucleoside levels (pmol/mg wet weight)*
Ado	High (15.9–23.9): cochlear nuclei, vestibular nuclei, cerebellar cortex, supraoptic nucleus, flocculonodular lobe Intermediate (8.0–15.8): spinal cord (ventral and dorsal horn)$^{+}$, amygdala^{+}, temporal^{+}, and prefrontal cortex^{+}, caudate nucleus^{+}, mediodorsal thalamic nucleus^{+} Low (1.4–7.9): frontal, somatosensory, cingulate, and entorhinal cortex; hippocampus, nuclei of diagonal band, septum, globus pallidus externa, ventral lateral nucleus, habenula, pulvinar, zona incerta, preoptic area, paraventricular nucleus, dorsomedial nucleus (hypothalamus), lateral hypothalamic area, substantia nigra, inferior colliculus, locus coeruleus, dorsal vagal nuclei, nucleus accumbens^{+}, spinal central gray[1]
Ino	High (107.7–161.5): cochlear nuclei, spinal trigeminal nucleus Intermediate (53.9–107.6): frontal, temporal, somatosensory^{+}, prefrontal^{+}, cingulate^{+}, and occipital cortex; caudate nucleus, substantia innominata, nucleus basalis, nucleus accumbens^{+}, reticular formation (medulla oblongata), amygdala^{+}, cerebellar nuclei, spinal cord (ventral and dorsal^{+} horn), mediodorsal thalamic nucleus^{+}, spinal cord (white matter) Low (29.8–53.8): entorhinal and parahippocampal cortex; hippocampus, nuclei of diagonal band, habenula, pulvinar, zona incerta, paraventricular nucleus, substantia nigra, inferior colliculus, locus coeruleus[1]
Guo	High (17.7–26.4): cochlear nuclei; temporal and occipital cortex; caudate nucleus, nucleus basalis, medial geniculate body, amygdala^{+} Intermediate (8.9–17.6): insular, prefrontal^{+}, entorhinal^{+}, cingulate^{+}, and somatosensory cortex^{+}; white matter (cerebral and cerebellar), nuclei of diagonal band, substantia innominata, lateral geniculate body, hippocampus^{+}, nucleus accumbens^{+}, cerebellar nuclei, mediodorsal thalamic nucleus^{+}, spinal cord (ventral and dorsal^{+} horn) Low (4.1–8.8): septum, habenula, pulvinar, zona incerta, paraventricular nucleus, lateral hypothalamic area, substantia nigra, superior colliculus, inferior colliculus, locus coeruleus, spinal cord (white matter)[1]
Urd	High (44.1–66.2): cochlear nuclei, temporal and occipital cortex, cerebellar cortex, amygdala^{+}, spinal central gray, spinal cord (ventral horn) Intermediate (22.1–44.0): cerebral and cerebellar white matter, somatosensory^{+}, prefrontal^{+}, cingulate^{+}, insular and entorhinal cortex; hippocampus^{+}, caudate nucleus, globus pallidus externa, anterior nuclei (thalamus), substantia nigra, inferior colliculus, nucleus accumbens^{+}, locus coeruleus, inferior olive, reticular formation (medulla oblongata), cerebellar nuclei, mediodorsal thalamic nucleus^{+}, spinal cord (white matter), spinal cord (dorsal horn)$^{+}$ Low (15.7–22.0): ventral anterior nucleus, zona incerta, preoptic area, motor facial nucleus[1]

(continued)

Table 29.1 (continued)

Metabolic enzymes of nucleosides in the CNS	
Enzymes	*Activity level*
5′NT	nmol/h/mg protein: High (749–1,123): temporal cortex, thalamus (medial and lateral), colliculus superior Intermediate (375–748): parietal lobe, cingulate cortex, insula, caudate nucleus, putamen, pallidum (internal), claustrum, thalamus (anterior), subthalamic nucleus, nucleus ruber, substantia nigra, amygdala, hypothalamus, midbrain (paramedian) Low (210–374): cerebellar cortex, lateral geniculate body, pallidum (external), centrum semiovale, corpus callosum, mamillary body, internal capsule[2]
ADA	nmol of ammonia/min/g of wet weight: High (387–579): white matter of frontal, orbital and temporal lobe Intermediate (194–386): gray matter of frontal, occipital, orbital, parietal and temporal lobe; pons, putamen, hippocampus, caudate nucleus, globus pallidus, thalamus, midbrain, cerebellar white matter, white matter of parietal, cingulate, and occipital lobe; corpus callosum Low (16–193): gray matter of cingulate cortex and cerebellum; hypothalamus, medulla oblongata, spinal cord[3]
ADK	nmol/min/g wet weight: High (16.4–19.4): hypothalamus, pons, hind brain Intermediate (13.1–16.3): cerebellum, temporal cortex, corpus callosum, occipital cortex Low (9.8–13.0): parietal lobe, frontal cortex[4]
PNP	Substrate transformed (μmol)/min/g wet weight: High (223–261): pons, midbrain, thalamus, white and gray matter of occipital lobe, amygdala Intermediate (183–222): caudate nucleus, white matter of cerebellum, medulla oblongata, white matter of frontal lobe, gray matter of temporal, parietal, and frontal lobe; corpus callosum Low (143–182): gray matter of cerebellum, white matter of temporal and parietal lobe, putamen, spinal cord[5]
GDA	Substrate transformed (μmol)/min/mg protein High (12.9–19.2): thalamus, mamillary body Intermediate (6.5–12.8): parietal cortex, caudate nucleus, putamen, pons (basis), hippocampus, substantia nigra Low (0.005–6.4): cerebellum, olivary nucleus, corpus callosum, lateral geniculate body[6]

Nucleoside transporters in the CNS	
Transporter family (gene)	*Relative density (by comparison of different brain areas with each other)*
ENT1 (SLC29A1)	High: frontal and parietal cortex Intermediate: temporal and occipital cortex, thalamus, midbrain, caudate nucleus, putamen, globus pallidus Low: medulla oblongata, pons, cerebellum, hippocampus[7]
ENT2 (SLC29A2)	High: midbrain, pons, cerebellum Intermediate: medulla oblongata, thalamus Low: frontal, occipital, temporal, and parietal cortex; hippocampus, caudate nucleus, putamen, globus pallidus[7]

(continued)

Table 29.1 (continued)

Nucleoside transporters in the CNS	
ENT3 (SLC29A3)	High: occipital and temporal lobe, corpus callosum, medulla oblongata, putamen Intermediate: frontal lobe, paracentral gyrus, pons, hippocampus, nucleus accumbens, thalamus, spinal cord, cerebellum (right) Low: parietal lobe, cerebellum (left), amygdala, caudate nucleus, substantia nigra, pituitary gland[8]
ENT4 (SLC29A4)	High: temporal lobe, paracentral gyrus, amygdala, caudate nucleus, hippocampus, medulla oblongata, putamen Intermediate: parietal and occipital lobe, pons, cerebellum (right), corpus callosum, thalamus, pituitary gland, spinal cord, substantia nigra, nucleus accumbens Low: frontal lobe, cerebellum (left)[9]
CNT1 (N2/cit; SLC28A1)	Uniform distribution[10]
CNT2 (N1/cif; SLC28A2)	High: cerebellum, putamen, hippocampus, medulla oblongata Intermediate/low: amygdala, cerebral cortex, frontal, occipital, and temporal lobe; substantia nigra, thalamus, spinal cord[10]
CNT3 (N3/cib; SLC28A3)	High: hippocampus, medulla oblongata, pituitary gland Intermediate/low: frontal, parietal and occipital lobe; corpus callosum, cerebellum, amygdala, caudate nucleus, putamen, thalamus, temporal lobe, paracentral gyrus, pons, substantia nigra, nucleus accumbens, spinal cord[11]

Adenosine receptors in the CNS	
Receptor type	*Relative density (by comparison of different brain areas with each other)*
A_1	High: frontal, parietal and occipital cortex; caudate nucleus, putamen, globus pallidus Intermediate: temporal cortex, thalamus, hippocampus Low: medulla oblongata, midbrain, pons, cerebellum[7]
A_{2A}	High: caudate nucleus, putamen, globus pallidus, nucleus accumbens[7,12] Intermediate/low: frontal, temporal, parietal, and occipital cortex; thalamus, hippocampus, medulla oblongata, midbrain, pons, cerebellum[7]
A_{2B}	Uniform distribution[13]
A_3	High: cerebellum, hippocampus Intermediate/low: other brain areas[13]

The levels of nucleosides in brain and spinal cord areas were compared to the grand average concentration values of the total brain and spinal cord areas (Kovács et al. 2010a). We also listed some brain and spinal cord areas, which are implicated in particular CNS diseases, even though their nucleoside levels did not differ from average values (these brain areas are labeled by "+")

References: [1]Kovács et al. 2010a; [2]Nagata et al. 1984; [3]Norstrand et al. 1984; [4]Phillips and Newsholme 1979; [5]Norstrand and Glantz 1980; [6]Dawson 1971; [7]Jennings et al. 2001; [8]Baldwin et al. 2005; [9]Barnes et al. 2006; [10]Pennycooke et al. 2001; [11]Ritzel et al. 2001; [12]Svenningsson et al. 1997; [13]Fredholm et al. 2001; *Abbreviations*: Nucleosides—*Ado* Adenosine, *Guo* Guanosine, *Ino* Inosine, *Urd* Uridine; Nucleoside metabolizing enzymes—*5′NT* 5′-Nucleotidase, *ADA* Adenosine deaminase, *ADK* Adenosine kinase, *GDA* Guanine deaminase, *PNP* Purine nucleoside phosphorylase. For nucleoside transporter and nucleoside receptor abbreviations, see text

29.3.2 Distribution of Nucleoside Metabolic Enzymes

Nucleoside metabolic enzymes form a complex network, including several alternative metabolic pathways (Ipata et al. 2011; Kovács et al. 2011) (Fig. 29.1). The distribution and activity of nucleoside metabolic enzymes are uneven in the human brain, reflecting spatial differences in the nucleoside metabolic network. The distribution of 5′NTs, ADA, ADK, PNP, and GDA (Fig. 29.1) activities in the human brain have been previously described (Dawson 1971; Nagata et al. 1984; Norstrand et al. 1984; Norstrand and Glantz 1980) (Table 29.1).

The activities of 5′NT, PNP, and GDA are high in the thalamus (Table 29.1). High to intermediate activity of 5′NT was found in several brain regions, including the temporal cortex, colliculus superior, basal ganglia, nucleus ruber, substantia nigra, amygdala, and hypothalamus. In contrast, the cerebellar cortex, lateral geniculate body, pallidum, corpus callosum, and mamillary body showed low activity levels of this enzyme.

Interestingly, the white matter of frontal, orbital, and temporal lobes contain the highest ADA activity, while only intermediate activity has been observed in the gray matter of these brain areas (Table 29.1). An intermediate ADA activity was also found, e.g., in the basal ganglia, pons, hippocampus, and thalamus. On the contrary, low ADA activity was measured in the cerebellum, hypothalamus, medulla oblongata, and spinal cord. Others have observed the highest level of ADA activity in the hypothalamus (Phillips and Newsholme 1979).

High ADK activity has been found in the hypothalamus, pons, and hind brain. The temporal and occipital cortices and cerebellum show intermediate ADK activity, whereas the parietal lobe and frontal cortex contain low levels of this enzyme (Table 29.1). GDA activity is also high in the mamillary body. Intermediate GDA activity was measured in the parietal cortex, basal ganglia, substantia nigra, and hippocampus, with the lowest activity in the cerebellum. High PNP activity was also revealed in the pons, midbrain and amygdala whereas low enzyme activity was demonstrated, e.g., in the putamen and spinal cord.

Spatial differences in the distribution of nucleosides are correlated with nucleoside metabolic enzyme activities and the neuron–glia ratio in the human brain (Kovács et al. 2010a). Nucleoside metabolism is different in neuronal and glial cells (Ceballos et al. 1994; Zoref-Shani et al. 1995). Consequently, alterations in the glia/neuron may cause regional differences in nucleoside levels. However, the correlation between the neuron–glia ratio and nucleoside levels in the human brain is weak (Kovács et al. 2010a). Importantly, the neuron–glia ratio is changed in some brain areas implicated in the development of major depressive and bipolar disorders, schizophrenia, Huntington's and Alzheimer's disease, and frontotemporal dementia (Bowley et al. 2002; Brauch et al. 2006; Harper et al. 2008; Öngür et al. 1998; Roos et al. 1985).

Table 29.1 shows that altered nucleoside metabolic enzyme activity may result in an uneven distribution of nucleosides and their metabolites in the human brain (Kovács et al. 2010a). For example, high or intermediate 5′NT activity and low or intermediate ADA/ADK activity can generate elevated Ado, Ino, and Guo levels

(Fig. 29.1) in the temporal cortex and caudate nucleus (Table 29.1). High 5′NT, PNP and GDA activities may result in low Guo levels in thalamic areas, such as the habenula, pulvinar, and zona incerta (Table 29.1).

Altogether, these results suggest that the uneven distribution of nucleoside levels may be due to complex interactions between regionally different glia–neuron ratios and nucleoside metabolic enzyme activities.

29.3.3 Distribution of Nucleoside Transporters

The distribution of nucleoside transporters in the human brain is uneven, and the regionally different distribution of nucleoside transporters reflects the functional significance of nucleoside neuromodulation in different brain areas (Baldwin et al. 2005; Barnes et al. 2006; Jennings et al. 2001; Pennycooke et al. 2001; Ritzel et al. 2001).

ENT1 expression is high in the frontal and the parietal cortices, whereas the occipital and temporal lobe shows the highest ENT3 activity and high or intermediate ENT4 activity (Table 29.1). Intermediate or low ENT3 and ENT4 density occurs in the frontal and parietal lobes. Low levels of ENT1 expression are found in the medulla oblongata and the pons, whereas these brain areas show high to intermediate ENT2 expression. ENT2 expression is low in cortical areas and the basal ganglia. All ENT transporters are expressed at intermediate levels in the thalamus. The hippocampus shows low or intermediate ENT levels with the exception of ENT4, which is expressed at high levels in this brain area.

CNT transporters are also widely distributed in the human brain. Relatively high expression of CNT subtypes (CNT1, CNT2, and CNT3) occurs in the cerebellum, putamen, hippocampus, and medulla oblongata (Table 29.1).

29.3.4 Distribution of Nucleoside Receptors

The distribution of adenosine receptors in the brain reflects the physiological activity and effects of Ado in brain structures, whereas changes in the density of adenosine receptors may indicate functional and pathological changes (Boison 2005; Fastbom et al. 1986, 1987; Jenner et al. 2009).

Adenosine receptors are unevenly distributed in the human brain (Fredholm et al. 2001; Jennings et al. 2001) (Table 29.1). High expression of A_1Rs has been measured in several cerebral cortical areas and the basal ganglia. The temporal cortex, thalamus, and hippocampus contain intermediate levels of A_1Rs, whereas the cerebellum, midbrain, pons, and medulla oblongata show low density of this adenosine receptor type. $A_{2A}Rs$ are expressed at high levels in the basal ganglia and high A_3R density occurs in the cerebellum and hippocampus. In other brain areas, A_{2A} and A_3Rs are expressed at lower levels. Uniform distribution of $A_{2B}Rs$, however, has been shown to occur.

29.4 Implications for Therapy

Drugs acting on the nucleoside system are widely used for therapeutic purposes (Table 29.2, Fig. 29.1). Nucleoside metabolic enzyme inhibitors are used in anticancer therapies and the treatment of gout. In addition, several different nucleoside transport inhibitors are used as coronary vasodilators. Drugs acting on adenosine receptors are also used as vasodilators and to treat cardiac arrhythmias, carcinomas, rheumatoid arthritis, acute renal failure, and asthma. In addition, some synthetic nucleosides (nucleoside drugs) are used in antiviral and anticancer therapies.

Some drugs acting on the adenosine system have already been tested for the potential to treat brain disorders (Table 29.2). Based on its distribution and physiological roles in the CNS, the adenosine system has much wider potential for the treatment of pain, movement and mood disorders, schizophrenia, epilepsy, drug addiction, insomnia, multiple sclerosis, dementias, and stroke. Guanosine and Ino may also be neuroactive purines with therapeutic potential (Deutsch et al. 2005; Schmidt et al. 2007). The recent discovery of pyrimidine nucleotide receptors and the emerging neural functions of Urd imply that this pyrimidine nucleoside could also have therapeutic applications in the future (Cansev 2006; Connolly and Duley 1999; Dobolyi et al. 2011).

In the following sections, we discuss several neurological disorders where drugs acting on the nucleoside system may have therapeutic potential (Table 29.2).

29.4.1 Movement Disorders

The initiation of movement is governed by the interaction of the motor cortex, the thalamus, and a circuit consisting of several members of the basal ganglia, including the striatum, globus pallidus, and substantia nigra. The underlying pathologies for Parkinson's and Huntington's diseases are loss of nigrostriatal dopaminergic cells and degeneration of GABA/enkephalin neurons projecting from the striatum to the external globus pallidus, respectively (Harris et al. 2009).

Nucleosides and nucleoside metabolic enzymes are found in brain areas involved in movement disorders (Dawson 1971; Kovács et al. 2010a; Nagata et al. 1984; Norstrand et al. 1984; Norstrand and Glantz 1980) (Table 29.1). Nucleoside transporters are present in the caudate nucleus, putamen, globus pallidus, and substantia nigra (Barnes et al. 2006; Jennings et al. 2001; Ritzel et al. 2001). Caudate nucleus, putamen, and globus pallidus contain high levels of A_1 and A_{2A}Rs (Jennings et al. 2001). In particular, striatopallidal GABAergic enkephalin-containing neurons in the basal ganglia show the highest expression of A_{2A}Rs (Durieux et al. 2011; Popoli et al. 2007). These A_{2A}Rs tightly interact structurally and functionally with the dopamine D2 receptor and have been suggested to drive striatopallidal output balance (Xu et al. 2005).

Table 29.2 Some of the drugs (licensed or in preclinical/clinical stages of development) that modulate the nucleoside system and their potential therapeutic applications in the brain

Nucleoside system and therapy

Based on inhibition of nucleoside metabolic emzymes

Enzyme inhibitor name: (pre)clinical or licensed	*Inhibited enzyme*	*Applications and ongoing clinical trials*	*Potential therapeutic applications of enzyme inhibitors in the brain*
Pentostatin (2′-deoxycoformycin; Nipent®)	ADA: adenosine deaminase	Anticancer therapy (e.g., hairy cell myelogenous leukemia; cutaneous T-cell lymphomas; psoriasis; cancer of colon and rectum)	Neuroprotective, antiepileptic, antinociceptive and antiinflammatic effects (ADK or XO inhibition)
Peldesine	PNP: purine nucleoside phosphorylase		
Raltitrexed (Tomudex®)	TS: thymidylate synthase		Antipsychotic effects (schizophrenia, mania; ADK, ADA, or XO inhibition)
Tiazofurin (Tiazole™)	IMPDH: IMP dehydrogenase		
Allopurinol (Aloprim®)	XO: xanthine oxidase	Gout	
GP-3269	ADK: adenosine kinase	Pain, epilepsy	

Based on nucleoside transporters

Nucleoside transport inhibition

Transporter inhibitor name: (pre)clinical or licensed	*Applications and ongoing clinical trials*	*Potential therapeutic applications of transport inhibitors in the brain*
Dipyridamole (Persantine®)	Coronary vasodilator; platalet aggregation inhibitor	Ischemic cerebral injury; psychosis; seizures; pain; insomnia; inflammatory diseases; potentiation of cytotoxic effects (in chemotherapy); drug addiction and alcoholism
Dilazep (Cormelian®)		
Cilostazol (Pletal®)		
Lidoflazine (Clinium®)		
Nimodipine (Nimotop®)	Prevention of vasospasm and ischemic damage	
Propentofylline	Alzheimer disease; vascular dementia	

Nucleoside drugs (synthetic nucleosides)

Nucleoside drug name:	*Nucleoside transporter (involved in uptake of drug)*	*Applications and ongoing clinical trials*	*Potential therapeutic applications of nucleoside drugs in the brain*
Didanosine (ddI; Videx®)	ENT1, ENT2, ENT3, CNT2, CNT3	Antiviral therapy (e.g., HIV, types 1 and 2; hepatitis B and C; influenza A and B; paramyxoviruses)	Multiple sclerosis (Cladribine)
Zidovudine (AZT; Retrovir®)	ENT2, ENT3, CNT1, CNT3		
Zalcitabine (ddC; Hivid®)	ENT1, ENT2, ENT3, CNT1, CNT3		
Stavudine (d4T; Zerit®)	CNT1		
Ribavirine (Virazole®)	ENT1, CNT2, CNT3		
Lamivudine (3TC; Epivir®)	ENT1, CNT1		
Cladribine (2-CdA; Leustatin®)	ENT1, ENT3, CNT2, CNT3	Anticancer therapy (e.g., hairy cell, acute and chronic lymphocytic leukemia; pancreatic, lung, bladder, and breast cancer)	
Cytarabine (Ara-C; Cytosar®)	ENT1, CNT1		
Fludarabine (F-AraA; Fludara®)	ENT1, CNT3		
Gemcitabine (dFdC; Gemzar®)	ENT1, ENT2, CNT1, CNT3		

Based on adenosine receptors

Agonists

Ado receptor type	*Receptor agonist name: (pre) clinical or licensed*	*Applications and ongoing clinical trials*	*Potential therapeutic applications of agonists in the brain*
A_1	Adenosine (Adenocard®)	Arrhythmia	Neuroprotection, ischemia; sleep disorders; multiple sclerosis; Huntington's disease; antiepileptic and antinociceptive effects
	Selodenoson		
	Tecadenoson		
	GW-493838	Pain and migraine	
A_{2A}	Regadenoson	Coronary vasodilator	Multiple sclerosis; infectious meningitis; sleep disorders; psychosis; cognitive disorders
	Binodenoson		
A_{2B}	LUF5835	–	Sepsis
A_3	IB-MECA	Colon carcinoma; rheumatoid arthritis	Neuroprotection, ischemia

(continued)

Table 29.2 (continued)

Based on adenosine receptors

Ado receptor type	*Receptor antagonist name: (pre) clinical or licensed*	*Applications and ongoing clinical trials*	*Potential therapeutic applications of antagonists in the brain*
Antagonists			
A_1	Adentri®	Acute renal failure	Dementia; cognitive and anxiety disorders; antimetastatic therapy
	FR194921	Dementia and anxiety disorders	
A_{2A}	Istradefylline (KW6002)	Parkinson's disease	Antidepressant effect; neuroprotection, ischemia; epilepsy; cocaine abuse; pain; migraine; Alzheimer's and Huntington's disease; sleep disorders
	Preladenant		
	BIIB014		
	ST-1535		
A_{2B}	Enprofylline	Asthma	Alzheimer's disease
A_3	MRE 3008F20	–	Neuroprotection, stroke

References—Based on inhibition of nucleoside metabolic emzymes: Akhondzadeh et al. 2006; Boison et al. 2012; Bzowska et al. 2000; De Clercq 2004; De Mattia and Toffoli 2009; Erion et al. 1997; Jacobson and Gao 2006; Kaiser and Quinn 1999; Kowaluk and Jarvis 2000; Lara et al. 2006; Lehman 2002; Marro et al. 2006; McGaraughty et al. 2005; Nabhan et al. 2004; Robak et al. 2009; Togha et al. 2007; Weber et al. 1990; Weber and Prajda 1994; Wiesner et al. 1999; Willis et al. 1978. *Nucleoside transport inhibition*: Baldwin et al. 1999; Ciccarelli et al. 2001; Griffith and Jarvis 1996; Hanley and Hampton 1983; King et al. 2006; Kittner et al. 1997; Lara et al. 2006; Li et al. 2007; Mangravite et al. 2003; Noji et al. 2004; Pearce et al. 2008; Podgorska et al. 2005; Tomassoni et al. 2008; Weyrich et al. 2009. *Nucleoside drugs (synthetic nucleosides)*: Arias-Menendez 2002; Baldwin et al. 1999; Benesch and Urban 2008; Breckenridge 2005; De Clercq 2004, 2009, 2011; Franklin and Blanden 2007; Galmarini et al. 2002; King et al. 2006; Linker et al. 2008; Mangravite et al. 2003; Nabhan et al. 2004; Pastor-Anglada et al. 2005; Podgorska et al. 2005; Rando and Nguyen-Ba 2000; Robak et al. 2009; Van Rompay et al. 2003; Warnke et al. 2010; Zapor et al. 2004. *Adenosine receptor agonists*: Barrett et al. 2005; Beukers et al. 2004; Blum et al. 2003; Cunha 2005; Elzein and Zablocki 2008; Fredholm et al. 2001; Haskó et al. 2005; Headrick et al. 2011; Hendel et al. 2005; Jacobson and Gao 2006; Kaiser and Quinn 1999; Moreau and Huber 1999; Müller 2003; Paul and Pfammatter 1997; Peterman and Sanoski 2005; Popoli et al. 2007; Ribeiro et al. 2003; Zaza 2002. *Adenosine receptor antagonists*: Blum et al. 2003; Cunha 2005; Dall'Igna et al. 2003; Ferré et al. 2007; Fredholm et al. 2001; Gottlieb et al. 2002; Hauser et al. 2003, 2011; Headrick et al. 2011; Jacobson 1998; Jacobson and Gao 2006; Kaiser and Quinn 1999; Lopes et al. 2011; Merighi et al. 2003; Moreau and Huber 1999; Müller 2003; Pinna 2009; Popoli et al. 2007; Ribeiro et al. 2003; Schwarzschild et al. 2006; Varani et al. 2000; Volpini et al. 2003; Wardas 2002; Xu et al. 2005

In cases of dopaminergic hypofunction, $A_{2A}R$ activation contributes to the overdrive of the indirect pathway (Schiffmann et al. 2007). $A_{2A}R$ antagonists (Table 29.2), therefore, have the potential to restore this inhibitor imbalance. Consequently, these drugs have therapeutic potential in diseases of dopaminergic hypofunction such as Parkinson's disease. Indeed, $A_{2A}R$ antagonists have been effective in a variety of animal models of Parkinson's disease (Bastia et al. 2005; Chen et al. 2001; Hodgson et al. 2010; Kanda et al. 1998, 2000). Furthermore, caffeine ameliorates the freezing of gait that occurs in Parkinson's disease patients (Kitagawa et al. 2007). A number of clinical trials are under way to evaluate the potential of $A_{2A}R$ antagonists in the treatment of Parkinson's disease (Table 29.2), and the modulation of A_1 and $A_{2A}Rs$ may be effective in the treatment of Huntington's disease as well (Blum et al. 2003; Chou et al. 2005; Popoli et al. 2007).

Uridine might also be potentially effective in the treatment for Parkinson's disease. Coadministration of uridine monophosphate (UMP) and docosahexaenoic acid known to increase Urd levels and synapse formation in the brain increased striatal dopamine levels and alleviated the behavioral effects of 6-hydroxydopamine injections in a rat model of Parkinson's disease (Cansev et al. 2008).

29.4.2 Addiction

Although the different classes of drugs of abuse influence numerous neurotransmitter systems within the brain, all either directly or indirectly enhance the activity of the mesolimbic dopaminergic system. Within this system, ascending dopaminergic fibers project from the ventral tegmental area to the prefrontal cortex and nucleus accumbens, areas that are involved in the rewarding effects of drugs of abuse (Lajtha and Sershen 2010; Willuhn et al. 2010).

Similar to the striatum, the level of $A_{2A}Rs$ is also particularly high in the nucleus accumbens (Ferré et al. 2007; Svenningsson et al. 1997), an area that contains low to intermediate levels of nucleosides (Kovács et al. 2010a) and ENT3, ENT4, and CNT3 transporters (Baldwin et al. 2005; Barnes et al. 2006; Ritzel et al. 2001) (Table 29.1). Based on the presence of nucleoside transporters in the nucleus accumbens, transport inhibitors might have therapeutic potential in the treatment of drug addiction and alcoholism (Table 29.2). Indeed, adenosine transport in the nucleus accumbens decreases following chronic administration of morphine to rats (Brundege and Williams 2002). Adenosine may inhibit the reward process via $A_{2A}Rs$ (Baldo et al. 1999). In animal models, $A_{2A}R$ agonists inhibit cocaine self-administration, while antagonists reinstate this behavior (Knapp et al. 2001; Weerts and Griffiths 2003). Furthermore, mice lacking the $A_{2A}R$ exhibit attenuated reward processes (Castane et al. 2006). Some novel human data also supports the involvement of the adenosine system in addiction. An elevated $A_{2A}R$ binding affinity was found in platelets of patients suffering from pathological gambling (Martini et al. 2011). Clinical trials based on these data are expected in the near future (Lopes et al. 2011).

29.4.3 Pain Management

Nociceptive impulses first reach the posterior horn of the spinal cord. From here, information is transmitted to several brain regions involved in nociception. The reticular formation regulates arousal reactions and autonomic reflexes to pain, and thalamic nuclei relay and differentiate the nociceptive stimuli. Specific nuclei of the hypothalamus mediate autonomic and neuroendocrine responses. The limbic system mediates the emotional and motivation-related aspects of nociception, while the somatosensory cortex is mainly responsible for pain differentiation and localization (Apkarian et al. 2005). Additional pathways descending from a handful of brain regions, including the periaqueductal gray, rostroventromedial medulla, lateral reticular nucleus, and some brainstem monoamine cell groups, modulate nociception (Heinricher et al. 2009).

Nucleosides and their metabolic enzymes and transporters have been observed in different regions of large anatomical structures such as the spinal cord, medulla oblongata, midbrain, thalamus, hypothalamus, and diencephalon (Table 29.1). However, there is little data on the presence of nucleoside system in the specific areas of nociceptive circuitry in the brain, and further studies are needed. Nevertheless, a significant expression of A_1Rs has been described in primary sensory neurons associated with nociceptive pathways (Lima et al. 2010).

There is a great body of evidence indicating that the activation of A_1Rs produces antinociception (Curros-Criado and Herrero 2005). Mice lacking the A_1R exhibit hyperalgesia (Johansson et al. 2001). Consequently, drugs that target the nucleoside system have potential for the treatment of pain. GP-3269, an adenosine kinase inhibitor (Fig. 29.1), and GW-493838, an A_1R agonist, may be useful in the treatment of pain and migraines (Elzein and Zablocki 2008; Erion et al. 1997; Kowaluk and Jarvis 2000; McGaraughty et al. 2005; Wiesner et al. 1999) (Table 29.2). Guanosine was also found to have an antinociceptive effect in mice (Schmidt et al. 2009), suggesting that it may also be a potential target for the treatment of pain.

29.4.4 Mood Disorders

Anxiety, panic disorder, mania, and different forms of depression do not involve major neuronal degeneration in any brain regions. Nevertheless, animal studies and various imaging techniques have identified a number of limbic brain regions that play a role in the etiology of mood disorders. These regions include the prefrontal and cingulate cortices, septohippocampal circuits, amygdala, hypothalamus, and central gray matter of the midbrain (Garakani et al. 2006; Kalia 2005). Neurons in the locus coeruleus and raphe nuclei are thought to modulate these systems, explaining the effects of noradrenergic and serotonergic drugs on mood disorders (Fava 2003).

Only some of these structures have been studied for the presence of the elements of the nucleoside system (Table 29.1). The amygdala is particularly rich in nucleosides (Kovács et al. 2010a). Intermediate or high activities of 5´NT and PNP occur in the amygdala and the frontal and cingulate cortices (Nagata et al. 1984; Norstrand and Glantz 1980). Intermediate/low CNT2 and CNT3 levels are also observed in the amygdala and frontal cortex (Ritzel et al. 2001). ENT4 is abundant only in the amygdala, while ENT1 is believed to be the major equilibrative nucleoside transporter subtype in the frontal cortex (Barnes et al. 2006; Jennings et al. 2001). A high level of A_1Rs is found in the frontal cortex. Similar to the stratium (Schiffmann et al. 2007), the cortex also contains adenosine receptors both pre- and postsynaptically (Kirmse et al. 2008). Some other important brain regions, including the periaqueductal gray and monoamine systems, however, have not been systematically investigated for the presence of nucleoside metabolic enzymes and transporters. Nevertheless, the nucleoside system is expected to be a target for new drugs to treat mood disorders (Boison et al. 2012). Caffeine, a competitive antagonist of the A_1 and A_{2A}Rs (Fredholm et al. 1999), promotes anxious behavior both in animal models and humans (Klein et al. 1991), and A_{2A}R polymorphisms are associated with increased incidence of panic disorder and depression (Hamilton et al. 2004; Lam et al. 2005; Tsai et al. 2006). In addition, mice lacking A_1 or A_{2A}Rs demonstrate anxiogenic-like behaviors (Gimenez-Llort et al. 2002; Johansson et al. 2001; Ledent et al. 1997). Indeed, the application of an A_1R antagonist might be an effective treatment strategy for patients with anxiety disorders (Table 29.2). Allopurinol has been found to elicit therapeutic effects in the treatment of mania (Akhondzadeh et al. 2006) (Table 29.2, Fig. 29.1). Chronically administrated Guo produced anxiolytic effects in mice (Vinadé et al. 2003), suggesting a potential role of this purine nucleoside in the management of anxiety.

29.4.5 Schizophrenia

Pharmacological studies indicate the involvement of dopaminergic and glutamatergic neurons in the etiology of schizophrenia. A leading current hypothesis is that schizophrenia arises due to abnormalities in the dopamine–glutamate system of the corticostriatal pallidothalamic circuit, including the prefrontal cortex, nucleus accumbens, ventral tegmental area, mediodorsal thalamic nucleus, and ventral pallidum. Some drugs inducing drug dependence, probably by increasing the level of dopamine in the nucleus accumbens, also cause hallucinations suggesting that surplus dopamine may be a common ethiological factor. In addition to abnormalities in the corticostriatal system, alterations in the ventral limbic circuits of the dopamine–glutamate system, including the hippocampus, enthorinal cortex, and basolateral amygdala, may also be involved (Ross et al. 2006).

There is an intermediate to high level of nucleosides in most of the brain regions implicated in schizophrenia (Kovács et al. 2010a) (Table 29.1), and the evidence suggests that schizophrenia is associated with a hypofunctioning adenosine system (Lara et al. 2006). Adenosine levels can be increased by inhibiting adenosine transporters or xanthine oxidase with dypiridamole or allopurinol, respectively (Fig. 29.1). Both of these treatments had beneficial antipsychotic effects in clinical trials when administered in combination with haloperidol (Akhondzadeh et al. 2000, 2005) (Table 29.2). Furthermore, psychotic symptoms in schizophrenic patients are worsened by caffeine (Lucas et al. 1990). An interaction between adenosine and the dopamine system (Ferré et al. 1997) or the glutamate system (De Mendonca et al. 1995; Gerevich et al. 2002) could be driving these effects. Indeed, $A_{2A}R$ agonists and antagonists may have therapeutic potential for different types of psychosis (Table 29.2).

In animal models, some of the effects of haloperidol were augmented by coadministration with Urd (Agnati et al. 1989; Myers et al. 1994). Chronic Urd administration was also found to increase stereotypy scores and catalepsy induced by an acute haloperidol injection (Agnati et al. 1989). Furthermore, chronic Urd treatment reduced expression of dopamine receptors and enhanced their turnover rate in the striatum (Farabegoli et al. 1988). These data suggest that Urd coadministration might enhance the antipsychotic actions of traditional neuroleptics.

Moreover, the neuroprotective and neurotrophic effects of Guo may also be advantageous for the treatment of schizophrenia; Guo was found to attenuate hyperlocomotion induced by dizocilpine, a pharmacological model of schizophrenia, in mice (Tort et al. 2004).

29.4.6 Epilepsy

Epilepsy is characterized by a variety of recurrent symptoms resulting from the synchronous or sustained discharge of a group of neurons. The pathophysiology of epilepsy is poorly understood, and so far, there is no clear association between the abnormal function of a specific group of neurons and the genesis of seizures. There is some evidence, however, that the impairment of inhibitory signals, often occurring in the neocortex and hippocampus, may be primarily involved (Bertram 2009).

In the human hippocampus, ADA and GDA have intermediate activity (Dawson 1971; Norstrand et al. 1984) (Table 29.1). Adenosine and Ino levels are low, but intermediate concentrations of Guo and Urd are present (Kovács et al. 2010a). Based on their abundance, ENT4, CNT2, and CNT3 are believed to be the major nucleoside transporters in the hippocampus (Barnes et al. 2006; Ritzel et al. 2001). Furthermore, an intermediate level of A_1 and intermediate/low level of $A_{2A}R$ is present (Jennings et al. 2001). Indeed, the interaction of Ado with the inhibitory A_1R has been shown to have anticonvulsant effects in animal models (Barraco et al. 1984; Fedele et al. 2006). As A_1R agonists have peripheral cardiac and central sedative side-effects, adenosine kinase inhibitors (Fig. 29.1) have been used to indirectly increase Ado levels (Boison 2008). These drugs were shown to have anticonvulsant properties (McGaraughty et al. 2005). In particular, GP-3269, an adenosine kinase inhibitor, was found to be useful for the treatment of epilepsy (Erion et al. 1997; Kowaluk and Jarvis 2000; McGaraughty et al. 2005; Wiesner et al. 1999) (Table 29.2).

Recently, the distribution of A_{2A}Rs in the brain has been found to be altered in an animal model of human absence epilepsy (Wistar Albino Glaxo/Rijswijk rat: WAG/Rij), both before and after appearance of absence seizures (D'Alimonte et al. 2009). A low density of A_1Rs was also found in the thalamic reticular nucleus in another animal model of human absence epilepsy (Genetic Absence Epilepsy Rat from Strasbourg: GAERS) when compared with control animals (Ekonomou et al. 1998). These results suggest that adenosine receptors might represent a novel target for the treatment of absence epilepsy.

The anticonvulsant effects of Urd have also been hypothesized; Urd was found to reduce penicillin- (Roberts 1973; Roberts et al. 1974), pentylenetetrazole- (Dwivedi and Harbison 1975), and electroconvulsion-induced (Piccoli et al. 1971) seizures in experimental rodent models of epilepsy. Indeed, Urd is released following depolarization and inhibits unit activity (Dobolyi et al. 1999, 2000). Recently, Urd has been found to act as an antiepileptogen in hippocampal kindling models (Zhao et al. 2006, 2008). In addition, Guo prevented seizures induced by quinolinic acid and other glutamatergic agents (De Oliveira et al. 2004; Schmidt et al. 2000). These data suggest that Urd and Guo also have antiepileptic potential.

29.4.7 Insomnia

EEG recordings and other evidence indicate that sleep affects most cortical areas. Sleep waves are generated by an interaction between cortical and thalamic circuits, including thalamic reticular and relay nuclei. Sleep states are regulated by specific brain centers, and dysfunction of these regions leads to insomnia. Serotonergic and noradrenergic projections ascending from the brainstem and histaminergic cells in the tuberomamillary nucleus promote consciousness, while the preoptic area of the hypothalamus and cholinergic neurons in the basal forebrain and tegmental nuclei of the pons promote sleep. Orexinergic cells in the lateral hypothalamus may also have important on/off functions regarding sleep states (Datta and Maclean 2007; Saper 2006). The involvement of Ado in regulating sleep has long been suspected due to the hypnotic effects of adenosine analogues (Radulovacki 1985). The distribution of Ado and its inhibitory A_1R and increases of Ado levels in metabolically challenged cells are relatively ubiquitous. Furthermore, caffeine and theophylline are widely used as stimulants of the CNS. Therefore, the hypothesis emerged that, during daytime activity, ATP is degraded to adenosine, which could induce sleep. Indeed, prolonged wakefulness is known to increase Ado levels in the basal forebrain that, in turn, may decrease the activity of cholinergic cells to promote sleep (Porkka-Heiskanen and Kalinchuk 2011). A selective decrease in CNT2 mRNA levels was demonstrated in the cerebral cortex of sleep-deprived rats (Guillén-Gómez et al. 2004). These data suggest that adenosine receptor agonists and nucleoside transport inhibitors might be effective in the treatment of sleep disorders (Table 29.2).

Uridine was identified as an active component of a sleep-promoting substance purified from the brainstem of sleep-deprived rats (Borbely and Tobler 1989; Inoue 1986). Infusion of Urd increased slow wave and paradoxical sleep (Honda et al. 1984). Intraperitoneally injected Urd resulted in a dose-dependent appearance of slow-wave sleep when administered shortly before onset of the dark period (Honda et al. 1985). Based on these data, drugs elevating Urd levels in the brain should be tested for the treatment of insomnia in future studies.

29.4.8 Dementia

Alzheimer's disease is a progressive, degenerative disease of the brain that is the most common cause of dementia in the elderly. Typical pathological features of Alzheimer's disease are neuritic plaques and neurofibrillary tangles occurring primarily in the cholinergic basal forebrain and the hippocampus, frontal, parietal, and temporal lobes of the cerebral cortex (Peskind 1996).

The cerebral cortex and the basal forebrain contain all elements of the nucleoside system (Table 29.1). Neuroprotection achieved by manipulating the brain nucleoside system could be beneficial in the treatment of dementia. Animal models implicate the involvement of A_{2A}Rs in the development Alzheimer's disease. Caffeine and A_{2A}R antagonists prevent beta-amyloid (25–35)-induced cognitive deficits in mice (Dall'Igna et al. 2007). Additionally, caffeine elevates alertness and improves cognition in humans (Eskelinen et al. 2009; Ritchie et al. 2007). These effects might be due to altered acetylcholine release by A_{2A}Rs (Cunha et al. 1995; Jin and Fredholm 1997). In addition to receptor antagonists, propentofylline, an inhibitor of "es" nucleoside transporters, has established neuroprotective effects (Kittner et al. 1997), and its administration to patients with Alzheimer disease and vascular dementia resulted in functional improvements in clinical trials (Mielke et al. 1998).

Low to intermediate Ado levels but intermediate to high A_1R density has been observed in brain areas implicated in Alzheimer disease (Table 29.1), and loss of human hippocampal A_1Rs has been shown in dementia patients (Deckert et al. 1998). Therefore, A_1 receptor antagonists are potential targets for the treatment of dementia and cognitive disorders (Table 29.2). Administration of a nucleoside–nucleotide mixture reduced memory deterioration in elderly senescence-accelerated mice (Chen et al. 2000). In addition, age-dependent alterations in the adenosine system have been found (Kovács et al. 2010b; Meyer et al. 2007). These findings suggest that Ado might participate in the pathophysiology of learning and memory disorders, as well as the normal aging process.

In animal studies, Urd was found to improve certain types of memory function (Holguin et al. 2008; Teather and Wurtman 2003, 2005, 2006). Therefore, increased Urd formation may mediate the positive effects of cytidine diphosphocholine (CDP-choline) on verbal memory in aging humans (Spiers et al. 1996). Consequently, CDP-choline and other nutritional components that increase brain Urd levels (Wurtman et al. 2000) may be important, especially during the early phases of Alzheimer's disease (Van der Beek and Kamphuis 2008; Wurtman et al. 2009).

Recently, Guo has been found to protect against beta-amyloid-induced apoptosis (Pettifer et al. 2004). This effect appeared to be mediated by the antiapoptotic properties of Guo (Di Iorio et al. 2004). Guanosine was also found to modulate memory processes: its pretraining administration impaired retention of inhibitory avoidance responses in rats (Roesler et al. 2000). Furthermore, the amnesic effects associated with GMP pretreatment are also dependent on its conversion to Guo (Saute et al. 2006) (Fig. 29.1).

29.4.9 Stroke

In stroke, tissue damage is most often caused by ischemia resulting from an occluded blood vessel (Dietrich 1998). Neuroprotection by manipulation of brain nucleoside system may be beneficial in stroke victims. Adenosine and other nucleosides are elevated during ischemia (Rudolphi et al. 1992). While adenosine released from neurons or accumulated by the extracellular degradation of released ATP could reach a concentration efficient for the activation of adenosine receptors, a pathophysiological release from neurons as well as glial cells occurs during an ischemic event (Latini and Pedata 2001). Agonist stimulation of the A_1R may inhibit excessive neuronal firing and may enhance local cerebral blood flow (O'Regan 2005), reducing brain damage following experimentally induced ischemia in animals. Indeed, lacking the A_1 receptor exhibited decreased hypoxic neuroprotection in mice (Johansson et al. 2001). Thus, Ado may be involved in ischemic preconditioning, an endogenous neuroprotective mechanism (Liu et al. 2009). Consequently, drugs that act on adenosine receptors, adenosine metabolizing enzymes, and nucleoside transporters (Table 29.2, Fig. 29.1) and increase EC Ado levels could be targets for the development of clinical therapeutics suitable for treatment of ischemic brain disorders (Stone 2002; Von Lubitz 2001). Importantly, the effectiveness of all of these potential therapies may vary between patients due to differences in the spatial distribution of the nucleoside system (Table 29.1). Indeed, the nucleoside system may be modulated differently in men and women (Kovács et al. 2010b). Changes in nucleoside levels in female brain cortical samples may serve as a protective mechanism against excitotoxic insults, suggesting that several normal and pathological brain functions are based on gender-dependent nucleoside microenvironments in humans.

Other nucleosides might also have neuroprotective functions in response to ischemic injury, and increasing their expression might be beneficial both during and after an ischemic attack. In animal models, Guo had neuroprotective effects in both in vitro and in vivo stroke models (Chang et al. 2008). Inosine was also shown to reduce ischemic brain injury in rats (Shen et al. 2005). Inosine and Guo preserved the viability of cultured astrocytes, neurons (Jurkowitz et al. 1998; Litsky et al. 1999), and brain slices maintained under hypoxic or hypoglycemic conditions (Frizzo et al. 2002). The potential neuroprotective effects of Guo are also supported by the finding that neuronal and astrocytic cell cultures are able to release Guo and Ado under both basal and ischemic conditions (Ciccarelli et al. 2001).

29.4.10 Multiple Sclerosis

Multiple sclerosis is characterized by multiple symptoms of brain and spinal cord dysfunction that reflect degeneration of particular areas of the nervous system that are involved. The affected regions vary between patients and are not specific to the disease. The pathological hallmark is inflammatory demyelination and axonal lesions. Inflammation is primarily driven by autoreactive lymphocytes, which recruit immune cells, such as macrophages, causing tissue damage (Hauser and Oksenberg 2006).

A synthetic nucleoside, cladribine, was shown to be effective in the treatment of multiple sclerosis (Table 29.2). The biologic activity of cladribine is dependent on the preferential accumulation of cladribine phosphates in cells with a high intracellular ratio of deoxycytidine kinase to 5′NT. Cladribine-phosphates incorporate into DNA, interfering with DNA synthesis and repair and inhibiting enzymes involved in DNA metabolism, such as DNA polymerase and ribonucleotide reductase. This, in turn, leads to DNA strand breaks and, ultimately, cell death (Leist and Weissert 2011). Because activated macrophages, but not neuronal and glial cells, have a high deoxycytidine kinase to 5′NT ratio (Ceruti et al. 2000; Nagata et al. 1984), cladribine can selectively inhibit the damaging inflammatory process that occurs in multiple sclerosis.

Inosine may also have beneficial effects in the treatment of multiple sclerosis (Markowitz et al. 2009). These data suggest that regional differences in nucleoside system may influence the pathological processes of multiple sclerosis, but further studies are needed to confirm this hypothesis.

In conclusion, the current data suggest that nucleoside system offers promising drug targets for the treatment of a variety of brain disorders, including Alzheimer's, Huntington's and Parkinson's diseases, epilepsy, and schizophrenia. Unfortunately, although the nucleoside system has been implicated in the development and treatment of a number of brain disorders, a systematic investigation of the nucleoside system in most brain areas has not yet been performed. These data are needed to elucidate therapeutic strategies driven by the anatomical distribution of nucleoside system. In addition, attention must be given to the effects of gender and age in future studies. These data are eagerly awaited and will help form the foundation for studies of the physiological and pathophysiological functions of nucleosides and for the development of effective treatments for several CNS diseases.

Acknowledgments This work was supported by the Scientific Foundation of NYME SEK/NYME SEK TTK (2010–2011) Hungary (Zsolt Kovács) and the OTKA NNF2 85612 Research Grant as well as the Bolyai János Grant of the Hungarian Academy of Sciences (Arpád Dobolyi).
Conflict of interest: All authors declare no conflicts of interest.

References

Agnati LF, Fuxe K, Ruggeri M, Merlo Pich E, Benfenati F, Volterra V, Ungerstedt U, Zini I (1989) Effects of chronic treatment with uridine on striatal dopamine release and dopamine related behaviours in the absence or the presence of chronic treatment with haloperidol. Neurochem Int 15:107–113

Akhondzadeh S, Shasavand E, Jamilian H, Shabestari O, Kamalipour A (2000) Dipyridamole in the treatment of schizophrenia: adenosine-dopamine receptor interactions. J Clin Pharm Ther 25:131–137

Akhondzadeh S, Safarcherati A, Amini H (2005) Beneficial antipsychotic effects of allopurinol as add-on therapy for schizophrenia: a double blind, randomized and placebo controlled trial. Prog Neuropsychopharmacol Biol Psychiatry 29:253–259

Akhondzadeh S, Milajerdi MR, Amini H, Tehrani-Doost M (2006) Allopurinol as an adjunct to lithium and haloperidol for treatment of patients with acute mania: a double-blind, randomized, placebo-controlled trial. Bipolar Disord 8:485–489

Apkarian AV, Bushnell MC, Treede RD, Zubieta JK (2005) Human brain mechanisms of pain perception and regulation in health and disease. Eur J Pain 9:463–484

Arias-Menendez L (2002) Targeting HIV: antiretroviral therapy and development of drug resistance. Trends Pharmacol Sci 23:381–388

Baldo BA, Koob GF, Markou A (1999) Role of adenosine A2 receptors in brain stimulation reward under baseline conditions and during cocaine withdrawal in rats. J Neurosci 19:11017–11026

Baldwin SA, Mackey JR, Cass CE, Young JD (1999) Nucleoside transporters: molecular biology and implications for therapeutic development. Mol Med Today 5:216–224

Baldwin SA, Yao SYM, Hyde RJ, Ng AML, Foppolo S, Barnes K, Ritzel MWL, Cass CE, Young JD (2005) Functional characterization of novel human and mouse equilibrative nucleoside transporters (hENT3 and mENT3) located in intracellular membranes. J Biol Chem 280:15880–15887

Barnes K, Dobrzynski H, Foppolo S, Beal PR, Ismat F, Scullion ER, Sun L, Tellez J, Ritzel MW, Claycomb WC, Cass CE, Young JD, Billeter-Clark R, Boyett MR, Baldwin SA (2006) Distribution and functional characterization of equilibrative nucleoside transporter-4, a novel cardiac adenosine transporter activated at acidic pH. Circ Res 99:510–519

Barraco RA, Swanson TH, Phillis JW, Berman RF (1984) Anticonvulsant effects of adenosine analogues on amygdaloid-kindled seizures in rats. Neurosci Lett 46:317–322

Barrett RJ, Lamson MJ, Johnson J, Smith WB (2005) Pharmacokinetics and safety of binodenoson after intravenous dose escalation in healthy volunteers. J Nucl Cardiol 12:166–171

Bastia E, Xu YH, Scibelli AC, Day YJ, Linden J, Chen JF, Schwarzschild MA (2005) A crucial role for forebrain adenosine A(2A) receptors in amphetamine sensitization. Neuropsychopharmacology 30:891–900

Bender E, Buist A, Jurzak M, Langlois X, Baggerman G, Verhasselt P, Ercken M, Guo HQ, Wintmolders C, Van den Wyngaert I, Van Oers I, Schoofs L, Luyten W (2002) Characterization of an orphan G protein-coupled receptor localized in the dorsal root ganglia reveals adenine as a signaling molecule. Proc Natl Acad Sci USA 99:8573–8578

Benesch M, Urban C (2008) Liposomal cytarabine for leukemic and lymphomatous meningitis: recent developments. Expert Opin Pharmacother 9:301–309

Berne RM, Rubio R, Curnish RR (1974) Release of adenosine from ischemic brain: effect on cerebral vascular resistance and incorporation into cerebral adenine nucleotides. Circ Res 35:262–271

Bertram EH (2009) Temporal lobe epilepsy: where do the seizures really begin? Epilepsy Behav 14(Suppl 1):32–37

Beukers MW, Chang LC, von Frijtag Drabbe Kunzel JK, Mulder-Krieger T, Spanjersberg RF, Brussee J, Ijzerman AP (2004) New, non-adenosine, high-potency agonists for the human adenosine A2B receptor with an improved selectivity profile compared to the reference agonist N-ethylcarboxamidoadenosine. J Med Chem 47:3707–3709

Bjerring PN, Hauerberg J, Jørgensen L, Frederiksen HJ, Tofteng F, Hansen BA, Larsen FS (2010) Brain hypoxanthine concentration correlates to lactate/pyruvate ratio but not intracranial pressure in patients with acute liver failure. J Hepatol 53:1054–1058

Blum D, Hourez R, Galas MC, Popoli P, Schiffmann SN (2003) Adenosine receptors and Huntington's disease: implications for pathogenesis and therapeutics. Lancet Neurol 2:366–374

Boison D (2005) Adenosine and epilepsy: from therapeutic rationale to new therapeutic strategies. Neuroscientist 11:25–36

Boison D (2008) The adenosine kinase hypothesis of epileptogenesis. Prog Neurobiol 84:249–262

Boison D (2011) Modulators of nucleoside metabolism in the therapy of brain diseases. Curr Top Med Chem 11:1068–1086

Boison D, Singer P, Shen HY, Feldon J, Yee BK (2012) Adenosine hypothesis of schizophrenia-Opportunities for pharmacotherapy. Neuropharmacology 62:1527–1543

Borbely AA, Tobler I (1989) Endogenous sleep-promoting substances and sleep regulation. Physiol Rev 69:605–670

Borrmann T, Abdelrahman A, Volpini R, Lambertucci C, Alksnis E, Gorzalka S, Knospe M, Schiedel AC, Cristalli G, Müller CE (2009) Structure-activity relationships of adenine and deazaadenine derivatives as ligands for adenine receptors, a new purinergic receptor family. J Med Chem 52:5974–5989

Bowley MP, Drevets WC, Öngür D, Price JL (2002) Low glial numbers in the amygdala in major depressive disorder. Biol Psychiatry 52:404–412

Brauch RA, Adnan El-Masri M, Parker JC Jr, El-Mallakh RS (2006) Glial cell number and neuron/glial cell ratios in postmortem brains of bipolar individuals. J Affect Disord 91:87–90

Breckenridge A (2005) Pharmacology of drugs for HIV. Medicine 33:30–31

Brundege JM, Williams JT (2002) Increase in adenosine sensitivity in the nucleus accumbens following chronic morphine treatment. J Neurophysiol 87:1369–1375

Burnstock G, Fredholm BB, Verkhratsky A (2011) Adenosine and ATP receptors in the brain. Curr Top Med Chem 11:973–1011

Bzowska A, Kulikowska E, Shugar D (2000) Purine nucleoside phosphorylases: properties, functions, and clinical aspects. Pharmacol Therapeut 88:349–425

Cansev M (2006) Uridine and cytidine in the brain: their transport and utilization. Brain Res Rev 52:389–397

Cansev M, Ulus IH, Wang L, Maher TJ, Wurtman RJ (2008) Restorative effects of uridine plus docosahexaenoic acid in a rat model of Parkinson's disease. Neurosci Res 62:206–209

Castane A, Soria G, Ledent C, Maldonado R, Valverde O (2006) Attenuation of nicotine-induced rewarding effects in A2A knockout mice. Neuropharmacology 51:631–640

Ceballos G, Tuttle JB, Rubio R (1994) Differential distribution of purine metabolizing enzymes between glia and neurons. J Neurochem 62:1144–1153

Ceruti S, Franceschi C, Barbieri D, Malorni W, Camurri A, Giammarioli AM, Ambrosini A, Racagni G, Cattabeni F, Abbracchio MP (2000) Apoptosis induced by 2-chloro-adenosine and 2-chloro-2'-deoxy-adenosine in a human astrocytoma cell line: differential mechanisms and possible clinical relevance. J Neurosci Res 60:388–400

Chang R, Algird A, Bau C, Rathbone MP, Jiang S (2008) Neuroprotective effects of guanosine on stroke models in vitro and in vivo. Neurosci Lett 431:101–105

Chen TH, Wang MF, Liang YF, Komatsu T, Chan YC, Chung SY, Yamamoto S (2000) A nucleoside-nucleotide mixture may reduce memory deterioration in old senescence-accelerated mice. J Nutr 130:3085–3089

Chen JF, Xu K, Petzer JP, Staal R, Xu YH, Beilstein M, Sonsalla PK, Castagnoli K, Castagnoli N Jr, Schwarzschild MA (2001) Neuroprotection by caffeine and A(2A) adenosine receptor inactivation in a model of Parkinson's disease. J Neurosci 21:RC143

Chou SY, Lee YC, Chen HM, Chiang MC, Lai HL, Chang HH, Wu YC, Sun CN, Chien CL, Lin YS, Wang SC, Tung YY, Chang C, Chern Y (2005) CGS21680 attenuates symptoms of Huntington's disease in a transgenic mouse model. J Neurochem 93:310–320

Ciccarelli R, Ballerini P, Sabatino G, Rathbone MP, D'Onofrio M, Caciagli F, Di Iorio P (2001) Involvement of astrocytes in purine-mediated reparative processes in the brain. Int J Dev Neurosci 19:395–414
Connolly GP, Duley JA (1999) Uridine and its nucleotides: biological actions, therapeutic potentials. Trends Pharmacol Sci 20:218–225
Cornfield LJ, Hu S, Hurt SD, Sills MA (1992) [3 H]2-phenylaminoadenosine ([3 H]CV 1808) labels a novel adenosine receptor in rat brain. J Pharmacol Exp Ther 263:552–561
Cunha RA (2005) Neuroprotection by adenosine in the brain: from A(1) receptor activation to A (2A) receptor blockade. Purinergic Signal 1:111–134
Cunha RA, Johansson B, Fredholm BB, Ribeiro JA, Sebastiao AM (1995) Adenosine A2A receptors stimulate acetylcholine release from nerve terminals of the rat hippocampus. Neurosci Lett 196:41–44
Curros-Criado MM, Herrero JF (2005) The antinociceptive effects of the systemic adenosine A1 receptor agonist CPA in the absence and in the presence of spinal cord sensitization. Pharmacol Biochem Behav 82:721–726
D'Alimonte I, D'Auro M, Citraro R, Biagioni F, Jiang S, Nargi E, Buccella S, Di Iorio P, Giuliani P, Ballerini P, Caciagli F, Russo E, De Sarro G, Ciccarelli R (2009) Altered distribution and function of A2A adenosine receptors in the brain of WAG/Rij rats with genetic absence epilepsy, before and after appearance of the disease. Eur J Neurosci 30:1023–1035
Dall'Igna OP, Porciúncula LO, Souza DO, Cunha RA, Lara DR (2003) Neuroprotection by caffeine and adenosine A2A receptor blockade of beta-amyloid neurotoxicity. Br J Pharmacol 138:1207–1209
Dall'Igna OP, Fett P, Gomes MW, Souza DO, Cunha RA, Lara DR (2007) Caffeine and adenosine A(2a) receptor antagonists prevent beta-amyloid (25-35)-induced cognitive deficits in mice. Exp Neurol 203:241–245
Datta S, Maclean RR (2007) Neurobiological mechanisms for the regulation of mammalian sleep-wake behavior: reinterpretation of historical evidence and inclusion of contemporary cellular and molecular evidence. Neurosci Biobehav Rev 31:775–824
Dawson DM (1971) Absence of guanine deaminase from cerebellum. Neurology 21:621–626
De Clercq E (2004) Antiviral drugs in current clinical use. J Clin Virol 30:115–133
De Clercq E (2009) Anti-HIV drugs: 25 compounds approved within 25 years after the discovery of HIV. Int J Antimicrob Agents 33:307–320
De Clercq E (2011) A 40-year journey in search of selective antiviral chemotherapy. Annu Rev Pharmacol Toxicol 51:1–24
De Mattia E, Toffoli G (2009) C677T and A1298C MTHFR polymorphisms, a challenge for antifolate and fluoropyrimidine-based therapy personalisation. Eur J Cancer 45:1333–1351
De Mendonca A, Sebastiao AM, Ribeiro JA (1995) Inhibition of NMDA receptor-mediated currents in isolated rat hippocampal neurones by adenosine A1 receptor activation. Neuroreport 6:1097–1100
De Oliveira DL, Horn JF, Rodrigues JM, Frizzo ME, Moriguchi E, Souza DO, Wofchuk S (2004) Quinolinic acid promotes seizures and decreases glutamate uptake in young rats: reversal by orally administered guanosine. Brain Res 1018:48–54
Deckert J, Abel F, Kunig G, Hartmann J, Senitz D, Maier H, Ransmayr G, Riederer P (1998) Loss of human hippocampal adenosine A_1 receptors in dementia: evidence for lack of specificity. Neurosci Lett 244:1–4
Deutsch SI, Long KD, Rosse RB, Mastropaolo J, Eller J (2005) Hypothesized deficiency of guanine-based purines may contribute to abnormalities of neurodevelopment, neuromodulation, and neurotransmission in Lesch-Nyhan syndrome. Clin Neuropharmacol 28:28–37
Di Iorio P, Ballerini P, Traversa U, Nicoletti F, D'Alimonte I, Kleywegt S, Werstiuk ES, Rathbone MP, Caciagli F, Ciccarelli R (2004) The antiapoptotic effect of guanosine is mediated by the activation of the PI 3-kinase/AKT/PKB pathway in cultured rat astrocytes. Glia 46:356–368
Dietrich WD (1998) Neurobiology of stroke. Int Rev Neurobiol 42:55–101
Dobolyi A, Szikra T, Kekesi AK, Kovacs Z, Juhasz G (1999) Uridine is released by depolarization and inhibits unit activity in the rat hippocampus. Neuroreport 10:3049–3053

Dobolyi A, Reichart A, Szikra T, Nyitrai G, Kekesi KA, Juhasz G (2000) Sustained depolarisation induces changes in the extracellular concentrations of purine and pyrimidine nucleosides in the rat thalamus. Neurochem Int 37:71–79

Dobolyi A, Juhasz G, Kovacs Z, Kardos J (2011) Uridine function in the central nervous system. Curr Top Med Chem 11:1058–1067

Durieux PF, Schiffmann SN, de Kerchove d'Exaerde A (2011) Targeting neuronal populations of the striatum. Front Neuroanat 5:40

Dwivedi C, Harbison RD (1975) Anticonvulsant activities of delta-8 and delta-9 tetrahydrocannabinol and uridine. Toxicol Appl Pharmacol 31:452–458

Eells JT, Spector R (1983) Purine and pyrimidine base and nucleoside concentrations in human cerebrospinal fluid and plasma. Neurochem Res 8:1451–1457

Ekonomou A, Angelatou F, Vergnes M, Kostopoulos G (1998) Lower density of A1 adenosine receptors in nucleus reticularis thalami in rats with genetic absence epilepsy. Neuroreport 9:2135–2140

Elzein E, Zablocki J (2008) A1 adenosine receptor agonists and their potential therapeutic applications. Expert Opin Investig Drugs 17:1901–1910

Erion MD, Ugarkar BG, Dare J, Castellino AJ, Fujitaki JM, Dixon R, Appleman JR, Wiesner JB (1997) Design, synthesis and anticonvulsant activity of the potent adenosine kinase inhibitor GP3269. Nucleos Nucleot 16:1013–1021

Eskelinen MH, Ngandu T, Tuomilehto J, Soininen H, Kivipelto M (2009) Midlife coffee and tea drinking and the risk of late-life dementia: a population-based CAIDE study. J Alzheimers Dis 16:85–91

Farabegoli C, Merlo Pich E, Cimino M, Agnati LF, Fuxe K (1988) Chronic uridine treatment reduces the level of [3 H]spiperone-labelled dopamine receptors and enhances their turnover rate in striatum of young rats: relationship to dopamine-dependent behaviours. Acta Physiol Scand 132:209–216

Fastbom J, Pazos A, Probst A, Palacios JM (1986) Adenosine A1 receptors is human brain: characterisation and autoradiographic visualization. Neurosci Lett 65:127–132

Fastbom J, Pazos A, Palacios JM (1987) The distribution of adenosine A1 receptors and 5'-nucleotidase is the brain of some commonly used experimental animals. Neuroscience 22:813–826

Fava M (2003) The role of the serotonergic and noradrenergic neurotransmitter systems in the treatment of psychological and physical symptoms of depression. J Clin Psychiatry 64(Suppl 13):26–29

Fedele DE, Li T, Lan JQ, Fredholm BB, Boison D (2006) Adenosine A1 receptors are crucial in keeping an epileptic focus localized. Exp Neurol 200:184–190

Fernández JR, Sweet ES, Welsh WJ, Firestein BL (2010) Identification of small molecule compounds with higher binding affinity to guanine deaminase (cypin) than guanine. Bioorg Med Chem 18:6748–6755

Ferré S, Fredholm BB, Morelli M, Popoli P, Fuxe K (1997) Adenosine-dopamine receptor-receptor interactions as an integrative mechanism in the basal ganglia. Trends Neurosci 20:482–487

Ferré S, Diamond I, Goldberg SR, Yao L, Hourani SM, Huang ZL, Urade Y, Kitchen I (2007) Adenosine A2A receptors in ventral striatum, hypothalamus and nociceptive circuitry implications for drug addiction, sleep and pain. Prog Neurobiol 83:332–347

Fields RD, Burnstock G (2006) Purinergic signalling in neuron-glia interactions. Nat Rev Neurosci 7:423–436

Firestein BL, Brenman JE, Aoki C, Sanchez-Perez AM, El-Husseini AE, Bredt DS (1999) Cypin: a cytosolic regulator of PSD-95 postsynaptic targeting. Neuron 24:659–672

Franklin A, Blanden RV (2007) Potential inhibition of somatic hypermutation by nucleoside analogues. Mol Immunol 44:666–669

Fredholm BB, Battig K, Holmen J, Nehlig A, Zvartau EE (1999) Actions of caffeine in the brain with special reference to factors that contribute to its widespread use. Pharmacol Rev 51:83–133

Fredholm BB, Ijzerman AP, Jacobson KA, Klotz KN, Linden J (2001) International Union of Pharmacology. XXV. Nomenclature and classification of adenosine receptors. Pharmacol Rev 53:527–552

Frizzo ME, Lara DR, Prokopiuk Ade S, Vargas CR, Salbego CG, Wajner M, Souza DO (2002) Guanosine enhances glutamate uptake in brain cortical slices at normal and excitotoxic conditions. Cell Mol Neurobiol 22:353–363
Galmarini CM, Mackey JR, Dumontet C (2002) Nucleoside analogues and nucleobases in cancer treatment. Lancet Oncol 3:415–424
Garakani A, Mathew SJ, Charney DS (2006) Neurobiology of anxiety disorders and implications for treatment. Mt Sinai J Med 73:941–949
Geiger JD, Nagy JI (1986) Distribution of adenosine deaminase activity in rat brain and spinal cord. J Neurosci 6:2707–2714
Gerevich Z, Wirkner K, Illes P (2002) Adenosine A2A receptors inhibit the N-methyl-D-aspartate component of excitatory synaptic currents in rat striatal neurons. Eur J Pharmacol 451:161–164
Gimenez-Llort L, Fernandez-Teruel A, Escorihuela RM, Fredholm BB, Tobena A, Pekny M, Johansson B (2002) Mice lacking the adenosine A1 receptor are anxious and aggressive, but are normal learners with reduced muscle strength and survival rate. Eur J Neurosci 16:547–550
Gottlieb SS, Brater DC, Thomas I, Havranek E, Bourge R, Goldman S, Dyer F, Gomez M, Bennett D, Ticho B, Beckman E, Abraham WT (2002) BG9719 (CVT-124), an A1 adenosine receptor antagonist, protects against the decline in renal function observed with diuretic therapy. Circulation 105:1348–1353
Gracia E, Cortés A, Meana JJ, García-Sevilla J, Herhsfield MS, Canela EI, Mallol J, Lluís C, Franco R, Casadó V (2008) Human adenosine deaminase as an allosteric modulator of human A(1) adenosine receptor: abolishment of negative cooperativity for [H](R)-pia binding to the caudate nucleus. J Neurochem 107:161–170
Griffith GA, Jarvis SM (1996) Nucleoside and nucleobase transport systems of mammalian cells. Biochim Biophys Acta 1286:153–181
Guillén-Gómez E, Calbet M, Casado J, de Lecea L, Soriano E, Pastor-Anglada M, Burgaya F (2004) Distribution of CNT2 and ENT1 transcripts in rat brain: selective decrease of CNT2 mRNA in the cerebral cortex of sleep-deprived rats. J Neurochem 90:883–893
Hagberg H, Andersson P, Lacarewicz J, Jacobson I, Butcher S, Sandberg M (1987) Extracellular adenosine, inosine, hypoxanthine, and xanthine in relation to tissue nucleotides and purines in rat striatum during transient ischemia. J Neurochem 49:227–231
Hamilton SP, Slager SL, De Leon AB, Heiman GA, Klein DF, Hodge SE, Weissman MM, Fyer AJ, Knowles JA (2004) Evidence for genetic linkage between a polymorphism in the adenosine 2A receptor and panic disorder. Neuropsychopharmacology 29:558–565
Hanley SP, Hampton JR (1983) Ventricular arrhythmias associated with lidoflazine: side-effects observed in a randomized trial. Eur Heart J 4:889–893
Harper DG, Stopa EG, Kuo-Leblanc V, McKee AC, Asayama K, Volicer L, Kowall N, Satlin A (2008) Dorsomedial SCN neuronal subpopulations subserve different functions in human dementia. Brain 131:1609–1617
Harris MK, Shneyder N, Borazanci A, Korniychuk E, Kelley RE, Minagar A (2009) Movement disorders. Med Clin North Am 93:371–388
Haskó G, Sitkovsky MV, Szabó C (2004) Immunomodulatory and neuroprotective effects of inosine. Trends Pharmacol Sci 25:152–157
Haskó G, Pacher P, Vizi ES, Illes P (2005) Adenosine receptor signaling in the brain immune system. Trends Pharmacol Sci 26:511–516
Hauser SL, Oksenberg JR (2006) The neurobiology of multiple sclerosis: genes, inflammation, and neurodegeneration. Neuron 52:61–76
Hauser RA, Hubble JP, Truong DD (2003) Randomized trial of the adenosine A(2A) receptor antagonist istradefylline in advanced PD. Neurology 61:286–287
Hauser RA, Cantillon M, Pourcher E, Micheli F, Mok V, Onofrj M, Huyck S, Wolski K (2011) Preladenant in patients with Parkinson's disease and motor fluctuations: a phase 2, double-blind, randomised trial. Lancet Neurol 10:221–229
Headrick JP, Peart JN, Reichelt ME, Haseler LJ (2011) Adenosine and its receptors in the heart: regulation, retaliation and adaptation. Biochim Biophys Acta 1808:1413–1428

Heinricher MM, Tavares I, Leith JL, Lumb BM (2009) Descending control of nociception: specificity, recruitment and plasticity. Brain Res Rev 60:214–225

Hendel RC, Bateman TM, Cerqueira MD, Iskandrian AE, Leppo JA, Blackburn B, Mahmarian JJ (2005) Initial clinical experience with regadenoson, a novel selective A2A agonist for pharmacologic stress single-photon emission computed tomography myocardial perfusion imaging. J Am Coll Cardiol 46:2076–2078

Hodgson RA, Bedard PJ, Varty GB, Kazdoba TM, Di Paolo T, Grzelak ME, Pond AJ, Hadjtahar A, Belanger N, Gregoire L, Dare A, Neustadt BR, Stamford AW, Hunter JC (2010) Preladenant, a selective A(2A) receptor antagonist, is active in primate models of movement disorders. Exp Neurol 225:384–390

Holguin S, Martinez J, Chow C, Wurtman R (2008) Dietary uridine enhances the improvement in learning and memory produced by administering DHA to gerbils. FASEB J 22:3938–3946

Honda K, Komoda Y, Nishida S, Nagasaki H, Higashi A, Uchizono K, Inoue S (1984) Uridine as an active component of sleep-promoting substance: its effects on nocturnal sleep in rats. Neurosci Res 1:243–252

Honda K, Okano Y, Komoda Y, Inoue S (1985) Sleep-promoting effects of intraperitoneally administered uridine in unrestrained rats. Neurosci Lett 62:137–141

Huang ZL, Urade Y, Hayaishi O (2011) The role of adenosine in the regulation of sleep. Curr Top Med Chem 11:1047–1057

Inoue S (1986) Sleep and sleep substances. Brain Dev 8:469–473

Ipata PL, Camici M, Micheli V, Tozzi MG (2011) Metabolic network of nucleosides in the brain. Curr Top Med Chem 11:909–922

Jacobson KA (1998) Adenosine A3 receptors: novel ligands and paradoxical effects. Trends Pharmacol Sci 19:184–191

Jacobson KA, Gao ZG (2006) Adenosine receptors as therapeutic targets. Nat Rev Drug Discov 5:247–264

Jenner P, Mori A, Hauser R, Morelli M, Fredholm BB, Chen JF (2009) Adenosine, adenosine A_{2A} antagonists, and Parkinson's disease. Parkinsonism Relat Disord 15:406–413

Jennings LL, Hao C, Cabrita MA, Vickers MF, Baldwin SA, Young J, Cass CE (2001) Distinct regional distribution of human equilibrative nucleoside transporter proteins 1 and 2 (hENT1 and hENT2) in the central nervous system. Neuropharmacology 40:722–731

Jin S, Fredholm BB (1997) Adenosine A2A receptor stimulation increases release of acetylcholine from rat hippocampus but not striatum, and does not affect catecholamine release. Naunyn Schmiedeberg's Arch Pharmacol 355:48–56

Johansson B, Halldner L, Dunwiddie TV, Masino SA, Poelchen W, Gimenez-Llort L, Escorihuela RM, Fernandez-Teruel A, Wiesenfeld-Hallin Z, Xu XJ, Hårdemark A, Betsholtz C, Herlenius E, Fredholm BB (2001) Hyperalgesia, anxiety, and decreased hypoxic neuroprotection in mice lacking the adenosine A1 receptor. Proc Natl Acad Sci USA 98:9407–9412

Jurkowitz MS, Litsky ML, Browning MJ, Hohl CM (1998) Adenosine, inosine, and guanosine protect glial cells during glucose deprivation and mitochondrial inhibition: correlation between protection and ATP preservation. J Neurochem 71:535–548

Kaiser SM, Quinn RJ (1999) Adenosine receptors as potential therapeutic targets. Drug Discov Today 4:542–551

Kalia M (2005) Neurobiological basis of depression: an update. Metabolism 54:24–27

Kanda T, Jackson MJ, Smith LA, Pearce RK, Nakamura J, Kase H, Kuwana Y, Jenner P (1998) Adenosine A2A antagonist: a novel antiparkinsonian agent that does not provoke dyskinesia in parkinsonian monkeys. Ann Neurol 43:507–513

Kanda T, Jackson MJ, Smith LA, Pearce RK, Nakamura J, Kase H, Kuwana Y, Jenner P (2000) Combined use of the adenosine A(2A) antagonist KW-6002 with L-DOPA or with selective D1 or D2 dopamine agonists increases antiparkinsonian activity but not dyskinesia in MPTP-treated monkeys. Exp Neurol 162:321–327

Kékesi KA, Kovács Z, Szilágyi N, Bobest M, Szikra T, Dobolyi Á, Juhász G, Palkovits M (2006) Concentration of nucleosides and related compounds in cerebral and cerebellar cortical areas and white matter of the human brain. Cell Mol Neurobiol 26:831–842

Kimura T, Ho IK, Yamamoto I (2001) Uridine receptor: discovery and its involvement in sleep mechanism. Sleep 24:251–260

King AE, Ackley MA, Cass CE, Young JD, Baldwin SA (2006) Nucleoside transporters: from scavengers to novel therapeutic targets. Trends Pharmacol Sci 27:416–425

Kirmse K, Dvorzhak A, Grantyn R, Kirischuk S (2008) Developmental downregulation of excitatory GABAergic transmission in neocortical layer I via presynaptic adenosine A(1) receptors. Cereb Cortex 18:424–432

Kitagawa M, Houzen H, Tashiro K (2007) Effects of caffeine on the freezing of gait in Parkinson's disease. Mov Disord 22:710–712

Kittner B, Rossner M, Rother M (1997) Clinical trials in dementia with propentofylline. Ann N Y Acad Sci 826:307–316

Klein E, Zohar J, Geraci MF, Murphy DL, Uhde TW (1991) Anxiogenic effects of m-CPP in patients with panic disorder: comparison to caffeine's anxiogenic effects. Biol Psychiatry 30:973–984

Knapp CM, Foye MM, Cottam N, Ciraulo DA, Kornetsky C (2001) Adenosine agonists CGS 21680 and NECA inhibit the initiation of cocaine self-administration. Pharmacol Biochem Behav 68:797–803

Kovács Z, Dobolyi A (2011) Functions and metabolism of brain nucleosides and their metabolites. Curr Top Med Chem 11:907–908

Kovács Z, Dobolyi Á, Szikra T, Palkovits M, Juhász G (1998) Uneven regional distribution of nucleotide metabolism in human brain. Neurobiology (Bp) 6:315–321

Kovács Z, Kékesi KA, Bobest M, Török T, Szilágyi N, Szikra T, Szepesi Z, Nyilas R, Dobolyi Á, Palkovits M, Juhász G (2005) Post mortem degradation of nucleosides in the brain: comparison of human and rat brains for estimation of in vivo concentration of nucleosides. J Neurosci Methods 148:88–93

Kovács Z, Dobolyi A, Juhász G, Kékesi AK (2010a) Nucleoside map of the human central nervous system. Neurochem Res 35:452–464

Kovács Z, Juhász G, Dobolyi A, Bobest M, Papp V, Takáts L, Kékesi KA (2010b) Gender- and age-dependent changes in nucleoside levels in the cerebral cortex and white matter of the human brain. Brain Res Bull 81:579–584

Kovács Z, Juhász G, Palkovits M, Dobolyi A, Kékesi KA (2011) Area, age and gender dependence of the nucleoside system in the brain: a review of current literature. Curr Top Med Chem 11:1012–1033

Kowaluk EA, Jarvis MF (2000) Therapeutic potential of adenosine kinase inhibitors. Expert Opin Investig Drugs 9:551–564

Lajtha A, Sershen H (2010) Heterogeneity of reward mechanisms. Neurochem Res 35:851–867

Lam P, Hong CJ, Tsai SJ (2005) Association study of A2a adenosine receptor genetic polymorphism in panic disorder. Neurosci Lett 378:98–101

Lara DR, Dall'Igna OP, Ghisolfi ES, Brunstein MG (2006) Involvement of adenosine in the neurobiology of schizophrenia and its therapeutic implications. Prog Neuropsychopharmacol Biol Psychiatry 30:617–629

Latini S, Pedata F (2001) Adenosine in the central nervous system: release mechanisms and extracellular concentrations. J Neurochem 79:463–484

Ledent C, Vaugeois JM, Schiffmann SN, Pedrazzini T, El Yacoubi M, Vanderhaeghen JJ, Costentin J, Heath JK, Vassart G, Parmentier M (1997) Aggressiveness, hypoalgesia and high blood pressure in mice lacking the adenosine A2a receptor. Nature 388:674–678

Lehman NL (2002) Future potential of thymidylate synthase inhibitors in cancer therapy. Expert Opin Investig Drugs 11:1775–1787

Leist TP, Weissert R (2011) Cladribine: mode of action and implications for treatment of multiple sclerosis. Clin Neuropharmacol 34:28–35

Li RW, Tse CM, Man RY, Vanhoutte PM, Leung GP (2007) Inhibition of human equilibrative nucleoside transporters by dihydropyridine-type calcium channel antagonists. Eur J Pharmacol 568:75–82

Lima FO, Souza GR, Verri WA Jr, Parada CA, Ferreira SH, Cunha FQ, Cunha TM (2010) Direct blockade of inflammatory hypernociception by peripheral A1 adenosine receptors: involvement of the NO/cGMP/PKG/KATP signaling pathway. Pain 151:506–515

Linden J, Rosin DL (2006) Purinergic systems. In: Siegel G, Albers RW, Brady S, Price D (eds) Basic neurochemistry: molecular, cellular and medical aspects, 7th edn. Elsevier, Academic Press Inc., New York, USA, pp 303–316

Linker RA, Kieseier BC, Gold R (2008) Identification and development of new therapeutics for multiple sclerosis. Trends Pharmacol Sci 29:558–565

Litsky ML, Hohl CM, Lucas JH, Jurkowitz MS (1999) Inosine and guanosine preserve neuronal and glial cell viability in mouse spinal cord cultures during chemical hypoxia. Brain Res 821:426–432

Liu XQ, Sheng R, Qin ZH (2009) The neuroprotective mechanism of brain ischemic preconditioning. Acta Pharmacol Sin 30:1071–1080

Lopes LV, Sebastiao AM, Ribeiro JA (2011) Adenosine and related drugs in brain diseases: present and future in clinical trials. Curr Top Med Chem 11:1087–1101

Lucas PB, Pickar D, Kelsoe J, Rapaport M, Pato C, Hommer D (1990) Effects of the acute administration of caffeine in patients with schizophrenia. Biol Psychiatry 28:35–40

Luthin DR, Linden J (1995) Comparison of A_4 and A_{2A} binding sites in striatum and COS cells transfected with adenosine A_{2A} receptors. J Pharmacol Exp Ther 272:511–518

Mangravite LM, Badagnani I, Giacomini KM (2003) Nucleoside transporters in the disposition and targeting of nucleoside analogs in the kidney. Eur J Pharmacol 479:269–281

Markowitz CE, Spitsin S, Zimmerman V, Jacobs D, Udupa JK, Hooper DC, Koprowski H (2009) The treatment of multiple sclerosis with inosine. J Altern Complement Med 15:619–625

Marro PJ, Mishra OP, Delivoria-Papadopoulos M (2006) Effect of allopurinol on brain adenosine levels during hypoxia in newborn piglets. Brain Res 1073–1074:444–450

Martini C, Daniele S, Picchetti M, Panighini A, Carlini M, Trincavelli ML, Cesari D, Da Pozzo E, Golia F, Dell'Osso L (2011) A(2A) adenosine receptor binding parameters in platelets from patients affected by pathological gambling. Neuropsychobiology 63:154–159

McGaraughty S, Cowart M, Jarvis MF, Berman RF (2005) Anticonvulsant and antinociceptive actions of novel adenosine kinase inhibitors. Curr Top Med Chem 5:43–58

Melani A, De Micheli E, Pinna G, Alfieri A, Corte LD, Pedata F (2003) Adenosine extracellular levels in human brain gliomas: an intraoperative microdialysis study. Neurosci Lett 346:93–96

Merighi S, Mirandola P, Varani K, Gessi S, Leung E, Baraldi PG, Tabrizi MA, Borea PA (2003) A glance at adenosine receptors: novel target for antitumor therapy. Pharmacol Ther 100:31–48

Meyer PT, Elmenhorst D, Boy C, Winz O, Matusch A, Zilles K, Bauer A (2007) Effect of aging on cerebral A_1 adenosine receptors: a [18 F]CPFPX PET study in humans. Neurobiol Aging 28:1914–1924

Mielke R, Moller HJ, Erkinjuntti T, Rosenkranz B, Rother M, Kittner B (1998) Propentofylline in the treatment of vascular dementia and Alzheimer-type dementia: overview of phase I and phase II clinical trials. Alzheimer Dis Assoc Disord 12(Suppl 2):S29–S35

Moreau JL, Huber G (1999) Central adenosine A_{2A} recetors: an overview. Brain Res Rev 31:65–82

Müller CE (2003) Medicinal chemistry of adenosine A3 receptor ligands. Curr Top Med Chem 3:445–462

Myers CS, Fisher H, Wagner GC (1994) Uridine potentiates haloperidol's disruption of conditioned avoidance responding. Brain Res 651:194–198

Nabhan C, Gartenhaus RB, Tallman MS (2004) Purine nucleoside analogues and combination therapies in B-cell chronic lymphocytic leukemia: dawn of a new era. Leuk Res 28:429–442

Nagata H, Mimori Y, Nakamura S, Kameyama M (1984) Regional and subcellular distribution in mammalian brain of the enzymes producing adenosine. J Neurochem 42:1001–1007

Noji T, Karasawa A, Kusaka H (2004) Adenosine uptake inhibitors. Eur J Pharmacol 495:1–16

Norstrand IF, Glantz MD (1980) Topographical distribution of purine nucleoside phosphorylase in the human neuraxis. Enzyme 25:118–122

Norstrand IF, Siverls VC, Libbin RM (1984) Regional distribution of adenosine deaminase in the human neuraxis. Enzyme 32:20–25

O'Regan M (2005) Adenosine and the regulation of cerebral blood flow. Neurol Res 27:175–181

Öngür D, Drevets WC, Price JL (1998) Glial reduction in the subgenual prefrontal cortex in mood disorders. Proc Natl Acad Sci USA 95:13290–13295
Parkinson FE, Damaraju VL, Graham K, Yao SYM, Baldwin SA, Cass CE, Young JD (2011) Molecular biology of nucleoside transporters and their distributions and functions in the brain. Curr Top Med Chem 11:948–972
Pastor-Anglada M, Cano-Soldado P, Molina-Arcas M, Lostao MP, Larráyoz I, Martinez-Picado J, Casado JF (2005) Cell entry and export of nucleoside analogues. Virus Res 107:151–164
Paul T, Pfammatter JP (1997) Adenosine: an effective and safe antiarrhythmic drug in pediatrics. Pediatr Cardiol 18:118–126
Pearce L, Ghosh J, Counsell A, Serracino-Inglott F (2008) Cilostazol and peripheral arterial disease. Expert Opin Pharmacother 9:2683–2690
Pennycooke M, Chaudary N, Shuralyova I, Zhang Y, Coe IR (2001) Differential expression of human nucleoside transporters in normal and tumor tissue. Biochem Biophys Res Commun 280:951–959
Peskind ER (1996) Neurobiology of Alzheimer's disease. J Clin Psychiatry 57(Suppl 14):5–8
Peterman C, Sanoski CA (2005) Tecadenoson: a novel, selective A1 adenosine receptor agonist. Cardiol Rev 13:315–321
Pettifer KM, Kleywegt S, Bau CJ, Ramsbottom JD, Vertes E, Ciccarelli R, Caciagli F, Werstiuk ES, Rathbone MP (2004) Guanosine protects SH-SY5Y cells against beta-amyloid-induced apoptosis. Neuroreport 15:833–836
Phillips E, Newsholme EA (1979) Maximum activities, properties and distribution of 5'-nucleotidase, adenosine kinase and adenosine deaminase in rat and human brain. J Neurochem 33:553–558
Piccoli F, Camarda R, Bonavita V (1971) The brain nucleotide pattern of the rat after injection of uracil, uridine and uridine phosphate. Acta Neurol (Napoli) 26:109–117
Pinna A (2009) Novel investigational adenosine A2A receptor antagonists for Parkinson's disease. Expert Opin Investig Drugs 18:1619–1631
Podgorska M, Kocbuch K, Pawelczyk T (2005) Recent advences in studies on biochemical and structural properties of equilibrative and concentrative nucleoside transporters. Acta Biochim Pol 52:749–758
Popoli P, Blum D, Martine A, Ledent C, Ceruti S, Abbracchio MP (2007) Functions, dysfunctions and possible therapeutic relevance of adenosine A2A receptors in Huntington's disease. Prog Neurobiol 81:331–348
Porkka-Heiskanen T, Kalinchuk AV (2011) Adenosine, energy metabolism and sleep homeostasis. Sleep Med Rev 15:123–135
Radulovacki M (1985) Role of adenosine in sleep in rats. Rev Clin Basic Pharm 5:327–339
Rando RF, Nguyen-Ba N (2000) Development of novel nucleoside analogues for use against drug rersisitant strains of HIV-1. Drug Discov Today 5:465–476
Ribeiro JA, Sebastiao AM, de Mendonca A (2003) Adenosine receptors in the nervous system: pathophysiological implications. Prog Neurobiol 68:377–392
Ritchie K, Carriere I, de Mendonca A, Portet F, Dartigues JF, Rouaud O, Barberger-Gateau P, Ancelin ML (2007) The neuroprotective effects of caffeine: a prospective population study (the Three City Study). Neurology 69:536–545
Ritzel MWL, Ng AML, Yao SYM, Graham K, Loewen SK, Smith KM, Ritzel RG, Mowles DA, Carpenter P, Chen XZ, Karpinski E, Hyde RJ, Baldwin SA, Cass CE, Young JD (2001) Molecular identification and characterization of novel human and mouse concentrative Na^+-nucleoside cotransporter proteins (hCNT3 and mCNT3) broadly selective for purine and pyrimidine nucleosides (system cib). J Biol Chem 276:2914–2927
Robak T, Korycka A, Lech-Maranda E, Robak P (2009) Current status of older and new purine nucleoside analogues in the treatment of lymphoproliferative diseases. Molecules 14:1183–1226
Roberts CA (1973) Anticonvulsant effects of uridine: comparative analysis of metrazol and penicillin induced foci. Brain Res 55:291–308
Roberts CA, Kreisman NR, Waltman M (1974) Uridine anticonvulsant effects: selective control of nucleoside incorporation in experimental epilepsy. Epilepsia 15:479–500

Roesler R, Vianna MR, Lara DR, Izquierdo I, Schmidt AP, Souza DO (2000) Guanosine impairs inhibitory avoidance performance in rats. Neuroreport 11:2537–2540
Roos RA, Bots GT, Hermans J (1985) Neuronal nuclear membrane indentation and astrocyte/neuron ratio in Huntington's disease. A quantitative electron microscopic study. J Hirnforsch 26:689–693
Ross CA, Margolis RL, Reading SA, Pletnikov M, Coyle JT (2006) Neurobiology of schizophrenia. Neuron 52:139–153
Rudolphi KA, Schubert P, Parkinson FE, Fredholm BB (1992) Adenosine and brain ischemia. Cerebrovasc Brain Metab Rev 4:346–369
Saper CB (2006) Staying awake for dinner: hypothalamic integration of sleep, feeding, and circadian rhythms. Prog Brain Res 153:243–252
Saute JA, da Silveira LE, Soares FA, Martini LH, Souza DO, Ganzella M (2006) Amnesic effect of GMP depends on its conversion to guanosine. Neurobiol Learn Mem 85:206–212
Schiffmann SN, Fisone G, Moresco R, Cunha RA, Ferré S (2007) Adenosine A2A receptors and basal ganglia physiology. Prog Neurobiol 83:277–292
Schmidt AP, Lara DR, de Faria Maraschin J, da Silveira Perla A, Souza DO (2000) Guanosine and GMP prevent seizures induced by quinolinic acid in mice. Brain Res 864:40–43
Schmidt AP, Lara DR, Souza DO (2007) Proposal of a guanine-based purinergic system in the mammalian central nervous system. Pharmacol Therapeut 116:401–416
Schmidt AP, Böhmer AE, Schallenberger C, Antunes C, Pereira MS, Leke R, Wofchuk ST, Elisabetsky E, Souza DO (2009) Spinal mechanisms of antinociceptive action caused by guanosine in mice. Eur J Pharmacol 613:46–53
Schulte G, Fredholm BB (2003) Signalling from adenosine receptors to mitogen-activated protein kinases. Cell Signal 15:813–827
Schwarzschild MA, Agnati L, Fuxe K, Chen JF, Morelli M (2006) Targeting adenosine A2A receptors in Parkinson's disease. Trends Neurosci 29:647–654
Shen H, Chen GJ, Harvey BK, Bickford PC, Wang Y (2005) Inosine reduces ischemic brain injury in rats. Stroke 36:654–659
Sperlágh B, Vizi ES (2011) The role of extracellular adenosine in chemical neurotransmission in the hippocampus and Basal Ganglia: pharmacological and clinical aspects. Curr Top Med Chem 11:1034–1046
Spiers PA, Myers D, Hochanadel GS, Lieberman HR, Wurtman RJ (1996) Citicoline improves verbal memory in aging. Arch Neurol 53:441–448
Stone TW (2002) Purines and neuroprotection. Adv Exp Med Biol 513:249–280
Svenningsson P, Hall H, Sedvall G, Fredholm BB (1997) Distribution of adenosine receptors in the postmortem human brain: an extended autoradiographic study. Synapse 27:322–335
Teather LA, Wurtman RJ (2003) Dietary cytidine (5')-diphosphocholine supplementation protects against development of memory deficits in aging rats. Prog Neuropsychopharmacol Biol Psychiatry 27:711–717
Teather LA, Wurtman RJ (2005) Dietary CDP-choline supplementation prevents memory impairment caused by impoverished environmental conditions in rats. Learn Mem 12:39–43
Teather LA, Wurtman RJ (2006) Chronic administration of UMP ameliorates the impairment of hippocampal-dependent memory in impoverished rats. J Nutr 136:2834–2837
Togha M, Akhondzadeh S, Motamedi M, Ahmadi B, Razeghi S (2007) Allopurinol as adjunctive therapy in intractable epilepsy: a double-blind and placebo-controlled trial. Arch Med Res 38:313–316
Tomassoni D, Lanari A, Silvestrelli G, Traini E, Amenta F (2008) Nimodipine and its use in cerebrovascular disease: evidence from recent preclinical and controlled clinical studies. Clin Exp Hypertens 30:744–766
Tort AB, Mantese CE, dos Anjos GM, Dietrich MO, Dall'Igna OP, Souza DO, Lara DR (2004) Guanosine selectively inhibits locomotor stimulation induced by the NMDA antagonist dizocilpine. Behav Brain Res 154:417–422
Traut TW (1994) Physiological concentrations of purines and pyrimidines. Mol Cell Biochem 140:1–22

Traversa U, Bombi G, Di Iorio P, Ciccarelli R, Werstiuk ES, Rathbone MP (2002) Specific [(3) H]-guanosine binding sites in rat brain membranes. Br J Pharmacol 135:969–976

Tsai SJ, Hong CJ, Hou SJ, Yen FC (2006) Association study of adenosine A2a receptor (1976 C>T) genetic polymorphism and mood disorders and age of onset. Psychiatr Genet 16:185

Tucker AL, Linden J (1993) Cloned receptors and cardiovascular responses to adenosine. Cardiovasc Res 27:62–67

Van der Beek EM, Kamphuis P (2008) The potential role of nutritional components in the management of Alzheimer's Disease. Eur J Pharmacol 585:197–207

Van Rompay AR, Johansson M, Karlsson A (2003) Substrate specificity and phosphorylation of antiviral and anticancer nucleoside analogues by human deoxyribonucleoside kinases and ribonucleoside kinases. Pharmacol Therapeut 100:119–139

Varani K, Merighi S, Gessi S, Klotz KN, Leung E, Baraldi PG, Cacciari B, Romagnoli R, Spalluto G, Borea PA (2000) [(3)H]MRE 3008 F20: a novel antagonist radioligand for the pharmacological and biochemical characterization of human A(3) adenosine receptors. Mol Pharmacol 57:968–975

Vinadé ER, Schmidt AP, Frizzo ME, Izquierdo I, Elisabetsky E, Souza DO (2003) Chronically administered guanosine is anticonvulsant, amnesic and anxiolytic in mice. Brain Res 977:97–102

Volonté C, D'Ambrosi N (2009) Membrane compartments and purinergic signalling: the purinome, a complex interplay among ligands, degrading enzymes, receptors and transporters. FEBS J 276:318–329

Volpini R, Costanzi S, Vittori S, Cristalli G, Klotz KN (2003) Medicinal chemistry and pharmacology of A2B adenosine receptors. Curr Top Med Chem 3:427–443

Von Lubitz DK (2001) Adenosine in the treatment of stroke: yes, maybe, or absolutely not? Expert Opin Investig Drugs 10:619–632

Wardas J (2002) Neuroprotective role of adenosine in the CNS. Pol J Pharmacol 54:313–326

Warnke C, Wiendl H, Hartung HP, Stüve O, Kieseier BC (2010) Identification of targets and new developments in the treatment of multiple sclerosis–focus on cladribine. Drug Des Devel Ther 4:117–126

Weber G, Prajda N (1994) Targeted and non-targeted actions of anti-cancer drugs. Adv Enzyme Regul 34:71–89

Weber G, Ichikawa S, Nagai M, Natsumeda Y (1990) Azidothymidine inhibition of thymidine kinase and synergistic cytotoxicity with methotrexate and 5-fluorouracil in rat hepatoma and human colon cancer cells. Cancer Commun 2:129–133

Weerts EM, Griffiths RR (2003) The adenosine receptor antagonist CGS15943 reinstates cocaine-seeking behavior and maintains self-administration in baboons. Psychopharmacology (Berl) 168:155–163

Weyrich AS, Skalabrin EJ, Kraiss LW (2009) Targeting the inflammatory response in secondary stroke prevention: a role for combining aspirin and extended-release dipyridamole. Am J Ther 16:164–170

Wiesner JB, Ugarkar BG, Castellino AJ, Barankiewicz J, Dumas DP, Gruber HE, Foster AC, Erion MD (1999) Adenosine kinase inhibitors as a novel approach to anticonvulsant therapy. J Pharmacol Exp Ther 289:1669–1677

Willis RC, Carson DA, Seegmiller JE (1978) Adenosine kinase initiates the major route of ribavirin activation in a cultured human cell line. Proc Natl Acad Sci USA 75:3042–3044

Willuhn I, Wanat MJ, Clark JJ, Phillips PE (2010) Dopamine signaling in the nucleus accumbens of animals self-administering drugs of abuse. Curr Top Behav Neurosci 3:29–71

Wurtman RJ, Regan M, Ulus I, Yu L (2000) Effect of oral CDP-choline on plasma choline and uridine levels in humans. Biochem Pharmacol 60:989–992

Wurtman RJ, Cansev M, Sakamoto T, Ulus IH (2009) Use of phosphatide precursors to promote synaptogenesis. Annu Rev Nutr 29:59–87

Xu K, Bastia E, Schwarzschild M (2005) Therapautic potential of adenosine A_{2A} receptor antagonists in Parkinson's disease. Pharmacol Ther 105:267–310

Yegutkin GG (2008) Nucleotide- and nucleoside converting ectoenzymes: important modulators of purinergic signalling cascade. Biochim Biophys Acta 1783:673–694

Zapor MJ, Cozza KL, Wynn GH, Wortmann GW, Armstrong SC (2004) Antiretrovirals, part II: focus on non-protease inhibitor antiretrovirals (NRTIs, NNRTIs, and fusion inhibitors). Psychosomatics 45:524–535

Zaza A (2002) CVT-510 (CV Therapeutics). Curr Opin Investig Drugs 3:96–100

Zhao Q, Marolewski A, Rusche JR, Holmes GL (2006) Effects of uridine in models of epileptogenesis and seizures. Epilepsy Res 70:73–82

Zhao Q, Shatskikh T, Marolewski A, Rusche JR, Holmes GL (2008) Effects of uridine on kindling. Epilepsy Behav 13:47–51

Zimmermann H (1996) Biochemistry, localization and functional roles of ecto-nucleotidases in the nervous system. Prog Neurobiol 49:589–618

Zoref-Shani E, Bromberg Y, Lilling G, Gozes I, Brosh S, Sidi Y, Sperling O (1995) Developmental changes in purine nucleotide metabolism in cultured rat astroglia. Int J Dev Neurosci 13:887–896

Index

S. Masino and D. Boison (eds.), *Adenosine: A Key Link between Metabolism and Brain Activity*, DOI 10.1007/978-1-4614-3903-5,

D

MIX
Papier aus verantwortungsvollen Quellen
Paper from responsible sources
FSC® C105338

If you have any concerns about our products,
you can contact us on
ProductSafety@springernature.com

In case Publisher is established outside the EU,
the EU authorized representative is:
Springer Nature Customer Service Center GmbH
Europaplatz 3, 69115 Heidelberg, Germany

Printed by Libri Plureos GmbH
in Hamburg, Germany